AF581908

Advances in Mineral Processing and Hydrometallurgy

Advances in Mineral Processing and Hydrometallurgy

Editors

Corby G. Anderson
Hao Cui

MDPI • Basel • Beijing • Wuhan • Barcelona • Belgrade • Manchester • Tokyo • Cluj • Tianjin

Editors
Corby G. Anderson
Colorado School of Mines
USA

Hao Cui
Colorado School of Mines
USA

Editorial Office
MDPI
St. Alban-Anlage 66
4052 Basel, Switzerland

This is a reprint of articles from the Special Issue published online in the open access journal *Metals* (ISSN 2075-4701) (available at: http://www.mdpi.com).

For citation purposes, cite each article independently as indicated on the article page online and as indicated below:

LastName, A.A.; LastName, B.B.; LastName, C.C. Article Title. *Journal Name* **Year**, *Volume Number*, Page Range.

ISBN 978-3-0365-2079-7 (Hbk)
ISBN 978-3-0365-2080-3 (PDF)

Contents

About the Editors

Prof. Dr. Corby G. Anderson is a licensed Professional Chemical Engineer with over 40 years of global experience in industrial operations, corporate level management, engineering, design, consulting, teaching, research and professional service. He is a native of Butte, America. His career includes positions with Thiokol Chemical Corporation, Key Tronic Corporation, Sunshine Mining and Refining Company, H. A. Simons Ltd., and at CAMP-Montana Tech. He holds a BSc in Chemical Engineering from Montana State University and an MSc from Montana Tech in Metallurgical Engineering, as well as a PhD from the University of Idaho in Mining Engineering—Metallurgy. He is a Fellow of both the Institution of Chemical Engineers and of the Institute of Materials, Minerals and Mining. He shares 14 international patents and 5 new patent applications, covering several innovative technologies, 2 of which were successfully reduced to industrial practice. He currently serves as the Harrison Western Professor in the Kroll Institute for Extractive Metallurgy, as part of both the Mining Engineering Department and the George S. Ansell Department of Metallurgical and Materials Engineering at the Colorado School of Mines. In 2009, he was honored by the Society for Mining Metallurgy and Exploration, with the Milton E. Wadsworth Extractive Metallurgy Award for his contributions in hydrometallurgical research. In 2015, he was awarded the International Precious Metals Institute's Tanaka Distinguished Achievement Award. In 2016, he received the Distinguished Member Award from the Society for Mining, Metallurgy and Exploration, the Outstanding Faculty Award from the George S. Ansell Department of Metallurgical and Materials Engineering at Colorado School of Mines, and became a Distinguished Member of the U of Idaho Academy of Engineering. In 2017, he received the EPD Distinguished Lecturer Award from The Minerals, Metals and Materials Society. In 2019, he was named as a Henry Krumb Distinguished SME Lecturer. In 2019, he was also appointed as a Visiting Faculty within the Minerals Engineering Department of Central South University in China, the largest Mineral Processing and Extractive Metallurgy program in the world. Lastly, in 2016 and in 2021 he received an Outstanding Faculty Award from the Colorado School of Mines.

Dr. Hao Cui's research expertise concentrates on refractory gold treatment, rare earth beneficiation, and the recycling of e-waste. Hao received his BS degree in mineral processing engineering from the University of Technology & Science, Beijing in 2012, and obtained his MS and PhD in metallurgical engineering from the Colorado School of Mines in 2015 and 2018, respectively. In 2017, he was awarded the International Precious Metals Institute's Gero Family Trust Bright Future Award. Hao worked as a metallurgist (2019–2021) at Nevada Gold Mines, LLC, looking at opportunities to improve gold recovery through the implementation of test programs and technical research, specifically pertaining to gold cyanide and thiosulfate leaching.

Preface to "Advances in Mineral Processing and Hydrometallurgy"

A Special Issue of *Metals* was commissioned that was devoted to aspects of Mineral Processing and Hydrometallurgy. The editors were Prof. Dr. Corby Anderson and Dr. Hao Cui. This issue was designed to include submissions for characterization, along with recycling and waste minimization, mineralogy, geometallurgy, thermodynamics, kinetics, comminution, classification, physical separations, liquid solid separations, leaching, solvent extraction, ion exchange, activated carbon, precipitation, reduction, process economics, and process control. Suggested application areas were gold, silver, PGMs, aluminum, copper, zinc, lead, nickel, and titanium. Critical metal articles on topics such as lithium, antimony, tellurium, gallium, germanium, cobalt, graphite, indium, and the rare earth metals were welcome. Thirty-one high-quality peer-reviewed articles from around the globe were selected and accepted for inclusion, along with an Editorial Review overview article.

As such, this Special Issue of *Metals* was well supported by diverse submissions and a final publication of high-quality peer-reviewed articles. It is anticipated that, due to this success, a new Special Issue ("Advances in Mineral Processing and Hydrometallurgy II", website: https://www.mdpi.com/journal/metals/special_issues/mineral_process_hydrometallurgy2) will be commissioned as a follow-up to accept global contributions from the hydrometallurgy and mineral processing community.

Corby G. Anderson, Hao Cui
Editors

Editorial

Advances in Mineral Processing and Hydrometallurgy

Corby G. Anderson * and Hao Cui *

Kroll Institute for Extractive Metallurgy, Mining Engineering Department & George S. Ansell Department of Metallurgical and Materials Engineering, Colorado School of Mines, Golden, CO 80401, USA
* Correspondence: cganders@mines.edu (C.G.A.); hcuimetallurgy@outlook.com (H.C.)

1. Introduction and Scope

A Special Issue of *Metals* was commissioned that was devoted to aspects of Mineral Processing and Hydrometallurgy. The editors were Prof. Dr. Corby Anderson and Dr. Hao Cui. This issue was designed to include submissions for characterization, along with recycling and waste minimization, mineralogy, geometallurgy, thermodynamics, kinetics, comminution, classification, physical separations, liquid solid separations, leaching, solvent extraction, ion exchange, activated carbon, precipitation, reduction, process economics, and process control. Suggested application areas were gold, silver, PGMs, aluminum, copper, zinc, lead, nickel, and titanium. Critical metal articles on topics such as lithium, antimony, tellurium, gallium, germanium, cobalt, graphite, indium, and the rare earth metals were welcome. Thirty-one high-quality peer-reviewed articles from around the globe were selected and accepted for inclusion.

2. Contributions

Wang et al. [1] contributed an article entitled "Comparison of Butyric Acid Leaching Behaviors of Zinc from Three Basic Oxygen Steelmaking Filter Cakes". The selective leaching of zinc from three different basic oxygen steelmaking (BOS) filter cakes by butyric acid was investigated to compare the leaching behaviors of zinc and to further establish the correlation between the zinc leaching performance and chemical composition. The effects of acid concentration and the acid to solid (L/S) stoichiometric ratio were studied, with different optimal leaching conditions being obtained. BOS-1 showed the lowest leachability, with only less than 10% of zinc removed by a 0.5 M acid concentration and 90% of the L/S stoichiometric ratio in 10 h. The best zinc selectivity was achieved with BOS-2 at 51.2% of the zinc leaching efficiency, with only 0.47% of iron loss under the optimal conditions of a 1.5 M acid concentration and a 70% stoichiometric ratio. BOS-3 showed the highest leaching of zinc, but the optimal conditions depended on priority consideration. Using 1.0 M acid and a 90% stoichiometric ratio for 10 h, the leaching efficiency of zinc was 84.6% with 20% iron loss. The filter cakes and the leaching residues were characterized. The results indicated different zinc and iron leaching behaviors, which were probably related to the storage conditions, zinc-containing phases, and the leaching parameters.

Toro et al. [2] provided an article entitled "Leaching of Pure Chalcocite in a Chloride Media Using Sea Water and Waste Water". Chalcocite is the most important and abundant secondary copper ore in the world, with a rapid dissolution of copper in an acid-chloride environment. In this investigation, the surface optimization methodology will be applied to evaluate the effect of three independent variables (time, concentration of sulfuric acid, and chloride concentration) in the leaching of pure chalcocite to extract the copper with the objective of obtaining a quadratic model that allows us to predict the extraction of copper. The kinetics of copper dissolution with regard to the function of temperature are also analyzed. An ANOVA indicated that the linear variables with the greatest influence are time and the chloride concentration. Moreover, the concentration of chloride–time exerts a significant synergic effect in the quadratic model. The ANOVA indicated that the quadratic model is representative and that the R^2 value of 0.92 is valid. The highest copper

Citation: Anderson, C.G.; Cui, H. Advances in Mineral Processing and Hydrometallurgy. *Metals* **2021**, *11*, 1393. https://doi.org/10.3390/met11091393

Received: 19 August 2021
Accepted: 26 August 2021
Published: 1 September 2021

extraction (67.75%) was obtained at 48 h leaching under the conditions of 2 mol/L H_2SO_4 and 100 g/L chloride. The XRD analysis shows the formation of a stable and non-polluting residue, such as elemental sulfur (S^0). This residue was obtained in a leaching time of 4 h at room temperature under the conditions of 0.5 mol/L H_2SO_4 and 50 g/L Cl^-.

Benavente et al. [3] submitted an article entitled "Copper Dissolution from Black Copper Ore under Oxidizing and Reducing Conditions". Black copper oxides are amorphous materials of copper-bearing phases of manganese. They are complex mineral compounds with difficult-to-recognize mineralogy and have slow dissolution kinetics in conventional hydrometallurgical processes. This study evaluates the effects of various leaching media on copper dissolution from black copper minerals. The leaching of a pure black copper sample from Lomas Bayas Mine and another from a regional mine were characterized by inductively coupled plasma atomic emission spectroscopy (ICP-AES), X-ray diffraction (XRD), scanning electron microscopy (SEM), and Qemscan and were mechanically prepared for acid leaching under standard, oxidizing, and reducing conditions through the addition of oxygen, iron sulfate, or sulfur dioxide, respectively. Standard and high-potential leaching (770 mV (SHE)) resulted in a copper dissolution rate of 70% and manganese dissolution rate of 2%. The addition of potential reducing agents ($FeSO_4$ or SO_2) decreased the redox potential to 696 and 431 mV, respectively, and favored the dissolution of manganese, thus increasing the overall copper extraction rate. The addition of SO_2 resulted in the lowest redox potential and the highest copper extraction rates of 86.2% and 75.5% for the Lomas Bayas and regional samples, respectively, which represent an increase of 15% over the copper extract rates under standard and oxidizing conditions.

Coello-Velázquez et al. [4] provided an article entitled "Use of the Swebrec Function to Model Particle Size Distribution in an Industrial-Scale Ni-Co Ore Grinding Circuit". Mathematical models of particle size distribution (PSD) are necessary in the modelling and simulation of comminution circuits. In order to evaluate the application of the Swebrec PSD model (SWEF) in the grinding circuit at the Punta Gorda Ni-Co plant, a sampling campaign was conducted with variations in the operating parameters. Subsequently, the fitting of the data to the Gates–Gaudin–Schumann (GGS), Rosin–Rammler (RRS), and SWEF PSD functions was evaluated under statistical criteria. The fitting of the evaluated distribution models showed that these functions are characterized as being sufficiently accurate, as the estimation error does not exceed 3.0% in any of the cases. In the particular case of the Swebrec function, the reproducibility for all of the products was high. Furthermore, its estimation error did not exceed 2.7% in any of the cases, with a correlation coefficient of the ratio between the experimental and simulated data greater than 0.99.

Soria-Aguilar et al. [5] contributed an article entitled "Oxidative Leaching of Zinc and Alkalis from Iron Blast Furnace Sludge". The sludge from a wet-off gas cleaning system of an iron blast furnace (BF) contains significant amounts of iron; however, this iron cannot be recycled due to its high content of zinc and alkalis. These compounds are detrimental to the optimal performance of iron and steelmaking furnaces. In this work, a comparative laboratory study to reduce zinc and alkali contained in the blast furnace sludge (BFS) is presented. The effect of leaching parameters, such as oxidants (i.e., ferric ion, oxygen, or ozone), aqueous solution media (i.e., 0.2 M NH_4Cl, 0.2 M HCl, and 0.1 M H_2SO_4), and temperature (i.e., 27 and 80 °C) on Zn and alkali (Na_2O and K_2O) removal were studied by applying an experimental design. The obtained results show that Zn and K_2O removals of 85% and 75% were achieved under the following conditions: ozone as an oxidant agent and 0.1 M H_2SO_4 as an aqueous medium; temperature had no significant effect. The results are supported by thermodynamic diagrams and the possible chemical reactions are mentioned. Although the results also indicate that leaching under the above conditions dissolves up to 9% of iron, this loss is much less than leaching without the oxidizing conditions generated by the ozone. The BFS obtained from this treatment could be recirculated to the iron or steelmaking processes to recover iron values.

Pérez et al. [6] contributed an article entitled "Extraction of Mn from Black Copper Using Iron Oxides from Tailings and Fe^{2+} as Reducing Agents in Acid Medium". Exotic-type deposits include several species of minerals, such as atacamite, chrysocolla, copper pitch, and copper wad. Among these, copper pitch and copper wad have considerable concentrations of manganese. However, their non-crystalline and amorphous structure makes it challenging to recover elements of interest (such as Cu or Mn) using conventional hydrometallurgical methods. For this reason, black copper ores are generally not incorporated into the extraction circuits or left unprocessed, whether in stock, leach pads, or waste. Therefore, to dilute MnO_2, the use of reducing agents is essential. In the present research, agitated leaching was performed to dissolve the Mn of black copper in an acidic medium, comparing the use of ferrous ions and tailings as reducing agents. Two samples of black copper were studied, of high and low grade of Mn, respectively, the latter with a high content of clays. The effect of the reducing agent/black copper ratio and the concentration of sulfuric acid in the system was evaluated. Better results in removing Mn were achieved using the highest-grade black copper sample when working with ferrous ions at a ratio of Fe^{2+}/black copper of 2/1 and 1 mol/L of sulfuric acid. Moreover, the low-grade sample induced significant H_2SO_4 consumption due to the high presence of gangue and clays.

Jeldres et al. [7] authored an article entitled "Viscoelasticity of Quartz and Kaolin Slurries in Seawater: Importance of Magnesium Precipitates". In this study, the viscoelastic properties of quartz and kaolin suspensions in seawater were analysed considering two distinct conditions: pH 8 and 10.7. Creep and oscillatory sweep tests provided the rheological parameters. An Anton Paar MCR 102 rheometer (ANAMIN Group, Santiago, Chile) was used with a vane-in-cup configuration, and the data were processed with RheoCompass^{TM} Light software (ANAMIN Group, Santiago, Chile). The outcomes were associated with the formation of solid species principally composed of magnesium precipitates. The magnesium in the solution reduced from 1380 to 1280 mg/L in the presence of quartz (68 wt %). Since the difference was not large regarding the solid-free seawater, the disposition of the solid complexes at pH 10.7 was expected to be similar. The jump in pH caused both yield stress and viscoelastic moduli to drop, suggesting that the solid precipitates diminished the strength of the particle networks that made up the suspension. For the kaolin slurries (37 wt %), the yield stress increased when the pH increased, but unlike quartz, there was a significant adsorption of magnesium cations. In fact, the concentration of magnesium in solution fell from 1380 to 658 mg/L. Dynamic oscillatory assays revealed structural changes in both pulps; in particular, the phase angle was greater at pH 8 than at pH 10.7, which indicates that at more alkaline conditions, the suspension exhibits a more solid-like character.

Sokić et al. [8] contributed a manuscript entitled "Kinetics of Chalcopyrite Leaching by Hydrogen Peroxide in Sulfuric Acid". In ores, chalcopyrite is usually associated with other sulfide minerals, such as sphalerite, galena, and pyrite, in a dispersed form with complex mineralogical structures. Concentrates obtained by the flotation of such ores are unsuitable for pyrometallurgical processing, owing to their poor quality and low metal recovery. This paper presents the leaching of chalcopyrite concentrate from the location of Rudnik, Serbia. The samples from the flotation plant were treated with hydrogen peroxide in sulfuric acid. The influences of temperature, particle size, and stirring speed as well as the concentrations of hydrogen peroxide and sulfuric acid were followed and discussed. Hence, the main objective was to optimize the relevant conditions and to determine the reaction kinetics. It was remarked that the increase in temperature, hydrogen peroxide content, and sulfuric acid concentration as well as the decrease in particle size and stirring speed contribute to the dissolution of chalcopyrite. The dissolution kinetics follow a model controlled by diffusion, and the lixiviant diffusion controls the rate of reaction through the sulfur layer. Finally, the main characterization methods used to corroborate the obtained results were X-ray diffraction (XRD) as well as the qualitative and quantitative light microscopy of the chalcopyrite concentrate samples and the leach residue.

Jeldres et al. [9] contributed their research entitled "Copper Tailing Flocculation in Seawater: Relating the Yield Stress with Fractal Aggregates at Varied Mixing Conditions". The implications of the physical conditions of the feedwell on the rheological properties of synthetic copper tailings flocculated in seawater were analyzed. The mixing intensity of flocculation was related to the structural characteristics of the aggregates, and the outcomes were linked to the yield stress of the pulp sediments. Tailing settling assays were conducted using a 30 mm turbine type stirrer with an in situ aggregate size characterization. The structural characteristics of the aggregates were determined using the focused beam reflectance measurement (FBRM). After mixing time between the pulp and the flocculant, the sample was allowed to settle for 2.5 h, where the variation of the sediment height was minimal. The sediment was gently removed and was subjected to rheological characterization. The yield stress was measured on an Anton Paar MCR 102 rheometer (ANAMIN Group, Santiago, Chile) with a vane-in-cup configuration. The mixing intensity was related to the characteristics of the aggregates, and the outcomes were linked to the yield stress of the flocculated pulp sediments. More aggressive hydrodynamics deteriorated the structure of the aggregates, promoting the reduction of both its size and the fractal dimension. This brought direct consequences to the rheological properties of the sediments: at a higher mixing level, the yield stress was lower. The explanation lies in the structural changes of the aggregates, where at a fixed mixing rate, the yield stress presented a seemingly exponential increase over the fractal dimension. Additionally, correlations were found between the rheological properties regarding settling rate and aggregate size.

Torres et al. [10] provide an article entitled "Leaching Manganese Nodules in an Acid Medium and Room Temperature Comparing the Use of Different Fe Reducing Agents". The deposits of Fe-Mn in the seabed of the planet are a good alternative source for the extraction of elements of interest. Among these are marine nodules, which contain approximately 24% manganese and may be a solution to the shortage of high-grade ores on the surface. In this investigation, an ANOVA analysis was performed to evaluate the time-independent variables and the MnO_2/reducing agent in the leaching of manganese nodules with the use of different Fe reducing agents (FeS_2, Fe^{2+}, Fe^0 and Fe_2O_3). Tests were also conducted for the different reducing agents evaluating the MnO_2/Fe ratio, in which the Fe^0 (FeC) proved to be the best reducing agent for the dissolution of Mn from marine nodules, achieving solutions of 97% in 20 min. In addition, it was discovered that, at low MnO_2/Fe ratios, the acid concentration in the system is not very relevant, and the potential and pH were in ranges of −0.4–1.4 V and −2–0.1, favoring the dissolution of Mn from MnO_2.

Karimov et al. [11] submitted two articles. The first is entitled "Leaching Kinetics of Arsenic Sulfide-Containing Materials by Copper Sulfate Solution". The overall decrease in the quality of mineral raw materials, combined with the use of arsenic-containing ores, results in large amounts of various intermediate products containing this highly toxic element. The use of hydrometallurgical technologies for these materials is complicated by the formation of multicomponent solutions and the difficulty of separating copper from arsenic. Previously, for the selective separation of As from copper–arsenic intermediates, a leaching method in the presence of Cu(II) ions was proposed. This paper describes the investigation of the kinetics of arsenic sulfide-containing materials leaching due to copper sulfate solution. After the leaching of arsenic trisulfide with a solution of copper sulfate, the cakes were described using methods such as X-ray diffraction spectrometry (XRD), X-ray fluorescence spectrometry (XRF), scanning electron microscopy (SEM), and energy-dispersive X-ray spectroscopy analysis (EDS). The effect of temperature (70–90 °C), the initial concentration of $CuSO_4$ (0.23–0.28 M), and the time on the As recovery into the solution were studied. The process temperature has the greatest effect on the kinetics, while an increase in copper concentration from 0.23 to 0.28 M effects an increase in As transfer into a solution from 93.2% to 97.8% for 120 min of leaching. However, the shrinking core model that best fits the kinetic data suggests that the process occurs via the intra-diffusion mode, with an average activation energy of 44.9 kJ/mol. Using the time-to-a-given-fraction kinetics analysis, it was determined that the leaching mechanism does not change during the reaction. The semi-empirical expression describing the reaction rate under the studied conditions can be written as follows: $1/3\ln(1-X) + [(1-X) - 1/3 - 1] = 4560000Cu^{3.61}e^{-44900/RT}t$.

Salinas-Rodriguez et al. [12] offered a publication entitled "Assessment of Silica Recovery from Metallurgical Mining Waste, by Means of Column Flotation". The generation of mining waste commonly leads to the use of spaces for its disposal. Challenges such as mitigating the damage to surrounding communities have promoted the need to reuse, recycle, and/or reduce their generation. Moreover, these residues may become the source of materials that are capable of being recovered and reused in several industries, minimizing their environmental impact. In the mining region of Pachuca, Mexico, waste from the mining industry has been generated for more than 100 years and has a high SiO_2 content that can be recovered for various industrial applications. This work aims to recover silica from a material from the Dos Carlos dam. A columnar system composed of two-stage of cleaning was used, considering a J_{LT} (surface liquid rate) value of 0.45 and 0.68 cm/s, respectively, while the J_g (surface gas rate) value was 0.30 cm/s for both stages. Similar bubble sizes in the range of J_g 0.10 to 0.30 cm/s, with values between 0.14 and 0.16 cm were found in the first stage, and those that were 0.05 to 0.06 cm in size were found in the second one. This provided a recovery of 75.10% for all of the allotropic phases of silica (quartz, trydimite, and cristobalite), leaving a concentration of 24.90% for the feldspathic phase (orthoclase) as flotation tails.

Torres et al. [13] authored a submission entitled "Leaching Chalcopyrite with High MnO_2 and Chloride Concentrations". Most copper minerals are found as sulfides, with chalcopyrite being the most abundant. However, this ore is refractory to conventional hydrometallurgical methods, so it has been historically exploited through froth flotation followed by smelting operations. This implies that the processing involves polluting activities, either by the formation of tailings dams or the emission of large amounts of SO_2 into the atmosphere. Given increasing environmental restrictions, it is necessary to consider new processing strategies that are compatible with the environment, and, if feasible, combine the reuse of industrial waste. In the present research, the dissolution of pure chalcopyrite was studied, considering the use of MnO_2 and wastewater with a high chloride content. Fine particles (−20 μm) generated an increase in the extraction of copper from the mineral. Moreover, it was discovered that while working at high temperatures (80 °C), large concentrations of MnO_2 become irrelevant. The biggest copper extractions of this work (71%) were achieved when operating at 80 °C, with a particle size of −47 + 38 μm, a $MnO_2/CuFeS_2$ ratio of 5/1, and 1 mol/L of H_2SO_4.

Roldán-Contreras et al. [14] contributed an article entitled "Leaching of Silver and Gold Contained in a Sedimentary Ore, Using Sodium Thiosulfate; A Preliminary Kinetic Study". Some sedimentary minerals have attractive contents of gold and silver, similar to the sedimentary exhalative ore available in the eastern Hidalgo in Mexico. The gold and silver contained represent an interesting opportunity for processing via non-toxic and aggressive leaching reagents such as thiosulfate. The preliminary kinetic study indicated that the leaching process was poorly affected by temperature and thiosulfate concentration. The reaction order was −0.61 for Ag, considering a thiosulfate concentration between 200–500 $mol \cdot m^{-3}$, while for Au, it was −0.09 for a concentration range between 32–320 $mol \cdot m^{-3}$. By varying the pH between 7–10, it was found that the reaction order was $n = 5.03$ for Ag, while for Au, the value was $n = 0.94$ when considering pH 9.5–11. The activation energy obtained during the silver leaching process was 3.15 $kJ \cdot mol^{-1}$ (298–328 K), which was indicative of the diffusive control of the process. On the other hand, during gold leaching, the obtained activation energy was 36.44 $kJ \cdot mol^{-1}$, which was indicative that this process was a mixed controlled process, first at low temperatures by diffusive control (298–313 K), and then by chemical control (318–323 K).

Rodríguez et al. [15] offered their research paper entitled "Leaching Chalcopyrite with an Imidazolium-Based Ionic Liquid and Bromide". The unique properties of ionic liquids (ILs) drive the growing number of novel applications in different industries. The main features of ILs are their high thermal stability, recyclability, low flash point, and low vapor pressure. This study investigated pure chalcopyrite dissolution in the presence of the ionic liquid 1-butyl-3-methylimidazolium hydrogen sulfate, $[BMIm]HSO_4$, and a bromide-like complexing agent. The proposed system was compared to acid leaching in sulfate media with the addition of chloride and bromide ions. The results demonstrated that the use of ionic liquid and bromide ions improved chalcopyrite leaching performance. The best operational conditions were at a temperature of 90 °C, with an ionic liquid concentration of 20% and 100 g/L of bromide.

Castellón et al. [16] contributed an article entitled "Depression of Pyrite in Seawater Flotation by Guar Gum". The application of guar gum for pyrite depression in seawater flotation was assessed through microflotation tests, focused beam reflectance measurements (FBRM), and particle vision measurements (PVM). Potassium amyl xanthate (PAX) and methyl isobutyl carbinol (MIBC) were used as a collector and frother, respectively. Chemical species on the pyrite surface were characterized by Fourier-transform infrared (FTIR) spectroscopy. The microflotation tests were performed at pH 8, which is the pH at the copper sulfide processing plants that operate with seawater. Pyrite flotation recovery was correlated with FBRM and PVM characterization to delineate the pyrite depression mechanisms by the guar gum. The high flotation recovery of pyrite with PAX was significantly lowered by guar gum, indicating that this polysaccharide could be used as an effective depressant in flotation with sea water. FTIR analysis showed that PAX and guar gum co-adsorbed on the pyrite surface, but the highly hydrophilic nature of the guar gum embedded the hydrophobicity due to the PAX. FBRM and PVM revealed that the guar gum promoted the formation of flocs whose size depended on the addition of guar gum and PAX. It is proposed that the highest pyrite depression occurred not only because of the hydrophilicity induced by the guar gum, but also due to the formation of large flocs, which could not be transported by the bubbles to the froth phase. Furthermore, it is shown that an overdose of guar gum hinders the depression effect due to the redispersion of the flocs.

Quezada et al. [17] submitted a work entitled "Describing Mining Tailing Flocculation in Seawater by Population Balance Models: Effect of Mixing Intensity". A population balance model (PBM) was used to describe the flocculation of particle tailings in seawater at pH 8 for a range of mixing intensities. The size of the aggregates is represented by the mean chord length, determined by the focused beam reflectance measurement (FBRM) technique. The PBM follows the dynamics of aggregation and breakage processes underlying flocculation and provides a good approximation to the temporal evolution of aggregate size. The structure of the aggregates during flocculation is described by a constant or time-dependent fractal dimension. The results revealed that the compensation between the aggregation and breakage rates leads to a correct representation of the flocculation kinetics of the tailings of particles in seawater and, in addition, that the representation of the flocculation kinetics in optimal conditions is equally good with a constant or variable fractal dimension. The aggregation and breakage functions and their corresponding parameters are sensitive to the choice of the fractal dimension of the aggregates, whether they are constant or time-dependent; however, under optimal conditions, a constant fractal dimension is sufficient. The model is robust and predictive with a few parameters and can be used to find the optimal flocculation conditions at different mixing intensities, and the optimal flocculation time can be used for a cost-effective evaluation of the quality of the flocculant used.

Matsumoto et al. [18] reported their research entitled "Selective and Mutual Separation of Palladium (II), Platinum (IV), and Rhodium (III) Using Aliphatic Primary Amines". The selective recovery of platinum-group metals (PGMs) remains a huge challenge. Although solvent extraction processes are generally used for PGM separation, the use of organic solvents is problematic because of their toxicity and environmental concerns. Here, we have developed a new PGM recovery method by means of precipitation from hydrochloric acid (HCl) solutions containing Pd(II), Pt(IV), and Rh(III), using aliphatic primary amines as precipitants. Pt(IV) was precipitated using the amines with alkyl chains longer than the hexyl, independent of HCl concentration. The precipitation of Pd(II) required longer alkyl amines than octyl, regardless of the HCl concentration. Rh(III) was recovered by precipitation at high HCl concentrations using the amines longer than hexyl. The mutual separation of Pt(IV), Rh(III), and Pd(II), in this order, was successfully achieved by changing the HCl concentrations and alkyl chain lengths of the amines. X-ray photoelectron spectroscopy and thermogravimetric analysis evidently showed that the metal-containing precipitates were ion-pair complexes composed of metal chloro-complex anions and ammonium cations.

Quezada et al. [19] submitted their work entitled "Reducing the Magnesium Content from Seawater to Improve Tailing Flocculation: Description by Population Balance Models". Experimental assays and mathematical models, through population balance models (PBM), were used to characterize the particle aggregation of mining tailings flocculated in seawater. Three systems were considered for the preparation of the slurries: (i) seawater at natural pH (pH 7.4), (ii) seawater at pH 11, and (iii) treated seawater at pH 11. The treated seawater had reduced magnesium content in order to avoid the formation of solid complexes, which damage the concentration operations. For this, the pH of seawater was increased with lime before being used in the process, generating solid precipitates of magnesium that were removed by means of vacuum filtration. The mean size of the aggregates was represented by the mean chord length obtained with the focused beam reflectance measurement (FBRM) technique, and their descriptions, obtained by the PBM, showed that an aggregation and a breakage kernel had evolved. The fractal dimension and permeability were included in the model in order to improve the representation of the irregular structure of the aggregates. Then, five parameters were optimized: three for the aggregation kernel and two for the breakage kernel. The results show that raising the pH from 8 to 11 was severely detrimental to the flocculation performance. Nevertheless, for pH 11, the aggregates slightly exceeded 100 μm, causing undesirable behaviour during the thickening operations. Interestingly, magnesium removal provided a suitable environment to perform the tailing flocculation at alkaline pH, creating aggregates with sizes that exceeded 300 μm. Only the fractal dimension changed between pH 8 and treated seawater at pH 11—as reflected in the permeability outcomes. The PBM fit well with the experimental data, and the parameters showed that the aggregation kernel was dominant at all-polymer dosages. The descriptive capacity of the model might have been utilized as a support in practical decisions regarding the best-operating requirements in the flocculation of copper tailings and water clarification.

Cui et al. [20] submitted a research effort entitled "Hydrometallurgical Treatment of Waste Printed Circuit Boards: Bromine Leaching". This paper demonstrates the recovery of valuable metals from shredded waste-printed circuit boards (WPCBs) by bromine leaching. The effects of the sodium bromide concentration, bromine concentration, leaching time, and inorganic acids were investigated. The most critical factors were sodium concentration and bromine concentration. It was found that more than 95% of copper, silver, lead, gold, and nickel could be dissolved simultaneously under the optimal conditions: a 50 g/L solid/liquid ratio, 1.17 M NaBr, 0.77 M Br_2, 2 M HCl, a 400 RPM agitation speed, and 23.5 °C for 10 h. The study shows that the dissolution of gold from waste-printed circuit boards in a Br_2-NaBr system is controlled by film diffusion and chemical reaction.

Prasetyo et al. [21] authored an innovative article entitled "Platinum Group Elements Recovery from Used Catalytic Converters by Acidic Fusion and Leaching". The recovery of platinum group elements (PGE (platinum group element coating); Pd, Pt, and Rh) from used catalytic converters using low energy and fewer chemicals was developed using a potassium bisulfate fusion pretreatment and were subsequently leached using hydrochloric acid. In the fusion pre-treatment, potassium bisulfate alone (without the addition of an oxidant) proved to be an effective and selective fusing agent. It altered PGE into a more soluble species and did not react with the cordierite support, based on X-ray diffraction (XRD) and metallographic characterization results. The fusion efficacy was due to the transformation of bisulfate into pyrosulfate, which is capable of oxidizing PGE. However, the introduction of potassium through the fusing agent proved to be detrimental in general since the potassium formed insoluble potassium PGE chloro-complexes during leaching (decreasing the recovery) and required a higher HCl concentration and a higher leaching temperature to restore solubility. Optimization of the fusion and leaching parameters resulted in 106% ± 1.7%, 93.3% ± 0.6%, and 94.3% ± 3.9% recovery for Pd, Pt, and Rh, respectively. These results were achieved at the following fusion conditions: temperature 550 °C, potassium bisulfate/raw material mass ratio 2.5, and fusion time within 30 min. The leaching conditions were a HCl concentration 5 M, a temperature of 80 °C, and a time within 20 min.

Torres et al. [10] contributed a revised article entitled "Leaching Manganese Nodules in an Acid Medium and Room Temperature Comparing the Use of Different Fe Reducing Agents". The deposits of Fe-Mn in the seabed of the planet are a good alternative source for the extraction of elements of interest. Among these are marine nodules, which have approximately 24% manganese and may be a solution to the shortage of high-grade ores on the surface. In this investigation, an ANOVA analysis was performed to evaluate the time-independent variables and the MnO_2/reducing agent in the leaching of manganese nodules with the use of different Fe reducing agents (FeS_2, Fe^{2+}, Fe^0 and Fe_2O_3). Tests were also conducted for the different reducing agents evaluating the MnO_2/Fe ratio, in which the Fe^0 (FeC) proved to be the best reducing agent for the dissolution of Mn from marine nodules, achieving solutions of 97% in 20 min. In addition, it was discovered that at low MnO_2/Fe ratios, the acid concentration in the system is not very relevant, and the potential and pH were in ranges of −0.4–1.4 V and −2–0.1, favoring the dissolution of Mn from MnO_2.

Castellón et al. [22] authored a submission entitled "An Alternative Process for Leaching Chalcopyrite Concentrate in Nitrate-Acid-Seawater Media with Oxidant Recovery". An alternative copper concentrate leaching process using sodium nitrate and sulfuric acid diluted in seawater followed by gas scrubbing to recover the sodium nitrate was evaluated. The work involved a leaching test conducted under various conditions with varying temperatures, leaching times, particle sizes, and concentrations of $NaNO_3$ and H_2SO_4. Regarding the amount of copper extracted from the chalcopyrite concentrate leached with seawater, 0.5 M of H_2SO_4 and 0.5 M of $NaNO_3$, copper recovery increased from 78% at room temperature to 91% at 45 °C after 96 h and 46 h of leaching, respectively. Gas scrubbing with the alkaline solution of NaOH was explored to recover part of the sodium nitrate. The dissolved salts were recovered by evaporation as sodium nitrate and sodium nitrite crystals.

Cho et al. [23] offered a novel concept paper entitled "Recovery of Gold from the Refractory Gold Concentrate Using Microwave Assisted Leaching". Microwave technology has been confirmed to be suitable for use in a wide range of mineral leaching processes. Compared to conventional leaching, microwave-assisted leaching has significant advantages. It is a proven process because of its short processing time and reduced energy. The purpose of this study was to enhance the gold content in a refractory gold concentrate using microwave-assisted leaching. The leaching efficiencies of metal ions (As, Cu, Zn, Fe, and Pb) and the recovery of gold from refractory gold concentrate were investigated via nitric acid leaching followed by microwave treatment. As the acid concentration increased, the metal ion leaching increased. In the refractory gold concentrate leaching experiments, nitric acid leaching at high temperatures could limit the decomposition of sulfide minerals because of the passive layer in the refractory gold concentrate. Microwave-assisted leaching experiments for gold recovery were conducted for the refractory gold concentrate. More extreme reaction conditions (nitric acid concentration > 1.0 M) facilitated the decomposition of passivation species derived from metal ion dissolution and the liberation of gangue minerals on the sulfide surface. The recovery rate of gold in the leach residue was improved with microwave-assisted leaching, with a gold recovery of ~132.55 g/t after 20 min of the leaching experiment (2.0 M nitric acid), according to fire assays.

Valeev et al. [24] produced a research article entitled "Acid and Acid-Alkali Treatment Methods of Al-Chloride Solution Obtained by the Leaching of Coal Fly Ash to Produce Sandy Grade Alumina". Sandy grade alumina is a valuable intermediate material that is mainly produced by the Bayer process and is used for manufacturing primary metallic aluminum. Coal fly ash is generated in coal-fired power plants as a by-product of coal combustion that consists of submicron ash particles and is considered to be a potentially hazardous technogenic waste. The present paper demonstrates that the Al-chloride solution obtained by leaching coal fly ash can be further processed to obtain sandy grade alumina, which is essentially suitable for metallic aluminum production. The novel process developed in the present study involves the production of amorphous alumina via the calcination of aluminium chloride hexahydrate obtained by salting-out from acid Al-Cl liquor. Following this, alkaline treatment with further Al_2O_3 dissolution and recrystallization as $Al(OH)_3$ particles is applied, and a final calcination step is employed to obtain sandy grade alumina with minimum impurities. The process does not require high-pressure equipment and reutilizes the alkaline liquor and gibbsite particles from the Bayer process, which allows the sandy grade alumina production costs to be significantly reduced. The present article also discusses the main technological parameters of the acid treatment and the amounts of major impurities in the sandy grade alumina obtained by the different (acid and acid-alkali) methods.

Prasetyo et al. [25] provide novel research in "Monosodium Glutamate as Selective Lixiviant for Alkaline Leaching of Zinc and Copper from Electric Arc Furnace Dust". The efficacy of monosodium glutamate (MSG) as a lixiviant for the selective and sustainable leaching of zinc and copper from electric arc furnace dust was tested. Batch leaching studies and XRD, XRF, and SEM-EDS characterization confirmed the high leaching efficiency of zinc (reaching 99%) and copper (reaching 86%), leaving Fe, Al, Ca, and Mg behind in the leaching residue. The separation factor (concentration ratio in pregnant leach solution) between zinc vs. other elements and copper vs. other elements in the optimum conditions could reach 11,700 and 250 times, respectively. The optimum conditions for the leaching scheme were a pH of 9, a MSG concentration of 1 M, and a pulp density of 50 g/L. Kinetic studies (leaching time and temperature) revealed that the saturation value of the leaching efficiency was attained within 2 h for zinc and within 4 h for copper. Modeling of the kinetic experimental data indicated that the role of temperature on the leaching process was minor. The study also demonstrated the possibility of MSG recycling from pregnant leach solutions by precipitation as glutamic acid (>90% recovery).

Karimov et al. [26] furthered the understanding of ammoniacal hydrometallurgy with their work entitled "Effect of Preliminary Alkali Desilication on Ammonia Pressure Leaching of Low-Grade Copper–Silver Concentrate". Ammonia leaching is a promising method for processing low-grade copper ores, especially those containing large amounts of oxidized copper. In this paper, we study the effect of Si-containing minerals on the kinetics of Cu and Ag leaching from low-grade copper concentrates. The results of experiments on the pressure leaching of the initial copper concentrate in an ammonium/ammonium carbonate solution with oxygen as an oxidizing agent are in good agreement with the shrinking core model in the intra-diffusion mode: in this case, the activation energies were 53.50 kJ/mol for Cu and 90.35 kJ/mol for Ag. Energy-dispersive X-ray spectroscopy analysis (EDX) showed that reagent diffusion to Cu-bearing minerals can be limited by aluminosilicate minerals of the gangue. The recovery rate for copper and silver increases significantly after a preliminary alkaline desilication of the concentrate, and the new shrinking core model is the most adequate, showing that the process is limited by diffusion through the product layer and interfacial diffusion. The activation energy of the process increases to 86.76 kJ/mol for Cu and to 92.15 kJ/mol for Ag. Using the time-to-a-given-fraction method, it has been shown that a high activation energy is required in the later stages of the process, when the most resistant sulfide minerals of copper and silver apparently remain.

Aracena et al. [27] also contributed further insights into ammoniacal hydrometallurgy with their contribution "Mechanism and Kinetics of Malachite Dissolution in an NH_4OH System". Copper oxide minerals composed of carbonates consume high quantities of leaching reagent. The present research proposes an alternative procedure for malachite leaching ($Cu_2CO_3(OH)_2$) through the use of only one compound, ammonium hydroxide (NH_4OH). Preliminary studies were also conducted for the dissolution of malachite in an acid system. The variables evaluated were solution pH, stirring rate, temperature, NH_4OH concentration, particle size, solid/liquid ratio, and different ammonium reagents. The experiments were conducted in a stirred batch system with controlled temperatures and stirring rates. For the acid dissolution system, sulfuric acid consumption reached excessive values (986 kg H_2SO_4/ton of malachite), invalidating the dissolution in these common systems. On the other hand, for the ammoniacal system, there was no acid consumption, and the results show that copper recovery was very high, reaching values of 84.1% for a concentration of 0.2 mol/dm^3 of NH_4OH and an experiment time of 7200 s. The theoretical/thermodynamic calculations indicate that the solution pH was a significant factor in maintaining the copper soluble as $Cu(NH_3)_4^{2+}$. This was validated by the experimental results and solid analysis by X-ray diffraction (XRD) from which the reaction mechanisms were obtained. A heterogeneous kinetic model was obtained from the diffusion model in a porous layer for the particles that begin the reaction as nonporous but that become porous during the reaction as the original solid splits and cracks to form a highly porous structure. The reaction order for the NH_4OH concentration was 3.2 and was inversely proportional to the square of the initial radius of the particle. The activation energy was calculated at 36.1 kJ/mol in the temperature range of 278 to 313 K.

Zhang et al. [28] provide a review of rare earth recycling entitled "Hydrometallurgical Recovery of Rare Earth Elements from NdFeB Permanent Magnet Scrap: A Review". NdFeB permanent magnet scrap is regarded as an important secondary resource that contains rare earth elements (REEs) such as Nd, Pr, and Dy. Recovering these valuable REEs from the NdFeB permanent magnet scrap not only increases economic potential, but also helps to reduce problems related to disposal and the environment. Hydrometallurgical routes are considered to be the primary choice for recovering the REEs because of the applications of higher REE recovery to all types of magnet compositions. In this paper, the authors first reviewed the chemical and physical properties of NdFeB permanent magnet scrap and then conducted an in-depth discussion on a variety of hydrometallurgical processes for recovering REEs from the NdFeB permanent magnet scrap. The methods mainly included selective leaching or complete leaching processes followed by precipitation, solvent extraction, or ionic liquid extraction processes. Particular attention is devoted to the specific technical challenge that emerges in the hydrometallurgical recovery of REEs from NdFeB permanent magnet scrap and to the corresponding potential measures for improving REE recovery by promoting processing efficiency. This summarized review will be useful for researchers who are developing processes for recovering REEs from NdFeB permanent magnet scrap.

Cho et al. [29] furthered our understanding of lithium refining by authoring "Application of Multistage Concentration (MSC) Electrodialysis to Concentrate Lithium from Lithium-Containing Waste Solution". In order to manufacture lithium carbonate to be used as a raw material for a secondary lithium battery, lithium sulfate solution is used as a precursor, and the concentration of lithium is required to be 10 g/L or more. Electrodialysis (ED) was used as a method to concentrate lithium in a low-concentration lithium sulfate solution, and multistage concentration (MSC) electrodialysis was used to increase the concentration ratio (%). When MSC was performed using a raw material solution containing a large amount of sodium sulfate, the process lead time was increased by 60 min. Furthermore, the concentration ratio (%) of lithium decreased as the number of concentration stages increased. In order to remove sodium sulfate, methanol was added to the raw material solution to precipitate sodium sulfate, and when it was added in a volume ratio of 0.4, lithium was not lost. Using a solution in which sodium sulfate was partially removed, fourth-stage concentration ED was performed to obtain a lithium sulfate solution with a lithium concentration of 10 g/L.

Medina and Anderson [30] provided an insightful document, "Review of the Cyanidation Treatment of Copper-Gold Ores and Concentrates". Globally, copper, silver, and gold orebody grades have been dropping, and the mineralogy surrounding them has become more diversified and complex. The cyanidation process for gold production has remained dominant for over 130 years because of its selectivity and feasibility in the mining industry. For this reason, the industry has been adjusting its methods for the extraction of gold by utilizing more efficient processes and technologies. Often, gold may be found in conjunction with copper and silver in ores and concentrates. Hence, the application of cyanide to these types of ores can present some difficulty, as the diversity of the minerals found within these ores can cause the application of cyanidation to become more complicated. This paper outlines the practices, processes, and reagents proposed for the effective treatment of these ores. The primary purpose of this review paper is to present the hydrometallurgical processes that currently exist in the mining industry for the treatment of silver, copper, and gold ores as well as for concentration treatments. In addition, this paper aims to present the most important challenges that the industry currently faces so that future processes that are both more efficient and feasible may be established.

3. Summary and Outlook

This Special Issue of *Metals* was well supported by diverse submissions and a final publication of thirty-one high-quality peer-reviewed articles. It is anticipated that due to this success, a new Special Issue ("Advances in Mineral Processing and Hydrometallurgy II", website: https://www.mdpi.com/journal/metals/special_issues/mineral_process_hydrometallurgy2) will be commissioned as a follow-up to accept global contributions from the hydrometallurgy and mineral processing community.

Conflicts of Interest: The authors declare no conflict of interest.

References

1. Wang, J.; Wang, Z.; Zhang, Z.; Zhang, G. Comparison of Butyric Acid Leaching Behaviors of Zinc from Three Basic Oxygen Steelmaking Filter Cakes. *Metals* **2019**, *9*, 417. [CrossRef]
2. Toro, N.; Briceño, W.; Pérez, K.; Cánovas, M.; Trigueros, E.; Sepúlveda, R.; Hernández, P. Leaching of Pure Chalcocite in a Chloride Media Using Sea Water and Waste Water. *Metals* **2019**, *9*, 780. [CrossRef]
3. Benavente, O.; Hernández, M.; Melo, E.; Núñez, D.; Quezada, V.; Zepeda, Y. Copper Dissolution from Black Copper Ore under Oxidizing and Reducing Conditions. *Metals* **2019**, *9*, 799. [CrossRef]
4. Coello-Velázquez, A.; Quijano Arteaga, V.; Menéndez-Aguado, J.; Pole, F.; Llorente, L. Use of the Swebrec Function to Model Particle Size Distribution in an Industrial-Scale Ni-Co Ore Grinding Circuit. *Metals* **2019**, *9*, 882. [CrossRef]
5. Soria-Aguilar, M.; Davila-Pulido, G.; Carrillo-Pedroza, F.; Gonzalez-Ibarra, A.; Picazo-Rodriguez, N.; Lopez-Saucedo, F.; Ramos-Cano, J. Oxidative Leaching of Zinc and Alkalis from Iron Blast Furnace Sludge. *Metals* **2019**, *9*, 1015. [CrossRef]
6. Pérez, K.; Toro, N.; Campos, E.; González, J.; Jeldres, R.; Nazer, A.; Rodriguez, M. Extraction of Mn from Black Copper Using Iron Oxides from Tailings and Fe^{2+} as Reducing Agents in Acid Medium. *Metals* **2019**, *9*, 1112. [CrossRef]
7. Jeldres, M.; Piceros, E.; Robles, P.; Toro, N.; Jeldres, R. Viscoelasticity of Quartz and Kaolin Slurries in Seawater: Importance of Magnesium Precipitates. *Metals* **2019**, *9*, 1120. [CrossRef]
8. Sokić, M.; Marković, B.; Stanković, S.; Kamberović, Ž.; Štrbac, N.; Manojlović, V.; Petronijević, N. Kinetics of Chalcopyrite Leaching by Hydrogen Peroxide in Sulfuric Acid. *Metals* **2019**, *9*, 1173. [CrossRef]
9. Jeldres, M.; Piceros, E.; Toro, N.; Torres, D.; Robles, P.; Leiva, W.; Jeldres, R. Copper Tailing Flocculation in Seawater: Relating the Yield Stress with Fractal Aggregates at Varied Mixing Conditions. *Metals* **2019**, *9*, 1295. [CrossRef]
10. Torres, D.; Ayala, L.; Saldaña, M.; Cánovas, M.; Jeldres, R.I.; Nieto, S.; Castillo, J.; Robles, P.; Toro, N. Leaching Manganese Nodules in an Acid Medium and Room Temperature Comparing the Use of Different Fe Reducing Agents. *Metals* **2019**, *9*, 1316; Correction in **2020**, *10*, 485. [CrossRef]
11. Karimov, K.; Rogozhnikov, D.; Kuzas, E.; Shoppert, A. Leaching Kinetics of Arsenic Sulfide-Containing Materials by Copper Sulfate Solution. *Metals* **2020**, *10*, 7. [CrossRef]
12. Salinas-Rodriguez, E.; Flores-Badillo, J.; Hernandez-Avila, J.; Cerecedo-Saenz, E.; Gutierrez-Amador, M.; Jeldres, R.; Toro, N. Assessment of Silica Recovery from Metallurgical Mining Waste, by Means of Column Flotation. *Metals* **2020**, *10*, 72. [CrossRef]
13. Torres, D.; Ayala, L.; Jeldres, R.; Cerecedo-Sáenz, E.; Salinas-Rodríguez, E.; Robles, P.; Toro, N. Leaching Chalcopyrite with High MnO_2 and Chloride Concentrations. *Metals* **2020**, *10*, 107. [CrossRef]
14. Roldán-Contreras, E.; Salinas-Rodríguez, E.; Hernández-Ávila, J.; Cerecedo-Sáenz, E.; Rodríguez-Lugo, V.; Jeldres, R.I.; Toro, N. Leaching of Silver and Gold Contained in a Sedimentary Ore, Using Sodium Thiosulfate; A Preliminary Kinetic Study. *Metals* **2020**, *10*, 159. [CrossRef]
15. Rodríguez, M.; Ayala, L.; Robles, P.; Sepúlveda, R.; Torres, D.; Carrillo-Pedroza, F.; Jeldres, R.; Toro, N. Leaching Chalcopyrite with an Imidazolium-Based Ionic Liquid and Bromide. *Metals* **2020**, *10*, 183. [CrossRef]
16. Castellón, C.; Piceros, E.; Toro, N.; Robles, P.; López-Valdivieso, A.; Jeldres, R. Depression of Pyrite in Seawater Flotation by Guar Gum. *Metals* **2020**, *10*, 239. [CrossRef]
17. Quezada, G.; Ayala, L.; Leiva, W.; Toro, N.; Toledo, P.; Robles, P.; Jeldres, R.I. Describing Mining Tailing Flocculation in Seawater by Population Balance Models: Effect of Mixing Intensity. *Metals* **2020**, *10*, 240. [CrossRef]

18. Matsumoto, K.; Sezaki, Y.; Yamakawa, S.; Hata, Y.; Jikei, M. Selective and Mutual Separation of Palladium (II), Platinum (IV), and Rhodium (III) Using Aliphatic Primary Amines. *Metals* **2020**, *10*, 324. [CrossRef]
19. Quezada, G.; Jeldres, M.; Toro, N.; Robles, P.; Jeldres, R. Reducing the Magnesium Content from Seawater to Improve Tailing Flocculation: Description by Population Balance Models. *Metals* **2020**, *10*, 329. [CrossRef]
20. Cui, H.; Anderson, C. Hydrometallurgical Treatment of Waste Printed Circuit Boards: Bromine Leaching. *Metals* **2020**, *10*, 462. [CrossRef]
21. Prasetyo, E.; Anderson, C. Platinum Group Elements Recovery from Used Catalytic Converters by Acidic Fusion and Leaching. *Metals* **2020**, *10*, 485. [CrossRef]
22. Castellón, C.; Hernández, P.; Velásquez-Yévenes, L.; Taboada, M. An Alternative Process for Leaching Chalcopyrite Concentrate in Nitrate-Acid-Seawater Media with Oxidant Recovery. *Metals* **2020**, *10*, 518. [CrossRef]
23. Cho, K.; Kim, H.; Myung, E.; Purev, O.; Choi, N.; Park, C. Recovery of Gold from the Refractory Gold Concentrate Using Microwave Assisted Leaching. *Metals* **2020**, *10*, 571. [CrossRef]
24. Valeev, D.; Shoppert, A.; Mikhailova, A.; Kondratiev, A. Acid and Acid-Alkali Treatment Methods of Al-Chloride Solution Obtained by the Leaching of Coal Fly Ash to Produce Sandy Grade Alumina. *Metals* **2020**, *10*, 585. [CrossRef]
25. Prasetyo, E.; Anderson, C.; Nurjaman, F.; Al Muttaqii, M.; Handoko, A.; Bahfie, F.; Mufakhir, F. Monosodium Glutamate as Selective Lixiviant for Alkaline Leaching of Zinc and Copper from Electric Arc Furnace Dust. *Metals* **2020**, *10*, 644. [CrossRef]
26. Kaŕimov, K.; Shoppert, A.; Rogozhnikov, D.; Kuzas, E.; Zakhar'yan, S.; Naboichenko, S. Effect of Preliminary Alkali Desilication on Ammonia Pressure Leaching of Low-Grade Copper–Silver Concentrate. *Metals* **2020**, *10*, 812. [CrossRef]
27. Aracena, A.; Pino, J.; Jerez, O. Mechanism and Kinetics of Malachite Dissolution in an NH_4OH System. *Metals* **2020**, *10*, 833. [CrossRef]
28. Zhang, Y.; Gu, F.; Su, Z.; Liu, S.; Anderson, C.; Jiang, T. Hydrometallurgical Recovery of Rare Earth Elements from NdFeB Permanent Magnet Scrap: A Review. *Metals* **2020**, *10*, 841. [CrossRef]
29. Cho, Y.; Kim, K.; Ahn, J.; Lee, J. Application of Multistage Concentration (MSC) Electrodialysis to Concentrate Lithium from Lithium-Containing Waste Solution. *Metals* **2020**, *10*, 851. [CrossRef]
30. Medina, D.; Anderson, C. A Review of the Cyanidation Treatment of Copper-Gold Ores and Concentrates. *Metals* **2020**, *10*, 897. [CrossRef]

Article

Comparison of Butyric Acid Leaching Behaviors of Zinc from Three Basic Oxygen Steelmaking Filter Cakes

Jingxiu Wang [1], Zhe Wang [2,*], Zhongzhi Zhang [3] and Guangqing Zhang [1]

1 School of Mechanical, Materials, Mechatronic and Biomedical Engineering, University of Wollongong, Wollongong, NSW 2522, Australia; jw071@uowmail.edu.au (J.W.); gzhang@uow.edu.au (G.Z.)
2 State Key Laboratory of Advanced Metallurgy, University of Science and Technology Beijing, Beijing 100083, China
3 State Key Laboratory of Heavy Oil Processing, Faculty of Chemical Engineering, China University of Petroleum, Beijing 102249, China; bjzzzhang@163.com
* Correspondence: zhewang@ustb.edu.cn; Tel.: +86-010-62334444

Received: 19 March 2019; Accepted: 5 April 2019; Published: 7 April 2019

Abstract: The selective leaching of zinc from three different basic oxygen steelmaking (BOS) filter cakes by butyric acid was investigated to compare the leaching behaviors of zinc and further to establish the correlation of the zinc leaching performances and the chemical compositions. The effects of acid concentration and the acid to solid (L/S) stoichiometric ratio were studied, with different optimal leaching conditions obtained. BOS-1 showed the lowest leachability with only less than 10% of zinc removed by 0.5 M acid concentration and 90% of the L/S stoichiometric ratio in 10 h. The best zinc selectivity was achieved with BOS-2 at 51.2% of zinc leaching efficiency, with only 0.47% of iron loss under optimal conditions of 1.5 M acid concentration and a 70% stoichiometric ratio. BOS-3 showed the highest leaching of zinc but the optimal conditions depend on the priority consideration. Using 1.0 M acid and 90% stoichiometric ratio for 10 h, the leaching efficiency of zinc was 84.6% with 20% iron loss. The filter cakes and the leaching residues were characterized. The results indicate different zinc and iron leaching behaviors, which were probably related to the storage conditions, zinc containing phases and the leaching parameters.

Keywords: BOS filter cakes; butyric acid; selective leaching; leaching behaviors; zinc; iron

1. Introduction

In the iron and steel industry, over 30 million tonnes of dusts are generated each year [1,2]. Basic oxygen furnace steelmaking (BOS) represents about 57% of the annual steel production worldwide and electric arc furnace (EAF) accounts for 27%, hence more dusts will be produced and this increasing trend is likely to continue [3]. These dusts are categorized as hazardous metallurgical waste in light of the heavy metals contained [4]. BOS plays a key role in steel production where hot metallic iron from blast furnace (BF) and scrap steel is converted to steel [3,5]. The BOS dust contains metal oxides, thus zinc mostly exists as ZnO and $ZnFe_2O_4$ while iron is mainly found as Fe, FeO, Fe_2O_3, Fe_3O_4 and $ZnFe_2O_4$. The zinc content ranges from 0.5% to 8% with 50–80% of zinc as ZnO and the rest as $ZnFe_2O_4$. The iron content in the dust varies between 50% and 75% [6–9]. Considering its high iron content, BOS dust is potentially a secondary iron resource for recovery. However, direct recycling is hindered by the problematic zinc and therefore BOS dust has to be stored or disposed of in landfills [10–13].

It has been demonstrated that the composition of the steelmaking dust varies widely depending on the quality of the scrap used, the type of steel produced, and the operating and the aging conditions [14,15]. EAF dust has a much higher content of non-ferrous metal oxides than BOS dust. The content of carbon in the BF dust is usually higher than that contained in the BOS dust while zinc content is correspondingly 2–6% lower. Also, BF dust is coarser than BOS dust due to the difference between the processes. It should be noted that the zinc content in EAF dust is the highest among these three steelmaking dusts.

The best way to solve this problem is to selectively remove zinc from such waste materials by using hydrometallurgy methods, which complies with the policy of sustainable development, helps to alleviate furnace damage, reduces the environmental load of stored waste and also reduces the costs of the replacement of raw materials. Considering the lack of selectivity owing to the large amount of iron dissolution produced using inorganic acids, organic acids are preferred due to having no secondary waste, being environmentally benign and showing indications of a potential bioleaching technology. After selective removal of zinc, the waste becomes an excellent iron-bearing material that can be used as a feed component to produce iron and steel [4,16]. Hence, it is feasible to selectively remove zinc from the BOS dusts.

It is presumed that zinc leaching from steelmaking dust is related to the fraction of Zn in the form of ZnO because $ZnFe_2O_4$ has a very stable spinel structure that is considerably refractory against leaching [17–20]. However, no specific relationship between the leaching efficiency of zinc and the ZnO fraction has been established. Steer and Griffiths investigated the leaching of zinc from BF slurry with 2.25 wt.% of Zn in the form of ZnO using 1 M prop-2-enoic acid, and 83.1% of zinc was removed [21]. Kelebek et al. found that the fine fraction of BOS sludge contained a greater proportion of $ZnFe_2O_4$ than the coarse fraction, and 81% of zinc removal was achieved for the coarse fraction by leaching with H_2SO_4 while only 29.2% removal was achieved for the fine fraction [7]. It was reported by Vereš et al. that BOS dust contained 9.37 wt.% zinc with 14.5% of the total zinc as $ZnFe_2O_4$ and the remaining portion as ZnO, and the zinc extraction was less than 50% even if the H_2SO_4 concentration was increased to 2.0 M at room temperature [22]. Using H_2SO_4 to leach zinc from BOS sludge with 2.74% of Zn, about 10% of zinc was removed at room temperature and low acid concentration (0.1 and 0.2 M) since only Zn in ZnO was leached out. However, Zn in $ZnFe_2O_4$ was further dissolved at higher temperatures and higher concentrations (1 M), giving increased zinc and iron leaching efficiencies over 50% [11]. So, the characteristics of the steelmaking dusts are closely related to the zinc leaching performance.

Butyric acid ($CH_3CH_2CH_2COOH$), as a 4-carbon short-chain fatty acid, is known to have many applications in the chemical, food, and pharmaceutical industries. It is partially dissociated weak acid, and the stabilization of $CH_3CH_2CH_2COO^-$ could potentially alter the extraction capability as follows.

$$CH_3CH_2CH_2COOH + H_2O = CH_3CH_2CH_2COO^- + H_3O^+ \quad (1)$$

In a previous study [23], butyric acid was shown to be highly efficient in selective leaching of zinc over iron from a BOS filter cake. However, each steelmaking dust is unique in terms of the chemical composition and herein it is necessary to study the leaching performances of different steelmaking wastes using butyric acid. This paper compares the leaching behaviors of zinc from three different BOS filter cakes using butyric acid to develop a better understanding of the relations to their chemical, physical, morphological and mineralogical characterization. This is vital to the design of alternative techniques for treating such BOS filter cakes. The optimal leaching conditions for maximum zinc removal and minimum iron removal from the three BOS filter cakes were determined by varying acid concentration and the acid to solid (L/S) stoichiometric ratio. The filter cakes and the corresponding leaching residues were characterized by X-ray fluorescence (XRF), X-ray diffraction (XRD), scanning electron microscopy with energy dispersive X-ray spectrometer (SEM/EDS), thermo-gravimetric/differential scanning calorimetry (TG/DSC).

2. Materials and Methods

2.1. Materials

The three different BOS filter cake wastes were collected from a steelwork in Australia, and the details are provided in Table 1. These samples were oven dried, crushed and sieved to 300–500 μm for leaching experiments.

Table 1. The information of the three different BOS filter cake samples.

Sample	Storage Time	Pretreatment	Color
BOS-1	Several years	Sintered in Stockpile	Red brown
BOS-2	Half a year	Heated at 130 °C for 40 days	Shiny black
BOS-3	Within one month	Freshly stored in fridge	Dull black

Butyric acid with 99% purity supplied by Sigma Aldrich, Australia, was used as a leaching reagent. Zinc and iron standard solutions were used for the calibration of inductively coupled plasma-optical emission spectrometry (ICP-OES 710, Agilent, Australia) analysis to determine the concentrations of zinc and iron. The deionized water used for dilution of different acids was purified using a water super-purification apparatus (Milli-Q, Millipore, North Ryde, Australia).

2.2. Leaching Experiments

The leaching experiments were performed in 250 mL conical flasks on a horizontally oscillating shaker (RM2, Ratek, Boronia, Australia) with 120 rpm to stir the samples for 10 h at ambient temperature. The effects of acid concentration and L/S stoichiometric ratio were investigated in the study. The acid concentrations tested are 0.5, 1.0, 1.5, and 2.0 M with the butyric acid solution fixed at 150 mL. At each acid concentration, the 10, 30, 50, 70 and 90% of L/S stoichiometric ratio [23] were investigated to express the solid weight added.

The sampling was taking under agitated conditions by removing 1.0 mL of leaching solution to be diluted to 10 mL using deionized water and filtered by 0.22-μm cellulose nitrate membrane using syringe. Further dilution was made by 2% nitric acid to appropriate Fe and Zn concentration ranges for the analysis by ICP-OES. The leaching efficiency was calculated from a mass balance as follows:

$$\text{Leaching efficiency } (\%) = \frac{\text{Mass of M in leachate}}{\text{Mass of M in filter cake added into flask}} \times 100 \quad (2)$$

where M is either Zn or Fe.

The pH of the leaching solutions was measured. After leaching, the final solid residues were washed with deionized water, filtered, dried and weighed for further characterization.

2.3. Characterization of the Filter Cakes and Leaching Residues

The chemical compositions of the BOS filter cakes were determined by X-ray fluorescence (XRF, AMETEK SPECTRO XEPOS ED-XRF, Kleve, Germany). To avoid loss of zinc when it was melted with fluxing material, all the samples were first heated slowly to 600 °C in an oxygen stream to remove their carbonaceous matter. The mineralogical compositions of the original filter cakes and leaching residues were analyzed by X-ray diffraction (XRD, GBC MMA, Braeside, Australia). The XRD patterns for quantitative analysis were obtained at 35 kV and 28.5 mA with monochromated Cu-Kα X-ray radiation (λ = 1.5406 Å) from 10° to 142° at a scanning speed of 0.5°/min with a step size of 0.014°. The patterns for qualitative analysis were from 15° to 85° at a scanning speed of 1.5°/min with a step size of 0.05°. The morphologies were studied by scanning electron microscopy with energy dispersive X-ray spectrometer (SEM/EDS, JSM-6490LV, JEOL, Tokyo, Japan). The thermo-gravimetric (TG) and differential scanning calorimetry (DSC) analysis of the thermal behavior of the filter cake samples was performed using the NETZSCH STA 449 F5 Jupiter (Selb, Germany) under air and argon atmospheres. Approximately 100 mg of a sample was loaded in an Al_2O_3 crucible on a pan of the microbalance, and scanned in the temperature range of 50–800 °C at a heating efficiency of 10 °C/min and a flow efficiency of 100 mL/min. An empty pan served as the reference.

3. Results and Discussion

3.1. Characterization of the BOS Filter Cakes

The chemical compositions of the three BOS filter cakes were listed in Table 2, and similar elements were observed. They mainly contained Fe and Zn together with small amounts of Ca, Mg and Si. Other metals exhibited very low concentrations below 1%. BOS-1 contained the lowest zinc level at 2.42 wt.%, followed by 6.53 wt.% in BOS-2, which is approximately half of that in BOS-3 at 13.77 wt.%. The corresponding iron content in the three filter cakes showed insignificant differences at 56.0 wt.%, 56.5 wt.% and 60.0 wt.%, respectively.

Table 2. The chemical compositions of the three BOS filter cakes (wt.%) based on the oxide content.

Sample	BOS-1	BOS-2	BOS-3
Fe	56.0	56.4	60.0
Zn	2.42	6.52	13.8
Ca	4.04	2.86	3.02
Mg	1.73	1.16	0.95
Si	0.72	0.65	1.01
Al	0.07	0.02	0.09
Mn	0.83	0.67	0.61
Ti	0.015	0.022	0.007
K	0.031	0.029	0.020
Pb	0.096	0.085	0.387
Cr	0.053	0.022	0.035
Ni	0.030	0.008	0.000
V	0.039	0.012	0.013
LOI [1]	4.73	1.89	3.56

[1] LOI-Loss on Ignition at 1050 °C.

Figure 1 presents the XRD patterns of the three BOS filter cakes. It can be found that zinc existed as ZnO and $ZnFe_2O_4$ in all the samples, but the contents of zinc in these two compounds varied widely. The quantitative analysis showed that the content of zinc in ZnO was significantly lower, equal to, while much higher than that in $ZnFe_2O_4$ for BOS-1, BOS-2, and BOS-3, respectively. For iron containing phases, FeO, Fe_3O_4 and Fe were detected in all these samples but Fe_2O_3 was only found in BOS-1. Fe_3O_4 and Fe_2O_3 were the predominant crystalline phases in BOS-1 while FeO and Fe were the major iron states in both BOS-2 and BOS-3, indicating that BOS-1 is more oxidized. Calcite was also confirmed in all these three samples.

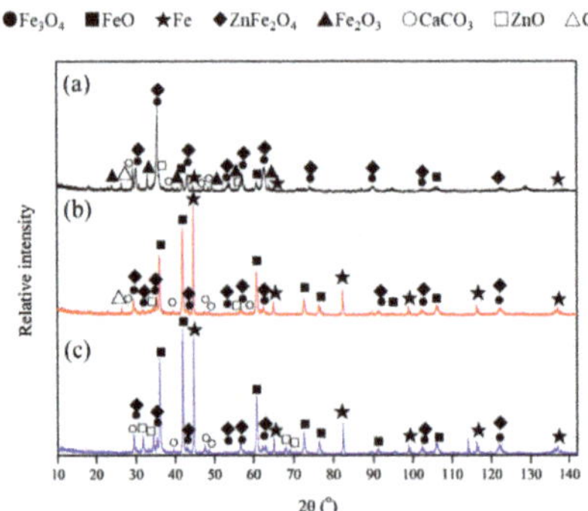

Figure 1. XRD patterns of the three BOS filter cakes: (**a**) BOS-1; (**b**) BOS-2; (**c**) BOS-3.

Figure 2 presents the morphology of the three BOS filter cakes. All the samples consist of fine grains with various shapes and sizes, but the grain size of BOS-1 is obviously larger than those of BOS-2 and BOS-3. BOS-1 contains non-spherical grains and clusters of submicron grains, but the majority of the grains are >1 μm. EDS analysis confirmed the presence of metal oxides that were mainly iron oxides, followed by the oxides of Zn, Al, Ca and Mg. For BOS-2 and BOS-3, the individual grains are predominantly spherical at submicron level, and large grains >1 μm are rare. The fine grains agglomerated together to form a porous structure of the filter cakes which favors the access of acid to zinc in leaching.

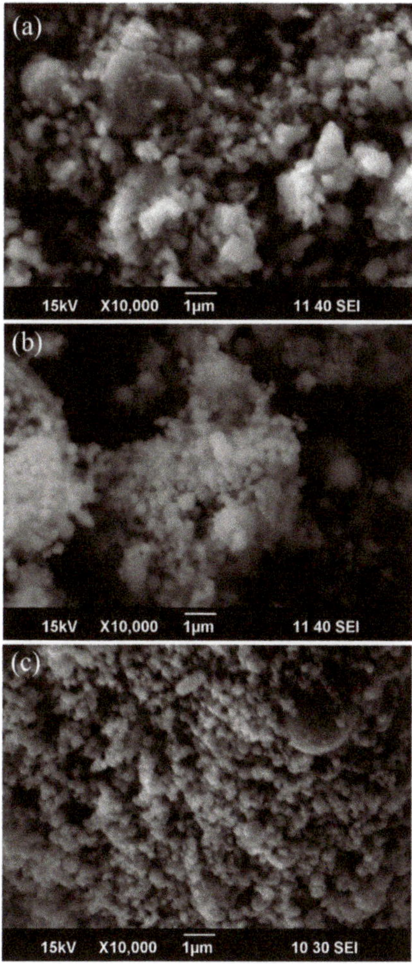

Figure 2. SEM micrographs of the three BOS filter cakes. (**a**) BOS-1; (**b**) BOS-2; (**c**) BOS-3.

Figure 3 presents the TG-DSC curves of the three BOS filter cakes, which were obtained in the temperature range of 50–800 °C both in air and in argon with a gas flow rate of 100 mL/min. Overall, the TG patterns of BOS-2 and BOS-3 show similar trends of weight increase in air atmosphere up to 500 °C, while mass loss in argon atmosphere. However, BOS-1 mainly was subjected to a weight loss with increasing temperature in both atmospheres. In air, the weight loss of BOS-1 was 5.15%. For filter cakes BOS-2 and BOS-3, there was 4.4% and 5.11% of total weight gain, respectively. The results indicated some oxidation and possible combustion reactions for both BOS-2 and BOS-3 with two strong exothermic peaks at about 340 °C and 430 °C. In argon atmosphere, the absence of free oxygen in the furnace led to no oxidation reactions in contrast to the analysis in air, and hence the mass loss became increasing monotonically as reflected by the TG curves. The weight loss recorded for BOS-1 was 5.96%, followed by 4.44% for BOS-2 and 3.36% for BOS-3.

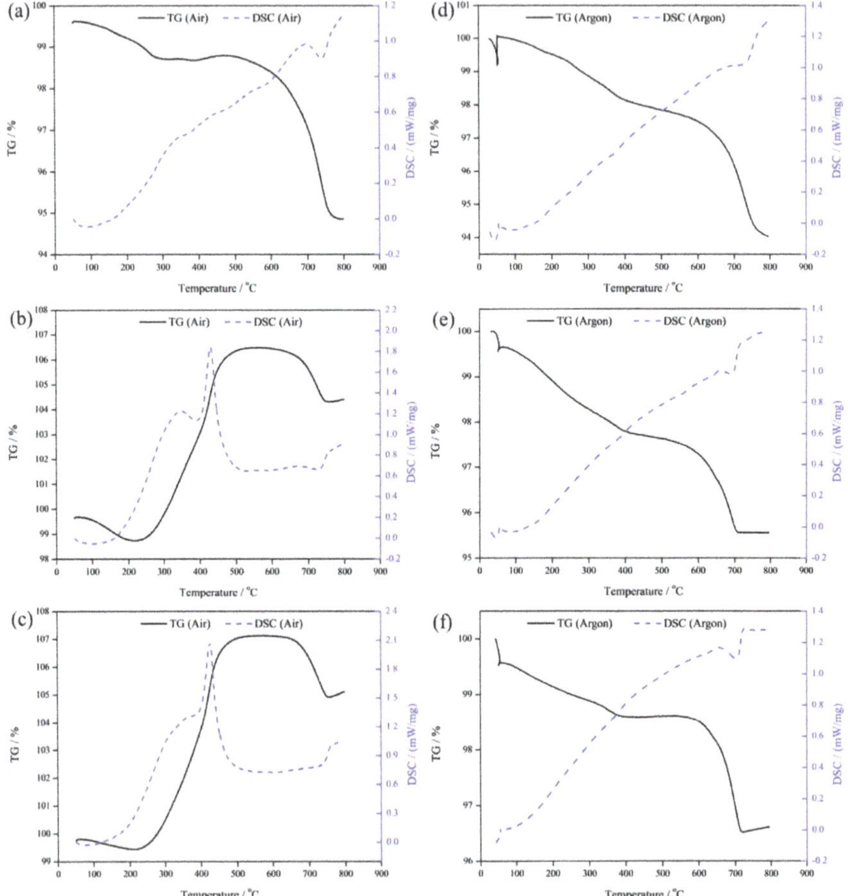

Figure 3. TG-DSC curves of the three BOS filter cakes in air and argon atmospheres: (**a**) BOS-1, in air; (**b**) BOS-2, in air; (**c**) BOS-3, in air; (**d**) BOS-1, in argon; (**e**) BOS-2, in argon; (**f**) BOS-3, in argon.

Figure 4 presents the XRD patterns of the filter cakes subjected to the TG-DSC analysis (reacted samples). The results confirmed that the weight changes are due to oxidation and reduction of the iron and its oxides in the BOS filter cakes. In air, all the three filter cakes are nearly fully oxidized during heating to 800 °C, since Fe_2O_3 and $ZnFe_2O_4$ (both containing Fe^{3+}) are the dominant phases in the reacted samples as seen in Figure 4a–c. However, the weight gain resulted from BOS-1 was less than the weight loss caused by the dehydration and dissociation of the calcium carbonate. This led to the total mass loss in air for BOS-1. On the other hand, BOS-2 and BOS-3 are converted mainly to FeO phase in addition to metallic Fe in the argon atmosphere (Figure 4e,f). The slight difference of mass loss for BOS-1 in air and argon can be attributed to the oxidation of low contents of Fe, FeO and Fe_3O_4 to Fe_2O_3 as shown in Figure 4a,b [24].

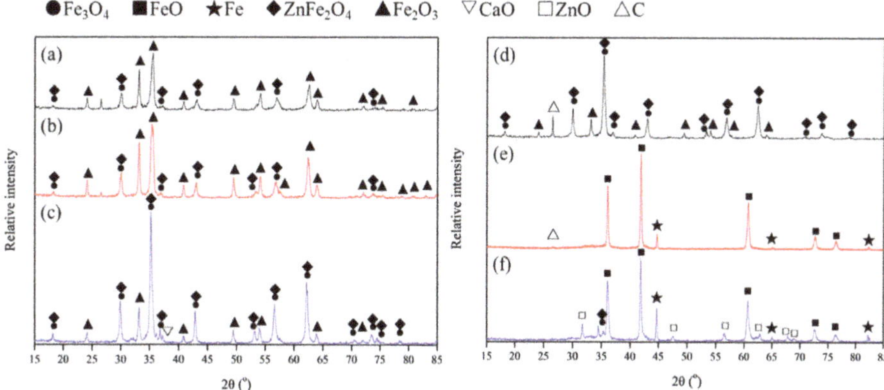

Figure 4. The XRD patterns of the BOS filter cakes subjected to TG-DSC analysis in air and argon atmospheres: (**a**) BOS-1, in air; (**b**) BOS-2, in air; (**c**) BOS-3, in air; (**d**) BOS-1, in argon; (**e**) BOS-2, in argon; (**f**) BOS-3, in argon.

3.2. The Leaching Performance of the Different BOS Filter Cakes

The three BOS filter cakes were leached by butyric acid under the same leaching conditions and the corresponding zinc and iron leaching efficiencies were obtained. Figure 5 compares the leaching efficiency of zinc at different L/S stoichiometric ratios and acid concentrations. BOS-3 filter cake shows the highest efficiency, followed by BOS-2, while less than 10% of zinc was dissolved from BOS-1 at all the above leaching conditions. BOS-3 shows a strong dependence on the L/S stoichiometric ratio, and increasing the ratio caused considerable increase in the leaching efficiency of zinc at all the acid concentrations. However, the acid concentration did not have a significant impact on the zinc leaching efficiency. For BOS-2, the increase in the L/S stoichiometric ratio from 10% to 50% increased zinc removal considerably, but further increases did not show an obvious increase in zinc removal. Like BOS-3, acid concentration also had a slight effect on the leaching efficiency. It should be noted that the maximum zinc leaching efficiency was limited at about 55%, and no larger increase can be achieved by increasing acid concentration and L/S stoichiometric ratios. BOS-1 showed no increasing trend at all the leaching conditions, indicating the poorest leaching performance of zinc in butyric acid solution.

Figure 6 shows the iron dissolution from the three BOS filter cakes in relation to L/S stoichiometric ratio at various acid concentrations. Like zinc leaching from BOS-1, iron dissolution was less than 1% at all the leaching conditions tested. The change pattern of the iron dissolution from BOS-2 was complex. At the low acid concentration of 0.5 M, the iron leaching efficiency increased significantly with an increase of the L/S stoichiometric ratio, and up to 30% of iron was dissolved at the ratio of 90%. Increasing the acid concentration to 1.0 M reduced the iron dissolution dramatically to less than 5%. A further increase in the acid concentration to 1.5 and 2.0 M caused the iron dissolution even below 1%, which seemed to be independent of the L/S stoichiometric ratio. It is inferred that acid concentration played a crucial role in controlling the iron dissolution from BOS-2. BOS-3 shows the same leaching behavior of iron with that of zinc, revealing the remarkable effect of the L/S stoichiometric ratio and the small impact of the acid concentration.

It can be concluded that the leaching of BOS-3 filter cake reached the best outcome with butyric acid concentration at 1.0 M. Under this concentration, the optimal leaching conditions can be determined for different objectives to achieve as indicated by Supplementary Figure S1. To achieve the maximum zinc removal, a leaching for 10 h with L/S stoichiometric ratio at 90% can be adopted which can reach 85% zinc removal and 20% iron loss. However, a better selectivity of zinc can be attained by decreasing the ratio to 70% or reducing the leaching time to 2 h.

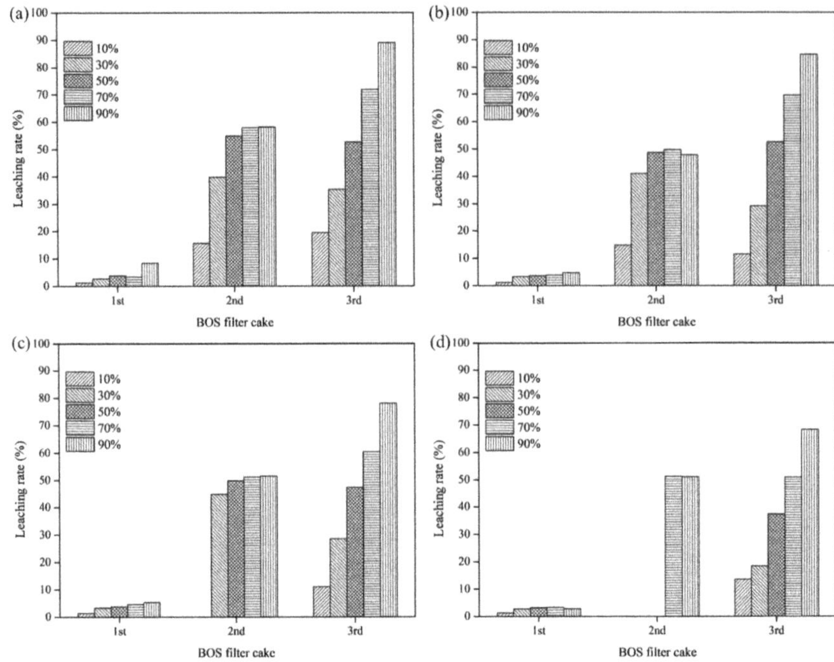

Figure 5. The leaching efficiency of Zn from three different BOS filter cakes at L/S stoichiometric ratios 10% to 90% and acid concentrations (**a**) 0.5 M, (**b**) 1.0 M, (**c**) 1.5 M and (**d**) 2.0 M.

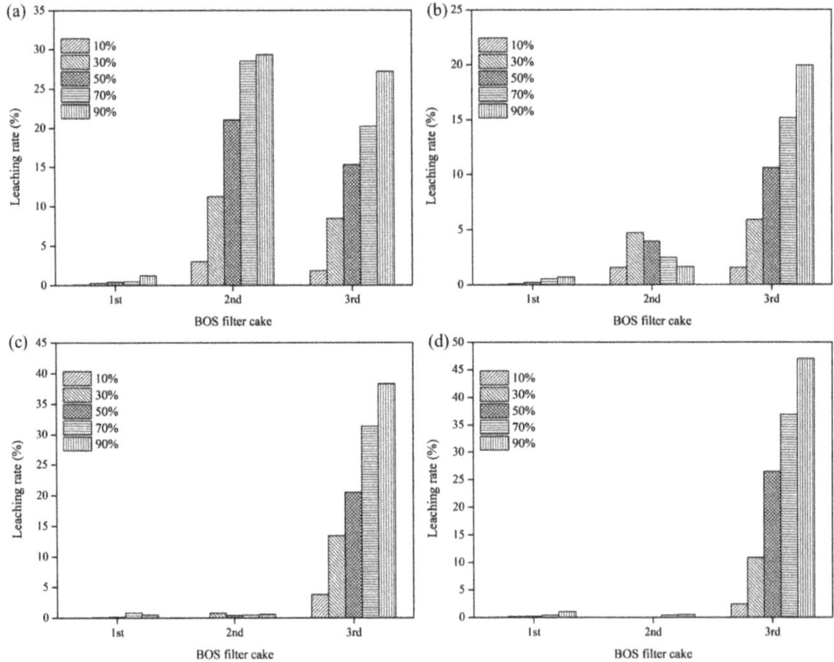

Figure 6. The leaching efficiency of Fe from three different BOS filter cakes at L/S stoichiometric ratio 10% to 90% and acid concentration (**a**) 0.5 M, (**b**) 1.0 M, (**c**) 1.5 M and (**d**) 2.0 M.

Overall, the three BOS filter cakes showed various leaching behaviors of zinc and iron. BOS-1 showed a poor leachability of zinc and iron in butyric acid solutions, which is likely related to the self-sinter behavior resulting from long term stockpile of the waste, and some exothermic oxidation-sintering reactions probably occurred [25]. The obtained sample generally has stronger physical properties with a harder surface, higher density and strength, and a larger grain size in comparison with the fresh filter cake, which can make the acid leaching very difficult. Another reason is the higher content of zinc in $ZnFe_2O_4$ than that in ZnO as demonstrated by quantitative XRD analysis, and $ZnFe_2O_4$ was very stable and insoluble in traditional acid leaching process [17–20]. The best leaching selectivity of zinc over iron was achieved from BOS-2, which depends on the increasing acid concentration to form the hydrophobic butyric film adsorbed on Fe-containing phases hindering the leaching of Fe. It was reported that zinc exhibits amphoteric character and is soluble in a wide range of pH from 0 to 6, while iron dissolution mainly dependent on pH (see Supplementary Table S1). By increasing the acid concentration to form a hydrophobic protective film for the adsorption of butyric acid by iron ions rather than zinc ions, iron dissolution can be effectively limited. However, the maximum zinc leaching can only be reached without the unleachable zinc contained in $ZnFe_2O_4$. In comparison, BOS-3 tended to be most easily dissolved in terms of zinc and iron. This could be ascribed to the adequate storage of this fresh BOS filter cake to avoid the exothermic oxidation reaction in stockpiles causing a higher stability of the zinc phases contained in the leaching material, and the binding force was not strong enough to resist acid leaching. To conclude, the optimum leaching conditions for the selectively leaching of zinc over iron from the three BOS filter cakes, the resulted leaching efficiencies of zinc and iron, and the selectivity as the Zn/Fe ratio are summarized in Table 3 (BOS-2 Data from [23]).

Table 3. The optimal conditions and the corresponding leaching efficiencies of the three BOS filter cakes.

Sample	Zn Removal (%)	Fe Removal (%)	Zn/Fe Ratio (*w*/*w*)	Optimal Parameters	
				Acid Concentration (M)	L/S Stoichiometric Ratio (%)
BOS-1	8.4	1.2	0.8	0.5	90
BOS-2	51.2	0.47	12.6	1.5	70
BOS-3	84.6	20.0	0.5	1.0	90

3.3. *Characterization of the Leached Residues*

Figure 7 presents the XRD patterns of the leaching residues obtained under the optimal leaching conditions. No obvious change of the phases were observed for the residue of BOS-1 except for the disappearance of ZnO and some peaks of Fe, and this matched well with the corresponding lower zinc and iron leaching efficiencies. The same phases of Fe, FeO and Fe_3O_4 were still detected from the residue of BOS-2 due to the 0.47% iron dissolution. Apparently, the residue of BOS-3 showed remarkable changes of phases and the relative contents. For iron-containing phases, Fe was mostly dissolved while no obvious dissolution of FeO and Fe_3O_4 was observed. The disappearance of ZnO peaks from all the filter cakes implied that it was completely dissolved, or the remaining amount was below the limit of the detection. As expected, there was insignificant variation of $ZnFe_2O_4$ peaks.

Figure 8 presents the morphology of the corresponding leaching residues and there were some changes in contrast with the original samples. The shape and grain size of the BOS-1 residue are almost the same as before leaching. The EDS analysis also does not show much difference. This corresponds to the low leaching efficiencies of zinc and iron by butyric acid. For the residue of BOS-2, although the morphology generally unchanged after leaching, the overall zinc content appears to be decreased based on the EDS results. Considering that up to 50% of zinc was removed by leaching with very low loss of iron, it can be recognized that iron oxides forming the basic structure of the filter cake were not affected by the removal of zinc. For the residue of BOS-3, it seems that the fine particles were leached, leaving the residue with relatively coarse grains (comparable to BOS-1) and higher porosity, which is consistent with the high zinc and iron leaching efficiencies.

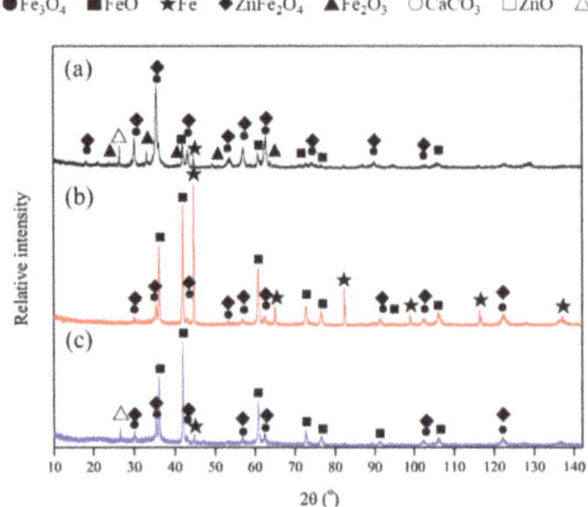

Figure 7. The XRD patterns of the corresponding leaching residues. (**a**) Residue of BOS-1; (**b**) Residue of BOS-2; (**c**) Residue of BOS-3.

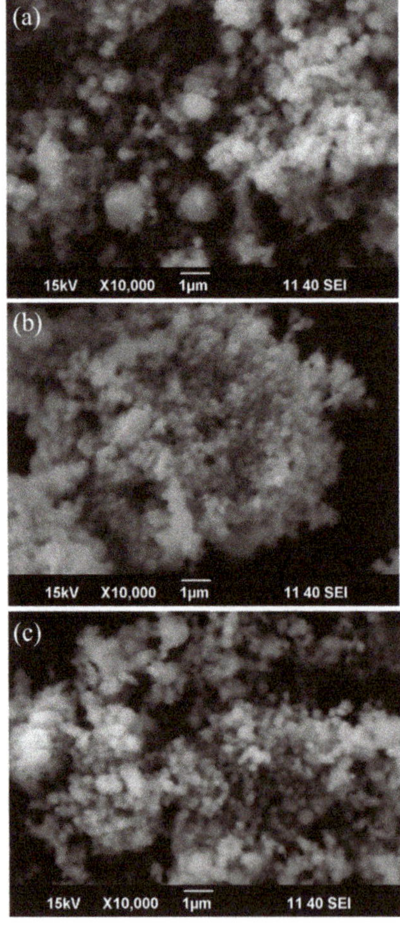

Figure 8. SEM micrographs of the corresponding leaching residues. (**a**) Residue of BOS-1; (**b**) Residue of BOS-2; (**c**) Residue of BOS-3.

4. Conclusions

Selective leaching of zinc over iron from three BOS filter cakes by butyric acid was investigated. The effects of acid concentration and L/S stoichiometric ratio were systematically examined. The following conclusions are obtained:

1. Among the three BOS filter cakes, BOS-3 showed the highest removal of zinc by butyric acid, while the best selectivity of zinc over iron was achieved with BOS-2. The BOS-1 filter cake showed the lowest leaching performance of zinc and iron.

2. The optimal leaching conditions for BOS-3 can be selected depending on the priority consideration. Considering both zinc removal and zinc selectivity, 90% L/S stoichiometric ratio and 1.0 M acid concentration for 10 h were chosen as the optimal conditions with 84.6% zinc removal and 20.0% iron loss.

3. BOS-1 probably have the self-sinter behavior resulting from long term stockpile with some exothermic oxidation reactions occurred, and those materials generally have stronger physical properties with harder surface, higher density and strength, and larger grain size which can make the acid leaching very difficult. The zinc leaching from BOS-2 was limited by the franklinite but can reach the maximum by increasing the acid concentration.

4. For the leaching behaviors of zinc and iron from the BOS filter cakes using butyric acid, mineral compositions played a more important role than the leaching parameters.

Supplementary Materials: The following are available online at http://www.mdpi.com/2075-4701/9/4/417/s1, Figure S1: Progress of leaching of (**a**) Zn and (**b**) Fe from BOS-3 using 1.0 M butyric acid with different L/S stoichiometric ratios, Table S1: The pH variations obtained from the leachates for the three BOS filter cakes under different leaching conditions.

Author Contributions: Conceptualization, investigation, writing the manuscript—J.W.; Data analysis, reviewing and editing—Z.W. and G.Z.; Supervision, project administration—G.Z. and Z.Z. All authors have discussed the results, read and approved the final manuscript.

Funding: This research received no external funding.

Acknowledgments: The first author is supported by a University of Wollongong-China Scholarship Council joint scholarships. The authors acknowledge the use of facilities within the UOW Electron Microscopy Centre. Paul Carr and José Abrantes assisted with sample preparation and XRF analysis. Linda Tie assisted in ICP-OES analysis.

Conflicts of Interest: The authors declare no conflict of interest.

References

1. Yang, S.; Zhao, D.; Jie, Y.; Tang, C.; He, J.; Chen, Y. Hydrometallurgical process for zinc recovery from C.Z.O. Generated by the steelmaking industry with ammonia–ammonium chloride solution. *Metals* **2019**, *9*, 83. [CrossRef]
2. Vereš, J.; Jakabský, Š.; Lovás, M. Zinc recovery from iron and steel making wastes by conventional and microwave assisted leaching. *Acta Montanistica Slovaca* **2011**, *16*, 185.
3. Jaafar, I. Chlorination for the Removal of Zinc from Basic Oxygen Steelmaking (BOS) by-product. Ph.D. Thesis, Cardiff University, Cardiff, UK, 2014.
4. Vereš, J.; Jakabský, Š.; Lovás, M.; Hredzák, S. Non-isothermal microwave leaching kinetics of zinc removal from basic oxygen furnace dust. *Acta Montanistica Slovaca* **2010**, *15*, 204–211.
5. Ma, N. Recycling of basic oxygen furnace steelmaking dust by in-process separation of zinc from the dust. *J. Clean. Prod.* **2016**, *112*, 4497–4504. [CrossRef]
6. Cantarino, M.V.; de Carvalho Filho, C.; Mansur, M.B. Selective removal of zinc from basic oxygen furnace sludges. *Hydrometallurgy* **2012**, *111*, 124–128. [CrossRef]
7. Kelebek, S.; Yörük, S.; Davis, B. Characterization of basic oxygen furnace dust and zinc removal by acid leaching. *Miner. Eng.* **2004**, *17*, 285–291. [CrossRef]
8. Gargul, K.; Boryczko, B. Removal of zinc from dusts and sludges from basic oxygen furnaces in the process of ammoniacal leaching. *Arch. Civil Mech. Eng.* **2015**, *15*, 179–187. [CrossRef]

9. Stefanova, A.; Aromaa, J. Alkaline Leaching of Iron and Steelmaking Dust. Research Report, Aalto University publication series Science + Technology 1/2012. Available online: https://aaltodoc.aalto.fi/handle/123456789/3570 (accessed on 24 April 2018).
10. Wang, J.; Wang, Z.; Zhang, Z.; Zhang, G. Zinc removal from basic oxygen steelmaking filter cake by leaching with organic acids. *Metall. Mater. Trans. B* **2019**, *50*, 480–490. [CrossRef]
11. Trung, Z.H.; Kukurugya, F.; Takacova, Z.; Orac, D.; Laubertova, M.; Miskufova, A.; Havlik, T. Acidic leaching both of zinc and iron from basic oxygen furnace sludge. *J. Hazard. Mater.* **2011**, *192*, 1100–1107.
12. Zeydabadi, B.A.; Mowla, D.; Shariat, M.; Kalajahi, J.F. Zinc recovery from blast furnace flue dust. *Hydrometallurgy* **1997**, *47*, 113–125. [CrossRef]
13. Jaafar, I.; Griffiths, A.J.; Hopkins, A.; Steer, J.M.; Griffiths, M.H.; Sapsford, D.J. An evaluation of chlorination for the removal of zinc from steelmaking dusts. *Miner. Eng.* **2011**, *24*, 1028–1030. [CrossRef]
14. Sofilić, T.; Rastovčan-Mioč, A.; Cerjan-Stefanović, Š.; Novosel-Radović, V.; Jenko, M. Characterization of steel mill electric-arc furnace dust. *J. Hazard. Mater.* **2004**, *109*, 59–70. [CrossRef]
15. Bakkar, A. Recycling of electric arc furnace dust through dissolution in deep eutectic ionic liquids and electrowinning. *J. Hazard. Mater.* **2014**, *280*, 191–199. [CrossRef]
16. Gargul, K.; Jarosz, P.; Malecki, S. Alkaline leaching of low zinc content iron-bearing sludges. *Arch. Metall. Mater.* **2016**, *61*, 43–50. [CrossRef]
17. Aromaa, J.; Kekki, A.; Stefanova, A.; Makkonen, H.; Forsén, O. New hydrometallurgical approaches for stainless steel dust treatment. *Miner. Process. Extr. Metall.* **2016**, *125*, 242–252. [CrossRef]
18. Miki, T.; Chairaksa-Fujimoto, R.; Maruyama, K.; Nagasaka, T. Hydrometallurgical extraction of zinc from CaO treated EAF dust in ammonium chloride solution. *J. Hazard. Mater.* **2016**, *302*, 90–96. [CrossRef]
19. Palimąka, P.; Pietrzyk, S.; Stępień, M.; Ciećko, K.; Nejman, I. Zinc recovery from steelmaking dust by hydrometallurgical methods. *Metals* **2018**, *8*, 547. [CrossRef]
20. Langová, Š.; Leško, J.; Matýsek, D. Selective leaching of zinc from zinc ferrite with hydrochloric acid. *Hydrometallurgy* **2009**, *95*, 179–182. [CrossRef]
21. Steer, J.M.; Griffiths, A.J. Investigation of carboxylic acids and non-aqueous solvents for the selective leaching of zinc from blast furnace dust slurry. *Hydrometallurgy* **2013**, *140*, 34–41. [CrossRef]
22. Vereš, J.; Jakabský, Š.; Lovás, M. Comparison of conventional and microwave assisted leaching of zinc from the basic oxygen furnace dust. *Miner. Slovaca* **2010**, *42*, 369–374.
23. Wang, J.; Wang, Z.; Zhang, Z.; Zhang, G. Removal of zinc from basic oxygen steelmaking filter cake by selective leaching with butyric acid. *J. Clean. Prod.* **2019**, *209*, 1–9. [CrossRef]
24. Mikhail, S.A.; Turcotte, A.-M. Thermal reduction of steel-making secondary materials: I. Basic-oxygen-furnace dust. *Thermochim. Acta* **1998**, *311*, 113–119. [CrossRef]
25. Longbottom, R.J.; Monaghan, B.J.; Zhang, G.; Chew, S.J.; Pinson, D.J. Characterisation of steelplant by-products to realise the value of Fe and Zn. In Proceedings of the 7th European Coke and Ironmaking Congress (ASME 2016), Linz, Austria, 12–14 September 2016.

Article

Leaching of Pure Chalcocite in a Chloride Media Using Sea Water and Waste Water

Norman Toro [1,2,*], Williams Briceño [3], Kevin Pérez [1], Manuel Cánovas [1], Emilio Trigueros [2], Rossana Sepúlveda [4] and Pía Hernández [5]

1 Departamento de Ingeniería en Metalurgia y Minas, Universidad Católica del Norte, Av. Angamos 610, Antofagasta 1270709, Chile
2 Department of Mining, Geological and Cartographic Department, Universidad Politécnica de Cartagena, Paseo Alfonso XIII N°52, 30203 Cartagena, Spain
3 Departamento de Ingeniería Industrial, Universidad Católica del Norte, Av. Angamos 610, Antofagasta 1270709, Chile
4 Departamento de Ingeniería en Metalurgia, Universidad de Atacama, Av. Copayapu 485, Copiapó 1531772, Chile
5 Departamento de Ingeniería Química y Procesos de Minerales, Universidad de Antofagasta, Avda. Angamos 601, Antofagasta 1240000, Chile
* Correspondence: ntoro@ucn.cl; Tel.: +56-552651021

Received: 25 June 2019; Accepted: 11 July 2019; Published: 12 July 2019

Abstract: Chalcocite is the most important and abundant secondary copper ore in the world with a rapid dissolution of copper in an acid-chloride environment. In this investigation, the methodology of surface optimization will be applied to evaluate the effect of three independent variables (time, concentration of sulfuric acid and chloride concentration) in the leaching of pure chalcocite to extract the copper with the objective of obtaining a quadratic model that allows us to predict the extraction of copper. The kinetics of copper dissolution in regard to the function of temperature is also analyzed. An ANOVA indicates that the linear variables with the greatest influence are time and the chloride concentration. Also, the concentration of chloride-time exerts a significant synergic effect in the quadratic model. The ANOVA indicates that the quadratic model is representative and the R^2 value of 0.92 is valid. The highest copper extraction (67.75%) was obtained at 48 h leaching under conditions of 2 mol/L H_2SO_4 and 100 g/L chloride. The XRD analysis shows the formation of a stable and non-polluting residue; such as elemental sulfur (S^0). This residue was obtained in a leaching time of 4 h at room temperature under conditions of 0.5 mol/L H_2SO_4 and 50 g/L Cl^-.

Keywords: chalcocite; sulphide leaching; copper; reusing water; desalination residue; ecological treatment

1. Introduction

Most of the copper minerals on the planet correspond to sulfur minerals and a smaller amount of oxidized minerals. A report by COCHILCO [1] mentions that the world copper production is currently 19.7 million tons. Seventy-five percent of this total comes from the pyrometallurgical processing of copper sulfide minerals processed in smelting plants [2], and 25% by the hydrometallurgical route [3].

There is a need to generate a new momentum that overcomes a certain stagnation in the growth capacity of the mining industry. Even in its role as a surplus generator, large-scale mining faces great challenges. These include an increase in costs due to various factors; such as the deterioration of laws and other elements associated with the aging of deposits and input costs to be compatible with sustainable development demands [4].

Sulfur minerals have been treated for decades with flotation and pyrometallurgical processes [5], which result in major environmental problems; such as tailings dams and the generation of acid drainage (sulfuric acid and oxides of iron) by the oxidation of sulfur minerals with a high presence of pyrite. This sulfide is one of the most common and abundant minerals in the world and is associated with hydrothermal mineralization [6]. On the other hand, foundries produce large emissions of sulfur dioxide (SO_2), which together, with NO_x and CO_2, can cause large problems; such as acid rain and increasing local pollution, therefore, the abatement of waste gases is an important task for the protection of the environment [7–9]. As a result, new hydrometallurgical alternatives are being developed in the mining industry, because they are more ecological and economic processes to recover copper [10,11].

Chalcocite is the most abundant copper sulfide mineral after chalcopyrite [5,12], and it is the copper sulfide the most easily treated by hydrometallurgical routes [13]. Several investigations have been carried out for the leaching of this mineral with the use of multiple additives and in different media such as; bioleaching [14–18], ferric sulfate solution [19], chloride media [20–23], pressure leaching for chalcocite [24] and synthetic chalcocite (white metal) [25].

When operating in sulphated or chloride media, the oxidative dissolution of the chalcocite occurs in two stages [13,19,20,23,25].

$$Cu_2S + 2Fe^{3+} = Cu^{2+} + 2Fe^{2+} + CuS \quad (1)$$

$$CuS + 2Fe^{3+} = Cu^{2+} + 2Fe^{2+} + S^0 \quad (2)$$

The first stage of leaching of the chalcocite is much faster than the second stage. This is controlled by diffusion of the oxidant on the surface of the ore at low values of activation energy (4–25 kJ mol^{-1}) [19]. The second stage is slower and can be accelerated depending on the temperature [13,26].

The investigations shown in Table 1 were obtained as a result of high extractions (90%), but these results were obtained with the application of high temperatures and/or with the addition of ferric or cupric ions as an oxidizing agent. In addition, the previous investigations were made with mixtures of copper sulfides, with the presence of gangue or with the use of synthetic chalcocite. It is emphasized that the present investigation will include a leaching of pure chalcocite in a chlorinated medium, without the addition of oxidizing agents (Fe^{3+}, Cu^{2+}, etc.) and at room temperature.

Table 1. Comparison of previous investigations of chalcocite with the use of Cl^-.

Investigation	Leaching Agent	Parameters Evaluated	Reference	Cu Ext (%)
The kinetics of leaching chalcocite (synthetic) in acidic oxygenated sulphate-chloride solutions	NaCl, H_2SO_4, HCl, HNO_3 and Fe^{3+}	Oxygen flow, stirring speed, temperature, sulfuric acid concentration, ferric ions concentration, chloride concentration and particle size.	[20]	97
The kinetics of dissolution of synthetic covellite, chalcocite and digenite in dilute chloride solutions at ambient temperatures	HCl, Cu^{2+} and Fe^{3+}	Potential effect, chloride concentration, acid concentration, temperature, dissolved oxygen and pyrite effect.	[13]	98
Leaching kinetics of digenite concentrate in oxygenated chloride media at ambient pressure	$CuCl_2$, HCl and NaCl	Effect of stirring speed, oxygen flow, cupric ion concentration, chloride concentration, acid concentration and temperature effect.	[27]	95
Leaching of sulfide copper ore in a NaCl–H_2SO_4–O_2 media with acid pre-treatment	NaCl and H_2SO_4	Chloride concentration, effect of agitation with compressed air, percentage of solids and particle size.	[22]	78

A chalcocite leaching is performed with the injection of O_2 at ambient pressure in a H_2SO_4-NaCl solution, where the leaching agents are Cu^{2+}, $CuCl^+$, $CuCl_2$ and $CuCl_3^-$, which are generated during leaching in a Cu^{2+}/Cl^- system. The general reaction of chalcocite leaching is as follows:

$$Cu_2S + 0.5O_2 + 2H^+ + 4Cl^- = 2CuCl_2^- + H_2O + S^0, \quad (3)$$

While the chalcocite leaching reactions occur in two stages, guiding us to Equation (3), the following occurs:

$$Cu_2S + 2Cl^- + H^+ + 0.25O_2 = CuCl_2^- + CuS + 0.5H_2O, \quad (4)$$

$$CuS + 2Cl^- + H^+ + 0.25O_2 = 0.5H_2O + CuCl_2^- + S^0, \quad (5)$$

The resulting products expected from this chalcocite leaching should be soluble copper; such as $CuCl_2^-$ and a solid residue of elemental sulfur (S^0) with covellite residues or copper polysulfides (CuS_2) that still contain valuable metals.

The $CuCl_2^-$ is the predominant soluble specie due to the complexation of Cu (I) with the presence of Cl^- at room temperature, in a system of high concentrations of chloride (greater than 1 M). This $CuCl_2^-$ is stable in a range of potentials between 0–500 mV and pH < 6–7 (depending on the chloride concentration in the system) [20,28].

The shortage of fresh water in arid areas is an economic, environmental and social problem [29]. The use of sea water has become increasingly important for mining in Chile, not only because of its positive effects on leaching processes due to its chloride content, but also as a strategic and indispensable resource. For example, some metallic and non-metallic mining companies in the north of Chile have deposits rich in copper, gold, silver, iron and minerals from salt lakes, which are found in hyper-arid zones and at high altitudes, which emphasizes the necessity of this resource [30]. In addition, it is important to mention that the Chilean authorities have indicated that large-scale mining projects involving the use of water from aquifers will not be authorized [31]. An attractive alternative is the use of waste water from desalination plants. These companies produce drinking water for the population, however, their disposal product pollutes the oceans, for this reason, it is necessary to think of possible alternatives to recycle this resource and at the same time optimize extraction processes in local mining.

In the present investigation, a statistical analysis will be carried out using the methodology of surface optimization (design of the central composite face) to sensitize independent parameters (time, sulfuric acid concentration and chloride concentration) in the leaching of a pure mineral of chalcocite in chlorinated media. In addition, the effect on chloride concentration in the system will be evaluated when operating with potable water, seawater and reusing waste water.

2. Materials and Methods

2.1. Chalcocite

The pure chalcocite mineral used for the present investigation was collected manually directly from the veins by expert geologists from Mina Atómica, located in the region of Antofagasta, Chile.

The pure chalcocite samples were checked by X-ray diffraction (XRD) analysis, using an automatic and computerized X-ray diffractometer Bruker model Advance D8 (Bruker, Billerica, MA, USA). Figure 1 shows the results of the XRD analysis, indicating the presence of 99.9% chalcocite. The chemical analysis was performed by atomic emission spectrometry via induction-coupled plasma (ICP-AES), the sample of chalcocite was digested using aqua regia and HF. Table 2 shows the chemical composition of the samples. The samples for XRD and ICP-OES were ground in a porcelain mortar to reach a size range between −147 + 104 µm. The procedures described were performed in the applied geochemistry laboratory of the department of geological sciences of the Universidad Católica del Norte.

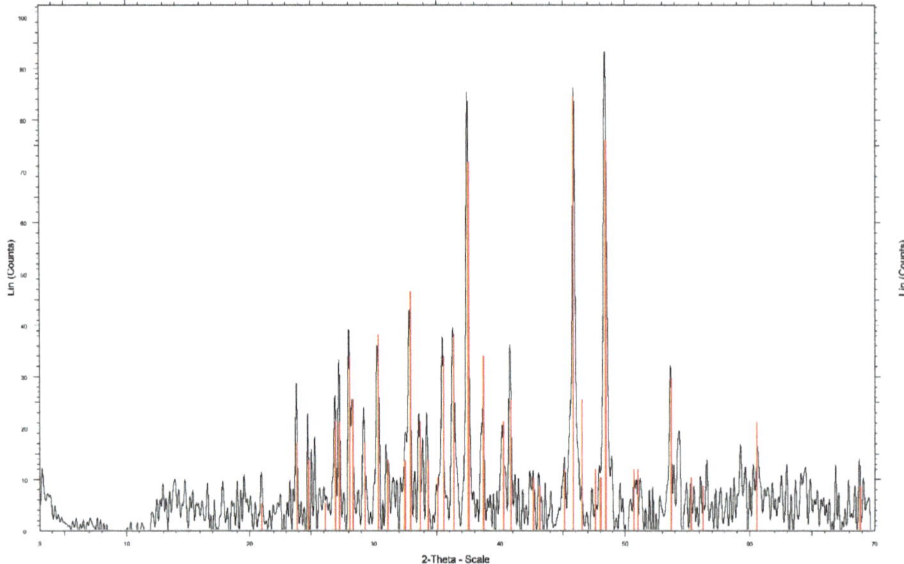

Figure 1. X-ray diffractogram for the chalcocite mineral.

Table 2. Chemical analysis of the chalcocite ore.

Component	Cu	S^0
Mass (%)	79.83	20.17

2.2. Leaching and Leaching Tests

The sulfuric acid used for the leaching tests was grade P.A., Merck brand, purity 95–97%, density 1.84 kg/L and molecular weight of 98.08 g/mol, though the tests also work with the use of sea water and waste water from the "Aguas Antofagasta" Desalination Plant. Table 3 shows the chemical composition of waste water.

Table 3. Chemical analysis of waste water.

Compound	Concentration (g/L)
Fluorine (F^-)	0.01
Calcium (Ca^{2+})	0.80
Magnesium (Mg^{2+})	2.65
Bicarbonate (HCO_3^-)	1.10
Chloride (Cl^-)	39.16
Calcium carbonate ($CaCO_3$)	13.00

Leaching tests were carried out in a 50 mL glass reactor with a 0.01 S/L ratio. A total of 200 mg of chalcocite ore in a size range between −147 + 104 µm and the addition of NaCl at different concentrations were maintained in agitation and suspension with the use of a 5-position magnetic stirrer (IKA ROS, CEP 13087-534, Campinas, Brazil) at a speed of 600 rpm and the temperature was controlled using an oil-heated circulator (Julabo, St. Louis, MO, USA). The temperature range tested in the experiments was 25 °C. Also, the tests were performed in duplicate and measurements (or analyzes) were carried out on 5 mL aliquot and diluted to a range of dilutions using atomic absorption spectrometry with a coefficient of variation ≤5% and a relative error between 5 to 10%. Measurements of pH and oxidation-reduction potential (ORP) of leach solutions were made using a pH-ORP meter (HANNA HI-4222, St. Louis, MO,

USA). The ORP solution was measured in a combination ORP electrode cell of a platinum working electrode and a saturated Ag/AgCl reference electrode.

2.3. Experimental Design

The effects of independent variables on Cu extraction rates from leaching of chalcocite were studied using the response surface optimization method [32–35]. The central composite face (CCF) design and a quadratic model were applied to the experimental design for Cu_2S leaching.

Twenty-seven experimental tests were carried out to study the effects of time, chloride and H_2SO_4 concentration as independent variables. Minitab 18 software (version 18, Pennsylvania State University, State College, PA, USA) was used for modeling and experimental design, which allowed the study of the linear and quadratic effects of the independent variables. The experimental data were fitted by multiple linear regression analysis to a quadratic model, considering only those factors that helped to explain the variability of the model. The empirical model contains coefficients of linear, quadratic, and two-factor interaction effects.

The general form of the experimental model is represented by:

$$Y = (overall\ constant) + (linear\ effects) + (interaction\ effects) + (curvature\ effects), \tag{6}$$

$$Y = b_0 + b_1x_1 + b_2x_2 + b_3x_3 + b_{12}x_1x_2 + b_{13}x_1x_3 + b_{23}x_2x_3 + b_{11}x_1^2 + b_{22}x_2^2 + b_{33}x_3^2, \tag{7}$$

Where, x_1 is time, x_2 is Chloride, x_3 is H_2SO_4 concentration, and b is the variable coefficients.

Table 4 presents the ranges of parameter values used in the experimental model. The variable values are codified in the model. The following Equation (8) is used for transforming a real value (Z_i) into a code value (X_i) according to the experimental design:

$$X_i = \frac{Z_i - \frac{Z_{high}+Z_{low}}{2}}{\frac{Z_{high}-Z_{low}}{2}}, \tag{8}$$

Z_{high} and Z_{low} are the highest and lowest levels of a variable, respectively [36].

Table 4. Experimental parameters for the central design of the composite face.

Experimental Parameters	Low	Medium	High
Time (h)	4	8	12
Concentration Cl^- (g/L)	20	50	100
Concentration H_2SO_4 (mol/L)	0.5	1	2
Codifications	−1	0	1

A factorial design was applied involving three factors, each one having three levels thus 27 experimental tests were carried out in Table 5, evaluating the effect of time and H_2SO_4 and chloride concentration.

The statistical R^2, R^2_{adj}, p-values and Mallows's Cp indicate whether the model obtained is adequate to describe Cu extraction under a given domain. The R^2 coefficient is a measure of the goodness of fit, which measures the proportion of total variability of the dependent variable with respect to its mean, which is explained by the regression model. The p-values represent statistical significance, which indicates whether there is a statistically significant association between the response variable and the terms. The predicted R^2 was used to determine how well the model predicts the response for new observations. Finally, Mallows's Cp is a precise measure in the model, estimating the true parameter regression [36].

Table 5. Experimental configuration and Cu extraction.

Exp. No.	Time (h)	Cl^- (g/L)	H_2SO_4 (mol/L)	Cu Ext. (%)
1	4	20	0.5	31.63
2	4	20	1	33.25
3	4	20	2	37.00
4	4	50	0.5	32.25
5	4	50	1	33.38
6	4	50	2	38.00
7	4	100	0.5	44.75
8	4	100	1	44.88
9	4	100	2	46.19
10	8	20	0.5	35.75
11	8	20	1	38.75
12	8	20	2	43.00
13	8	50	0.5	48.13
14	8	50	1	49.50
15	8	50	2	50.63
16	8	100	0.5	51.50
17	8	100	1	53.00
18	8	100	2	54.88
19	12	20	0.5	52.25
20	12	20	1	52.75
21	12	20	2	52.63
22	12	50	0.5	53.13
23	12	50	1	53.13
24	12	50	2	53.00
25	12	100	0.5	53.25
26	12	100	1	53.88
27	12	100	2	55.63

3. Results

3.1. ANOVA

An ANOVA analysis (Table 6) showed F-value and p-value for the model.

Table 6. ANOVA (analysis of variance) Cu extraction.

Source	F-Value	p-Value
Regression	22.73	0
Time	123.15	0
Cl^-	45.25	0
H_2SO_4	5.44	0.03
Time × Time	2.06	0.17
$Cl^- \times Cl^-$	0.13	0.72
$H_2SO_4 \times H_2SO_4$	0.00	0.97
Time × Cl^-	10.27	0.01
Time × H_2SO_4	1.18	0.29
$Cl^- \times H_2SO_4$	0.31	0.59

In the contour plot in Figure 2, it is observed that Cu extraction increases at long times, high chloride concentration, and high H_2SO_4 concentration.

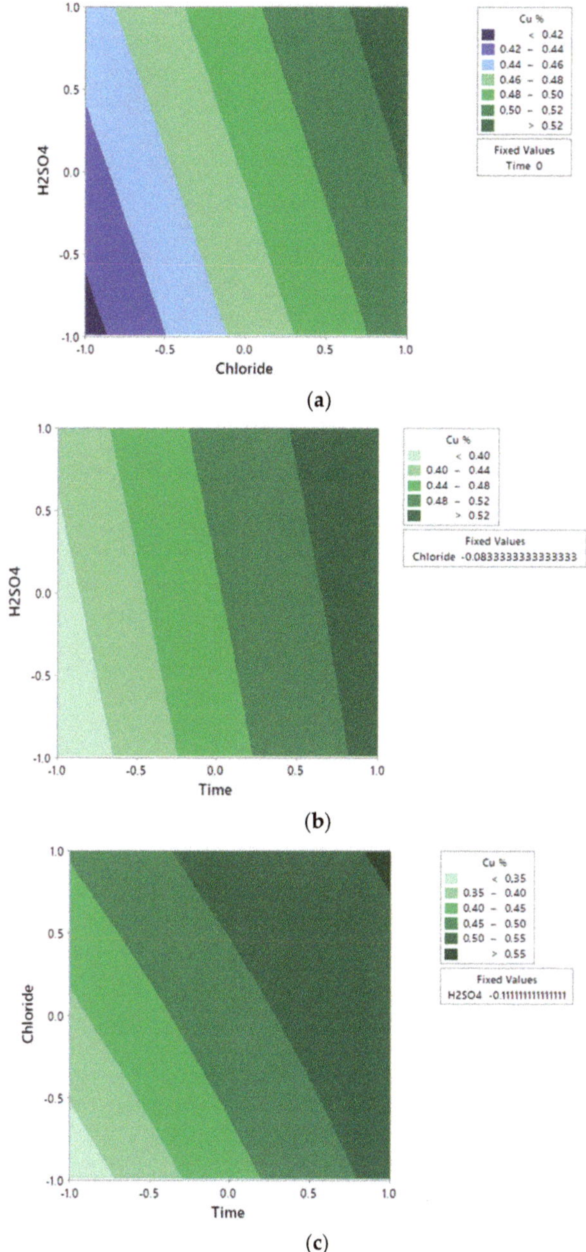

Figure 2. Experimental contour plot of independent variables of Chloride and H_2SO_4 concentration (**a**); Time and H_2SO_4 concentration (**b**); and Time and Chloride concentration (**c**) on the dependent variable Cu extraction.

Table 6 shows ANOVA analysis. There is no significant effect ($p > 0.05$) of the interactions concentration of chloride concentration of H_2SO_4 and time-concentration of H_2SO_4 in copper extraction, complying with the theory that the increase in sulfuric acid concentration does not have a great influence on the leaching of chalcocite above 0.02 mol/L [19,22]. Rather, it is only the time-concentration interaction of chloride that must be considered in the model. Additionally, the effects of curvature of the variable chloride concentration and H_2SO_4 concentration do not contribute significantly to explaining the variability of the model. On the other hand, the linear effects of chloride time and concentration contribute to explaining the experimental model, as shown in the contour plot of Figure 2.

Figures 3 and 4 show that time, chloride and H_2SO_4 concentration, as well as the interaction of time-H_2SO_4 and Cl-H_2SO_4 affected Cu extraction.

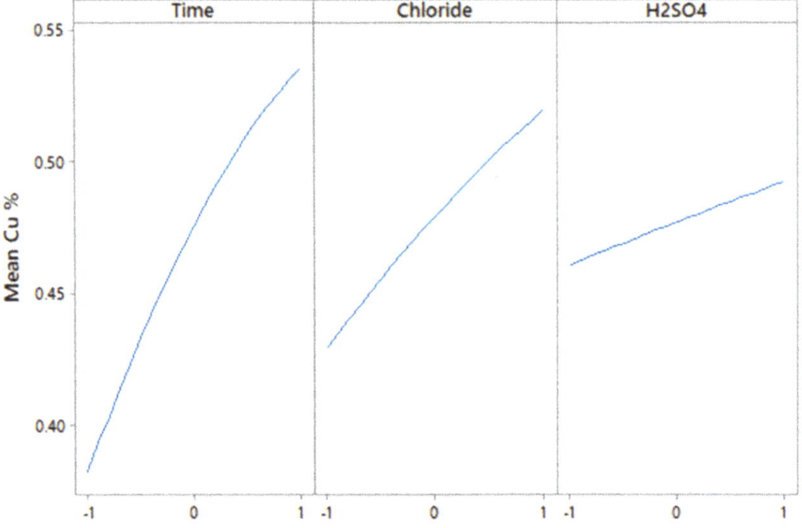

Figure 3. Linear effect plot for Cu extraction.

In Figure 3, the linear effects demonstrate what has been said by several authors [13,20,37], with respect to the effect of the concentration of chloride present in the leaching media and the effect of sulfuric acid concentration. The concentration of chloride has a great impact on the dissolution of copper from a sulfide; such as chalcocite. According to Velásquez-Yévenes et al. [38], the chloride ions present in the media increase the rate of oxidation of cuprous ions, while Cheng and Lawson [20,39] proposed that the effect of chloride ions promotes the formation of long sulfide crystals that allow the reactants to penetrate the sulfide layer, since in their tests they noticed that, in the absence of chloride ions, the kinetics of dissolution decreased considerably and that covellite did not dissolve. This with time was supported in the research of Nicol and Basson [37], without the presence of chloride ions or with a very low concentration of ions, the potential needed to dissolve covellite is very high.

Figure 4 shows the mean Cu extraction at different combinations of factor levels. In the interaction time-chloride, the lines are not parallel, and the plot indicates that there is an interaction between the factors. On the other hand, the interaction between time-H_2SO_4 and chloride-H_2SO_4 is low.

Equation (9) presents the Cu extraction model over the range of experimental conditions after eliminating the non-significant coefficients.

$$\%\ \text{Extraction} = 0.47782 + 0.07472\, x_1 + 0.04462\, x_2 + 0.01568\, x_3 - 0.0163\, x_1^2 - 0.02546\, x_1 x_2, \quad (9)$$

Where x_1, x_2 and x_3 are codified variables that respectively represent time, chloride and H_2SO_4 concentration.

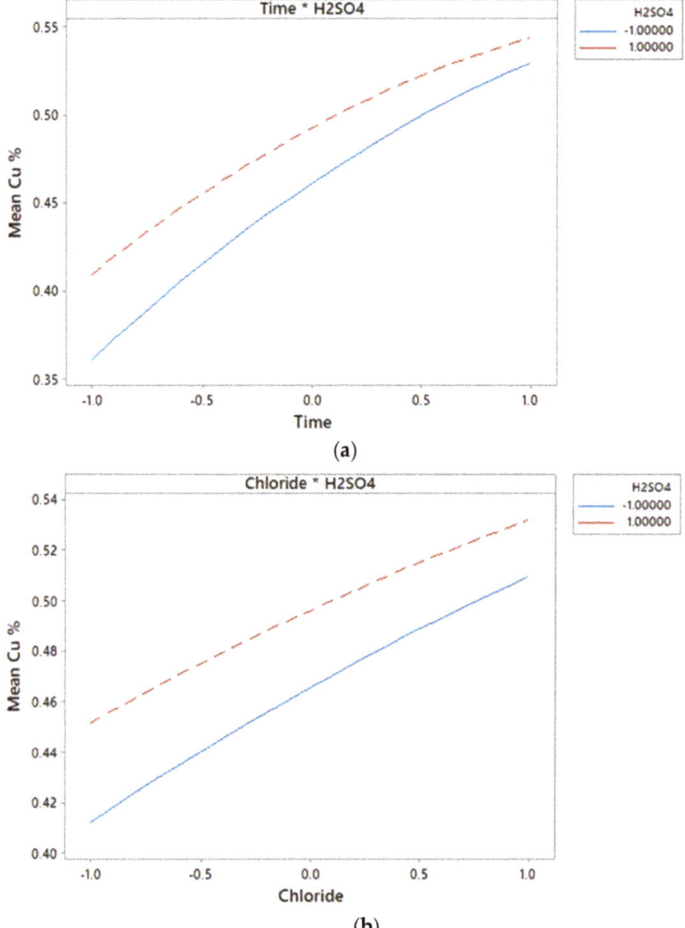

Figure 4. Interaction effect plot of independent variables Time and H2SO4 concentration (**a**); and Chloride and H2SO4 concentration (**b**) on the dependent variable Cu extraction.

An ANOVA test indicated that the quadratic model adequately represented Cu extraction from Cu_2S under the established parameter ranges. The model did not require adjustment and it was validated by the R^2 value (0.92) and R^2_{adj} value (0.90). The ANOVA analysis showed that the factors indicated influence Cu extraction from Cu_2S ($F_{Regression}$ (22.73) > $F_{T,95\% \text{ confidence level}} = F_{5,21}$ (2.68)). On the other hand, the p-value of the model (Equation (9)) is lower than 0.05, indicating that the model is statistically significant.

The Mallows's Cp = 3.62 (constant + 5 predictors) indicated that the model was accurate and did not present bias in estimating the true regression coefficients. This value of Cp of Mallows allows comparison with other models and establishes that the model found is the one that is most adjustable, due to the Cp closest to the number of constants and predictors.

In addition, all variance inflation factors (VIF) values are close to one, which ensures that there is no multicollinearity.

It also allows for prediction with an acceptable future forecast margin of error of R^2_{pred} = 0.8684.

Finally, from the adjustment of the ANOVA analysis, it was found that the factors considered, after analysis of the main components, explained the variation in the response. The difference between the R^2 and R^2_{pred} of the model was minimal, thus reducing the risk that the model was over adjusted. That means, the probability that the model fits only in the sample data is lower. The ANOVA analysis indicated that time, chloride concentration, H_2SO_4 concentration and the interaction of time-chloride are the factors that explain to a greater extent the behavior of the system for the sampled data set.

Table 5 shows that the increase in sulfuric acid concentration does not affect copper dissolution, obtaining similar results under similar conditions of leaching, only a minimum amount of sulfuric acid is needed in the leaching system. This result is consistent with other investigations, since according to Cheng and Lawson [20], a concentration of sulfuric acid of 0.02 mol/L, is sufficient to perform a leaching of chalcocite and its subsequent phases as it is the djurleite, digenite ore [22]. After this value its effect is null.

On the other hand, it is shown that at a leaching time of 12 h, the values of copper extraction do not vary regardless of the concentration of chloride and sulfuric acid. This could be explained with Equations (4) and (5); Equation (4) is the rapid reaction of the transformation of chalcocite to covellite, in which a low activation energy is required to achieve its transformation [13]. When covellite is formed (Equation (5)), it needs more energy (about 72 kJ approximately) to achieve its dissolution and later become a copper polysulfide (CuS_2), what it requires is even more demanding conditions to achieve its complete dissolution [37].

3.2. Effect on the Chloride Concentration

It has been known since the 1970s that it is beneficial to work with chloride ions in the leaching of sulfide minerals [23,40]. In Figure 5a, when operating at higher chloride concentrations, higher copper recoveries are obtained. When operating with the highest chloride concentrations (100 g/L), the highest recovery (68.82%) is obtained at 48 h. However, a large difference in copper recoveries cannot be seen when operating at chloride concentrations between 20 and 50 g/L. At 48 h and 20 g/L, a recovery of 63.58% of Cu was obtained and for chloride concentrations of 50 g/L, 65.45% was obtained. On the other hand, in Figure 5b it is observed that with the use of waste water (39.16 g/L of Cl^-), results similar to those presented in Figure 5a were obtained in a Cl^- concentration of 50 g/L, so it is noted that the presence of calcium ions, fluorine, magnesium and calcium carbonate did not affect the dissolution of copper from the chalcocite. In the tests carried out with seawater, which has approximately a concentration of 20 g/L Cl^-, obtained copper extractions of up to 63.4% at 48 h with a concentration of 0.5 mol/L of sulfuric acid. In previous investigations [13,20], it has been determined that leaching is independent of a chloride concentration between 0.5 and 2 mol/L, but a greater kinetic of dissolution is observed in the first minutes and then the difference decreases as a function of time and behavior similar to that of Figure 5.

Figure 6 shows a residue analysis performed under the conditions of 50 g/L Cl^- and 0.5 mol/L H_2SO_4, in a leaching time of 4 h. The result of this XRD is useful to understand the behavior of the chalcocite in a short time and in low reagent conditions, and to observe which mineralogical species are forming. The results show a high formation of synthetic covellite (77.34 wt %), early formation of elemental sulfur (20.20 wt %) and a remaining chalcocite (4.46 wt %), which still does not dissolve. From this, it follows that the transformation of chalcocite to covellite is faster than the transformation of covellite to elemental sulfur, which is similar to that observed in Equations (4) and (5), also, according to Figure 5, the slope of the curve is decreasing slowly, which means less kinetics of copper dissolution as a function of time. In the investigation of Senanayake [28], it is reported that the dissolution of chalcocite in a chloride-iron-water system at 25 °C occurs at potentials greater than 500 mV with a pH < 4, while in the research of Miki et al. [13] it is reported that the chalcocite dissolution occurs rapidly at a potential of 500 mV but stops when it reaches 50% copper extraction. When the potential increases to 550 mV, this extraction increases again because once it reaches 50% copper extraction, the mineral present is mainly covellite, which has a dissolution kinetics lower than the chalcocite and that needs potentials greater than 600 mV to dissolve.

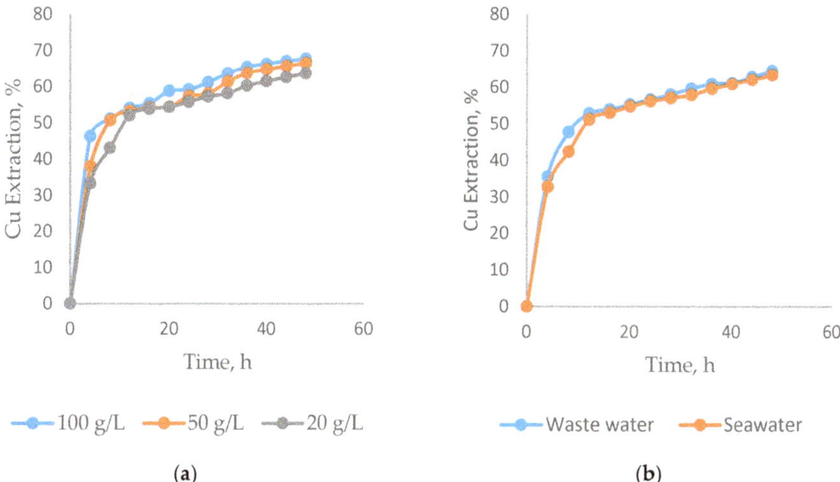

Figure 5. Effect of chloride concentration on Cu extraction from chalcocite ($T = 25°C$. H_2SO_4 = 0.5 mol/L); (**a**) Cl^- added by NaCl; (**b**) Cl^- added by waste water and seawater.

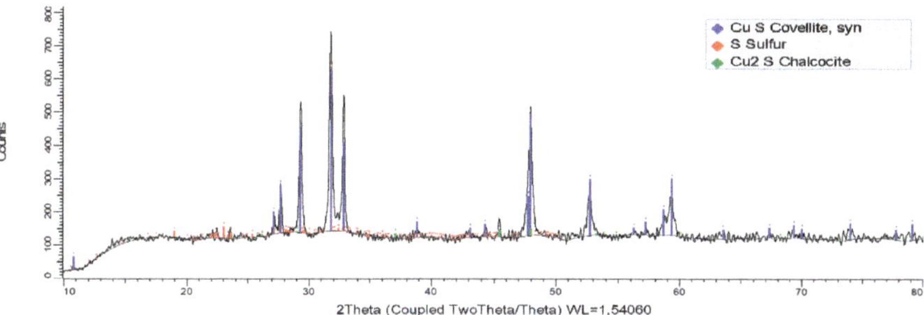

Figure 6. X-ray diffractogram for solid residues (chalcocite mineral) after being leached at 25 °C in a time of 4 h with 0.5 mol/L H_2SO_4 and 50 g/L Cl^-.

4. Conclusions

The present investigation shows the experimental results necessary to dissolve Cu from a chalcocite mineral in chloride media. The findings of this study were:

1. The linear variables with the greatest influence in the model are: time, chloride concentration and sulfuric acid concentration, respectively.
2. Under normal pressure and temperature conditions, only the chloride-time concentration exerts a significant synergistic effect on the extraction of copper from a chalcocite mineral.
3. The ANOVA analysis indicates that the presented quadratic model is adequate to represent the copper extractions and the value of R^2 (0.92) validates it.
4. The highest copper extraction is achieved under conditions of low concentration of sulfuric acid (0.5 mol/L), high concentrations of chloride (100 g/L) and a prolonged leaching time (48 h) to obtain an extraction of 67.75% copper.
5. The XRD analysis shows the formation of a stable and non-polluting residue; such as elemental sulfur (S^0). This residue was obtained in a leaching time of 4 h at room temperature under conditions of 0.5 mol/L H_2SO_4 and 50 g/L Cl^-.

Author Contributions: N.T. and K.P. contributed in project administration, investigation and wrote paper, W.B. contributed in the data curation and software, M.C. and E.T. contributed in validation and supervision and R.S. and P.H. performed the experiments, review and editing.

Funding: This research received no external funding.

Acknowledgments: The authors are grateful for the contribution of the Scientific Equipment Unit- MAINI of the Universidad Católica del Norte for aiding in generating data by automated electronic microscopy QEMSCAN® and for facilitating the chemical analysis of the solutions. We are also grateful to the Altonorte Mining Company for supporting this research and providing slag for this study, and we thank to Marina Vargas Aleuy and María Barraza Bustos of the Universidad Católica del Norte for supporting the experimental tests.

Conflicts of Interest: The authors declare they have no conflict of interest.

References

1. Comisión Chilena del Cobre. Sulfuros Primarios: Desafíos y Oportunidades. Registro Propiedad Intelectual N° 2833439. 2017. Available online: https://www.cochilco.cl/ListadoTemtico/sulfurosprimarios_desafíosyoportunidades.pdf (accessed on 25 April 2019).
2. Navarra, A.; Oyarzun, F.; Parra, R.; Marambio, H.; Mucciardi, F. System dynamics and discrete event simulation of copper smelters. *Miner. Metall. Process.* **2017**, *34*, 96–106. [CrossRef]
3. International Copper Study Group. *The World Copper Factbook 2017*; International Copper Study Group: Lisbon, Portugal, 2017.
4. CESCO. La Minería como plataforma para el desarrollo: Hacia una relación integral y sustentable de la industria minera en Chile. Available online: http://www.cesco.cl/wp-content/uploads/2018/06/Resumen-Position-Paper.pdf (accessed on 12 July 2019).
5. Schlesinger, M.E.; King, M.J.; Sole, K.C.; Davenport, W.G. *Extractive Metallurgy of Copper*, 5th ed.; Elsevier: Amsterdam, Netherlands, 2011.
6. Oyarzun, R.; Oyarzún, J.; Lillo, J.; Maturana, H.; Higueras, P. Mineral deposits and Cu-Zn-As dispersion-contamination in stream sediments from the semiarid Coquimbo Region, Chile. *Environ. Geol.* **2007**, *53*, 283–294. [CrossRef]
7. Dijksira, R.; Senyard, B.; Shah, U.; Lee, H. Economical abatement of high-strength SO2off-gas from a smelter. *J. South. African Inst. Min. Metall.* **2017**, *117*, 1003–1007. [CrossRef]
8. Serbula, S.M.; Milosavljevic, J.S.; Radojevic, A.A.; Kalinovic, J.V.; Kalinovic, T.S. Extreme air pollution with contaminants originating from the mining—Metallurgical processes. *Sci. Total Environ.* **2017**, *586*, 1066–1075. [CrossRef] [PubMed]
9. Zagoruiko, A.N.; Vanag, S.V. Reverse-flow reactor concept for combined SO2 and co-oxidation in smelter off-gases. *Chem. Eng. J.* **2014**, *238*, 86–92. [CrossRef]
10. Baba, A.A.; Balogun, A.F.; Olaoluwa, D.T.; Bale, R.B.; Adekola, F.A.; Alabi, A.G.F. Leaching kinetics of a Nigerian complex covellite ore by the ammonia-ammonium sulfate solution. *Korean J. Chem. Eng.* **2017**, *34*, 1133–1140. [CrossRef]
11. Pradhan, N.; Nathsarma, K.C.; Rao, K.S.; Sukla, L.B.; Mishra, B.K. Heap bioleaching of chalcopyrite: A review. *Miner. Eng.* **2008**, *21*, 355–365. [CrossRef]
12. Mindat. Copper: The mineralogy of Copper. Available online: https://www.mindat.org/element/Copper (accessed on 8 July 2019).
13. Miki, H.; Nicol, M.; Velásquez-Yévenes, L. The kinetics of dissolution of synthetic covellite, chalcocite and digenite in dilute chloride solutions at ambient temperatures. *Hydrometallurgy* **2011**, *105*, 321–327. [CrossRef]
14. Leahy, M.J.; Davidson, M.R.; Schwarz, M.P. A model for heap bioleaching of chalcocite with heat balance: Mesophiles and moderate thermophiles. *Hydrometallurgy* **2007**, *85*, 24–41. [CrossRef]
15. Lee, J.; Acar, S.; Doerr, D.L.; Brierley, J.A. Comparative bioleaching and mineralogy of composited sulfide ores containing enargite, covellite and chalcocite by mesophilic and thermophilic microorganisms. *Hydrometallurgy* **2011**, *105*, 213–221. [CrossRef]
16. Palencia, I.; Romero, R.; Mazuelos, A.; Carranza, F. Treatment of secondary copper sulphides (chalcocite and covellite) by the BRISA process. *Hydrometallurgy* **2002**, *66*, 85–93. [CrossRef]

17. Xingyu, L.; Biao, W.; Bowei, C.; Jiankang, W.; Renman, R.; Guocheng, Y.; Dianzuo, W. Bioleaching of chalcocite started at different pH: Response of the microbial community to environmental stress and leaching kinetics. *Hydrometallurgy* **2010**, *103*, 1–6. [CrossRef]
18. Ruan, R.; Zhou, E.; Liu, X.; Wu, B.; Zhou, G.; Wen, J. Comparison on the leaching kinetics of chalcocite and pyrite with or without bacteria. *Rare Met.* **2010**, *29*, 552–556. [CrossRef]
19. Niu, X.; Ruan, R.; Tan, Q.; Jia, Y.; Sun, H. Study on the second stage of chalcocite leaching in column with redox potential control and its implications. *Hydrometallurgy* **2015**, *155*, 141–152. [CrossRef]
20. Cheng, C.Y.; Lawson, F. The kinetics of leaching chalcocite in acidic oxygenated sulphate-chloride solutions. *Hydrometallurgy* **1991**, *27*, 249–268. [CrossRef]
21. Herreros, O.; Quiroz, R.; Viñals, J. Dissolution kinetics of copper, white metal and natural chalcocite in Cl2/Cl^- media. *Hydrometallurgy* **1999**, *51*, 345–357. [CrossRef]
22. Herreros, O.; Viñals, J. Leaching of sulfide copper ore in a NaCl-H2SO4-O2 media with acid pre-treatment. *Hydrometallurgy* **2007**, *89*, 260–268. [CrossRef]
23. Senanayake, G. Chloride assisted leaching of chalcocite by oxygenated sulphuric acid via Cu(II)-OH-Cl. *Miner. Eng.* **2007**, *20*, 1075–1088. [CrossRef]
24. Muszer, A.; Wódka, J.; Chmielewski, T.; Matuska, S. Covellinisation of copper sulphide minerals under pressure leaching conditions. *Hydrometallurgy* **2013**, *137*, 1–7. [CrossRef]
25. Ruiz, M.C.; Abarzúa, E.; Padilla, R. Oxygen pressure leaching of white metal. *Hydrometallurgy* **2007**, *86*, 131–139. [CrossRef]
26. Petersen, J.; Dixon, D. Principles, mechanisms and dynamics of chalcocite heap bioleaching. In *Microbial Processing of Metal Sulfides*; Springer: Dordrecht, The Netherlands, 2007; pp. 193–218.
27. Ruiz, M.C.; Honores, S.; Padilla, R. Leaching kinetics of digenite concentrate in oxygenated chloride media at ambient pressure. *Metall. Mater. Trans. B Process Metall. Mater. Process. Sci.* **1998**, *29*, 961–969. [CrossRef]
28. Senanayake, G. A review of chloride assisted copper sulfide leaching by oxygenated sulfuric acid and mechanistic considerations. *Hydrometallurgy* **2009**, *98*, 21–32. [CrossRef]
29. Tundisi, J.G. Water resources in the future: problems and solutions. *Estud. Avançados* **2008**, *22*, 7–16. [CrossRef]
30. Cisternas, L.A.; Gálvez, E.D. The use of seawater in mining. *Miner. Process. Extr. Metall. Rev.* **2018**, *39*, 18–33. [CrossRef]
31. MCH. Agua en la Minería. Agua en la Minería. 2018. Available online: https://www.mch.cl/columnas/agua-la-mineria/# (accessed on 3 June 2019).
32. Aguirre, C.L.; Toro, N.; Carvajal, N.; Watling, H.; Aguirre, C. Leaching of chalcopyrite (CuFeS2) with an imidazolium-based ionic liquid in the presence of chloride. *Miner. Eng.* **2016**, *99*, 60–66. [CrossRef]
33. Bezerra, M.A.; Santelli, R.E.; Oliveira, E.P.; Villar, L.S.; Escaleira, L.A. Response surface methodology (RSM) as a tool for optimization in analytical chemistry. *Talanta* **2008**, *76*, 965–977. [CrossRef]
34. Dean, A.; Voss, D.; Draguljic, D. Response Surface Methodology. In *Design and Analysis of Experiments*; Springer Texts in Statistics: Cham, Switzerland, 2017; pp. 565–614.
35. Toro, N.; Herrera, N.; Castillo, J.; Torres, C.; Sepúlveda, R. Initial Investigation into the Leaching of Manganese from Nodules at Room Temperature with the Use of Sulfuric Acid and the Addition of Foundry Slag—Part I. *Minerals* **2018**, *8*, 565. [CrossRef]
36. Montgomery, D.C. Cap. 3, 6, 7 and 10. In *Design and Analysis of Experiments*, 8th ed.; Wiley: New York, NY, USA, 2012.
37. Nicol, M.; Basson, P. The anodic behaviour of covellite in chloride solutions. *Hydrometallurgy* **2017**, *172*, 60–68. [CrossRef]
38. Velásquez-Yévenes, L.; Nicol, M.; Miki, H. The dissolution of chalcopyrite in chloride solutions: Part 1. The effect of solution potential. *Hydrometallurgy* **2010**, *103*, 108–113. [CrossRef]
39. Cheng, C.Y.; Lawson, F. The kinetics of leaching covellite in acidic oxygenated sulphate-chloride solutions. *Hydrometallurgy* **1991**, *27*, 269–284. [CrossRef]
40. Dutrizac, J.E. The leaching of sulphide minerals in chloride media. *Hydrometallurgy* **1992**, *29*, 1–45. [CrossRef]

Article

Copper Dissolution from Black Copper Ore under Oxidizing and Reducing Conditions

Oscar Benavente [1], María Cecilia Hernández [1], Evelyn Melo [1,*], Damián Núñez [1], Víctor Quezada [1] and Yuri Zepeda [2]

1 Laboratorio de Procesos Hidrometalúrgicos, Departamento de Ingeniería Metalúrgica y Minas, Universidad Católica del Norte, Avenida Angamos 0610, 1270709 Antofagasta, Chile

2 Compañía Minera Lomas Bayas, General Borgoño 934, 1270242 Antofagasta, Chile

* Correspondence: emelo@ucn.cl; Tel.: +565521012

Received: 4 July 2019; Accepted: 15 July 2019; Published: 19 July 2019

Abstract: Black copper oxides are amorphous materials of copper-bearing phases of manganese. They are complex mineral compounds with difficult to recognize mineralogy and have slow dissolution kinetics in conventional hydrometallurgical processes. This study evaluates the effects of various leaching media on copper dissolution from black copper minerals. Leach of a pure black copper sample from Lomas Bayas Mine and another from a regional mine were characterized by inductively coupled plasma atomic emission spectroscopy (ICP-AES), X-ray diffraction (XRD), scanning electron microscopy (SEM), Qemscan and mechanically prepared for acid leaching under standard, oxidizing and reducing conditions through the addition of oxygen, iron sulfate or sulfur dioxide, respectively. Standard and high potential leaching (770 mV (SHE)) results in a copper dissolution rate of 70% and manganese dissolution rate of 2%. The addition of potential reducing agents ($FeSO_4$ or SO_2) decreases the redox potential to 696 and 431 mV, respectively, and favors the dissolution of manganese, thus increasing the overall copper extraction rate. The addition of SO_2 results in the lowest redox potential and the highest copper extraction rates of 86.2% and 75.5% for the Lomas Bayas and regional samples, respectively, which represent an increase of 15% over the copper extract rates under standard and oxidizing conditions.

Keywords: leaching; black copper ore; copper wad; copper pitch

1. Introduction

There are numerous copper porphyry ore deposits with oxidization zones in the central Chilean Andes that have generated enriched surface oxide ores and sometimes resulted in the development of neighboring bodies of exotic copper mineralization. These "exotic" deposits vary widely in size and grade. The Mina Sur deposit contains of 3.63 million tons of fine Cu [1]. Black copper ore is most often found in exotic deposits [2], including in those at Mina Sur, Spence, Lomas Bayas and Centinela [3]. Black copper ore is a dark colored mineral compound that is difficult to recognize due to its mineralogical complexity and the presence of polymetallic connections and characterized by a resistance to dissolution and slow dissolution kinetics in hydrometallurgical systems conventionally applied to copper oxide ores [4]. Black copper ores are heterogeneous mixtures of amorphous materials, traditionally referred as copper wad ($CuO{\cdot}MnO_2{\cdot}7H_2O$), copper pitch ($MnO(OH)CuSiO_2{\cdot}nH_2O$) and other copper-bearing manganese phases [5]. However, samples from Mina Sur show that there is no significant difference between copper pitch and copper wad. Furthermore, their chemical composition varies widely, showing black copper content of mainly Cu (1% to 54%), Si, Mn, Fe and Al, with trace quantities of Ca, Na, K, Mg, S, P, Cl, Mo, Co, Ni, As, Zn, Pb, U and V [4].

Resources associated with black copper ores are generally not incorporated into extraction circuits or are left untreated, either in stock, leach pads or residues [6]. Little research has been performed on black copper ores because research has largely been aimed at the treatment of copper sulfides [7]. Consequently, black copper ore is now becoming an important subject of geological and metallurgical interest because it is found close to or in existing mines and extraction facilities, and represents an important resource that can extend operational life. In particular, the Lomas Bayas mine has mineral deposits with the presence of black copper ore. Lomas Bayas operates with a head grade of 0.30% total copper (0.26% soluble and 0.04% insoluble copper). Mineralogical analyses estimate that 30% of the insoluble copper is black copper.

There have been few studies on the treatment of black copper. Its non-crystalline character, variable composition, with Cu, Mn, Fe, Al and Si as major elements, and its resistance to dissolution are similar to the characteristics of marine polymetallic nodules and manganese minerals. Minerals associated with manganese have been studied from a metallurgical point of view and provide a precedent for leaching black copper [8,9]. Manganese dioxide minerals like pyrolusite (MnO_2) and oxides with Mn and Cu content in marine nodules are stable under acidic or alkaline oxidizing conditions. Manganese can be extracted through the medium of sulfuric acid under reductive conditions, incorporating SO_2 [10–12], oxalic acid [13], hydrogen peroxide [14,15], iron sulfate, $FeSO_4$ [12,16,17], iron metal [17] or through the incorporation of pyrite in hydrochloric acid media [18].

Among the few works on the metallurgical behavior of black copper minerals [19] is a study on the classification of copper types associated with copper wad and the leaching behavior of minerals that contain copper wad mixed with secondary sulfurs, which results in copper yields of over 90%. Sulfuric acid leaching of copper wad and chalcocite is proposed according to the following reaction:

$$2(CuO.MnO_2.7H_2O) + 6H_2SO_4 + Cu_2S \rightarrow 4CuSO_4 + 2MnSO_4 + S + 20H_2O \quad (1)$$

There is an intermediary reaction that generates and consumes iron ions, which favors the overall reaction [19]. Copper, nickel, cobalt, zinc and other metals can be extracted from leach solutions by several proven techniques, among them direct electrowinning, solvent extraction–electrowinning, crystallization, cementation with iron (for copper), and sulfide precipitation. The choice of one method over another depends on a number of factors, including solution chemistry, the pregnant leach solution (PLS) flow rate and concentration, the shape of the metal product, input availability and costs, and capital costs. In general, high metal mass flows (solution flow and grade) are necessary to justify the capital expenditures of solvent extraction and electrowinning. The metallurgical efficiency and cost effectiveness of these processes can be reduced as feed metal concentrations decline. The present study investigated the effects of different methods of acid leaching on the dissolution of copper from black copper ore.

2. Materials and Methods

To determine the behavior of black copper ore under acid leaching with different modifying potential agents, a program of eight agitation leach test was performed. Two mineral samples were used, the first from Minera Lomas Bayas (LB) and the second from an Antofagasta Region mine (RA). A standard test was performed in an acid medium, followed by a test in oxidizing medium through the addition of oxygen. Finally, two tests were carried out in reducing media, one in iron sulfate ($FeSO_4$) and the other in SO_2. Both mineral samples were subjected to the same four tests. Feed samples and leaching residues were characterized.

2.1. Experimental Procedure

2.1.1. Sample Characterization

The feed samples used were identified as LB and RA. Samples were crushed and milled to a particle size 100% below 295 µm. The chemical composition was determined using inductively coupled plasma atomic emission spectroscopy (ICP-AES). The mineralogy of the samples was determined by X-ray diffraction (XRD). For XRD analysis, the sample was ground in an agate mortar to a size of less than 45 μm and analyzed in an automatic and computerized X-ray diffractometer (Siemens model D5600, Bruker, Billerica, MA, USA), with an analysis time of one hour. The ICDD (International Center for Diffraction Data) database was used to identify the species present. Qemscan analysis was performed using a Model Zeiss EVO Series, with Bruker AXS XFlash 4010 detectors and iDiscover 5.3.2.501 software. Qemscan analyses of exotic Cu deposits yielded more accurate and precise Cu mineralogy and deportment [20].

The morphology was characterized by scanning electron microscopy (SEM) using JEOL 6360-LV equipment with a coupled energy-dispersive X-ray spectroscopy (EDS) microanalysis system and operated at 30 kV under high vacuum conditions. Mineral samples were metallized with a thin carbon layer to improve their conductivity.

2.1.2. Leaching Experiments

Eight leaching tests were conducted in the present work, four for each of the two samples: under standard conditions, with an oxidizing media, with SO_2 gas and iron sulfate ($FeSO_4$) used separately as reductive media, all the tests were performed in duplicate. The results of the standard leaching test were compared to those with an oxidizing or reductive medium. The standard conditions were a particle size below 295 μm, a temperature of 25 °C, a leaching time of 240 min, 10 g/L of H_2SO_4 (pH = 0.9), 300 min^{-1} of mechanical agitation and 4.8 g of mineral added to 250 mL solution. The oxidizing condition is achieved by adding technical grade oxygen (1 L/min) under standard conditions. The reductive conditions are generated by adding iron sulfate (1.2 g/L) or SO_2 (1 L/min) under standard conditions. O_2 or SO_2 gasses are injected with a porous frit. The aliquots obtained at different leaching times (15, 30, 60, 120 and 240 min) were filtered (0.2 μm) and the metal concentrations (Cu^{2+}, Mn^{2+} and Fe_t) in the filtrate were determined using inductively coupled plasma atomic emission spectroscopy (PerkinElmer, model Optima 2000 DV). The solid residues after the leaching experiment were analyzed by X-ray diffraction to determine the behavior of solids during leaching under different experimental conditions. SEM analysis was also applied. Figure 1 shows a representation of the system.

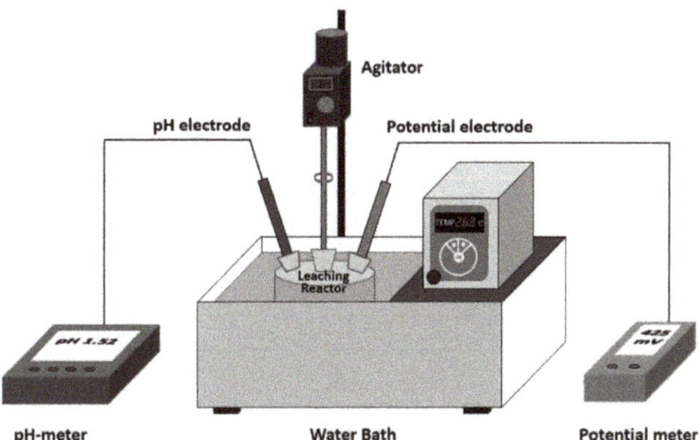

Figure 1. Experimental setup.

3. Results and Discussion

3.1. Samples Characterization

The feed samples, identified as LB and RA, were characterized by inductively coupled plasma atomic emission spectroscopy (ICP-AES), X-ray diffraction (XRD), and Qemscan analysis. Table 1 shows the ICP-AES results, with an evident difference in copper content.

Table 1. Chemical composition of black oxide samples.

Sample	$Cu_{(T)}$ (%)	$Fe_{(T)}$ (%)	$Mn_{(T)}$ (%)
RA	22.8	0.36	8.68
LB	2.59	2.56	2.48

The XRD analysis of the RA and LB samples yielded spectrums with high noise levels, characteristic of amorphous or low crystallinity minerals like black copper minerals. An analysis of the low concentration compounds detected the presence of copper silicates and crystalline types of manganese. Sample RA contained birnessite, $K_{0,46}Mn_{1,54}Mn_{0,46}O_4(H_2O)$; plancheite, $Cu_8(Si_4O_{11})_2(OH)_4H_2O$; crednerite, $CuMnO_2$; pyrite, (FeS_2); pyrochroite, $Mn(OH)_2$ and lithiodionite, $CuNaKSi_4O_{10}$, while sample LB contained albite, $NaAlSi_3O_8$; bixbyte, $FeMnO_3$; chlorite, $(Mg,Fe)_6(Si,Al)_4O_{10}(OH)_8$; quartz, SiO_2; illite, $(K,H_3O)Al_2Si_3AlO_{10}(OH)_2$; microcline, $KAlSi_3O_8$; molysite, $FeCl_3$; montmorillonite, $Na_{0.3}(Al,Mg)_2Si_4O_{10}(OH)_2{\cdot}xH_2O$; natrojarosite, $NaFe_3(SO_4)_2(OH)_6$ and nontronite, $Na_{0.3}Fe_2Si_4O_{10}(OH)_2{\cdot}4H_2O$.

The EDS (Figure 2) and semi-quantitative analyses of sample LB (Table 2) indicated the presence of oxidized Si, Mn, Cu, and Fe. SEM analysis of sample LB showed characteristic particles, permitting a qualitative analysis of not only the elements present, but also of the associations among them. It was possible to determine the presence of oxidized phases formed by the association of Cu-Mn-Fe and K, as well as quartz and iron oxides.

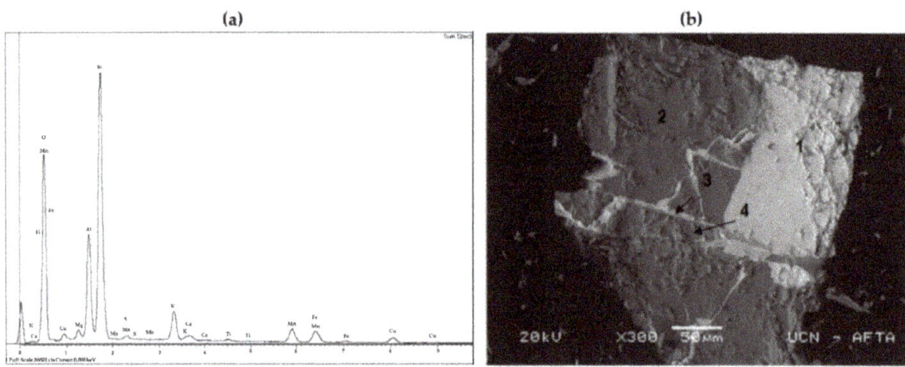

Figure 2. EDS analysis (**a**) and scanning electron microscopy (SEM) image (X 300) of sample LB (**b**). Association of (1) Cu-Mn-Fe-K-O (2) Si-O (3) Fe-O (4) Al-Fe-O.

Table 2. Semi-quantitative analysis of sample LB.

Element	O	Mg	Al	Si	S	K	Ca	Ti	Mn	Fe	Cu	Mo
% (wt)	51.9	0.93	8.31	23.6	0.30	3.38	0.56	0.35	3.77	3.41	3.02	0.48

Qemscan analysis showed that the black copper in the two samples was of the copper wad type, without other types present like copper pitch or copper oxides (Fe-Cu) (Figure 3). Copper wad is the term given to a subgroup of black copper, specifically hydroxides of Cu and Mn, with traces of other elements such as Co, Ca, Fe, Al and Mg. More than 90% of sample RA was copper bearing minerals. The 10% of gangue was composed principally of feldspar and kaolinite. Sample LB was approximately 20% copper-bearing minerals, while the gangue was mainly quartz, feldspar, muscovite/sericite and kaolinite.

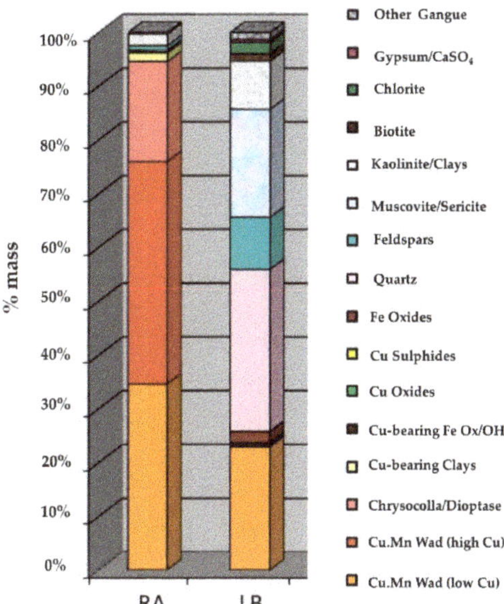

Figure 3. Mineralogical analysis by Qemscan of samples RA and LB.

Figure 4 shows the distribution of copper in the samples. Low-content copper wad was 70% of the copper in sample RA, while the remaining 30% was mainly chrysocolla/dioptase. Low-content copper wad accounted for 95% of the copper in sample LB.

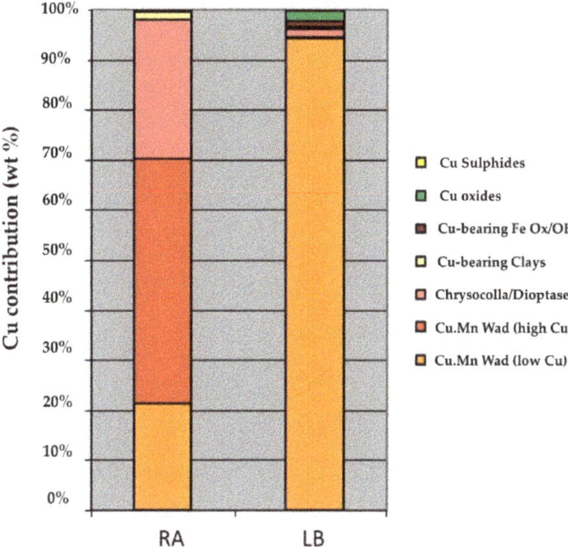

Figure 4. Cu distribution in samples RA and LB.

Figure 5 shows the copper wad associations in the samples, 42% of which was associated with chrysocolla/dioptase, while 53% was free from association (sample RA). About 45% of sample LB was free from association, while about 20% was associated with feldspar, muscovite/sericite and kaolinite/clays.

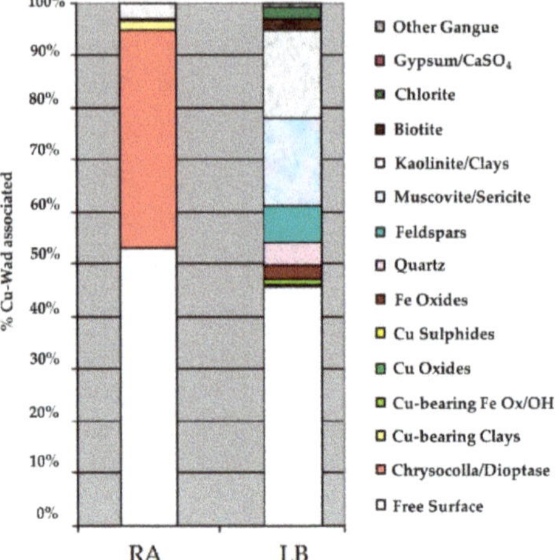

Figure 5. Cu wad associated in samples RA and LB.

3.2. Leaching Test

Figures 6–9 show the leaching kinetics with samples RA and LB in acid media under standard conditions and with the addition of oxidizing (O_2) or reducing agents ($FeSO_4$ or SO_2). The Cu extraction rate with the RA sample was around 70% under standard conditions (♦) with a high redox potential of 773 mV due to the small quantity of iron in the sample. The rate was not greatly altered by the addition of the oxidant (▲) (777 mV, practically the same as the rate under standard condition), which was due to the highly oxidized nature of black copper mineral and it's not affected for this condition. When reducing agents are added, the extraction increases by 15%, showing a greatest effect the SO_2 addition reaching up to 86% of copper extraction (■). The maximum extraction rates were reached in every test in less than 120 min (Figure 6).

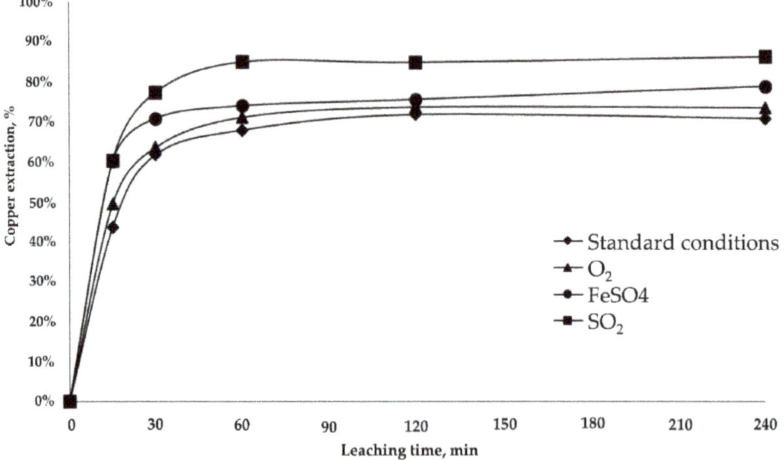

Figure 6. Copper extraction from the RA sample.

The Mn extraction rate for the RA sample was practically zero under standard and oxidizing conditions (♦ and ▲). This was confirmed by tests performed by [21], who obtained a manganese dissolution rate of around 1% in an acid medium at room temperature, indicating that manganese oxides are relatively insoluble in a conventional acid medium [12,21].

The addition of the reducing agents had notable effects, in particular SO_2 (■), which increased Mn extraction to 80% in less than 30 min (Figure 7). These results concur with those of [12], who indicated that SO_2 is a useful reagent. The benefit of SO_2 is due to its rapid and selective dissolution, yielding manganese extraction rates from MnO_2 of over 95% under moderate leaching conditions.

The Mn dissolution rate increased as a consequence of the addition of a reducing agent (solution potential of 431 mV (SHE)), which exposed the copper to acid and increased overall Cu dissolution from sample RA. Figure 7 shows that the test with SO_2 had an average solution potential of 430 mV (SHE), which generated a favorable reducing acid medium for dissolving black copper ore.

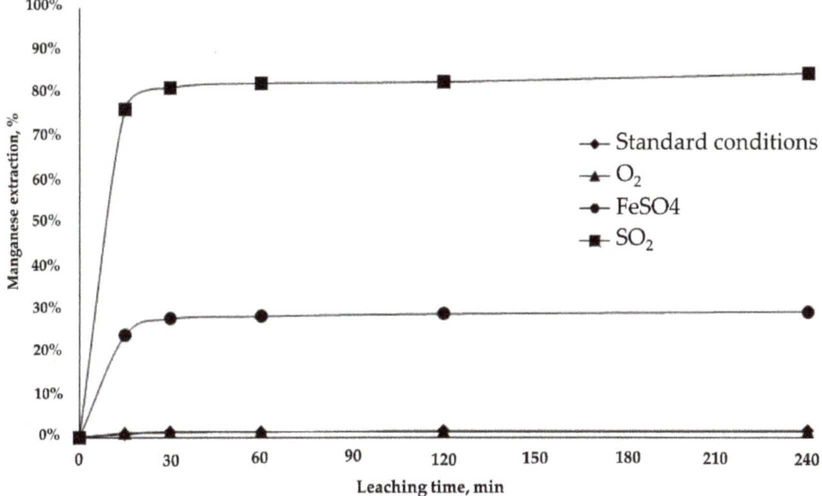

Figure 7. Manganese extraction from RA sample.

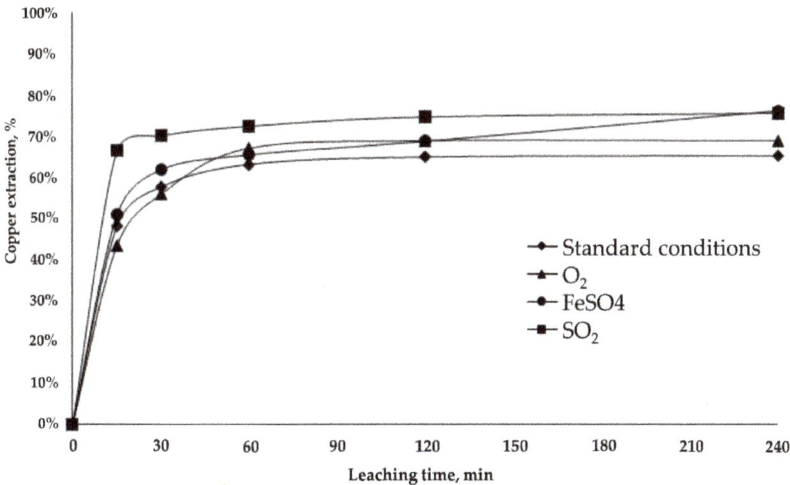

Figure 8. Copper extraction from LB sample.

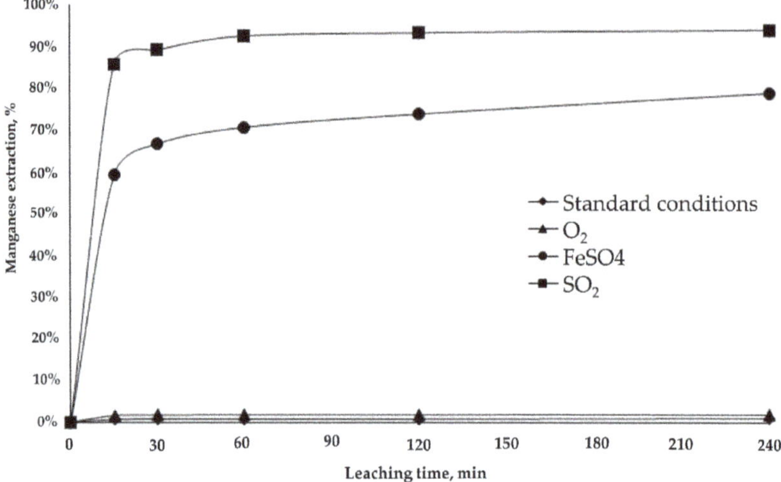

Figure 9. Manganese extraction from LB sample.

Ferrous sulfate allowed a manganese extraction of 30% in less than 30 min. The dissolution of Mn with this reducing agent is affected by the concentration of acid according to the following reactions, [12].

With ferrous sulfate solution and small amount of sulfuric acid,

$$MnO_2 + 2FeSO_4 + 2H_2SO_4 \rightarrow MnSO_4 + Fe(OH)SO_4 \quad (2)$$

With ferrous sulfate solution and excess of sulfuric acid,

$$MnO_2 + 2FeSO_4 + 2H_2SO_4 \rightarrow MnSO_4 + Fe(SO_4)_3 + 2H_2O \quad (3)$$

The Mn dissolution with ferrous sulfate is temperature and sulfuric acid concentration dependent. In a study performed by [12] The authors exposed Mn dissolution above 90% at 90 °C and 147 g/L H_2SO_4. Furthermore, [22] achieved a 90% Mn dissolution at 60 °C and at a molar ratio H_2SO_4/MnO_2 of 2.0. This can explain the low Mn dissolution (less than 80%) for RA and LB samples. In this study, 10 g/L H_2SO_4 was used at room temperature.

The copper extraction rates were similar for the two samples. Under standard (♦) and oxidant (▲) conditions, the copper extraction rates were between 65% and 67%, and increased by 10% with the reducing agents (■) (●) (Figure 8). The smaller increase in extraction is associated with the mineralogy of this sample, which contains more low-grade copper wad than does the RA sample, the latter being 30% chrysocolla.

The Mn extraction rate is associated directly with the presence of reducing agents in the leaching process, and its dissolution favors Cu extraction (Figure 9). Mn extraction increased from 2% to 95% with the reducing agents. The use of SO_2 stands out by the rapid dissolution of Cu and the Mn that it produces, reaching its maximum value at 30 min of leaching (■).

According to the results and the literature, black copper resists leaching under standard and oxidizing conditions. The addition of a reducing agent increases copper and manganese extraction rates because of the rupture of the Cu-Mn matrix, which increases copper extraction from copper wad. Our results concur with those of [12], who indicated that SO_2 plays an important role in the dissolution of Mn from MnO_2 in accordance with the following reaction

$$MnO_{2(s)} + SO_{2(aq)} \rightarrow Mn^{2+}{}_{(aq)} + SO_4^{2-}{}_{(aq)} \quad (4)$$

In addition to this reaction, there are intermediate and secondary reactions that depend on the pH of the medium [12].

The result obtained opens the possibility of recovering copper from minerals and residues with black copper content. This process can be through primary or secondary leaching, modifying the redox potential of the irrigation solution with a reducing agent decreasing its value to 460 mV.

3.3. Residues Characterization

The solid residues from the tests under standard conditions and with $FeSO_4$ as a reducing agent were characterized by SEM and mapping analysis. The elements shown by mapping analysis are: copper (red), manganese (green) and silica (blue).

Figure 10a shows an abundance of manganese and silica, reflecting the low manganese extraction rate under standard leaching conditions (2% of Mn extracted). The presence of copper is still evident considering the high level in the initial sample (22.8% Cu). Figure 10a shows the Cu extraction rate at 70%, with no direct relationship with Mn extraction, due to the presence of chrysocolla or dioptase in the feed. Figure 10b shows a significant decrease of Mn in the residue when $FeSO_4$ is used as a reducing agent (Figure 10b). The above had a direct effect in terms of increasing the copper extraction rate under these conditions (80%), thus giving way to the dissolution of species associated in the Cu–Mn wad that did not react. Figure 11 shows the approach to a particle from the leaching residue obtained using $FeSO_4$. There is a notable presence of manganese that did not react. A low presence of copper is also evident. It thus seems that there are Cu–Mn phases (with a low presence of copper) without reacting.

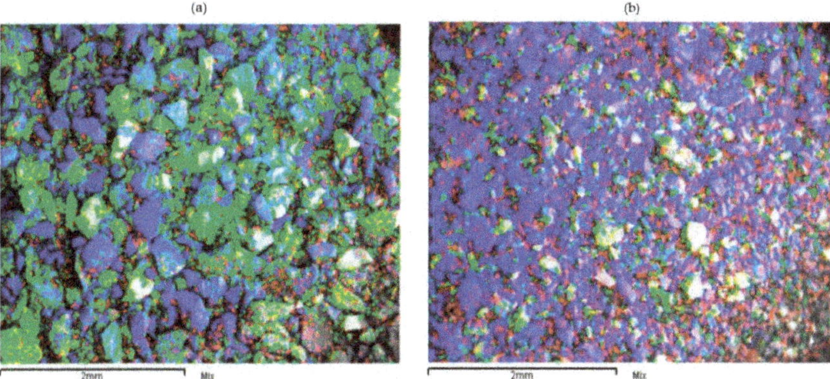

Figure 10. SEM image shows RA residues obtained under standard conditions (**a**) and with $FeSO_4$ as a reducing agent (**b**). Manganese (green), copper (red) and silica (blue).

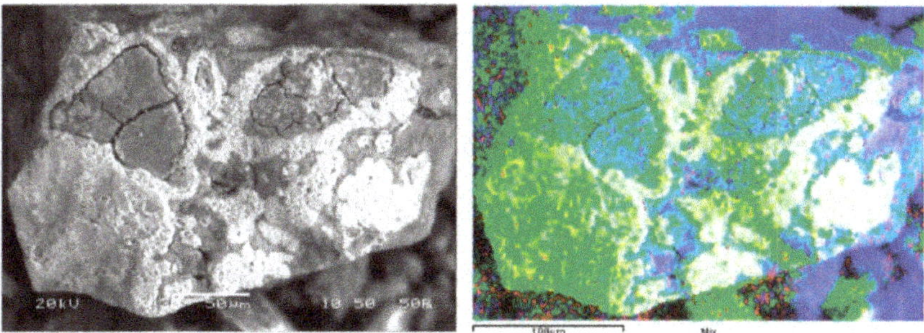

Figure 11. SEM image shows RA residues obtained under reducing conditions using $FeSO_4$. Manganese (green), copper (red) and silica (blue).

Figure 12 shows the comparative results of SEM mapping analysis for residues obtained from the LB sample under standard leaching conditions (12a) and under reducing conditions using $FeSO_4$ (12b). An abundance of manganese and silica can be observed in Figure 12a, mainly resulting from Cu–Mn wad and quartz, respectively. This abundance is due to almost no Mn dissolution under standard leaching conditions (1% of Mn dissolved). The residue obtained under reducing conditions using $FeSO_4$ was considerably less in the presence of manganese, but remained largely unchanged in the presence of quartz (12b). The low presence of manganese was due to the 80% dissolution rate obtained under these conditions. Finally, the presence of copper was due to the fact that 30% of copper could not be dissolved.

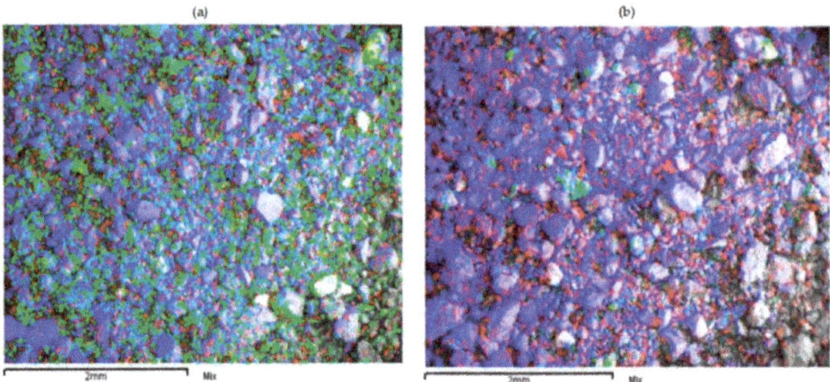

Figure 12. SEM image shows LB residues obtained under standard conditions (**a**) and with $FeSO_4$ as a reducing agent (**b**). Manganese (green), copper (red) and silica (blue).

4. Conclusions

Despite the difference in grade, the black copper from both the RA and LB samples can be classified as copper wad and behaved similarly in the leaching tests under all studied conditions.

The lowest copper extraction rates are under standard leaching conditions, and are in the range of 70%, while manganese dissolution rates are less than 2%. The redox potential is high and in the order of 770 mV.

The results from leaching under oxidizing conditions (produced by adding oxygen) are analogous to those obtained under standard conditions, with an extraction of 2% for manganese and 73% for copper. Therefore, the addition of an oxidant does not improve the black copper dissolution rate.

The addition of reducing agents ($FeSO_4$ or SO_2) decreases the redox potential to 696 and 431 mV, respectively, and favors the dissolution of manganese. This increases the overall copper extraction rate in minerals with high copper wad content. The use of SO_2 resulted in the highest extraction rates of 86.2% for sample RA and 75.5% for LB, which is 15% higher than the rates obtained under standard conditions.

Author Contributions: Conceptualization, O.B., M.H., and Y.Z.; methodology, O.B., M.H., and E.M.; validation, O.B., M.H., and Y.Z.; formal analysis, O.B., E.M., V.Q., and D.N.; investigation, O.B., and M.H.; resources, O.B., M.H., and Y.Z.; data curation, E.M., V.Q., and D.N.; writing—original draft preparation, O.B., E.M., and V.Q.; writing—review and editing, O.B., E.M., V.Q.; visualization, E.M., V.Q.; supervision, O.B., M.H., D.N., and Y.Z.; project administration, O.B., and M.H.; funding acquisition, O.B., M.H., and Y.Z.

Funding: This research was funded by CORFO INNOVA (No. 08CM01-09).

Acknowledgments: The authors are grateful for the financial support provided by the CORFO INNOVA Program and Glencore Lomas Bayas to the project "Geometallurgical Modelling and Design of a Specific Process for the Recovery of Ores with Black Copper Content".

Conflicts of Interest: The authors declare no conflict of interest. The funders had no role in the design of the study, the collection, analysis, or interpretation of data, the writing of the manuscript, or the decision to publish the results.

References

1. Münchmeyer, C. Exotic deposits-products of lateral migration of supergene solutions from porphyry copper deposits. In *Andean Copper Deposits: New Discoveries, Mineralization, Styles and Metallogeny;* Society of Economic Geologists, Special Publication: Tulsa, OK, USA, 1996; pp. 43–58.
2. Pinget, M.; Dold, B.; Fontboté, L. Exotic mineralization at Chuquicamata, Chile: Focus on the copper wad enigma. In Proceedings of the 10th Swiss Geoscience Meeting, Bern, Swiss, 16–17 November 2012; pp. 88–89.
3. Riquelme, R.; Tapia, M.; Campos, E.; Mpodozis, C.; Carretier, S.; González, R.; Muñoz, S.; Fernández-Mort, A.; Sanchez, C.; Marquardt, C. Supergene and exotic Cu mineralization occur during periods of landscape stability in the Centinela Mining District, Atacama Desert. *Basin Res.* **2018**, *30*, 395–425. [CrossRef]
4. Pincheira, M.; Dagnini, A.; Kelm, U.; Helle, S. Copper Pitch Y Copper Wad: Contraste Entre Las Fases Presentes En Las Cabezas Y En Los Ripios En Pruebas De Mina sur, Chuquicamata. In Proceedings of the X Congreso Geológico Chileno, Concepción, Chile, 6–10 October 2003; p. 10.
5. Mote, T.I.; Becker, T.A.; Renne, P.; Brimhall, G.H. Chronology of exotic mineralization at El Salvador, Chile, by 40Ar 39Ar dating of copper wad and supergene alunite. *Econ. Geol.* **2001**, *92*, 351–366. [CrossRef]
6. Helle, S.; Pincheira, M.; Jerez, O.; Kelm, U. Sequential extraction to predict the leaching potential of refractory black copper ores. In Proceedings of the XV Balkan Mineral Processing Congress, Sozopol, Bulgaria, 12–16 June 2013; pp. 109–111.
7. Velásquez-Yévenes, L.; Quezada-Reyes, V. Influence of seawater and discard brine on the dissolution of copper ore and copper concentrate. *Hydrometallurgy* **2018**, *180*, 88–95. [CrossRef]
8. Mukherjee, A.; Raichur, A.M.; Modak, J.M.; Natarajan, K.A. Bioprocessing of polymetallic Indian Ocean nodules using a marine isolate. *Hydrometallurgy* **2004**, *73*, 205–213. [CrossRef]
9. Acharya, R.; Ghosh, M.K.; Anand, S.; Das, R.P. Leaching of metals from Indian ocean nodules in SO2-H_2O-H_2SO_4-$(NH_4)_2SO_4$ medium. *Hydrometallurgy* **1999**, *53*, 169–175. [CrossRef]
10. Senanayake, G. A mixed surface reaction kinetic model for the reductive leaching of manganese dioxide with acidic sulfur dioxide. *Hydrometallurgy* **2004**, *73*, 215–224. [CrossRef]
11. Naik, P.K.; Sukla, L.B.; Das, S.C. Aqueous SO_2 leaching studies on Nishikhal manganese ore through factorial experiment. *Hydrometallurgy* **2000**, *54*, 217–228. [CrossRef]
12. Sinha, M.K.; Purcell, W. Reducing agents in the leaching of manganese ores: A comprehensive review. *Hydrometallurgy* **2019**, *187*, 168–186. [CrossRef]
13. Sahoo, R.N.; Naik, P.K.; Das, S.C. Leaching of manganese from low-grade manganese ore using oxalic acid as reductant in sulphuric acid solution. *Hydrometallurgy* **2001**, *62*, 157–163. [CrossRef]
14. Jiang, T.; Yang, Y.; Huang, Z.; Qiu, G. Simultaneous leaching of manganese and silver from manganese-silver ores at room temperature. *Hydrometallurgy* **2003**, *69*, 177–186. [CrossRef]
15. Jiang, T.; Yang, Y.; Huang, Z.; Zhang, B.; Qiu, G. Leaching kinetics of pyrolusite from manganese-silver ores in the presence of hydrogen peroxide. *Hydrometallurgy* **2004**, *72*, 129–138. [CrossRef]
16. Das, S.C.; Sahoo, P.K.; Rao, P.K. Extraction of manganese from low-grade manganese ores by $FeSO_4$ leaching. *Hydrometallurgy* **1982**, *8*, 35–47. [CrossRef]
17. Bafghi, M.S.; Zakeri, A.; Ghasemi, Z.; Adeli, M. Reductive dissolution of manganese ore in sulfuric acid in the presence of iron metal. *Hydrometallurgy* **2008**, *90*, 207–212. [CrossRef]
18. Kanungo, S.B. Rate process of the reduction leaching of manganese nodules in dilute HCl in presence of pyrite. Part II: Leaching behavior of manganese. *Hydrometallurgy* **1999**, *52*, 331–347. [CrossRef]
19. Helle, S.; Kelm, U.; Barrientos, A.; Rivas, P.; Reghezza, A. Improvement of mineralogical and chemical characterization to predict the acid leaching of geometallurgical units from Mina Sur, Chuquicamata, Chile. *Miner. Eng.* **2005**, *18*, 1334–1336. [CrossRef]
20. Menzies, A.; Campos, E.; Hernández, V.; Sola, S.; Riquelme, R. Understanding Exotic-Cu Mineralisation Part II: Characterization of "Black Copper"ore (Cobre Negro). In Proceedings of the 13th SGA Biennial Meeting, Nancy, France, 24–27 August 2015; pp. 3–6.

21. Han, K.; Fuerstenau, D. Acid leaching of ocean floor manganese at elevated temperature. *Int. J. Miner. Process.* **1975**, *2*, 163–171. [CrossRef]
22. Zakeri, A.; Bafghi, M.S.; Shahriari, S.; Das, S.C.; Sahoo, P.K.; Rao, P.K. Dissolution kinetics of manganese dioxide ore in sulfuric acid in the presence of ferrous ion. *Hydrometallurgy* **2007**, *8*, 22–27.

Article

Use of the Swebrec Function to Model Particle Size Distribution in an Industrial-Scale Ni-Co Ore Grinding Circuit

Alfredo L. Coello-Velázquez [1], Víctor Quijano Arteaga [1], Juan M. Menéndez-Aguado [2,*], Francisco M. Pole [3,4] and Luis Llorente [5]

1 CETAM, Universidad de Moa Dr. Antonio Núñez Jiménez, Bahía de Moa 83300, Cuba
2 Escuela Politécnica de Mieres, Universidad de Oviedo, Mieres 33600, Spain
3 Escola Superior Politécnica da Lunda Sul, Universidade Lueji A'Nkonde, Saurimo 161, Angola
4 Sociedade Mineira de Catoca Lda., Luanda 10257, Angola
5 Empresa Comandante Ernesto Ché Guevara., Punta Gorda 83310, Cuba
* Correspondence: maguado@uniovi.es; Tel.: +34-985-458-033

Received: 9 July 2019; Accepted: 8 August 2019; Published: 10 August 2019

Abstract: Mathematical models of particle size distribution (PSD) are necessary in the modelling and simulation of comminution circuits. In order to evaluate the application of the Swebrec PSD model (SWEF) in the grinding circuit at the Punta Gorda Ni-Co plant, a sampling campaign was carried out with variations in the operating parameters. Subsequently, the fitting of the data to the Gates-Gaudin-Schumann (GGS), Rosin-Rammler (RRS) and SWEF PSD functions was evaluated under statistical criteria. The fitting of the evaluated distribution models showed that these functions are characterized as being sufficiently accurate, as the estimation error does not exceed 3.0% in any of the cases. In the particular case of the Swebrec function, reproducibility for all the products is high. Furthermore, its estimation error does not exceed 2.7% in any of the cases, with a correlation coefficient of the ratio between experimental and simulated data greater than 0.99.

Keywords: comminution; grinding circuit; Swebrec function; size distribution models; modelling; lateritic ore

---

1. Introduction

The particle size distribution (PSD) in granular materials is one of the quality indicators of many transformation processes. In the particular case of mining, it is a measure of effectiveness [1] from blasting operations through to comminution. It is a crucial element of control in the processing of materials. In the specific case of the Punta Gorda milling plant (Moa, Cuba), P80 governs the quality of the final product, together with P95 and P70, product sizes which should correspond to 149 μm, 74 μm and 44 μm, respectively.

The Punta Gorda industrial plant processes lateritic minerals from deposits in northeastern Cuba by means of the ammonium-carbonate technology to obtain nickel and cobalt concentrates. Grinding plays an important role in this technological context, ensuring an adequate contact surface of the ore to allow the reduction of nickel and cobalt to their metallic phases and the subsequent solution of these elements in the leaching process [2–5]

The mathematical representation of the PSD is a mandatory tool in the modelling and simulation of processes, in addition to being crucial for controlling efficiency in comminution circuits. In most existing PSD models, the particle size is plotted against the cumulative undersize [6], p representing the probability that a fragment is smaller than x, the particle size.

In ore processing, PSD can be obtained directly through sieve analysis, or indirectly via the processing of a digital image and the use of sensors based on artificial neural networks [7–9].

Ouchterlony [10,11] argues that these techniques based on image processing have many drawbacks, one of which is the tendency to provide incorrect shapes of the size distribution curves. Therefore, the integration of PSD models in the different ways in which particle sizes are experimentally determined and the determination of the parameters of these functions is both a straightforward and satisfactory means of characterization, control and automation in comminution processes.

In comminution and mechanical separation processes, the PSD is limited to the determination of the F80 feed size and P80 product size diameters, as likewise occurs in the standard tests to determine the Bond index [12]. In many cases, the process engineer uses the graphic method for this purpose. This behaviour can have serious consequences in decision-making in the operation and control of comminution and mechanical concentration processes.

The aim of the present study is to present the potential advantages of applying the Swebrec (SWEF) function [10] in the modelling of the PSD of the products of the dry grinding circuit of lateritic minerals.

1.1. Models of Particle Size Distribution

In practice, the most widespread PSD models in ore processing are two-parameter mathematical functions, one parameter being the size coefficient and the other, the distribution coefficient [13].

The Rosin-Rammler model (RRS) is usually considered the best size distribution model for ore processing applications, although the Gates-Gauss-Schumann distribution (GGS) is preferred in certain applications such as coal processing, especially in North America [9]. Álvarez Rodríguez et al. [12] concluded that it is preferable to use the RRS model to determine the F80 and P80 diameters in the Bond ball mill test.

For the characterization of the PSD of blast fragmentation, Blair [14] obtained excellent results with the two-parameter logarithmic function for particles smaller than 0.1 mm; however, the worst results were obtained using the RRS function. For this case, the fragmentation function that provided the best fit was based on the Kuz-Ram model [15]. Other models exist, such as the two-parameter model proposed by Djordjevic [16] and the crush zone model [17], which were developed for the particle size characterization of blast fragmentation [6]. There are no known applications of these models in ore processing.

The main drawbacks of the larger models in ore processing (GGS and RRS) reside in the respective expansions and contractions of certain regions of the curve.

1.2. Swebrec Function

The Swebrec function (SWEF), proposed by Ouchterlony [10,18], is given by Equation (1):

$$P(x) = \frac{1}{1 + \left[\frac{\ln\left(\frac{x_{max}}{x}\right)}{\ln\left(\frac{x_{max}}{x_{50}}\right)}\right]^{b}} \quad (1)$$

where:

- $P(x)$, cumulative undersize, u
- x_{max}, maximum size of ore particles, mm.
- x, size, mm.
- x_{50}, sieve size that retains 50% of the material, mm.

Unlike the most commonly used models in ore processing, this is a three-parameter size distribution model. The parameters are: x_{max}, x_{50} and b. When $b = 1$, the inflection point tends towards $x = x_{max}$ and when b increases to 2, the inflection point tends towards $x = x_{50}$.

2. Methodology

Data were obtained by carrying out a full factorial experiment in the dry grinding circuit at the Punta Gorda plant. In this experiment, eight trials were devised with different operating conditions. Three operational variables were studied: ore feed flow (t/h), air flow rate to the mill (m^3/h) and the position of the air separator vanes (u), with a 45% ball load in all cases. The base levels of the variables were: 100 t/h; 68700 m^3, and 5, respectively.

Primary samples were taken every 15 min in each operating mode, which had a duration of 8 h. The samples were subjected to sieving analysis to determine the particle size composition: up to 1 mm dry, and below 1 mm wet. The following sieves were used for the fresh feed material: 40.0, 25.0, 18.0, 12.0, 10.0, 6.0, 4.75, 3.5, 2.0, 1.25, 1.0, 0.6, 0.4, 0.2, 0.1, and 0.074 mm. For the recirculated product in the mill, the product discharged from the mill and final product of the circuit, the following sieve sizes were used: 0.149; 0.074, and 0.044 mm.

The resulting data were subjected to the Coello procedure [19] in order to avoid the effect of fluctuations in the characteristics and properties of the fresh material, and errors and anomalous variations in the size class balance. Results are shown in Appendix A (Tables A1–A3).

The experimentally obtained PSDs of the different products of the milling circuit were fitted to the GGS, RRS and SWEF models according to the procedures set out in [6,9,11]. The coefficient of determination, the estimation error and Cochran's criterion were used as quality criteria for the fitting. The procedures were run with the help of Excel software templates prepared for such purposes. Considering the results of the values of the Cochran and Chi-Square criteria, the average values of the experiment were presented for each size distribution function.

3. Results and Discussion

3.1. Balance of the Size Class Distribution of the Grinding Circuit Minimizing Residual Errors

A simplified closed circuit scheme of lateritic ore comminution similar to that used in the grinding plant at the Punta Gorda nickel processing plant is presented in [19,20].

The results of calculating the circulating load per equation by means of the conventional equation appear in Table A1. As a result of the inherent fluctuations in the technological process, above all at the scale at which the sampling took place, the values of the residuals vary considerably (Table A2), by both excess and defect.

Table A3 shows the recalculated values of the circulating load and the PSD whose residual in the class balance is zero. This means that the newly found values of the particle size distribution better satisfy the calculation of the circulating load and hence the general balance of the particle size distribution of the size classes.

3.2. Fresh Feed to the Mill

Figure 1 shows the results of the particle size distribution of the fresh feed product to the mill.

In general, the experimental PSD occupies an intermediate position between the RRS and SWEF distributions. A more detailed analysis shows that for particles larger than 10 mm, the experimental data is closer to the RRS distribution; i.e., this function provides a better fit to the experimental data in this region of the curve. The mass fraction corresponding to this coarse class (>10 mm) is approximately 8%.

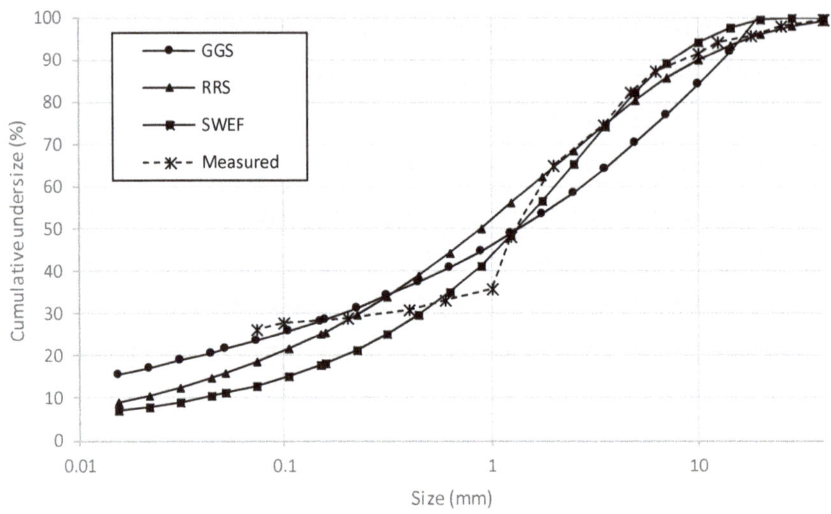

Figure 1. Size distribution of the ore particles of the fresh feed to the mill.

The region of the experimental distribution between sizes below 6–8 mm behaves very differently, however. This region is closer to the distribution of the SWEF, and accounts up to 87% of particles. In short, the RRS function best describes the coarse size ranges, while the SWEF better fits the experimental data in the fine fractions.

According to the R^2 values, the SWEF provides the best fit, the value of its coefficient of determination, R^2, being lower than those presented by Ouchterlony [1] for products of fragmentation by blasting and crushing. It should be noted that the fresh feed to the mill did not undergo any comminution operation after it was mined, although this product was influenced by the operations of mining, loading, transport, unloading and drying. Therefore, it may be considered a run-of-mine material. As to the value of coefficient b, in this case it does fit the range of values reported by Sanchidrián et al. [6], although it is slightly higher than the values of b reported for crushing.

It can be seen in Figure 2 that only four values are separated to some extent from the zero error line, which is an indicator of a good fit of the experimental data to the Swebrec function.

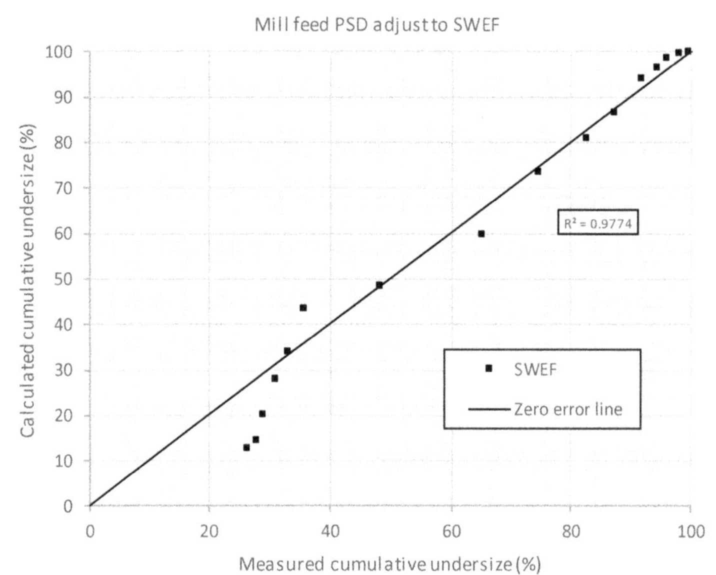

Figure 2. Zero error graph for the SWEF.

3.3. *Mill Discharge*

Figure 3 shows the particle size distribution of the mill discharge fitted to the PSD models studied in this paper. As can be seen in the figure, the SWEF occupies an intermediate position between the GGS and RRS distributions for sizes between 35 and 7 microns. For sizes below 7 μm and between 35 μm and 149 μm, the SWEF distribution respectively approaches the GGS and RRS distribution functions.

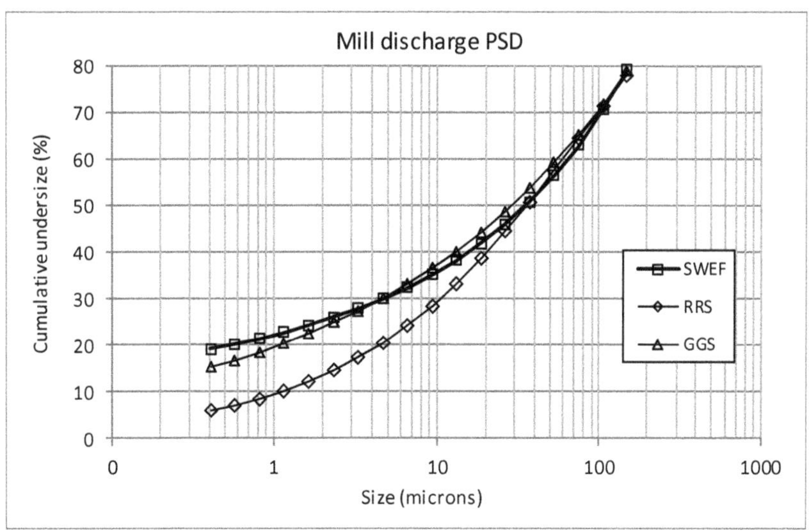

Figure 3. Particle size distribution simulated by the models under study.

The reported value of b (Table 1) falls within the range of values reported in [6], although it is somewhat lower than the value of b reported for fresh feed to the mill. The R^2 is higher than that obtained for the latter product. Considering the value of the estimation error (σ), the best fit corresponds to the SWEF model. The coefficient of determination for the experimental data and those simulated by the SWEF model is 0.9999 (Figure 4).

Table 1. Studied model parameters, mill discharge.

MDTP	Model Parameters					R^2	σ
	Slope, m	K (*)	X_{max}	X_{50}	b		
GGS	0.2792	0.3461	-	-	-	0.9867	2.14
RRS	0.5425	0.0703	-	-	-	0.9867	3.04
SWEF	-	-	0.3461	0.0353	1.3354	0.9927	1.77

(*) $d_{62.8}$—for RRS and d_{100}—for GGS.

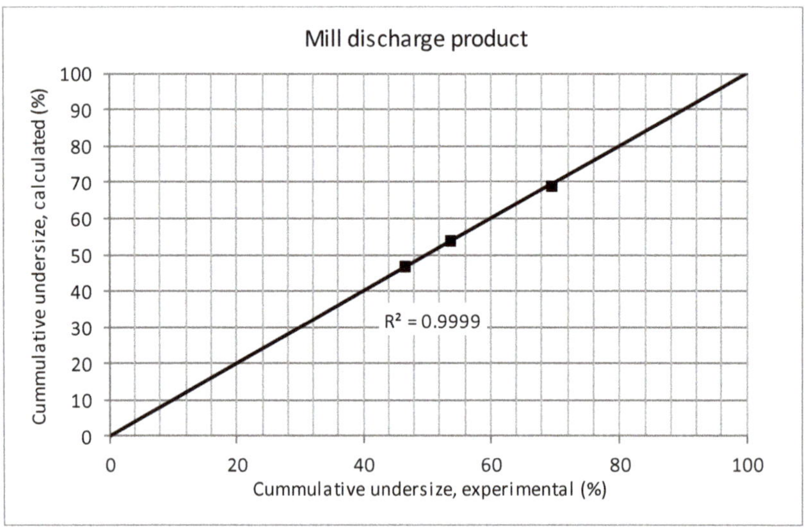

Figure 4. Zero error graph for the mill discharge.

The calculated Cochran criterion (0.2858) is lower than the critical criterion (0.6798), which indicates good homogeneity in the error made in the experimentation and the absence of significant differences between the series of experimental data and the data simulated by this function. It was similar for the other distributions under study.

3.4. Recirculated Product in the Mill

Figure 5 shows the size distribution of the product recirculated in the grinding circuit at the Punta Gorda plant fitted to the GGS, RRS and SWEF models.

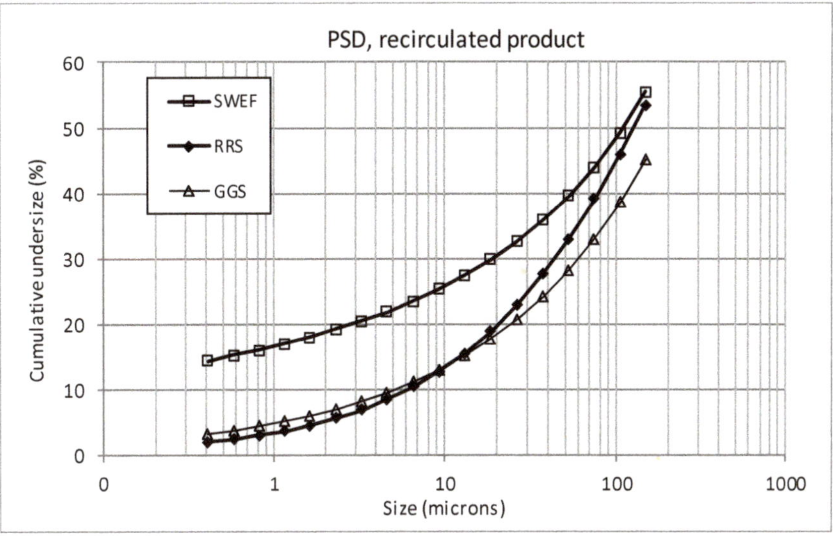

Figure 5. Particle size distribution of the product recirculated in the grinding circuit.

The configuration of the particle size distribution curve is similar to that of the mill discharge, the value of $b = 1.3206$ being very similar (see Tables 1 and 2).

Table 2. Parameters of the distribution models under study, recirculated product.

MDTP	Model Parameters					R^2	σ
	Slope, m	K *	X_{max}	X_{50}	b		
GGS	0.4507	0.8732	-	-	-	0.9605	1.9277
RRS	0.6224	0.2299	-	-	-	0.9538	2.1163
SWEF	-	-	0.7787	0.1104	1.3206	0.9743	2.6468

(*) $d_{62.8}$—for RRS and d_{100}—for GGS.

The curve of the Swebrec function is always above the other distribution models, tending to approach the curve of the RRS distribution. This function occupies precisely the intermediate position.

For this product, the SWEF distribution shows the best-fit parameter value, although the estimation error is greater than in the other two distributions. However, the error is lower than 2.7% and it may be stated that the SWEF characterizes the particle size distribution of the mill discharge with sufficient accuracy.

Figure 6 shows the zero error graph for the case of the recirculated product. It includes the coefficient of determination, which in this case is 0.9925, thus confirming the model's suitability. The calculated Cochran criterion is lower than the critical criterion (0.6088 versus 0.6798).

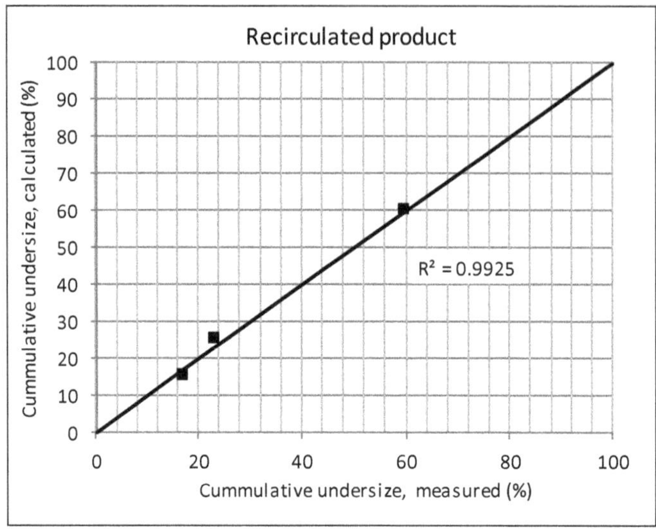

Figure 6. Zero error graph, recirculated product fitted to the SWEF model.

3.5. Final Product of the Circuit

The fitting of the size distribution functions evaluated in this paper is presented in Figure 7. The SWEF occupies an intermediate position between the other two studied distributions, tending to parallel the GGS model to a certain extent. Here, the RRS distribution behaves quite differently from the other two models towards the finer fractions, although all of the functions converge from 37 μm to the coarser classes. The estimation error does not exceed 2.5% for any of the functions.

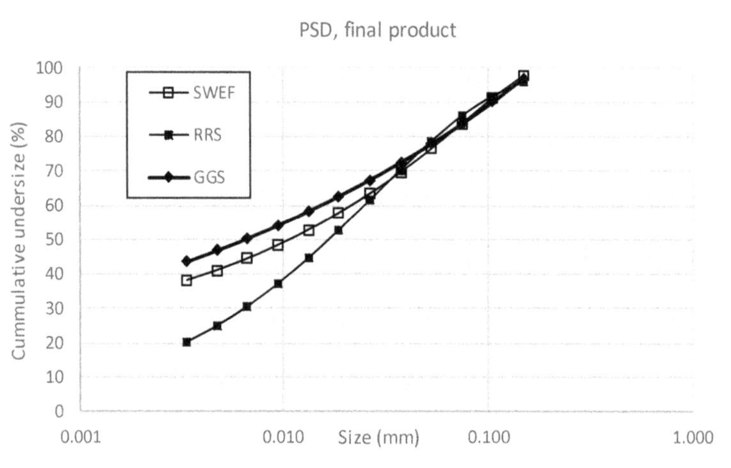

Figure 7. Distribution models fitted to the experimental data.

The value of coefficient *b* is somewhat greater than coefficient *b* of the size distributions of the products discharged from and recirculated in the mill. This justifies the similar behaviour of the SWEF distribution model for these products.

As expected, the maximum size in the distribution is lower than that of the other products. Something similar occurs with the x_{50} size, which constitutes the cut-off size of the pneumatic separator.

The zero error graph (Figure 8) confirms the reproducibility of the SWEF model. The coefficient of determination of the experimental data and the data simulated by this model is 0.9925. The values of the calculated Cochran criterion and the critical criterion (0.2858 versus 0.6798) confirm the high quality of the model.

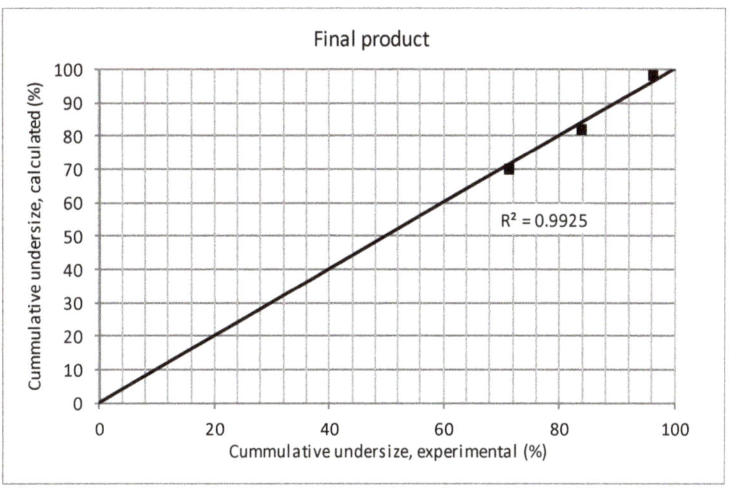

Figure 8. Zero error graph of the experimental data and the data simulated by the SWEF.

Figure 9 shows the PSDs of the different products of the grinding circuit simulated by the Swebrec function for normalized sizes with a sieving scale of $\sqrt{2}$. The shape of the distribution for the products inside the circuit is similar, the tendency following the logic of the x_{max} and x_{50} sizes reported in Tables 1–4.

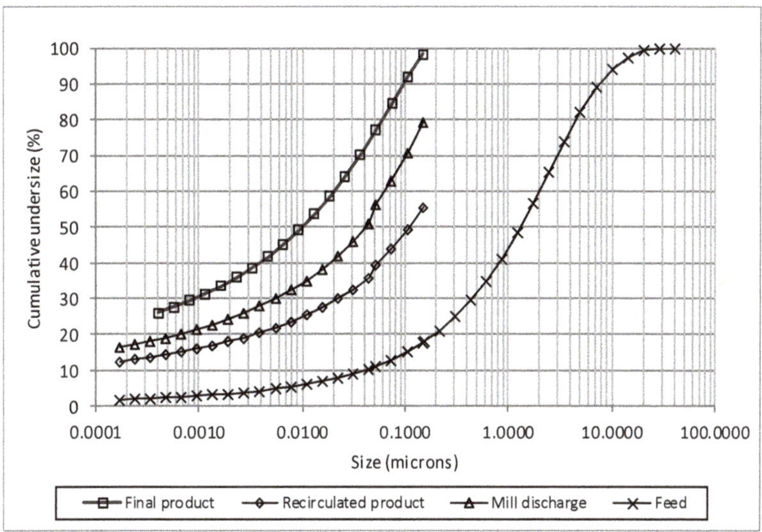

Figure 9. Simulation of the particle size distribution for products in the grinding circuit.

Table 3. Parameters of the distribution models under study.

Model	Model Parameters					R^2
	Slope, m	K (*)	X_{max}	X_{50}	b	
GGS	0.2493	0.429	-	-	-	0.900
RRS	0.3352	1.857	-	-	-	0.954
SWEF	-	-	40.0	1.331	3.10	0.977

(*) $d_{62.8}$ for RRS and d_{100} for GGS.

Table 4. Parameters of the distribution models for the final product of the circuit.

MDTP	Model Parameters					R^2	σ
	Slope, m	K (*)	X_{max}	X_{50}	b		
GGS	0.2026	0.1690	-	-	-	0.9763	1.4051
RRS	0.7489	0.0284	-	-	-	0.9869	1.1234
SWEF	-	-	0.1749	0.0099	1.401	0.9855	2.2193

(*) $d_{62.8}$ for RRS and d_{100} for GGS.

4. Conclusions

The results of applying the Coello procedure [19] allowed us to obtain the particle size distribution of the different products of the milling circuit suitably balanced with zero residual values.

For all the size distributions, no significant differences were found between the experimental data and the data simulated by the size distribution models. In combination with the values obtained from the coefficient of determination and the estimation error, this means that all the studied models are reproducible. The Cochran criterion and Chi-square values confirm the above: the calculated values are always better than the critical values of the considered statistics.

The Swebrec function characterizes the particle size distribution of the products of the grinding circuit very accurately. The coefficient of determination for the experimental data and those simulated by the model is greater than 0.9900. The average error of its fitting to the size distribution of the products of the grinding circuit does not exceed 3%, demonstrating the versatility of its use. The coefficient of determination values for the experimental data and those simulated by this function are predominantly very high.

The b value for the products inside the grinding circuit ranges between 1.32 and 1.4. The b value for the fresh feed (crushed product) coincides with the values reported by Sanchidrián et al. [6] for crushing.

Author Contributions: Conceptualization, J.M.M.-A., A.L.C.-V.; Formal analysis, V.Q.A. and F.M.P.; Investigation, F.M.P. and R.M.A.; Methodology, A.L.C.-V. and J.M.M.-A.; Resources, V.Q.A. and F.M.P.; Supervision, V.Q.A. and A.L.C.-V.; Validation, L.L.; Writing - original draft, A.L.C.-V.;Writing—review & editing, J.M.M.-A.

Funding: This research received no external funding.

Acknowledgments: The authors are grateful to Alexis García for his support during the carrying out of the industrial tests and to the Technical Board of the ECCG company for the trust placed in the research team.

Conflicts of Interest: The authors declare no conflict of interest

Appendix A

Table A1. Calculation of the circulating load in the grinding circuit.

Flow of Material, m^3/h	Air Flow, m^3/h	Slope Angle	Size Class, mm	Content of the Class, %			Circulating Load, $C(i)$
				Mill Discharge, f'	Recirculated Product, c	Fine Product, p	
			−0.149 + 0.074	87.58	54.66	95.66	0.25
+	+	+	−0.074 + 0.044	79.22	39.70	85.08	0.15
			−0.044 + 0.0	64.13	39.20	75.71	0.46
			−0.149 + 0.074	67.67	64.00	97.09	8.01
+	−	−	−0.074 + 0.044	50.41	44.66	86.92	6.35
			−0.044 + 0.0	44.91	39.03	79.38	5.86
			−0.149 + 0.074	81.03	62.71	97.67	0.91
−	+	−	−0.074 + 0.044	61.31	48.54	88.16	2.10
			−0.044 + 0.0	50.2	42.06	80.34	3.70
			−0.149 + 0.074	85.28	49.405	94.06	0.24
−	−	+	−0.074 + 0.044	69.21	38.41	87.3	0.59
			−0.044 + 0.0	60.56	37.93	73.56	0.57
			−0.149 + 0.074	85.34	65.60	97.41	0.61
+	+	−	−0.074 + 0.044	76.89	46.12	87.17	0.33
			−0.044 + 0.0	53.49	38.73	78.77	1.71
			−0.149 + 0.074	70.63	61.04	95.92	2.64
+	−	+	−0.074 + 0.044	52.58	46.46	84.2	5.16
			−0.044 + 0.0	46.51	39.94	71.33	3.78
			−0.149 + 0.074	81.93	51.51	96.29	0.47
−	+	+	−0.074 + 0.044	66.17	43.86	86.9	0.93
			−0.044 + 0.0	59.77	33.46	74.84	0.57
			−0.149 + 0.074	79.46	60.08	97.02	0.91
−	−	−	−0.074 + 0.044	61.37	22.37	85.1	0.61
			−0.044 + 0.0	53.13	17.18	78.77	0.71

Table A2. Calculation of residuals.

Recalculated Circulating Load					Residual, ε	Lagrange Multiplier, λ
p-f	f-c	$(f\text{-}c)^2$	(p-f)*(f-c)	C (i)		
8.08	32.92	1083.73	265.99		−0.16	−0,06
5.86	39.53	1562.23	231.62		3.65	1.41
11.58	24.93	621.50	288.69		−5.58	−2.15
Sum Total		3267.46	786.30	0.24		
29.42	3.68	13.51	108.12		−5.84	−0,06
36.51	5.75	33.06	209.93		0.39	0.00
34.47	5.88	34.57	202.68		3.27	0.03
Sum Total		81.14	520.73	6.42		
16.64	18.33	335.81	304.93		12.32	1.21
26.85	12.77	163.07	342.87		−6.67	−0.66
30.14	8.14	66.26	245.34		−17.28	−1.70
Sum Total		565.14	893.14	1.58		
8.78	35.875	1287.02	314.98		6.45	2.01
18.09	30.8	948.64	557.17		−5.02	−1.56
13	22.63	512.12	294.19		-3.39	−1.06
Sum Total		2747.77	1166.34	0.42		
12.07	19.74	389.67	238.26		−0.29	−0.07
10.28	30.775	947.10	316.37		8.09	2.07
25.28	14.76	217.86	373.13		−16.47	−4.22
Sum Total		1554.63	927.76	0.60		

Table A2. *Cont.*

Recalculated Circulating Load					Residual, ε	Lagrange Multiplier, λ
p-f	f-c	$(f\text{-}c)^2$	(p-f)*(f-c)	C (*i*)		
25.29	9.59	91.97	242.53		8.00	0.24
31.62	6.125	37.52	193.67		−10.36	−0.31
24.82	6.57	43.16	163.07		−2.02	−0.06
Sum Total		172.65	599.27	3.47		
14.36	30.42	925.38	436.83		4.27	1.07
20.73	22.315	497.96	462.59		−7.06	−1.78
15.07	26.315	692.48	396.57		1.05	0.26
Sum Total		2115.81	1295.99	0.61		
17.56	19.385	375.78	340.40		−4.26	−0.99
23.73	39.005	1521.39	925.59		3.02	0.70
25.64	35.955	1292.76	921.89		−0.98	−0.23
Sum Total		3189.93	2187.88	0.69		

Table A3. Recalculated values of the particle size distribution for the grinding circuit products.

Recalculated Class Content, %			Recalculated Residual, ε
Mill Discharge, f.	Recirculated Product, c	Fine Product, p	
87.66	54.65	95.60	0
77.48	40.03	86.49	0
66.80	38.68	73.56	0
68.1	63.6	97.03	0
50.4	44.7	86.92	0
44.7	39.2	79.41	0
77.9	64.6	98.88	0
63.0	47.5	87.50	0
54.6	39.4	78.64	0
82.4	50.3	96.07	0
71.4	37.7	85.74	0
62.1	37.5	72.50	0
85.5	65.6	97.34	0
73.6	47.4	89.24	0
60.2	36.2	74.55	0
69.5	61.9	96.16	0
54.0	45.4	83.89	0
46.8	39.7	71.27	0
80.2	52.2	97.36	0
69.0	42.8	85.12	0
59.3	33.6	75.10	0
81.1	59.4	96.03	0
60.2	22.8	85.80	0
53.5	17.0	78.54	0

References

1. Ouchterlony, F. Fragmentation characterization; the Swebrec function and its use in blast engineering. In Proceedings of the 9th International Symposium on Rock Fragmentation by Blasting—Fragblast 9, Granada, Spain, 13–17 August 2009; pp. 3–22.
2. Coello-Velázquez, A.L. Sovershenstvovanie tecnologii izmilchenii lateritobij pud na zabode "Punta Gorda". Ph.D. Thesis, IMS, Saint Petersburg, Russia, 1993.
3. Aguado, J.M.M.; Velázquez, A.L.C.; Tijonov, O.N.; Díaz, M.A.R. Implementation of energy sustainability concepts during the comminution process of the Punta Gorda nickel ore plant (Cuba). *Powder Technol.* **2006**, *170*, 153–157. [CrossRef]
4. Coello Velázquez, A.L.; Menéndez-Aguado, J.M.; Brown, R.L. Grindability of lateritic nickel ores in Cuba. *Powder Technol.* **2008**, *182*, 113–115. [CrossRef]

5. Chalkley, M.E.; Collins, M.J.; Iglesias, C.; Tuffrey, N.E. Effect of magnesium on pressure leaching of Moa Laterite ore. *Can. Metall. Q* **2010**, *49*, 227–234. [CrossRef]
6. Sanchidrián, J.A.; Ouchterlony, F.; Moser, P.; Segarra, P.; López, L.M. Performance of some distributions to describe rock fragmentation data. *Int. J. Rock Mech. Min. Sci.* **2012**, *53*, 18–31. [CrossRef]
7. Sbarbaro, D.; Ascencio, P.; Espinoza, P.; Mujica, F.; Cortes, G. Adaptive soft-sensors for on-line particles estimation in wet grinding circuits. *Control Eng. Practice.* **2008**, *16*, 171–178. [CrossRef]
8. Ko, Y.-D.; Shang, H. A neural network-based soft sensor for particle size distribution using image analysis. *Powder Technol.* **2011**, *212*, 359–366. [CrossRef]
9. Wills, B.A.; Finch, J.A. *Wills' Mineral Processing Technology: An Introduction to the Practical Aspects of Ore Treatment and Mineral Recovery*, 8th ed.; Elsevier: Amsterdam, The Netherlands, 2016; p. 498.
10. Ouchterlony, F. The Swebrec function: Linking fragmentation by blasting and crushing. Institution of Mining and Metallurgy. *Min. Technol.* **2005**, *114*, 29–44. [CrossRef]
11. Ouchterlony, F.; Moser, P. Likenesses and differences in the fragmentation of full-scale and model-scale blasts. In Proceedings of the 8th International Symposium on Rock Fragmentation by Blasting–Frag blast 8, Santiago, Chile, 7–11 May 2006; pp. 207–220.
12. Álvarez Rodríguez, B.; González García, G.; Coello-Velázquez, A.L.; Menéndez-Aguado, J.M. Product size distribution function influence on interpolation calculations in the Bond ball mill grindability test. *Int. J. Miner. Process.* **2016**, *157*, 16–20. [CrossRef]
13. Kelly, E.G.; Spottiswood, D.J. *Introduction to Mineral Processing*; John Willey and Sons: Hoboken, NJ, USA, 1982.
14. Blair, D.P. Curve-fitting schemes for fragmentation data. *Fragblast. Int. J. Blast. Fragment.* **2004**, *8*, 137–150.
15. Kuznetsov, V.M. The mean diameter of the fragments formed by blasting rock. *Sov. Min. Sci.* **1973**, *9*, 144–148. [CrossRef]
16. Djordjevic, N. Two-component model of blast fragmentation. In Proceedings of the 6th International Symposium on Rock Fragmentation by Blasting, Johannesburg, South Africa, 8–12 August 1999; pp. 213–219.
17. Kanchibotla, S.S.; Valery, W.; Morell, S. Modelling fines in blast fragmentation and its impact on crushing and grinding. In Proceedings of the Conference on Rock Breaking, Kalgoorlie, WA, USA, 7–11 November 1999; pp. 137–144.
18. Osorio, A.M.; Menéndez-Aguado, J.M.; Bustamante, O.; Restrepo, G.M. Fine grinding size distribution analysis using the Swebrec function. *Powder Technol.* **2014**, *258*, 206–208. [CrossRef]
19. Coello-Velázquez, A.L. Procedimiento para la determinación de la carga circulante en circuitos cerrados de trituración y molienda. *Minería y Geología* **2015**, *31*, 66–79.
20. Menéndez-Aguado, J.M.; Coello, A.L.; Tikjonov, O.N.; Rodríguez, M. Implementation of sustainable concepts during comminution process in Punta Gorda nickel ore plant. *Powder Technol.* **2006**, *170*, 153–157. [CrossRef]

Article

Oxidative Leaching of Zinc and Alkalis from Iron Blast Furnace Sludge

Ma. de Jesus Soria-Aguilar [1], Gloria Ivone Davila-Pulido [2], Francisco Raul Carrillo-Pedroza [1,*], Adrian Amilcare Gonzalez-Ibarra [2], Nallely Picazo-Rodriguez [1], Felipe de Jesus Lopez-Saucedo [2] and Juan Ramos-Cano [1]

[1] Facultad de Metalurgia, Universidad Autónoma de Coahuila, Carretera 57 Km 5, C.P. 25710 Monclova, Coahuila, Mexico; ma.soria@uadec.edu.mx (M.d.J.S.-A.); nallely_pikzo@hotmail.com (N.P.-R.); jramos@uadec.edu.mx (J.R.-C.)

[2] Escuela Superior de Ingenieria, Universidad Autónoma de Coahuila, Blvd. Adolfo López Mateos s/n C.P. 26800 Nueva Rosita, Coahuila, Mexico; gloriadavila@uadec.edu.mx (G.I.D.-P.); gonzalez-adrian@uadec.edu.mx (A.A.G.-I.); felipe.lopez@uadec.edu.mx (F.d.J.L.-S.)

* Correspondence: raul.carrillo@uadec.edu.mx; Tel.: +52-866-639-0330

Received: 20 August 2019; Accepted: 5 September 2019; Published: 18 September 2019

Abstract: The sludge from a wet-off gas cleaning system of the iron blast furnace (BF) contains significant amounts of iron; however, they cannot be recycled due to their high content of zinc and alkalis. These compounds are detrimental to the optimal performance of iron and steelmaking furnaces. In this work, a comparative laboratory study to reduce zinc and alkali contained in the blast furnace sludge (BFS) is presented. The effect of leaching parameters such as oxidant (i.e., ferric ion, oxygen or ozone), aqueous solution media (i.e., 0.2 M NH_4Cl, 0.2 M HCl and 0.1 M H_2SO_4) and temperature (i.e., 27 and 80 °C) on Zn and alkalis (Na_2O and K_2O) removal were studied by applying an experimental design. The results obtained show that Zn and K_2O removal of 85% and 75% were achieved under the following conditions: Ozone as an oxidant agent and 0.1 M H_2SO_4 as an aqueous medium, temperature had no significant effect. The results are supported by thermodynamic diagrams and the possible chemical reactions are mentioned. Although the results also indicate that leaching under the above conditions dissolves up to 9% of iron, this loss is much less than leaching without the oxidizing conditions generated by the ozone. The BFS obtained from this treatment could be recirculated to the iron or steelmaking processes to recover iron values.

Keywords: iron and steelmaking wastes; zinc; alkalis; aqueous treatment

---

1. Introduction

The iron and steelmaking processes have been always related to the generation of dust/sludge, slag and other wastes or byproducts. Dust and sludge from the gas cleaning systems of the blast furnace (BF) and the basic oxygen furnace (BOF) contains significant amounts of iron, carbon and other elements. Such materials cannot be recycled because of its impurities (nonferrous metals, mainly zinc) and represent a loss of 2% of the iron and coal contained in the raw material. In 2018, the world mine production was of 2.5 billion tons of iron ore; then, dust and sludge can be representing a loss of 15 to 20 million tons of Fe units annually) [1,2]. The composition of the blast furnace and basic oxygen furnace dusts varies significantly depending on the process conditions. Blast furnace sludge (BFS) containing 0.77–5 wt.% Zn has been reported [3,4]. A high content of zinc in the raw material might damage the blast furnace refractory materials and, consequently, shorten its campaign life [5,6]. In the case of alkalis (Na and K), they increase coke consumption due to the transfer of heat to higher levels in the blast furnace, causing catastrophic swelling and disintegration of ores, pellets and coke. It also causes deterioration of the furnace lining. In brief, zinc and alkalis cause irregularities in the operation process and decrease the production, so only it is recycled up to 20% of the dust and sludge generated [7–9].

As mentioned, not recycling of BFS represents an economical lost, due to the relatively high contents of iron (up to 50 wt.% Fe). This also represents an environmental issue, because this sludge contains leachable compounds of lead, cadmium and other elements that are classified as hazardous wastes [10]. This "waste" material is pile stocked on the steelmaking sites, which is an environmental hazard and occupies valuable land. To recycle BFS to the process, it is necessary to reduce their zinc and alkalis content to at least 0.03 wt.% and 0.12wt.%, respectively [11,12].

Pyrometallurgical and hydrometallurgical processes have been developed to recycle dust and/or to recover zinc from it. Pyrometallurgical routes are considered the primary choice because of its high potential metal recovery and among the most important processes are: Waelz, rotary heart furnace, PRIMUS, OXYCUP, coke-packed bed, Ausmelt, electric smelting reduction furnace, Plasamadust, plasma-arc, Elkem multi-purpose furnace, submerged plasma, etc. [13,14]. It is worth noting that in the metallurgical industry, the energy cost and pollution problems produced by pyrometallurgical processes (e.g., combustion gas emissions) have led to a more intensive search of hydrometallurgical alternatives [3].

The hydrometallurgical processes developed for dust treatment might be classified as leaching in acid or alkaline media [15,16]. The efforts have been focused on conventional technology, based on H_2SO_4, as is shown in Table 1. As can be observed, almost all of studies have been realized for dust from the electric arc furnace steelmaking process (EAF). It is interesting to observe the relativity high H_2SO_4 concentration used (>1 M) and temperature (>50 °C). In the case of the treatment time, it varies from 15 min (microwave heating leaching assisted) to 1.5 h [17–19].

Table 1. Leaching studies for zinc removal base in H_2SO_4 media.

Sample Type	Optimal Condition	Zn Removal, %	Fe Loss, %	Reference
EAF dust (17.05% Zn)	1 M H_2SO_4, 80 °C, 1 h.	87	up to 40%	[5]
EAF dust (26.95% Zn)	2.35 M H_2SO_4, 25 °C, 1 h.	79	4	[16]
BOF sludge (2.74% Zn)	1 M H_2SO_4, 80 °C, 15 min.	70	up to 60%	[17]
BOF sludge (0.77% Zn)	1 M H_2SO_4, 80 °C, 15 min (microwave assisted).	86	4	[18]
EAF dust (20.32% Zn)	3 M H_2SO_4, 60 °C, 1.5 h.	80	45	[10]
EAF dust (29.1% Zn)	1 M H_2SO_4, 50 °C, 1 h.	72	-	[19]
EAF dust (20.9% Zn)	1.2 M H_2SO_4, 80 °C, 1 h.	80	20	[15]

Table 1 shows that Zn removal is >70% but the iron loss has a great variability (4% to 60%), which could be due to the different composition between each material. Kukurugya et al. [5] studied the behavior of various elements (i.e., zinc, iron and calcium) during the leaching of electric arc furnace dust in sulfuric acid solutions. The authors proposed that the rate limiting step for zinc and calcium was diffusion, while for iron it was the chemical reaction. This confirms the high solubility of some iron oxide compounds, but the iron loss will depend of the type of iron oxides contained in the material.

Other acid media has been studied to steelmaking dust. Shawabkeh [19] studied the zinc extraction from a Jordanian electric arc furnace dust by testing different acids (i.e., nitric, hydrochloric and sulphuric) at different concentrations. The highest zinc extraction (i.e., 72%) was obtained with H_2SO_4 and 50 °C. Although 10% nitric acid concentration can dissolve approximately 33% of zinc [19], this leaching agent must be discarded due to the large amount of iron dissolved (it is preferable that the iron remain in the dust to recycle it as raw material).

Novel leaching agents has been used for the treatment of EAF, BF and BOF dust as is shown in Table 2. Organic acid as butyric, citric or iminodiacetic have been studied to extract zinc from basic oxygen sludges or dust and electric arc furnace dust by using the coordination reaction between the organic ligand and zinc ions [20–23]. According to results shown in the table, in some case high Zn removal are obtained (up to 100%) with low Fe loss. The table also shows the alkaline leaching (hydroxide and carbonate solutions) that has been studied due to the insolubility of the ferric in these medium [24,25]. However, pre-treatment or high concentration of lixiviant is necessary to obtain a good Zn extraction (>70%).

Table 2. Leaching studies for Zn removal using different aqueous medium.

Sample Type	Leached Media	Optimal Condition	Zn Removal, %	Fe Loss, %	Reference
BOS filter cake (6.52% Zn)	Butyric acid	1.5 M butyric acid	66	<1	[11]
EAF dust (33.2% Zn)	Citric acid (previous NaOH roasting)	0.8 M citric acid, 40 °C, 1 h	100	8	[3]
EAF dust (33.2% Zn)	Leaching with different acid and alkaline media	1.2 M HCl or 90% aqua regia; room temperature, 168 h	>90	<5	[23]
BOF dust (5.1% Zn)	Iminodiacetic acid	0.2 M iminodiacetic acid, 20 °C, 2 h	63	6	[21]
EAF dust (24.24% Zn)	Sodium hydroxide (CaO pre-treatment)	CaO pre-treatment: 1100 °C, 5 h, air atm. Leaching: 2M NaOH; 70 °C, 2 h	>95	-	[24]
EAF dust (29% Zn)	Ammonium carbonate	1 M $(NH_4)_2CO_3$, 20 °C, 2 h	49	-	[25]
EAF dust (12.2% Zn)	Sodium hydroxide	6 M NaOH; 90 °C, 4 h	74	-	[6]

It is worth mentioning that some processes based on chlorides were developed in the past by the industry. In the Hoogovens Ijmuiden and Delft process [26,27] the EAF dust is leached under oxidant conditions at approximately 140 °C with HCl to dissolve zinc and a small quantity of iron. This process only considers the Zn dissolution from zinc oxides, but not the presence of zinc ferrite. The limited amount of iron dissolved can be achieved controlling the pH of the solution to limit the attack of the ferric oxides and leaching with O_2 to get enough oxidation conditions. A similar process was developed by Terra de Gaia, which is based on the dissolution of ZnO and $ZnFe_2O_4$ by $FeCl_3$–HCl at 175 °C [28]. The dust is mixed with the solution of chloride, reacting with the dust of iron and producing gas chloride, then it is injected into the autoclave at 175 °C. The other one was the Ezinex process de Engite Impianti [29] that uses ammonium chloride as aqueous media at 80 °C. The zinc oxide in the dust was dissolved but the zinc ferrite was not perceptibly attacked.

Most of the studies mentioned so far, have been focused on the removal and recovery of zinc, mainly in the electric arc furnace. Fewer studies have been done to treat basic oxygen furnace and blast furnace dust/sludge. For the case of the alkalis, the problem has been addressed since the operation conditions of the blast furnace perspective. Practically no study addresses the simultaneous removal and recovery of zinc and alkalis, respectively.

Considering the previous developments and the most recent studies, it could be concluded that the issue of the reduction of zinc and alkalis in iron and steelmaking dusts is still relevant for the industry due to the iron-containing values and the environment aspects. The aim of the present investigation was to evaluate the elimination of zinc and alkali in a BFS using H_2SO_4, HCl and NH_4Cl solutions and employing different oxidizing agents (i.e., ferric chloride, ferric sulfate, oxygen and ozone) at 27 and 80 °C. Eh–pH diagrams were added to support the experimental results obtained.

2. Materials and Methods

2.1. Material

Samples of blast furnace sludge (BFS) were obtained from the wet-off gas cleaning system of an iron blast furnace (the collection system of the blast furnace dusts consists of three devices: Cyclone, baghouse and wet-off gas cleaning or wet scrubber; the samples were obtained from the latter). The sample was dried (110 °C by 24 h) and crumbled manually. The particle size was determined by wet sieving, using a Tyler sieve.

The BFS sample was analyzed by atomic absorption spectroscopy (AAS) using a Thermo Electron Solaar S4 spectrophotometer (Thermo Fisher Scientific Inc., Waltham, MA, USA); Ca, K and Na (reported as CaO, K_2O and Na_2O, respectively) (Sigma-Aldrich Inc., San Luis, MO, USA), a 0.2% *w*/*v* Cl_3La and 0.1% *w*/*v* CsCl (Sigma-Aldrich Inc., San Luis, MO, USA)were added to prevent reduction of the signal by aluminum and as an ionization buffer, respectively. Mg and Al (reported as MgO and Al_2O_3) were measured under more oxidative conditions using N_2O/acetylene mixed gases. SiO_2 retained in the solid was determined by gravimetry (oxidation with $HClO_4$). Determination of sulphur and total carbon was carried out using a LECO CS-244 analyzer (LECO, St. Joseph, MI, USA). Complementary chemical analyses were performed using X-Ray fluorescence (XRF) using a Brucker AXS S4 PIONEER spectrophotometer (Brucker, Billerica, MA, USA), following the Li-tetraborate fused bead technique. X-ray diffraction analysis was performance using a Brucker D8 diffractometer (Brucker, Billerica, MA, USA) and monochromatized CuKa radiation. Scanning electron microscopy analysis (Hitachi 5500 microscope; Hitachi, Tokyo, Japan) was realized using BFS powder dispersed in ethanol, sonicated and deposited on the carbon coated grid.

2.2. Experimental Procedure

Experimental tests were performed using a reactor (glass vessel) (Pyrex glass Vessel; Corning Inc., New York, NY, USA) with mechanic agitation. The BFS was mixed with a solution previously prepared, which could be a strong acid (0.1 M H_2SO_4 and 0.2 M HCl, pH = 0.9 and 1.1, respectively) or weak acid (0.2 M NH_4Cl, pH = 5), prepared in deionized pure water (conductivity = 0.06 µS). The material and solution were poured in the reactor and were mixed using an agitation rate of 500 rpm. Solutions 0.1 M of ferric ion as ferric chloride ($FeCl_3$ reagent grade, 97%, Sigma-Aldrich or ferric sulphate ($Fe_2(SO_4)_3$, reagent grade 97%, Sigma-Aldrich), oxygen (industrial grade, 99.5%) and ozone (1% *v*/*v* O_3 in O_2, generated by a Pacific Ozone L22; Pacific Ozone Technology, Benicia, CA, USA) were used to have an oxidant condition in the aqueous media. Oxygen and ozone were injected to the solution through a bubbler positioned at the bottom of the reactor. The treatment duration was of 60 minutes, at room temperature (27 °C) and 80 °C.

Table 3 shows the experimental design used in this work. The variables studied were three different leaching media: (1) NH_4Cl, (2) HCl and (3) H_2SO_4; three different oxidant agents: (1) Ferric ion ($FeCl_3$ for NH_4Cl and HCl, and $Fe_2(SO_4)_3$ for H_2SO_4, respectively), (2) oxygen and (3) ozone; and two different temperature levels: (1) 27 °C and (2) 80 °C.

Samples of the solid material were withdrawn as well as samples of the aqueous solution. Chemical analyses were carried out in the collected samples to determine the final composition of solid residues and the barren solution.

Table 3. Experimental design implemented to study the different aqueous media, oxidant agents and temperature.

Test	Leaching Media	Oxidant	Temperature
1	1	1	1
2	1	1	2
3	1	2	1
4	1	2	2
5	1	3	1
6	1	3	2
7	2	1	1
8	2	1	2
9	2	2	1
10	2	2	2
11	2	3	1
12	2	3	2
13	3	1	1
14	3	1	2
15	3	2	1
16	3	2	2
17	3	3	1
18	3	3	2

3. Results

3.1. BFS Characterization

Table 4 shows the chemical composition and the main mineral species contained in the BFS. This reconstruction was based in the chemical analysis and the X-ray diffraction analysis (Figure 1) and elemental analysis by Scanning Electron Microscopy/Energy Dispersive X-Ray Spectroscopy (SEM/EDS) (Hitachi 5500 microscope, Hitachi, Tokyo, Japan), Figure 2.

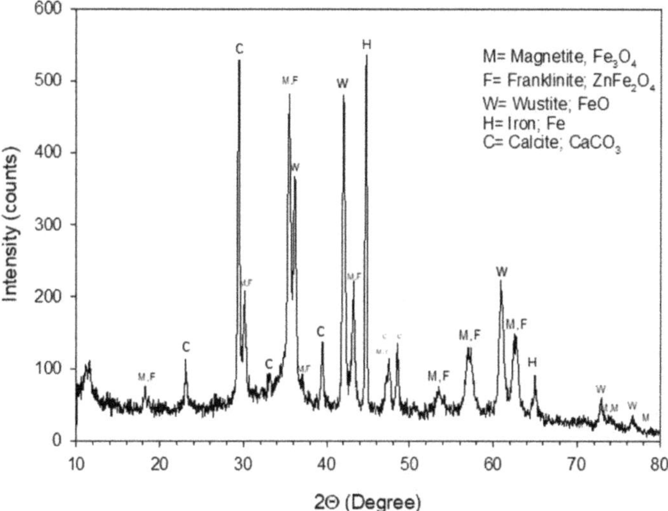

Figure 1. X-Ray diffraction pattern obtained for the BFS.

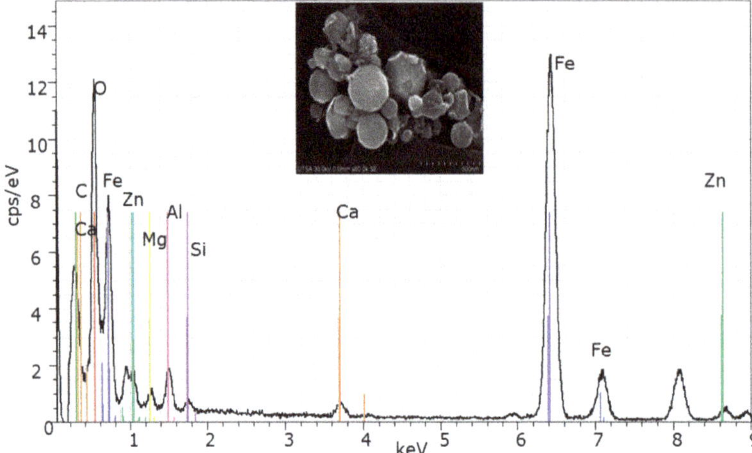

Figure 2. SEM/EDS micrograph of the BFS.

Table 4. Chemical analysis of the blast furnace sludge (BFS) and mineralogical reconstruction of the Fe compounds.

Element	Fe	Zn	Na_2O	K_2O	SiO_2	Al_2O_3	CaO	MgO	S	C
wt. %	41.55	0.45	0.98	0.28	6.36	0.27	29.90	1.12	0.50	3.94

Compounds		Weight (%)
Metallic Iron	Fe	42
Wustite	FeO	35
Magnetite	Fe_3O_4	18
Franklinite	$ZnFe_2O_4$	2

Figure 1 shows the XRD pattern of the initial sample, indicating that the residue is formed by multiple compounds. We want to show only those of Fe to determine its phases and their possible applications. As can be seen, iron metallic, wustite (FeO) and magnetite (Fe_3O_4) the main phases observed.

Figure 2 presents the SEM image showing the size and morphology of BFS particles. These microstructures exhibit spherical-shaped particles and some side-striped eroded particles with octahedral shape, which result due to intense erosion that occurs during transportation through the dust collection system. On the other hand, can be observed that SEM shows a particle size below one micron.

Element mapping by scanning electron microscopy shown in Figure 3 indicate that zinc and alkalis are integrated to iron compounds (metallic and iron oxides). It is clear that BFS can have a homogeneous chemical composition, which suggests they are formed by multi-metallic oxides complex where iron, silicon, zinc, alkalis and minor elements could have joined by sintering due to high temperatures and atmospheric conditions prevailing during the extraction process. It is worth noting that liberated franklinite or sodium and potassium oxides were not observed.

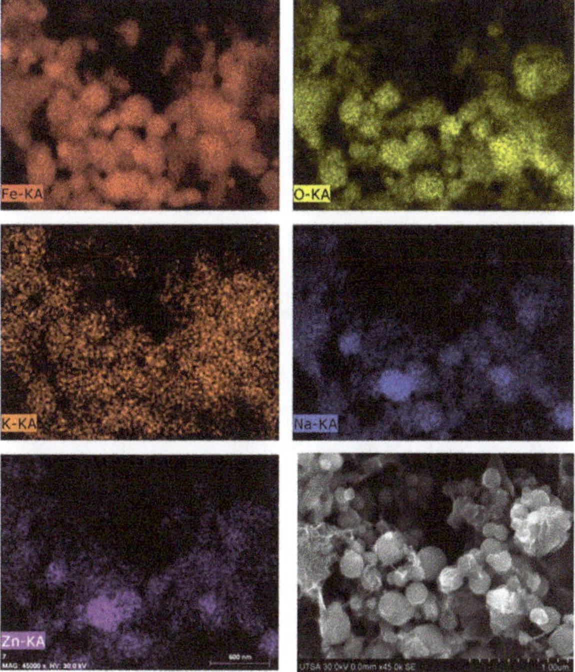

Figure 3. SEM element mapping of BFS.

3.2. Removal of Zinc and Alkali from Blast Furnace Dust by Different Experimental Conditions

The following results show the removal of zinc and alkali obtained treating the steelmaking process BFS at different experimental conditions. Figure 4 shows the extraction of Zn and alkalis (as Na_2O and K_2O) for each of the tests carried out. The extraction was obtained from the balance of the chemical analysis of the aqueous solutions and solid residues sampling in the tests. The results showed in the figure indicate that there are great differences in the zinc and alkalis contents obtained with each experimental treatment.

Figure 4 also shows that the best results for zinc removal (>70%) were obtained when ozone and sulphuric acid were used. When other oxidants were used, it was not observed a significant decrease in the extraction of zinc. Exception case was obtained when NH_4Cl was used, with a zinc extraction of 40%, due possibly to the dissolution of zinc oxides. It is interesting to note that the zinc removal was not observed when ferric ions were used. Possibly, the quantity of iron contained in the sample, mainly composed by hematite, under acid conditions have already conditions oxidant, but that are not enough for it to separate and to remove the zinc ferrites contained in the BFS sample. These results are in accordance with the conclusions made by other researchers [16,17], although the Zn/Fe ratio of the samples was different. It is important bear in mind that a smaller Zn/Fe ratio diminishes the effect of the ion ferric as an oxidant.

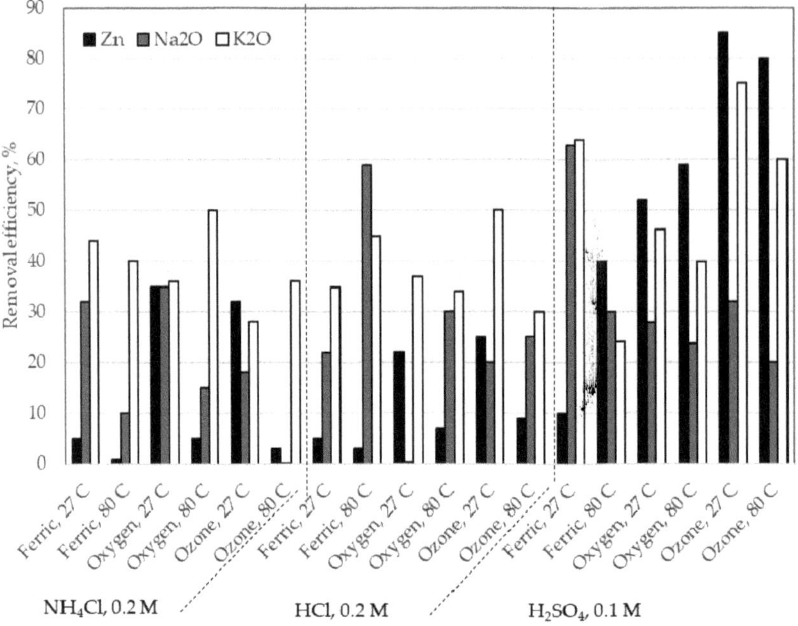

Figure 4. Zinc and alkalis (Na_2O and K_2O) removal from BFS at different experimental tests.

Figure 5 shows the remaining zinc and alkalis ($Na_2O + K_2O$) content after the BFS leaching tests. In this figure it can be appreciated that the removal of zinc varied greatly depending on the conditions of the aqueous media and the oxidant employed, as well as with the temperature. The results indicate that it was possible to obtain a zinc removal of 0.45% to less than 0.1%, when the sulfuric acid and the ozone were used. In these tests, a removal of alkalis of 75% and 60% for K_2O and 30% and 20% for Na_2O, respectively, were obtained.

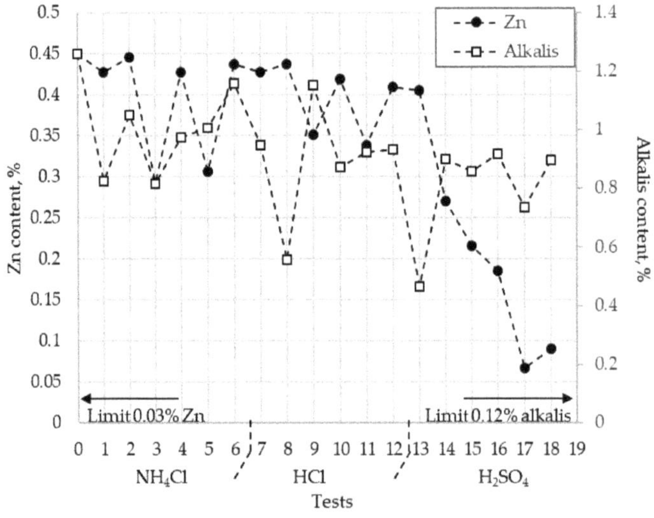

Figure 5. Zn and alkalis content in the BFS at different experimental conditions.

It is interesting to note in Figure 5 that there were conditions allowing at the same time the removal of zinc and a significant decreasing of the alkali content. They were precisely those performed in an aqueous media that allowed dissolution of the sulphate-type compounds.

However, the results shown in Figure 6 indicate that there is a reduction in the iron containing in BFS, of 41.6 (original sample) to a minimum of 36%. However, this loss in iron is smaller than the one that would be presented if the oxidizing conditions were not present. These conditions allow the maintenance of the stability of the most acid soluble iron oxides species in the aqueous medium, also considering the low solubility of the metallic iron (the main element of the Fe compounds containing in the BFS in the low acid sulphuric concentration media). In any case, the recovery of iron of the solution is possibly increasing the pH of the aqueous solution, also under oxidant conditions, to obtain the precipitation of ferric oxide, previous separation of the zinc and alkalis dissolved in solution.

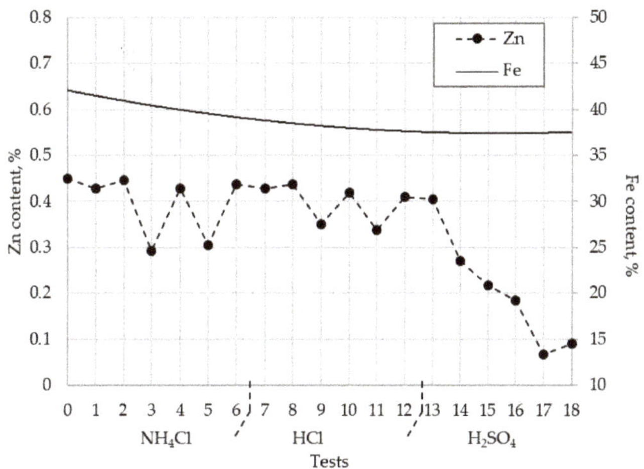

Figure 6. Zn and Fe content in blast furnace sludge at a different experimental test.

3.3. 3 × 3 × 2 Experimental Design and Analysis of Variance

The effect of the individual experimental variables on the Zn, Na_2O and K_2O removal was examined using the statistical analysis of data. Analysis of variance (ANOVA) of the experimental tests data at different conditions was used to evaluate the effect of each individual variable.

Table 5 shows the main effects, as determined by ANOVA, using NCSS software (NCSS 2000 software, NCSS, LLC, Kaysville, UT, USA) [30]. The table indicates the values of degree freedom (DF), sum of squares (SS), media of squares (MS), Fisher ratio (F), probability level (Prob Level) and the probability that a false null hypothesis can be rejected (Power) with a 95% level of confidence ($\alpha = 0.05$). According to F, Prob Level and Power values, ANOVA shows that under the studied conditions, leaching media and the type of oxidant were the most important factors during the extraction of zinc. Results also indicated that within the analyzed range, the three variables studied had the lower effect in Na_2O and K_2O removal.

Table 5. Analysis of variance (ANOVA) for Zn, Na_2O and K_2O removal from blast furnace sludge.

ANOVA for Zn						
Parameter	**DF**	**SS**	**MS**	**F**	**Prob Level**	**Power**
Leaching media	2	6952.778	3476.389	15.43	0.000481 *	1.0
Oxidant	2	2515.111	1257.556	5.58	0.019320 *	0.751343
Temperature	1	227.5556	227.5556	1.01	0.334684	0.152623
S	12	2703	225.25	-	-	-
Total (Adjusted)	17	12398.44	-	-	-	-
Total	18	-	-	-	-	-
ANOVA for Na_2O						
Parameter	**DF**	**SS**	**MS**	**F**	**Prob Level**	**Power**
Leaching media	2	621.1822	310.5911	1.32	0.303421	0.231437
Oxidant	2	972.4589	486.2294	2.07	0.169474	0.342345
Temperature	1	77.95842	77.95842	0.33	0.5756	0.082851
S	12	2824.788	235.399	-	-	-
Total (Adjusted)	17	4496.387	-	-	-	-
Total	18	-	-	-	-	-
ANOVA for K_2O						
Parameter	**DF**	**SS**	**MS**	**F**	**Prob Level**	**Power**
Aqueous media	2	651.17	325.585	1.98	0.180997	0.329381
Oxidant	2	116.95	58.47501	0.36	0.708176	0.094407
Temperature	1	175.7813	175.7813	1.07	0.321842	0.158614
S	12	1975.63	164.6358	-	-	-
Total (Adjusted)	17	2919.531	-	-	-	-
Total	18	-	-	-	-	-

* $\alpha = 0.05$.

These results are graphically shown in Figure 7, which shows the box graph for the individual effects of the experimental parameters or variables. Each plot shows the box with the minor and major values, the average value and percentiles for each parameter. Furthermore, the comparison between different oxidants shows that for the case of zinc, the ozone could generate slightly more favorable conditions for increased extraction of zinc BFS. For alkalis, although the average removal of these compounds between the three oxidants was similar, under certain conditions the ferric ion allowed a better removal. These observations are discussed below.

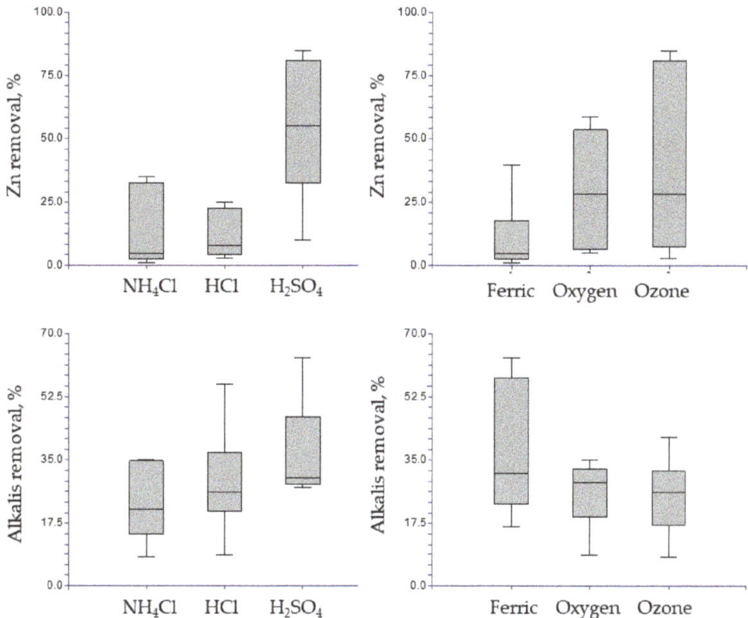

Figure 7. Box graphs for the individual effects of leaching media and oxidant in Zn and alkalis removal.

Figure 8 shows the average response curves for the individual parameters to evaluate the effect of the ozone (considering that the ferric ion has the least effect in the Zn removal). The graphs were constructed with data from the tests that used the ozone and oxygen. The most significant comparative results were found for the oxidant (see Figure 8a), which Zn removal increased from 16% to 40% using oxygen and the ozone, respectively. However, alkalis removal was not affected by the change of oxidant. On the other hand, Fe loss increased from 4% to 9% when the ozone was employed. In the case of the type of aqueous solution, Figure 8b shows that the best result was obtained with H_2SO_4 where Zn removal increased from 20% and 18% to 80% using NH_4Cl, HCl and H_2SO_4 respectively. Alkalis removal and Fe loss also increased when sulfuric acid was employed. In the case of temperature (see Figure 8c), Zn removal decreased by increasing temperature from 27 °C to 80 °C, while alkalis and Fe increased. Generally, in a leaching process, at higher temperature the velocity of dissolution is increased. However, an increment of temperature diminishes the absorption of ozone in the solution. In this case, it is possible that the effect of temperature in the reaction of dissolution were annulled, by which a change in the removal of zinc was observed.

According to results, the best conditions to remove up to 70% of Zn and alkalis are using a solution of 0.1 M H_2SO_4 and ozone (1% *v*/*v*) as the oxidant, at 25 °C. Comparing with conditions shown in Table 1, the sulphuric acid concentration used in the present work was very low. Besides the additional advantage of using ambient temperature, it also has low loss of iron during leaching. Obtaining a zinc extraction, like those obtained in the studies cited in the table.

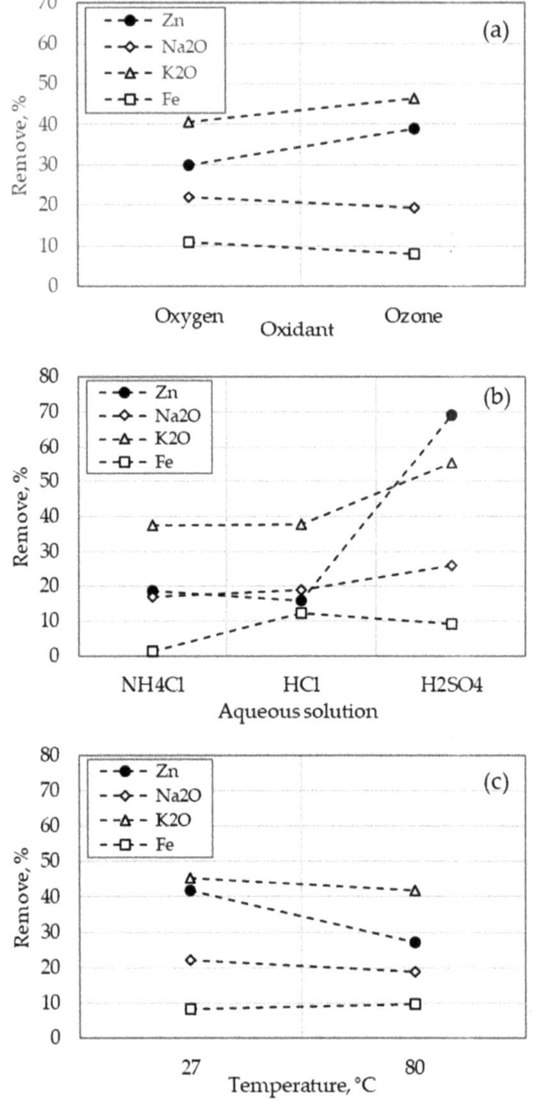

Figure 8. Response curves of individual effects: (**a**) Oxidant, (**b**) type of aqueous solution and (**c**) temperature.

4. Discussion

The fundament of zinc and alkalis dissolution into strong or weak acid media using ozone or oxygen is that aqueous media react with the oxides of zinc and alkalis, incorporating them with the media as aqueous ions, just like it is shown by the average chemical analysis of the BFS with and without ozone treatment.

The presence of an oxidizing agent, like oxygen or ozone, allows the creation of severe oxidizing conditions in aqueous media, keeping stable and practically insoluble the ferric oxides. This oxidizing environment also creates the necessary conditions to transform oxides or ferrites of zinc and alkalis, as sodium and potassium oxides, to sulfates, chloride or chlorates, compounds with high solubility in an oxidizing aqueous media.

Eh–pH diagram [31] for a Zn–O–S system (see Figure 9a), shows that zinc, which is in the form of zinc ferrite in the BFS, reacts in acid aqueous medium (pH < 5) under oxidant conditions (Eh > 0) to produce soluble species. According to the results discussed above, the aqueous medium appropriate is the sulfuric acid and ozone as an oxidizing agent. The following reactions can be possible:

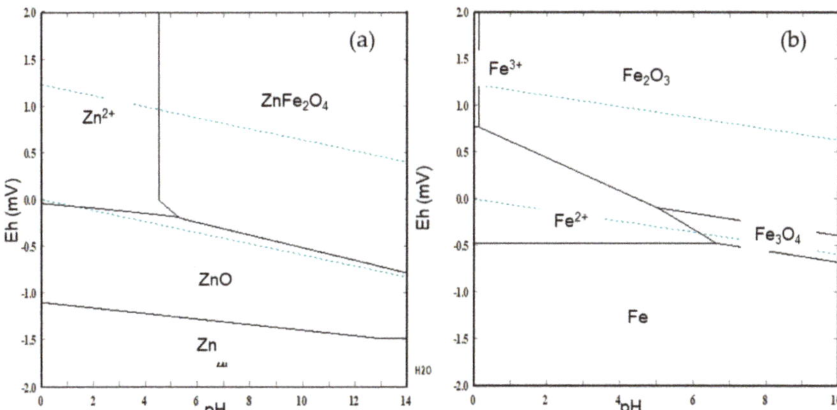

Figure 9. Eh–pH diagrams. (**a**) Zn–O–S system and (**b**) Fe–O–S system; under standard conditions at 25 °C [31].

In the absence of the ozone:

$$ZnFe_2O_4 + H_2SO_4 \rightarrow Fe_2O_3 + ZnSO_4(a) + H_2O \rightarrow \Delta G^{\circ}_{20\,^{\circ}C} = -107 \text{ kJ}. \quad (1)$$

In the presence of the ozone:

$$ZnFe_2O_4 + H_2SO_4 + 2/3O_3(g) \rightarrow Fe_2O_3 + ZnSO_4(a) + H_2O + O_2(g) \rightarrow \Delta G^{\circ}_{20\,^{\circ}C} = -215 \text{ kJ}. \quad (2)$$

$\Delta G^{\circ}_{20\,^{\circ}C}$ values were obtained from HSC Chemistry 6 [31].

According to the Eh–pH diagram for a Fe system (Figure 9b), under very acid and oxidant conditions, iron oxides could be dissolved. Then, low concentrations of acid (pH > 5) decrease iron leaching. In the case of iron metallic or iron species as magnetite or wustite, they could dissolve to a pH between 1 and 5 (weak acid conditions). However, an oxidant acid media (Eh potential > 0.5 for pH > 2) could inhibit iron dissolution.

The alkalis contained in BFS are found as ferrites or sodium (potassium)-iron jarosites. These compounds could react in different aqueous medium to obtain soluble species. When an acid medium and oxidant agent is used, the alkalis can be dissolved and separated from the BFS. The possible reactions involved are:

In the absence of the ozone:

$$[K]Na_2O{*}Fe_2O_3 + H_2SO_4 \rightarrow [K]Na_2SO_4(a) + Fe_2O_3 + H_2O \rightarrow \Delta G^{\circ}_{20\,^{\circ}C} = -277 \text{ kJ}. \quad (3)$$

In the presence of the ozone:

$$[K]Na_2O{*}Fe_2O_3 + H_2SO_4 + 2/3O_3 \rightarrow [K]Na_2SO_4(a) + Fe_2O_3 + H_2O + O_2 \rightarrow \Delta G^{\circ}_{20\,^{\circ}C} = -385 \text{ kJ}. \quad (4)$$

According to the results, the removal of alkalis only was possible in the test where high oxidation conditions were employed, obtaining the best results when the ozone was employed as the oxidizing agent. The use of other oxidizing agents, as oxygen or ferric chloride, did not decrease the zinc and alkalis content in the blast furnace sludge.

The Zn/Fe and [K]Na/Fe ratios in the BFS allow zinc and alkalis to be as stable species (e.g., oxides/ferrites) under alkaline and acid non-oxidizing conditions, being necessary to increase the potential of oxidation of the aqueous medium to obtain the dissolution of these compounds. The use of the Fe^{3+} ion could inhibit the Zn dissolution. This is possible if some Fe dissolution from zinc ferrite can be necessary to zinc leaching. Then, an initial Fe concentration in solution could decrease the ferrite dissolution due to equilibrium between the Fe concentration in products. In the case of oxygen or, better, the ozone, the oxidant condition for zinc ferrite and alkalis removal is possible.

5. Conclusions

The high content of undesirable elements (alkalis and zinc) in iron blast furnace sludge (with respect to the 0.03% Zn and 0.12% alkalis required by iron and steelmaking industry) generated a loss of valuable units of iron, additionally to the potential environment hazards and the cost by handling and disposal/confinement. In accordance with the results presented, the BFS recycling by hydrometallurgical process is possible. The best conditions were obtained when the BFS was subjected to an acid leach process with sulfuric acid (0.1 M H_2SO_4), employing ozone (1% *v*/*v* O_3) to obtain high oxidizing conditions. Only under these conditions, an important removal of zinc (up to 85%) and alkalis (up to 35% Na_2O and 75% K_2O) was possible to decrease the content of these compounds in the BFS. The fact that similar results were obtained under ambient temperature and at 80 °C indicates a compensation of two factors; the increment in the iron dissolution rate under high temperature and the diminishing of the oxidizing conditions due to the ozone decomposition. Although the zinc and alkalis concentration are still high according to the maximum values recommended, their lower content would allow it to increase the percentage of BFS added to the raw material of the blast furnace. On the other hand, the Fe loss (up to 13%) is still important, if blast furnace sludge volume is considered. Then, a Fe recovery treatment of the leaching solutions (e.g., precipitation) will be required.

Author Contributions: Data curation, F.R.C.P.; Formal analysis, F.R.C.P.; Funding acquisition, M.d.J.S.-A.; Investigation, M.d.J.S.-A., N.P.-R. and J.R.-C.; Methodology, F.R.C.P. and J.R.-C.; Project administration, M.d.J.S.-A.; Supervision, F.d.J.L.-S.; Validation, G.I.D.-P. and A.A.G.-I.; Visualization, F.d.J.L.-S.; Writing—original draft, F.R.C.P.; Writing—review & editing, G.I.D.-P. and A.A.G.-I.

Funding: This research received no external funding.

Acknowledgments: The authors would like to thank to International Center for Nanotechnology and Advanced Materials–Kleberg Advanced Microscopy Center–at the University of Texas at San Antonio (ICNAM-UTSA) for their technical assistance with the SEM characterization.

Conflicts of Interest: The authors declare no conflict of interest.

References

1. Trinkel, V.; Mallow, O.; Aschenbrenner, P.; Rechberger, H.; Fellner, J. Characterization of blast furnace sludge with respect to heavy metal distribution. *Ind. Eng. Chem. Res.* **2016**, *55*, 5590–5597. [CrossRef]
2. Geological Survey, U.S. *Mineral Commodity Summaries*; United States Government Printing Office: Washington, DC, USA, 2019; p. 89.
3. Halli, P.; Hamuyuni, J.; Leikola, M.; Lundström, M. Developing a sustainable solution for recycling electric arc furnace dust via organic acid leaching. *Miner. Eng.* **2018**, *124*, 1–9. [CrossRef]
4. Sedlakova, Z.; Havlik, T. Appearance of non-ferrous metals in iron and steel making plant and their possible treatment. *Acta Metall. Slovaca* **2006**, *12*, 209–218.
5. Kukurugya, F.; Vindt, T.; Havlík, T. Behavior of zinc, and calcium from electric arc furnace (EAF) dust in hydrometallurgical processing in sulfuric acid solutions: thermodynamic and kinetic aspects. *Hydrometallurgy* **2015**, *154*, 20–32. [CrossRef]

6. Dutra, A.J.B.; Paiva, P.R.P.; Tavares, L.M. Alkaline leaching of zinc from electric arc furnace steel dust. *Miner. Eng.* **2006**, *19*, 478–485. [CrossRef]
7. Williams, P.J.; Cloete, T.E. The production and use of cictric acid for the removal of potassium from the iron ore concentrate of the Sishen Iron Ore Mine, South Africa. *S. Afr. J. Sci.* **2010**, *106*, 1–5. [CrossRef]
8. Yusfin, Y.S.; Chernousov, P.I.; Garte, V.; Karpov, Y.A.; Petelin, A.L. The role of alkalis and conserving resources in blast-furnca smelting. *Metallurgist* **1999**, *43*, 54–58. [CrossRef]
9. Aydin, S.; Dikec, F. The removal of alkalis from iron ores by chloride volatilization. *Can. Metall. Q.* **1990**, *29*, 213–216. [CrossRef]
10. Oustadakis, P.; Tsakiridis, P.E.; Katsiapi, A.; Agatzini-Leonardou, S. Hydrometallurgical process for zinc recovery from electric arc furnace dust (EAFD): Part I: Characterization and leaching by diluted sulphuric acid. *J. Hazard. Mater.* **2010**, *179*, 1–7. [CrossRef]
11. Wang, J.; Wang, Z.; Zhang, Z.; Zhang, G. Removal of zinc from basic oxygen steelmaking filter cake by selective leaching with butyric acid. *J. Cleaner Prod.* **2019**, *209*, 1–9. [CrossRef]
12. Leclerc, N.; Meux, E.; Lecuire, J.M. Hydrometallurgical recovery of zinc and lead from electric arc furnace dust using mononitrilotricetate anion and hexahydrated ferric chloride. *J. Hazard. Mater.* **2002**, *91*, 257–270. [CrossRef]
13. Lin, X.; Peng, Z.; Yan, J.; Li, Z.; Hwang, J.Y.; Zhang, Y.; Li, G.; Jiang, T. Pyrometallurgical recycling of electric arc furnace dust. *J. Cleaner Prod.* **2017**, *149*, 1079–1100. [CrossRef]
14. She, X.; Wang, J.; Wang, G.; Xue, Q.; Zhang, X. Removal mechanism of Zn, Pb and alcalis from metallurgical dusts in direct reduction process. *J. Iron. Steel Res. Int.* **2014**, *21*, 488–495. [CrossRef]
15. Havlik, T.; Turzakova, M.; Stopic, S.; Friedrich, B. Atmospheric leaching of EAF dust with diluted sulphuric acid. *Hydrometallurgy* **2005**, *77*, 41–50. [CrossRef]
16. Kul, M.; Oskay, K.O.; Simsir, M.; Subutay, H.; Kirgezen, H. Optimization of selective leaching of Zn from electric arc furnace steelmaking dust using response surface methodology. *Trans. Nonferrous Met. Soc. China* **2015**, *25*, 2753–2762. [CrossRef]
17. Trung, Z.H.; Kukurugya, F.; Takacova, Z.; Orac, D.; Laubertova, M.; Miskufova, A.; Havlik, T. Acidic leaching both of zinc and iron from basic oxygen furnace sludge. *J. Hazard. Mater.* **2011**, *192*, 1100–1107. [CrossRef] [PubMed]
18. Veres, J.; Lovás, M.; Jakabský, S.; Sepelák, V.; Hredzák, S. Characterization of blast furnace slufge and removal of zinc by microwave assisted extraction. *Hydrometallurgy* **2012**, *129–130*, 67–73. [CrossRef]
19. Shawabkeh, R.A. Hydrometallurgical extraction of zinc from Jordanian electric arc furnace dust. *Hydrometallurgy* **2010**, *104*, 61–65. [CrossRef]
20. Ciba, J.; Zolotajkin, M.; Kluczka, J.; Loska, K.; Cebula, J. Comparison of methods for leaching heavy metals from composts. *Waste Manage.* **2003**, *23*, 897–905. [CrossRef]
21. Zhang, D.; Zhang, X.; Yang, T.; Rao, S.; Hu, W.; Liu, W.; Chen, L. Selective leaching of zinc from blast furnace dust with mono-ligand and mixed-ligand complex leaching systems. *Hydrometallurgy* **2017**, *169*, 219–228. [CrossRef]
22. Omran, M.; Fabrotius, T. Improved removal of zinc from blast furnace sludge by particle size separation and microwave heating. *Miner. Eng.* **2018**, *127*, 265–276. [CrossRef]
23. Halli, P.; Hamuyuni, J.; Revitzer, H.; Lundström, M. Selection of leaching media for metal dissolution from electric arc furnace dust. *J. Cleaner Prod.* **2017**, *164*, 265–276. [CrossRef]
24. Chaijraksa-Fujimoto, R.; Maruyama, K.; Miki, T.; Nagasaka, T. The selective alkaline leaching of zinc oxide from electric arc furace dust pre-treated with calcium oxide. *Hydrometallurgy* **2016**, *159*, 120–125. [CrossRef]
25. Ruiz, O.; Clemente, C.; Alonso, M.; Alguacil, F.J. Recycling of an electric arc furnace flue dust to obtain high grade ZnO. *J. Hazard. Mater.* **2007**, *141*, 33–36. [CrossRef] [PubMed]
26. Genstskens, R. Pressure leaching of zinc-bearing blast furnace dust. In Proceedings of the Lead-Zinc '90, Warrendale, PA, USA, 18–21 February 1990; Minerals, Metals & Materials Society: Warrendale, PA, USA, 1990; pp. 529–545.
27. Van Weert, G.; Van Sandwijk, A.; Kat, W.; Honingh, S. The treatment of iron making blast furnace dust by chloride hydrometallurgy. In Proceedings of the Hydrometallurgy Fundamentals, Technology and Innovation, Littleton, CO, USA, 1–5 August 1993; Hiskey, J.B., Warren, G.W., Eds.; Society for Mining, Metallurgy and Exploration: Englewood, IL, USA, 1993; pp. 931–946.

28. Mcelroy, R.O.; Murray, W. Developments in the Terra Gaia process for the treatment of EAF dust. In Proceedings of the Iron Control and Disposal, Montreal, QC, Canada, 20–23 October 1996; Dutrizac, J.E., Harris, G.B., Eds.; Canadian Institute of Mining, Metallurgy and Petroleum: Montreal, QC, Canada, 1996; pp. 505–517.
29. Olper, M. Zinc extraction from EAF dust with Ezinex Preocess. In Proceedings of the Recycling of Metals and Engineered Materials Minerals, Warrendale, PA, USA, 12–16 November 1995; Queneau, P.B., Peterson, R.D., Eds.; The Metals & Materials Society: Pittsburgh, PA, USA, 1995; pp. 563–578.
30. Hintze, J. NCSS 2000. Available online: www.ncss.com (accessed on 19 April 2019).
31. Roine, A. HSC Chemistry 6. Available online: www.outotec.com/hsc (accessed on 22 March 2019).

Article

Extraction of Mn from Black Copper Using Iron Oxides from Tailings and Fe^{2+} as Reducing Agents in Acid Medium

Kevin Pérez [1], Norman Toro [1,2,*], Eduardo Campos [3], Javier González [1], Ricardo I. Jeldres [4], Amin Nazer [5] and Mario H. Rodriguez [6]

1 Departamento de Ingeniería en Metalurgia y Minas, Universidad Católica del Norte, Antofagasta 1270709, Chile; kps003@alumnos.ucn.cl (K.P.); javier.gonzalez@ucn.cl (J.G.)
2 Department of Mining, Geological and Cartographic Department, Universidad Politécnica de Cartagena, Murcia 30203, Spain
3 Departamento de Ciencias Geológicas, Universidad Católica del Norte, Antofagasta 1270709, Chile; edcampos@ucn.cl
4 Departamento de Ingeniería Química y Procesos de Minerales, Universidad de Antofagasta, Antofagasta 1270300, Chile; ricardo.jeldres@uantof.cl
5 Departamento de Construcción, Universidad de Atacama, Copiapó 1531772, Chile; amin.nazer@uda.cl
6 Laboratorio de Metalurgia Extractiva y Síntesis de Materiales (MESiMat-ICB-UNCUYO-CONICET-FCEN, Padre Contreras 1300, Parque Gral. San Martín), CP 5500 Mendoza, Argentina; mrodriguez@uncu.edu.ar
* Correspondence: ntoro@ucn.cl; Tel.: +56-55-2651-021

Received: 13 September 2019; Accepted: 16 October 2019; Published: 18 October 2019

Abstract: Exotic type deposits include several species of minerals, such as atacamite, chrysocolla, copper pitch, and copper wad. Among these, copper pitch and copper wad have considerable concentrations of manganese. However, their non-crystalline and amorphous structure makes it challenging to recover the elements of interest (like Cu or Mn) by conventional hydrometallurgical methods. For this reason, black copper ores are generally not incorporated into the extraction circuits or left unprocessed, whether in stock, leach pads, or waste. Therefore, to dilute MnO_2, the use of reducing agents is essential. In the present research, agitated leaching was performed to dissolve Mn of black copper in an acidic medium, comparing the use of ferrous ions and tailings as reducing agents. Two samples of black copper were studied, of high and low grade of Mn, respectively, the latter with a high content of clays. The effect on the reducing agent/black copper ratio and the concentration of sulfuric acid in the system were evaluated. Better results in removing Mn were achieved using the highest-grade black copper sample when working with ferrous ions at a ratio of Fe^{2+}/black copper of 2/1 and 1 mol/L of sulfuric acid. Besides, the low-grade sample induced a significant consumption of H_2SO_4 due to the high presence of gangue and clays.

Keywords: waste treatment; reducing agent; manganese

---

1. Introduction

Copper mining is Chile's most important economic activity, accounting for 10% of the gross national product (GNP) [1]. According to the latest figures from the Chilean Copper Commission, 5.83 million metric tons of copper were produced in 2018, making Chile the leading copper producer, accounting for 27.7% of global copper production. Experts from the Chilean Association of Geologists have stated that Chile has the largest copper deposits in the world [2], with a total copper reserve of 170 million metric tons [3].

Porphyry minerals in deposits like pyrite oxidize when submitted to geological agents. When pyrite reacts with water, it generates sulfuric acid, promoting the mobility of metals like copper that can be transported under certain potential and pH conditions, precipitating downstream and forming what are termed exotic deposits [4–8].

These deposits are composed of different copper containing phases such as chrysocolla, atacamite, copper pitch, and copper wad [6,9]. The latter two are defined as mineraloids because they crystalize amorphously [2]. They are also termed silicates rich in Si-Fe-Cu-Mn [10].

Some examples of exotic deposits in Chile are Mina Sur in Chuquicamata [11], Damiana in El Salvador [7], Huanquintipa in Collahuasi [12], and La Cascada, Lomas Bayas Spence, El Tesoro [2], and Angélica in Tocopilla [13]. The copper and manganese of this type of deposit are often associated with oxidized minerals, mainly chrysocolla, which, in turn, are associated with gangue that can negatively affect leaching [11]. Silicates and aluminosilicates, like mica and clay minerals, have the capacity to consume some of the acid generated by oxidization [14]. Clay minerals, like montmorillonite, kaolinite, and smectite, easily absorb acid [15]. Other minerals, like chlorites and biotite, also consume large amounts of acid over the long term [15]. Helle et al. [11] studied the effect of gangue and clay minerals on the leaching of copper oxides such as atacamite, chrysocolla, and malachite. The copper oxides were treated with a strong solution of sulfuric acid (265 g/L) in small columns at ambient temperature (18 to 21 °C), with the addition of synthetic rocks composed of 57% quartz, 1% phase mineral, and 42% reactive gangue. The authors concluded that copper retention and acid consumption were the result of the presence of smectite, mordenite gangue, kaolinite, illite, and quartz.

Researchers have indicated that it is not possible to recover copper associated with these silicates using conventional hydrometallurgical methods for oxidized copper because of their non-crystalline or amorphous structure [16]. However, recent studies on techniques for extracting manganese have found that silicates can be recovered by treating them in a similar manner to treatment for manganese, owing to the similarity in their metallurgical behavior [17].

It has been demonstrated that a reducing agent is required to extract Mn from MnO_2 in acid media [18,19]. Other studies have obtained good results dissolving MnO_2 with different reducing agents like H_2SO_3 [20], SO_2 [21], wastewater from producing molasses-based alcohol [22], and various iron-based reducing agents [20,23,24]. Iron, which is abundant and inexpensive, has proven to be a good alternative when working with MnO_2 in acid media.

Zakeri et al. [25] obtained an Mn extraction rate of 90% in 20 min at ambient temperature with the addition of ferrous ions to the system, with an Fe^{2+}/MnO_2 molar ratio of 3.0 and H_2SO_4/MnO_2 ratio of 2.0. They proposed the following series of reactions for MnO_2 dissolution:

$$MnO_2 + 4H^+ + 2e^- = Mn^{2+} + 2H_2O \quad (1)$$

$$2Fe^{2+} = 2Fe^{3+} + 2e^- \quad (2)$$

$$MnO_2 + 2Fe^{2+} + 4H^+ = Mn^{2+} + 2Fe^{3+} + 2H_2O \quad (3)$$

Toro et al. [26] leached Mn nodules using tailings with high Fe_3O_4 contents (58.52%) from slag flotation for the recovery of Cu from the Alto Norte Foundry Plant and optimized the working parameters (Fe_2O_3/MnO_2 ratio and H_2SO_4 concentration). They found that for short periods of time (5 to 20 min), the optimal MnO_2/Fe_2O_3 ratio is 1/3, with H_2SO_4 concentration of 0.1 mol/L, giving Mn extraction rates of approximately 70%. The authors proposed the following reactions to dissolve MnO_2 with the addition of iron oxides:

$$Fe_2O_{3(s)} + 3\ H_2SO_{4(aq)} = Fe_2(SO_4)_{3(s)} + 3\ H_2O_{(l)} \quad (4)$$

$$Fe_3O_{4(s)} + 4H_2SO_{4(l)} = FeSO_{4(aq)} + Fe_2(SO_4)_{3(s)} + 4\ H_2O_{(l)} \quad (5)$$

$$2\ FeSO_{4(aq)} + 2\ H_2SO_{4(aq)} + MnO_{2(s)} = Fe_2(SO_4)_{3(s)} + 2\ H_2O_{(l)} + MnSO_{4(aq)} \quad (6)$$

The following reactions are proposed to dissolve manganese from black copper:

$$(CuO \times MnO_2 \times 7H_2O)_{(s)} + 3\,H_2SO_{4(aq)} + 2\,FeSO_{4(aq)} = Fe_2(SO_4)_{3(aq)} + MnSO_{4(aq)} + CuSO_{4(aq)} + 10\,H_2O_{(l)} \quad (7)$$

$$(CuO \times MnO_2 \times 7H_2O)_{(s)} + 11\,H_2SO_{4(aq)} + 3\,Fe_3O_{4(s)} = 3\,Fe_2(SO_4)_{3(aq)} + MnSO_{4(aq)} + CuSO_{4(aq)} + 18\,H_2O_{(l)} \quad (8)$$

Equation (5) gives the reaction of magnetite with sulfuric acid forming ferrous sulfate, which is a good reducing agent for the leaching of MnO_2. This is shown in Equation (6), where Mn^{4+} is reduced to Mn^{2+}. In Equation (7), the solution of manganese from black copper (copper wad) is proposed, using ferrous sulfate expressed in Equation (5). In general, Equation (8) represents the dissolution of manganese with iron oxide as a reducing agent, which demands high concentrations of sulfuric acid to first form $FeSO_4$ from Fe_3O_4 and then continues to dissolve manganese until a manganese sulfate solution is obtained.

In Chile, big copper mining poses new challenges and needs. It seeks to diversify the extractions of other elements (besides the Cu) in order to boost the export of commodities and raise employment. Black copper ores are resources that are generally not incorporated into the extraction circuits or left untreated, whether in stock, leach pads, or waste [27]. These exotic minerals have considerable amounts of Mn (approximately 29%), which represent a commercial appeal. Besides, according to the study conducted by Benavente et al. [27], by dissolving black copper ores in a reducing condition, the decrease in redox potential favors the dissolution of manganese. This would allow the subsequent extraction of the Cu present in black copper, given the potential commercial value of these "wastes".

This work aimed to study the dissolution of MnO_2 from black copper in acid media comparing the use of iron and iron oxide tailings as reducing agents.

2. Methodology

2.1. Black Oxide Samples

Two samples of black copper, obtained from different mines in northern Chile, were used in this investigation. One sample, black copper sample-1 (BCS-1), was from a high-grade vein and was almost 100% pure, while the other, black copper sample-2 (BCS-2), was low-grade and taken from the mine dumpsite. The black oxides ores were ground in a porcelain mortar to sizes ranging from −173 to +147 μm. Chemical composition was determined by inductively coupled plasma atomic emission spectrometry (ICP-AES). Table 1 shows the chemical composition of the samples. A QEMSCAN analysis was applied, which is an electronic scanning microscope that was modified both in hardware and software. This performed the identification and automated quantification of ranges of elementary definitions that can be associated with inorganic solid phases (minerals, alloys, slags, etc.). To determine the mineralogical composition, the samples were mounted on briquettes and polished. The identification, mapping of 2-D distribution, and quantification of inorganic phases, was done by combining the emissions of retro-dispersed electrons (BSE) with a Zeiss EVO series, a Bruker AXS XFlash 4010 detector (Bruker, Billerica, MA, USA) and the iDiscover 5.3.2.501 software (FEI Company, Brisbane, Australia). The QEMSCAN analyses are based on the automated obtaining of EDS spectra (dispersed energy from X-rays) in hundreds of thousands or millions of collected analysis points, each in a time of milliseconds. The classification of mineralogical phases is done by classifying each EDS spectrum in a hierarchical and descending compositional list known as the "SIP List". The BSE image is used to discriminate between resin and graphite in the sample, to specify entries in the SIP list, and to establish thresholds for acceptance or rejection of particles. As a result, pixelated, 2-D and false color images of a specimen or a representative subsample of particles are obtained. Each pixel retains its elementary and BSE brightness information, which allows subsequent offline data processing. Through software, customized filters are generated that allow the quantification of ore and gangue species, mineral release, associations between inorganic phases, and the classification of particles according to criteria of shape, size, texture, etc. Figure 1 shows the chemical species to black oxides using QEMSCAN.

Table 1. Chemical composition of black oxide samples.

Sample	Mn (%)	Fe (%)
Black Copper Sample-1	22.01	7.92
Black Copper Sample-2	0.51	3.88

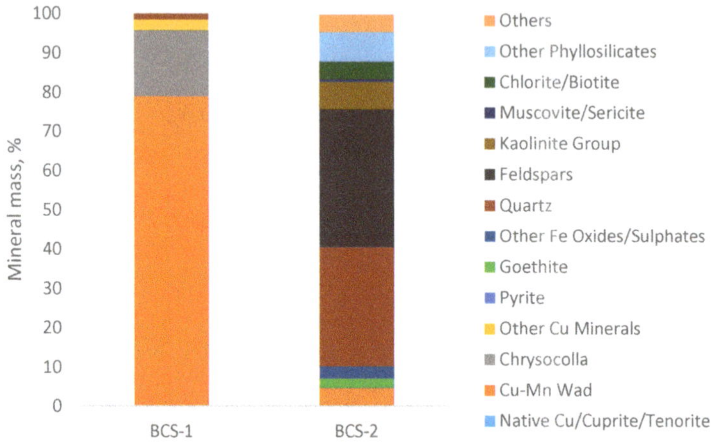

Figure 1. Detailed modal mineralogy.

Table 2 shows the mineralogical composition of the black copper samples. Copper wad refers to a subgroup of copper composed of manganese and copper hydroxides, as well as also traces of other elements such as Co, Ca, Fe, Al, Si, and Mg.

Table 2. The mineralogical composition of the black copper samples as determined by QEMSCAN.

Mineral (% Mass)	Black Copper Sample-1	Black Copper Sample-2
Native Cu/Cuprite/Tenorite	0.12	0.00
Cu-Mn Wad	78.90	4.64
Chrysocolla	16.72	0.03
Other Cu Minerals	2.69	0.03
Pyrite	0.00	0.01
Goethite	0.01	2.39
Other Fe Oxides/Sulphates	0.00	3.15
Quartz	1.41	30.20
Feldspars	0.02	35.11
Kaolinite Group	0.01	7.08
Muscovite/Sericite	0.01	0.67
Chlorite/Biotite	0.01	4.45
Montmorillonite	0.00	4.56
Others	0.09	7.35
Total	100	100

2.2. Ferrous Ions

The ferrous ions used for this investigation ($FeSO_4 \times 7H_2O$) were WINKLER brand, with a molecular weight of 278.01 g/mol.

2.3. Iron Oxide Tailings

The iron oxide tailings used were from the Altonorte Smelting Plant. The particle sizes were in a range between −75 to +53 µm. The methods used to determine its chemical and mineralogical composition were the same as those used in black copper ores. Table 3 shows the minerals (and chemical formulas) from QEMSCAN analysis, noting that several iron-containing phases were present from which the Fe content was estimated at 41.9%. As the Fe was mainly in the form of magnetite, the most appropriate method of extraction was the same as that used by Toro et al. [26].

Table 3. Mineralogical composition of tailings, as determined by QEMSCAN.

Mineral	Amount % (*w/w*)
Chalcopyrite/Bornite ($CuFeS_2/Cu_5FeS_4$)	0.47
Tennantite/Tetrahedrite ($Cu_{12}As_4S_{13}/Cu_{12}Sb_4S_{13}$)	0.03
Other Cu Minerals	0.63
Cu–Fe Hydroxides	0.94
Pyrite (FeS_2)	0.12
Magnetite (Fe_3O_4)	58.52
Specular Hematite (Fe_2O_3)	0.89
Hematite (Fe_2O_3)	4.47
Ilmenite/Titanite/Rutile ($FeTiO_3/CaTiSiO_5/TiO_2$)	0.04
Siderite ($FeCO_3$)	0.22
Chlorite/Biotite $(Mg)_3(Si)_4O_{10}(OH)_2(Mg)_3(OH)_6/K(Mg)_3AlSi_3O_{10}(OH)_2$	3.13
Other Phyllosilicates	11.61
Fayalite (Fe_2SiO_4)	4.59
Dicalcium Silicate (Ca_2SiO_4)	8.30
Kirschsteinite ($CaFeSiO_4$)	3.40
Forsterite (Mg_2SiO_4)	2.30
Barite ($BaSO_4$)	0.08
Zinc Oxide (ZnO)	0.02
Lead Oxide (PbO)	0.01
Sulfate (SO_4)	0.20
Others	0.03
Total	100.00

2.4. Reagent and Leaching Test

The sulfuric acid used for the leaching tests was grade P.A., with 95–97% purity, a density of 1.84 kg/L, and a molecular weight of 98.80 g/mol. The leaching tests were carried out in a 50 mL glass reactor with a 0.01 solid/liquid ratio. A total of 200 mg of black oxide ore was maintained in suspension with the use of a five-position magnetic stirrer (IKA ROS, CEP 13087-534, Campinas, Brazil) at a speed of 600 rpm. The tests were conducted at a room temperature of 25 °C, while variations were iron additives, particle size, and leaching time. The tests were performed in duplicate and measurements (or analyses) were carried out on 5 mL undiluted samples using atomic absorption spectrometry with a coefficient of variation ≤ 5% and a relative error between 5 to 10%. The measurements of pH and oxidation-reduction potential (ORP) of the leach solutions were made using a pH-ORP meter (HANNA HI-4222 (HANNA instruments, Woonsocket, Rhode Island, USA)). The solution ORP was measured in a combination ORP electrode cell composed of a platinum working electrode and a saturated Ag/AgCl reference electrode.

2.5. The Effect of the Fe/MnO$_2$ Ratio

Other investigations have shown that variables of particle size and stirring speed do not have significant effects when working with a high Fe/MnO_2 ratio [26,28]. Given this result, we decided to work with the following parameters: Fe/MnO_2 ratios of 1/1, 2/1 and 3/1, a particle size range of −75–+53 µm, a stirring speed of 600 rpm, 1 mol/L sulfuric acid, and room temperature (25 °C).

2.6. The Effect of the Acid Concentration on the System

The present research studied the effect of the sulfuric acid concentration on the system, working with H_2SO_4 concentrations of 0.5, 1, 2, and 3 mol/L under the following operating conditions: Reducing agent/black copper ratio of 1/2, particle size range of −75 + 53 μm, stirring speed of 600 rpm, and a temperature of 25 °C.

3. Results

3.1. The Effect of the Fe^{2+}/MnO_2 Ratio

Figure 2a,b show the results for the dissolution of two black copper samples using Fe^{2+} in acid media. As can be seen, better results were achieved with the sample BCS-1, which was due to the high presence of clay in sample BSC-2. It was observed that high Mn extraction rates can be achieved in short periods of time using MnO_2/Fe^{2+} ratios of 1/2 or less, achieving dissolution rates of over 78% in 5 min with sample BCS-1, and 65% in 5 min with sample BCS-2. The results shown in Figure 2a are similar to the 90% recovery obtained by Zakeri et al. [25] in 20 min leaching MnO_2 from manganese nodules with an Fe^{2+}/MnO_2 ratio of 3, and an H_2SO_4/MnO_2 molar ratio of 2/1. A 1/1 Fe^{2+}/MnO_2 ratio resulted in a lower MnO_2 dissolution kinetics, with an extraction of 40% in 5 min with sample A and 31% in 5 min with sample B. In general, Mn dissolution rates were similar with a longer period (30 min). However, the dissolution kinetics were slower for the sample BCS-2.

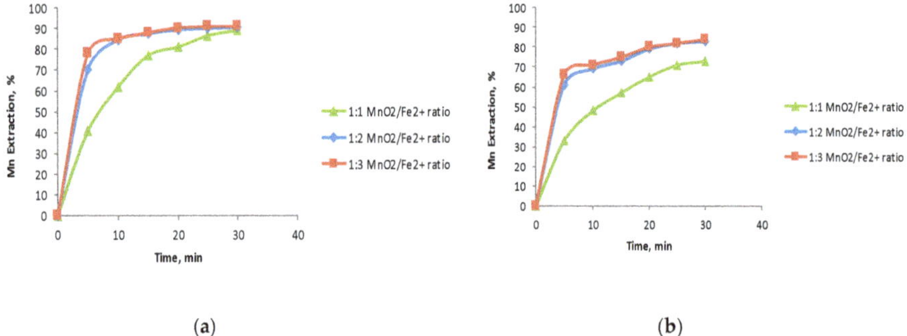

Figure 2. The effect of the Fe^{2+} concentration on MnO_2 dissolution (**a**) Black copper sample-1 (BSC-1); (**b**) Black copper sample-2 (BCS-2), 25 °C, particle size range of −75–+53 μm, 1 mol/L H_2SO_4.

3.2. The Effect of the Fe_2O_3/MnO_2 Ratio

Figure 3a,b show Mn dissolution with two black copper samples using Fe_2O_3 in acid media. As in earlier investigations by Toro et al. [26,28], working with an Fe_2O_3/MnO_2 ratio of 2/1 or higher significantly increased MnO_2 dissolution kinetics. There was little difference in the Mn extraction rates working with Fe_2O_3/MnO_2 ratios of either 2/1 or 3/1, while the Mn extraction fell significantly when the quantity of Fe_2O_3 was reduced. Potential and pH levels were respectively in the ranges of −0.5 to 1.3 V and −1.5 to 0.4 in all the tests in this study.

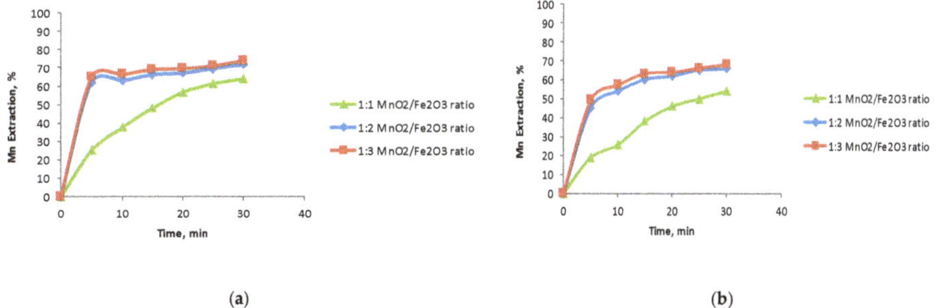

Figure 3. The effect of the iron oxide tailings concentration on MnO_2 dissolution (**a**) BSC-1; (**b**) BSC-2, 25 °C, particle size range of −75–+53 µm, 1 mol/L H_2SO_4.

3.3. The Effect of the H_2SO_4 Concentration

Figure 4 shows the effect of the acid concentration on dissolving Mn dissolution from the two black copper samples with the addition of high concentrations of iron oxides from tailings or ferrous ions. It can be seen from Figure 4a,c that for sample BCS-1, the sulfuric acid concentration was not significant in either case when working with high concentrations of the reducing agent. Differences in the effect of the acid concentration could only be noted with very low concentrations of iron oxide tailings (0.5 mol/L). The above concurs with findings of previous studies by Toro et al. [24,26] on extraction of MnO_2 from manganese nodules. The Mn extraction rate from the BCS-2 sample increased with higher concentrations of H_2SO_4, possibly owing to the high consumption of acid generated by the presence of mineral impurities in this sample, mainly montmorillonite, kaolinite, and chlorite. This is consistent with what was previously found by Helle and Kelm [29], where the leaching of exotic Cu minerals (atacamite, chrysocolla, and malachite) required higher acid consumption by incorporating reactive bargains into the system. This was driven by smectites, mordenite bargain, and the presence of kaolinite, illite, and quartz.

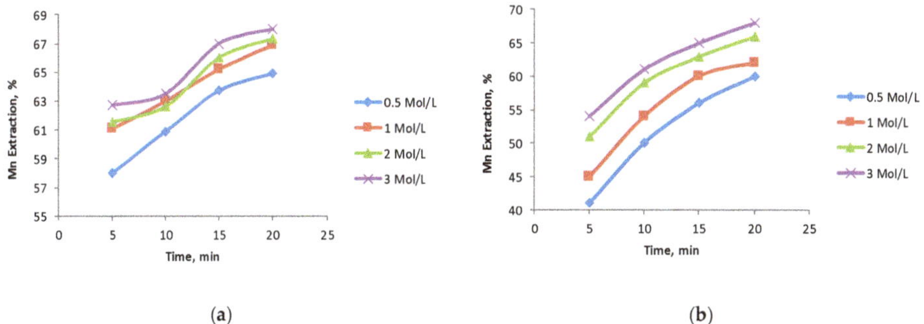

Figure 4. *Cont.*

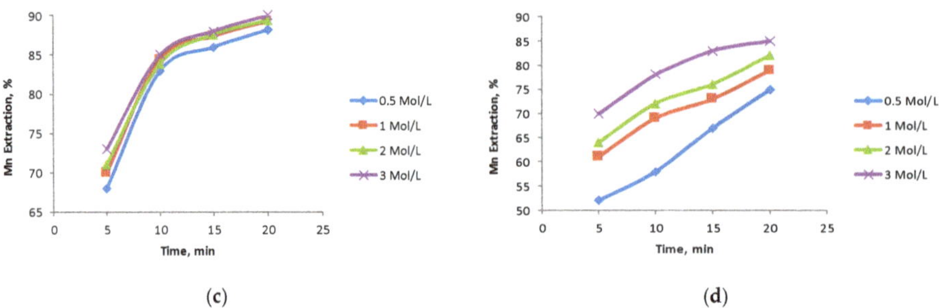

Figure 4. The effect of the sulfuric acid concentration on the system (**a**) BSCe-1; (**b**) BSC-2, MnO_2/Fe_2O_3 ratio of 1/2; (**c**) BSC-1; (**d**) BSC-2, MnO_2/Fe^{2+} ratio of 1/2.

4. Conclusions

This study presents the results obtained for dissolving Mn from black copper using iron oxides (and specifically magnetite) from tailings and Fe^{2+} as reducing agents in acid media. Both reducing agents yielded good results with the two samples studied. Similar behavior was observed with the two samples in relation to Mn extraction, with the best results obtained in all the experiments with the BCS-1 sample. These encouraging results give new options to extract the Cu present in these exotic minerals, which are considered as industrial waste today. The main findings are the following:

(1) The ferrous ions were a better reducing agent than iron oxides to dissolve MnO_2 in black copper.
(2) The optimal reducing agent/black copper ratio was 2:1 for the studied reducing agents studied.
(3) High concentrations of H_2SO_4 had a positive effect on the dissolution of Mn with the BCS-2 sample owing to the high content of clay (montmorillonite and kaolinite) and gangue (chlorite), which consume significant amounts of acid. The acid concentration was not significant with the BCS-1 sample.
(4) The best results in this study were obtained working with the sample with fewer impurities (BCS-1), with an Fe^{2+}/black copper ratio of 2:1, and 1 mol/L of sulfuric acid.

Despite the good results obtained with BCS-1, BCS-2 was more like the mineralogy found at the industrial scale. It should be noted that although lower Mn extraction rates are obtained using tailings instead of ferrous ions, tailings can be a more attractive additive for leaching black copper because they are an industrial waste with no economic value. Given the above results, future investigations should aim to optimize operational parameters for leaching black copper minerals with high gangue content using industrial waste or wastewater as reducing agents, with the aim of taking this process to the industrial scale.

Author Contributions: K.P. contributed in research and wrote paper, N.T. and R.I.J. contributed in project administration, E.C. and A.N. contributed resources, J.G. contributed in review and editing and M.H.R. contributed in data curing.

Funding: This research received no external funding.

Acknowledgments: The authors are grateful for the contribution of the Scientific Equipment Unit- MAINI of the Universidad Católica del Norte for aiding in generating data by automated electronic microscopy QEMSCAN® and for facilitating the chemical analysis of the solutions. We are also grateful to the Altonorte Mining Company for supporting this research and providing slag for this study, and we thank to Marina Vargas Aleuy and María Barraza Bustos of the Universidad Católica del Norte for supporting the experimental tests. Also, we Conicyt Fondecyt 11,171,036 and Centro CRHIAM Project Conicyt/Fondap/15130015.

Conflicts of Interest: The authors declare they have no conflict of interest.

References

1. Servicio Nacional de Geología y Minería (SERNAGEOMIN). *Anuario de la Minería de Chile*; Servicio Nacional de Geología y Minería: Santiago, Chile, 2018; p. 274.
2. Menzies, A.; Campos, E.; Hernández, V.; Sola, S.; Riquelme, R. Understanding Exotic-Cu Mineralisation Part II: Characterization of 'Black Copper' ore ('Cobre Negro'). In Proceedings of the 13th SGA Biennial Meeting, Nancy, France, 24–27 August 2015; pp. 3–6.
3. U.S. Geological Survey and U.S. Department of the Interior, Mineral. *Commodity Summaries*; U.S. Geological Survey: Reston, Virginia, 2018.
4. Ossandon, C.G.; Freraut, C.R.; Gustafson, L.B.; Lindsay, D.D.; Zentilli, M. Geology of the Chuquicamata Mine: A Progress Report. *Econ. Geol.* **2001**, *96*, 249–270. [CrossRef]
5. Mote, T.I.; Becker, T.A.; Renne, P.; Brimhall, G.H. Chronology of Exotic Mineralization at El Salvador, Chile, by 40Ar/39Ar Dating of Copper Wad and Supergene Alunite. *Econ. Geol.* **2001**, *96*, 351–366. [CrossRef]
6. Cuadra, C.P.; Rojas, S.G. Oxide mineralization at the Radomiro Tomic porphyry copper deposit, Northern Chile. *Econ. Geol.* **2001**, *96*, 387–400.
7. Mora, R.; Artal, J.; Brockway, H.; Martinez, E.; Muhr, R. El Tesoro exotic copper deposit, Antofagasta region, northern Chile. *Econ. Geol. Spec. Publ.* **2004**, *11*, 187–197.
8. Pinget, M.; Dold, B.; Fontboté, L. Exotic mineralization at Chuquicamata, Chile: Focus on the copper wad enigma. In Proceedings of the 10th Swiss Geoscience Meeting, Bern, Switzerland, 16–17 November 2012; pp. 88–89.
9. Kojima, S.; Astudillo, J.; Rojo, J.; Tristá, D.; Hayashi, K.-I. Ore mineralogy, fluid inclusion, and stable isotopic characteristics of stratiform copper deposits in the coastal Cordillera of northern Chile. *Miner. Deposita* **2003**, *38*, 208–216. [CrossRef]
10. Pincheira, M.; Dagnini, A.; Kelm, U.; Helle, S. *Copper Pitch Y Copper Wad: Contraste Entre Las Fases Presentes En Las Cabezas Y En Los Ripios En Pruebas De Mina sur, Chuquicamata*; X Congreso Geológico Chileno: Concepción, Chile, 2003; p. 10.
11. Hellé, S.; Kelm, U.; Barrientos, A.; Rivas, P.; Reghezza, A. Improvement of mineralogical and chemical characterization to predict the acid leaching of geometallurgical units from Mina Sur, Chuquicamata, Chile. *Miner. Eng.* **2005**, *18*, 1334–1336. [CrossRef]
12. García, C.; Garcés, J.P.; Rojas, C.; Zárate, G. Efecto sinérgico del tratamiento de mezcla de minerales conteniendo copper wad y sulfuros secundarios. In Proceedings of the IV International Copper Hydrometallurgy Workshop, Viña del Mar, Chile, 16–18 May 2007.
13. Zambra, J.; Kojima, S.; Espinoza, S.; Definis, A. Angélica Copper Deposit: Exotic Type Mineralization in the Tocopilla Plutonic Complex of the Coastal Cordillera, Northern Chile. *Resour. Geol.* **2007**, *57*, 427–434. [CrossRef]
14. Consejo Minero. *Guía Metodológica sobre Drenaje en la Industria Minera*; Subsecretaría de economía Consejo Nacional de Producción Limpia: Santiago, Chile, 2002.
15. Sequeira, R. A note on the consumption of acid through cation exchange with clay minerals in atmospheric precipitation. *Atmos. Environ. Part. A Gen. Top.* **1991**, *25*, 487–490. [CrossRef]
16. Helle, S.; Pincheira, M.; Jerez, O.; Kelm, U. Sequential extraction to predict the leaching potential of refractory. In Proceedings of the XV Balkan Mineral Processing Congress, Sozopol, Bulgaria, 12–16 June 2013; pp. 109–111.
17. Hernández, M.C.; Benavente, O.; Melo, E.; Núñez, D. Copper leach from black copper minerals. In Proceedings of the 6th International Seminar on Copper Hydrometallurgy, Viña del Mar, Chile, 6–8 July 2011; pp. 1–10.
18. Randhawa, N.S.; Hait, J.; Jana, R.K. A brief overview on manganese nodules processing signifying the detail in the Indian context highlighting the international scenario. *Hydrometallurgy* **2016**, *165*, 166–181. [CrossRef]
19. Saldaña, M.; Toro, N.; Castillo, J.; Hernández, P.; Trigueros, E.; Navarra, A. Development of an Analytical Model for the Extraction of Manganese from Marine Nodules. *Metals* **2019**, *9*, 903. [CrossRef]
20. Kanungo, S. Rate process of the reduction leaching of manganese nodules in dilute HCl in presence of pyrite. *Hydrometallurgy* **1999**, *52*, 313–330. [CrossRef]
21. Kanungo, S.; Das, R. Extraction of metals from manganese nodules of the Indian Ocean by leaching in aqueous solution of sulphur dioxide. *Hydrometallurgy* **1988**, *20*, 135–146. [CrossRef]

22. Su, H.; Liu, H.; Wang, F.; Lu, X.; Wen, Y.-X. Kinetics of Reductive Leaching of Low-grade Pyrolusite with Molasses Alcohol Wastewater in H2SO4. *Chin. J. Chem. Eng.* **2010**, *18*, 730–735. [CrossRef]
23. Bafghi, M.S.; Zakeri, A.; Ghasemi, Z.; Adeli, M. Reductive dissolution of manganese ore in sulfuric acid in the presence of iron metal. *Hydrometallurgy* **2008**, *90*, 207–212. [CrossRef]
24. Toro, N.; Saldaña, M.; Gálvez, E.; Cánovas, M.; Trigueros, E.; Castillo, J.; Hernández, P.C. Optimization of Parameters for the Dissolution of Mn from Manganese Nodules with the Use of Tailings in An Acid Medium. *Minerals* **2019**, *9*, 387. [CrossRef]
25. Zakeri, A.; Bafghi, M.; Shahriari, S. Dissolution Kinetics of Manganese Dioxide Ore in Sulfuric Acid in the Presence of Ferrous Ion. *Iran. J. Mater. Sci. Eng.* **2007**, *4*, 22–27.
26. Toro, N.; Saldaña, M.; Castillo, J.; Higuera, F.; Acosta, R. Leaching of Manganese from Marine Nodules at Room Temperature with the Use of Sulfuric Acid and the Addition of Tailings. *Minerals* **2019**, *9*, 289. [CrossRef]
27. Benavente, O.; Hernández, M.C.; Melo, E.; Núñez, D.; Quezada, V.; Zepeda, Y. Copper Dissolution from Black Copper Ore under Oxidizing and Reducing Conditions. *Metals* **2019**, *9*, 799. [CrossRef]
28. Toro, N.; Herrera, N.; Castillo, J.; Torres, C.M.; Sepúlveda, R. Initial Investigation into the Leaching of Manganese from Nodules at Room Temperature with the Use of Sulfuric Acid and the Addition of Foundry Slag—Part, I. *Minerals* **2018**, *8*, 565. [CrossRef]
29. Helle, S.; Kelm, U. Experimental leaching of atacamite, chrysocolla and malachite: Relationship between copper retention and cation exchange capacity. *Hydrometallurgy* **2005**, *78*, 180–186. [CrossRef]

Article

Viscoelasticity of Quartz and Kaolin Slurries in Seawater: Importance of Magnesium Precipitates

Matías Jeldres [1], Eder Piceros [2], Pedro A. Robles [3], Norman Toro [4,5,*] and Ricardo I. Jeldres [1]

1 Departamento de Ingeniería Química y Procesos de Minerales, Facultad de Ingeniería, Universidad de Antofagasta, Antofagasta 1240000, Chile; mjeldresvalenzuela@gmail.com (M.J.); ricardo.jeldres@uantof.cl (R.I.J.)

2 Faculty of Engineering and Architecture, Universidad Arturo Pratt, P.O. Box 121, Iquique 1100000, Chile; edpicero@unap.cl

3 Escuela de Ingeniería Química, Pontificia Universidad Católica de Valparaíso, Valparaíso 2340000, Chile; pedro.robles@pucv.cl

4 Departamento de Ingeniería Metalúrgica y Minas, Universidad Católica del Norte, Antofagasta 1270709, Chile

5 Department of Mining, Geological and Cartographic Department, Universidad Politécnica de Cartagena, 30202 Murcia, Spain

* Correspondence: ntoro@ucn.cl; Tel.: +56-552-651-021

Received: 14 September 2019; Accepted: 17 October 2019; Published: 19 October 2019

Abstract: In this study, the viscoelastic properties of quartz and kaolin suspensions in seawater were analysed considering two distinct conditions: pH 8 and 10.7. Creep and oscillatory sweep tests provided the rheological parameters. An Anton Paar MCR 102 rheometer (ANAMIN Group, Santiago, Chile) was used with a vane-in-cup configuration, and the data were processed with RheoCompass^{TM} Light software (ANAMIN Group, Santiago, Chile). The outcomes were associated with the formation of solid species principally composed of magnesium precipitates. The magnesium in solution reduced in the presence of quartz (68 wt %), from 1380 to 1280 mg/L. Since the difference was not large regarding the solid-free seawater, the disposition of solid complexes at pH 10.7 was expected to be similar. The jump in pH caused both yield stress and viscoelastic moduli to drop, suggesting that the solid precipitates diminished the strength of the particle networks that made up the suspension. For the kaolin slurries (37 wt %), the yield stress raised when the pH increased, but unlike quartz, there was significant adsorption of magnesium cations. In fact, the concentration of magnesium in solution fell from 1380 to 658 mg/L. Dynamic oscillatory assays revealed structural changes in both pulps; in particular, the phase angle was greater at pH 8 than at pH 10.7, which indicates that at more alkaline conditions, the suspension exhibits a more solid-like character.

Keywords: viscoelasticity; quartz; kaolin; seawater; magnesium precipitates

---

1. Introduction

Diverse mining companies have their deposits in arid areas, especially in northern Chile, southern Peru, Australia, Africa and Asia [1,2]. In these regions, water availability may be a serious challenge and is the primary purpose for plants to reduce their water demand while avoiding significant alterations in production. A current strategy implemented in several plants is the use of seawater (see Table 1), either through prior desalination treatment by reverse osmosis, where all salts are eliminated, or by simply applying it directly, without altering the salinity [3–5]. The direct use of seawater entails advantages associated with environmental impact and operational costs that involve the building of a desalination plant [6]; however, the complexity of such an undertaking is the subject of constant discussion and research for engineers and scientists [3]. For example, there is the issue of the location of plants that are far from the coast and at a high altitude. This may require a great amount of investment in pumping, but it is the only option in some cases. In this scenario, the efficient use of hydric resources is inevitable, which means advancing effective closures of seawater circuits. Here, it is essential to acquire a deep rheological understanding of the mineral slurries, which restrict performance at the dewatering stages [7].

Table 1. Examples of seawater use in mining (adapted from Cisternas and Gálvez [3]).

Plant	Country	Metal	Technology
El Boleo Proyect	Mexico	Copper, cobalt, zinc, manganese	Leaching
Mount Keith (*)	Australia	Nickel	Flotation
Sierra Gorda SCM	Chile	Copper, molybdenum	Flotation, leaching
Black Angel	Greenland	Lead–zinc	Flotation
Batu Hijau	Indonesia	Copper-gold	Flotation
Beverly Uranium Mine (*)	Australia	Uranium	Leaching in situ
Minera Michilla	Chile	Copper	Leaching
Antucoya Proyect	Chile	Copper	Leaching
Minera Las Luces	Chile	Copper	Flotation
Minera Algorta Norte S.A.	Chile	Iodine	Leaching

* Saline water.

The high concentration of electrolytes in solution can hinder distinct stages of mineral processing, such as flotation and tailings management [8,9]. Ions alter the interactions between the particles that constitute the slurries, as well as the behaviour of several reagents which are commonly employed in solid–liquid separation. Consequently, the industry is being driven to implement new methods or technologies that can adapt to the challenges posed by a highly saline environment such as seawater [10,11].

Understanding the fundamentals that govern the rheology of mineral pulps requires an in-depth knowledge of the surface interactions that rule the stability of colloidal systems. It is generally considered that the total interaction energy corresponds to the addition of van der Waals forces with the electric double-layer forces. This is the origin of the classical Derjaguin-Landau-Vervey-Overbeek (DLVO) theory. However, additional interactions may arise in highly saline media that are induced by the solvation of ions and structural changes in the association of water molecules that interact with mineral surfaces such as metal oxides [12]. An essential issue for solvation phenomena is the type of ions dissolved in the solution, wherein maker ions such as Na^+, Mg^{2+} and Ca^{2+} are small in size and generate a strong electric field that causes the surrounding water molecules to become highly structured. Otherwise, breaker ions such as K^+ and Cl^- are larger than the maker ions and, therefore, produce a smaller electric field on their neighbouring molecules, disorganising the layers of associated water molecules around the ions [13–15].

Copper concentration operations in freshwater are normally carried out under highly alkaline conditions (above pH 10.5) to depress pyrite and prevent contamination of the concentrate [16–18]; however, it is challenging to operate in seawater at this condition and, generally, the processes are restricted to the natural pH (approx. pH 8), requiring further reagents to deal with the quality of concentrates [5,19]. At a high pH, the seawater divalent cations (Mg^{2+} and Ca^{2+}) can hydrolyse to form soluble complexes (e.g., $Mg(OH)^+$ and $Ca(OH)^+$) and varied solid substances (e.g., $Mg(OH)_2$, $Ca(OH)_2$, etc.). These species may adsorb onto the mineral's surface, which includes the main component of copper tailings such as kaolin and silica. Some reports indicate that the presence of complexes reduces the magnitude of the anionic zeta potential of particles, while solid precipitates can even assign them cationic values [20]. This leads to a substantial impact on the interaction between particles, and consequently, the effect on the rheological properties can be remarkable [21,22].

The tailings produced in copper processing operations mainly contain a significant amount of clays and quartz, which are consistently the focus of study since these determine the rheological behaviour of the tailings [23]. Teh et al. [24] analysed the yield stress and zeta potential at different pH levels in kaolin pulps. The results confirmed that, in most cases, the yield-stress–DLVO model is obeyed, which means that the highest yield stress occurs at low magnitudes of zeta potential. Kaolinite, the molecular formula of which is $[Al_4(Si_4O_{10})(OH)_8]$, has three crystallographically different surfaces: a silica face, an alumina face and the edges. Previous studies revealed that, at a low pH, the alumina face and the edges have cationic charges, while the silica face is anionic. This promotes the formation of strong bonds between particles and, consequently, high yield stress. However, at a high pH, the faces and edges have anionic charges, which causes the dispersion of the particles and thus leads to lower yield stress [25]. Avadiar et al. [26] studied the rheological behaviour of alumina, silica and kaolin slurries in the presence of calcium and magnesium salts. It was detected that the adsorption of $Ca(OH)^+$ and the precipitation of $Ca(OH)_2$ induced a stronger association of particles than that caused by adsorption of $Mg(OH)^+$ and solids of $Mg(OH)_2$. The authors ascribed this to the larger size of the calcium complexes and precipitates, and due to hydration/solvation phenomena, these possess a lower enthalpy of hydration than $Mg(OH)^+$ and $Mg(OH)_2$, respectively. Therefore, the former should adsorb more readily on the particles' surface [27].

The yield stress is the most used rheological parameter to characterise the flow properties of mining tailings, indicating the stress that must be overcome by the pulps so that they begin to flow [28–30]. However, extending to a better rheological characterisation can provide more insight into the strength of the particle networks that make up the suspensions, giving further control opportunities in tailings handling, especially in the discharge of the underflow by rakes in thickeners. In this case, mineral pulps can exhibit viscoelastic behaviours, which are described by a viscous component, represented by the storage modulus (G'), and an elastic part, represented by the loss modulus (G'') [31]. The viscoelastic modulus can be obtained by means of dynamic methods of oscillatory rheology, which are carried out by subjecting the material to an oscillatory strain $\gamma(t) = \gamma_0 \sin(\omega t)$, where the resulting stress in relation to time is $\tau(t) = \gamma_0(G'(\omega)\sin(\omega t) + G''(\omega)\cos(\omega t))$, where G' is a measure of the stored energy of the material and, therefore, is related to molecular events of an elastic nature, while G'' is a measure of energy dissipated as heat, associated with molecular events of a viscous character [32]. Additionally, creep tests consist of applying permanent stress to a specific material, measuring the temporal evolution of its shear strain. The tests can be used as a descriptor of the strength of the particle networks since stronger structures deform less [33,34]. Jeldres et al. [15] studied the effect of the type of salt on the viscoelastic behaviour of silica suspensions prepared in monovalent brines. By increasing the size of the cations, higher values of yield stress, elastic modulus and complex viscosity were obtained, while the shear strain after application of fixed stress was lower. The authors explained that silica has a greater tendency to agglomerate in the presence of larger ions such as K^+, forming stronger particle networks compared with smaller salts such as Na^+ and Li^+.

Several studies have analysed rheological behaviour in saline media, with respect to the primary minerals that compose mining tailings, such as quartz and clays. The reported investigations covered interpretation in monovalent [15,35,36] and divalent salt solutions [26,37]. However, no systematic study has explained the viscoelastic behaviour of these minerals in seawater, especially at a high pH (pH $>$ 10.5), where interference by complexes of calcium and magnesium may appear. These conditions are potentially attractive to carry out the concentration stages, particularly in the copper industry, when the ores have a high content of pyrite [38]. In this sense, the objective of this study was to examine the influence of the principal ionic species present in seawater on the viscoelastic behaviour of quartz and kaolin slurries. The varied pH levels were linked to the formation of soluble complexes and solid precipitates of divalent cations, and their remarkable impact on rheological properties were studied through creep and oscillatory rheology tests.

2. Methodology

2.1. Materials

Kaolin was acquired from Ward's Science, and quantitative X-ray diffraction (XRD) analysis showed that it contained 84 wt % kaolinite ($Al_2Si_2O_5(OH)_4$) and 16 wt % halloysite ($Al_2Si_2O_5(OH)_4{\cdot}2H_2O$) (Figure 1). A D5000 X-ray diffractometer (Siemens S.A., Lac Condes, Chile) was used and the data were processed with Total Pattern Analysis Software (TOPAS) (Siemens S.A., Lac Condes, Chile). The FTIR spectrum (Figure 2) showed a double peak at 3696 and 3654 cm^{-1}, characteristic of the kaolin group. Three absorption bands at 3696, 3654 and 3621 cm^{-1} reflected the high structural order of the samples by the OH stretching of the inner-surface hydroxyl groups. There was Si–O stretching at 1115, 1032 and 1009 cm^{-1}. Al–O–Si deformation appeared at 539 cm^{-1}, Si–O–Si deformation appeared at 471 cm^{-1} and Si–O deformation appeared at 431 cm^{-1}. The zeta potential at pH 8 in distilled water was −40 mV. This was measured using a Zetameter System 4.0 (Zeta-meter, Staunton, VA, USA) following the methodology of Jeldres et al. [15].

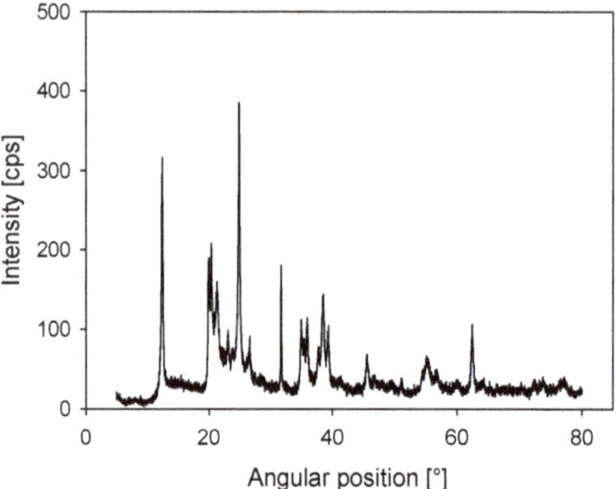

Figure 1. X-ray diffraction (XRD) for kaolin powder.

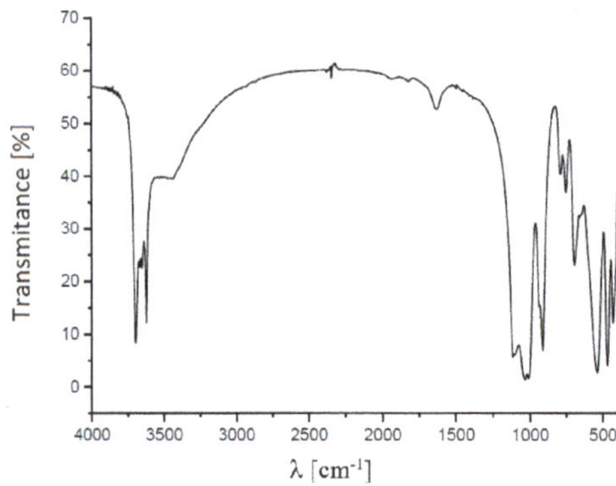

Figure 2. Fourier Transform Infrared Spectroscopy (FTIR) spectrum for kaolin powder.

Quartz was acquired from a local Chilean store, where the SiO_2 content detected by quantitative XRD was over 99 wt % (see Figure 3) and the zeta potential in distilled water at pH 8 was −45 mV.

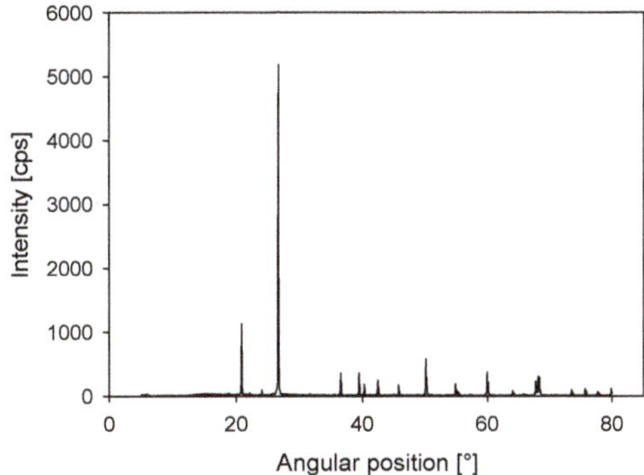

Figure 3. X-ray diffraction (XRD) for quartz powder.

The volume weighted particle size distribution (PSD) was obtained using a Microtrac S3500 laser diffraction particle size analyser (Verder Scientific, Newtown, PA, USA). The PSD of the quartz and kaolin samples are shown in Figure 4, where it was found that 10% of the particles were smaller than d_{10} = 1.8 and 3.8 µm in the samples of kaolin and quartz, respectively.

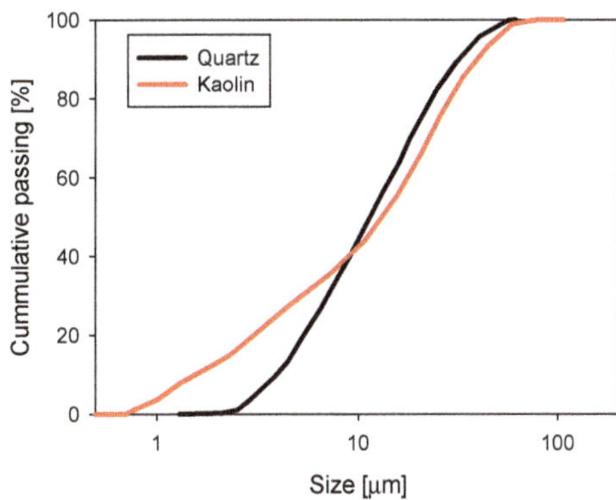

Figure 4. Particle size distribution for quartz and kaolin in distilled water at natural pH.

Seawater was obtained from the San Jorge Bay in Antofagasta, Chile. The water was filtered at 1 µm using a UV filter system to eliminate the bacterial activity. The cation concentrations were determined by atomic absorption spectrophotometry: Na^+: 10.9 g/L; Mg^{2+}: 1.38 g/L; Ca^{2+}: 0.4 g/L; K^+: 0.39 g/L. By the argentometric method, Cl^- was found to be 19.6 g/L. By acid-base volumetry, HCO^{3-} was found to be 0.15 mg/L. The conductivity was 50.4 mS/cm at 25 °C and natural pH.

The pH was raised with sodium hydroxide (NaOH) of analytical grade (over 98%), with the presence of sodium carbonate less than 0.5%. The pulps were mixed with a mechanical stirrer, and the reagent was added gradually. The pH was recorded with a PHS-3BW pH meter (Bante Instrument, Shanghai, China).

2.2. Rheology

Stock slurries of silica and kaolin were prepared at 68 and 37 wt %, respectively, by mixing the solid with seawater overnight with magnetic stirring. The pH was adjusted with sodium hydroxide either to pH 8 or 10.7. Once prepared, a 60 mL aliquot was extracted to perform the rheological assays, while the remaining pulp continued under mixing. Each aliquot was used for a single rheological test and then discarded.

The experiments were carried out on an Anton Paar MCR 102 rheometer, and the data were processed with Rheocompass software. A vane-in-cup configuration was used to minimise the wall-slip effects. The diameter of the vane was 2.2 cm and that of the cup was 4.2 cm.

Creep tests were performed by applying constant stress for 1200 s, which generated an increase in the strain angle (θ) as a function of time. Then, the shear strain (γ) was obtained by Equation (1), considering that the material has linear viscoelastic behaviour:

$$\gamma = \frac{2\theta}{\left(1 - (d_v/d_c)^2\right)} \tag{1}$$

where θ is the strain angle, d_v is the vane diameter and d_c is the cup diameter. The apparent compliance (J) is the apparent shear strain divided by the applied shear stress as follows:

$$J = \frac{\gamma}{\tau_{\text{creep}}}. \tag{2}$$

If J remains constant concerning the applied stress, the slurry exhibits a linear viscoelastic behaviour; otherwise, it is a nonlinear viscoelastic material. Equation (1) is valid only for linear viscoelastic materials, but errors due to nonlinear viscoelasticity are expected to be small because the main interest of this study was to capture only the trend of transition behaviour. The yield stress was determined as the average between the stress that causes a maximum strain (critical strain) and the minimum stress necessary for the pulp to start flowing instantly.

The oscillatory tests were carried out for a frequency sweep between 0.1 and 100 rad/s, keeping the shear strain at 0.5%, previously selected by a sweep amplitude. This ensured that the test was performed in the linear viscoelastic regime, wherein the deformation is proportional to the applied stress, and the viscoelastic moduli have physical meaning.

3. Results

This section analyses the viscoelastic behaviour of silica and kaolin suspensions in seawater, using the results of the creep and oscillatory sweeps tests. Different rheological parameters were obtained, such as apparent critical strain, yield stress, viscoelastic modulus and phase angle. These parameters were associated with the variations of pH and the appearance of magnesium precipitates, which are made in seawater at alkaline conditions, and were distinguished by speciation graphs and the consumption of sodium hydroxide.

3.1. Formation of Magnesium Precipitates

The principal chemical reactions that make soluble magnesium complexes and solid precipitates in seawater are the following:

$$CO_{2(g)} + H_2O_{(l)} \leftrightarrow H_2CO_3$$

$$H_2CO_3 \leftrightarrow HCO_3^- + H^+$$

$$HCO_3^- \leftrightarrow CO_3^{2-} + H^+$$

$$Mg^{2+} + HCO_3^- \leftrightarrow MgHCO_3^+$$

$$Mg^{2+} + OH^- \leftrightarrow MgOH^+$$

$$Mg^{2+} + CO_3^{2-} \leftrightarrow MgCO_3$$

$$Mg^{2+} + 2OH^- \leftrightarrow Mg(OH)_2$$

The formation of magnesium precipitates and complexes generates a buffer effect, which can be identified in a range where a high amount of reagent is required to raise the pH slightly. Nevertheless, the presence of particles (concentrated slurries) interferes with the concentration of ions dissolved in solution, modifying precipitation reactions and the buffer effect. Indeed, a chemical analysis by Atomic Absorption Spectroscopy (AAS) determined that the magnesium in seawater reduced from 1380 to 1280 mg/L in the quartz pulps, while it dropped from 1380 to 658 mg/L in the kaolin suspension.

The amount of sodium hydroxide used (g/L) to raise the pH of seawater, quartz pulp and kaolin pulp, normalised by the amount required to obtain pH 12.2, is shown in Figure 5. A marked buffer effect was observed for pure seawater and quartz slurry between pH 10 and 11, which coincided with the formation of magnesium precipitates, where carbonate precipitation started at pH 9.3 and ended at pH 10, while hydroxide deposition started at pH 10.3 and completed at pH 11 [39–41]. This effect was not clear in the kaolin pulp, indicating a lower formation of magnesium precipitates, which is compatible with the loss of magnesium in solution (from 1380 to 659 mg/L). Considering the low cationic exchange capacity of kaolinite [42,43], it is expected that magnesium cations are attached to the anionic sites of the clay's surface by electrostatic attraction. The adsorption of magnesium ions reduces its concentration in solution, reducing the formation of solid complexes at a high pH.

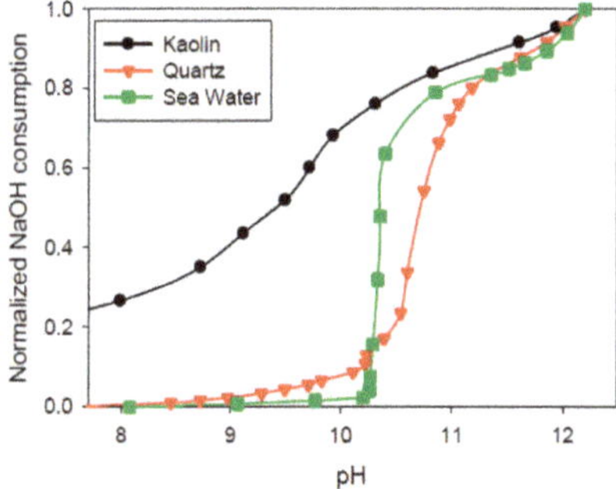

Figure 5. Sodium hydroxide (NaOH) required to modify the pH of (i) seawater, (ii) quartz slurry in seawater and (iii) kaolin slurry in seawater. NaOH is mass (g) consumed per 1 L of solution, normalised by the concentration at pH 12.2 (seawater: 6.1 g/L; quartz: 5.7 g/L; kaolin: 9.1 g/L).

3.2. Creep Tests

Figure 6a–d display the creep tests for quartz and kaolin slurries prepared with seawater at pH 8 and 10.7. The temporal evolution of the apparent strain was measured under fixed stress for 1200 s, and in all cases, a maximum strain was observed that remained constant at high times. When the applied stress was greater than a critical value, the pulp started to flow, and its deformation diverged. An apparent instantaneous strain was observed for the quartz pulps when the stress was lower than half of the yield stress (Figure 6a,b). This effect was not observed for kaolin pulps, which exhibited a gradual increase in their apparent strains (Figure 6c,d).

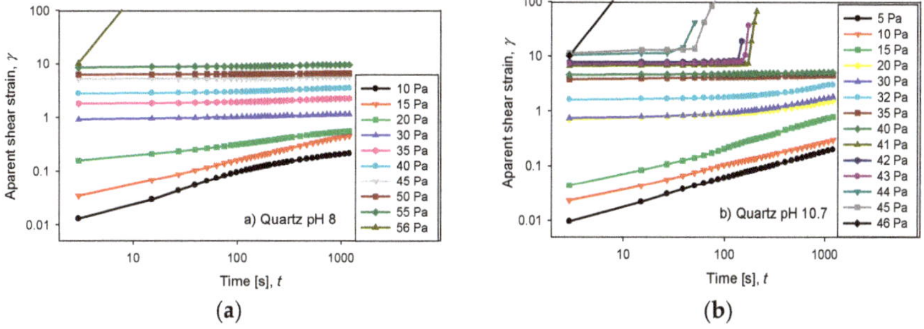

(a) (b)

Figure 6. *Cont.*

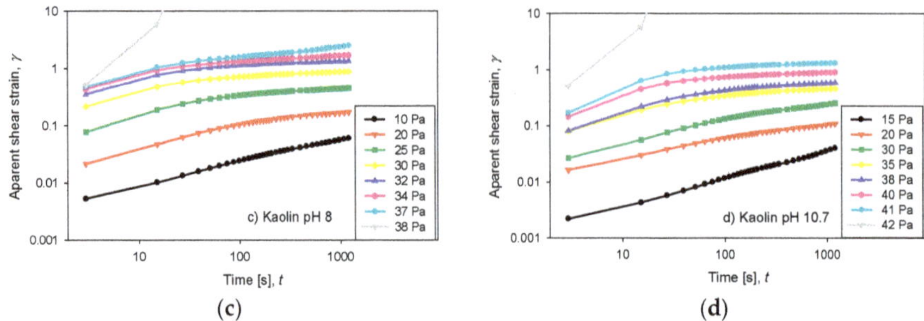

(c) (d)

Figure 6. Creep test for quartz and kaolin slurries at varied applied stresses: (**a**) quartz at pH 8, (**b**) quartz at pH 10.7, (**c**) kaolin at pH 8 and (**d**) kaolin at pH 10.7.

The yield stress for each pulp examined in this study is shown in Figure 7. The quartz slurry at pH 8 reached a yield stress of 56 Pa, which was significantly higher than the 43 Pa obtained at pH 10.7. Interestingly, the decrease of yield stress by increasing the pH (from pH 8 to 10.7) coincided with the formation of solid complexes in seawater. In the literature, it has been reported that these products can adsorb onto the surface of the particles, having different consequences on the rheological properties [22]. The solid complexes may form bonds between quartz particles, mainly due to the reduction of zeta potential [20], but the strength of these bonds would be weaker compared with the cationic bridges generated by the divalent Mg^{2+} ions. For this reason, it is expected that at pH 8, where the concentration of Mg^{2+} ions in solution is higher, the yield stress is higher. On the other hand, the kaolin suspension at pH 8 had a yield stress of 38 Pa, while at pH 10.7, it increased to 42 Pa. This behaviour is contrary to the quartz pulp and can be justified by the low presence of solid precipitates.

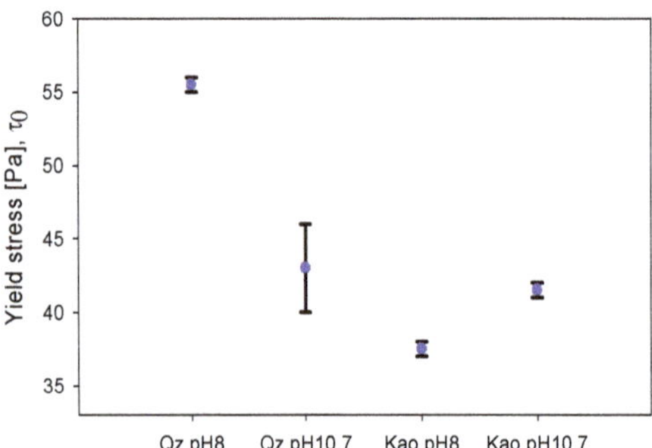

Figure 7. Yield stress for quartz and kaolin slurries at different pH levels.

Critical shear strain is an attractive property of materials, which is directly associated with the strength of the bonds that form the particle networks just before they begin to flow [44]. Figure 8 shows the critical strain, where for both minerals, the shear strain decreased as there was an increase in pH.

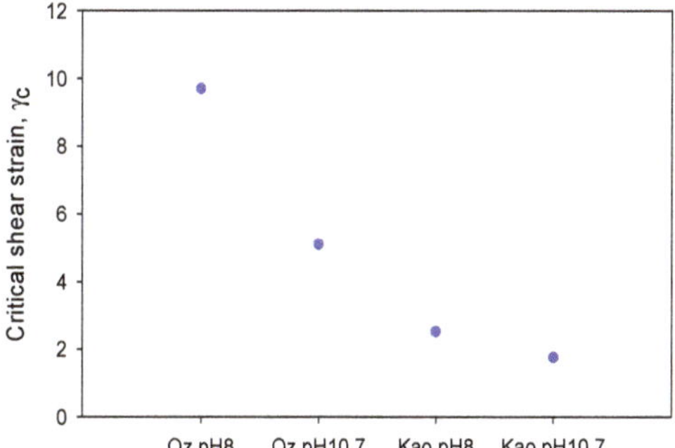

Figure 8. Critical shear strain for quartz and kaolin slurries at different pH levels.

An exceptional contrast of the system that held precipitates (quartz at pH 10.7) with respect to the rest of the slurries is that the yield stress was displayed in a range of stress (40–46 Pa) instead of a single point. For that matter, the pulp began to flow depending on both the applied stress and the time of stress application (Figure 9). Otherwise, for pulps in which there were no solid precipitates, the yield stress was a precise value: quartz pH 8, 55 Pa; kaolin pH 8, 38 Pa; and kaolin pH 10.7, 42 Pa.

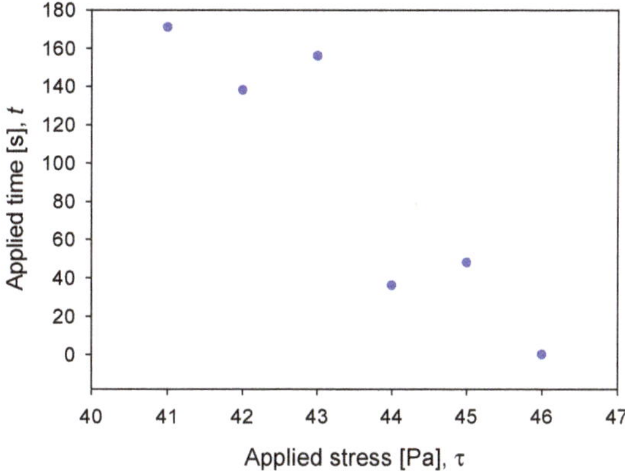

Figure 9. Relationship between yield stress and applied time for quartz slurries at pH 10.7.

Mineral pulps may exhibit linear or nonlinear viscoelastic behaviour, or both, depending on the applied stress. A linear response implies that compliance remains constant, independent of the stress which was employed. This characteristic is shown in Figure 10, considering the compliance at the end of the experiment (1200 s). The kaolin pulp at pH 8 suggested linear viscoelastic behaviour when the stress was lower than 20 Pa, while at pH 10.7, the linear response limit was up to the stress of 35 Pa. On the other hand, the quartz slurry at pH 8 had a linear behaviour up to 30 Pa, while at pH 10.7, the system of which was affected by the presence of complexes and solid precipitates, the linearity decreased considerably to a stress lower than 10 Pa.

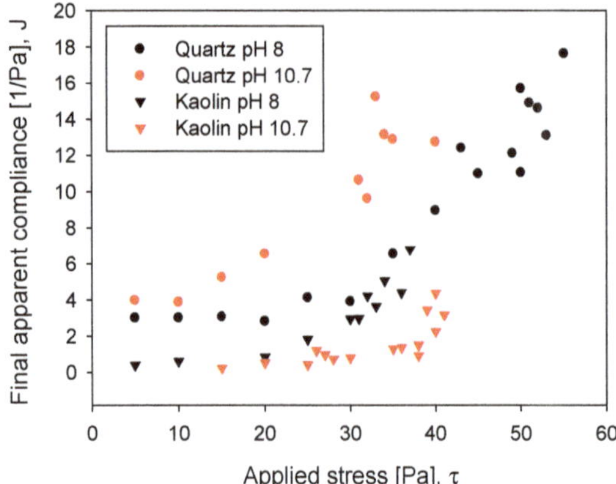

Figure 10. Final apparent compliance as a function of stress for quartz and kaolin slurries.

3.3. Oscillatory Rheology

Oscillatory rheology tests are a good descriptor of pulp strength. In this case, a frequency sweep was performed considering a constant amplitude of 0.5%, previously defined by an amplitude sweep test. This ensured that the pulp was in a linear viscoelastic regime. Figure 11 shows the response of both moduli G' and G'' concerning the angular frequency for the silica suspension. Viscoelastic moduli had a higher value when silica was at pH 8, meaning a greater resistance of the particle networks, while at pH 10.7, both moduli were diminished. Interestingly, for both pH levels, the moduli G' and G'' crossed within the range of frequencies examined. The crossover point was an indicator of the relaxation time, using the reciprocal of frequency under conditions where both moduli intersected. While at pH 8 the relaxation time was 8.6 s, at pH 10.7, the value increased to 10.8 s. For the kaolin pulps, the behaviour was different (Figure 12), and for both pH values (8 and 10.7), the storage modulus (G') was higher than the loss modulus (G''). It can be argued that both moduli had a weak dependence on frequency, which is a typical gel behaviour [32]. The phase angle, as shown in Figure 13, was higher at pH 8 than at pH 10.7, for both minerals. This suggests that the pH changed the internal structure of the pulps, adopting a more solid-like character when the pH was more alkaline. In the silica suspension, it was observed that at frequencies greater than 30 rad/s, the phase angle was independent of the applied frequency, while for kaolin pulp, this happened at over 10 rad/s.

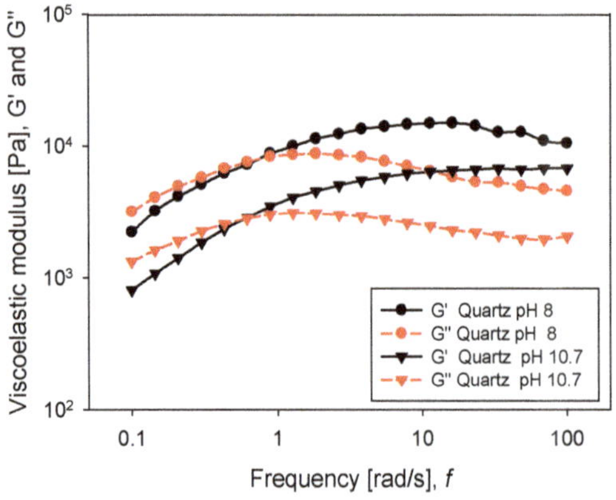

Figure 11. Frequency sweep for quartz slurries at different pH levels.

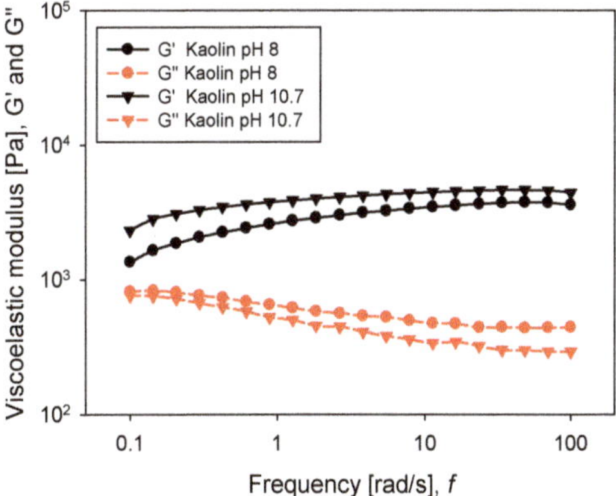

Figure 12. Frequency sweep for kaolin slurries at different pH levels.

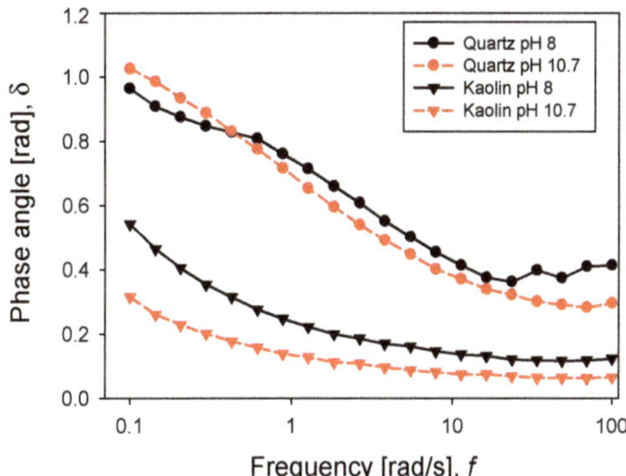

Figure 13. Phase angle for quartz and kaolin slurries at varied pH levels.

4. Conclusions

The viscoelastic properties of quartz and kaolin suspensions in seawater were analysed, examining two different conditions: (i) pH 8, which resembles that used by copper concentrating plants, and (ii) pH 10.7, which is attractive since it may enhance the quality of concentrates. Nevertheless, the latter is characterised by speciation reactions of divalent seawater cations, which lead to the formation of solid precipitates, mainly magnesium complexes. This feature changes for slurries according to the mineral species because the particles' surfaces can adsorb a portion of the cations, reducing their content in solution. Here, quartz pulp reduced magnesium from 1380 to 1280 mg/L. Since it was not a large difference from the solid-free seawater, the disposition of solid complexes at pH 10.7 was similar. As rheological consequences, the increase in pH caused both yield stress and viscoelastic moduli to drop, suggesting that the solid precipitates diminished the strength of the particle networks that made up the suspension. For the kaolin slurries, the yield stress raised when the pH increased, but unlike quartz, there was significant adsorption of magnesium cations. In fact, the concentration of magnesium in solution fell from 1380 to 658 mg/L.

The dynamic oscillatory assays revealed structural changes for both pulps; in particular, the phase angle was higher at pH 8 than at pH 10.7, showing that the more alkaline suspension exhibited a more solid-like character.

Author Contributions: All of the authors contributed to analysing the results and writing the paper.

Funding: This research was funded by Conicyt Fondecyt 11171036 and Centro CRHIAM Project Conicyt/Fondap/15130015.

Acknowledgments: Ricardo I. Jeldres thanks Conicyt Fondecyt 11171036 and Centro CRHIAM Project Conicyt/Fondap/15130015. The authors are grateful for the contribution of the Scientific Equipment Unit—MAINI of the Universidad Católica del Norte for facilitating the XRD analysis. Pedro Robles thanks the Pontificia Universidad Católica de Valparaíso for the support provided.

Conflicts of Interest: The authors declare no conflict of interest.

References

1. Northey, S.A.; Mudd, G.M.; Werner, T.T.; Jowitt, S.M.; Haque, N.; Yellishetty, M.; Weng, Z. The exposure of global base metal resources to water criticality, scarcity and climate change. *Glob. Environ. Chang.* **2017**, *44*, 109–124. [CrossRef]
2. Shao, D.; Li, X.; Gu, W. A method for temporary water scarcity analysis in humid region under droughts condition. *Water Resour. Manag.* **2015**, *29*, 3823–3839. [CrossRef]
3. Cisternas, L.A.; Gálvez, E.D. The use of seawater in mining. *Miner. Process. Extr. Metall. Rev.* **2018**, *39*, 18–33. [CrossRef]
4. Li, W.; Li, Y.; Wei, Z.; Xiao, Q.; Song, S. Fundamental studies of SHMP in reducing negative effects of divalent ions on molybdenite flotation. *Minerals* **2018**, *8*, 404. [CrossRef]
5. Jeldres, R.I.; Forbes, L.; Cisternas, L.A. Effect of seawater on sulfide ore flotation: A review. *Miner. Process. Extr. Metall. Rev.* **2016**, *37*, 369–384. [CrossRef]
6. Lattemann, S.; Höpner, T. Environmental impact and impact assessment of seawater desalination. *Desalination* **2008**, *220*, 1–15. [CrossRef]
7. McFarlane, A.J.; Bremmell, K.E.; Addai-Mensah, J. Optimising the dewatering behaviour of clay tailings through interfacial chemistry, orthokinetic flocculation and controlled shear. *Powder Technol.* **2005**, *160*, 27–34. [CrossRef]
8. Lucay, F.; Cisternas, L.A.; Gálvez, E.D.; López Valdivieso, A. Study of the natural floatability of molybdenite fines in saline solutions and effect of gypsum precipitation. *Miner. Metall. Process.* **2015**, *32*, 203–208. [CrossRef]
9. Ramos, O.; Castro, S.; Laskowski, J.S. Copper–molybdenum ores flotation in sea water: Floatability and frothability. *Miner. Eng.* **2013**, *53*, 108–112. [CrossRef]
10. Cruz, C.; Reyes, A.; Jeldres, R.I.; Cisternas, L.A.; Kraslawski, A. Using partial desalination treatment to improve the recovery of copper and molybdenum minerals in the Chilean mining industry. *Ind. Eng. Chem. Res.* **2019**. [CrossRef]
11. Jeldres, R.I.; Arancibia-Bravo, M.P.; Reyes, A.; Aguirre, C.E.; Cortes, L.; Cisternas, L.A. The impact of seawater with calcium and magnesium removal for the flotation of copper-molybdenum sulphide ores. *Miner. Eng.* **2017**, *109*, 10–13. [CrossRef]
12. Israelachvili, J. *Intermolecular and Surface Forces*, 3rd ed.; Elsevier: Santa Barbara, CA, USA, 2011.
13. Hofmeister, F. Zur Lehre von der Wirkung der Salze. *Arch. Exp. Pathol. Pharmcol.* **1888**, *24*, 247–260. [CrossRef]
14. Morag, J.; Dishon, M.; Sivan, U. The governing role of surface hydration in ion specific adsorption to silica: An AFM-based account of the hofmeister universality and its reversal. *Langmuir* **2013**, *29*, 6317–6322. [CrossRef] [PubMed]
15. Jeldres, R.I.; Piceros, E.C.; Leiva, W.H.; Toledo, P.G.; Quezada, G.R.; Robles, P.A.; Valenzuela, J. Analysis of silica pulp viscoelasticity in saline media: The effect of cation size. *Minerals* **2019**, *9*, 216. [CrossRef]
16. Mu, Y.; Peng, Y.; Lauten, R.A. The depression of pyrite in selective flotation by different reagent systems—A Literature review. *Miner. Eng.* **2016**, *96–97*, 143–156. [CrossRef]
17. Jeldres, R.I.; Uribe, L.; Cisternas, L.A.; Gutierrez, L.; Leiva, W.H.; Valenzuela, J. The effect of clay minerals on the process of flotation of copper ores—A critical review. *Appl. Clay Sci.* **2019**, *170*, 57–69. [CrossRef]
18. Zanin, M.; Lambert, H.; du Plessis, C.A. Lime use and functionality in sulphide mineral flotation: A review. *Miner. Eng.* **2019**, *143*, 105922. [CrossRef]
19. Castro, S. Challenges in flotation of Cu-Mo Sulfide Ores in Sea Water I. In *Water in Mineral Processing*; Society for Mining, Metallurgy, and Exploration: Englewoog, NJ, USA, 2012; pp. 29–40.
20. Atesok, G.; Somasundaran, P.; Morgan, L.J. Adsorption properties of Ca^{2+} on Na-kaolinite and its effect on flocculation using polyacrylamides. *Colloids Surf.* **1988**, *32*, 127–138. [CrossRef]

21. De Kretser, R.G.; Scales, P.J.; Boger, D.V. Surface chemistry-rheology inter-relationships in clay suspensions. *Colloids Surf. A* **1998**, *137*, 307–318. [CrossRef]
22. Mpofu, P.; Addai-Mensah, J.; Ralston, J. Influence of hydrolyzable metal ions on the interfacial chemistry, particle interactions, and dewatering behavior of kaolinite dispersions. *J. Colloid Interface Sci.* **2003**, *261*, 349–359. [CrossRef]
23. Castillo, C.; Ihle, C.F.; Jeldres, R.I. Chemometric optimisation of a copper sulphide tailings flocculation process in the presence of clays. *Minerals* **2019**, *9*, 582. [CrossRef]
24. Teh, E.J.; Leong, Y.K.; Liu, Y.; Fourie, A.B.; Fahey, M. Differences in the rheology and surface chemistry of kaolin clay slurries: The source of the variations. *Chem. Eng. Sci.* **2009**, *64*, 3817–3825. [CrossRef]
25. Ndlovu, B.; Becker, M.; Forbes, E.; Deglon, D.; Franzidis, J.-P. The influence of phyllosilicate mineralogy on the rheology of mineral slurries. *Miner. Eng.* **2011**, *24*, 1314–1322. [CrossRef]
26. Avadiar, L.; Leong, Y.K.; Fourie, A. Physicochemical behaviors of kaolin slurries with and without cations-Contributions of alumina and silica sheets. *Colloids Surf. A* **2015**, *468*, 103–113. [CrossRef]
27. Colic, M.; Fisher, M.L.; Franks, G.V. Influence of ion size on short-range repulsive forces between silica surfaces. *Langmuir* **1998**, *14*, 6107–6112. [CrossRef]
28. De Kretser, R.; Scales, P.J.; Boger, D.V. Improving clay-based tailings disposal: Case study on coal tailings. *AIChE J.* **1997**, *43*, 1894–1903. [CrossRef]
29. Scales, P.J.; Johnson, S.B.; Healy, T.W.; Kapur, P.C. Shear yield stress of partially flocculated colloidal suspensions. *AIChE J.* **1998**, *44*, 538–544. [CrossRef]
30. Sofrá, F.; Boger, D.V. Environmental rheology for waste minimisation in the minerals industry. *Chem. Eng. J.* **2002**, *86*, 319–330. [CrossRef]
31. Barnes, H.A.; Hutton, J.F.; Walters, K. *An Introduction to Rheology. Rheology Series (NL)*; Elsevier Science: Amsterdam, The Netherlands, 1989.
32. Lin, Y.; Phan-Thien, N.; Lee, J.B.P.; Khoo, B.C. Concentration dependence of yield stress and dynamic moduli of kaolinite suspensions. *Langmuir* **2015**, *31*, 4791–4797. [CrossRef]
33. Stickland, A.D.; Kumar, A.; Kusuma, T.E.; Scales, P.J.; Tindley, A.; Biggs, S.; Buscall, R. The effect of premature wall yield on creep testing of strongly flocculated suspensions. *Rheol. Acta* **2015**, *54*, 337–352. [CrossRef]
34. Jeldres, R.I.; Toledo, P.G.; Concha, F.; Stickland, A.D.; Usher, S.P.; Scales, P.J. Impact of seawater salts on the viscoelastic behavior of flocculated mineral suspensions. *Colloids Surf. A* **2014**, *461*, 295–302. [CrossRef]
35. Oncsik, T.; Trefalt, G.; Borkovec, M.; Szilagyi, I. Specific ion effects on particle aggregation induced by monovalent salts within the Hofmeister series. *Langmuir* **2015**, *31*, 3799–3807. [CrossRef] [PubMed]
36. Jeldres, R.I.; Piceros, E.C.; Leiva, W.H.; Toledo, P.G.; Herrera, N. Viscoelasticity and yielding properties of flocculated kaolinite sediments in saline water. *Colloids Surf. A* **2017**, *529*, 1009–1015. [CrossRef]
37. Abu-Jdayil, B. Rheology of sodium and calcium bentonite–water dispersions: Effect of electrolytes and aging time. *Int. J. Miner. Process.* **2011**, *98*, 208–213. [CrossRef]
38. Li, Y.; Chen, J.; Kang, D.; Guo, J. Depression of pyrite in alkaline medium and its subsequent activation by copper. *Miner. Eng.* **2012**, *26*, 64–69. [CrossRef]
39. El-Manharawy, S.; Hafez, A. Study of seawater alkalization as a promising RO pretreatment method. *Desalination* **2003**, *153*, 109–120. [CrossRef]
40. Ayoub, G.M.; Zayyat, R.M.; Al-Hindi, M. Precipitation softening: A pretreatment process for seawater desalination. *Environ. Sci. Pollut. Res.* **2014**, *21*, 2876–2887. [CrossRef]
41. Kapp, E.M. The precipitation of calcium and magnesium from sea water by sodium hydroxide. *Biol. Bull.* **1928**, *55*, 453–458. [CrossRef]
42. Ma, C.; Eggleton, R.A. Cation exchange capacity of kaolinite. *Clays Clay Miner.* **1999**, *47*, 174–180.
43. Zhou, Z.; Gunter, W.D. The nature of the surface charge of kaolinite. *Clays Clay Miner.* **1992**, *40*, 365–368. [CrossRef]
44. Buscall, R.; Mills, P.D.A.; Stewart, R.F.; Sutton, D.; White, L.R.; Yates, G.E. The rheology of strongly-flocculated suspensions. *J. Non-Newtonian Fluid Mech.* **1987**, *24*, 183–202. [CrossRef]

Article

Kinetics of Chalcopyrite Leaching by Hydrogen Peroxide in Sulfuric Acid

Miroslav Sokić [1,*], Branislav Marković [1], Srđan Stanković [1], Željko Kamberović [2], Nada Štrbac [3], Vaso Manojlović [2] and Nela Petronijević [1]

1 Institute for Technology of Nuclear and Other Mineral Raw Materials, 11000 Belgrade, Serbia; b.markovic@itnms.ac.rs (B.M.); s.stankovic@itnms.ac.rs (S.S.); n.petronijevic@itnms.ac.rs (N.P.)
2 Faculty of Technology and Metallurgy, University of Belgrade, 11000 Belgrade, Serbia; kamber@tmf.bg.ac.rs (Ž.K.); v.manojlovic@tmf.bg.ac.rs (V.M.)
3 Technical Faculty, University of Belgrade, 19210 Bor, Serbia; nstrbac@tf.bor.ac.rs
* Correspondence: m.sokic@itnms.ac.rs; Tel.: +381-11-3691-722

Received: 4 September 2019; Accepted: 26 October 2019; Published: 30 October 2019

Abstract: In ores, chalcopyrite is usually associated with other sulfide minerals, such as sphalerite, galena, and pyrite, in a dispersed form, with complex mineralogical structures. Concentrates obtained by flotation of such ores are unsuitable for pyrometallurgical processing owing to their poor quality and low metal recovery. This paper presents the leaching of chalcopyrite concentrate from the location "Rudnik, Serbia". The samples from the flotation plant were treated with hydrogen peroxide in sulfuric acid. The influences of temperature, particle size, stirring speed, as well as the concentrations of hydrogen peroxide and sulfuric acid were followed and discussed. Hence, the main objective was to optimize the relevant conditions and to determine the reaction kinetics. It was remarked that the increase in temperature, hydrogen peroxide content, and sulfuric acid concentration, as well as the decrease in particle size and stirring speed, contribute to the dissolution of chalcopyrite. The dissolution kinetics follow a model controlled by diffusion, and the lixiviant diffusion controls the rate of reaction through the sulfur layer. Finally, the main characterization methods used to corroborate the obtained results were X-ray diffraction (XRD) as well as qualitative and quantitative light microscopy of the chalcopyrite concentrate samples and the leach residue.

Keywords: chalcopyrite concentrate; hydrogen peroxide; sulfuric acid; leaching kinetics

1. Introduction

The global demand for copper is increasing while its content in mined ores decreases. Chalcopyrite, as the most abundant copper mineral in the earth's crust, has a very stable crystal structure, and usually co-exists with other sulfide minerals such as sphalerite, galena, pyrite, and gangue minerals. Copper producers are forced to cope with low-grade ores dominated by chalcopyrite [1,2]. Due to the complex mineralogy and/or presence of toxic chemical elements, such as arsenic, some ores are not suitable for concentration by froth flotation. Therefore, it is convenient to treat these ores using some of the hydrometallurgical processes. Leaching of copper from chalcopyrite requires the presence of the oxidants in an acidic environment; some of the frequently present oxidants are: the ferric ions [3–9], the cupric ions [10–12], some acidophilic bacteria [13–18], the nitrate ions [19,20], the nitrite ions [21,22], the dichromate ions [23,24], the manganese ions [25,26], the permanganate ions [27], the chlorate ions [28], the oxygen ions [29–34], ozone [35], the silver ions [36,37], and the use of microwaves [38,39].

Hydrogen peroxide was used as a relatively cheap and strong oxidizing agent, with a redox potential of +1.77 mV in the acidic medium [40]. The application of hydrogen peroxide for the leaching of chalcopyrite in sulfuric acid solution has previously been examined by several authors [40–53].

Oxidative activity of the hydrogen peroxide is based on its dissociation to the reactive hydroxyl anion group (OH^-) and hydroxyl radical ($HO^{\cdot}$), dissociation is catalyzed by ferric ion, this reaction occurs according to the chemical Equation (1) [40,54]:

$$H_2O_2 + Fe^{2+} \rightarrow HO^{\cdot} + OH^- + Fe^{3+} \quad (1)$$

A dominant mechanism of chalcopyrite dissolution by the hydrogen peroxide in acidic solution is given by the Equation (2) [42]:

$$2CuFeS_2 + 17H_2O_2 + 2H^+ \rightarrow 2Cu^{2+} + 2Fe^{3+} + 4SO_4{}^{2-} + 18H_2O \quad (2)$$

At the same time, a small part of sulfide sulfur is oxidized to its elemental form, this is confirmed by XRD analysis of the leach residue and a leaching degree of 55%, the reaction is presented by Equation (3):

$$2CuFeS_2 + 5H_2O_2 + 10H^+ \rightarrow 2Cu^{2+} + 2Fe^{3+} + 4S^0 + 10H_2O \quad (3)$$

Agacayak et al. obtained similar results [46]. They did not confirm the formation of elemental sulfur in the leaching process, as confirmed by the XRD analysis of the leach residues. Simultaneously, they found a pH decrease in the solution, which confirms that sulfide sulfur transforms to sulfate during the leaching. On the other hand, Misra and Fuerstenau [43] have shown that most of the sulfide sulfur was transformed into its elemental form.

Adebayo et al. [40] examined the leaching kinetics of chalcopyrite and found that increasing the concentration of sulfuric acid and hydrogen peroxide could increase the copper leaching rate; also, the chalcopyrite dissolution reaction takes place by a shrinking-core model with the surface reaction as the rate-controlling step with an activation energy of 39 kJ/mol.

Hydrogen peroxide is unstable and quickly decomposes in the leaching process of sulfide minerals in the presence of iron and copper ions, mineral particles, and different impurities. Intensive depletion of hydrogen peroxide by metal cations and discontinuation of chalcopyrite leaching after 60 min were confirmed by Olubambi and Potgieter [40], and Agacayak et al. [46]. They also confirmed the decomposition of hydrogen peroxide at elevated temperatures.

To reduce the decomposition of peroxide and reduce the decomposition of peroxide, polar organic solvents can be used as its stabilizers. Mahajan et al. [51] confirmed that in the presence of ethylene glycol, chalcopyrite dissolution was possible at low peroxide concentrations. They also confirmed the presence of elemental sulfur at the chalcopyrite surface in the form of separate crystalline particles instead of a continuous sulfur layer; these particles cause the passivation of the chalcopyrite surface. in their experiments, the activation energy was 30 kJ/mol, and the leaching rate was controlled by the surface reaction. Similar results of chalcopyrite leaching in the presence of ethylene glycol and methanol were obtained by Ruiz-Sanchez and Lapidus [52], and Solís Marcial et al. [53].

This work is related to the study of kinetics and the mechanism of dissolution of chalcopyrite from its concentrate by hydrogen peroxide in sulfuric acid solution. The main reason to perform this study was significant differences (~50%) observed in leaching rates of eleven different chalcopyrite samples from different localities under various leaching conditions [55]. The same concentrate was used for the previous study by Sokić et al. [56], where he studied the influence of chalcopyrite structure on leaching by sodium nitrate in sulfuric acid.

In order to optimize parameters for copper leaching, effects of the particle size, temperature, leaching time, concentrations of hydrogen peroxide on copper extraction yield were measured. The obtained results were used to select and to determine the adequate reaction kinetic model.

2. Materials and Methods

2.1. Experimental Procedure

All leaching experiments were performed in the glass reactor of 1.2 dm^3 equipped with the Heidolph RZR 2021 model mechanic stirrer (Heidolph Instruments GmbH & CO. KG, Schwabach, Germany) with a Teflon-covered propeller, the condenser to prevent evaporation, the thermometer, the funnel for adding solid samples, and the sampling device. The temperature in the glass reactor was controlled by a heating mantle station (Electrothermal, Staffordshire, UK). The 600 mL sulfuric acid-hydrogen peroxide solution of a definite concentration was poured into the reactor and heated to the set temperature. A 1.2 g concentrate samples were added to the reactor in all experiments when the desired temperature was reached. Samples of the leaching solution were collected in the selected time intervals, and then analyzed on dissolved copper, zinc and iron by atomic absorption spectrophotometry. The solid residues obtained by leaching were washed, dried, and characterized.

Analytical grade chemicals (sulfuric acid, hydrogen peroxide, and distilled water) (MOSS & HEMOSS, Belgrade, Serbia) were used for preparation leaching solutions.

In the leaching experiments, the influence of the stirring speed up to 300 rpm, the temperature in the range 25–45 °C, the hydrogen peroxide concentration in the range of 0.2–2.0 M, and the particle size range was −37, +37–50, +50–75, and +75 μm were investigated.

2.2. Methods

In order to determine the characteristics of the chalcopyrite concentrate, leaching solutions and leach residues, detailed chemical and qualitative and quantitative mineralogical examinations were done. The following methods were used: atomic absorption spectrophotometry (AAS), X-ray diffraction (XRD), and qualitative and quantitative light microscopy.

For chemical analysis, samples were dissolved in aqua regia, and then the concentration of the selected chemical elements was measured by the atomic absorption spectrophotometer PERKIN ELMER 703 model (Perkin Elmer, Norwalk, CT, USA).

The mineralogical compositions of the concentrate and leaching residue were determined using Philips PW-1710 X-ray diffractometer (Philips, Eindhoven, Netherlands). Data obtained by X-ray diffraction were analyzed by Powder Cell computer software [57]. Qualitative and quantitative microscopic investigations were performed in reflected light on Carl Zeiss-Jena, JENAPOL-U microscope (Carl Zeiss, Jena, Germany).

2.3. Material

The chalcopyrite concentrate samples were provided by the "Rudnik" flotation plant (Rudnik, Serbia), and their granulometric and chemical composition for each size fraction are presented in Table 1.

Table 1. Granulometric and chemical composition of the chalcopyrite concentrate samples.

Class, μm	Mass Fraction, %	Content (%)				
		Cu	Zn	Pb	Fe	S
+75	7.32	23.38	3.43	4.08	22.25	28.31
−75 + 50	21.15	26.55	4.28	1.70	24.43	30.19
−50 + 37	5.18	26.95	4.36	1.85	24.75	30.92
−37	66.35	27.08	4.15	2.28	25.12	31.12

The chemical analysis showed that chalcopyrite concentrate contains significant amounts of zinc, lead, and iron.

X-ray diffraction (XRD) analysis shows that the main phase in concentrate is chalcopyrite, with small amounts of sphalerite (Figure 1).

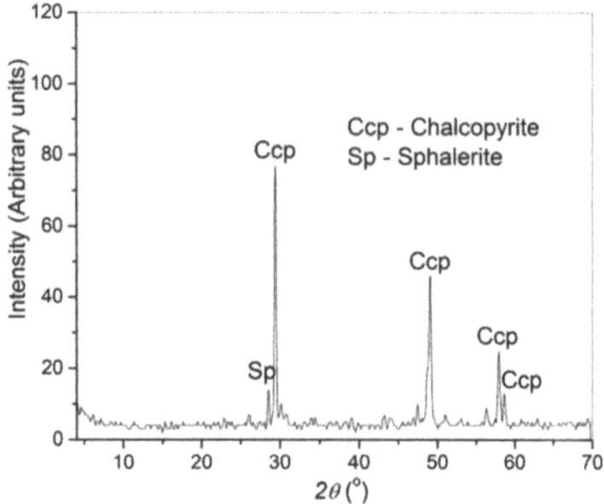

Figure 1. X-ray diffraction analysis of the chalcopyrite concentrate sample [56].

The complete mineralogical composition of the sample is shown in Table 2; the qualitative and quantitative analyses are done using light microscopy, with corrections of Cu, Zn, and Pb content using AAS. Qualitative and quantitative mineralogical analyses show that total sulfides content was 88.7%, while the share of the liberated suplhide grains was 91% and liberated chalcopyrite grains were about 95%.

Table 2. Mineral composition of the chalcopyrite concentrate [56].

Mineral	Mass %
Chalcopyrite	78.247
Sphalerite	6.249
Galena	2.633
Pyrrhotite	1.161
Pyrite	0.099
Arsenopyrite	0.211
Covellite	0.081
Native bismuth	0.031
Limonite	0.118
Gangue	11.170
Total:	100.000

The reflected light microscopic examination of chalcopyrite concentrate is shown in Figure 2. Figure 2 shows the presence of chalcopyrite in the form of free grains and simple intergrowths with sphalerite.

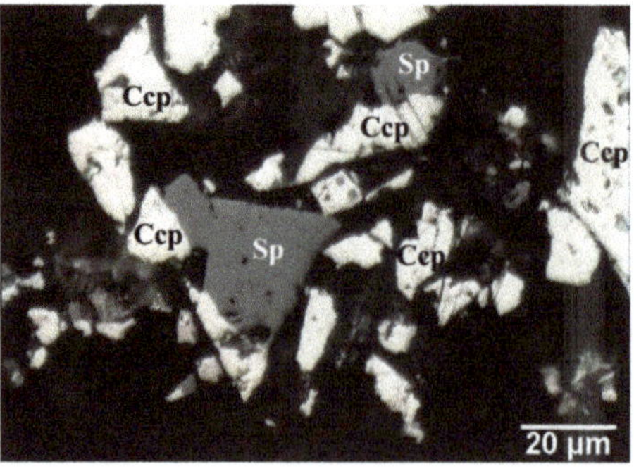

Figure 2. Optical micrograms of the chalcopyrite concentrate sample [56].

3. Results and Discussion

3.1. Effect of Particle Size

The influence of the particle size on the leaching degree of copper was determined under the following conditions: the temperature—40 °C, stirring speed–100 rpm, concentrations of H_2SO_4—1.5 M, H_2O_2—1 M, and solid/liquid ratio—2 gdm^{-3}. The results of these experiments are presented in Figure 3.

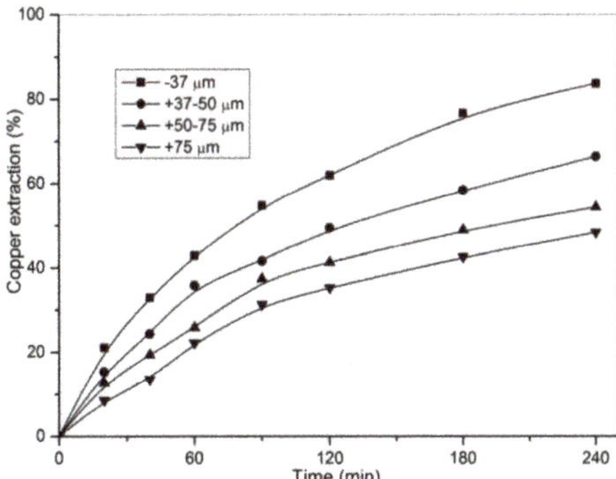

Figure 3. Influence of the particle size on the copper extraction yield (temperature—40 °C, stirring speed—100 rpm, solid/liquid ratio—2 g/dm^3, —1.5 M H_2SO_4, —1 M H_2O_2).

As can be seen from Figure 3, with decreasing particle size, copper extraction under the same conditions increases. The similar results were obtained by other authors [40,46,50]. The contact surface between oxidant and chalcopyrite increases, for smaller particles compared to large ones, significantly improving copper extraction; thus, particles with size −37 μm was used for testing other parameters.

3.2. Effect of Temperature and Leaching Time

Influences of the temperature and leaching time on a copper extraction yields from chalcopyrite concentrate was measured at temperatures 25, 30, 35, 40, and 45 °C and leaching time from 0 to 240 min. The grain size was 100% −37 μm, the stirring speed was 100 rpm, and the were of H_2SO_4 and H_2O_2 1.5 M, and 1 M, respectively. The amount of concentrate in the solution was 2 gdm^{-3}. The leaching degree of copper as a function of temperature and leaching time is shown in Figure 4. The copper extraction yield sharply increased with temperature up to 40 °C. After 240 min, at 25 °C, 33.21% of the copper was extracted, compared to 83.64% at 40 °C. At temperatures above 40 °C, copper leaching yields decreased due to a reduction in the amount of H_2O_2 from the solution, caused by its decomposition, as confirmed by Antonijevic et al. [41,42] and Mahajan et al. [51]. Agacayak et al. [46] found that during the leaching with H_2O_2, increasing of the temperature above 50 °C decreases the copper extraction yield due to the promoted decomposition of the hydrogen peroxide. They also noticed that the copper extraction increased with increasing temperatures until the leaching time reached 90 min. After that time, Cu extraction does not change significantly at temperatures above 50 °C. Accordingly, a temperature of 40 °C was used in other experiments.

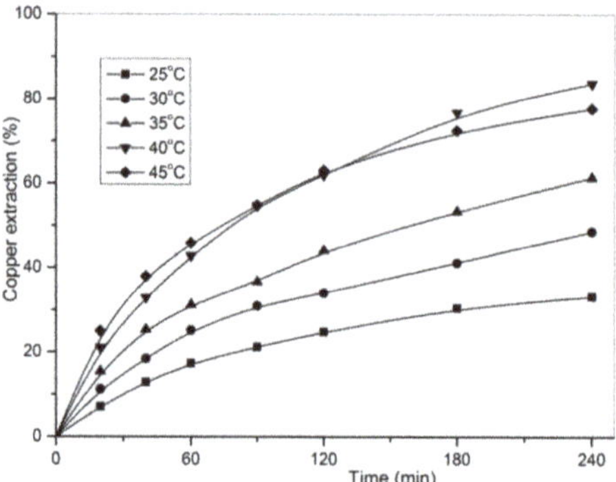

Figure 4. Influence of the temperature and time on the copper extraction yield (particle size −37 μm, stirring speed—100 rpm, solid/liquid ratio—2 g/dm^3, 1.5 M H_2SO_4, 1 M H_2O_2).

3.3. *Effect of Stirring Speed*

The influence of the stirring speed was examined on the sample with the particle size 100% −37 μm, temperature of 40 °C, concentrations of H_2SO_4 1.5 M, and H_2O_2 1 M, and solid/liquid ratio—2 gdm^{-3}. These experimental results are shown in Figure 5.

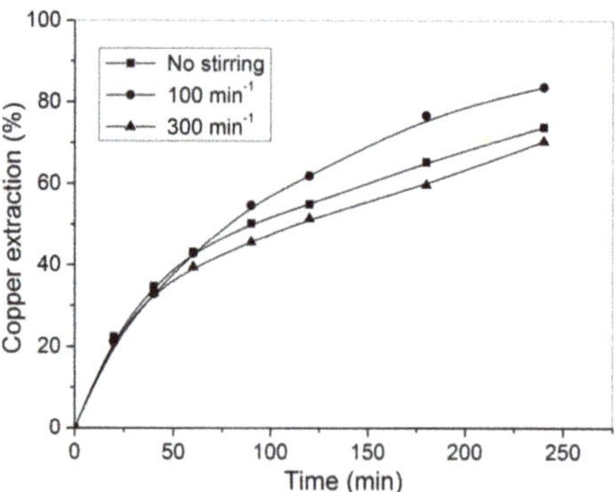

Figure 5. Influence of the stirring speed on copper extraction yield (temperature—40 °C, particle size -37 μm, solid/liquid ratio—2 g/dm^3, 1.5 M H_2SO_4, 1 M H_2O_2).

Maximal leaching degree was achieved at the stirring speed of 100 rpm, which is in line with the results published by Mahajan et al. [51]. Accordingly, the stirring speed of the 100 rpm was optimal, and it has been used in other experiments.

A similar result was obtained by Hu et al. [50], who confirmed that the copper dissolution increased slowly with increasing stirring speed up to 200 rpm, after which it does not change.

The increase of steering speed possibly interferes with the adsorption of hydrogen peroxide on the chalcopyrite and sulfure surface and facilitates the decomposition of the hydrogen peroxide [41]. Adebayo et al. [40] observed the highest dissolution rate when there was no mechanical mixing, indicating that better contacts were made. The decomposition of H_2O_2 is associated with a stirring speed by the evolution of molecular oxygen adsorbed at the particle surface, thus aggravating particle/peroxide contact. In addition, peroxide decomposition was catalyzed by cations as well as Cu^{2+}, Fe^{3+}, and Fe^{2+} [40,52]. On the other hand, Antonijević et al. [42] and Agacayak et al. [46] concluded that stirring speed does not affect the yield of copper during the leaching with hydrogen-peroxide. During the chalcopyrite leaching with potassium dichromate in sulfuric acid solution, Aydogan et al. [24] showed that chalcopyrite dissolution increased up to 400 rpm stirring speed, and after that, the dissolution rate significantly decreased. The same result was obtained by Petrović et al. [48] during chalcopyrite leaching with peroxide in hydrochloric acid solution.

3.4. Effect of Hydrogen Peroxide Concentration

The influence of the concentration of hydrogen peroxide was examined on the sample size 100% −37 μm, temperature 40 °C, the concentration of H_2SO_4 1.5 M, and concentration of solid phase 2 gdm^{-3}. The experimental results are shown in Figure 6.

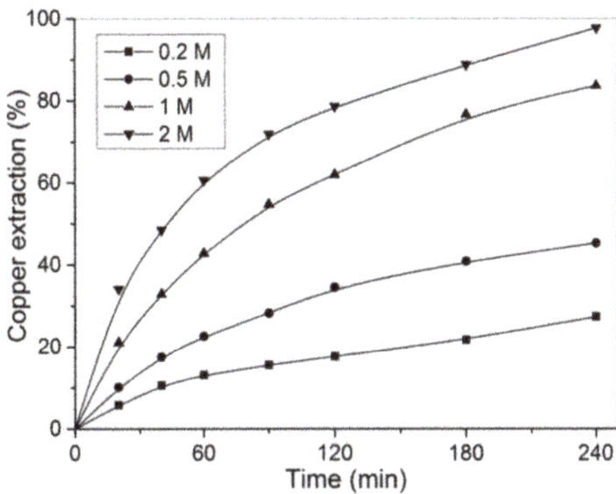

Figure 6. Influence of H_2O_2 concentration on the copper extraction yield (temperature—40 °C, particle size −37 μm, stirring speed—100 rpm, solid/liquid ratio—2 g/dm^3, —1.5 M H_2SO_4).

Increasing the hydrogen peroxide concentration in the solution has a significant influence on the copper leaching rate. The copper extraction yield increased from 27.35% to 97.96% after 240 min, while the concentration of H_2O_2 was increased from 0.2 to 2.0 M, which shows that the increase of peroxide concentration had a positive effect on the chalcopyrite leaching. The influence of peroxide concentration is larger as the concentration increased from 0.2 to 1.0 M, while the smaller influence was noticed in increasing the peroxide concentration from 1.0 to 2.0 M. These results are in agreement with the earlier results that the decomposition rate of peroxide was proportional to its concentration [40,41,48]. Because of that, the concentration of 1.0 M hydrogen peroxide was used for the investigation of other parameters.

3.5. Characterization of Leach Residue

The chemical analysis of the leach residue, obtained at 40 °C, grain size −37 μm, stirring speed—100 rpm, time—240 min, concentrations of H_2SO_4—1.5 M and H_2O_2—1 M, and a concentration of solid phase of 2 gdm^{-3} is shown in Table 3. As can be seen in Table 3, the zinc, copper, and iron content in the residue decreases, while the lead content increases. This confirms that in the leaching process, copper, zinc, and iron dissolve, while lead in the form of poorly soluble anglesite remains in the residue. The sulfur content is increased, which confirms that the part of the sulfide sulfur oxidized to its elemental form.

Table 3. Chemical composition of the leach residue.

Mass of the Leach Residue (g)	Content (%)				
	Cu	Zn	Pb	Fe	S
0.38	21.80	0.14	4.95	20.12	41.34

The XRD analysis of the leach residue is shown in Figure 7. The presence of chalcopyrite and elemental sulfur as the major phases was determined.

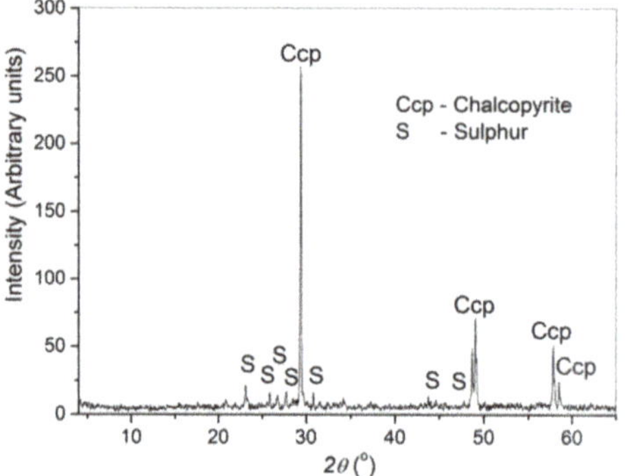

Figure 7. X-ray diffraction analysis of the leach residue sample.

The leach residue was analyzed using light microscopy, and mineral composition is shown in Table 4; the total sulfides content was 64.8%, with liberated sulfide grains of about 87% and liberate chalcopyrite grains of about 89%.

Table 4. Mineral composition of leach residue.

Mineral	Mass %
Chalcopyrite	63.002
Sphalerite	0.222
Galena	0.417
Pyrrhotite	1.215
Anglesite	6.754
Elemental sulfur	18.887
Quartz	9.503
Total:	100.000

The reflected light microscopic examination of leach residues is shown in Figure 8. The leach residue obtained at 25 °C, as seen from Figure 8a, displays corroded chalcopyrite grains and elemental sulfur crystals, while the leach residue obtained at 40 °C as seen from Figure 8b only contains elemental sulfur.

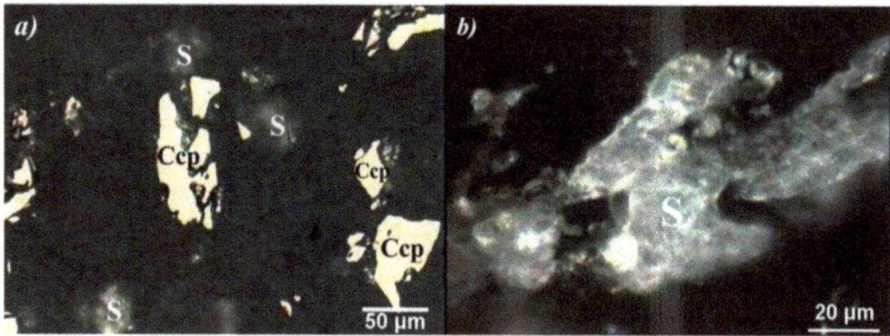

Figure 8. Optical micrograms of the leach residue for different temperatures of leaching (particle size −37 µm, stirring speed—100 rpm, time—240 min, solid/liquid ratio—2 g/dm^3, —1.5 M H_2SO_4, —1 M H_2O_2): (**a**) temperature 25 °C, (**b**) temperature 40 °C.

3.6. Leaching Kinetics

The process of copper leaching from chalcopyrite concentrate with H_2SO_4 + H_2O_2 solution is a complex heterogeneous process. The kinetic model applied for linearization of the experimental results was selected based on Sharp's method, including the reduced reaction half time [19,50,58,59]. The reduced reaction half time is given in Equation (4):

$$F(\alpha) = A(\tau/\tau_{0.5}) \quad (4)$$

where: α is the extraction degree, τ is the extraction time, $\tau_{0.5}$ is the time required to reach $\alpha = 50\%$, while constant A depends on a function $F(\alpha)$. For the isotherms presented in Figure 4, it is possible to determine the values of the reaction half time and time for different copper extraction yields for each temperature. According to the data obtained by such an analysis, the experimental results ($\tau/\tau_{0.5}$) and values for different kinetic models may be compared, as shown in Figure 9.

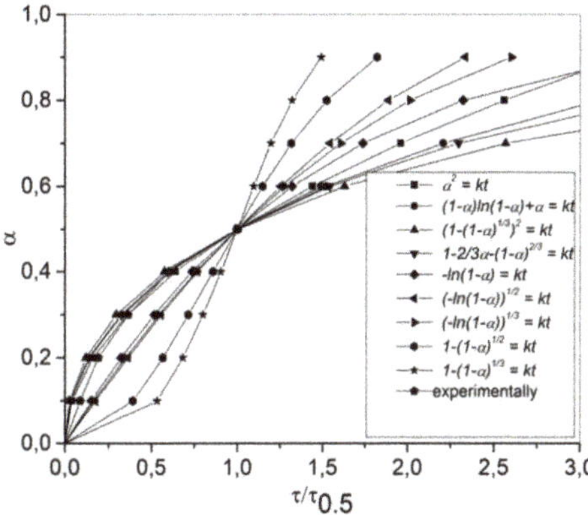

Figure 9. Selection of the kinetic model for linearization of the experimental results.

From Figure 9 it can be seen that the experimental data could be approximated with values that are presented by the kinetic Equation (5)

$$\alpha^2 = kt \quad (5)$$

where k is the rate constant and t is the reaction time. This kinetic model is used for a description of the process controlled by diffusion. The rate of reaction may be limited by diffusion of the reactants through the sulfur layer, precipitated at the surfaces of the particles.

In order to determine the activation energy, Equation (5) was used for the linearization of the kinetic curves from Figure 4. The variation in $\alpha^2 = kt$ with time at different temperatures is shown in Figure 10.

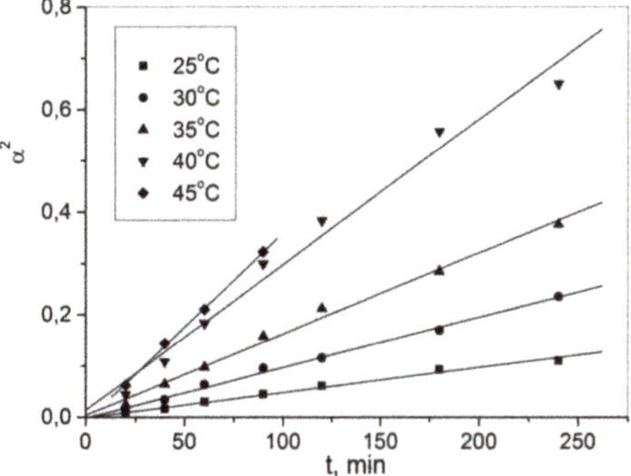

Figure 10. The plot of α^2 vs time for Cu extraction at different temperatures.

The activation energy was calculated from the slopes of the lines in Figure 10; for the chalcopyrite oxidation in the H_2SO_4-H_2O_2-H_2O system, activation energy is 80 kJ/mol (Arrhenius equation $k = A{\cdot}e^{-Ea/RT}$ was used—Figure 11).

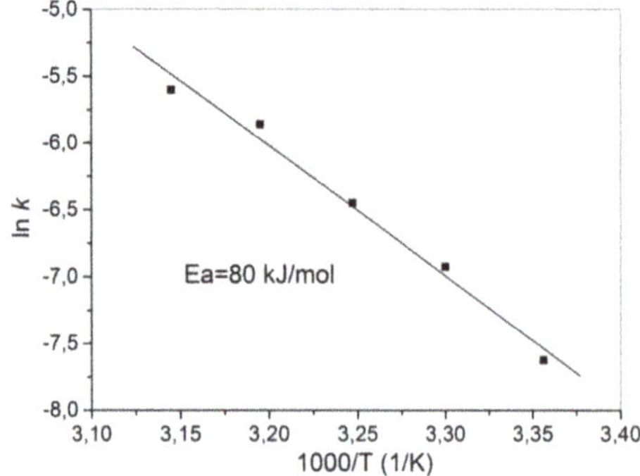

Figure 11. Arrhenius plot of the chalcopyrite oxidation in the H_2SO_4-H_2O_2-H_2O system.

Relatively high activation energy suggests that the chemical surface reaction might control the kinetics of the reaction. At the same time, some of the reported results indicate high activation energy in the reactions controlled by the diffusion mechanism [3,60].

Considering that the experimental data obtained at the temperature range from 25–45 °C best fit to the linearization equation α^2 = kt; it can be concluded that the rate of the chemical reaction was most likely controlled by diffusion of reactants through the compact layer of the elemental sulfur. The chemical surface reaction mechanism is probably important only throughout the initial stages of the process.

Other authors also confirmed the existence of elemental sulfur at the surface of the leach residue. Thus, Misra and Fuerstenau [43], and Mahajan et al. [51] have shown that most sulfide sulfur is transformed into its elemental form. Also, Antonijevic et al. [42] confirmed the presence of elemental sulfur in the leach residue and found an activation energy value of 59 kJ/mol, using a kinetic model for the process controlled by chemical reaction. Unlike them, Olubambi and Potgeiter [44] proved that elemental sulfur on the mineral surface was oxidized by hydrogen peroxide to sulfates, which increased the permeability of the passivation layer and thus promoted oxidation of the chalcopyrite. Adebayo et al. [40] found that the dissolution kinetics follow a shrinking-core model, controlled by surface chemical reaction mechanism as the rate-determining step, and they have reported an activation energy value of 39 kJmol^{-1}. Hu et al. [50] examined chalcopyrite leaching with hydrogen peroxide in 1-hexyl-3-methyl-imidazolium hydrogen sulfate solution and found that the process was chemically controlled with an activation energy of 52 kJ/mol and with the formation of sulfur as the main leaching product located near unleached chalcopyrite. Similar results with a chemically controlled mechanism, activation energy of 39.9 kJ/mol, and elemental sulfur in the residues were obtained by Ahn et al. [49].

4. Conclusions

Chalcopyrite concentrate from the "Rudnik" polymetallic ore, containing 27.08% of copper can be effectively leached using the hydrogen peroxide, as an oxidant, in the sulfuric acid medium. The copper extraction yield increased with increasing sulfuric acid and hydrogen peroxide concentrations and with smaller particles. The maximal reaction rate was obtained with a stirring speed of 100 rpm. Copper extraction yield increased with temperatures up to 40 °C; after that, the yield begins to decline due to the decomposition of hydrogen peroxide at higher temperatures. The highest yield of copper, 97.69% was obtained under the following conditions: particle size 100% −37 µm, temperature of 40 °C, leaching time of 240 min, stirring speed of 100 rpm, concentrations of H_2SO_4 1.5 M and H_2O_2 2.0 M, and a concentration of solid phase at 2 gdm^{-3}. The kinetic data for the copper leaching process show a good fit for the model controlled by diffusion, and the rate of reaction controlled by the lixiviant diffusion through the sulfur layer. This sulfur layer is formed as the reaction product during the leaching process. The calculated activation energy value is 80 kJ/mol. The formation of elemental sulfur was confirmed by XRD and mineralogical analysis, showing sulfur precipitation at the particle surfaces.

Author Contributions: Conceptualization, M.S. and B.M.; methodology, M.S. and V.M.; formal analysis, S.S., N.Š. and V.M., investigation, M.S., N.P. and B.M.; writing—original draft preparation, S.S.; writing—review and editing, M.S. and Ž.K.; visualization, B.M. and V.M.; supervision, M.S. and Ž.K.; project administration, M.S. and B.M.

Funding: This work was realized in the frame of projects TR 34023 and TR 34033, supported by the Ministry of Education, Science and Technological Development of the Republic of Serbia.

Conflicts of Interest: The authors declare no conflict of interest.

References

1. Watling, H.R. Chalcopyrite hydrometallurgy at atmospheric pressure: 1. Review of acidic sulfate, sulfate-chloride and sulfate-nitrate process options. *Hydrometallurgy* **2013**, *140*, 163–180. [CrossRef]
2. Li, Y.; Kawashima, N.; Li, J.; Chandra, A.P.; Gerson, A.R. A review of the structure, and fundamental mechanisms and kinetics of the leaching of chalcopyrite. *Adv. Colloid Interface Sci.* **2013**, *197*, 1–32. [CrossRef] [PubMed]
3. Dutrizac, J.E. The dissolution of chalcopyrite in ferric sulfate and ferric chloride media. *Metall. Mater. Trans. B* **1981**, *12*, 371–378. [CrossRef]
4. Hackl, R.P.; Dreisinger, D.B.; Peters, E.; King, J.A. Passivation of chalcopyrite during oxidative leaching in sulfate madia. *Hydrometallurgy* **1995**, *39*, 25–48. [CrossRef]
5. Maurice, D.; Hawk, J.A. Simultaneous autogenous milling and ferric chloride leaching of chalcopyrite. *Hydrometallurgy* **1999**, *51*, 371–377. [CrossRef]
6. Klauber, C. A critical review of the surface chemistry of acidic ferric sulfate dissolution of chalcopyrite with regards to hindered dissolution. *Int. J. Miner. Process.* **2008**, *86*, 1–17. [CrossRef]
7. Cordoba, E.M.; Munoz, J.A.; Blazquez, M.L.; Gonzalez, F.; Ballester, A. Leaching of chalcopyrite with ferric iron: Part I: General aspects. *Hydrometallurgy* **2008**, *93*, 81–87. [CrossRef]
8. Yang, C.; Jiao, F.; Qin, W. Leaching of chalcopyrite: An emphasis on effect of copper and iron ions. *J. Cent. South Univ.* **2018**, *25*, 2380–2386. [CrossRef]
9. Nikoloski, A.N.; O'Malley, G.P. The acidic ferric sulfate leaching of primary copper sulfides under recycle solution conditions observed in heap leaching. Part 1. Effect of standard conditions. *Hydrometallurgy* **2018**, *178*, 231–239. [CrossRef]
10. Lundstrom, M.; Aromaa, J.; Forsen, O.; Hyvarinen, O.; Barker, M.H. Leaching of chalcopyrite in cupric chloride solution. *Hydrometallurgy* **2005**, *77*, 89–95. [CrossRef]
11. Winand, R. Chloride hydrometallurgy. *Hydrometallurgy* **1991**, *27*, 285–316. [CrossRef]
12. Skrobian, M.; Havlik, T.; Ukasik, M. Effect of NaCl concentration and particle size on chalcopyrite leaching in cupric chloride solution. *Hydrometallurgy* **2005**, *77*, 109–114. [CrossRef]
13. Zhang, R.; Sun, C.; Kou, J.; Zhao, H.; Wei, D.; Xing, Y. Enhancing the Leaching of Chalcopyrite Using *Acidithiobacillus ferrooxidans* under the Induction of Surfactant Triton X-100. *Minerals* **2019**, *9*, 11. [CrossRef]
14. Martins, F.L.; Patto, G.B.; Leão, V.A. Chalcopyrite bioleaching in the presence of high chloride concentrations. *J. Chem. Technol. Biotechnol.* **2019**, *94*, 2333–2344. [CrossRef]
15. Sandstrom, A.; Petersson, S. Bioleaching of a complex sulfide ore with moderate thermophilic and extreme thermophilic microorganisms. *Hydrometallurgy* **1997**, *46*, 181–190. [CrossRef]
16. Dreisinger, D. Copper leaching from primary sulfides: options for biological and chemical extraction of copper. *Hydrometallurgy* **2006**, *83*, 10–20. [CrossRef]
17. Watling, H.R. The bioleaching of sulfide minerals with emphasis on copper sulfides-A review. *Hydrometallurgy* **2006**, *84*, 81–108. [CrossRef]
18. Pradhan, N.; Nathsarma, K.C.; Srinivasa Rao, K.; Sukla, L.B.; Mishra, B.K. Heap bioleaching of chalcopyrite: A review. *Miner. Eng.* **2008**, *21*, 355–365. [CrossRef]
19. Sokić, M.; Marković, B.; Živković, D. Kinetics of chalcopyrite leaching by sodium nitrate in sulfuric acid. *Hydrometallurgy* **2009**, *95*, 273–279. [CrossRef]

20. Sokić, M.; Matković, V.; Marković, B.; Štrbac, N.; Živković, D. Passivation of Chalcopyrite During the Leaching with Sulfuric Acid Solution in Presence of Sodium Nitrate. *Hemijska Industrija* **2010**, *64*, 343–350.
21. Anderson, C.G. Treatment of copper ores and concentrates with industrial nitrogen species catalized pressure laching and non-cyanide precious metal recovery. *JOM* **2003**, *55*, 32–36. [CrossRef]
22. Anderson, C.G.; Harrison, K.D.; Krys, L.E. Theoretical considerations of sodium nitrite oxidation and fine grinding in refractory precious metals concentrate pressure leaching. *Trans. Soc. Min. Metall. Explor.* **1996**, *299*, 4–11. [CrossRef]
23. Antonijević, M.; Janković, Z.; Dimitrijević, M. Investigation of the kinetics of chalcopyrite oxidation by potassium dichtomate. *Hydrometallurgy* **1994**, *35*, 187–201. [CrossRef]
24. Aydogan, S.; Ucar, G.; Canbazoglu, M. Dissolution kinetiks of chalcopyrite in acidic potassium dichromate solution. *Hydrometallurgy* **2006**, *81*, 45–51. [CrossRef]
25. Devi, N.B.; Madhuchhanda, M.; Rath, P.C.; Srinivasa Rao, K.; Paramguru, R.K. Simultaneous leaching of a deep-sea manganese nodule and chalcopyrite in hydrochloric acid. *Metall. Mater. Trans. B* **2001**, *32B*, 777–784. [CrossRef]
26. Havlik, T.; Laubertova, M.; Miskufova, A.; Kondas, J.; Vranka, F. Extraction of copper, zinc, nickel and cobalt in acid oxidative leaching of chalcopyrite at the presence of deep-sea manganese nodules as oxidant. *Hydrometallurgy* **2005**, *77*, 51–59. [CrossRef]
27. Sandstrom, A.; Schukarev, A.; Paul, J. XPS characterization of chalcopyrite chemically and bioleached at high and low redox potential. *Miner. Eng.* **2005**, *18*, 505–515. [CrossRef]
28. Kariuki, S.; Moore, C.; McDonald, A.M. Chlorate based oxidative hydrometallurgical extraction of copper and zinc from copper concentrate sulfide ores using mild acidic conditions. *Hydrometallurgy* **2009**, *96*, 72–76. [CrossRef]
29. Zhong, S.; Li, Y. An improved understanding of chalcopyrite leaching kinetics and mechanisms in the presence of NaCl. *J. Mater. Res. Technol.* **2019**, *8*, 3487–3494. [CrossRef]
30. Padilla, R.; Pavez, P.; Ruiz, M.C. Kinetics of copper dissolution from sulfidized chalcopyrite at high pressures in H_2SO_4–O_2. *Hydrometallurgy* **2008**, *91*, 113–120. [CrossRef]
31. McDonald, R.G.; Muir, D.M. Pressure oxidation leaching of chalcopyrite. Part I. Comparison of high and low temperature reaction kinetics and products. *Hydrometallurgy* **2007**, *86*, 191–205. [CrossRef]
32. Akcil, A.; Ciftci, H. Metals recovery from multimetal sulfide concentrates ($CuFeS_2$–PbS–ZnS): combination of thermal process and pressure leaching. *Int. J. Miner. Process.* **2003**, *71*, 233–246. [CrossRef]
33. Cháidez, J.; Parga, J.; Valenzuela, J.; Carrillo, R.; Almaguer, I. Leaching chalcopyrite concentrate with oxygen and sulfuric acid using a low-pressure reactor. *Metals* **2019**, *9*, 189. [CrossRef]
34. Benavente, O.; Hernández, M.C.; Melo, E.; Núñez, D.; Quezada, V.; Zepeda, Y. Copper Dissolution from Black Copper Ore under Oxidizing and Reducing Conditions. *Metals* **2019**, *9*, 799. [CrossRef]
35. Carillo Pedroza, F.R.; Sanchez-Castillo, M.A.; Soria-Aguilar, M.J.; Martinez-Luevanos, A.; Gutierrez, E.C. Evaluation of of acid leaching of low grade chalcopyrite using ozone by statistical analysis. *Can. Metall. Q.* **2010**, *49*, 219–226. [CrossRef]
36. Gomez, C.; Roman, E.; Blazquez, M.L.; Ballester, A. SEM and AES studies of chalcopyrite bioleaching in the presence of catalytic ions. *Miner. Eng.* **1997**, *10*, 825–835. [CrossRef]
37. Hiroyoshi, N.; Arai, M.; Miki, H.; Tsunekawa, M.; Hirajima, T. A new reaction model for the catalytic effect of silver ions on chalcopyrite leaching in sulfuric acid solution. *Hydrometallurgy* **2002**, *63*, 257–267. [CrossRef]
38. Al-Harahsheh, M.; Kingman, S.W. Microwave-assisted leaching – A review. *Hydrometallurgy* **2004**, *73*, 189–203. [CrossRef]
39. Al-Harahsheh, M.; Kingman, S.; Hankins, N.; Somerfield, C.; Bradshaw, S.; Louw, W. The influence of microwaves on the leaching kinetics of chalcopirite. *Miner. Eng.* **2005**, *18*, 1259–1268. [CrossRef]
40. Adebayo, A.O.; Ipinmorti, K.O.; Ajayi, O.O. Dissolution kinetics of chlacopyrite with hydrogen peroxide in sulfuric acid medium. *Chem. Biochem. Eng. Q.* **2003**, *17*, 213–218.
41. Antonijević, M.; Dimitrijević, M.; Janković, Z. Leaching of pyrite with hydrogen peroxide in sulfuric acid. *Hydrometallurgy* **1997**, *46*, 71–83. [CrossRef]
42. Antonijević, M.M.; Janković, Z.D.; Dimitrijević, M.D. Kinetics of chalcopyrite dissolution by hydrogen peroxide in sulfuric acid. *Hydrometallurgy* **2004**, *71*, 329–334. [CrossRef]
43. Misra, M.; Fuerstenau, M.C. Chalcopyrite leaching at moderate temperature and ambient pressure in the presence of nanosize silica. *Miner. Eng.* **2005**, *18*, 293–297. [CrossRef]

44. Olubambi, P.A.; Potgieter, J.H. Investigations on the machanisms of sulfuric acid leaching of chlacopyrite in the presence of hydrogen peroxide. *Miner. Process. Extr. Metall. Rev.* **2009**, *30*, 327–345. [CrossRef]
45. Turan, M.D.; Altundogan, H.S. Leaching of chalcopyrite concentrate with hydrogen peroxide and sulfuric acid in an autoclave system. *Metall. Mater. Trans. B* **2013**, *44B*, 809–819. [CrossRef]
46. Agacayak, T.; Aras, A.; Aydogan, S.; Erdemoglu, M. Leaching of chlacopyrite concentrate in hydrogen peroxide solution. *Physicochem. Prob. Miner. Process.* **2014**, *50*, 657–666.
47. Wu, J.; Ahn, J.; Lee, J. Comparative leaching study on conichalcite and chalcopyrite under different leaching systems. *Korean J. Met. Mater.* **2019**, *57*, 245–250. [CrossRef]
48. Petrović, S.J.; Bogdanović, G.D.; Antonijević, M.M. Leaching of chalcopyrite with hidrogen peroxide in hydrochloric acid solution. *Trans. Nonferrous Met. Soc. China* **2018**, *28*, 1444–1455. [CrossRef]
49. Ahn, J.; Wu, J.; Lee, J. Investigation on chalcopyrite leaching with methanesulfonic acid (MSA) and hydrogen peroxide. *Hydrometallurgy* **2019**, *187*, 54–62. [CrossRef]
50. Hu, J.; Tian, G.; Zi, F.; Hu, X. Leaching of chalcopyrite with hydrogen peroxide in 1-hexil-3-methil-imidazolium hydrogen sulfate ionic liquid aqueous solution. *Hydrometallurgy* **2017**, *169*, 1–8. [CrossRef]
51. Mahajan, V.; Misra, M.; Zhong, K.; Fuerstenau, M.C. Enhanced leaching of copper from chalcopyrite in hydrogen peroxide-glycol system. *Miner. Eng.* **2007**, *20*, 670–674. [CrossRef]
52. Ruiz-Sanchez, A.; Lapidus, G.T. Study of chalcopyrite leaching from a copper concentrate with hydrogen peroxide in aqueous ethylene glycol media. *Hydrometallurgy* **2017**, *169*, 192–200. [CrossRef]
53. Solís Marcial, O.J.; Nájera Bastida, A.; Bañuelos, J.E.; Valdés Martínez, O.U.; Luevano, L.A.; Serrano Rosales, B. Chalcopyrite Leaching Kinetics in the Presence of Methanol. *Int. J. Chem. Reactor Eng.* **2019**, in press. [CrossRef]
54. Lin, H.K.; Luong, H.V. Column leaching for simultaneous heap and in-situ soli remediation with metallic Fenton reaction. *J. Miner. Mater. Charact. Eng.* **2004**, *3*, 33–39.
55. Dutrizac, J. Ferric ion leaching of chalcopyrites from different localities. *Metall. Mater. Trans. B* **1982**, *13B*, 303–309. [CrossRef]
56. Sokić, M.; Radosavljević, S.; Marković, B.; Matković, V.; Štrbac, N.; Kamberović, Ž.; Živković, D. Influence of chalcopyrite structure on their leaching by sodium nitrate in sulfuric acid. *Metall. Mater. Eng.* **2014**, *20*, 53–60.
57. Craus, W.; Nolze, G. POWDER CELL-a program for the representation and manipulation of crystal structures and calculation of the resulting X-ray powder patterns. *J. Appl. Crystallogr.* **1996**, *29*, 301–303. [CrossRef]
58. Sharp, J.H.; Brindley, G.W.; Achar, B.N.N. Numerical data for some commonly used solid state reaction equations. *J. Am. Ceram. Soc.* **1996**, *49*, 379–382. [CrossRef]
59. Sokić, M.; Marković, B.; Matković, V.; Živković, D.; Štrbac, N.; Stojanović, J. Kinetics and mechanism of sphalerite leaching by sodium nitrate in sulfuric acid solution. *J. Min. Metall. Sect. B.* **2012**, *48*, 185–195. [CrossRef]
60. Ammou-chokroom, M.; Canbazoglu, M.; Steinmetz, D. Oxydation ménagée de la chalcopyrite en solution acide: analyse cinétique des réactions. II. Modèles diffusionnels. *Bulletin de Minéralogie* **1997**, *100*, 161–177. [CrossRef]

Article

Copper Tailing Flocculation in Seawater: Relating the Yield Stress with Fractal Aggregates at Varied Mixing Conditions

Matías Jeldres [1], Eder C. Piceros [2], Norman Toro [3,4,*], David Torres [3,4], Pedro Robles [5], Williams H. Leiva [1] and Ricardo I. Jeldres [1,*]

1 Departamento de Ingeniería Química y Procesos de Minerales, Facultad de Ingeniería, Universidad de Antofagasta, Antofagasta 1240000, Chile; mjeldresvalenzuela@gmail.com (M.J.); wleivajeldres@gmail.com (W.H.L.)
2 Faculty of Engineering and Architecture, Universidad Arturo Pratt, Iquique 1100000, Chile; edpicero@unap.cl
3 Departamento de Ingeniería Metalúrgica y Minas, Universidad Católica del Norte, Antofagasta 1270709, Chile; David.Torres@sqm.com
4 Department of Mining, Geological and Cartographic Department, Universidad Politécnica de Cartagena, 30202 Murcia, Spain
5 Escuela de Ingeniería Química, Pontificia Universidad Católica de Valparaíso, Valparaíso 2340000, Chile; pedro.robles@pucv.cl
* Correspondence: ntoro@ucn.cl (N.T.); ricardo.jeldres@uantof.cl (R.I.J.); Tel.: +56-552-651-021 (N.T.); +56-552-637-901 (R.I.J.)

Received: 12 November 2019; Accepted: 29 November 2019; Published: 1 December 2019

Abstract: The implications of physical conditions of the feedwell on the rheological properties of synthetic copper tailings, flocculated in seawater, were analysed. The mixing intensity of flocculation was related to the structural characteristics of the aggregates, and the outcomes were linked to the yield stress of the pulp sediments. Tailings settling assays were conducted by using a 30 mm turbine type stirrer with an in-situ aggregate size characterisation. The structural characteristics of the aggregates were determined by using the focused beam reflectance measurement (FBRM). After a mixing time between the pulp and the flocculant, the sample was allowed to settle for 2.5 h, where the variation of the sediment height was minimal. The sediment was gently removed and subjected to rheological characterisation. The yield stress was measured on an Anton Paar MCR 102 rheometer (ANAMIN Group, Santiago, Chile), with a vane-in-cup configuration. The mixing intensity was related to the characteristics of the aggregates, and the outcomes were linked to the yield stress of the flocculated pulp sediments. More aggressive hydrodynamics deteriorated the structure of the aggregates, promoting the reduction of both its size and the fractal dimension. This brought direct consequences on the rheological properties of the sediments: at higher mixing level, the yield stress was lower. The explanation lies in the structural changes of the aggregates, where at a fixed mixing rate, the yield stress presented a seemingly exponential increase over the fractal dimension. Additionally, correlations were found between the rheological properties with settling rate and aggregate size.

Keywords: seawater; copper tailings; rheology; fractal aggregates; thickening

---

1. Introduction

The recovery and succeeding reuse of processed water is essential to guarantee the sustainability of the copper mining industry, and it is directly associated with operational and environmental costs of the mineral processing. Water recovery is principally accomplished in thickeners, where chemical reagents (i.e., flocculants) are used to stimulate the decantation of solid particles, generating a clarified supernatant liquid [1]. The discharge is carried out by a lower cone (underflow), and it must be expelled with the help of rakes that move with the application of mechanical torque. The magnitude of the rheological properties is critical, and several studies show a qualitative improvement in dewaterability as a result of raking the suspension [2,3]. Ultimately, the thickened tailings are transported through pipes to a disposal area for subsequent dehydration and consolidation [4].

Tailings produced in the copper mining operations generally have clays contents like kaolinite, where their colloidal size and amorphous structure cause varied challenges in thickening. Usually, the settling rate of the flocculated slurries drop [5–7], and the rheological parameters can rise to acquire non-Newtonian behaviours [8,9]. Several studies have examined the adsorption of polyacrylamides on the surface of kaolinite, standing out as the main flocculants and most common clay that appear in the copper industry. It has been concluded that polymer adsorption occurs mostly on the kaolinite edge, that is, on the broken bonds of the aluminol (Al-OH) and silanol (Si-OH) groups, through hydrogen bonds [10,11].

Depending on the physical and chemical conditions throughout the flocculation process, the aggregates may confer significant structural differences (size, porosity, shape, etc.). Several studies reveal the influence of mixing intensity on the structure of the flocculated aggregates [12,13]. Overall, by increasing the slurry shearing, the initial mixing and aggregation rates may enhance, but this ultimately leads to a reduced final aggregate size due to the extended breakage. Vaezi et al. [14] flocculated kaolinite suspensions with A-PAM (Magnafloc® 1011) at constant mixing intensity (145 s^{-1}) and varied mixing time. Initially, large porous aggregates were obtained, with sizes that even reached 1000 µm. As the flocculation time elapsed, the aggregates reduced in size due to their fragmentation. The authors noted a monotonous trend between aggregate porosity (or effective density) and its diameter. The diversity of structures that flocculated mineral aggregates may take directly impacts the rate at which particles settle. For example, spherical agglomerates frequently deposit faster than non-spherical ones, although those that are needle-shaped settle quicker if the major axis remains vertical [15]. Porous aggregates have high hydrodynamic profiles. Then, the more significant drag considerably lessens their sedimentation rate compared to solid particles of the same size [16]. However, aggregates still deposit faster than individual fine particles.

A central feature to discern the performance of thickening operations is the rheological behaviour of the thickened slurries [17]. The knowledge and adequate control of the rheological properties allow optimising the energy consumption required to drain the thickener underflow, and this is directly related to the torque that must be applied to the rakes so that they work. The yield stress, which corresponds to the minimum stress required to cause a material to flow, is determined by the solid concentration and the strength of the particle networks. In this sense, Zhou et al. [18] found a direct relationship between yield stress with the adhesion forces between the mineral surfaces, which was determined by a Nanoscope III atomic adsorption microscope. The authors analysed flocculated silica surfaces for combination of polymer type and dosage, and the bond strength proved to be strongly influenced by the chemical conditions of the mineral surface and solution like pH, salinity and polymer characteristics. This agrees directly with the work reported by Johnson et al. [19], who interpreted the results of the yield stress for concentrated pulps of alumina, zirconia and kaolin, considering different types of interaction forces. The authors reviewed the classic sum of van der Waals' interaction with the double electrical layer; however, the surface complexity of the minerals led to incorporate non-DLVO (Derjaguin, Landau, Verwey, Overbeek) interactions such as hydration, hydrophobic and steric forces. Additionally, Neelakantan et al. [20] studied the effect of the mixing intensity on the structure of the kaolinite aggregate and the yield stress of the sediment when the particles were flocculated with PAM. The researchers noted that rising the agitation rate of the slurry causes a considerable reduction in yield stress as well as the size of the aggregates. It was explained that the aggregates are possibly squeezed under intense mixing, diminishing their apparent volume in solution.

It is interesting to comprehend the interplay of concentrated slurries in saline environments since the use of seawater is an option that is being increasingly used by the copper mining industry. Salinity alone is not a significant problem for concentration operations; when working at natural pH (pH 8 approx.), impressive recoveries of copper and molybdenum can be achieved in flotation stages [21–23]. It is noteworthy considering that a desalination plant by reverse osmosis leads to diverse financial and environmental concerns. For example, reverse osmosis (RO) plants commonly return hypersaline water to the sea, where the consequences to the marine life can be significant considering the high concentrations of ions and precipitating complexes ($CaCO_3$, $CaSO_4$, $BaSO_4$). Additionally, the high temperature of the stream (30–40 °C) increases the toxicity of metals and chemical [24,25].

Recently, Reyes et al. [26] proved the notable implication of seawater on the yield stress of magnetite tailings. For this, they used mixtures of seawater with process water, and the results were associated with measurements of zeta potential. Jeldres et al. [27] studied the yield stress and viscoelasticity of kaolin sediments of slurries flocculated with an anionic polyelectrolyte of high molecular weight. The authors found that pulps fixed with low salinity led to a rise in sediment yield stress, while increasing the NaCl concentration the yield stress diminished. This behaviour was interpreted in terms of the DLVO theory, where compressions of the ionic cloud that surround the particles surfaces and functional groups of the polymer produced aggregates with a weaker structure at high salinity. This phenomenon was shown later by molecular dynamics simulations, considering the interaction between a polyacrylamide with a quartz surface, varying the salt concentrations [28].

To date, several studies show the influence of mixing degree during flocculation on suspension dewatering [29–31] and, consequently, the relationship between the structure of the aggregates with the settling rate [32,33]. However, few systematic studies deepen in the relationship between the microscopic properties over the rheological behaviour of the underflow in thickening stages. The most relevant work was recently developed by Benn et al. [34], who performed turbulent flocculation in a large-scale pipe reactor (slurry flows 25 L min^{-1}). The authors analysed the yield stress profile from the bottom to the top of the sediment and related the outcomes to the solids concentration. Additional information was presented on the influence of flocculant management. In this context, our research aims to analyse the implications of physical conditions of the feedwell on the rheological properties of the underflow of flocculated clay-based tailings in seawater. The mixing intensity was related to the structural characteristics of the aggregates, and the outcomes were linked to the yield stress of the flocculated pulp sediment. The results provided in this study are of interest to the copper mining industry, especially those that use seawater in their operations.

2. Methodology

2.1. Materials

Quartz particles purchased from DondeCapo (Santiago, Chile) were used, where the SiO_2 content estimated by X-Ray diffraction (XRD) analysis was greater than 99 wt.%. A D5000 X-ray diffractometer (Siemens S.A., Lac Condes, Chile) was used and the data were processed with Total Pattern Analysis Software (TOPAS) (Siemens S.A., Lac Condes, Chile). The mean square chord length, determined by the focused beam reflectance measurement (FBRM) probe, was 24 μm. The kaolin particles were acquired from Ward's Science (Rochester NY, USA), and the XRD indicated a composition of 84% Kaolinite and 16% Halloysite. The mean square chord length was 12 μm.

Seawater was collected from San Jorge Bay in Antofagasta (Chile) and was filtered at 1 μm using a UV filter system (BIO&NAM, Santiago, Chile), to defeat its bacterial activity. The ionic composition was determined with varied chemical methods depending on the ion type. By atomic absorption spectrometry: Na^+ 10.9 g/L, Mg^{2+} 1.38 g/L, Ca^{2+} 0.4 g/L, K^+ 0.39 g/L. The composition of Cl^- was determined by argentometric method and it was 19.6 g/L. The concentration of HCO^{3-} was determined by acid–base volumetry, whose value was 0.15 mg/L. Seawater showed a conductivity of 50.4 mS/cm at 25 °C. The pH of the slurries was fixed at pH 8 with sodium hydroxide.

2.2. Flocculation Test: Aggregate Size and Settling

Flocculation experiments were conducted using a PTFE, 30 mm-diameter turbine type impeller at the end of vertical shaft (diameter 4 mm) within a 1 L capacity, 100 mm diameter vessel (see Figure 1). The base of the impeller was positioned 20 mm above the vessel's bottom surface. A volume of liquid (232.2 mL) was transferred to the vessel, with known masses of the solid phases then added as dry powders to give mixtures containing 92.3 wt.% quartz and 7.7 wt.% kaolin at the required total solid concentration. The suspension was preconditioned by mixing at 450 rpm with a laboratory stirring IKA EUROSTAR 60 Control (WReichmann, Santiago, Chile) for 30 min. After this period, the mixing rate was reduced according to the test (from 100 to 300 rpm), and diluted flocculant was added to give the required dosage, with the mixing maintained for a further required flocculation time (from 10 to 80 s). The flocculated suspension was then carefully transferred into a stoppered cylinder (300 cm^3 volume, 35 mm internal diameter), which was inverted twice by hand (the entire cylinder rotation process taking ~4 s in all cases). The fall of the mudline, defined as the suspension-supernatant interface, was monitored for one hour and the settling rate taken as the initial linear slope of a plot of the mudline height against time.

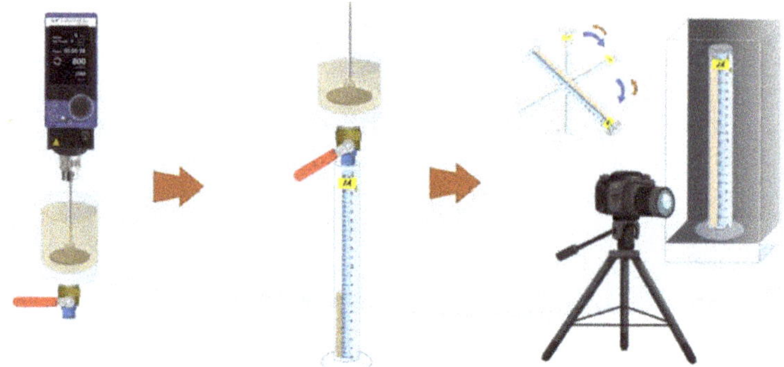

Figure 1. Schematic representation of experimental setup for settling assays.

The size distribution of the primary and aggregate particles was recorded using the Particle Track G400 equipment that uses Mettler Toledo's focused beam reflectance measurement (FBRM) technology. The measurement began with the tracking of the primary particles of the copper tailings and culminated three minutes after the polymer addition.

2.3. Yield Stress of Sediment

After 2.5 h of sedimentation, the supernatant liquid was gently removed and the lower cylinder separated (containing the sediment of the column). The yield stress was determined by employing the vane-in-cup configuration, with the stress-strain method. The measurements were carried out on an Anton Paar MCR 102 rheometer (ANAMIN Group, Santiago, Chile) and the RheocompassTM Light version software processed the data (ANAMIN Group, Santiago, Chile). The diameter of the vane was 2.2 cm and that of the cup was 4.2 cm.

2.4. Fractal Dimension

The aggregates usually have scale-invariant properties, allowing them to be characterised by a simple parameter: the fractal dimension. This parameter is directly linked to the porosity of the aggregates, and its value may range from $D_f = 1$ (for a one-dimensional line) up to 3 (solid body of three dimensions). As for flocculation, an aggregate with fractal dimension $D_f = 3$ would match to a solid sphere. The calculation of the fractal dimension is made with Equation (1) [33] and requires entering the values of the hindered settling rate and aggregate size. Both 'input' variables were obtained through the experiments described in Section 2.2.

$$U_h = \frac{\overline{d_{agg}^2}\, g(\rho_s - \rho_l)\left(\frac{\overline{d_{agg}}}{\overline{d_p}}\right)^{D_f-3}}{18\,\mu}\left(1 - \varphi_s\left(\frac{\overline{d_{agg}}}{\overline{d_p}}\right)^{3-D_f}\right)^{4.65} \quad (1)$$

U_h: Hidered settling rate, m/s
d_{agg}: Diameter of aggregate, m
d_p: Diameter of prymere particle, m
ρ_s and ρ_l: Density of solid and liquid phase, respectively, kg m^{-3},
g: Gravity acceleration constant, (9.81 m/s^2)
μ: Fluid viscosity, Ns/m^2
$\varnothing_s$: Volumetric solid concentration
D_f: Fractal dimension of mass length.

3. Results

3.1. Aggregate Size

Figure 2 shows the chord length distributions (CLD) of aggregates composed of synthetic tailings in seawater at pH 8. The square-weighted distributions reflect a volume-weighting and will be more influenced by the contribution of any aggregates formed. FBRM provides a distribution of particle chords, unlike a conventional distribution based on diameters, however Heath et al. [35] proved that the mean or mode averages of the square-weighted chord lengths are comparable to conventional sizing over the range from 50 to 400 µm. The onset of fine particle aggregation from flocculation would typically be expected to induce a shift to the right in CLDs; however, Figure 2 displayed a variation to the left while for increased mixing time. This is explained because of the irreversible fragmentation of the aggregates while they are subjected to the hydrodynamics shearing, and the finite duration that flocculants remain adsorbed on the particle surfaces. It is expected that the breakage will be more intense in larger structures; therefore, the differences are attenuated after a certain period when few large-aggregates remain in the system. For this reason, the shift to the left of the size distribution is attenuated as time progresses.

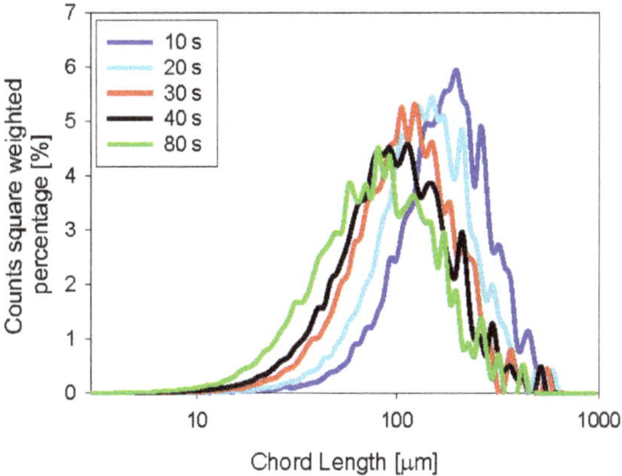

Figure 2. Square-weighted chord length distribution of flocculated copper tailings in seawater at varied flocculation time (pH 8, 225 rpm).

3.2. Analysis of Yield Stress

The vigour at which the slurry is mixed with the polymers is determined by the agitation rate, quantified by the impeller revolutions per minute (rpm) and the mixing time. These variables are crucial for the aggregation since they allow to ensure particle–particle collisions, and they form the basis for the proper design of thickener feedwells [36]. Figure 3 presents the effect of the mixing intensity through flocculation on the yield stress of the sediment separated after 2.5 h that the tailings sediment. The trend given is clear: higher mixing intensity leads to lower yield stress values. For example, the yield stress taken at 100 rpm for 10 s was higher than 300 Pa, while after an intensive stirring of 300 rpm for 80 s, the value was close to 150 Pa. Interestingly, no relevant differences in the solid concentration of sediment were distinguished concerning the various disturbances imposed to the system and mostly oscillates between 42–44 wt. % (see Table 1), without showing clear trends regarding the mixing variables. Accordingly, the performance of the yield stress observed in Figure 3 is explained by the structural changes of the particle aggregates, as a product of changing the mixing intensity. Stronger agitations cause a more significant breaking of bonds between the particle/aggregate networks that make up the slurry, before the sedimentation process. This causes the aggregate structures to be more fragile and open that is consistent with the characterisation of the fractal dimension, which is described later.

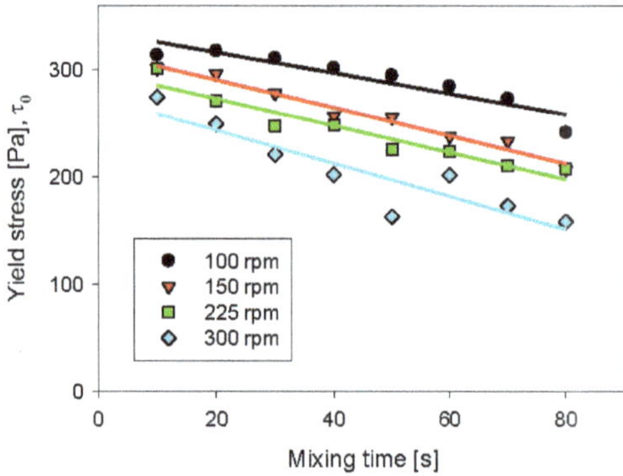

Figure 3. Yield stress of tailing sediment in seawater at varied flocculation time and mixing rate (flocculant dose: 14 g/tonne, pH 8, sedimentation time: 2.5 h). The experimental data are extracted from Table 1.

Figure 4 displays the relationship between the yield stress of the sediment and the average size of the aggregates that are formed after the flocculant addition. For a fixed mixing rate, a larger aggregate size leads to higher yield stress. The trend is similar for the all mixing rates examined in this research (from 100 to 300 rpm). However, care should be taken when relating only the aggregate size, since it is essential to consider the physical conditions in which they were formed. For example, aggregate with a mean chord length of 100 microns causes higher yield stress if it is built at a mixing rate of 300 rpm instead of 150 rpm. The reagent dosage is also relevant: at a higher amount of polymer, the yield stress is larger. This is illustrated in Figures 7 and 8.

The relationship between aggregate size and yield stress coincides with that proposed by Neelakantal et al. [20], who studied the effect of shearing kaolinite sediments, flocculated with polyacrylamide copolymers. The authors based their explanation on the elimination of retained water within the porous aggregates as a result of the fragmentation caused by shearing. This would create a dramatic reduction in the apparent volume of solids in the suspension. However, unlike the assays executed by Neelakantal et al., our work based on studying the mixing intensity on flocculation, that is, before the sedimentation begins; therefore, the structural behaviour between both studies differs.

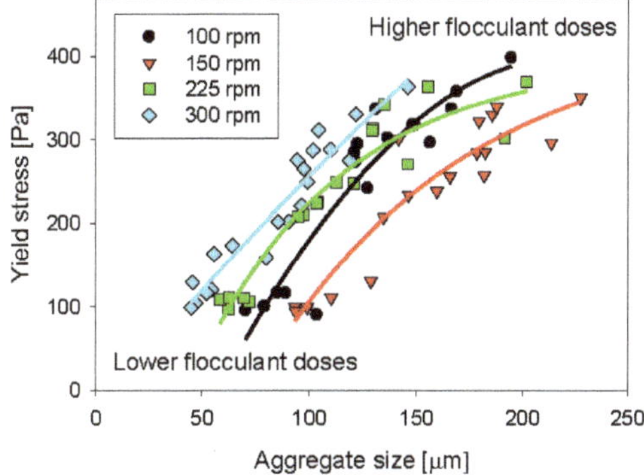

Figure 4. Relationship between yield stress of tailing sediment in seawater and aggregate size (mean chord length). The experimental data are extracted from Table 1.

Figure 5 gives the connection between fractal dimension and yield stress when the flocculation is carried out at different mixing rates. An apparent decrease in the fractal dimension is generated with the increase of the mixing, which is agreeable with that declared by Deng and Davè [37], who state that when aggregates are produced with a higher shear rate, they grow irregularly, resulting in lower fractal dimensions. On the other hand, more aggressive hydrodynamics deteriorates the structure of the aggregates, promoting the reduction of the fractal dimension.

At a fixed mixing rate, the yield stress presented a seemingly exponential increase over the fractal dimension. This agreement with the proposed by Deng and Davè [37], who established that a lower fractal dimension leads to a lower mechanical strength of the agglomerate, which was determined by the model proposed by Kendall and Stainton [38]. The authors considered the porosity, adhesion work and particle diameter. Zhou et al. presented two interesting works that directly links the rheological properties with the interaction forces [18,39]. The authors used an atomic force microscope to record the interaction forces between silica surfaces in solutions with cationic polyelectrolytes of different electric charge. The magnitude of adhesion force was related to the strength of the particle networks that make up a silica slurry. Interestingly, it was found that the strength of aggregates has a direct relationship with the rheological properties of the pulps.

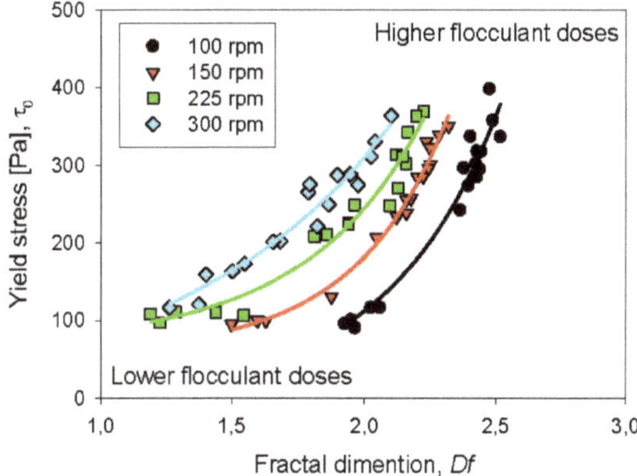

Figure 5. Relationship between yield stress of tailing sediment in seawater and fractal dimension of aggregates. The experimental data are extracted from Table 1.

The sedimentation rate of a suspension is strongly determined by the size and structure of the aggregates that compose it. For the same size, more compact structures have higher settling rates than open and porous have [40]. In this sense, it is expected that aggregates with larger fractal dimensions settle faster. In parallel, the structural aggregate features define the rheological behaviour of a suspension, so it is assumed that there is a close correlation among the settling rate and the rheology of the sediment, as shown in Figure 6: An increase in the settling rate was accompanied by higher yield stress. This result has a significant practical utility since there are continuous attempts in the mining industry to optimise the recovery of water from thickening stages, whose search generally lies in raising the sedimentation rate while maintaining low yield stress values in the thickener underflow. Then, to sustain proper thickener control, it is crucial to understand that both parameters are closely linked and the decisions that operators make inevitably impact both.

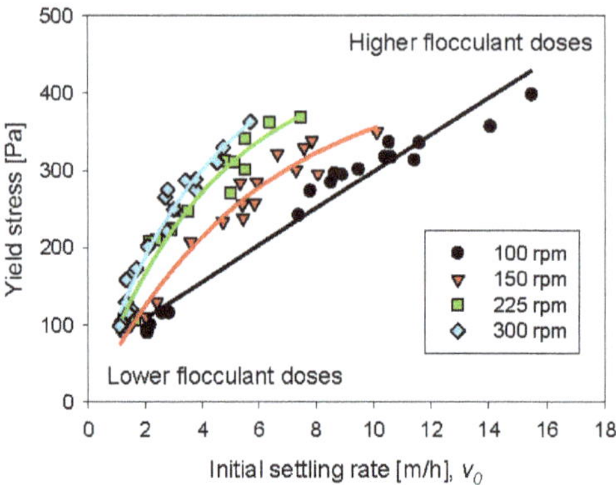

Figure 6. Relationship between yield stress of tailing sediment in seawater and settling rate. The experimental data are extracted from Table 1.

The flocculant dose was a key part of all the cases examined. An extract of the experimental data developed in the present work is shown in Figure 7, which shows a clear tendency of the rheology regarding the dosage of the polymer. A higher amount of polymer molecules leads to a greater amount of bonds that bind the particles, generating higher yield stress. On the other hand, the higher polymer dosage produces aggregates with a more robust structure, causing an increase in its fractal dimension (Figure 8), as previously reported by Ofori et al. [41].

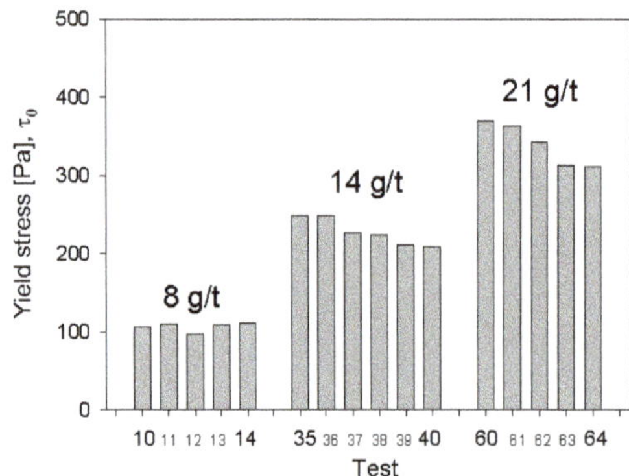

Figure 7. Influence of flocculant dose on yield stress of tailing sediment in seawater. The tests numbered in abscissa are according Table 1.

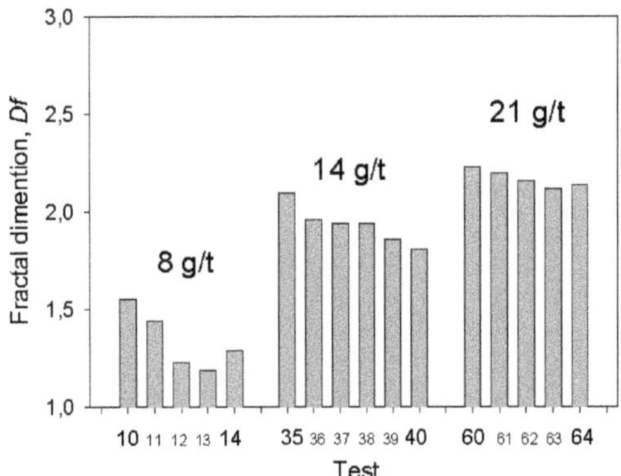

Figure 8. Influence of flocculant dose on fractal dimension of aggregates. The tests numbered in abscissa are according Table 1.

Table 1. Experimental condition at which the tests were performed.

Test	Flocculant Doses (g/t)	Mixing Rate (rpm)	Mixing Time (s)	D_f	Agregate Size (μm)	Initial Settling Rate (m/h)	Yield Stress (Pa)	Solid Concentrate (%)
1	8	100	15	2.03	85	2.6	116	41.9
2	8	100	30	1.96	104	2.1	90.0	41.9
3	8	100	45	2.06	89	2.8	116	43.0
4	8	100	60	1.95	79	2.2	100	42.5
5	8	100	75	1.93	70	2.1	95.0	43.3
6	8	150	15	1.87	129	2.4	130	43.2
7	8	150	45	1.63	99	1.5	99.1	43.8
8	8	150	60	1.60	94	1.5	99.6	43.1
9	8	150	75	1.50	94	1.1	94.6	43.2
10	8	225	15	1.55	72	1.7	106	45.6
11	8	225	30	1.44	70	1.4	110	45.2
12	8	225	45	1.23	62	1.2	97.0	45.0
13	8	225	60	1.19	59	1.3	108	45.0
14	8	225	75	1.29	63	1.3	111	44.4
15	8	300	15	1.38	55	1.5	120	45.6
16	8	300	30	1.26	52	1.4	116	45.2
17	14	100	10	3.95	48	11.4	314	43.5
18	14	100	20	2.43	148	10.4	318	44.2
19	14	100	30	2.40	167	10.5	337	42.9
20	14	100	40	2.42	137	9.5	302	43.9
21	14	100	50	2.44	123	8.9	295	43.7
22	14	100	60	2.43	121	8.5	285	43.1
23	14	100	70	2.40	121	7.8	273	42.9
24	14	100	80	2.36	127	7.4	242	42.8
25	14	150	10	2.25	142	7.3	301	44.1
26	14	150	20	2.24	214	8.1	296	43.3
27	14	150	30	2.24	186	7.6	330	42.7
28	14	150	40	2.18	182	5.9	257	43.5
29	14	150	50	2.16	166	5.4	256	42.8
30	14	150	60	2.16	160	5.4	238	44.0
31	14	150	70	2.12	146	4.7	234	42.7
32	14	150	80	2.04	135	3.6	207	44.1
33	14	225	10	2.16	192	5.5	302	43.3
34	14	225	20	2.13	146	5.0	271	44.6
35	14	225	30	2.10	121	3.5	248	44.0
36	14	225	40	1.96	113	3.0	249	44.5
37	14	225	50	1.94	104	2.9	226	44.6
38	14	225	60	1.94	104	2.9	224	44.8
39	14	225	70	1.86	98	2.4	211	45.5
40	14	225	80	1.81	95	2.2	208	43.4

Table 1. *Cont.*

Test	Flocculant Doses (g/t)	Mixing Rate (rpm)	Mixing Time (s)	D_f	Agregate Size (μm)	Initial Settling Rate (m/h)	Yield Stress (Pa)	Solid Concentrate (%)
41	14	300	10	1.97	119	3.8	275	44.8
42	14	300	20	1.86	99	3.0	250	44.2
43	14	300	30	1.82	96	2.8	221	44.6
44	14	300	40	1.68	91	2.1	202	44.1
45	14	300	50	1.50	86	1.5	163	42.5
46	14	300	60	1.66	86	2.1	202	45.6
47	14	300	70	1.55	84	1.7	173	44.2
48	14	300	80	1.40	80	1.3	159	42.8
49	21	100	10	2.77	133	23.0	538	41.2
50	21	100	20	2.48	195	15.5	398	42.0
51	21	100	30	2.49	169	14.1	357	41.8
52	21	100	45	2.38	157	8.7	296	44.1
53	21	100	60	2.44	149	10.6	317	42.7
54	21	100	75	2.52	131	11.6	336	42.9
55	21	150	15	2.32	228	10.1	350	41.7
56	21	150	30	2.28	188	7.9	339	41.5
57	21	150	45	2.25	180	6.7	322	43.4
58	21	150	60	2.20	179	5.3	284	42.8
59	21	150	75	2.22	183	5.9	285	45.9
60	21	225	15	2.23	202	7.5	369	42.7
61	21	225	30	2.20	156	6.4	363	42.6
62	21	225	45	2.16	135	5.5	342	43.7
63	21	225	60	2.12	130	4.8	313	41.6
64	21	225	75	2.14	129	5.1	311	43.7
65	21	300	15	2.10	146	5.7	363	44.1
66	21	300	30	2.04	122	4.7	330	42.3
67	21	300	40	1.95	110	3.8	288	43.3
68	21	300	50	2.02	104	4.5	311	43.2
69	21	300	60	1.90	102	3.4	287	43.1
70	21	300	70	1.79	98	2.7	265	42.4
71	21	300	80	1.79	95	2.8	275	44.2

4. Conclusions

This research aims to analyse the implications of physical conditions of the feedwell on the rheological properties of the underflow of flocculated clay-based tailings in seawater. The mixing intensity was related to the structural characteristics of the aggregates, and the outcomes were linked to the yield stress of the flocculated pulp sediment. Stronger agitations cause a more significant breaking of bonds between the particle/aggregate networks that make up the slurry, before the sedimentation process. This induces the aggregate structures to be more fragile and open; then, higher mixing intensity leads to lower yield stress values. In this context, at a fixed mixing rate, the yield stress presented a seemingly exponential increase over the fractal dimension. Additionally, a direct relationship between the yield stress and the settling rate was found since both parameters depend on the structural characteristic of the particle aggregates. The outcomes provided in this study are of interest to the copper mining industry, especially those that use seawater in their operations.

Author Contributions: All of the authors contributed to analysing the results and writing the paper.

Funding: This research was funded by Conicyt Fondecyt 11171036 and Centro CRHIAM Project Conicyt/Fondap/15130015.

Acknowledgments: Ricardo I. Jeldres thanks Conicyt Fondecyt 11171036 and Centro CRHIAM Project Conicyt/Fondap/15130015. The authors are grateful for the contribution of the Scientific Equipment Unit-MAINI of the Universidad Católica del Norte for supporting the experimental tests. Pedro Robles thanks the Pontificia Universidad Católica de Valparaíso for the support provided.

Conflicts of Interest: The authors declare no conflict of interest.

References

1. Kane, J.C.; La Mer, V.K.; Linford, H.B. The effect of solid content on the adsorption and flocculation behavior of silica suspensions. *J. Phys. Chem.* **1964**, *68*, 3539–3544. [CrossRef]

2. Rudman, M.; Simic, K.; Paterson, D.A.; Strode, P.; Brent, A.; Šutalo, I.D. Raking in gravity thickeners. *Int. J. Miner. Process.* **2008**, *86*, 114–130. [CrossRef]
3. Gladman, B.R.; Rudman, M.; Scales, P.J. The effect of shear on gravity thickening: Pilot scale modelling. *Chem. Eng. Sci.* **2010**, *65*, 4293–4301. [CrossRef]
4. De Kretser, R.; Scales, P.J.; Boger, D.V. Improving clay-based tailings disposal: Case study on coal tailings. *AIChE J.* **1997**, *43*, 1894–1903. [CrossRef]
5. Castillo, C.; Ihle, C.F.; Jeldres, R.I. Chemometric optimisation of a copper sulphide tailings flocculation process in the presence of clays. *Minerals* **2019**, *9*, 582. [CrossRef]
6. Du, J.; Morris, G.; Pushkarova, R.A.; Roger, S.C.S. Effect of surface structure of kaolinite on aggregation, settling rate, and bed density. *Langmuir* **2010**, *26*, 13227–13235. [CrossRef]
7. Liu, D.; Edraki, M.; Berry, L. Investigating the settling behaviour of saline tailing suspensions using kaolinite, bentonite, and illite clay minerals. *Powder Technol.* **2018**, *326*, 228–236. [CrossRef]
8. Jeldres, R.I.; Piceros, E.C.; Wong, L.; Leiva, W.H.; Herrera, N.; Toledo, P.G. Dynamic moduli of flocculated kaolinite sediments: Effect of salinity, flocculant dose, and settling time. *Colloid Polym. Sci.* **2018**, *296*, 1935–1943. [CrossRef]
9. Teh, E.J.; Leong, Y.K.; Liu, Y.; Fourie, A.B.; Fahey, M. Differences in the rheology and surface chemistry of kaolin clay slurries: The source of the variations. *Chem. Eng. Sci.* **2009**, *64*, 3817–3825. [CrossRef]
10. Nabzar, L.; Pefferkorn, E.; Varoqui, R. Polyacrylamide-sodium kaolinite interactions: Flocculation behavior of polymer clay suspensions. *J. Colloid Interface Sci.* **1984**, *102*, 380–388. [CrossRef]
11. Lee, L.T.; Rahbari, R.; Lecourtier, J.; Chauveteau, G. Adsorption of polyacrylamides on the different faces of kaolinites. *J. Colloid Interface Sci.* **1991**, *147*, 351–357. [CrossRef]
12. Heath, A.R.; Bahri, P.A.; Fawell, P.D.; Farrow, J.B. Polymer flocculation of calcite: Experimental results from turbulent pipe flow. *AIChE J.* **2006**, *52*, 1284–1293. [CrossRef]
13. Spicer, P.T.; Pratsinis, S.E. Shear-induced flocculation: The evolution of floc structure and the shape of the size distribution at steady state. *Water Res.* **1996**, *30*, 1049–1056. [CrossRef]
14. Vaezi, G.F.; Sanders, R.S.; Masliyah, J.H. Flocculation kinetics and aggregate structure of kaolinite mixtures in laminar tube flow. *J. Colloid Interface Sci.* **2011**, *355*, 96–105. [CrossRef]
15. Kousaka, Y.; Okuyama, K.; Payatakes, A. Physical meaning and evaluation of dynamic shape factor of aggregate particles. *J. Colloid Interface Sci.* **1981**, *84*, 91–99. [CrossRef]
16. Gabitto, J.; Tsouris, C. Drag coefficient and settling velocity for particles of cylindrical shape. *Powder Technol.* **2008**, *183*, 314–322. [CrossRef]
17. Nguyen, Q.D.; Boger, D.V. Application of rheology to solving tailings disposal problems. *Int. J. Miner. Process.* **1998**, *54*, 217–233. [CrossRef]
18. Zhou, Y.; Yu, H.; Wanless, E.J.; Jameson, G.J.; Franks, G.V. Influence of polymer charge on the shear yield stress of silica aggregated with adsorbed cationic polymers. *J. Colloid Interface Sci.* **2009**, *336*, 533–543. [CrossRef]
19. Johnson, S.B.; Franks, G.V.; Scales, P.J.; Boger, D.V.; Healy, T.W. Surface chemistry–rheology relationships in concentrated mineral suspensions. *Int. J. Miner. Process.* **2000**, *58*, 267–304. [CrossRef]
20. Neelakantan, R.; Vaezi, G.F.; Sanders, R.S. Effect of shear on the yield stress and aggregate structure of flocculant-dosed, concentrated kaolinite suspensions. *Miner. Eng.* **2018**, *123*, 95–103. [CrossRef]
21. Suyantara, G.P.W.; Hirajima, T.; Miki, H.; Sasaki, K. Floatability of molybdenite and chalcopyrite in artificial seawater. *Miner. Eng.* **2018**, *115*, 117–130. [CrossRef]
22. Jeldres, R.I.; Arancibia-Bravo, M.P.; Reyes, A.; Aguirre, C.E.; Cortes, L.; Cisternas, L.A. The impact of seawater with calcium and magnesium removal for the flotation of copper-molybdenum sulphide ores. *Miner. Eng.* **2017**, *109*, 10–13. [CrossRef]
23. Castro, S. Physico-chemical factors in flotation of Cu-Mo-Fe ores with seawater: A critical review. *Physicochem. Probl. Miner. Process.* **2018**, *54*, 1223–1236. [CrossRef]
24. Cambridge, M.L.; Zavala-Perez, A.; Cawthray, G.R.; Statton, J.; Mondon, J.; Kendrick, G.A. Effects of desalination brine and seawater with the same elevated salinity on growth, physiology and seedling development of the seagrass Posidonia australis. *Mar. Pollut. Bull.* **2019**, *140*, 462–471. [CrossRef]
25. Uddin, S. Environmental impacts of desalination activities in the Arabian gulf. *Int. J. Environ. Sci. Dev.* **2014**, *5*, 114–117. [CrossRef]

26. Reyes, C.; Álvarez, M.; Ihle, C.F.; Contreras, M.; Kracht, W. The influence of seawater on magnetite tailing rheology. *Miner. Eng.* **2019**, *131*, 363–369. [CrossRef]
27. Jeldres, R.I.; Piceros, E.C.; Leiva, W.H.; Toledo, P.G.; Herrera, N. Viscoelasticity and yielding properties of flocculated kaolinite sediments in saline water. *Colloids Surf. A Physicochem. Eng. Asp.* **2017**, *529*, 1009–1015. [CrossRef]
28. Quezada, G.R.; Jeldres, R.I.; Fawell, P.D.; Toledo, P.G. Use of molecular dynamics to study the conformation of an anionic polyelectrolyte in saline medium and its adsorption on a quartz surface. *Miner. Eng.* **2018**, *129*, 102–105. [CrossRef]
29. Gladman, B.; de Kretser, R.G.; Rudman, M.; Scales, P.J. Effect of shear on particulate suspension dewatering. *Chem. Eng. Res. Des.* **2005**, *83*, 933–936. [CrossRef]
30. Bubakova, P.; Pivokonsky, M.; Filip, P. Effect of shear rate on aggregate size and structure in the process of aggregation and at steady state. *Powder Technol.* **2013**, *235*, 540–549. [CrossRef]
31. Biggs, C. Activated sludge flocculation: On-line determination of floc size and the effect of shear. *Water Res.* **2000**, *34*, 2542–2550. [CrossRef]
32. Kyoda, Y.; Costine, A.D.; Fawell, P.D.; Bellwood, J.; Das, G.K. Using focused beam reflectance measurement (FBRM) to monitor aggregate structures formed in flocculated clay suspensions. *Miner. Eng.* **2019**, *138*, 148–160. [CrossRef]
33. Heath, A.R.; Bahri, P.A.; Fawell, P.D.; Farrow, J.B. Polymer flocculation of calcite: Relating the aggregate size to the settling rate. *AIChE J.* **2006**, *52*, 1987–1994. [CrossRef]
34. Benn, F.A.; Fawell, P.D.; Halewood, J.; Austin, P.J.; Costine, A.D.; Jones, W.G.; Francis, N.S.; Druett, D.C.; Lester, D. Sedimentation and consolidation of different density aggregates formed by polymer-bridging flocculation. *Chem. Eng. Sci.* **2018**, *184*, 111–125. [CrossRef]
35. Heath, A.R.; Fawell, P.D.; Bahri, P.A.; Swift, J.D. Estimating average particle size by focused beam reflectance measurement (FBRM). *Part. Part. Syst. Charact.* **2002**, *19*, 84. [CrossRef]
36. Fawell, P.D.; Nguyen, T.V.; Solnordal, C.B.; Stephens, D.W. Enhancing gravity thickener feedwell design and operation for optimal flocculation through the application of computational fluid dynamics. *Miner. Process. Extr. Metall. Rev.* **2019**, 1–15. [CrossRef]
37. Deng, X.; Davé, R.N. Breakage of fractal agglomerates. *Chem. Eng. Sci.* **2017**, *161*, 117–126. [CrossRef]
38. Kendall, K.; Stainton, C. Adhesion and aggregation of fine particles. *Powder Technol.* **2001**, *121*, 223–229. [CrossRef]
39. Zhou, Y.; Gan, Y.; Wanless, E.J.; Jameson, G.J.; Franks, G.V. Interaction forces between silica surfaces in aqueous solutions of cationic polymeric flocculants: Effect of polymer charge. *Langmuir* **2008**, *24*, 10920–10928. [CrossRef]
40. Tang, P.; Greenwood, J.; Raper, J.A. A model to describe the settling behavior of fractal aggregates. *J. Colloid Interface Sci.* **2002**, *247*, 210–219. [CrossRef]
41. Ofori, P.; Nguyen, A.V.; Firth, B.; McNally, C.; Ozdemir, O. Shear-induced floc structure changes for enhanced dewatering of coal preparation plant tailings. *Chem. Eng. J.* **2011**, *172*, 914–923. [CrossRef]

Article

Leaching Manganese Nodules in an Acid Medium and Room Temperature Comparing the Use of Different Fe Reducing Agents

David Torres [1,2], Luís Ayala [3], Manuel Saldaña [1], Manuel Cánovas [1], Ricardo I. Jeldres [4], Steven Nieto [4], Jonathan Castillo [5], Pedro Robles [6] and Norman Toro [1,3,*]

1 Departamento de Ingeniería Metalúrgica y Minas, Universidad Católica del Norte, Av. Angamos 610, Antofagasta 1270709, Chile; david.Torres@sqm.com (D.T.); manuel.saldana@ucn.cl (M.S); manuel.canovas@ucn.cl (M.C.)
2 Department of Mining, Geological and Cartographic Department, Universidad Politécnica de Cartagena, 30203 Cartagena, Spain
3 Faculty of Engineering and Architecture, Universidad Arturo Pratt, Almirante Juan José Latorre 2901, Antofagasta 1244260, Chile; luisayala01@unap.cl
4 Departamento de Ingeniería Química y Procesos de Minerales, Universidad de Antofagasta, Antofagasta 1240000, Chile; ricardo.jeldres@uantof.cl (R.I.J.); stevennietomejia@gmail.com (S.N.)
5 Departamento de Ingeniería en Metalurgia, Universidad de Atacama, Av. Copayapu 485, Copiapó 1531772, Chile; jonathan.castillo@uda.cl
6 Escuela de Ingeniería Química, Pontificia Universidad Católica de Valparaíso, Valparaíso 2340000, Chile; pedro.robles@pucv.cl
* Correspondence: ntoro@ucn.cl; Tel.: +56-552651021

Received: 23 October 2019; Accepted: 4 December 2019; Published: 6 December 2019

Abstract: The deposits of Fe-Mn, in the seabed of the planet, are a good alternative source for the extraction of elements of interest. Among these are marine nodules, which have approximately 24% manganese and may be a solution to the shortage of high-grade ores on the surface. In this investigation, an ANOVA analysis was performed to evaluate the time independent variables and MnO_2/reducing agent in the leaching of manganese nodules with the use of different Fe reducing agents (FeS_2, Fe^{2+}, Fe^0 and Fe_2O_3). Tests were also carried out for the different reducing agents evaluating the MnO_2/Fe ratio, in which the Fe^0 (FeC) proved to be the best reducing agent for the dissolution of Mn from marine nodules, achieving solutions of 97% in 20 min. In addition, it was discovered that at low MnO_2/Fe ratios the acid concentration in the system is not very relevant and the potential and pH were in ranges of −0.4–1.4 V and −2–0.1 favoring the dissolution of Mn from MnO_2.

Keywords: MnO_2; acid media; ANOVA; dissolution

1. Introduction

Deposits of ferromanganese (Fe-Mn) are found in the oceans around the world [1–4]. These deposits contain ferromanganese crusts, as well as cobalt-rich crusts and manganese nodules [5–7]. These marine resources are found mainly in the Pacific, Atlantic and Indian Ocean [8], and are formed by precipitation processes of Mn and Fe oxides around a nucleus, which is commonly composed of a fragment of an older nodule [9]. Manganese nodules also called polymetallic nodules because they are associated with large reserves of metals, such as Cu, Ni, Co, Fe and Mn, the latter being the most abundant, with an average content of around 24% [10]. In addition to the aforementioned elements, considerable quantities of Te, Ti, Pt and rare earths can also be found [11]. These nodules might be good source of manganese in the industry for high demand in steel production [12–14].

To extract Mn and other metals of interest from marine nodules, the use of a reducing agent is necessary [15,16]. Studies have used different reducing agents, such as, wastewater from the manufacture of alcohol from molasses [17], coal [18], H_2SO_3 [19,20], pyrite [21], sponge iron [22] and cast iron slag magnetite [23]. Iron has shown to be a good reducing agent for manganese extraction, from those, due to its low cost and abundance [23]. Several studies have been carried out to evaluate the effect of iron as a reducing agent in leaching in acid media of marine nodules [21,24]. For studies in acidic media and iron, it has been reported that the best results for extracting manganese are obtained by increasing the amounts of Fe in the Mn/Fe ratio and working at low acid concentrations [22,23].

In the studies by Kanungo [21,25], an acid leaching (HCl) was conducted at different temperatures with the addition of pyrite as a reducing agent achieving 50% manganese extractions. The author concluded that, in a moderately acidic medium, pH of 1.5, the Fe (II) and Fe (III) ratio in the system remains essentially constant up to 50 min above, which the ratio tends to increase exponentially. From this, it is suggested that the reduction of MnO_2 by ferrous ions occurs at a faster rate than the oxidation of pyrite generating ferric ions. For the dissolution of Mn with the use of pyrite in acidic media, the following series of reactions is proposed [21]:

$$3FeS_2 + 4H_2SO_4 = 3FeSO_4 + 4H_2O + 7S \tag{1}$$

$$6FeSO_4 + 4H_2SO_4 = 3Fe_2(SO_4)_3 + 4H_2O + S \tag{2}$$

$$2FeS_2 + 4H_2SO_4 = Fe_2(SO_4)_3 + 4H_2O + 5S \tag{3}$$

$$15MnO_2 + 2FeS_2 + 14H_2SO_4 = Fe_2(SO_4)_3 + 15MnSO_4 + 14H_2O \tag{4}$$

For the use of ferrous ions, Zakeri et al. [24] indicated that when working in a molar ratio of Fe^{2+}/MnO_2 of 3/1, a molar ratio of H_2SO_4/MnO_2 of 2/1 and a mineral particle size of −60 + 100 Tyler mesh, 90% extractions of Mn can be obtained in less than 20 min at a temperature of 20 °C. In their work they proposed the following series of reactions:

$$MnO_2 + 4H^+ + 2e^- = Mn^{2+} + 2H_2O \tag{5}$$

$$2Fe^{2+} = 2Fe^{3+} + 2e^- \tag{6}$$

$$MnO_2 + 2Fe^{2+} + 4H^+ = Mn^{2+} + 2Fe^{3+} + 2H_2O \tag{7}$$

Subsequently, Bafghi et al. [22] conducted a similar experiment but with the use of Fe sponge, where he compared the results reported by Zakeri et al. [24] and indicated that under the same operating conditions, sponge Fe delivers better results than the addition of ferrous ions, because the metal of Fe allows us to have a high activity ratio through the regeneration of ferrous ions. For the dissolution of Mn with the use of Fe (s), the following reactions are presented [22]:

$$Fe\ (s) + 2H^+ = Fe + H_2\ (g) \tag{8}$$

$$Fe\ (s) + 2Fe^{3+} = 3Fe^{2+} \tag{9}$$

$$MnO_2\ (s) + 2Fe^{2+} + 4H^+ = Mn^{2+} + 2Fe^{3+} + 2H_2O \tag{10}$$

$$MnO_2\ (s) + 2Fe\ (s) + 8H^+ = Mn^{2+} + 2Fe^{3+} + 2H_2O + 2\ H_2\ (g) \tag{11}$$

$$MnO_2\ (s) + Fe\ (s) + 4H^+ = Mn^{2+} + Fe^{2+} + 2H_2O \tag{12}$$

$$MnO_2\ (s) + 2/3Fe\ (s) + 4H^+ = Mn^{2+} + 2/3Fe^{3+} + 2H_2O \tag{13}$$

In the studies carried out by Toro et al. [23,26] smelting slag was used, taking advantage of the Fe_2O_3 presented in these to reduce MnO_2 in an acid medium. It was concluded that the ratios of MnO_2/Fe = 1/2 and 1 M H_2SO_4 significantly shorten the dissolution time of manganese (from 30 to

5 min). In addition, the authors indicated that the particle size is not as significant in Mn solutions as in the concentration of H_2SO_4. For the dissolution of Mn with the use of Fe_2O_3 in acid media, the following series of reactions is presented:

$$Fe_2O_3\ (s) + 3H_2SO_4 = Fe_2(SO_4)_3\ (s) + 3H_2O \quad (14)$$

$$Fe_3O_4\ (s) + 4H_2SO_4 = FeSO_4 + Fe_2(SO_4)_3\ (s) + 4H_2O \quad (15)$$

$$Fe_2(SO_4)_3\ (s) + 6H_2O = 2Fe(OH)_3\ (s) + 3H_2SO_4 \quad (16)$$

$$2FeSO_4 + 2H_2O = 2Fe\ (s) + 2H_2SO_4 + O_2\ (g) \quad (17)$$

$$2FeSO_4 + 2H_2SO_4 + MnO_2\ (s) = Fe_2(SO_4)_3\ (s) + 2H_2O + MnSO_4 \quad (18)$$

It is imperative to create innovative methods for the treatment of minerals that involve industrial waste reusing. Big mining companies are promoting recycling to generate a more sustainable sector. An example is the iron industry in China, where it is sought to reduce pollution by adding scrap in steelmaking [27]. Another example is mining in Chile, where companies like Collahuasi have recycling programs, in which they annually recover 3000 tons of scrap metal, 4 thousand kilos of electronic waste, 182 thousand units of plastic bottles and 680 kg of paper and cardboard [28]. Regarding steel scrap, the copper mining industry generates large amounts of this waste in the milling processes, but the steel balls or bars are discarded [29].

In this research, the leaching of MnO_2 to recover manganese with the use of different types of Fe reducing agents (pyrite, ferric ions, steel and magnetite) working under the same operating conditions was studied. The objective of this work is to find the most suitable iron reducing agent to extract manganese when working in an acidic environment and room temperature, with the novelty of testing the use of steel. A statistical analysis was conducted performed to evaluate the performance of the different selected reducers. Finally, the obtained results were compared in leaching tests over time, indicating which allow obtaining the best results.

2. Methodology

2.1. Manganese Nodule

The marine nodules used in this research were collected in the 1970s from the Blake Plateau in the Atlantic Ocean. The sample was reduced in size using a porcelain mortar and classified by mesh sieves until reaching a range between −140 + 100 µm. Later, it was analyzed chemically by atomic emission spectrometry via induction-coupled plasma (ICP-AES), developed in the Applied Geochemistry Laboratory of the Department of Geological Sciences of the Catholic University of the North, and its chemical composition was 0.12% of Cu, 0.29% Co and 15.96% Mn. Its mineralogical composition is presented in Table 1. Micro X-ray fluorescence spectrometry (Micro-XRF) is a method for elementary analysis of non-homogeneous or irregularly shaped samples, as well as small samples or even inclusions. The sample material was analyzed in a Bruker® M4-Tornado µ-FRX table (Fremont, CA, USA). This spectrometer consists of an X-ray tube (Rh-anode), and the system features a polycapillary X-ray optic, which concentrates the radiation of the tube in minimal areas, allowing a point size of 20 µm for Mo-K. The elementary maps created with the built-in software of the M4 Tornado ™ (Fremont, CA, USA), ESPRIT, indicate that the nodules were composed of fragments of pre-existing nodules that formed its nucleus, with concentric layers that precipitated around the core in later stages.

Table 1. Mineralogical analysis of the manganese nodule.

Component	MgO	Al_2O_3	SiO_2	P_2O_5	SO_3	K_2O	CaO	TiO_2	MnO_2	Fe_2O_3
Mass (%)	3.54	3.69	2.97	7.20	1.17	0.33	22.48	1.07	25.24	26.02

2.2. FeS_2

For this study, a cubic pyrite crystal obtained from the Navajún Mine (La Rioja, Spain) was used. This sample was reduced in size with the use of a cone crusher at laboratory level and later a sprayer. It was then classified through meshes sieves until reaching a size range of −75 + 53 μm. It was then analyzed chemically by atomic emission spectrometry via induction-coupled plasma (ICP-AES), developed in the Applied Geochemistry Laboratory of the Department of Geological Sciences of the Catholic University of the North. Table 2 shows the chemical composition of the samples.

Table 2. Chemical composition.

Component	Fe	S_2
Mass (%)	46.63	53.37

X-ray diffraction analyses (XRD) of the pyrite were performed on a Bruker D8 ADVANCE diffractometer (Billerica, MA, US) with Cu λ = 1.5406 Å radiation generated at 40 kV and 30 mA. The analysis and identification of the crystalline phases were obtained using the DIFFRAC.EVA V4.2.1 program, with the Powder Diffraction File of ICDD database (PDF-2 (2004)) (Billerica, MA, US). According to the initial qualitative analysis of XRD, the primary mineral phase in the samples was pyrite, whose main peaks are at 33.153°, 37.121° and 40.797°. These peaks correspond to those given in the reference pattern PDF 01-1295 (ICDD, 2004). As seen in Figure 1, the analysis showed the sample has a purity of 99.40%.

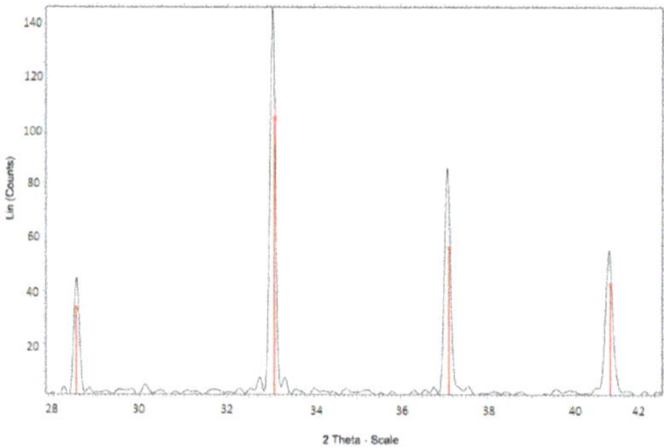

Figure 1. X-ray diffractogram for the pyrite mineral.

2.3. Fe_2O_3

The Fe_2O_3 used is found in tailings from the Altonorte Smelting Plant. Its size is in a range of −75 + 53 μm. The methods used to determine its chemical and mineralogical composition are the same as those used in marine nodules. Figure 2 and Table 3 shows the chemical species that use QEMSCAN (QEMSCAN has a database, which has the elemental composition, and density of the minerals that are detected. With this information, it is possible to obtain the elementary contribution of the measured sample), and several iron-containing phases are presented, while the Fe content is estimated at 41.9%.

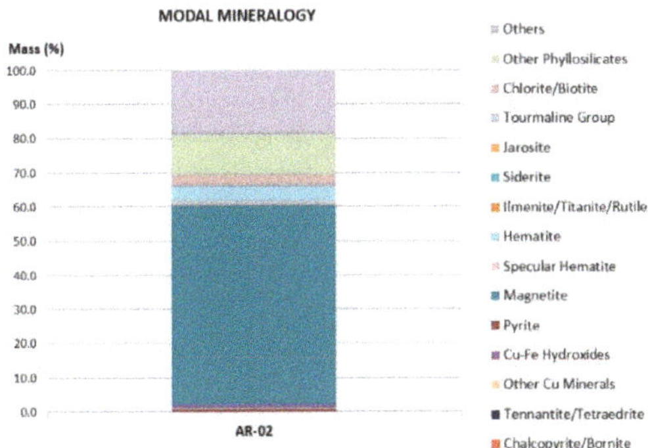

Figure 2. Detailed modal mineralogy.

Table 3. Shows the mineralogical composition of the tailings. The Fe in it was mainly in the form of magnetite.

Mineral	Amount % (w/w)
Chalcopyrite/Bornite $CuFeS_2/Cu_5FeS_4$	0.47
Tennantite/Tetrahedrite ($Cu_{12}As_4S_{13}/Cu_{12}Sb_4S_{13}$)	0.03
Other Cu Minerals	0.63
Cu–Fe Hydroxides	0.94
Pyrite (FeS_2)	0.12
Magnetite (Fe_3O_4)	58.52
Specular Hematite (Fe_2O_3)	0.89
Hematite (Fe_2O_3)	4.47
Ilmenite/Titanite/Rutile ($FeTiO_3/CaTiSiO_3/TiO_2$)	0.04
Siderite ($FeCO_3$)	0.22
Chlorite/Biotite ($Mg_3Si_4O_{10}(OH)_2(Mg)_3(OH)_6/K(Mg)_3AlSi_3O_{10}(OH)_2$)	3.13
Other Phyllosilicates	11.61
Fayalite (Fe_2SiO_4)	4.59
Dicalcium Silicate (Ca_2SiO_4)	8.3
Kirschsteinite ($CaFeSiO_4$)	3.4
Forsterita (Mg_2SiO_4)	2.3
Baritine ($BaSO_4$)	0.08
Zinc Oxide (ZnO)	0.02
Lead Oxide (PbO)	0.01
Sulfate (SO_4)	0.2
Others	0.03
Total	100

2.4. Steel (FeC)

A low carbon steel sheet (FeC; 0.25% C) from the steel supplier company Salomon Sack was used. This sample was reduced in size with the use of a cone crusher at laboratory level and later a pulverizer until reaching a size range between −75 + 53 μm.

2.5. Ferrous Ions

The ferrous ions used for this investigation ($FeSO_4 \times 7H_2O$) were the WINKLER brand (Santiago, Chile), with a molecular weight of 278.01 g/mol.

2.6. Reactor and Leaching Tests

The sulfuric acid used for the leaching tests was grade P.A., with 95–97% purity, a density of 1.84 kg/L and a molecular weight of 98.8 g/mol. The leaching tests were carried out in a 50 mL glass reactor with a 0.01 solid/liquid ratio in leaching solution. A total of 200 mg of Mn nodules were maintained in agitation and suspension with the use of a 5 position magnetic stirrer (IKA ROS, CEP 13087-534, Campinas, Brazil) at a speed of 600 rpm. The tests were conducted at a room temperature of 25 °C, with variations in additives, particle size and leaching time. The tests performed in duplicate, measurements (or analyses) carried on 5 mL undiluted samples using atomic absorption spectrometry with a coefficient of variation ≤5% and a relative error between 5% and 10%. Measurements of pH and oxidation-reduction potential (ORP) of leach solutions were made using a pH-ORP meter (HANNA HI-4222, St. Louis, MO, USA). The solution ORP was measured in a combination ORP electrode cell composed of a platinum working electrode and a saturated Ag/AgCl reference electrode.

2.7. Estimation of Linear and Interaction Coefficients for Factorial Designs of Experiments of 2^3

Two independent variables were chosen for the factorial design of 36 experiments, where: time and ratio MnO_2/reducing agent represent the independent variables that explain the extraction of Mn for a certain type of reducing agent. The analysis through a factorial design allowed us to study the effect of the factors and their levels in a response variable, helping to understand which factors are the most relevant [30,31]. Four factorial designs were carried out that involved two factors with three levels each, with a total of 36 experimental tests (Table 4). The Minitab 18 software (version 18, Pennsylvania State University, State College, PA, USA) was used for modeling, experimental design and adjustment of a multiple regression [32].

Table 4. Experimental conditions.

Parameters/Values	Low	Medium	High
Time (min)	10	20	30
MnO_2/Reducing agent	2/1	1/1	1/2
Codifications	−1	0	1

The expression of the response variable according to the linear effect of the variables of interest and considering the effects of interaction and curvature, is shown in Equation (19).

$$Cu\ Recovery(\%) = \alpha + \sum_{i=1}^{n} \beta_i \times x_i + \sum_{i=1}^{n} \beta_i^2 \times x_i^2 + \beta_{1,2} \times x_1 \times x_2, \tag{19}$$

where α is the overall constant, x_i is the value of the level "i" of the factor, β_i is the coefficient of the linear factor x_i, β_i^2 is the coefficient of the quadratic factors, $\beta_{1,2}$ is the coefficient of the interaction, n are the levels of the factors and Mn recovery is the dependent variable.

Table 4 shows the values of the levels for each factor, while Table 5 shows the recovery obtained for each configuration.

Table 5. Experimental configuration and Mn extraction data.

Exp. No.	Time (min)	MnO_2/Reducing Agent Ratio	Mn Recovery (%; Reducing Agent)			
			FeS_2	Fe^{2+}	FeC	Fe_2O_3
1	10	2/1	4.12	20.52	22.31	33.33
2	10	1/1	8.51	40.69	44.00	50.23
3	10	1/2	10.66	80.27	87.13	71.00
4	20	2/1	8.34	27.80	30.22	39.22
5	20	1/1	12.69	63.11	67.43	57.32
6	20	1/2	19.21	90.18	97.00	73.21
7	30	2/1	15.84	40.32	41.99	42.55
8	30	1/1	19.11	70.00	74.33	72.96
9	30	1/2	26.32	93.50	97.34	75.14

3. Results

3.1. Statistical Analysis

From the analysis of the main components, the time and ratio factors MnO_2/Reducing agent showed a main effect, since the variation between the different levels affected the response in a different way, as shown in Figure 3.

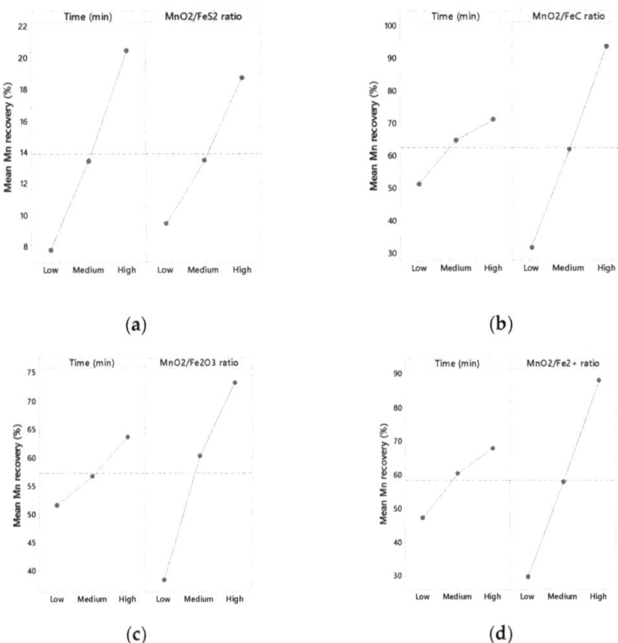

Figure 3. Main effect plots of Mn extraction in function of Time (min) and MnO_2/Reductant agent ratio for (**a**) FeS_2, (**b**) FeC, (**c**) Fe_2O_3 and (**d**) Fe^{2+} agents.

By developing the ANOVA test and the multiple linear regression adjustment for each of the configurations, it is necessary to recover the Mn as a function of the time predictor variables, and MnO_2/reducing agent, which is given by:

$$\text{Mn Extraction (\%) } [FeS_2] = 13.867 + 6.330 \times \text{Time} + 4.648 MnO_2/FeS_2 \times \text{ratio}. \tag{20}$$

$$\text{Mn Extraction (\%) } [Fe^{2+}] = 58.49 + 10.39 \times \text{Time} + 29.22\ MnO_2/Fe^{2+} \times \text{ratio}. \tag{21}$$

$$\text{Mn Extraction (\%) } [FeC] = 62.42 + 10.04 \times \text{Time} + 31.16\ MnO_2/FeC \times \text{ratio}. \tag{22}$$

$$\text{Mn Extraction (\%) } [Fe_2O_3] = 57.22 + 6.01 \times \text{Time} + 17.37\ MnO_2/Fe_2O_3 \times \text{ratio}. \tag{23}$$

The time and ratio MnO_2/reducing agents were coded according to low and medium high levels. From the adjustment of multiple regression models, the interactions of the factors together with the curvature of the time factor and MnO_2/reducing agent did not contribute to explain the variability in any of the adjusted models.

From Equations (20)–(23) and from the main effect graphs in Figure 3, the factor that had showed a higher marginal contribution in Mn recovery was the MnO_2/Reducing agent ratio for the experimental design whose reducing agent was Fe^{2+}, FeC and Fe_2O_3, while in case of using FeS_2 as a reducing agent, the factor that has a greater impact on recovery is time.

The ANOVA test indicates that the models adequately represent Mn extraction for the set of sampled values. The model does not require additional adjustments and is validated by the goodness-of-fit statistics shown in Table 6. The *p* value ($p < 0.05$) and the significance tests F ($F_{Regression} >> (F_{Table} = F_{2,6}(5.1432))$ for a level of significance of $\alpha = 0.05$ (95% confidence level) indicate that all models generated for the representation of the experimental tests were statistically significant. The normality tests indicate that the assumption of normality of the residuals was met. The low values of the S statistic indicate that there were no large deviations between the experimental data and the values of the adjusted model.

Table 6. Goodness of fit statistics.

Response	F-Value	*p*-Value	S	R^2	R^2 (Pred)
Mn Extraction (%) [FeS_2]	102.13	0.000	1.34602	97.15%	92.65%
Mn Extraction (%) [Fe^{2+}]	145.76	0.000	4.44890	97.98%	95.63%
Mn Extraction (%) [FeC]	116.48	0.000	5.25341	97.49%	94.35%
Mn Extraction (%) [Fe_2O_3]	42.02	0.000	4.91305	93.34%	84.37%

The value of the R^2 statistics was greater than 90%, which indicates that a large part of the total variability was explained by the models, while the similarity between the R^2 and R^2 predictive statistics indicates that the model could adequately predict the response to new observations.

3.2. *Effect on MnO_2/Reducing Agent Ratio*

In Figure 4, results are presented for the dissolution of Mn with the use of different Fe reducing agents at different ratios of Mn/Fe. For all the cases presented (Figure 4a–d), when working at low Mn/Fe ratios the highest recoveries of Mn were obtained. Ratios of 1/2 proved to be an optimum in Figure 4b–d. While for the Figure 4a in ratios of 1/3, the increase in the dissolution of Mn continued. The best results were obtained in Figure 4c when working with FeC because it allowed a high activity ratio through the regeneration of ferrous ions, favoring the dissolution of Mn and allowing better results to the use of Fe^{2+} in a direct way that is presented in Figure 4b. Using Fe_2O_3 shows good results when working with MnO_2/Fe_2O_3 ratios of 1/2, although it is lower than those presented when using Fe^{2+} and FeC. However, this may be an attractive proposal due to the reuse of tailings that are an environmental responsibility. For the use of pyrite, the lowest Mn solutions could be observed in this study. In previous studies [2,22,23,33], it has been indicated that it is not necessary to work at high concentrations of H_2SO_4 in the system to obtain high Mn solutions from marine nodules, but that if it is important to have low Mn/Fe ratios. The results presented in Figure 4a show a progressive increase in the Mn dissolution when increasing the amounts of FeS_2 in the system, however, it may be necessary to increase the acid concentration or temperature because of the kinetics of dissolution of ferrous ions from the pyrite ore.

For the performed tests, the values of potential and pH for the different reducing agents used for Mn/Fe ratios of 1/2 are presented in Figure 5. Senanayake [13] indicated that dissolving Mn from marine nodules requires to work in potential ranges between −0.4 and 1.4 V and pH between −2 and 0.1. With this, it is possible to avoid the precipitation of the Mn through the oxidation-reduction reaction, due to the presence of ferrous and ferric ions [34]. The outcomes met the operational condition mentioned above, which is due to the high concentrations of reducing agent. The lowest potential values were obtained with Fe^{2+} and FeC, wherein the iron (FeC) favored the regeneration of ferrous ions, which allows maintaining low potential ranges [22].

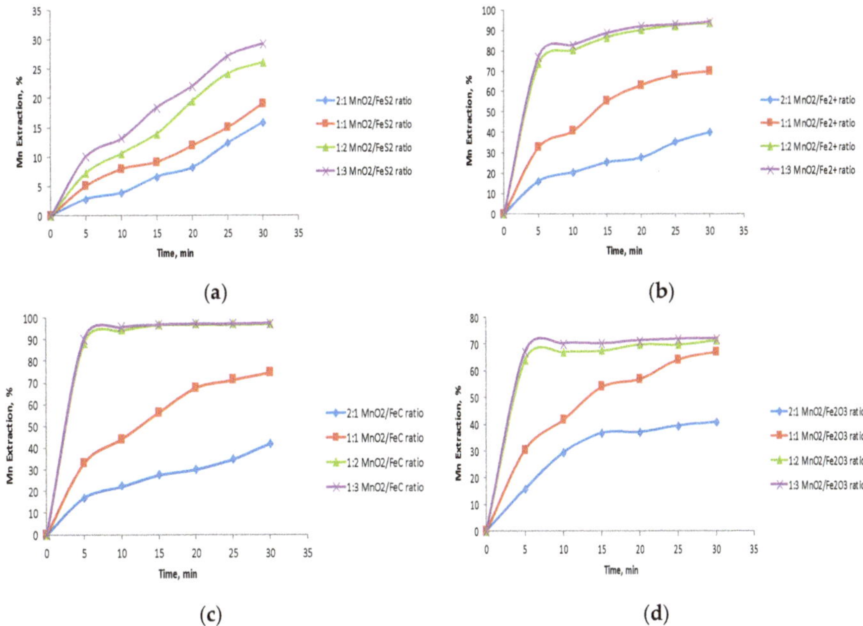

Figure 4. Effect on the ratio of MnO_2/reducing agent at room temperature (25 °C), 0.1 mol/L H_2SO_4, 600 rpm and particle size of –75 + 53 µm (reducing agent: (**a**) FeS_2, (**b**) Fe^{2+}, (**c**) FeC and (**d**) Fe_2O_3).

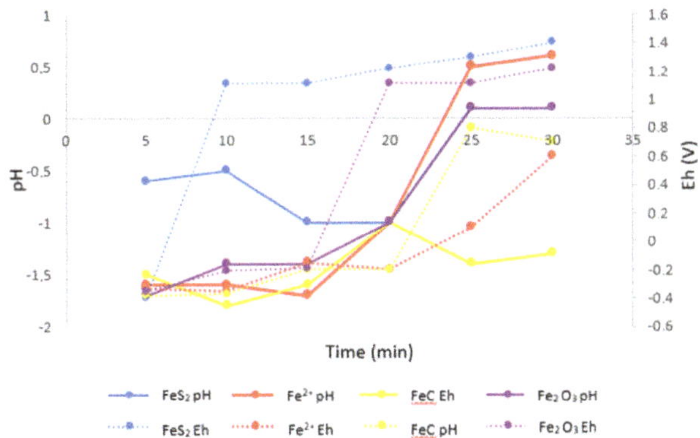

Figure 5. Effect of the potential and pH in the solution of Mn with different reducing agents (MnO_2/Fe_2O_3 ratio of 1/2, 25 °C, 600 rpm, −75 + 53 µm, acid concentration to 0.1 mol/L).

3.3. *Effect on the Concentration of H_2SO_4*

Figure 6 shows the effect of sulfuric acid concentration when working at Mn/Fe ratios of 1/2 with the use of different Fe reducing agents. Figure 6b,c shows that the concentration of H_2SO_4 was irrelevant in the extraction of Mn when working at low ratios of Mn/Fe with the use of Fe^{2+} and FeC. This is compatible with previous studies conducted by Zakeri et al. [24] and Bafghi et al. [22]. The researchers indicated that working at high concentrations of ferrous ions, variables like acid concentration and particle size were irrelevant. For the case shown in Figure 6d, it was observed that when working with the use of Fe_2O_3 there was a slight increase in Mn solutions when working above 0.1 mol/L, although it was observed that there were no differences between 0.5 and 1 mol/L, which reaffirms what was raised by Saldaña et al. [2], where they indicated that when working on acid-reducing leaching of MnO_2 using tailings, the acid concentration only influenced the extractions of Mn when it was not operated in high levels of Fe or no temperature increase. Finally, it can be seen in Figure 6a that when working with pyrite, the concentration of acid in the system was important. This was consistent with the results obtained by Kanungo et al. [21], which states that in an acid solution of marine nodules with the use of pyrite as the acidity of the medium decreases, the rate of reduction of MnO_2 decreases.

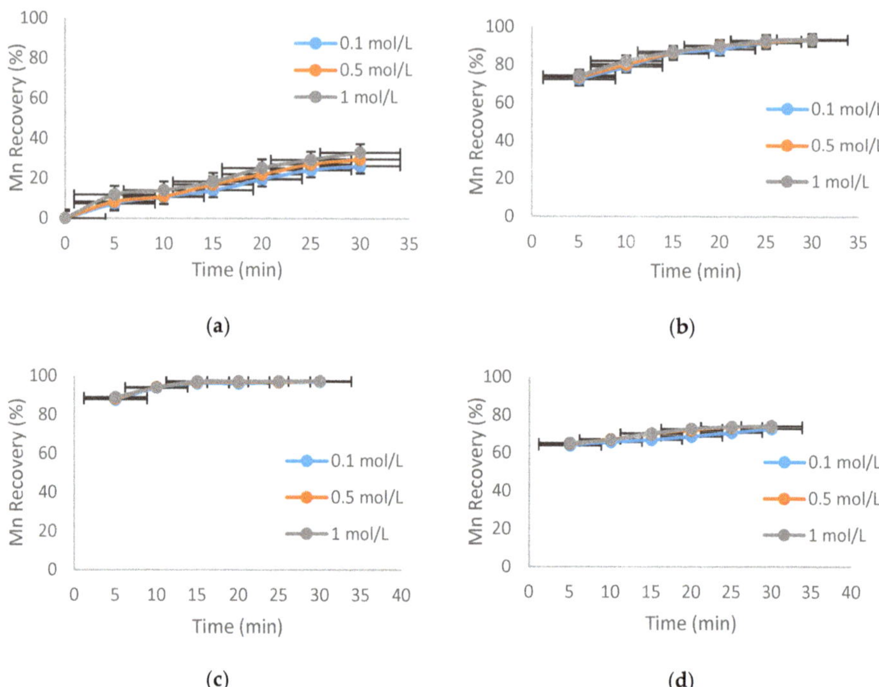

Figure 6. Effect on the concentration of H_2SO_4 at room temperature of (25 °C), ratio of MnO_2/reducing agent of 1/2, 600 rpm and particle size of −75 + 53 µm (reducing agent: (**a**) FeS_2, (**b**) Fe^{2+}, (**c**) FeC and (**d**) Fe_2O_3).

4. Conclusions

The Fe presented in the different additives proved to be a good reducing agent, increasing the dissolution of MnO_2. The main findings of this study were the following:

(1) Fe^0 (FeC) proved to be the best reducing agent for the dissolution of Mn from marine nodules since the direct contact of Fe in the liquid solution kept the regeneration of ferrous ions, due to high levels of ferrous and ferric ions.

(2) When working with Fe^{2+}, FeC and Fe_2O_3, and having high concentrations of reducing agent (MnO_2 ratios/reducing agent 1/2 or lower), low potential values were maintained, which allowed working at low acid concentrations (0.1 mol/L). However, for FeS_2, better results were achieved at higher ratios of MnO_2/FeS_2 (1/3) and acid levels of 1 mol/L, which was possibly due to the refractoriness of pyrite.

(3) For the tests carried out in this study with the different Fe reducing agents, the potential and pH ranges were from −0.4 to 1.4 V and −2 to 0.1, favoring the dissolution of Mn from marine nodules, and avoiding the formation of precipitates of the Fe.

(4) The best results of this research (97% of Mn) were obtained at MnO_2/FeC ratios of 1/2, 0.1 mol/L of H_2SO_4, in a time of 20 min.

In future work, other industrial iron wastes, generated in large industries, should be evaluated to create novel acid-reducing processes of MnO_2. Subsequently, to recover the manganese present in the solution, zero-valent iron (ZVI) is a good alternative. Zero valence iron can be reused, from scraps of the metal finishing industry.

The authors wish to make the following corrections to this paper:

We worked at a temperature of 60 °C for all the tests carried out in which FeS_2 was added as a reducing agent, whereas in the other experiments, other added Fe reducing agents were worked at room temperature (25 °C). This affects the results presented in Table 5 (results when working with FeS2), Figures 4a and 6a. By accident and through carelessness, we did not indicate this important detail in the work methodology. For this reason, we must correct it for the readers, otherwise the reproduction of the results of our experiments will not be possible due to incorrect working parameters. However, we confirm that this error does not affect the conclusions of the manuscript.

We must indicate that it is unlikely that the following series of reactions that were presented in the document could occur at room temperature:

$$3FeS_2 + 4H_2SO_4 = 3FeSO_4 + 4H_2O + 7S \quad (1)$$

$$6FeSO_4 + 4H_2SO_4 = 3Fe_2(SO_4)_3 + 4H_2O + S \quad (2)$$

$$2FeS_2 + 4H_2SO_4 = Fe_2(SO_4)_3 + 4H_2O + 5S \quad (3)$$

$$15MnO_2 + 2FeS_2 + 14H_2SO_4 = Fe_2(SO_4)_3 + 15MnSO_4 + 14H_2O \quad (4)$$

We will update the article and the original version will remain available on the article webpage.

Author Contributions: N.T. and R.I.J. contributed in project administration, M.C., S.N., L.A. and J.C. contributed in investigation and D.T. and N.T wrote paper, M.S. contributed in the data curation and software, P.R. contributed in validation and supervision and review and editing.

Funding: This research received no external funding.

Acknowledgments: The authors are grateful for the contribution of the Scientific Equipment Unit- MAINI of the Universidad Católica del Norte for aiding in generating data by automated electronic microscopy QEMSCAN®and for facilitating the chemical analysis of the solutions. We are also grateful to the Altonorte Mining Company for supporting this research and providing slag for this study, and we thank to Marina Vargas Aleuy of the Universidad Católica del Norte for supporting the experimental tests. Pedro Robles thanks the Pontificia Universidad Católica de Valparaíso for the support provided.

Conflicts of Interest: The authors declare no conflict of interest.

References

1. Marino, E.; González, F.J.; Somoza, L.; Lunar, R.; Ortega, L.; Vázquez, J.T.; Reyes, J.; Bellido, E. Strategic and rare elements in Cretaceous-Cenozoic cobalt-rich ferromanganese crusts from seamounts in the Canary Island Seamount Province (northeastern tropical Atlantic). *Ore Geol. Rev.* **2017**, *87*, 41–61. [CrossRef]
2. Saldaña, M.; Toro, N.; Castillo, J.; Hernández, P.; Trigueros, E.; Navarra, A. Development of an Analytical Model for the Extraction of Manganese from Marine Nodules. *Metals* **2019**, *9*, 903. [CrossRef]
3. Hein, J.R. The Geology of Cobalt-rich Ferromanganese Crusts. In *Deep Sea Minerals: Cobalt-Rich Ferromanganese Crusts, A Physical, Biological, Environmental, and Technical Review*; Secretariat of the Pacific Community (SPC): Noumea, New Caledonia, 2013; pp. 7–14.
4. Hein, J.R.; Mizell, K.; Koschinsky, A.; Conrad, T.A. Deep-ocean mineral deposits as a source of critical metals for high- and green-technology applications: Comparison with land-based resources. *Ore Geol. Rev.* **2013**, *51*, 1–14. [CrossRef]
5. González, F.J.; Somoza, L.; León, R.; Medialdea, T.; de Torres, T.; Ortiz, J.E.; Lunar, R.; Martínez-Frías, J.; Merinero, R. Ferromanganese nodules and micro-hardgrounds associated with the Cadiz Contourite Channel (NE Atlantic): Palaeoenvironmental records of fluid venting and bottom currents. *Chem. Geol.* **2012**, *310*, 56–78. [CrossRef]

6. Josso, P.; Pelleter, E.; Pourret, O.; Fouquet, Y.; Etoubleau, J.; Cheron, S.; Bollinger, C. A new discrimination scheme for oceanic ferromanganese deposits using high fi eld strength and rare earth elements. *Ore Geol. Rev.* **2017**, *87*, 3–15. [CrossRef]
7. Koschinsky, A.; Heinrich, L.; Boehnke, K.; Cohrs, J.C.; Markus, T.; Shani, M.; Singh, P.; Stegen, K.S.; Werner, W. Deep-sea mining: Interdisciplinary research on potential environmental, legal, economic, and societal implications. *Integr. Environ. Assess. Manag.* **2018**, *14*, 672–691. [CrossRef]
8. Ghosh, M.K.; Barik, S.P.; Anand, S. Sulphuric Acid Leaching Of Polymetallic Nodules Using Paper As A Reductant. *Trans. Indian Inst. Met.* **2008**, *61*, 477–481. [CrossRef]
9. Hein, J.R. Manganese nodules. In *Encyclopedia of Marine Geosciences*; Springer: Dordrecht, The Netherlands, 2016; pp. 408–412.
10. Sharma, R. Environmental Issues of Deep-Sea Mining. *Procedia Earth Planet. Sci.* **2015**, *11*, 204–211. [CrossRef]
11. Usui, A.; Nishi, K.; Sato, H.; Nakasato, Y.; Thornton, B.; Kashiwabara, T. Continuous growth of hydrogenetic ferromanganese crusts since 17 Myr ago on Takuyo-Daigo Seamount, NW Paci fi c, at water depths of 800—5500 m. *Ore Geol. Rev.* **2017**, *87*, 71–87. [CrossRef]
12. Jana, R.K.; Pandey, B.D. Ammoniacal leaching of roast reduced deep-sea manganese nodules. *Hydrometallurgy* **1999**, *53*, 45–56. [CrossRef]
13. Senanayake, G. Acid leaching of metals from deep-sea manganese nodules—A critical review of fundamentals and applications. *Miner. Eng.* **2011**, *24*, 1379–1396. [CrossRef]
14. Toro, N.; Pérez, K.; Saldaña, M.; Jeldres, R.I.; Jeldres, M.; Cánovas, M. Dissolution of pure chalcopyrite with manganese nodules and waste water. *J. Mater. Res. Technol.* **2019**, in press. [CrossRef]
15. Randhawa, N.S.; Hait, J.; Jana, R.K. A brief overview on manganese nodules processing signifying the detail in the Indian context highlighting the international scenario. *Hydrometallurgy* **2016**, *165*, 166–181. [CrossRef]
16. Pérez, K.; Toro, N.; Campos, E.; González, J.; Jeldres, R.I.; Nazer, A.; Rodriguez, M.H. Extraction of Mn from Black Copper Using Iron Oxides from Tailings and Fe^{2+} as Reducing Agents in Acid Medium. *Metals* **2019**, *9*, 1112. [CrossRef]
17. Su, H.; Liu, H.; Wang, F.; Lü, X.; Wen, Y. Kinetics of reductive leaching of low-grade pyrolusite with molasses alcohol wastewater in H_2SO_4. *Chin. J. Chem. Eng.* **2010**, *18*, 730–735. [CrossRef]
18. Kanungo, S.B.; Jena, P.K. Reduction leaching of manganese nodules of Indian Ocean origin in dilute hydrochloric acid. *Hydrometallurgy* **1988**, *21*, 41–58. [CrossRef]
19. Khalafalla, S.E.; Pahlman, J.E. Selective Extraction of Metals from Pacific Sea Nodules with Dissolved Sulfur Dioxide. *JOM J. Miner. Met. Mater. Soc.* **1981**, *33*, 37–42. [CrossRef]
20. Han, K.N.; Fuerstenau, D.W. Extraction behavior of metallic elements from deep-sea manganese nodules in reducing medium. *Mar. Min.* **1986**, *2*, 155–169.
21. Kanungo, S.B. Rate process of the reduction leaching of manganese nodules in dilute HCl in presence of pyrite. Part I. Dissolution behaviour of iron and sulphur species during leaching. *Hydrometallurgy* **1999**, *52*, 313–330. [CrossRef]
22. Bafghi, M.S.; Zakeri, A.; Ghasemi, Z.; Adeli, M. Reductive dissolution of manganese ore in sulfuric acid in the presence of iron metal. *Hydrometallurgy* **2008**, *90*, 207–212. [CrossRef]
23. Toro, N.; Herrera, N.; Castillo, J.; Torres, C.; Sepúlveda, R. Initial Investigation into the Leaching of Manganese from Nodules at Room Temperature with the Use of Sulfuric Acid and the Addition of Foundry Slag—Part I. *Minerals* **2018**, *8*, 565. [CrossRef]
24. Zakeri, A.; Bafghi, M.S.; Shahriari, S.; Das, S.C.; Sahoo, P.K.; Rao, P.K. Dissolution kinetics of manganese dioxide ore in sulfuric acid in the presence of ferrous ion. *Hydrometallurgy* **2007**, *8*, 22–27.
25. Kanungo, S.B. Rate process of the reduction leaching of manganese nodules in dilute HCl in presence of pyrite. Part II: Leaching behavior of manganese. *Hydrometallurgy* **1999**, *52*, 331–347. [CrossRef]
26. Toro, N.; Saldaña, M.; Castillo, J.; Higuera, F.; Acosta, R. Leaching of Manganese from Marine Nodules at Room Temperature with the Use of Sulfuric Acid and the Addition of Tailings. *Minerals* **2019**, *9*, 289. [CrossRef]
27. El Problema Global de la Chatarra de Mineral de Hierro se Agrava|Minería en Línea. 2018. Available online: https://mineriaenlinea.com/2018/11/el-problema-global-de-la-chatarra-de-mineral-de-hierro-se-agrava/ (accessed on 26 November 2019).

28. MCH, Reciclaje Minero: En Busca de un Sector Sustentable—Minería Chilena. 2013. Available online: http://www.mch.cl/2013/01/28/reciclaje-minero-en-busca-de-un-sector-sustentable/# (accessed on 25 November 2019).
29. Campos, C. EyN: Reciclaje minero: En busca de un sector sustentable. 2013. Available online: http://www.economiaynegocios.cl/noticias/noticias.asp?id=105240 (accessed on 26 November 2019).
30. Bezerra, M.A.; Santelli, R.E.; Oliveira, E.P.; Villar, L.S.; Escaleira, L.A. Response surface methodology (RSM) as a tool for optimization in analytical chemistry. *Talanta* **2008**, *76*, 965–977. [CrossRef] [PubMed]
31. Montgomery, D.C. *Montgomery: Design and Analysis of Experiments*, 8th ed.; John Wiley & Sons: New York, NY, USA, 2012.
32. Mathews, P.G. *Design of Experiments with MINITAB*; William, A., Ed.; ASQ Quality Press: Milwaukee, WI, USA, 2005; ISBN 0873896378.
33. Toro, N.; Briceño, W.; Pérez, K.; Cánovas, M.; Trigueros, E.; Sepúlveda, R.; Hernández, P. Leaching of Pure Chalcocite in a Chloride Media Using Sea Water and Waste Water. *Metals* **2019**, *9*, 780. [CrossRef]
34. Komnitsas, K.; Bazdanis, G.; Bartzas, G.; Sahinkaya, E.; Zaharaki, D. Removal of heavy metals from leachates using organic/inorganic permeable reactive barriers. *Desalin. Water Treat.* **2013**, *51*, 3052–3059. [CrossRef]

Article

Leaching Kinetics of Arsenic Sulfide-Containing Materials by Copper Sulfate Solution

Kirill A. Karimov, Denis A. Rogozhnikov, Evgeniy A. Kuzas and Andrei A. Shoppert *

Department of Non-Ferrous Metals Metallurgy, Ural Federal University, 620002 Yekaterinburg, Russia; kirill_karimov07@mail.ru (K.A.K.); darogozhnikov@yandex.ru (D.A.R.); e.kuzas@ya.ru (E.A.K.)

* Correspondence: andreyshop@list.ru; Tel.: +7-922-024-3963

Received: 25 November 2019; Accepted: 16 December 2019; Published: 19 December 2019

Abstract: The overall decrease in the quality of mineral raw materials, combined with the use of arsenic-containing ores, results in large amounts of various intermediate products containing this highly toxic element. The use of hydrometallurgical technologies for these materials is complicated by the formation of multicomponent solutions and the difficulty of separating copper from arsenic. Previously, for the selective separation of As from copper–arsenic intermediates a leaching method in the presence of Cu(II) ions was proposed. This paper describes the investigation of the kinetics of arsenic sulfide-containing materials leaching by copper sulfate solution. The cakes after leaching of arsenic trisulfide with a solution of copper sulfate were described using methods such as X-ray diffraction spectrometry (XRD), X-ray fluorescence spectrometry (XRF), scanning electron microscopy (SEM) and energy-dispersive X-ray spectroscopy analysis (EDS). The effect of temperature (70–90 °C), the initial concentration of $CuSO_4$ (0.23–0.28 M) and the time on the As recovery into the solution was studied. The process temperature has the greatest effect on the kinetics, while an increase in copper concentration from 0.23 to 0.28 M effects an increase in As transfer into solution from 93.2% to 97.8% for 120 min of leaching. However, the shrinking core model that best fits the kinetic data suggests that the process occurs by the intra-diffusion mode with the average activation energy of 44.9 kJ/mol. Using the time-to-a-given-fraction kinetics analysis, it was determined that the leaching mechanism does not change during the reaction. The semi-empirical expression describing the reaction rate under the studied conditions can be written as follows: $1/3\ln(1 - X) + [(1 - X) - 1/3 - 1] = 4560000Cu^{3.61}e^{-44900/RT}\ t$.

Keywords: arsenic; trisulfide; copper sulfate; kinetics; leaching; shrinking core model; time-to-a-given-fraction kinetics analysis

---

1. Introduction

The need to increase the production of copper and other non-ferrous metals compels the industry to engage various primary and man-made low-grade polymetallic materials such as copper–zinc, copper–lead–zinc, poor arsenic-containing copper concentrates, enrichment products, etc. [1]. The overall decrease in the quality of mineral raw materials, combined with the use of arsenic-containing ores, results in large amounts of various intermediate products containing this highly toxic element [2].

Currently, copper concentrates and other copper-containing sulfide intermediates are most often processed by pyrometallurgical methods, while most impurities such as arsenic, lead, zinc, antimony are sublimated and passed to fine dusts during gas cleaning [3–10]. When such arsenic-containing materials are processed along with ore raw materials, arsenic is redistributed and accumulated in target products (matte, crude metal, etc.) [11] as well as intermediates (dust, sublimates, etc.) [12].

The currently used methods of separate processing of arsenic-containing materials only allow concentrating arsenic in separate products, but fail to address the environmental safety of the generated waste containing active arsenic [13–15], and fail to produce effluent solutions with the concentration recommended by the World Health Organization (WHO) and the United States Environmental Protection Agency (USEPA) [16,17].

The use of hydrometallurgical technologies for these materials is complicated by the formation of multicomponent solutions and the difficulty of separating copper from arsenic [18–21].

Precipitation is a widely used method of removing arsenic from solutions [6,13,22–25]. The lowest soluble As compounds are arsenic sulfides [26], calcium arsenite, calcium arsenate [27,28] and iron arsenate [29–33]. Each of these precipitates exhibits relatively low solubility in the corresponding pH range, but these methods lead to loss of copper with arsenic-containing precipitates due to Cu co-precipitation, and do not yield high-quality respective copper products [20,29,34].

Arsenic precipitation as sulfide has the advantage of a high degree of Cu and As separation and water extraction due to the high specific gravity of As_2S_3. However, the residues are amorphous and heavily filtered. As_2S_3 is stable under acidic and reducing conditions (pH < 4), but is susceptible to atmospheric and bacterial oxidation. The solubility of arsenic sulfide is 1 mg/dm^3 at pH < 4.

In the works of Ke et al., and Lundström et al. [35,36] a method was proposed for the selective leaching of As from copper–arsenic intermediates in the presence of Cu(II) ions, based on their exchange reaction with As ions. This reaction was first described for the Nikkelverk process [35,37], which aimed at the separation of copper and nickel ions. Later it was described in the work of Lundström et al. [36].

In Japan, the application of this method is patented by Sumitomo [38,39]. The method allows efficient extraction of arsenic into the solution and its subsequent precipitation in the form of a well-filtered high-purity arsenic oxide. The solid residue of the leaching, consisting mainly of copper sulfide, can be further used in smelting.

According to the patent [40], arsenic cake is leached with an acidic solution with the addition of copper chloride. As a result, arsenic passes into solution, and copper precipitates as sulfide.

The work [21] describes the study to determine a method for processing mining and metallurgical acid effluents containing As and Cu. The proposed two-stage scheme includes precipitation of solutions containing up to 15% arsenic and up to 80% copper as sulfides by treating the solutions with hydrogen sulfide as a first stage, and selective leaching of arsenic from the obtained residue with the initial solution (which involves sulfide copper precipitation) as a second stage.

The mentioned works demonstrate the interest in this method from many researchers, since the need exists to process various arsenic-containing materials such as copper–arsenic, iron–arsenic gold-containing concentrates, dusts and various intermediates of copper and lead production, and many others. Loss of copper with arsenic waste and arsenic poisoning of target products are inevitable at various stages of processing such raw materials.

In our studies of hydrochemical processing of such materials [41,42], difficulties arise when separating copper and arsenic from the leachates. Therefore, in this study, we considered the physico-chemical aspects and mechanisms of the reactions during the leaching of arsenic trisulfide-based raw materials with water solutions of copper sulfate.

2. Materials and Methods

2.1. Materials and Reagents

The chemicals used in this study were of analytical grade; the water was distilled using a GFL-manufactured device (GFL mbH, Burgwedel, Germany). $CuSO_4{\cdot}5H_2O$ was obtained from AO Vekton (S-Petersburg, Russia).

The As_2S_3 containing material was obtained from an As-containing solution by reacting it with sodium hydrosulfide according to the method described elsewhere [43]. The chemical compositions of arsenic trisulfide-based raw material on a dry weight are presented in Table 1. The moisture content of the raw material was 60.2%.

Table 1. Chemical composition of arsenic trisulfide-based raw material, wt%.

Cu	As	S	Zn	Fe	Pb
0.14	40.20	44.86	0.42	0.23	1.24

2.2. Analysis

The chemical analysis of the original material and the resulting solid products of the studied processes was performed using an Axios MAX X-ray fluorescence spectrometer (XRF) (Malvern Panalytical Ltd., Almelo, The Netherlands). The chemical analysis of the obtained solutions was performed by mass spectrometry with inductively coupled plasma (ICP-MS) using an Elan 9000 instrument (PerkinElmer Inc., Waltham, MA, USA). The phase analysis was performed on an XRD 7000 Maxima diffractometer (Shimadzu Corp., Tokyo, Japan).

Scanning electron microscopy (SEM) was performed using a JSM-6390LV microscope (JEOL Ltd., Tokyo, Japan) equipped with an energy-dispersive X-ray spectroscopy analysis (EDS) module INCA Energy 450 X-Max 80 (Oxford Instruments, High Wycombe, UK).

Samples from each experiment were taken at the selected time intervals and analyzed using inductively coupled plasma mass spectrometry (ICP-MS–NexION 300S quadrupole mass spectrometer, PerkinElmer Inc., Waltham, MA, USA).

2.3. Experiments

Laboratory experiments on $CuSO_4$ leaching were carried out using an apparatus consisting of a 1 dm^3 borosilicate glass round bottom reactor (Lenz Laborglas GmbH & Co. KG, Wertheim, Germany), with openings for reagent injection, as well as for temperature control and removal of water vapor through a water-cooled reflux condenser. The reactor was thermostated. The materials were stirred using an overhead mixer at 400 rpm, which ensured uniform density of the pulp. A portion of the raw material weighing 60 g was added to a prepared solution of $CuSO_4$ of a required concentration. At the end of the experiment, the leaching pulp was filtered in a Buchner funnel (ECROSKHIM Co., Ltd., St. Petersburg, Russia); the solutions were sent for ICP-MS analysis; the leaching cake was washed with distilled water, dried at 100 °C to constant weight, weighed and sent for XRF analysis. All the experiments were performed twice and the mean values are presented here.

Eh-pH diagrams of the Cu-As-S-H_2O system were drawn using the HSC Chemistry Software Version 6.0 (Outokumpu Research Oy, Finland).

3. Results and Discussion

3.1. Thermodynamics of Leaching Arsenic Trisulfide with Copper-Containing Solution

Earlier, the thermodynamic features of the Cu-As-S-H_2O system were studied by Ke et al. [35], demonstrating that the temperature has practically no effect on the phase composition. So, Figure 1 shows the Pourbaix diagram of the Cu-As-S-H_2O system only at 80 °C. The concentration of Cu and As ions was 0.5 mol/dm^3, the concentration of S ions was 1 mol/dm^3. The diagram shows that the zone of existence of copper sulfide is in the pH range of 0 to 13 with Eh values of −0.7 to 0.4 V. So, under acidic conditions and with typical oxidation potentials, copper in the presence of S^{2-} or elemental S will be mainly in the form of copper sulfide. And according to [35] such conditions are actually created when arsenic-containing cakes are leached by copper cations. Arsenic trisulfide can also serve as a sulfur source. Under these conditions, it will either transform into arsenic trioxide or transfer to the solution as AsO_3^{3-} or AsO_4^{3-}. When the solution is supersaturated in arsenic and the system Eh is sufficiently high, the latter precipitates as oxide.

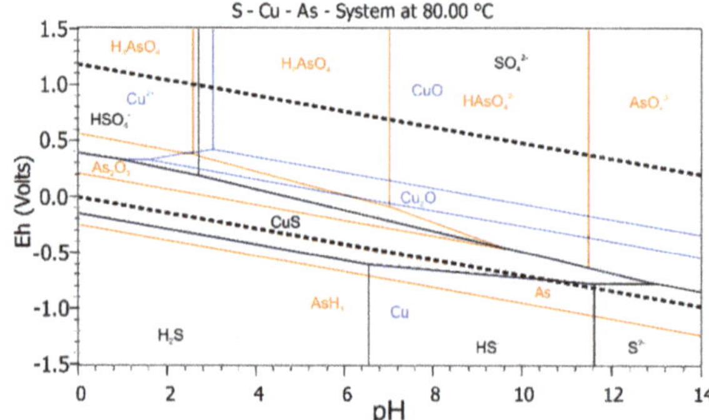

Figure 1. Pourbaix diagram of Cu-As-S-H_2O system at 80 °C: yellow—arsenic-containing phases; blue—copper-containing phases; black—sulfur-containing phases.

Figure 2 shows the X-ray diffraction patterns of the arsenic-containing raw materials and products obtained after leaching with different degrees of arsenic recovery: after 20 min of leaching at 80 °C (60% of As recovery) and after 240 min of leaching at 80 °C (more than 98% of As recovery). Liquid to solid ratio in both experiments was 10:1, copper concentration in initial solution—0.26 M.

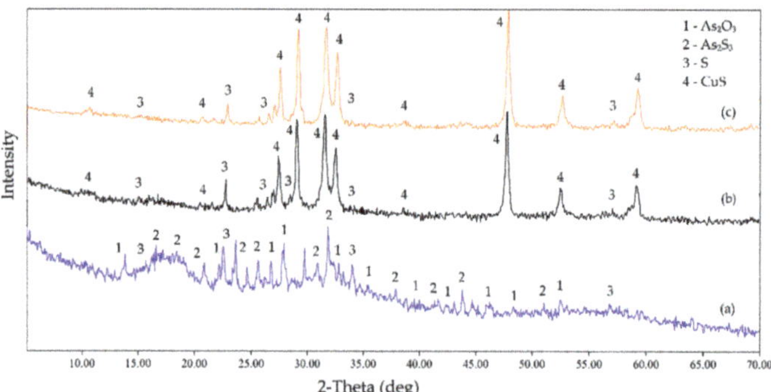

Figure 2. X-ray diffraction pattern of arsenic-containing raw material (**a**) and products of its leaching with copper sulfate at 60% recovery of As (**b**) and at more than 98% recovery of As (**c**).

Figure 2 shows that the arsenic-containing raw material contains amorphous material, as well as arsenic trisulfide and trioxide along with elemental S. While the leachates are mainly represented by copper sulfide and elemental S, higher degrees of the process completion result in smaller peaks for elemental sulfur, near complete reaction of the amorphous phase and a new crystalline phase, CuS.

Thus, the X-ray diffraction patterns confirm the previous [35] and our conclusions made regarding the thermodynamics of the process. Furthermore, the interaction of elemental S with copper cations is much slower than with arsenic trisulfide. A previous study assumed that the interaction of arsenic trisulfide with copper cations must proceed through the intermediate stages of elemental S and H_2S formation, since more negative Gibbs energies were obtained for these reactions [35]. Therefore, we also calculated the change in Gibbs free energy for the following chemical reactions (Equations (1)–(6)):

$$3CuSO_4(ia) + As_2S_3 + 6H_2O = 3CuS + 2H_3AsO_3(ia) + 3H_2SO_4(ia) \quad (1)$$

$$3CuSO_4(ia) + 4S + 4H_2O = 3CuS + 4H_2SO_4(ia) \quad (2)$$

$$CuSO_4(ia) + H_2S(ia) = CuS + H_2SO_4(ia) \quad (3)$$

$$As_2S_3 + H_2SO_4(ia) + 2H_2O = 2H_3AsO_3(a) + 4S \quad (4)$$

$$As_2S_3 + 3H_2O + 1.5O_2(g) = 2H_3AsO_3(a) + 3S \quad (5)$$

$$As_2S_3 + 6H_2O + 2O_2(g) = 2H_2S(ia) + 2H_3AsO_3(ia) + H_2SO_4(ia) \quad (6)$$

where ia—ionic./aqueous phase, g—gaseous phase.

The calculation results are presented in Figure 3, which shows the effect of temperature on the change in Gibbs free energy for each of the listed reactions.

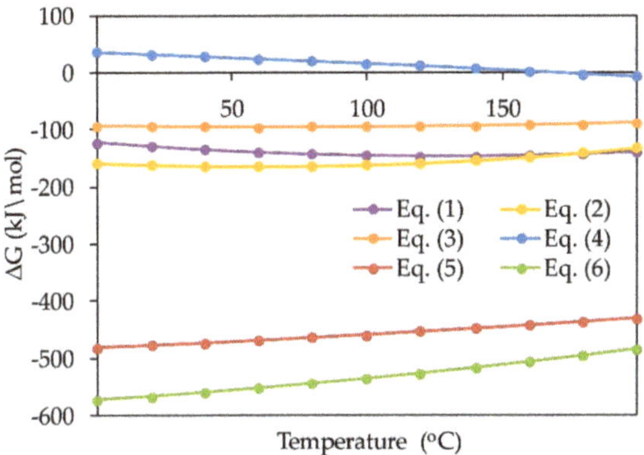

Figure 3. The calculation of the change in Gibbs free energy for Equations (1)–(6).

The calculations of the change in Gibbs free energy allow us to conclude that the probabilities of copper sulfate reaction with arsenic trisulfide, S, or H_2S (Equations (1)–(3)) are almost equal, although the change in Gibbs energy is lower in the presence of oxygen for the oxidation of arsenic trisulfide with the formation of elemental S and H_2S (Equations (5) and (6)). However, elemental S is still found in the cake, while arsenic almost completely passes into the solution (Figure 2). Therefore, it is likely that reaction 2 has a lower kinetic than reaction 1.

To determine if the reaction with elemental sulfur was kinetically controlled, we performed experiments on the effect of temperature on the Cu:S molar ratio of the product. The results of the experiments are presented in Figure 4. Notably, the degree of As extraction at all temperatures exceeded 95%, which means that the lower the Cu:S molar ratio in the leachate, the lower the reaction degree of the elemental sulfur.

Obviously, the Cu:S molar ratio of the leaching product approaches 1 as the temperature increases, which is consistent with the complete conversion of S into copper sulfide via reaction 2. Therefore, to accelerate this reaction, leaching under pressure at a temperature above 120 °C is necessary. In addition, reaction 2 slows down significantly at temperatures below 80 °C. The obtained data suggest that some of the As_2S_3 may generate elemental sulfur via reaction (5) and the reaction (2) of freshly formed sulfur also could have some kinetic difficulties.

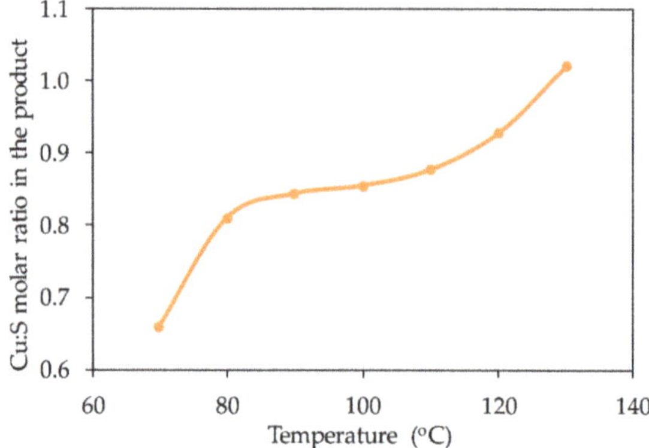

Figure 4. The effect of temperature on the Cu:S molar ratio in the leachate after 2 h of leaching with a 2-fold excess of copper in solution from the stoichiometry.

3.2. *Effect of Process Parameters on As Recovery from Trisulfide*

3.2.1. Effect of Temperature

As shown earlier, the initial material is highly amorphous and consists mainly of arsenic trisulfide. The effect of temperature and copper concentration on the leaching process was evaluated by As recovery in the solution. Figure 5 shows the dependence of As recovery on the duration and temperature. Liquid to solid ratio in all experiments was 10:1, copper concentration in initial solution—0.26 M.

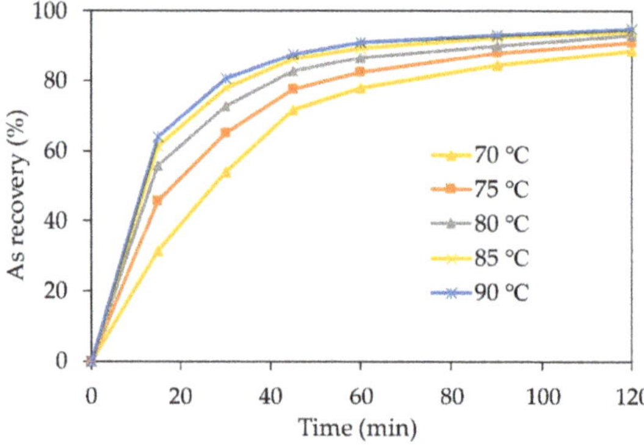

Figure 5. Dependence of the degree of As recovery from arsenic trisulfide on temperature and time.

Figure 5 shows a significant effect of temperature on the As recovery in the solution at the starting point of the process, from 0 to 60 min. At 90 °C, up to 80.7% of arsenic trisulfide is leached within 30 min. A decrease in temperature to 70 °C leads to a decrease in the degree of arsenic recovery to solution to 54.0% in a similar time.

An increase in the duration of leaching from 60 to 120 min leads to a decreased effect of temperature on the leaching of arsenic. Up to 88.6% of arsenic dissolves at 70 °C, and 94.9% at 90 °C within 120 min of the process. The data show that the temperature is the major factor in As recovery from arsenic trisulfide, which indicates kinetic control. However, a study of the kinetics is required to determine the limiting step.

3.2.2. Effect of Copper Concentration

The effect of copper concentration was evaluated in the concentration range 0.23–0.28 M, which corresponds to 90–110% of its stoichiometric requirement according to Equation (1). Liquid to solid ratio in all experiments was 10:1, temperature −80 °C.

The results in Figure 6 show the positive effect of copper concentration and leaching temperature on the dissolution of arsenic. An increase in copper concentration from 0.23 to 0.28 M effects an increase in As transfer into solution from 75.8% to 84.7% for 30 min of leaching, and from 93.2% to 97.8% for 120 min. Increasing the leaching time from 30 to 120 min leads to a decrease in the effect of copper concentration on the transition of arsenic to solution. A similar result was observed when studying the effect of temperature. The results also indicate that the influence of copper concentration has a lower value than temperature. However, the interval of concentration change was quite low.

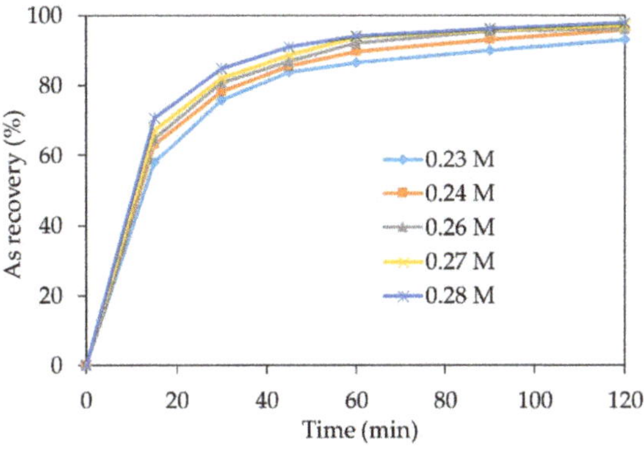

Figure 6. Dependence of As recovery on the concentration of copper in solution.

3.3. *Characteristics of Residue*

The cakes after leaching of arsenic trisulfide with a solution of copper sulfate were analyzed by scanning electron microscopy to assess the changes in the morphology of the sample. SEM images of the initial material and the cake after 240 min of leaching at 80 °C (more than 98% of As recovery) are presented in Figure 7.

According to the SEM images, the initial material is represented by 5–20 μm porous agglomerates of different shapes, with rough surfaces (Figure 7b). The apparent low level of crystallinity in this material is consistent with the X-ray phase analysis data shown in Figure 2. The interaction of arsenic sulfide with the copper ions in the solution forms agglomerates with lamellar, needle-like particles, the size of which are in the range 2–20 μm (Figure 7d,f).

SEM images of the surface of the formed particles with a magnification of 10,000 show the formation of 1–1.7 μm long and 100–130 nm thick plates on the surface of agglomerates (Figure 8).

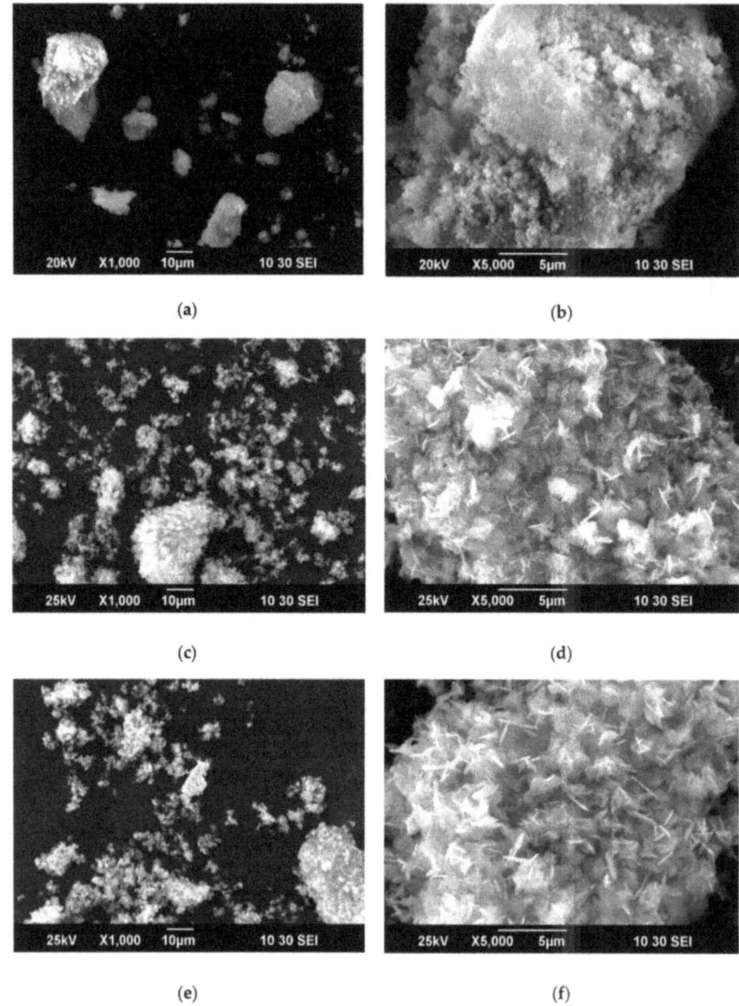

Figure 7. Scanning electron microscopy (SEM) images of the initial material (**a**,**b**), filter cake after leaching of 60% arsenic (**c**,**d**), filter cake after complete extraction (**e**,**f**).

Figure 8. SEM images of filter cake after leaching of 60% of arsenic (**a**) and filter cake after complete loosening (**b**).

The images show that the surface of the cake particles after 60% As recovery and after complete extraction are similar, but the plates are thicker after the leaching is complete (Figure 8a,b). Figure 9 shows energy dispersive spectroscopy microphotographs with points at which compositions were determined for the initial material and for cakes with varying degrees of As recovery (Table 2).

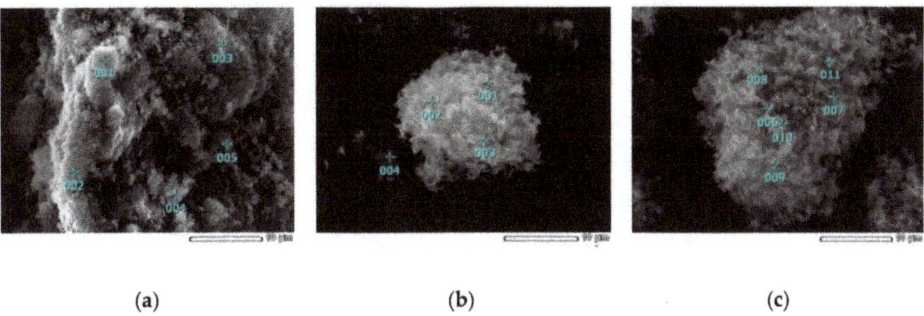

(a) (b) (c)

Figure 9. SEM images with points for determining by energy-dispersive X-ray spectroscopy analysis (EDS) the composition of the initial material (**a**), after 60% As recovery (**b**), full recovery (**c**).

Sulfur and arsenic are distributed unevenly, which is due to the presence of elemental sulfur in the initial material.

Arsenic trisulfide interacts with copper ions in the solution to form copper sulfide CuS, which is confirmed by the results of EDS and X-ray phase analysis (Figures 2 and 10). The copper content is 49–56% at 60% recovery of As, and increases to 57–64% when As is fully recovered.

Table 2. The results of energy dispersive spectroscopy, weight %.

Element	S	Fe	Cu	Zn	As	Total
Figure 9a. Point 001	52.8	0.4	2.8	2.8	41.2	100.0
Figure 9a. Point 002	39.6	0.4	4.7	3.9	51.4	100.0
Figure 9a. Point 003	61.4	0.3	2.9	2.6	32.7	100.0
Figure 9a. Point 004	51.2	0.2	3.1	3.2	42.3	100.0
Figure 9a. Point 005	64.6	0.3	2.7	2.2	30.2	100.0
Figure 9b. Point 001	44.4	0.1	51.4	0.5	3.6	100.0
Figure 9b. Point 002	40.7	0.2	56.0	0.4	2.7	100.0
Figure 9b. Point 003	45.2	0.1	48.5	1.1	5.1	100.0
Figure 9b. Point 004	39.4	0.0	54.3	1.0	5.3	100.0
Figure 9c. Point 006	39.8	0.0	60.2	0.0	0.0	100.0
Figure 9c. Point 007	41.6	0.1	57.4	0.3	0.6	100.0
Figure 9c. Point 008	41.1	0.0	58.2	0.3	0.4	100.0
Figure 9c. Point 009	35.6	0.0	64.4	0.0	0.0	100.0
Figure 9c. Point 010	41.5	0.1	58.0	0.2	0.1	100.0
Figure 9c. Point 011	42.7	0.1	56.9	0.0	0.3	100.0

Figure 10 shows that the copper content on the surface of the particles increases in the course of leaching. In the initial material (Figure 10b), copper is practically absent, only minor inclusions are found, as the copper was present in the solution from which the As_2S_3 containing raw material was precipitated. In the process of As recovery, the amount of copper sulfide on the surface increases (Figure 10b,e,h).

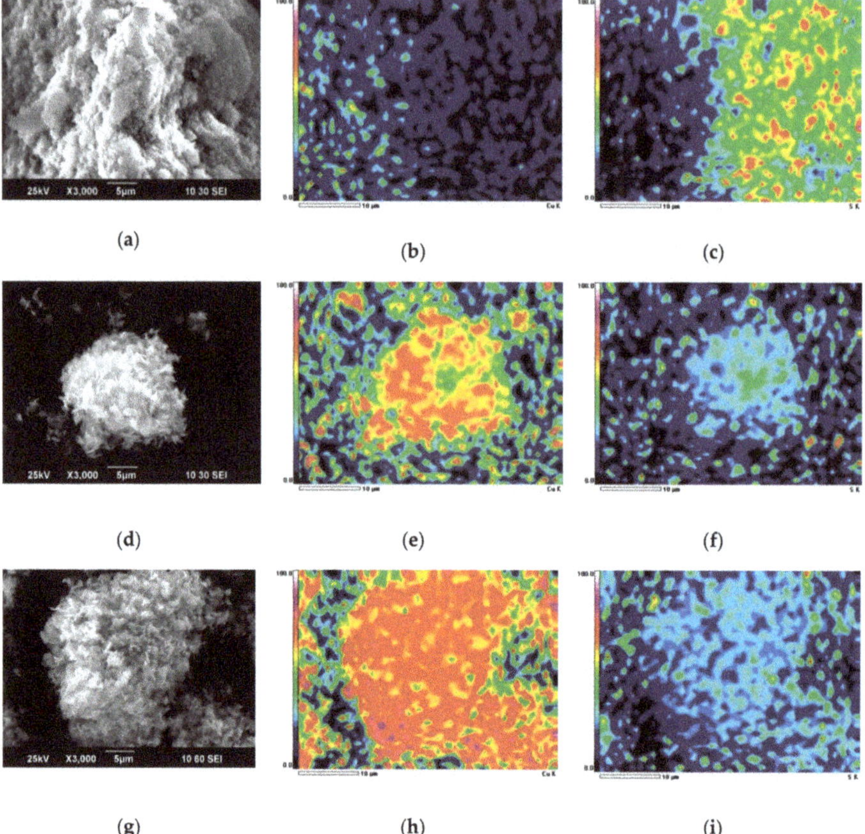

Figure 10. Microphotographs of the initial trisulfide (**a**), filter cake after leaching of 60% arsenic (**d**), filter cake after full recovery (**g**) and EDS mapping for copper (**b**,**e**,**h**) and sulfur (**c**,**f**,**i**).

Two theories exist to explain the mechanism of interaction of sulfides with copper ions [44]:

1. Interaction occurs immediately on the surface of the raw material, between the solid sulfur–sulfides and copper ions in the solution.
2. The process involves an intermediate stage of H_2S production when sulfides contact with acidified solutions. The H_2S then interacts with copper ions to form covellite.

SEM images, EDS, and EDS mapping suggest that in the process of arsenic extraction, the plates formed on the surface thicken, with a parallel increase in the copper content. The presence of plates or needle-shaped crystals on the surface (the predominant growth of crystals in one direction) and their thickening indicates that sulfide ions are being transported to the surface of existing particles and precipitation is initiated on the existing surfaces. So, arsenic trisulfide also could react with copper ions via the second mechanism.

3.4. Kinetic Model

The shrinking core model (SCM) is used to describe the kinetics of heterogeneous reactions, and suggests interaction of a substance with an external reagent only on the surface of the particles. The reaction zone gradually progresses inside the particles, leaving behind a converted product—an inert part of the particle. The core of the particle, containing the active component that has not yet reacted, gradually decreases during the reaction. The change in unreacted core size is assumed to be linear, the kinetics of the process is also considered linear. This model assumes that the rate of the process is limited either by diffusion of the agent to the surface through the diffusion layer, diffusion through the product layer (film diffusion through the surface layer of the shrinking sphere), or chemical reaction for particles of constant or decreasing size. Thus, the shrinking core model not only describes how the process proceeds, as it can determine the basic kinetic characteristics, but explains the mechanism of the process as well.

The slowest stage with the highest resistance will be the rate limiting one, and its intensification will effect an increase the leaching efficiency.

Levenspiel [45] proposes several variants of the SCM. Our previous studies [46] showed that a new variant of SCM proposed by Dickenson and Heal [47] accurately describes the kinetics of leaching reactions regulated by interfacial transfer and diffusion through the product layer.

The correlation coefficients (R^2) obtained by modeling the leaching of arsenic trisulfide by copper sulfate solutions using the SCM equations are shown in Table 3.

Table 3. SCM equations fitting.

#	Limiting Step	Equation	R^2				
			70 °C	75 °C	80 °C	85 °C	90 °C
1	Diffusion through the product layer (sp)	$1 - 3(1 - X)^{2/3} + 2(1 - X)$	0.955	0.937	0.873	0.810	0.774
2	Diffusion through the product layer (pp)	X^2	0.899	0.815	0.696	0.606	0.557
3	Diffusion through the product layer (cp)	$X + (1 - X)\ln(1 - X)$	0.948	0.905	0.823	0.749	0.708
4	Diffusion through the liquid film (sp)	X	0.691	0.539	0.414	0.341	0.305
5	Surface chemical reactions (cp)	$1 - (1 - X)^{1/2}$	0.828	0.727	0.625	0.558	0.523
6	Surface chemical reactions (sp)	$1 - (1 - X)^{1/3}$	0.869	0.789	0.699	0.639	0.607
7	New shrinking core model	$1/3\ln(1 - X) + [(1 - X)^{-1/3} - 1]$	0.968	0.985	0.993	0.991	0.987

Sp—spherical particles, pp—prismatic particles, cp—conus particles, X—the degree of As recovery into the solution.

According to the table data, the new SCM version (Equation (7) in Table 3) describes the derived dependences better than all other equations and has the highest correlation coefficients for the studied temperatures (R^2), which confirms that leaching proceeds in the intra-diffusion area.

To calculate the activation energy, we plotted the dependence of $\ln k_c$ vs. $1/T$, where k_c is the slope of each straight line obtained by substituting the experimental data into the SCM Equation (7) in Table 3, as can be seen in Figure 11a. Based on Equation (7), derived from Arrhenius law, we used the slope of the line to determine the apparent activation energy 44.9 kJ/mol. As shown before [48], a high value of the activation energy is not always representative of a kinetically controlled reaction.

$$\ln k_c = \ln A - E_a/RT. \quad (7)$$

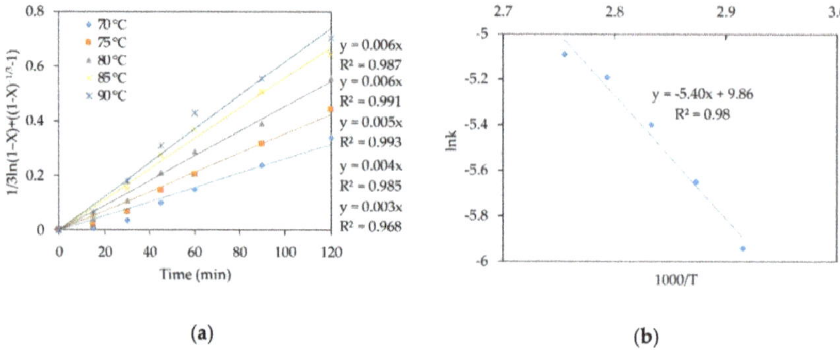

Figure 11. Calculating angular coefficients k_c (**a**), reverse dependence of $\ln k_c$ on temperature (**b**).

The Eyring equation (Equations (8) and (9)) (transition state theory) was used to calculate other kinetic characteristics, including enthalpies and entropies [49,50].

$$k_c = \frac{k_b T}{h} e^{\Delta G^{++}/RT}. \tag{8}$$

$$T * \ln\frac{k_c}{T} = T * \left(\ln\frac{k_b}{h} + \frac{\Delta S^{++}}{R}\right) - \frac{\Delta H^{++}}{R} \tag{9}$$

where $k_b = 1.381 \times 10^{-23}$ J/K (Boltzmann constant) and $h = 6.626 \times 10^{-34}$ J·s (Planck constant).

The values ΔH^{++} and ΔS^{++} can be used to calculate the free energy as per Equation (10).

$$\Delta G^{++} = \Delta H^{++} - T\Delta S^{++}. \tag{10}$$

Activation enthalpy and entropy were calculated graphically by plotting the temperature dependence in the reference frame $T \cdot \ln(k_c/T)$ (Figure 12).

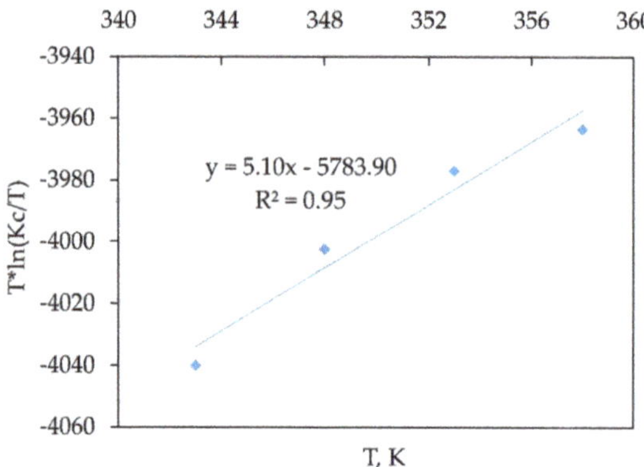

Figure 12. Eyring plot for arsenic leaching.

The enthalpy and activation entropy values obtained from the plot were 48.1 kJ/mol and −155.1 J/(mol·K). The small difference between the activation energies and the enthalpies is due to the low temperatures of the process (Equation (11)).

$$E_a = \Delta H^{++} + RT. \tag{11}$$

The negative values of entropy and positive values of free energy, enthalpy, indicate a non-spontaneous reaction in the entire range of temperatures studied.

Using the slope of the straight lines obtained by substituting the new SCM model with copper concentrations of 0.23–0.28 M and a temperature of 8 °C, a graph in the $\ln k_c$ vs. lnCu frame of reference was obtained to determine the reaction order with respect to Cu. The resulting empirical order with respect to Cu was 3.61 (Figure 13).

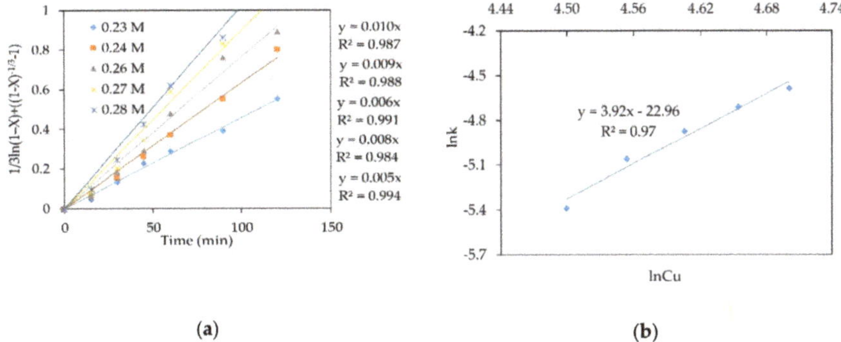

(a) (b)

Figure 13. Dependence of the new SCM equation on time (**a**), and $\ln k_c$ vs. lnCu for determining the order with respect to copper concentration (**b**).

By substituting the Arrhenius equation into the equation of the new shrinking core model, we can obtain Equation (12):

$$1/3\ln(1 - X) + [(1 - X)^{-1/3} - 1] = k_o e^{-E_a/RT}\ t. \quad (12)$$

In Equation (12), k_o depends on the initial concentration of copper in the solution; Equation (12) can be represented as follows (Equation (13)):

$$1/3\ln(1 - X) + [(1 - X)^{-1/3} - 1] = k_o Cu^n e^{-E_a/RT}\ t \quad (13)$$

where n is the order with respect to copper concentration.

Based on the earlier results, we can derive the following equation for leaching of arsenic trisulfide with copper sulfate solutions (Equation (14)):

$$1/3\ln(1 - X) + [(1 - X)^{-1/3} - 1] = k_o Cu^{3.61} e^{-44900/RT}\ t. \quad (14)$$

Graphs were plotted for all temperatures and copper concentrations, which allowed us to determine a fixed angle of inclination a = 4,560,000. The value of a obtained graphically and the corresponding value of the correlation coefficient R^2 are shown in Figure 14. The obtained value of coefficient "a" corresponds to k_o.

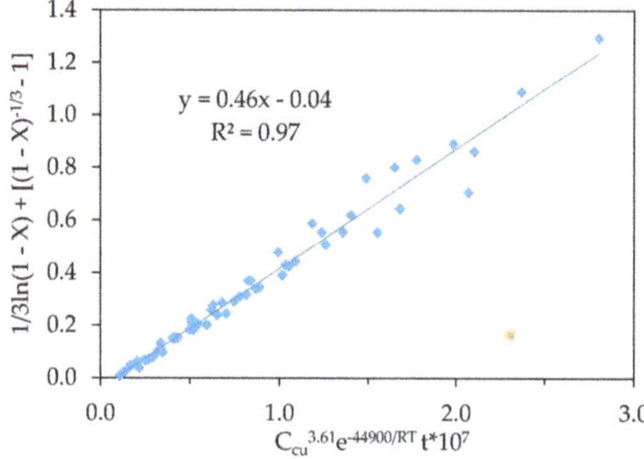

Figure 14. Graph for determining k_0.

Based on the data in Figure 14, k_o = 4,560,000, and therefore the kinetic Equation (15) is as follows:

$$1/3\ln(1 - X) + [(1 - X)^{-1/3} - 1] = 4560000Cu^{3.61}e^{-44900/RT}\,t. \quad (15)$$

Although an SCM is often used in hydrometallurgical kinetics to determine the activation energy of processes and to derive empirical equations, it has some drawbacks. For example, this method only allows us to determine the average values of the activation energy, which may lead to the omission of certain features of the reactions. It is therefore necessary to determine the kinetic characteristics during the process, which may change, e.g., due to the slower kinetics for the reaction of copper cations with elemental S. For this purpose, the time-to-a-given-fraction method [51] was used, allowing us to calculate the apparent activation energy at different points of the leaching process. The time required to achieve a certain degree of leaching and the apparent activation energy E_a are related as per Equation (16).

$$\ln t_x = \text{const} - \ln A + E_a/RT. \quad (16)$$

The slope of the graph plotted in the reference frame $\ln t_x$ vs. 1/T allowed us to calculate the apparent activation energy.

Figure 15 shows the graphical calculation of the apparent activation energy for different degrees of arsenic recovery.

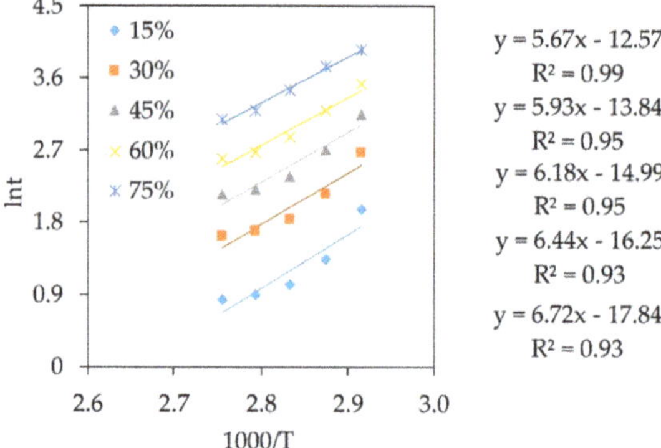

Figure 15. Dependence of $\ln t_x$ on 1/T for different degrees of As recovery in solution.

As seen in Figure 15, the values of the apparent activation energy calculated using the time-to-a-given-fraction method vary from 47.1 to 55.9 kJ/mol as the degree of arsenic recovery in solution increases from 15% to 75%. These values coincide with the previously obtained value of the SCM apparent activation energy, and confirm that the reaction mechanism does not change during leaching. However, a high activation energy at the later stage of leaching may be necessary for the reaction of elemental S (Figure 4), which may envelope arsenic sulfide and prevent more complete leaching of As.

4. Conclusions

This work studied the kinetics of the selective separation of copper and arsenic products by leaching arsenic trisulfide with a solution of copper sulfate. The study provides new data on the behavior of the initial components in the leaching process, using new approaches to studying the reaction kinetics. The conclusions are as follows:

1. The original arsenic sulfide material contains a substantial amorphous phase and, thermodynamically, an exchange reaction to form copper sulfide and various arsenic oxides should take place in the presence of copper sulfate in the entire studied temperature range. However, the leaching mechanisms may vary, since at the starting point arsenic sulfide is likely to be oxidized to form elemental sulfur or H_2S, which only then go on to react with copper cations.
2. Achieving the stoichiometric composition of copper sulfide in the final product is possible only with a large excess of copper cations and at increased temperature, which is due to the slow kinetics of the reaction of elemental sulfur with copper cations.
3. SEM images, EDS, and EDS mapping demonstrate that solid sulfur–sulfides react with copper ions. H_2S also could be generated and can react with the copper.
4. At 90 °C, up to 80.7% of arsenic trisulfide is leached within 30 min. A decrease in temperature to 70 °C leads to a decrease in the degree of arsenic recovery to solution to 54.0% in a similar time. Up to 88.6% of arsenic dissolves at 70 °C, and 94.9% at 90 °C within 120 min. This shows a significant effect of temperature on the As recovery. An increase in copper concentration from 0.23 to 0.28 M effects an increase in As transfer into solution from 75.8% to 84.7% for 30 min of leaching, and from 93.2 to 97.8% for 120 min.
5. The process temperature has the greatest effect on the kinetics. However, the shrinking core model that best fits the data suggests the process occurs by the intra-diffusion mode with the average activation energy of 44.9 kJ/mol.
6. Using the time-to-a-given-fraction kinetics analysis, it was determined that the leaching mechanism does not change during the reaction. The value of activation energy during the reaction increases from 47.1 kJ/mol at 15% arsenic recovery to 55.9 kJ/mol at 75%. This may be due to the slow kinetics of elemental sulfur reacting with copper cations.
7. Based on the findings, a semi-empirical equation was obtained, which allows us to describe the kinetics of the leaching of arsenic-containing cake by copper cations with a great accuracy: $1/3\ln(1-X) + [(1-X)^{-1/3} - 1] = 4560000Cu^{3.61}e^{-44900/RT}\,t$.

Author Contributions: Conceptualization, D.A.R.; methodology, K.A.K. and A.A.S.; validation, E.A.K..; formal analysis, K.A.K., D.A.R.; investigation, A.A.S. and E.A.K.; resources, E.A.K., D.A.R.; data curation, D.A.R.; writing—original draft preparation, K.A.K.; writing—review and editing, A.A.S., D.A.R.; visualization, A.A.S., K.A.K.; supervision, A.A.S.; project administration, D.A.R.; funding acquisition, D.A.R. All authors have read and agreed to the published version of the manuscript.

Funding: The research was funded by the Russian Science Foundation, grant number 18-19-00186. The SEM/EDS analyses were funded by State Assignment, grant number 10.7347.2017/8.9.

Acknowledgments: Technicians at Ural Branch of the Russian Academy of Sciences are acknowledged for their assistance with XRD, XRF, SEM, EDS, and ICP-MS analysis.

Conflicts of Interest: The authors declare no conflict of interest. The funders had no role in the design of the study; in the collection, analyses, or interpretation of data; in the writing of the manuscript, or in the decision to publish the results.

References

1. Celep, O.; Alp, İ.; Deveci, H. Improved gold and silver extraction from a refractory antimony ore by pretreatment with alkaline sulphide leach. *Hydrometallurgy* **2011**, *105*, 234–239. [CrossRef]
2. Bissen, M.; Frimmel, F.H. Arsenic–A review. Part I: Occurrence, Toxicity, Speciation, Mobility. *Acta Hydroch. Hydrob.* **2003**, *31*, 9–18. [CrossRef]
3. Shibayama, A.; Takasaki, Y.; William, T.; Yamatodani, A.; Higuchi, Y.; Sunagawa, S.; Ono, E. Treatment of smelting residue for arsenic removal and recovery of copper using pyro–hydrometallurgical process. *J. Hazard. Mater.* **2010**, *181*, 1016–1023. [CrossRef] [PubMed]
4. Long, G.; Peng, Y.; Bradshaw, D. A review of copper–arsenic mineral removal from copper concentrates. *Miner. Eng.* **2012**, *36–38*, 179–186. [CrossRef]
5. Dalewski, F. Removing Arsenic from Copper Smelter Gases. *JOM* **1999**, *51*, 24–26. [CrossRef]
6. Vircikova, E.; Havlik, M. Removing As from Converter Dust by a Hydrometallurgical Method. *JOM* **1999**, *51*, 20–23. [CrossRef]
7. Piret, N.L. The Removal and Safe Disposal of Arsenic in Copper Processing. *JOM* **1999**, *51*, 16–17. [CrossRef]
8. Cheng, R.; Ni, H.; Zhang, H.; Zhang, X.; Bai, S. Mechanism research on arsenic removal from arsenopyrite ore during a sintering process. *Int. J. Min. Met. Mater.* **2017**, *24*, 353–359. [CrossRef]
9. Safarzadeh, M.S.; Miller, J.D. Reaction of Enargite (Cu_3AsS_4) in Hot Concentrated Sulfuric Acid under an Inert Atmosphere. Part II: High-quality Enargite. *Int. J. Miner. Process.* **2014**, *128*, 79–85. [CrossRef]
10. Liu, J.; Chi, R.; Xu, Z.; Zeng, Z.; Liang, J. Selective arsenic-fixing roast of refractory gold concentrate. *Metall. Mater. Trans. B* **2000**, *31*, 1163–1168. [CrossRef]
11. Safarzadeh, M.S.; Miller, J.D. Reaction of Enargite (Cu_3AsS_4) in Hot Concentrated Sulfuric Acid under an Inert Atmosphere. Part I: Enargite Concentrate. *JOM* **2014**, *128*, 68–78. [CrossRef]
12. Ritcey, G.M. Tailings management in gold plants. *Hydrometallurgy* **2005**, *78*, 3–20. [CrossRef]
13. Riveros, P.A.; Dutrizac, J.E.; Spencer, P. Arsenic disposal practices in the metallurgical industry. *Can. Metall. Q.* **2001**, *40*, 395–420. [CrossRef]
14. Smedley, P.L.; Kinniburgh, D.G. A review of the source, behaviour and distribution of arsenic in natural waters. *Appl. Geochem.* **2002**, *17*, 517–568. [CrossRef]
15. Yao, L.; Min, X.; Xu, H.; Ke, Y.; Liang, Y.; Yang, K. Hydrothermal Treatment of Arsenic Sulfide Residues from Arsenic-Bearing Acid Waste Water. *J. Environ. Res. Public Health* **2018**, *15*, 1863. [CrossRef]
16. Smith, A.H.; Smith, M.M.H. Arsenic drinking water regulations in developing countries with extensive exposure. *Toxicology* **2004**, *198*, 39–44. [CrossRef]
17. Shoppert, A.; Loginova, I.; Rogozhnikov, D.; Karimov, K.; Chaikin, L. Increased As Adsorption on Maghemite-Containing Red Mud Prepared by the Alkali Fusion-Leaching Method. *Minerals* **2019**, *9*, 60. [CrossRef]
18. Nazari, A.M.; Radzinski, R.; Ghahreman, A. Review of arsenic metallurgy: Treatment of arsenical minerals and the immobilization of arsenic. *Hydrometallurgy* **2017**, *174*, 258–281. [CrossRef]
19. Ruiz, M.C.; Vera, M.V.; Padilla, R. Mechanism of enargite pressure leaching in the presence of pyrite. *Hydrometallurgy* **2011**, *105*, 290–295. [CrossRef]
20. Anderson, C.G.; Twidwell, L.G. Hydrometallurgical processing of gold-bearing copper enargite concentrates. *Can. Metall. Q.* **2008**, *47*, 337–346. [CrossRef]
21. Jiang, G.M.; Bing, P.E.N.G.; Chai, L.Y.; Wang, Q.W.; Shi, M.Q.; Wang, Y.Y.; Hui, L.I.U. Cascade sulfidation and separation of copper and arsenic from acidic wastewater via gas−liquid reaction. *Trans. Nonferrous Met. Soc. China* **2017**, *27*, 925–931. [CrossRef]
22. Jiang, J.Q. Research progress in the use of ferrate(VI) for the environmental remediation. *J. Hazard. Mater.* **2007**, *146*, 617–623. [CrossRef] [PubMed]
23. Prucek, R.; Tuček, J.; Kolařík, J.; Filip, J.; Marušák, Z.; Sharma, K.; Zbořil, R. Ferrate(VI)-induced arsenite and arsenate removal by in situ structural incorporation into magnetic iron(III) oxide nanoparticles. *Environ. Sci. Technol.* **2013**, *47*, 3283–3292. [CrossRef] [PubMed]

24. Lee, Y.; Um, I.H.; Yoon, J. Arsenic(III) oxidation by iron(VI) (ferrate) and subsequent removal of arsenic(V) by iron(III) coagulation. *Environ. Sci. Technol.* **2003**, *37*, 5750–5756. [CrossRef]
25. Gomez, M.A.; Becze, L.; Blyth, R.I.R.; Cutler, J.N.; Demopoulos, G.P. Molecular and structural investigation of yukonite (synthetic & natural) and its relation to arseniosiderite. *Geochim. Cosmochim. Acta* **2010**, *74*, 5835–5851. [CrossRef]
26. Guo, L.; Du, Y.; Yi, Q.; Li, D.; Cao, L.; Du, D. Efficient removal of arsenic from "dirty acid" wastewater by using a novel immersed multi-start distributor for sulphide feeding. *Sep. Purif. Technol.* **2015**, *142*, 209–214. [CrossRef]
27. Moon, D.H.; Dermatas, D.; Menounou, N. Arsenic immobilization by calcium–arsenic precipitates in lime treated soils. *Sci. Total Environ.* **2004**, *330*, 171–185. [CrossRef]
28. Zhu, Y.N.; Zhang, X.H.; Xie, Q.L.; Wang, D.Q.; Cheng, G.W. Solubility and stability of calcium arsenates at 25 °C. *Water Air Soil Pollut.* **2006**, *169*, 221–238. [CrossRef]
29. Gomez, M.A. The effect of copper on the precipitation of scorodite ($FeAsO_4 \cdot 2H_2O$) under hydrothermal conditions: Evidence for a hydrated copper containing ferric arsenate sulfate-short lived intermediate Original Research Article. *J. Colloid Interface Sci.* **2011**, *360*, 508–518. [CrossRef]
30. Fujita, T.; Taguchi, R.; Abumiya, M.; Matsumoto, M.; Shibata, E.; Nakamura, T. Effects of zinc, copper and sodium ions on ferric arsenate precipitation in a novel atmospheric scorodite process. *Hydrometallurgy* **2008**, *93*, 30–38. [CrossRef]
31. Yang, B.; Zhang, G.L.; Deng, W.; Ma, J. Review of arsenic pollution and treatment progress in nonferrous metallurgy industry. *Adv. Mater. Res.* **2013**, *634–638*, 3239–3243. [CrossRef]
32. Caetano, M.L.; Ciminelli, S.; Rocha, S.D.; Spitale, M.C.; Caldeira, C.L. Batch and continuous precipitation of scorodite from dilute industrial solutions. *Hydrometallurgy* **2009**, *95*, 44–52. [CrossRef]
33. Monhemius, A.J.; Swash, P.M. Removing and stabilizing as from copper refining circuits by hydrothermal processing. *JOM* **1999**, *51*, 30–33. [CrossRef]
34. Morales, A.; Cruells, M.; Roca, A.; Bergó, R. Treatment of copper flash smelter flue dusts for copper and zinc extraction and arsenic stabilization. *Hydrometallurgy* **2010**, *105*, 148–154. [CrossRef]
35. Ke, Y.; Shen, C.; Min, X.B.; Shi, M.Q.; Chai, L.Y. Separation of Cu and As in Cu-As-containing filter cakes by Cu^{2+}-assisted acid leaching. *Hydrometallurgy* **2017**, *172*, 45–50. [CrossRef]
36. Lundström, M.; Liipo, J.; Taskinen, P.; Aromaa, J. Copper precipitation during leaching of various copper sulfide concentrates with cupric chloride in acidic solutions. *Hydrometallurgy* **2016**, *166*, 136–142. [CrossRef]
37. Crundwell, F.K.; Moats, M.; Ramachandran, V. *Hydrometallurgical Production of High-Purity Nickel and Cobalt, Extractive Metallurgy of Nickel, Cobalt and Platinum Group Metals*; Elsevier: Oxford, UK, 2011; pp. 281–299. [CrossRef]
38. Toshimasa, I.; Yoshio, M. Method for Separating and Recovering Arsenious Acid from Substance Containing Arsenic Sulfide. J.P. Patent S59107923, 22 June 1984.
39. Tadashi, N.; Yoshio, M.; Naoki, K. Recovery of Arsenious Acid from Substance Containing Arsenic Sulfide. J.P. Patent S5869723, 26 April 1983.
40. Gong, D.; Gong, H.; Li, C.H. Method for Recovering Simple Substance Arsenic from Arsenic Sulfide Slag. C.N. Patent 101386915, 18 March 2009.
41. Karimov, K.A.; Naboichenko, S.S. Sulfuric Acid Leaching of High-Arsenic Dust from Copper Smelting. *Metallurgist* **2016**, *60*, 456–459. [CrossRef]
42. Karimov, K.A.; Naboichenko, S.S.; Kritskii, A.V.; Tret'yak, M.A.; Kovyazin, A.A. Oxidation Sulfuric Acid Autoclave Leaching of Copper Smelting Production Fine Dust. *Metallurgist* **2019**, *62*, 1244–1249. [CrossRef]
43. Mamyachenkov, S.V.; Anisimova, O.S.; Kostina, D.A. Improving the precipitation of arsenic trisulfide from washing waters of sulfuric-acid production of copper smelteries. *Russ. J. Non Ferr. Met.* **2017**, *58*, 212–217. [CrossRef]
44. Nabojčenko, S.S. (Ed.) *Avtoklavnaja Gidrometallurgija Cvetnych Metallov*; GOU UGTU-UPI: Ekaterinburg, Russia, 2002; ISBN 978-5-321-00065-6. (In Russia)
45. Levenspiel, O. *Chemical Reaction Engineering*, 3rd ed.; Wiley: New York, NY, USA, 1999; ISBN 978-0-471-25424-9.
46. Rogozhnikov, D.A.; Shoppert, A.A.; Dizer, O.A.; Karimov, K.A.; Rusalev, R.E. Leaching Kinetics of Sulfides from Refractory Gold Concentrates by Nitric Acid. *Metals* **2019**, *9*, 465. [CrossRef]

47. Dickinson, C.F.; Heal, G.R. Solid-liquid diffusion controlled rate equations. *Thermochim. Acta* **1999**, *340*, 89. [CrossRef]
48. Gok, O.; Anderson, C.G.; Cicekli, G.; Cocen, E.I. Leaching kinetics of copper from chalcopyrite concentrate in nitrous-sulfuric acid. *Physicochem. Probl. Mi.* **2014**, *50*, 399–413. [CrossRef]
49. Hidalgo, T.; Kuhar, L.; Beinlich, A.; Putnis, A. Kinetic study of chalcopyrite dissolution with iron(III) chloride in methanesulfonic acid. *Miner. Eng.* **2018**, *125*, 66–74. [CrossRef]
50. Lente, G.; Fábián, I.; Poë, A.J. A common misconception about the Eyring equation. *New J. Chem.* **2005**, *29*, 759. [CrossRef]
51. Hidalgo, T.; Kuhar, L.; Beinlich, A.; Putnis, A. Kinetics and mineralogical analysis of copper dissolution from a bornite/chalcopyrite composite sample in ferric-chloride and methanesulfonic-acid solutions. *Hydrometallurgy* **2019**, *188*, 140–156. [CrossRef]

Article

Assessment of Silica Recovery from Metallurgical Mining Waste, by Means of Column Flotation

Eleazar Salinas-Rodriguez [1], Javier Flores-Badillo [1], Juan Hernandez-Avila [1,*], Eduardo Cerecedo-Saenz [1,*], Ma. del Pilar Gutierrez-Amador [2], Ricardo I. Jeldres [3] and Normam Toro [4,5]

1 Area Academica de Ciencias de la Tierra y Materiales, Universidad Autonoma del Estado de Hidalgo, Carretera Pachuca-Tulancingo km 4.5, Mineral de la Reforma, Hidalgo C.P. 42184, Mexico; salinasr@uaeh.edu.mx (E.S.-R.); jfb.cmat@gmail.com (J.F.-B.)

2 Escuela Superior de Apan, Universidad Autonoma del Estado de Hidalgo. Carr. Apan-Calpulalpan km. 8, Apan, Hidalgo C.P. 43920, Mexico; gutierrezam@yahoo.com

3 Departamento de Ingenieria Quimica y Procesos de Minerales, Facultad de Ingenieria, Universidad de Antofagasta, Antofagasta 1240000, Chile; ricardo.jeldres@uantof.cl

4 Departamento de Ingenieria Metalurgica y Minas, Universidad Catolica del Norte, Antofagasta 1270709, Chile; ntoro@ucn.cl

5 Department of Mining, Geological and Cartographic Department, Universidad Politecnica de Cartagena, 30202 Murcia, Spain

* Correspondence: herjuan@uaeh.edu.mx (J.H.-A.); mardenjazz@yahoo.com.mx (E.C.-S.); Tel.: +52-771-125-8798 (J.H.-A.); +52-771-195-6006 (E.C.-S.)

Received: 20 November 2019; Accepted: 27 December 2019; Published: 2 January 2020

Abstract: The generation of mining waste commonly led to the use of spaces for its disposal. Challenges like mitigating the damage to surrounding communities have promoted the need to reuse, recycle and/or reduce their generation. Besides, these residues may become a source of materials, which are capable of being recovered and reused in several industries, minimizing the environmental impact. In the mining region of Pachuca, Mexico, waste from the mining industry have been generated for more than 100 years, which have a high SiO_2 content that can be recovered for various industrial applications. This work aims to recover silica from a material of the Dos Carlos dam. A columnar system composed of two-stage of cleaning was used, considering a J_{LT} (surface liquid rate) value of 0.45 and 0.68 cm/s, respectively; while the J_g (surface gas rate) value was 0.30 cm/s for both stages. Similar bubble sizes in the range of J_g 0.10 to 0.30 cm/s, with values between 0.14 and 0.16 cm in the first stage, and 0.05 to 0.06 cm in the second one. This provided a recovery of 75.10% for all the allotropic phases of silica (quartz, trydimite, and cristobalite) leaving a concentration of 24.90% of a feldspathic phase (orthoclase), as flotation tails.

Keywords: silica recovery; column flotation; mining waste; waste reprocessing

---

1. Introduction

Since the pre–Hispanic era to date, Mexico has been a mining country, and this activity has pointed more on precious metals ores. Particularly in the mining district of "Pachuca-Real del Monte" in the State of Hidalgo, the mineral processing has included several technologies ranging from the benefit of "Patio", froth flotation, and cyaniding, generating over 460 years of tailings that is deposited in four mining dams. "Dos Carlos" is one of the oldest [1]. The principal species found in this kind of residues is Si with content from 56% to 70%, and the presence of Ag and Au, which has given added value to these waste [1,2].

Due to the above pointed, some works have devoted to recovering the metallic values involved in these waste [1,3,4], leading so a second residue with important amounts of SiO_2 and feldspars, which could also be of commercial interest.

As was pointed, the "Dos Carlos" dam is one of the oldest in the Pachuca-Real del Monte Mining District, established since 1912. Consequently, this dam has an approximate amount of waste (coming from the processes of grinding, flotation and cyanidation) of more than 14.3 million tons [1]. These residues have a chemical composition corresponding to 70.43% SiO_2, 7.32% Al_2O_3, 2.32% Na_2O, 0.08% K_2O, 0.69% CaO, 0.54% MgO, 0.73% MnO, 2.80% Fe_2O_3, 2.41% FeO, 0.53% TiO_2, 0.12% P_2O_5, 55 ppm of Ag, and 0.58 ppm of Au [5,6]. Similarly, the described residues show a coarse granulometry of the order of 60% accumulated up to the 270 mesh (53 μm) [5].

These residues constitute an important source of materials that can be effectively reused, considering that the contents of silica and feldspars can be recovered; the silver and gold and finally the residual oxides. Therefore, the recovery and the obvious market for precious metals [7] results in direct profits, while the demand for silica and feldspars includes the ceramics, chemistry, glass, abrasives, silicate acid refractories, additives [8], and finally the oxides for the cement and glass industry.

Regarding the flotation of material from natural siliceous sand deposits, there are different methods and uses of reagents, depending on the impurities contained in the siliceous sand. For feldspathic sands with quartz, fatty acids are mainly used as collectors in a neutral or slightly basic medium during the primary phase of elimination of iron and titanium oxides. In the secondary stage, it is necessary the elimination of micas and feldspars, which is achieved from primary amine collectors in acid medium, by the addition of sulfuric acid. In addition, in the removal of feldspars, ammonium bifluorofuroide is used as a depressant agent, achieving that the silica law reaches 98% of purity, with minimum additions of the order of 0.97% Al_2O_3 and 0.05% Fe_2O_3 [9].

Moreover, the separation during the second phase is achieved by acidic pH of the order of 2 to 2.5, using H_2SO_4 or failing with HF, in a concentration of 2000–2500 ppm. The latter is also a depressant of the feldspathic portion, providing that the silica particle can be cleaned superficially from metal oxides; this modified process generated a silica content of 98.36% [10].

Besides, it is also possible to use cationic reagents based on complex sulfates, for the flotation of the silica from sands containing quartz and feldspar; by using silica activating agents such as Ca^{2+}, Ba^{2+} and Sr^{2+}, thus achieving an adequate flotation of the silica without the use of HF [11].

In a nutshell, the glass and chemical industries require a high standard of silica quality that means composition over 99.8%, with impurities less than 0.02% of hematite and alumina, as well as values lower than 0.1% of lime and magnesia [12]. Nevertheless, comparing with the previous investigations, the silica recoveries are still far from the amount expected to place it in the industries mentioned above.

Since flotation is a process widely used for mineral concentration, the study of size and quantity of bubbles, the particle sizes and hydrophobic conditions, and many other factors are of great importance [13]. According to this, mineral extraction from tailing is a technological challenge in mining and metallurgical operations. It needs an in-depth study of technical and economic feasibility [14], being the flotation an adequate technique for the concentration and recovery of minerals from ores and waste. The flotation in column also represents an appropriate option for the recovery and or concentration of minerals, where the model studied is related to a stable state behavior, taking in account variables such as dynamic mass balance, type of pulp, markers, bubble distribution and flux, bubble size, cations effect, and bubble coalescence, among others [15–17].

In this context, the principal aim of this work is to recover silica by column flotation, using the tailings form the "Dos Carlos" dam. This proposal offers significant advantages since it reuses a material considered waste. Thus it will be given a higher value, and unlike the processes used commercially, as described in previous paragraphs, it will be more advantageous and viable.

2. Materials and Methods

Firstly, the material used in this work was waste coming from the mining industry of the mining district of Pachuca-Real del Monte, which was sampled selectively at the "Dos Carlos" dam, using the channel technique. The sample so collected was homogenized by quartering and then characterized to disclose the mineral species involved and its chemical composition. The whole characterization was carried out by X-ray diffraction (XRD), using an Equinox 2000 diffractometer (INEL, Artenay, France) with CoKα_1 radiation (λ = 1.78901 Å), at 30 kV, 20 mA and a scanning speed of 22 θ/min. Phase identification was based on the COD Inorganics 2015 and Cements 2014 databases included in the Crystallography Open Database; Match! software used in this procedure (v.1.10, Crystal Impact, Bonn, Germany).

In addition, the shape and size of waste particles were determined by a scanning electron microscopy, in conjunction with energy dispersive spectrometry of X-Rays (SEM-EDS). In this case was used a JEOL scanning electron microscope, model JSM-IT300 (JEOL Ltd., Tokyo, Japan) which has an OXFORD X-ray detector (OXFORD Instruments, Oxfordshire, UK), employing an acceleration voltage of 30 kV.

Secondly, the chemical determination was executed by atomic absorption spectroscopy (AAS) using a spectrophotometer Perkin Elmer model 2380. In addition, X-ray fluorescence (XRF) chemical analysis was run by a SIEMENS spectrophotometer belonging to the ceramic company Morgan SA of CV, located in Pachuca, Mexico. Samples were mixed with lithium hexaborate in a 5:1 ratio and then fused in a furnace at 1473 K to obtain a button, which was irradiated with an Rh anticathode using the following crystals for their evaluation; LiF200, Ge111, and T1AP. Calibration was done using different oxides of known composition.

On the other hand, the recovery of the silica was carried out in a flotation column with two cleaning stages (Figure 1), using a concentration of 3% solids, and 50 ppm of analytical grade dodecylamine of the brand Sigma-Aldrich ($C_{12}H_{25}$-NH_2). The frother reagent was methyl isobutylcarbonyl or MIBC in a concentration of 60 ppm (Alkemin), adjusting to a J_g value of 0.3 cm/s, a J_{LT} of 0.45 and 0.68 cm/s, respectively, and without the addition of any depressant.

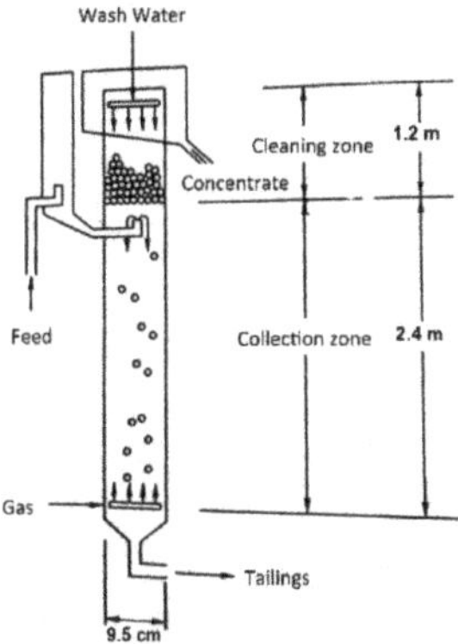

Figure 1. Schematic representation of the flotation column used in this study.

The respective values of ε_g (retained gas), D_b (particle size), and S_b (surface bubble rate) were determined from the experiments; taking samples in each J_g value, in the feed streams, tails and concentrate.

Finally, the concentrates resulting from the batch tests were characterized by XRD. The samples were sieved to obtain sizes between 74 and 53 μm (Tyler® 200 and 270 series meshes, respectively). Once this range of sizes was obtained, 1 g of sample was weighed and compacted in an aluminum sample holder for further analysis by XRD, determining the mineral phases present.

3. Results

The results obtained by SEM-EDS are shown in Figure 2, where can be seen the general morphology of samples (Saning Electron Microscopy–Secondary Electrons (SEM-SE)) and its punctual composition obtained by EDS (a) before flotation stage, and (b) after flotation process. It can be noted that after the flotation of material, the residue resting has been enriched with SiO_2.

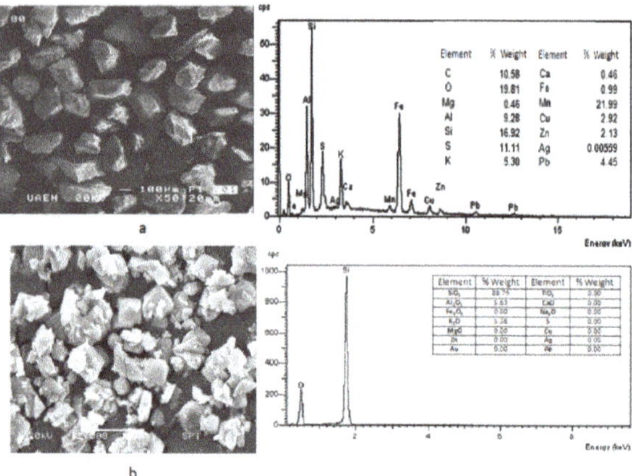

Figure 2. SEM-SE images and energy dispersive spectrometry (EDS) analysis executed on waste; (**a**) before flotation, and (**b**) after flotation process.

In the case of the X-ray fluorescence, the obtained results are shown in Table 1. It is appreciated that the dominant species contained were silica, alumina, hematite, and potassium oxide, having also minor species such as magnesium oxide, calcium oxide, lead, zinc copper, silver, and gold.

Table 1. X-ray fluorescence (XRF) chemical analysis of material used for the flotation of silica.

Element	Waste Prior Flotation%	Waste after Flotation%
SiO_2	73.3	91.2
Al_2O_3	6.5	4.6
Fe_2O_3	2.8	0.0
MgO	0.6	0.0
MnO	0.7	0.0
CaO	0.7	0.0
Na_2O	0.08	0.0
K_2O	0.08	4.2
P_2O_5	0.1	0.0
SO_3	0.9	0.0
ZnO	0.05	0.0
Ag	51 (ppm)	0.0
Au	0.6 (ppm)	0.0

Figure 3 (1st stage line) displays the gas retained in the columnar flotation system, for the recovery in the batch system during the first stage of cleaning, with a flow of liquid in the stream of tails (J_{LT}) of 0.45 cm/s, and applying a gas flow (J_g) of 0.3 cm/s. Additionally, the lowest concentration of gas retained appeared in the central area of the column (manometers 2 and 3), because in this zone, the bubble pack had a higher rate of ascent. On the other hand, in the upper part of the column, an increase of gas retained in the system was observed, with a value of 3.3%, decreasing the rising speed of the bubble package. The percentage of global retained gas was 2.3%.

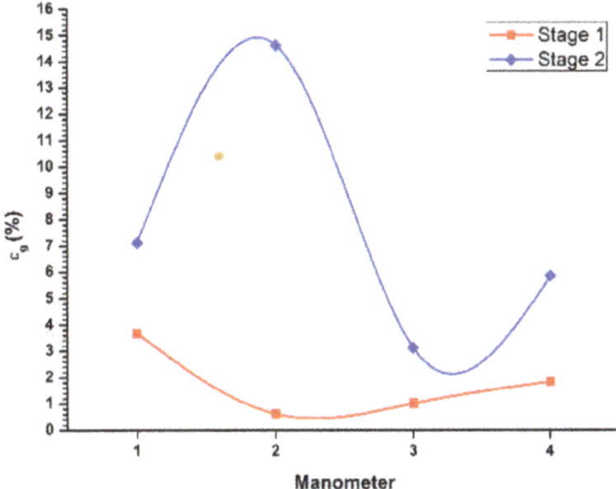

Figure 3. Determination of ε_g, in the bath flotation column of the material of the "Dos Carlos" dam, for J_{IT} of 0.446, 50 ppm of $C_{12}H_{25}NH_2$ as collector, and 60 ppm of MIBC as foaming agent.

Likewise, same Figure 3 (2nd stage line) shows the determination of retained gas (ε_g) in the columnar flotation system, for the recovery in batch system during the second cleaning stage, with a flow of liquid in the queuing stream (J_{LT}) of 0.45 cm/s, applying the gas flow (J_g) of 0.3 cm/s. In the central zone of the column (manometers 2 and 3), the highest concentration of gas retained in the system was presented, where the bubble packing has a lower acceleration. Besides, the values of gas retained at the ends of the column were similar at the top, which were slightly higher. The global percentage of gas retained was set at 15.2%.

Similarly, Figure 4 shows the bubble size (D_b) generated in the disperser, for the recovery in a batch system, with a flow of liquid in the tail stream (J_{LT}) of 0.45 cm/s, applying the gas flow (J_g) of 0.3 cm/s. The bubble size decreased between the two cleaning stages, thus generating an increase in gas retained in the system, as well as an increase in the superficial flow of bubbles. This, in turn, decreases the rising speed of the bubble pack, causing a lower recovery in the second stage.

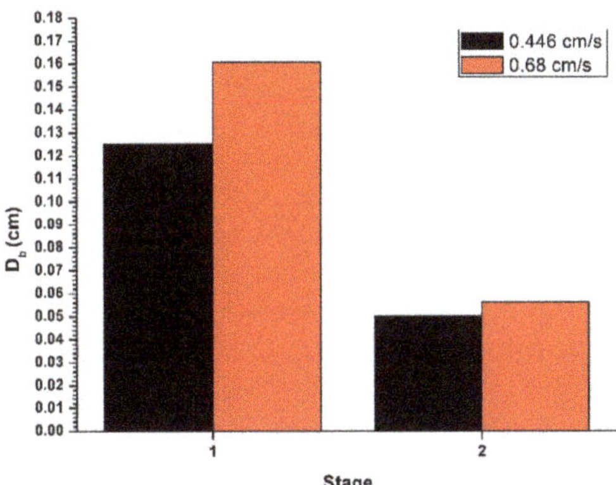

Figure 4. Determination of D_b in the batch flotation column of the material of the Dos Carlos dam, for J_{IT} of 0.446 & J_{IT} of 0.68 cm/s, 50 ppm $C_{12}H_{25}NH_2$ as a collector, and 60 ppm MIBC as foaming agent.

In the same Figure 4, it is shown the determination of the bubble size (D_b) formed in the disperser, for the recovery in a batch system, with a flow in the tail stream (J_{LT}) of 0.68 cm/s, applying the flow of gas (J_g) of 0.3 cm/s. The bubble size decreased among the two cleaning stages, thus causing an increase in gas retained in system, as well as an increment in the superficial flow of bubbles. This caused a reduction in the bubble rising and hence a lower recovery in the second stage.

Figure 5 (1st stage line) shows the determination of gas retained in the columnar flotation cell, for the recovery in a batch system through the first cleaning stage, with a flow of liquid in the tail flow (J_{LT}) of 0.68 cm/s, applying the gas flow (J_g) of 0.3 cm/s. There was a higher concentration of gas retained in the lower part of the system (manometers 1 and 2), thus creating a low ascent velocity of the bubble pack. Likewise, the middle and upper parts, presented an increase in the speed of ascent of the bubble pack, the middle section being the fastest rise. This makes the bubble pack disperse, increasing the possibility of recovery. Regarding the global retained gas, a value of 1.84% was found, generally giving a higher ascent rate, although with a low bubble concentration that recovered the mineral and with other bubbles of considerable size.

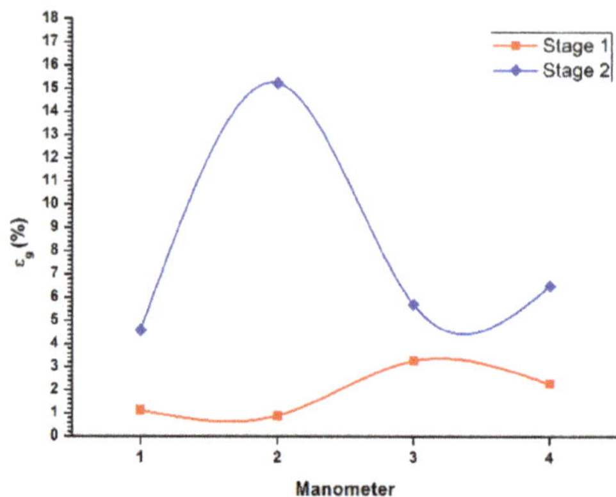

Figure 5. Determination of ε_g, in the bath flotation column of the material of the "Dos Carlos" dam, for J_{IT} of 0.68, 50 ppm of $C_{12}H_{25}NH_2$ as collector and 60 ppm of MIBC as foaming agent.

Similarly, Figure 5 (stage 2 line) illustrates the determination of retained gas (ε_g) in the columnar flotation system, for the recovery in the batch system during the second cleaning stage, with a flow of liquid in the tail stream (J_{LT}) of 0.68 cm/s, applying the gas flow (J_g) of 0.3 cm/s. In the middle zone of the flotation column (manometers 2 and 3), there was an increase in the concentration of the retained gas. This decelerated the ascent speed, and the recovery decreased, overloading the surface of the bubble with mineral. On the other hand, at the end of the flotation column, low concentrations of retained gas were present, and, therefore, higher rising rates of the bubble package. This result was more evident in the upper part of the column. Concerning global retained gas, a value of 5.9% is shown.

Figure 6 (stage 1) shows the determination of the bubble surface flow (S_b) from ε_g and D_b, for the recovery in batch system, with a flow of liquid in the tail stream (J_{LT}) of 0.45 cm/s, applying the gas flow (J_g) of 0.3 cm/s. The bubble surface flow increased between the two cleaning stages, thus generating an increase in gas retained in the system. Such effect was provided by a decrease in the initial bubble size from the first stage to a final one, in the second stage. Moreover, in Figure 6 (stage 2) is exposed the determination of the bubble surface flow (S_b) from ε_g and D_b, for the recovery in a batch system, with a flow in the stream of tails (J_{LT}) of 0.68 cm/s, applying the gas flow (J_g) of 0.3 cm/s. The bubble surface flow increased within the two cleaning stages, thus causing an increase in gas retained in the system. Such effect was provided by a decrease in the initial bubble size from the first stage to a final one, in the second stage. This, in turn, caused a reduction in the rising rate the bubble pack and, accordingly, a lower recovery in the second stage.

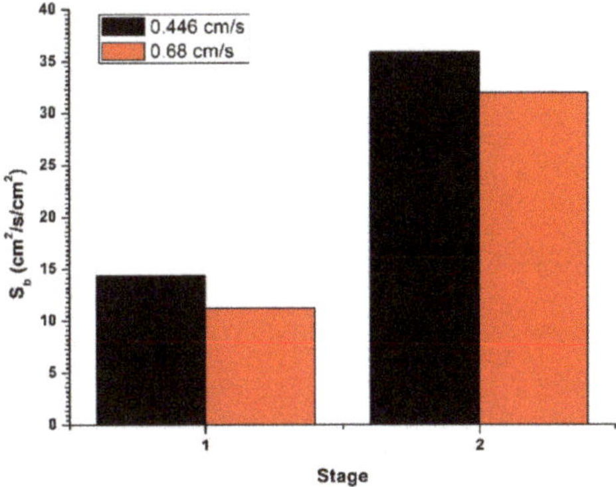

Figure 6. Determination of S_b in the batch flotation column of the material of the Dos Carlos dam, for J_{IT} of 0.446 & J_{IT} of 0.68 cm/s, 50 ppm $C_{12}H_{25}NH_2$, as a collector and 60 ppm MIBC as foaming agent.

Figure 7a illustrate the X-ray diffractogram corresponding to the first flotation stage at a J_{LT} = 0.446 cm/s. The major phases are Quartz, Orthoclase, and Tridymite, being the Cristobalite the remaining phase.

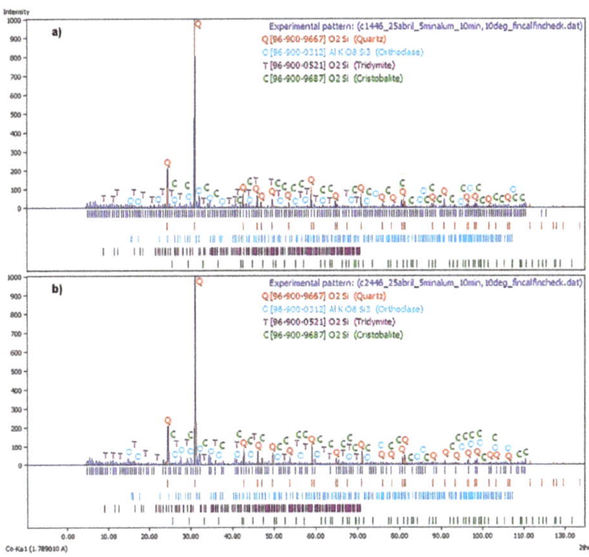

Figure 7. X-ray diffractogram of the material of the "Dos Carlos" dam recovered from the (**a**) first flotation stage at J_{IT} = 0.47 & cm/s, and (**b**) second stage of flotation at J_{IT} = 0.45 cm/s.

In Figure 7b, the X-ray diffractogram corresponds to the second flotation stage at a J_{LT} = 0.45 cm/s. It is observed that the principal phases were quartz, orthoclase, and tridymite. The cristobalite was a remaining phase. This spectrum displayed similar behavior to the previous.

On the other hand, Figure 8a shows the X-ray diffractogram corresponding to the first flotation stage at a J_{LT} = 0.68 cm/s. The principal phases were quartz, orthoclase, and berlinite, where the albite was the remaining phase. Note that due to an increase in liquid flow rates, there were more feldspathic phases and a phase analogous to quartz, although the major one was still quartz, concerning the same stage and at a J_{LT} = 0.47 cm/s.

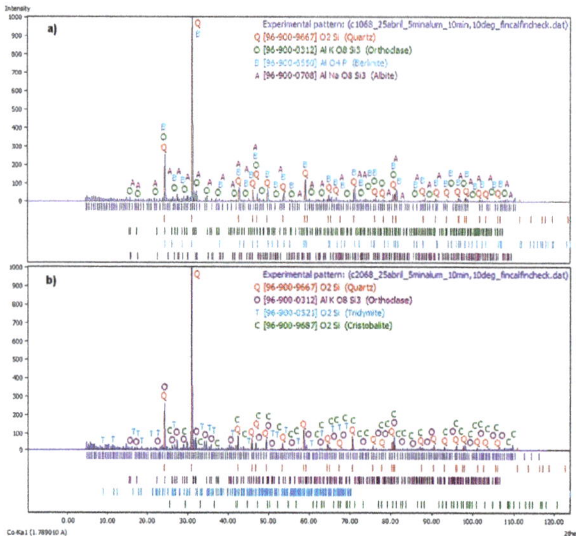

Figure 8. X-ray diffractogram of the material of the "Dos Carlos" dam recovered from the (**a**). First flotation stage at J_{lT} = 0.68 cm/s, and (**b**). Second stage of flotation at J_{lT} = 0.68 cm/s.

Figure 8B shows the X-ray diffractogram corresponding to the second flotation stage at a J_{LT} = 0.45 cm/s. It is observed that the major phases were quartz, orthoclase, and tridymite, while cristobalite was the remaining phase. The behavior was similar to the two stages of the batch process J_{LT} = 0.45 cm/s.

Regarding the percentages by weight of each phase, it is recognized in Table 2 that at a lower liquid velocity, a higher concentration of quartz was present. In the second stage at a J_{LT} = 0.68 cm/s, there was an increase in the level of quartz and orthoclase, cristobalite and tridymite also appearing. This might have happened due to the elimination of albite and berlinite. For flotation at lower liquid velocity, as well as a value of J_{LT} = 0.45 cm/s, a similar behavior was found between the two stages, since it increased the concentration of allotropic phases of the silica, decreasing the concentration of orthoclase.

Table 2. Present phases and their percentage in weight, obtained in the two stages of batch flotation, at values of J_{LT} = 0.68 and 0.45 cm/s, respectively. The phases were obtained by means of XRD.

Entry Number	Phase	Formula	Flotation J_{LT} = 0.68 cm/s		Flotation J_{LT} = 0.45 cm/s	
			1st Stage	2nd Stage	1st Stage	2nd Stage
96-900-9667	Quartz	SiO_2	56.5%	61.2%	69.9%	69.6%
96-900-9687	Cristobalite	SiO_2	0.0%	1.6%	0.8%	1.6%
96-900-0521	Tridymite	SiO_2	0.0%	7.6%	3.5%	3.9%
96-900-0312	Orthoclase	$KAlSi_3O_8$	27.2%	29.6%	31.8%	24.9%
96-900-0707	Albite	$NaAlSi_3O_8$	2.7%	0.0%	0.0%	0.0%
96-900-6550	Berlinite	$AlPO_4$	13.6%	0.0%	0.0%	0.0%

4. Discussion

In this work, a flotation column was used for the separation of silica and feldspars from mining waste, since flotation columns have advantages over conventional cells [18], because they do not use mechanical agitation, they operate with bubbles much smaller than common cells, and with long residence times, an improvement in the degree of recovery is achieved. Similarly, another additional advantage is that flotation columns can treat a large volume of ore, occupying a small area, as well as very fine minerals, at very attractive investment and operation costs [19,20].

Consequently, the main disadvantage of the flotation columns lies in their low capacity to recover thick particles [21], but this is not the case where there is a fine grain size.

Therefore, the use of flotation columns for the recovery of silica and feldspar, offers many advantages, especially in the improvement in grade and recovery for an isochemical mineral like the one studied here.

Due to the need of understanding what happens during quartz flotation, precious studies related to this topic using DDA, showed that flotation radically decrease with sizes from hydrophobic aggregate, which in turn increases with the length of the collector chain [22]. Similarly, flotation depends more significant on particle and bubble sizes. For instance, when there are small bubbles, high flotation rates can be reached at very low power intensities. On the other hand, when larger bubbles are present, flotation rate increases with increasing of the surface area's intensity, when both fine and intermediate particle sizes are treated. In addition, an optimal intensity of power was reached for thicker particles, just as is the case found in this work [23]. In the same way, flotation was not affected by the particle size, although we did the study with fine and thick particle sizes.

For SiO_2 flotation, the influence of ions (type, concentration, and their combination) on the bubble coalescence disclosed that the hydrophobic attraction was reduced with the increase of collector concentration. Due to the decrease in the solubility of the gas phase, the change of the superficial tension and the rheology of the thin liquid film in the gas–water and ion–water interfaces, depending on the hydrodynamic condition (static condition) [16]. In the case of this study, a low concentration of retained gas was generated. Therefore, a continuous recovery was obtained, due to the existence of a good flux rate, which was generated for the bubble pack, and also due to the minimum concentration of sulphur species and the Fe_2O_3, when using $C_{12}H_{25}NH_2$ as collector, because at low values of J_g, especially 0.3 cm/s, a concentration of 81.66% was reached joint to a decrease to zero concentration of Fe_2O_3.

Similarly, some researchers [24] did conventional flotation experiments with a mixture of feldspars and quartz in diluted solutions of hydrofluoric acid (HF), obtaining important recoveries for quartz (95.4% quartz recovery with a global recovery of 88.6%), and a concentration of feldspar of 99.9% (95.9% for albite and 95.5 for microcline). In this work during the batch flotation in column a 75.13% of recovery was reached for all the allotropic phases of silica (quartz, tridymite and cristobalite), and of 24.87% of a feldspathic phase (orthoclase), which is favourable to environment because there is no use of acids, activators and depressants, combined with a low concentration of collecting agents and foaming agents.

Consequently, as is shown in this work, this new concept could be used to produce a quartz concentrate, which also could be of great value as an important stage of cleaning to remove the last remnants of feldspars from the product. In the same way, this process has also probed for recycling porpoises. Some researches [14] have pointed that the used model for the flotation, are efficient for this kind of processes used for iron minerals waste. Larsen and Kleiv [24], found that silica was recovered quite successfully from a slag coming from the production of silicon in an electric reduction smelting furnace.

5. Conclusions

In the medium dodecylamine ($C_{12}H_{25}$-NH_2), it was studied the recovery of silica from the waste of the "Dos Carlos" dam, in a batch flotation system with two cleaning stages. The experimental condition considered J_g of 0.3 cm/s, in both cleaning stages, and J_{LT} of 0.45 and 0.68 cm/s, for each one. Similar bubble sizes were obtained for the range of J_g 0.1 to 0.3 cm/s of the effect of JL, generating a similar retained gas concentration and therefore promoting a steady recovery, since there was a comparable rising speed of the bubble pack.

In the characterization by XRD, a particular phenomenon was presented since at a lower surface velocity of the liquid and especially at a J_{LT} value of 0.45 cm/s, an area of the higher collection was introduced in the lower half of the column. So in the batch column flotation, consisting of two stages, 75.13% of recovery was achieved for all the allotropic phases of the silica (quartz, tridymite and cristobalite). This leaves a concentration of 24.87% of a feldspathic phase (orthoclase) as a remnant. Finally, according the obtained results, we can propose the use of the silica obtained in the cement industry, principally.

Feldspar residue with varying amounts of calcium, potassium or sodium [25,26], has a low melting point 758 K (485 °C), and it is used in the manufacture of glass, since alumina gives it hardness, durability and resistance to chemical corrosion. The second use of feldspar is in the ceramic industry, where its use as a flux for ceramic vitrification. Similarly, it is known that according to the content of potassium, the greater the amount of it, the material has a better performance as a cementing agent [27].

The advantage of using feldspar obtained as a byproduct of an epithermal reservoir like the one studied here, offers advantages over one of residual origin; since feldspar of isochemical primary origin is a pure material [25]. Additionally, the bounded particle size is of importance for the ceramic industry, because the ideal size should be greater than 170–200 meshes (74–88 microns), according to the US Geological Survey [27,28]. For this case, the waste material recovered here, exceeds 170 meshes (88 microns), and due to the comminution of the waste is more competent, mechanically. Likewise, due to surface exposure, the secondary feldspar usually has hematite alterations that affect the color of the ceramics and the flux function is modified because the iron is high temperature.

Consequently, recovered feldspars can be considered mixed because of their sodium, potassium and calcium contents. Therefore, its use could be in the elaboration of construction material (tiles, decoration material), abrasives (sandpaper for wood), seed coating, scrubbing soaps. However, this feldspar could be used in white paste and fine porcelain ceramics, but a second flotation would have to be done to remove residual impurities such as iron, since a maximum of 1% Fe is required for this use.

On the other hand, quartz, tridymite, and cristobalite, since they are of primary origin, are usually isochemicals of greater purity, unlike the supergenic quartz of areas of residual sand, due to the great variety of quartz components that are the result of their secondary alteration, such as el, jasper, chalcedony, chert, opal, or other siliceous varieties. This is due to the fact that the chemistry of supergenic quartz is varied and depends essentially on the type of rock affected. Industrial sand and gravel, often called "silica," "silica sand," and "quartz sand," include sands and gravels that are high in silicon dioxide (SiO_2). These arenas are used in the manufacture of glass; for hydraulic fracture, fracture, and abrasive applications, mainly.

The chemical composition and physical properties of the silica recovered from the "Dos Carlos" dam, show that it can be used as a raw material for the ceramic industry. Similarly, it can also be an alternative for the cement industry because of its silica content [29], and for the manufacture of building materials, as well as a substitute for mixed feldspars, in some common applications and uses.

Similarly, silica sand is used as a granular filter medium in the treatment of drinking and wastewater. Because of its hardness, it can be used in the manufacture of sandpaper, industrial abrasives, and blasting sand, in formulas of detergents, paints, concrete, and special mortars, in addition to being the basis for the manufacture of refractories based on silica [30–33].

Author Contributions: E.S.-R., J.H.-A., and E.C.-S. contributed in the conceptualization and develop of methodology, J.F.-B. contributed in methodology and data curing, M.d.P.G.-A. contributing in analysis of data, R.I.J. and N.T. contributed resources and review, J.F.-B. and E.S.-R. contributing in wrote paper, R.I.J. and N.T. contributing in funding acquisition for publishing fees. All authors have read and agreed to the published version of the manuscript.

Funding: This research received no external funding for experimental and analytical work.

Acknowledgments: Authors thank to the PRODEP–SEP for the financial support granted in the publication of this paper. Also thanks go to the UAEH for the technical support given for the realization of this work.

Conflicts of Interest: The authors declare no conflict of interest.

References

1. Salinas-Rodríguez, E.; Hernández-Ávila, J.; Cerecedo-Sáenz, E.; Arenas-Flores, A.; Reyes-Valderrama, M.I.; Roldán-Contreras, E.; Rodríguez-Lugo, V. Leaching of Silver Contained in Mining Tailings: A Comparative Study of Several Leaching Reagents. In *Silver Recovery from Assorted Spent Sources: Toxicology of Silver Ions*; Syeb, S., Ed.; King Saudi University: Riyadh, Saudi Arabia, 2018.
2. Geyne, A.R.; Fries, C. Geology and mineral deposits of the Pachuca–Real del Monte district, State of Hidalgo. In *Consejo de Recursos Minerales no Renovables*; Real del Monte y Pachuca Mining Company: Pachuca, México, 1963.
3. Hernández, A.J.; Rivera, L.I.; Patiño, C.F.; Juárez, T.J. Characterization and kinetics of the grinding of the Dos Carlos burrows in the State of Hidalgo. In Proceedings of the 31st Annual Conference of the International Precious Metals Institute 2007 and Petroleum Refining Seminar, Miami, FL, USA, 9–12 June 2007.
4. Salinas, R.E.; Patiño, C.F.; Hernández, A.J.; Hernández, C.L.; Rivera, L.I. *La problemática de los Jales en el Estado de Hidalgo*; V Congreso Nacional de Metalurgia y Materiales: Monclova, México, 2006; pp. 237–328. (In Spanish)
5. Flores, B.J.; Hernández, A.J.; Salinas, R.E.; Pérez, L.M.; Rivera, L.I.; Mireles, G.I.; Cerecedo, S.E. Preparation of blocks from tailings. In *Engineering Solutions for Sustainability. Materials and Resources II*, 1st ed.; Jeffrey, W.F., Brajendra, M., Dayan, A., Emily, A.S., Neale, R.N., Eds.; Wiley and Sons/TMS: Orlando, FL, USA, 2015; Volume 1, pp. 127–134. ISBN 978-1-119-17984-9.
6. Salinas, R.E.; Flores, B.J.; Hernández, A.J.; Vargas, R.M.; Flores, H.J.A.; Rodríguez, L.V.; Cerecedo, S.E. Design and production of a new constriction material (Bricks) using mining tailings. *J. Eng. Sci. Res. Technol.* **2017**, *6*, 225–238. [CrossRef]
7. Salinas-Rodríguez, E.; Hernández-Ávila, J.; Rivera-Landero, I.; Cerecedo-Sáenz, E.; Reyes-Valderrama, M.I.; Correa-Cruz, M.; Rubio-Mihi, D. Leaching of silver contained in mining tailings, using sodium thiosulfate: A kinetic study. *Hydrometallurgy* **2016**, *160*, 6–11. [CrossRef]
8. Galán, E.; Aparicio, P. Materias primas para la industria ceramic. In *Seminarios de la Sociedad Española de Mineralogía*, 1st ed.; Universidad de Alicante & Sociedad Española de Mineralogia: Alicante, Spain, 2006; Volume 2, pp. 31–48. (In Spanish)
9. Klyachin, V.V.; Stepanova, M.V. Production of high-quality quartz concentrate from quartz-feldspar sands. *Glas. Ceram.* **1982**, *39*, 515–516. [CrossRef]
10. Klyachin, V.V. The selective flotation of quartz? feldspar sands by a new method. *Glas. Ceram.* **1969**, *26*, 174–176. [CrossRef]
11. El-Salmawy, M.; Nakahiro, Y.; Wakamatsu, T. The role of alkaline earth cations in flotation separation of quartz from feldspar. *Miner. Eng.* **1993**, *6*, 1231–1243. [CrossRef]
12. Stulov, A. (Universidad de Concepción, Instituto de Investigaciones Tecnológicas, Concepción, Región del Bío Bío, Chile). Personal communication, 1963.
13. Alvarez-Silva, M.; Vinnett, L.; Langlois, R.; Waters, K. A comparison of the predictability of batch flotation kinetic models. *Miner. Eng.* **2016**, *99*, 142–150. [CrossRef]
14. Büttner, P.; Osbahr, I.; Zimmermann, R.; Leißner, T.; Satge, L.; Gutzmer, J. Recovery potential of flotation tailings assessed by spatial modelling of automated mineralogy data. *Miner. Eng.* **2018**, *116*, 143–151. [CrossRef]
15. Bouchard, J.; Desbiens, A.; Del Villar, R. Column flotation simulation: A dynamic framework. *Miner. Eng.* **2014**, *55*, 30–41. [CrossRef]
16. Yianatos, J.; Vinnett, L.; Panire, I.; Alvarez-Silva, M.; Díaz, F. Residence time distribution measurements and modelling in industrial flotation columns. *Miner. Eng.* **2017**, *110*, 139–144. [CrossRef]
17. Filippov, L.O.; Javor, Z.; Piriou, P.; Filippova, I.V. Salt effect on gas dispersion in flotation column—Bubble size as a fucntion of turbulent intensity. *Miner. Eng.* **2018**, *127*, 6–14. [CrossRef]
18. Bergh, L.; Yianatos, J. Control alternatives for flotation columns. *Miner. Eng.* **1993**, *6*, 631–642. [CrossRef]
19. Suárez, C.G.A.; García, R.E.; Amariz, B.J.J.J. Column flotation as technique for the benefit of fine minerals. *Tecnura* **2005**, *9*, 4–15. (In Spanish)
20. Dávila, P.G.O. Producción de Concentrados de Hierro de Alta Ley, Haciendo Uso de Columnas de Flotación. *Rev. Met. UTO* **2010**, *29*, 19–28. (In Spanish)

21. Azañero, O.A.; Núnez, J.P.A.; León, D.E.; Morales, V.M.; Jara, I.J.; Rendón, L.J.L. Avances en flotación columnar. *Rev. Inst. Invest. Fac. Ing. Geo. Min. Met. Geogr.* **2012**, *6*, 82–90. (In Spanish)
22. Roberia, S.D.M.A.; Magalhães, B.C.A. Importance of collector chain length in flotation of fine particles. *Miner. Eng.* **2018**, *122*, 179–184. [CrossRef]
23. Massey, W.; Harris, M.; Deglon, D. The effect of energy input on the flotation of quartz in an oscillating grid flotation cell. *Miner. Eng.* **2012**, *36*, 145–151. [CrossRef]
24. Larsen, E.; Kleiv, R. Flotation of quartz from quartz-feldspar mixtures by the HF method. *Miner. Eng.* **2016**, *98*, 49–51. [CrossRef]
25. Guilbert, J.M.; Park, C.F. *The Geology of Ore Deposits*, 1st ed.; Waveland Press, Inc.: New York, NY, USA, 2007; p. 985.
26. Harben, P.W.; Kuzvart, M. *Industrial Minerals. A Global Geology*; Metall Bulletin PLC: London, UK, 1996; pp. 168–174.
27. *U.S. Geological Survey Mineral Commodity Summaries*; USGS: Reston, VA, USA, 2018; p. 200.
28. Lefond, S.J. *Industrial Minerals and Rocks*, 5th ed.; American Institute of Mining, Metallurgical, and Petroleum Engineers: New York, NY, USA, 1983; pp. 709–722.
29. Mexican Official Standard Number C-414-ONNCCE-1999. retrieved 03/03/17. Available online: http://www.imcyc.com/cemento/ (accessed on 20 November 2019).
30. Norma ASTM-C-24-56 retrieved 05/03/15. Available online: https://es.scribd.com/document/328102093/Normas-Astm-c-resumen (accessed on 20 November 2019).
31. *Ceramic DataBook*; Gordon and Breach Science Publishers: Montreux, Switzerland, 1987; p. 105.
32. ASTM Standard C-24-56. Available online: https://books.google.com.mx/books?id=WRg_AAAAQBAJ&pg=PA748&lpg=PA748&dq=ASTM+Standard+C-24-56&source=bl&ots=7-Bi25hrkR&sig=ACfU3U3HO2rBDgkGowHdS7HV5j4P39K7Ow&hl=es&sa=X&ved=2ahUKEwjkzZid_t3mAhUQIKwKHcSBBbcQ6AEwAHoECAgQAQ#v=onepage&q=ASTM%20Standard%20C-24-56&f=false (accessed on 20 November 2019).
33. Chang, L.L.Y. *Industrial Mineralogy: Materials, Processes, and Uses*; Prentice-Hall: New Jersey, NJ, USA, 2001; pp. 107–113.

Article

Leaching Chalcopyrite with High MnO_2 and Chloride Concentrations

David Torres [1,2], Luís Ayala [3], Ricardo I. Jeldres [4], Eduardo Cerecedo-Sáenz [5], Eleazar Salinas-Rodríguez [5], Pedro Robles [6] and Norman Toro [1,2,*]

1 Departamento de Ingeniería Metalúrgica y Minas, Universidad Católica del Norte, Antofagasta 1270709, Chile; David.Torres@sqm.com
2 Department of Mining, Geological and Cartographic Department, Universidad Politécnica de Cartagena, 30203 Murcia, Spain
3 Faculty of Engineering and Architecture, Universidad Arturo Prat, Almirante Juan José Latorre 2901, Antofagasta 1244260, Chile; luisayala01@unap.cl
4 Departamento de Ingeniería Química y Procesos de Minerales, Universidad de Antofagasta, Antofagasta 1270300, Chile; ricardo.jeldres@uantof.cl
5 Área Académica de Ciencias de la Tierra y Materiales, Universidad Autónoma del Estado de Hidalgo, Carretera Pachuca—Tulancingo km. 4.5, C.P. 42184, Mineral de la Reforma, Hidalgo C.P. 42184, Mexico; mardenjazz@yahoo.com.mx (E.C.-S.); salinasr@uaeh.edu.mx (E.S.-R.)
6 Escuela de Ingeniería Química, Pontificia Universidad Católica de Valparaíso, Valparaíso 2340000, Chile; pedro.robles@pucv.cl
* Correspondence: ntoro@ucn.cl; Tel.: +56-552651021

Received: 5 December 2019; Accepted: 4 January 2020; Published: 9 January 2020

Abstract: Most copper minerals are found as sulfides, with chalcopyrite being the most abundant. However; this ore is refractory to conventional hydrometallurgical methods, so it has been historically exploited through froth flotation, followed by smelting operations. This implies that the processing involves polluting activities, either by the formation of tailings dams and the emission of large amounts of SO_2 into the atmosphere. Given the increasing environmental restrictions, it is necessary to consider new processing strategies, which are compatible with the environment, and, if feasible, combine the reuse of industrial waste. In the present research, the dissolution of pure chalcopyrite was studied considering the use of MnO_2 and wastewater with a high chloride content. Fine particles (−20 μm) generated an increase in extraction of copper from the mineral. Besides, it was discovered that working at high temperatures (80 °C); the large concentrations of MnO_2 become irrelevant. The biggest copper extractions of this work (71%) were achieved when operating at 80 °C; particle size of −47 + 38 μm, $MnO_2/CuFeS_2$ ratio of 5/1, and 1 mol/L of H_2SO_4.

Keywords: dissolution; $CuFeS_2$; chloride media; manganese nodules

1. Introduction

The most abundant type of copper mineral is chalcopyrite [1–5]. Chalcopyrite has traditionally been treated by conventional pyrometallurgical techniques [6], which consist of flotation, smelting and refining, and electrorefining [7]. These techniques yield approximately 19 million tonnes per annum [8]. Despite the high level of copper production, there is concern about the environmental contamination resulting from the application of these techniques owing to SO_2 atmospheric emissions [9,10]. Because of this, it is necessary to study more environmentally friendly hydrometallurgical alternatives [11]. The slow copper extraction rate of conventional leaching from chalcopyrite in sulfur media makes commercial scale leaching economically unfeasible [12]. This may be due to the formation of a passive layer that forms on the surface of the mineral [13–15]. There have been numerous studies on dissolving copper from chalcopyrite [16–18]. However, none of these studies have obtained positive results working at ambient temperature and atmospheric pressure [19].

The polymetallic nodules are rock concretions formed by concentric layers of hydroxides [20]. Their high content of base, critical and rare metals makes them commercially interesting [21–23]. Their metal content includes high concentrations of Co, Ni, Te, Ti, and Pt, as well rare earth elements [24, 25].

There have been few studies on acid leaching of chalcopyrite using marine nodules (MnO_2) as an oxidizing agent [26–29]. These studies showed that good copper dissolution rates of chalcopyrite can be obtained at room temperature, provided that the $MnO_2/CuFeS_2$ rate is high. Devi et al. [26,27] indicated that this is due to the galvanic interaction between chalcopyrite and MnO_2, the action of Fe^{3+}/Fe^{2+} ratio, and the formation of chlorine gas through the reaction between MnO_2 and HCl. Havlik et al. [28] showed that 4 mol/L of HCl and a 4/1 de $MnO_2/CuFeS_2$ ratio is optimal conditions to obtain good results at ambient temperature (54% of copper in 90 min).

The proposed reaction for chalcopyrite leaching with magnesium nodules is expressed as follows [29]:

$$CuFeS_{2(s)} + 2\,MnO_{2(s)} + 8\,H^+_{(aq)} + 5\,Cl^-_{(aq)} = 2\,Mn^{2+}_{(aq)} + 4\,H_2O_{(aq)} + CuCl^-_{2(aq)} + FeCl_{3(s)} + 2\,S^0_{(s)}\ \Delta G^0 = -202.6\text{ kJ} \quad (1)$$

$$Ca^{2+}_{(aq)} + SO^{2-}_{4(aq)} = CaSO_{4(s)}\ \Delta G^0 = -28.0\text{ kJ} \quad (2)$$

Equation (1) represents dissolving copper in a sulfur-chloride medium, owing to the use of sulfuric acid and the high presence of chloride (wastewater) in the system. Among the advantages of leaching in a chlorinated rather than sulphated environment is increased leaching kinetics, the generation of elemental sulfur and cupric and/or cuprous ions are stable in the form of chloride complexes. The Gibbs free energy of Equation (1), which negative, is spontaneous under normal conditions and forms a stable copper product and a non-polluting elemental sulfur residue. While the calcium in wastewater and the manganese nodules reacts with the sulfate in the system, forming Equation (2) which is spontaneous and more likely to occur under normal conditions with the elements present (higher affinity of sulfate for calcium than magnesium and manganese in solution), the calcium sulfate formed is insoluble because calcium precipitates when it comes in contact with sulfate, nitrates and other elements. Equation (1) shows a 2/1 $MnO_2/CuFeS_2$ ratio for leaching copper using manganese nodules as an oxidizing agent, which was initially proposed by Toro et al. [29], but the best conditions to leach copper is at a 4/1 $MnO_2/CuFeS_2$ ratio. The values of the Gibbs free energy were calculated using the software HSC 5.1.

Other investigations have reported the positive effect of the chloride concentration on chalcopyrite leaching [18,30–32]. Velasquez et al. [33] indicated that chloride ions play an important role in oxidizing copper and iron. The copper dissolution is improved with high chloride concentrations.

The level of energy consumed in industrial scale operations related to comminution processes, reactor design, and leaching residence time largely depend on the particle size of the working material [19]. Studies have found a positive effect of smaller particle size on chalcopyrite leaching owing to the large area of contact for leaching [34,35]. Skrobian et al. [36] conducted chalcopyrite leaching tests in agitating reactors, with the addition of 300 g/L of NaCl to all the reactors, but with different particle sizes (−40 µm, −80 + 60 µm and −200 + 100 µm) and a temperature of 100 °C. Their results indicate that particle size has a negligible effect on the copper dissolution rate from chalcopyrite.

Different researchers agree on the positive effects of higher temperature on copper dissolution from chalcopyrite in terms of substantially increasing dissolution velocity [37]. Ruiz et al. [17] used sulfate–chloride media for dissolve chalcopyrite of a particle size 12.3 µm, 20 g/L of acid, 35.5 g/L of chloride, a stirring rate of 1000 rpm and 0.3 L/min O_2 and obtained a copper dissolution rate of 90% in 180 min, with. Other studies of chalcopyrite leaching in chloride media and using oxidizing agents like cupric ions [30] and nitrates [38] have also reported good results in copper extraction at higher temperatures.

The scarcity of fresh water in arid zones is an economic, environmental and social problem [39,40]. The availability of water resources and the quality of potable water have decreased significantly owing to human activity, whose effects at the small-scale are significant for the entire basin [41]. Because of this situation the mining industry is driven to conserve the water it uses and minimize water discharges [41,42]. As well, conventional water resources that mining companies and communities compete for are limited [43]. Seawater has been shown to be a good alternative for mining, not only because of its positive effects on leaching owing to its chloride content, but also as a strategic and indispensable resource [40]. Another attractive alternative is using wastewater from desalination plants, not only because of the economic benefits, but also to avoid contaminating ocean waters [44].

There are few studies for the dissolution of chalcopyrite incorporating MnO_2 and chloride in the system [26–29], achieving positive results in the extraction of Cu at room temperature, mainly evaluating the concentration of MnO_2 in the system. Previously, Toro et al. [29] conducted an investigation in which they evaluated the use of wastewater with high chloride, seawater and manganese nodules contents, for the dissolution of chalcopyrite in an acidic medium. In this investigation, the effect on the concentration of MnO_2, chloride and agitation speed in the system was evaluated. The authors found that high levels of MnO_2 (4/1 and 5/1) allow potential values to be between 580 and 650 mV, favouring the dissolution of $CuFeS_2$, and preventing the formation of a passivating layer. However, no other fundamental variables have been evaluated to favor the dissolution of $CuFeS_2$. In the present research, we evaluated the use of wastewater with high chloride content, and MnO_2 present in manganese nodules as an oxidizing agent in leaching chalcopyrite. Also, wastewater with high chloride levels from a desalination plant was reused. The particle size and temperature were optimized.

2. Methodology

2.1. Chalcopyrite Sample

The chalcopyrite sample used in this study was the same as that used in the first part, published in Toro et al. [29]. The sample was taken selectively from a copper deposit (800 g) and then crushed in a porcelain mortar to avoid contamination. We removed the impurities by hand (with the help of a microscope). The homogenization of the material was done by sampling techniques, selecting a representative fraction of 40 g (20 g for chemical analysis and 20 g for mineralogical analysis). Through a mineralogical analysis using a Bruker brand X-ray diffractometer (Bruker, Billerica, MA, USA), automatic and computerized model of D8 determined that the sample has a purity of 99.9% as can be seen in Figure 1. Finally, a chemical analysis performed by means of an atomic emission spectrometry via induction-coupled plasma (ICP-AES) (AMETEK, SPECTRO, Boschstraße, Germany) determined 33.89% of Cu, 30.62% of Fe and 35.49% of S (See Table 1).

Table 1. Chemical analysis of chalcopyrite.

Component	Cu	Fe	S
Mass (%)	33.89	30.62	35.49

In addition, the sample was analyzed mineralogically using a Bruker brand X-ray diffractometer, automatic and computerized model of D8. In Figure 1, you can see the results of the analysis, from which it was obtained that the chalcopyrite mineral has a purity of 99.90%.

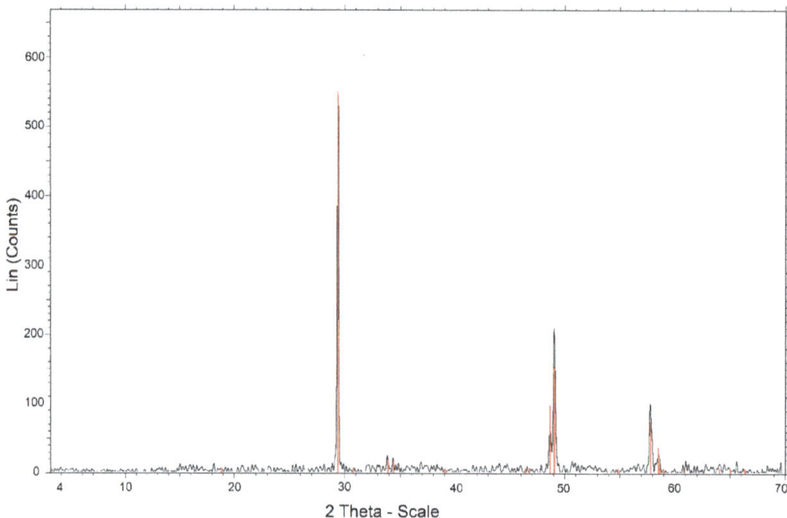

Figure 1. X-ray diffractogram for the chalcopyrite.

2.2. MnO_2 (Manganese Nodules)

The manganese nodules used in this study are the same as those used in the study by Toro et al. [29]. This sample was reduced in size with the use of a porcelain mortar until reaching a size range between −75 + 53 μm. This sample contains 15.96% of Mn. Table 2 shows the mineralogical composition. The sample was analyzed with a Bruker®tabletop M4-Tornado μ-FRX (Fremont, CA, USA). The interpretation of the μ-XRF data shows that the nodules were composed of fragments of preexisting nodules that formed their nucleus, with concentric layers that precipitated around the nucleus in later stages.

Table 2. Mineralogical Analysis of the Manganese Nodule.

Component	MgO	Al_2O_3	SiO_2	P_2O_5	SO_3	K_2O	CaO	TiO_2	MnO_2	Fe_2O_3
Mass (%)	3.54	3.69	2.97	7.20	1.17	0.33	22.48	1.07	29.85	26.02

2.2.1. Reagent and Leaching Test

The sulfuric acid used for the leaching tests was P.A. grade (Merck, Darmstadt, Germany), purity 95–97%. We also work with the use of waste water from the Aguas Antofagasta Desalination Plant, which has a concentration of 39.16 g/L of chloride. Tables 3 and 4 shows the chemical composition of waste water and sea water.

Table 3. Chemical composition of waste water.

Compound	Concentration (g/L)
Fluoride (F^-)	0.002
Calcium (Ca^{2+})	0.8
Magnesium (Mg^{2+})	2.65
Bicarbonate (HCO_3^-)	1.1
Chloride (Cl^-)	39.16
Calcium carbonate ($CaCO_3$)	13

Table 4. Reference composition of seawater, with principal ions (Modified from Cisternas and Gálvez, [45]).

Solute	g/kg of solution
Na^+	10.781
Mg^{2+}	1.283
Ca^{2+}	0.412
K^+	0.399
Cl^-	19.353
SO_4^{2-}	2.712
HCO_3^-	0.105
Br^-	0.067
CO_3^{2-}	0.014
Total	35.146

Leaching tests were carried out in a 50 mL glass reactor with a 0.01 S/L ratio. A total of 200 mg of chalcopyrite ore, with the addition of different concentrations of MnO_2 (manganese nodules), was maintained in agitation and suspension with a 5 position magnetic stirrer (IKA ROS, CEP 13087-534, Campinas, Brazil) at a speed of 800 rpm. Temperature was controlled using an oil-heated circulator (Julabo). The temperature range tested in the experiments was 25 to 80 °C. The tests were performed in triplicate, and measurements (or analyzes) were carried out on 5 mL undiluted samples using atomic absorption spectrometry with a coefficient of variation ≤5% and a relative error between 5 to 10%. The pH levels and oxidation-reduction potential (ORP) of leaching solutions were measured with a pH-ORP meter (HANNA HI-4222 (HANNA instruments, Woonsocket, RI, USA)). The ORP solution was measured in a combination ORP electrode cell of a platinum working electrode and a saturated Ag/AgCl reference electrode.

In the previous study (Toro et al. [29]) the ratio of $MnO_2/CuFeS_2$, agitation rate, H_2SO_4 concentration, and chloride concentration were evaluated. Besides, the obtained residues were analyzed, but the formation of contaminating elements was not observed. However, no other fundamental variables were evaluated, and the performance in the extraction of Fe and Mn was not mentioned. For the reasons discussed above, the leaching of $CuFeS_2$ with MnO_2 and wastewater in the present investigation continues, evaluating the particle size and temperature.

2.2.2. Effect of Particle Size

In previous studies conducted by Devi et al. [27] and Havlik et al. [28], it was shown that high MnO_2 concentrations favour the kinetics of chalcopyrite dissolution. Recently, Toro et al. [29] indicated that when working on $MnO_2/CuFeS_2$ ratios of 5/1, attractive results were obtained for short periods. Based on previous background, the effect of the chalcopyrite particle size was evaluated by adding MnO_2 at different sulfuric acid concentrations over time under the conditions shown in Table 5.

Table 5. Experimental conditions for the study of the effect of chalcopyrite particle size.

Parameters	Values
Particle size (μm)	−75 + 53, −47 + 38, −20
Time (min)	5, 20, 40, 60, 80
H_2SO_4 (mol/L)	1, 2, 3
$MnO_2/CuFeS_2$ ratio	5/1

2.2.3. Effect of Temperature

In the study conducted by Toro et al. [29], positive results were obtained when working at high ratios of $MnO_2/CuFeS_2$ (5/1). However, the effect of temperature was not evaluated to shorten leaching times or decrease MnO_2 concentrations.

This study investigated the effect of temperature (in which interval 25–80 °C) on the copper dissolution rate from chalcopyrite with the addition of MnO_2, working with a particle size of −47 + 38 μm, $MnO_2/CuFeS_2$ ratios of 2/1 and 5/1, 1 mol/L of sulfuric acid, 39.16 g/L of chloride (wastewater) and at a stirring speed of 800 rpm.

3. Results

3.1. The Effect of Particle Size on $CuFeS_2$ Dissolution

Figure 2 shows the effect on $CuFeS_2$ dissolution of particle size with the addition of MnO_2 (5/1 $MnO_2/CuFeS_2$ ratio, at different concentrations of H_2SO_4 and wastewater. It can be observed that no differences in copper dissolution rates can be obtained at particle size of −75 + 53 to −47 + 38 μm (Figure 2a,b). At small increases in the dissolution rate can be obtained by increasing the sulfuric acid concentration, with the best results obtained at 3 mol/L. This concurs with Skrobian et al. [36], who stated that particle size is irrelevant in chalcopyrite leaching in stirring reactors. However, the copper extraction rate increases with smaller particles (−20 μm), which could be due to the mechanical-chemical activation of the mineral resulting from extended milling [19], which Juhász and Opoczky [46] termed electrochemical activation. At this size, the concentration of H_2SO_4 is irrelevant.

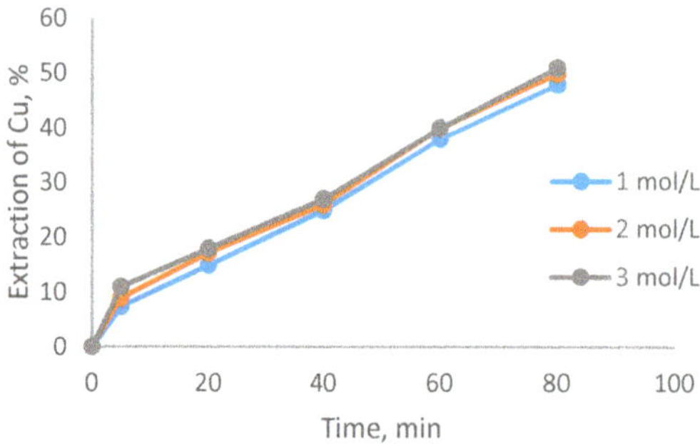

(a)

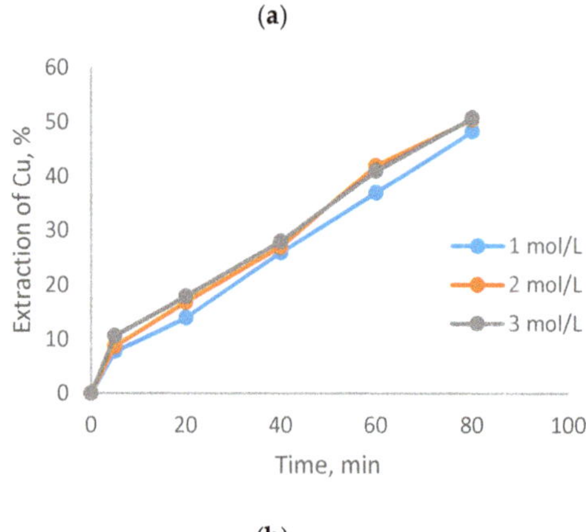

(b)

Figure 2. *Cont.*

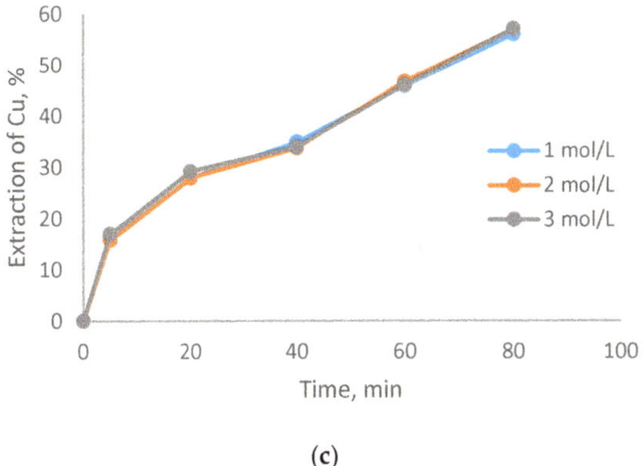

(**c**)

Figure 2. Effect of particle size and sulfuric acid concentration on copper dissolution. Particle size of: (**a**) −75 + 53 µm; (**b**) −47 + 38 µm; (**c**) −20 µm (acid concentration of 1, 2 and 3 mol/L and 25 °C).

3.2. The Effect of Temperature on $CuFeS_2$ Dissolution

Figure 3 shows the effect of temperature and $MnO_2/CuFeS_2$ ratios on $CuFeS_2$ dissolution. Dutrizac [35] stated that it is difficult to precisely determine the effect of temperature on copper dissolution from chalcopyrite in chloride media, owing to the presence of small amounts of secondary copper mineralization that can affect data interpretation. However, this problem was avoided in this study by using pure chalcopyrite. It can be seen from Figure 3 that at high temperatures (80 °C), the extraction of copper in the system is greater, with similar results obtained with $MnO_2/CuFeS_2$ ratios of 2/1 and 5/1. It can also be seen that at ambient temperature there is a significant difference in Cu extraction (About 30%) at ambient temperature between $MnO_2/CuFeS_2$ ratios of 2/1 and 5/1. The potential values for the tests at room temperature were between 540 and 590 mV, which is within the potential range where the dissolution rate of the chalcopyrite is linear (550 and 620 mV), as Velásquez-Yévenes et al. [37] noted. The potential values in the tests at temperatures of 50 and 80 °C were between 610 and 660 mV, and yielded higher copper dissolution rates. This is because high concentrations of chloride can raise the range of potential values [34]. The pH levels ranged between −0.5 and 1.4 in all the tests.

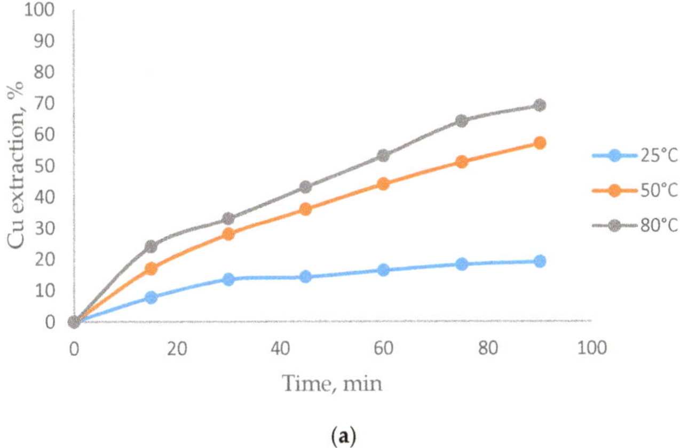

(**a**)

Figure 3. *Cont.*

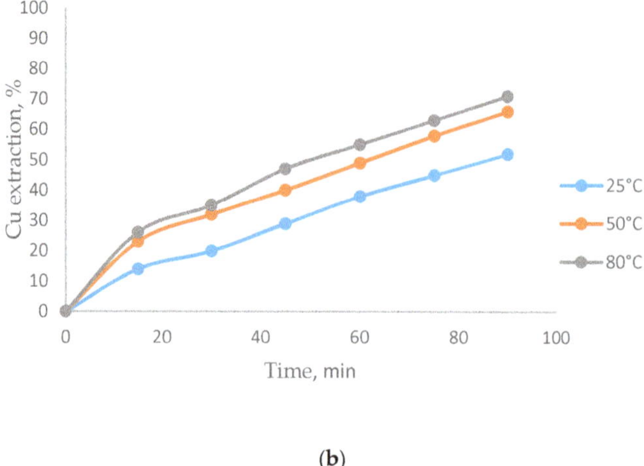

(**b**)

Figure 3. Effect of the temperature on the dissolution of Cu from chalcopyrite at different ratios of $MnO_2/CuFeS_2$ (particle size of −47 + 38 μm, ratio $MnO_2/CuFeS_2$ of 2/1 (**a**) and 5/1 (**b**), H_2SO_4 concentration to 1 mol/L and 39.19 g/L of chloride).

Manganese nodules are composed of 29.85% MnO_2 and 26.02% Fe_2O_3, which are dissolved in the acidic environment. Figure 4a shows how the MnO_2 used as an oxidising agent for the dissolution of copper, where the Mn (IV) is reduced to Mn (II). Manganese has a high extraction at a temperature of 80 °C. The manganese dissolved in the PLS can be present in two forms, such as $MnSO_4$ or as $MnCl_2$, due to the effect of sulfuric acid and/or chloride, respectively. In Figure 4b, the iron (II) present in the chalcopyrite oxidises and slowly dissolves in an environment of high concentration of sulfuric acid and high level of chloride forming ferric chloride, which is positive since it is a compound that helps the copper solution from the chalcopyrite, and the kinetics of the iron solution increases with temperature. Regarding the pH, it can be seen that lowering the acidity in the Mn solution it does not decrease the dissolution of this element. While in iron, more positive results are presented in a more acidic environment.

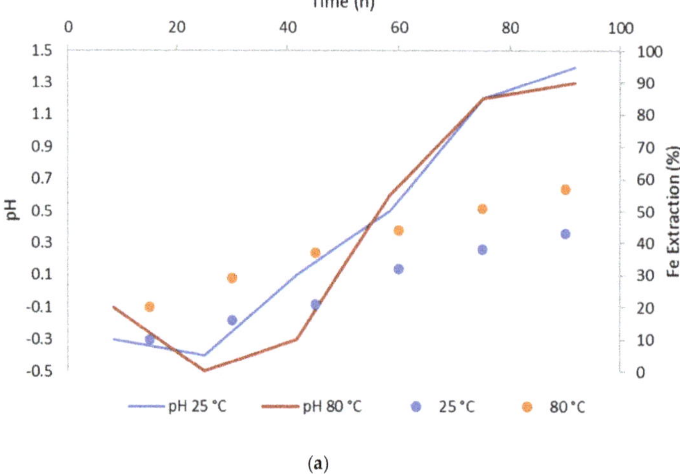

(**a**)

Figure 4. *Cont.*

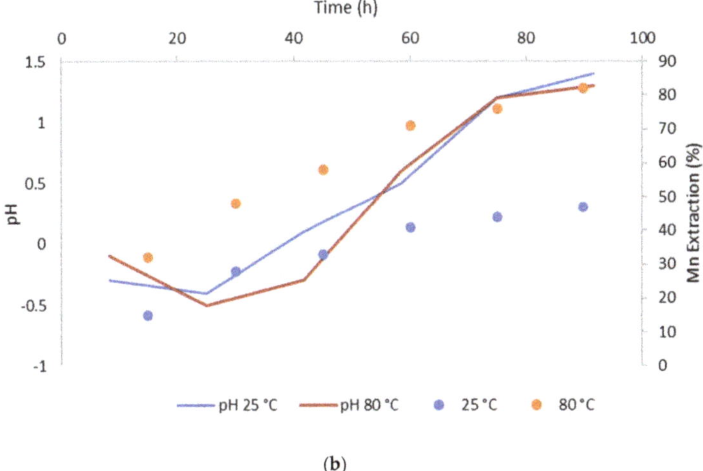

(**b**)

Figure 4. Dissolution of Fe and Mn over time from $CuFeS_2$ at room temperature (25 °C) and high temperature (80 °C) (**a**): Dissolution of Fe and its behavior at changes in pH; (**b**): Dissolution of Mn and its behavior at changes in pH).

In the big copper mining in Chile, the Fe and Mn present in the pregnant leaching solution (PLS) are considered as impurities, this because in the electro-obtaining process, the Fe reduces the efficiency of current. At the same time, the Mn generates corrosion in lead anodes. However, both impurities are controlled in metallurgical plants minimizing the problems that they could produce. Manganese and iron can be transferred to the electro-obtaining stage through physical drag generated by solvent extraction. To reduce the physical transfer of impurities, the plants must optimize the equipment to retain water trawlers in organic (A/O), in addition to adopting some operational practices like maintaining good organic quality through the treatment of organic with clay; avoid over agitation in the mixers; surfactant addition control; and maintain design parameters within the recommended range (linear speed, specific decantation flow, etc.). For the reasons stated, it is possible to apply this process at the industrial level through the conventional hydrometallurgical route (leaching, solvent extraction and electro-obtaining), since solvent extraction processes in Chile have solved this problem. Also, it works in several miners with the use of seawater or adding high concentrations of chloride in synthetic form.

4. Conclusions

This research presents the results of dissolving copper from chalcopyrite by adding MnO_2 as an oxidizing agent (manganese nodules) in a chloride medium (wastewater). As previously concluded by Devi et al. [26]; Devi et al. [27]; Havlik et al. [28] and Toro et al. [29], the addition of MnO2 and chloride in high concentrations generate a positive effect on the chalcopyrite solution. The main findings of this study are:

- There were no differences in copper dissolution rates at particle sizes between −75 + 53 and −47 + 38 µm. at different H_2SO_4 concentrations.
- Small particle size (−20 µm) increases $CuFeS_2$ dissolution kinetics, due to the mechanical-chemical activation of the mineral.
- Temperatures of 80 °C positively affect $CuFeS_2$ dissolution, while the MnO_2 concentration did not have a significant effect in the system.
- The biggest copper extractions in this research (71%) was obtained working at 80 °C, a particle size of −47 + 38 µm, a 5/1 $MnO_2/CuFeS_2$ ratio, and 1 mol/L of H_2SO_4.

Author Contributions: All of the authors contributed to analyzing the results and writing the paper. All authors have read and agreed to the published version of the manuscript.

Funding: This research received no external funding

Acknowledgments: The authors are grateful for the contribution of the Scientific Equipment Unit-MAINI of the Universidad Católica del Norte for facilitating the chemical analysis of the solutions. Pedro Robles thanks the Pontificia Universidad Católica de Valparaíso for the support provided. Also, we thanks Conicyt Fondecyt 11171036 and Centro CRHIAM Project Conicyt/Fondap/15130015.

Conflicts of Interest: The authors declare no conflict of interest.

References

1. Aguirre, C.L.; Toro, N.; Carvajal, N.; Watling, H.; Aguirre, C. Leaching of chalcopyrite (CuFeS2) with an imidazolium-based ionic liquid in the presence of chloride. *Miner. Eng.* **2016**, *99*, 60–66. [CrossRef]
2. Cerda, C.P.; Taboada, M.E.; Jamett, N.E.; Ghorbani, Y.; Hernández, P.C. Effect of pretreatment on leaching primary copper sulfide in acid-chloride media. *Minerals* **2018**, *8*, 1. [CrossRef]
3. Lu, D.; Wang, W.; Chang, Y.; Xie, F.; Jiang, K. Thermodynamic analysis of possible chalcopyrite dissolution mechanism in sulfuric acidic aqueous solution. *Metals* **2016**, *6*, 303. [CrossRef]
4. Beiza, L.; Quezada, V.; Melo, E.; Valenzuela, G. Electrochemical behaviour of chalcopyrite in chloride solutions. *Metals* **2019**, *9*, 67. [CrossRef]
5. Sokić, M.; Marković, B.; Kamberović, Ž.; Stanković, S. Leaching of chalcopyrite concentrate by hydrogen peroxide in sulphuric acid solution. *Tehnika* **2019**, *74*, 66–70. [CrossRef]
6. Choubey, P.K.; Lee, J.C.; Kim, M.S.; Kim, H.S. Conversion of chalcopyrite to copper oxide in hypochlorite solution for selective leaching of copper in dilute sulfuric acid solution. *Hydrometallurgy* **2018**, *178*, 224–230. [CrossRef]
7. Li, Y.; Kawashima, N.; Li, J.; Chandra, A.P.; Gerson, A.R. A review of the structure, and fundamental mechanisms and kinetics of the leaching of chalcopyrite. *Adv. Coll. Interface Sci.* **2013**, *197–198*, 1–32. [CrossRef]
8. International Copper Study Group. *The World Copper Factbook 2017*; International Copper Study Group: Lisbon, Portugal, 2017.
9. Hiroyoshi, N.; Arai, M.; Miki, H.; Tsunekawa, M.; Hirajima, T. A new reaction model for the catalytic effect of silver ions on chalcopyrite leaching in sulfuric acid solutions. *Hydrometallurgy* **2002**, *63*, 257–267. [CrossRef]
10. Velásquez-yévenes, L.; Torres, D.; Toro, N. Hydrometallurgy Leaching of chalcopyrite ore agglomerated with high chloride concentration and high curing periods. *Hydrometallurgy* **2018**, *181*, 215–220. [CrossRef]
11. Baba, A.A.; Ayinla, K.I.; Adekola, F.A.; Ghosh, M.K.; Ayanda, O.S. A Review on Novel Techniques for Chalcopyrite Ore Processing. *Int. J. Min. Eng. Miner. Process.* **2012**, *1*, 1–16. [CrossRef]
12. Padilla, R.; Pavez, P.; Ruiz, M.C. Kinetics of copper dissolution from sulfidized chalcopyrite at high pressures in H2SO4-O2. *Hydrometallurgy* **2008**, *91*, 113–120. [CrossRef]
13. Dutrizac, J.E. Elemental sulphur formation during the ferric chloride leaching of chalcopyrite. *Hydrometallurgy* **1990**, *23*, 153–176. [CrossRef]
14. Hackl, R.P.; Dreisinger, D.B.; Peters, E.; King, J.A. Passivation of chalcopyrite during oxidative leaching in sulfate media. *Hydrometallurgy* **1995**, *39*, 25–48. [CrossRef]
15. Torres, C.M.; Ghorbani, Y.; Hernández, P.C.; Justel, F.J.; Aravena, M.I.; Herreros, O.O. Cupric and Chloride Ions: Leaching of Chalcopyrite Concentrate with Low Chloride Concentration Media. *Minerals* **2019**, *9*, 639. [CrossRef]
16. Carneiro, M.F.C.; Leão, V.A. The role of sodium chloride on surface properties of chalcopyrite leached with ferric sulphate. *Hydrometallurgy* **2007**, *87*, 73–82. [CrossRef]
17. Ruiz, M.C.; Montes, K.S.; Padilla, R. Chalcopyrite leaching in sulfate-chloride media at ambient pressure. *Hydrometallurgy* **2011**, *109*, 37–42. [CrossRef]
18. Lu, J.; Dreisinger, D. Copper leaching from chalcopyrite concentrate in Cu (II)/Fe (III) chloride system. *Miner. Eng.* **2013**, *45*, 185–190. [CrossRef]
19. Tanda, B.C.; Eksteen, J.J.; Oraby, E.A.; Connor, G.M.O. The kinetics of chalcopyrite leaching in alkaline glycine/glycinate solutions. *Miner. Eng.* **2019**, *135*, 118–128. [CrossRef]

20. Toro, N.; Herrera, N.; Castillo, J.; Torres, C.; Sepúlveda, R. Initial Investigation into the Leaching of Manganese from Nodules at Room Temperature with the Use of Sulfuric Acid and the Addition of Foundry Slag—Part I. *Minerals* **2018**, *8*, 565. [CrossRef]
21. Toro, N.; Saldaña, M.; Gálvez, E.; Cánovas, M.; Trigueros, E.; Castillo, J.; Hernández, P.C. Optimization of Parameters for the Dissolution of Mn from Manganese Nodules with the Use of Tailings in An Acid Medium. *Minerals* **2019**, *9*, 387. [CrossRef]
22. Saldaña, M.; Toro, N.; Castillo, J.; Hernández, P.; Trigueros, E.; Navarra, A. Development of an Analytical Model for the Extraction of Manganese from Marine Nodules. *Metals* **2019**, *9*, 903. [CrossRef]
23. Cronan, D.S. *Handbook of Marine Mineral Deposits*; CRC Press: Boca Raton, FL, USA, 1999.
24. Usui, A.; Nishi, K.; Sato, H.; Nakasato, Y.; Thornton, B.; Kashiwabara, T. Continuous growth of hydrogenetic ferromanganese crusts since 17 Myr ago on Takuyo-Daigo Seamount, NW Paci fi c, at water depths of 800–5500 m. *Ore Geol. Rev.* **2017**, *87*, 71–87. [CrossRef]
25. Toro, N.; Saldaña, M.; Castillo, J.; Higuera, F.; Acosta, R. Leaching of Manganese from Marine Nodules at Room Temperature with the Use of Sulfuric Acid and the Addition of Tailings. *Minerals* **2019**, *9*, 289. [CrossRef]
26. Devi, N.B.; Madhuchhanda, M.; Rao, K.S.; Rath, P.C.; Paramguru, R.K. Oxidation of chalcopyrite in the presence of manganese dioxide in hydrochloric acid medium. *Hydrometallurgy* **2000**, *57*, 57–76. [CrossRef]
27. Devi, N.B.; Madhuchhanda, M.; Rath, P.C.; Rao, K.S.; Paramguru, R.K. Simultaneous Leaching of a Deep-Sea Manganese Nodule and Chalcopyrite in Hydrochloric Acid. *Metall. Mater. Trans. B* **2001**, *32*, 777–784. [CrossRef]
28. Havlik, T.; Laubertova, M.; Miskufova, A.; Kondas, J.; Vranka, F. Extraction of copper, zinc, nickel and cobalt in acid oxidative leaching of chalcopyrite at the presence of deep-sea manganese nodules as oxidant. *Hydrometallurgy* **2005**, *77*, 51–59. [CrossRef]
29. Toro, N.; Pérez, K.; Saldaña, M.; Jeldres, R.; Jeldres, M.; Cánovas, M. Dissolution of pure chalcopyrite with manganese nodules and waste water. *J. Mater. Res. Technol.* **2019**, *9*. [CrossRef]
30. Lundstrom, M.; Aromaa, J.; Forsén, O.; Hyvarinen, O.; Barker, M. Leaching of chalcopyrite in cupric chloride solution. *Hydrometallurgy* **2005**, *77*, 89–95. [CrossRef]
31. Bobadilla-Fazzini, R.A.; Pérez, A.; Gautier, V.; Jordan, H.; Parada, P. Primary copper sulfides bioleaching vs. chloride leaching: Advantages and drawbacks. *Hydrometallurgy* **2017**, *168*, 26–31. [CrossRef]
32. Velásquez-Yévenes, L.; Miki, H.; Nicol, M. The dissolution of chalcopyrite in chloride solutions. Part 2: Effect of various parameters on the rate. *Hydrometallurgy* **2010**, *103*, 80–85. [CrossRef]
33. Velásquez-Yévenes, L.; Nicol, M.; Miki, H. The dissolution of chalcopyrite in chloride solutions: Part 1. The effect of solution potential. *Hydrometallurgy* **2010**, *103*, 108–113. [CrossRef]
34. Nicol, M.; Miki, H.; Velásquez-yévenes, L. Hydrometallurgy The dissolution of chalcopyrite in chloride solutions Part 3. Mechanisms. *Hydrometallurgy* **2010**, *103*, 86–95. [CrossRef]
35. Dutrizac, J.E. The Dissolution of Chalcopyrite in Ferric Sulfate and Ferric Chloride Media. *Metall. Trans. B* **1981**, *12*, 371–378. [CrossRef]
36. Skrobian, M.; Havlik, T.; Ukasik, M. Effect of NaCl concentration and particle size on chalcopyrite leaching in cupric chloride solution. *Hydrometallurgy* **2005**, *77*, 109–114. [CrossRef]
37. Yévenes, L.V. *The Kinetics of the Dissolution of Chalcopyrite in Chloride Media*; Murdoch University: Perth, Australia, 2009.
38. Hernández, P.C.; Taboada, M.E.; Herreros, O.O.; Graber, T.A.; Ghorbani, Y. Leaching of chalcopyrite in acidified nitrate using seawater-based media. *Minerals* **2018**, *8*, 238. [CrossRef]
39. Tundisi, J.G. Water resources in the future: Problems and solutions. *Estudos Avançados* **2008**, *22*, 7–16. [CrossRef]
40. Toro, N.; Briceño, W.; Pérez, K.; Cánovas, M.; Trigueros, E.; Sepúlveda, R.; Hernández, P. Leaching of Pure Chalcocite in a Chloride Media Using Sea Water and Waste Water. *Metals (Basel)* **2019**, *9*, 780. [CrossRef]
41. Peters, N.E.; Meybeck, M. Water quality degradation effects on freshwater availability: Impacts of human activities. *Water Int.* **2000**, *25*, 185–193. [CrossRef]
42. Ridoutt, B.G.; Pfister, S. A revised approach to water footprinting to make transparent the impacts of consumption and production on global freshwater scarcity. *Glob. Environ. Chang.* **2010**, *20*, 113–120. [CrossRef]

43. Cruz, C.; Reyes, A.; Jeldres, R.I.; Cisternas, L.A.; Kraslawski, A. Using Partial Desalination Treatment To Improve the Recovery of Copper and Molybdenum Minerals in the Chilean Mining Industry. *Ind. Eng. Chem. Res.* **2019**, *58*, 8915–8922. [CrossRef]
44. MCH. Agua en la Minería. 2018. Available online: https://www.mch.cl/columnas/agua-la-mineria/# (accessed on 3 June 2019).
45. Cisternas, L.A.; Gálvez, E.D. The use of seawater in mining. *Miner. Process. Extr. Metall. Rev.* **2018**, *39*, 18–33. [CrossRef]
46. Juhász, A.; Opoczky, L. *Mechanical Activation of Minerals by Grinding Pulverizing and Morphology of Particles*; Akademia Kiado: Budapest, Hungary, 1990.

Article

Leaching of Silver and Gold Contained in a Sedimentary Ore, Using Sodium Thiosulfate; A Preliminary Kinetic Study

Edmundo Roldán-Contreras [1], Eleazar Salinas-Rodríguez [1,*], Juan Hernández-Ávila [1], Eduardo Cerecedo-Sáenz [1,*], Ventura Rodríguez-Lugo [1], Ricardo I. Jeldres [2] and Norman Toro [3,4]

1 Área Académica de Ciencias de la Tierra y Materiales, Universidad Autónoma del Estado de Hidalgo, Carretera Pachuca-Tulancingo km 4.5, Mineral de la Reforma, Hidalgo 42184, Mexico; ed.roldan@hotmail.com (E.R.-C.); herjuan@uaeh.edu.mx (J.H.-Á.); ventura.rl65@gmail.com (V.R.-L.)

2 Departamento de Ingeniería Química y Procesos de Minerales, Facultad de Ingeniería, Universidad de Antofagasta, Antofagasta 1240000, Chile; ricardo.jeldres@uantof.cl

3 Departamento de Ingeniería Metalúrgica y Minas, Universidad Católica del Norte, Antofagasta 1270709, Chile; ntoro@ucn.cl

4 Departament of Mining, Geological and Cartographic Department, Universidad Politécnica de Cartagena, 30202 Murcia, Spain

* Correspondence: salinasr@uaeh.edu.mx (E.S.-R.); mardenjazz@yahoo.com.mx (E.C.-S.); Tel.: +52-771-207-4171 (E.S.-R.)

Received: 11 December 2019; Accepted: 15 January 2020; Published: 21 January 2020

Abstract: Some sedimentary minerals have attractive contents of gold and silver, like a sedimentary exhalative ore available in the eastern of Hidalgo in Mexico. The gold and silver contained represent an interesting opportunity for processing by non-toxic and aggressive leaching reagents like thiosulfate. The preliminary kinetic study indicated that the leaching process was poorly affected by temperature and thiosulfate concentration. The reaction order was −0.61 for Ag, considering a thiosulfate concentration between 200–500 mol·m^{-3}, while, for Au, it was −0.09 for a concentration range between 32–320 mol·m^{-3}. By varying the pH 7–10, it was found that the reaction order was n = 5.03 for Ag, while, for Au, the value was n = 0.94, considering pH 9.5–11. The activation energy obtained during the silver leaching process was 3.15 kJ·mol^{-1} (298–328 K), which was indicative of a diffusive control of the process. On the other hand, during gold leaching, the activation energy obtained was of 36.44 kJ·mol^{-1}, which was indicative that this process was mixed controlled process, first at low temperatures by diffusive control (298–313 K) and then by chemical control (318–323 K).

Keywords: thiosulfate; gold leaching; silver leaching; kinetic analysis; sedimentary ore; diffusion control; mixed control

1. Introduction

The mining activity in Mexico has focused mainly on gold and silver ores, the processing of which entails grinding, froth flotation, and cyanidation stages. While the latter involves low operation costs, the reagent utilized is highly toxic, which has even been outlawed in several countries worldwide [1–3]. Moreover, particular operational challenges may arise, wherein the lasting cyanidation process can require up to 24 h [3]. This impairs the leaching efficiency since refractory minerals can be produced that encapsulate in small pyritic and quartz-type particles [4,5], and some cyanide-consuming minerals may also appear, hampering the suitable extraction of metallic contents [5–7].

To date, several studies have been carried out to find leaching reagents that can replace cyanide, falling into chloride, thiourea, and thiosulfate [8,9]. Chloride is corrosive and can cause hazardous working conditions, together with low selectivity during the extraction process [4,6,8]. Processing with thiourea is expensive, and the background indicated that it is a potential carcinogen reagent [10,11]. However, leaching using thiosulfate is an economical and promising method for ore treatment. It offers high reaction selectivity during the process, reduced environmental risks, and low corrosive solutions, low price, offering also an efficient dissolution medium for refractory ores [12–19].

Several studies have displayed good performance on the use of thiosulfate instead of cyanide to extract gold and silver [20–23]. However, a particular challenge persists that is mostly related to the low stability of thiosulfate ions, where proper alternative considers the solution of copper–ammonium–thiosulfate. The Cu^{2+} ions may oxidize the gold and silver, while the thiosulfate makes quite stable complexes with them, which might allow a suitable extraction from ores and waste. Also, the ammonia ions can also form a stable complex with copper ions, avoiding their precipitation [9]. Some studies have shown their performance in the presence of additives and electrolytes, considering the effect of ligands and oxidants reagents [24–27].

Recent studies have found that during the kinetic of silver leaching using thiosulfate, the overall process is controlled by the mass transfer of oxygen to the solid-liquid interface [28]. Additionally, other similar studies done for silver leaching have concluded that for different ranges of concentrations of ammonia and thiosulfate, silver complexes preferentially with thiosulfate [29–32]. On the other hand, the only company that at present is using the thiosulfate solutions for gold leaching is Barrick Gold Corporation (Elko, NV, USA), after an acidic or alkaline pressure oxidation pretreatment [33,34]. The carbonaceous gold ore cannot be treated with cyanide due to the "pre-robbing" phenomena, which does not occur during the leaching using thiosulfate solutions. The weak affinity of carbonaceous material for the gold thiosulfate complex forwards this stage [35]. Some authors have gotten extractions of 11% for gold and 21% for silver using both ammonium thiosulfate with the addition of H_2O_2 and cupric ions [36]. On the other hand, thiosulfate easily can be decomposed by some factors, like the Cu(II) content and the presence of different minerals like pyrite and hematite, promoting also high amounts of thiosulfate consumption during leaching process, generating diverse polythionates and at the end, S^0, S^{2-}, and $SO_3{}^{2-}$, which can be deposited on the surface of mineral, passivating the dissolution of the metals of interest [9]. Therefore, this work aimed to analyze the dissolution kinetics of gold and silver from a sedimentary ore, where silver can be presented as metallic and/or sulfur, and gold is joint to carbonaceous material. Air-$Na_2S_2O_3{}^{2-}$ solutions were used without adding cupric ions, which allowed evaluating how thiosulfate could extract these metallic values. The ore considered in this study might have some trace copper contents (less than 10 ppm, perhaps like chalcopyrite) that acted as an oxidizing reagent, improving the gold and silver leaching. Finally, the mechanisms that control the chemical reactions of both metals were discussed.

2. Materials and Methods

2.1. Materials

An ore, located at the northeast of the State of Hidalgo, Mexico, was collected selectively [37], taking 50,000 g of each sample in 4 different points of the mineralized zone of the outcrop. Samples collected were mixed and quartered, taking a representative sample to carry out the kinetics leaching study. The mineralogical characterization was executed to get accurate data of the phases present, for which an analysis of general phases was proposed through X-Ray diffraction (XRD). The samples were ground up to get an average particle size, less than 78 μm, and kept in an Equinox 2000 X-ray Diffractometer (INEL, Artenary, France, located at UAEH) with $CoK\alpha_1$ radiation. The identification of present phases was executed using the COD Inorganics 2015 databases, which is included in the crystallography open database match software (v.1.10, Crystal Impact, Bonn, Germany).

The scanning electron microscopy (SEM) identified texture, particle sizes, and morphology of the detected phases. The semi-quantitative and punctual analysis was done by energy dispersive spectrometry of X-ray (EDS) (OXFORD Instruments, Oxford, UK). It determined the punctual and semi-quantitative composition of the previously identified particles, using a JEOL scanning electron microscope JSM-IT300 (JEOL Ltd., Tokyo, Japan, located at UAEH, Apan, Mexico)) and an OXFORD X-ray detector (OXFORD Instruments, Oxford, UK) with 30 kV of acceleration voltage. The analysis of samples was done using powders of the sample placed in uniform layers, where punctual semi-quantitative routines were done on scanning areas about 4.5 mm^2.

An inductively coupled plasma spectrometry (ICP-MS) analysis was performed by Actlabs (Activation Laboratories Ltd., Ancaster, ON, Canada); this determined the average total rock composition of the mineralized phase, where the positive anomalies of the light rare earth and mineral contents of the platinum group (PGE) were found. In this case, the samples were fused and then diluted and analyzed by a Perkin Elmer Sciex ELAN 9000 ICP-MS spectrometer (Located at Actlabs, Ancaster, ON, Canada). Fused blank was run in triplicate every 10 samples, and then the instrument was recalibrated after every 44 samples.

An analysis by copelation determined the Au and Ag contents. This was executed in an oven EMISON brand, CL Series (Located at UAEH, Pachuca, Mexico). The sample preparation was done utilizing borax, PbO, bone ash, sodium carbonate as flux. The melting temperature was of 1273 K (1000 °C) during 90 min. In the slag separation, a button contained the values of Ag and Au. The release of these metals was executed in porcelain crucibles on a heating plate at 313 K (40 °C), adding 15% nitric acid to obtain a solution of silver nitrate. Then, the release of Au and Ag was done using regal water, adding 10% of hydrochloric acid in test tubes. The determination of Au and Ag contents was done using an ICP-OES Varian brand 735ES ICP (Located at UAEH, Pachuca, Mexico), where samples were analyzed with a minimum of 10 certified reference materials, all prepared with sodium peroxide fusion. Every 10 samples were analyzed by duplicate, and the blank was renewed after every 30 samples measurement. For the kinetics study, the samples were ground during 360 s at a working speed of 150 s^{-1} with a ball charge of 10,230 g, a pulp charge of 2100 g, and a volume of 9.5×10^{-4} m^3 of water in the ball mill "Denver" (Located at the UAEH-Mexico). Then, wet sieve determined the particle size distribution. Finally, the samples were separated for the next stage of kinetics leaching.

2.2. Experimental Procedure

The experiments for the kinetics study were executed in a 0.001 m^3 flat bottom glass reactor mounted on a hot plate having a magnetic stirring system and coupled to a pH meter. The pH was monitored and adjusted continuously with NaOH solution at a concentration of 200 mol·m^{-3}. A thermocouple attached to the hot plate controlled the temperature. All assays were done at an open atmosphere with vigorous mechanical stirring (500 s^{-1}). The chemical reagents were of analytical grade, and distilled water was used for the preparation of solutions. Thus, leaching reagent was added like sodium thiosulfate from Sigma MEYER brand with an essay quality of [$Na_2S_2O_3{\cdot}5H_2O$] of 99.5–100%.

For the leaching of silver, the kinetics leaching s using thiosulfate solutions was done using the following experimental conditions: concentration of sodium thiosulfate [$Na_2S_2O_3$], 200 to 500 mol·m^{-3}; temperature range, 298 to 328 K; mineral weight, 40 g·m^3; pH range, 7 to 10; volume of reaction, 0.0005 m^3; stirring rate, 500 s^{-1}. This allowed comparing with previous results on mining waste [38]. For the leaching of gold, the kinetics study was executed under the following conditions: concentration of [$Na_2S_2O_3$], 32 to 320 mol·m^{-3}; temperature range, 298 to 323 K; mineral weight, 40 g·m^{-3}, pH range, 9.5 to 11; volume of reaction, 0.0005 m^3; stirring rate, 500 s^{-1}. This followed recommendations stated previously [39,40].

The progress of all experiences was followed by sampling at pre-set times (0–14,400 s) throughout the experiment, and then the dissolved Au and/or Ag were analyzed by ICP and AAS. Mathematical calculations corrected variations in the mass balance in the sampling addition of a reagent.

3. Results

3.1. Mineral Characterization

The XRD analysis (Figure 1) showed that the mineral species involved were characteristics of a sedimentary exhalative ore [37]. This was principally composed of quartz, ilmenite, and monazite, with some contents of precious metals, such as Pt, Pd, Au, and Ag. Some light rare earths could be present (Table 1), and also trace contents of base metals sulfurs of Pb, Zn, Cu, chalcopyrite, and pyrite, associated with organic material (carbonaceous substance) with attractive contents of gold (5 g/ton) and silver (25 g/ton), determined by ICP, AAS, and cupellation test (Figure 2). All the above gave additional value to this ore.

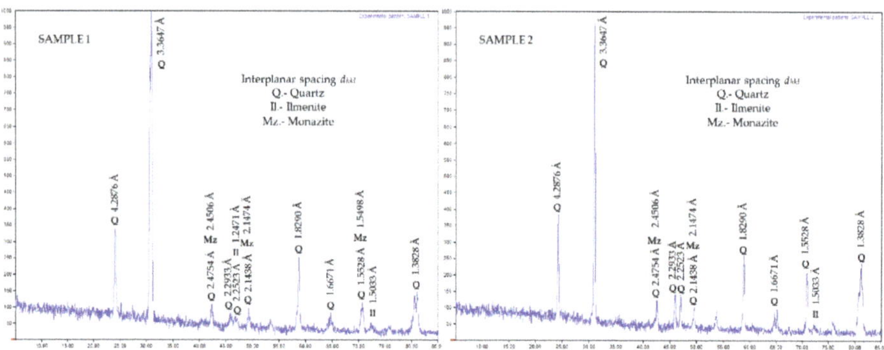

Figure 1. XRD spectra of sedimentary ore, used for kinetic study of gold and silver leaching.

Table 1. Average chemical composition of mineral executed by ICP-OES/MS, XRF, SEM-EDS, and AAS.

Element	% wt
S	34.3
Fe	32.6
O	16.6
C	2.4
Si	0.8
Mg	0.9
Al	0.6
Na	0.6
K	0.05
Ti	0.006
Cu	0.0007
Pt	0.0002
Pd	0.0007
Au	0.0005
Ag	0.0025

The morphology obtained by SEM-SE showed irregular shape particles, typical of the mineral species mentioned above, that presented irregular faces in similar dimensional style, having sizes that vary from 30 to 354 µm. Figure 3 shows a general image of ore particles used in this study, also showing the size distribution of particles of this material and the anhedral type of morphology. A detailed zone from where an EDS was executed, revealing the presence of Au and Ag, is also shown in Figure 3. The particle size distribution is shown in Table 2, where the predominant particle (77.6%) had 44 µm of diameter.

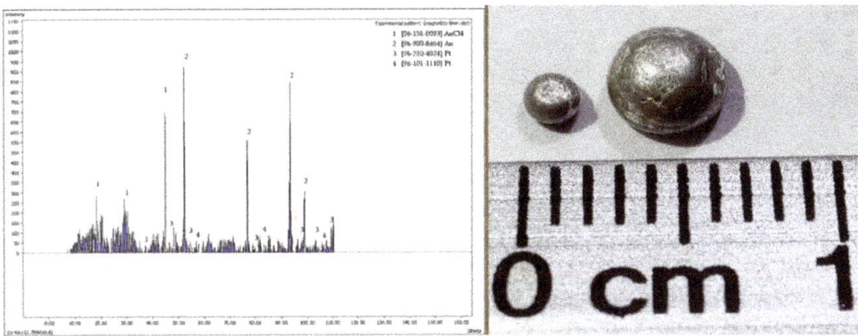

Figure 2. Image of the button obtained by copelation and the corresponding XRD spectra.

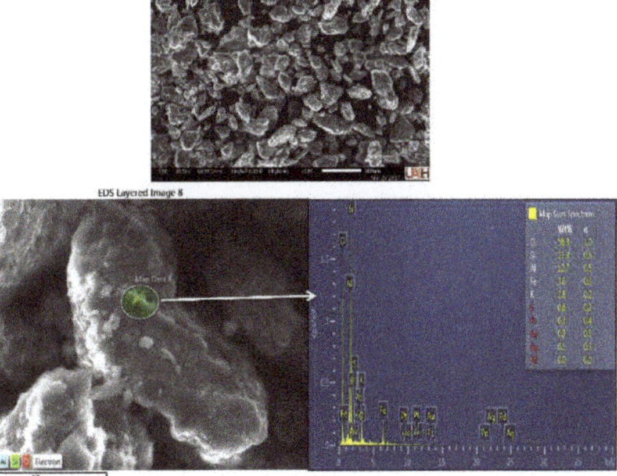

Figure 3. The general image of the ore particle size distribution, and detailed zone with an SEM-EDS analysis (SEM-SE), similar for all samples used in this study.

Table 2. Ore particle size distribution.

D (μm)	Partial Weight (pct)	Accumulative Weight (pct)
37	3.8	3.8
44	77.6	81.4
53	8.5	89.9
63	3.5	93.4
105	2.7	96.1
125	2.2	98.3
149	1.5	99.8
354	0.2	100

3.2. *Kinetic Study of Gold and Silver Leaching*

3.2.1. Nature and Stoichiometry of Reactions

The experimental conditions for silver leaching were: concentration of [$Na_2S_2O_3$], 500 mol·m^{-3}; temperature, 298 K; pH, 9; mineral weight, 40 g·m^{-3}; stirring rate 500 s^{-1}; the volume of reaction, 0.0005 m^3, reaching a maximum silver recovery of 80%. For gold leaching, the experimental conditions were the following: concentration of [$Na_2S_2O_3$], 130 mol·m^{-3}; temperature, 298 K; pH, 9.5; mineral weight, 40 g·m^{-3}; stirring rate 500 s^{-1}; the volume of reaction, 0.0005 m^3, with a maximum recovery of

20%. Figure 4 shows a representation of the leached element (Ag or Au) versus time. For silver, there was no induction period, and the reaction started immediately, describing the progressive conversion until the end of the reaction, where it appeared a stabilization zone (Figure 4A). For gold leaching (Figure 4B), the graph showed a curve type "S" with a small period of induction, then a period of progressive conversion, a stabilization zone. The existence of a short induction period could be caused by the presence of pyrite, which could influence thiosulfate decomposition, leading to a slow gold dissolution.

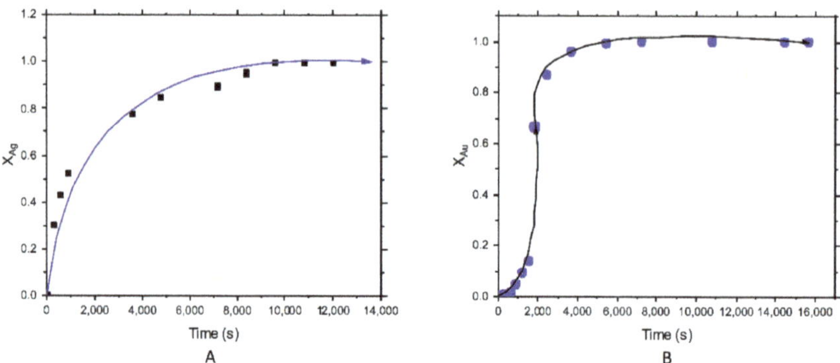

Figure 4. Graph of the curve type "S", showing the period of progressive conversion for; (**A**) silver leaching and (**B**) gold leaching, in thiosulfate solutions.

The results showed in Figure 4 were managed according to the core model for diffusive (Equation (1)) and chemical control (Equation (3)) [41–43], determining which kinetics leaching model fit better.

$$[1 - \frac{2}{3}X - (1 - X)^{\frac{2}{3}}] = k_{exp} \times t \tag{1}$$

where

$$k_{exp} = \frac{2V_M\, D_c\, c_A}{r_0^2} \tag{2}$$

$$[1 - (1 - X)^{\frac{1}{3}}] = k_{exp} \times t \tag{3}$$

where

$$k_{exp} = \frac{V_M\, k_q\, c_A^n}{r_0} \tag{4}$$

X is the reacted fraction of Ag or Au, V_M is the molar volume of the mineral, c_A is the concentration of leaching reactant (in this case thiosulfate), D_e is the diffusion coefficient through the product layer, k_q is the kinetic coefficient, r_0 is the initial radius of particle (in average), k_{exp} is the experimental constant, and, finally, n is the order of reaction.

Figure 5 shows the results for silver leaching, considering the fit of experimental data to (A) core model for diffusive control and (B) core model for chemical control. A better representation of the diffusive control model could be appreciated. The same behavior occurred for gold leaching in the thiosulfate solution, and then the kinetics study was analyzed according to the core model for diffusive control.

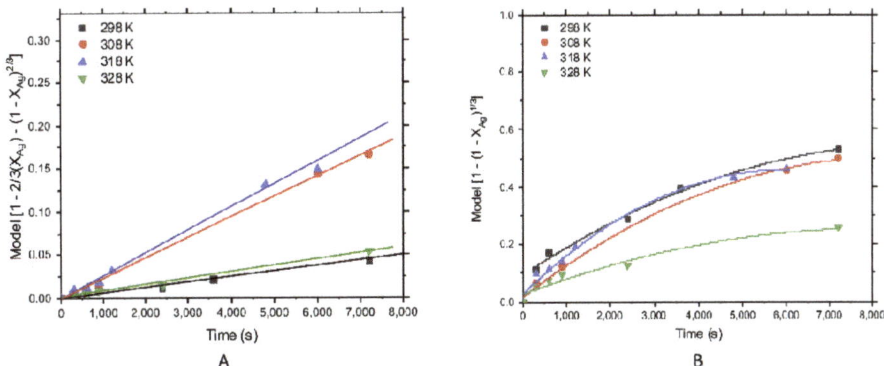

Figure 5. Treatment of silver leached data with kinetic models; (**A**) diffusive control and (**B**) chemical control.

3.2.2. Stoichiometry of Leaching of Gold and Silver

Mineralogical complexities prevented the determination of stoichiometry for gold and silver leaching system under study. However, a theoretical estimation based on the characterization results determined the presence of gold and silver in the sedimentary ore. These appeared in native form for both metals and like sulfur for the case of silver.

For the case of metallic gold:

$$4Au(s) + 8(S_2O_3)aq^{-2} + O_2g + 2H_2Oaq \rightarrow 4Au(S_2O_3)aq^{-2} + 4OH(aq)^{-} \quad (5)$$

For the metallic silver:

$$2Ag_{(s)} + 4(S_2O_3)^{-2}_{(aq)} + O_{2(g)} + 2H_2O_{(aq)} \rightarrow 2Ag(S_2O_3)^{-3}_{(aq)} + 4OH^{-}_{(aq)} \quad (6)$$

When silver is like silver sulfide:

$$Ag_2S_{(s)} + 4(S_2O_3)^{-2}_{(aq)} + O_{2(g)} + 2H_2O_{(aq)} \rightarrow 2Ag(S_2O_3)^{-3}_{(aq)} + S^{+2}_{(s)}4OH^{-}_{(aq)} \quad (7)$$

3.2.3. Effect of the Concentration of $[Na_2S_2O_3]$

The study of gold and silver extraction contained in a sedimentary ore, using the $Na_2S_2O_3$-Air system, was done to establish the effect of the $[Na_2S_2O_3]$ concentration, temperature, and pH.

Figure 6A shows the leached fraction of Ag that was analyzed by the core model for diffusive control [41–43], which was $[1 - 2/3X_{Ag} - (1 - X_{Ag})^{2/3}]$. Straight lines were obtained, and their slopes represented the experimental rate constant (k_{exp}). Thiosulfate concentration had no effect on the rate of reaction. Since all experiences were done at a high stirring rate, the oxygen input could be enough to maintain a stoichiometric excess during the progress of the reaction. This caused a limited effect on the leaching of silver with respect to thiosulfate concentration, getting a low order of the reaction, n = −0.61 (Figure 6B).

Figure 7A shows the leached fraction of Au, which was evaluated by the same core model for diffusive control $[1 - 2/3X_{Au} - (1 - X_{Au})^{2/3}]$. Similar to silver leaching, thiosulfate concentration did not affect the leaching reaction rate. In this case, the presence of pyrite and Ag_2S could promote the generation of $\mathbf{S}^0_{(s)})$ and/or $\mathbf{S}^{+2}_{(s)}$, which would be deposited on the gold surface, passivating its dissolution. This also provided a low order of reaction, n = −0.09 (Figure 7B).

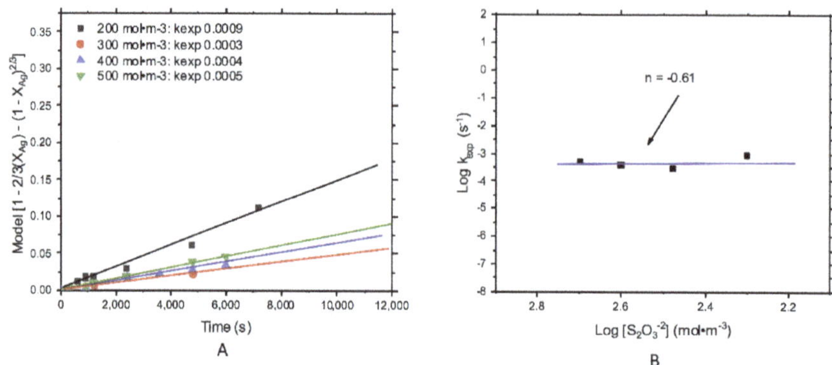

Figure 6. Kinetic study of silver leaching; effect of the thiosulfate concentration: (**A**) k_{exp} and (**B**) order of reaction n = −0.61.

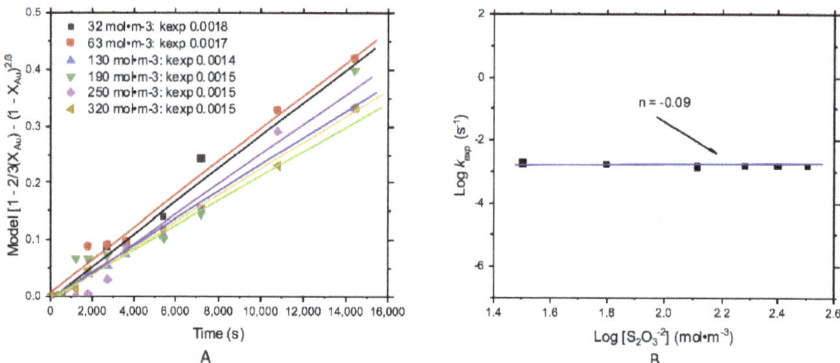

Figure 7. Kinetic study of gold leaching; effect of the thiosulfate concentration: (**A**) k_{exp} and (**B**) order of reaction n = −0.09.

3.2.4. Effect of the Temperature

The effect of temperature for silver leaching in thiosulfate solutions is shown in Figure 8A,B. For low and high temperatures (298–328 K), the values were quite similar, giving low reaction rates. In conclusion, the influence of temperature was limited.

The activation energy was calculated by plotting the natural logarithm of k_{exp} against the reciprocal of temperature. The slope ($m = -(E_a/R)$) of the linear curve represented the activation energy (E_a) divided by the negative value of the universal gas constant (Figure 8B). For silver leaching, the calculated energy of activation was of 3.15 kJ/mol, which was representative of a diffusive control [41–43]. This was also observed by the fitting obtained with the diffusive control model. The obtained order of reaction was low, and the curve type "S" had no induction period, then the process was not dependent on both thiosulfate and temperature, and the slow diffusion of products from the particle's surfaces to the deep of solution controlled the overall process.

For gold leaching, the effect of temperature according to the diffusive control model is shown in Figure 9A. The rate constants were quite similar in the range of temperature from 298 to 313 K (0.0011–0.0014), but he slight increase at higher temperatures (318 and 323 K) could indicate that the process was carried out by a combination of both chemical and diffusive control. Besides, the activation energy that is shown in Figure 9B was 36.44 kJ/mol, which, according to the literature, is representative of mixed control [41–43].

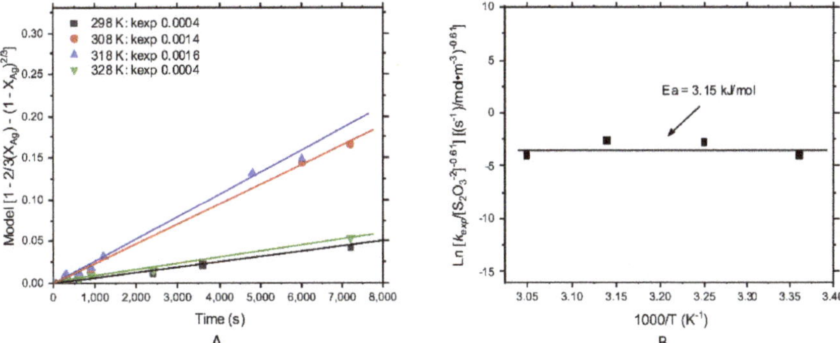

Figure 8. Kinetic study of silver leaching; effect of the temperature: (**A**) k_{exp} and (**B**) energy of activation, Ea = 3.15 kJ/mol.

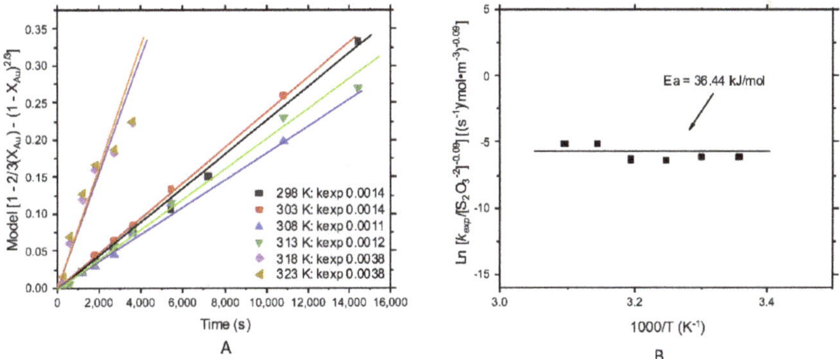

Figure 9. Kinetic study of gold leaching; effect of the temperature: (**A**) k_{exp} and (**B**) energy of activation, Ea = 36.44 kJ/mol.

According to the obtained results, the kinetic expressions for silver (Equation (7)) and gold (Equation (8)) leaching in $S_2O_3^{-2}$ medium were:

For silver, diffusive control

$$\frac{r_0^2}{V_M}\left[1-\frac{2}{3}X_{Ag}-\left(1-X_{Ag}\right)^{\frac{2}{3}}\right]=2D_e\left[S_2O_3^{-2}\right]\times t \quad (8)$$

For gold, when control could be by chemical reaction

$$\frac{r_0^2}{V_M}\left[1-\left(1-X_{Au}\right)^{\frac{1}{3}}\right]=3.736\,x10^3\,exp^{\frac{-36,440}{RT}}\left[S_2O_3^{-2}\right]^{-0.09}\times t \quad (9)$$

For gold, when control could be by diffusion of products through the product layer

$$\frac{r_0}{V_M}\left[1-\frac{2}{3}X_{Au}-\left(1-X_{Au}\right)^{\frac{2}{3}}\right]=2D_e\left[S_2O_3^{-2}\right]\times t \quad (10)$$

where $V_M = 1.44 \times 10^{-9}$ m^3·mol^{-1} for silver, and $V_M = 6.22 \times 10^{-9}$ m^3·mol^{-1} for gold, $R = 8.31$ J·mol^{-1}·K^{-1}, r_0 in m, D_e is the diffusion coefficient through the product layer, T in Kelvin, $[S_2O_3{}^{-2}]$ is in mol·m^{-3}, and t is in seconds.

3.2.5. Effect of the pH

The effect of the pH for silver leaching is shown in Figure 10A,B. Figure 10A represents the treatment of results with the diffusive control model and the obtained experimental rates. The rates decreased with increasing the pH, which was faster at pH 7, having a poor effect at a higher pH. Figure 10B shows that the influence of pH over the overall reaction rate was significant, finding an order of reaction of n = 5.03.

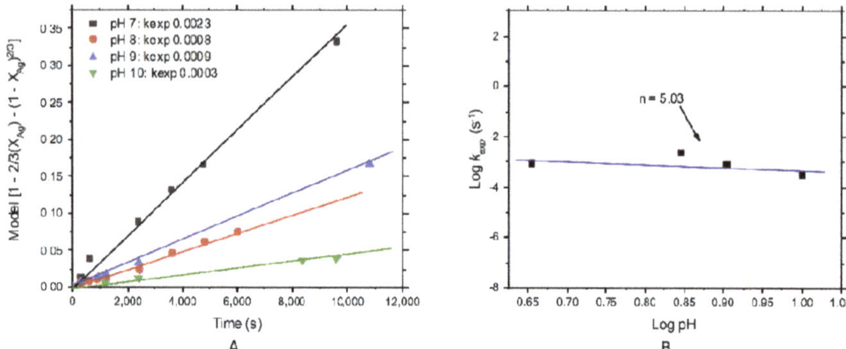

Figure 10. Kinetic study of silver leaching; effect of the pH: (**A**) k_{exp} and (**B**) order of reaction n = 5.03.

To conclude, for gold leaching, the pH effect is shown in Figure 11A,B. Unlike what happened with the leaching of silver, the experimental rate constants were similar for all the analyzed pH values. Consequently, a poor effect of this variable over the overall rate of the process was observed (Figure 11A). However, all experimental rate constants were low. This might be explained by the presence of small amounts of sulfur minerals since their formation is inevitable even in small quantities, promoting a partial degradation of thiosulfate. The deposition of this formed sulfur on the surface of the gold particles promoted the decrease in its dissolution rate. Finally, Figure 11B displays the effect of pH over the overall gold leaching reaction rate. The order of reaction was n = 0.94, concluding that the pH had no effect on gold leaching under the thiosulfate concentration and temperatures considered.

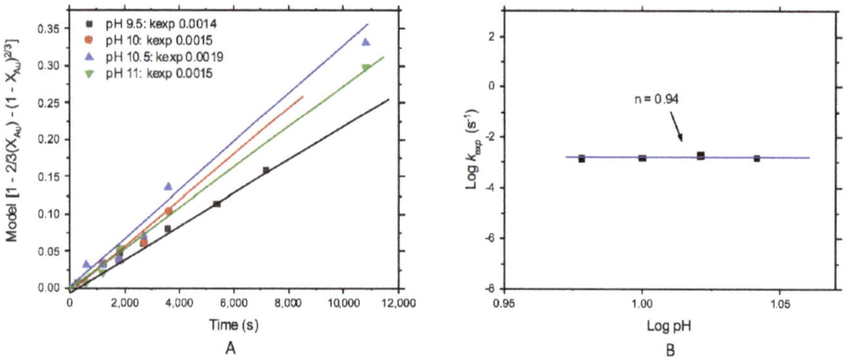

Figure 11. Kinetic study of gold leaching; effect of the pH: (**A**) k_{exp} and (**B**) order of reaction n = 0.94.

4. Discussion

The mineral studied here presented significant amounts of Ag and Au, including light rare earths contents, which increased its commercial importance. However, this research aimed to analyze the leaching kinetics of gold and silver in thiosulfate solutions without adding Cu(II), which allowed verifying the ability of the leaching reagent according to the nature and composition of the ore. Although leaching using thiosulfate solutions is considered a viable alternative to cyanidation [16–18,39], nowadays, it is not used at a big scale due to the high reagent consumption and the difficulty in recovering metallic gold and silver [9]. Still, it works appropriately with metallic ores containing carbonaceous materials [35], like the ore here studied (having about a 2% C).

The use of thiosulfate as a leaching reagent needs more attention since some problems can arise, the reagent is unstable, and it can self-decomposed or reduced to S^0, S^-, and S_3^{-2}. These species may deposit on the surface of the metal, hindering its dissolution [25]. Additionally, the reaction is prolonged without the addition of Cu(II) and ammonia [44]. However, for this study, we decided to use sodium thiosulfate solely since the characteristics of mineral gave a chance to get promising results. Yen et al. [45] concluded that high concentrations of thiosulfate, high dissolved oxygen, and high temperatures increased the consumption of thiosulfate. Consequently, low metallic recoveries were obtained. In parallel, dilute concentrations of thiosulfate, low oxygen concentration, and low temperatures could reduce the rate of gold dissolution. However, the results of this study were encouraging, getting 80% of silver recovery using 500 mol·m^3 of sodium thiosulfate, pH 7, the temperature of 298 K, stirring rate of 500 s^{-1}, and using air in low concentration (just by incorporation during mechanical stirring). In the case of gold, the maximum dissolution was of 20% using 130 mol·m^3 of sodium thiosulfate, pH 9, the temperature of 298 K, stirring rate of 500 s^{-1}, and using air in low concentration (just by incorporation during mechanical stirring). As known, the experimental conditions constantly change during leaching, and it could be difficult to control each of them with adequate precision. Similarly, the use of relatively low leaching reagent concentrations with limited oxygen supply is an easy method to avoid high thiosulfate consumption during the leaching process [9].

Some authors [28,38] reported interesting results for the silver leaching with thiosulfate solutions, concluding that the diffusion of oxygen controls the process through the product layer. The order of the reaction was similar to that found here, which was $n = -0.61$ for thiosulfate concentrations 200–500 mol·m^{-3}. Other cases displayed values like $n = 0.074$ for thiosulfate concentrations 100–500 mol·m^{-3}, for silver leaching contained in mining burrows [38], $n = 0.41$ for thiosulfate concentration 25–200 mol·m^{-3} [28], and $n = 0$ for thiosulfate concentration 200–600 mol·m^{-3}, for leaching of metallic silver [28]. Thiosulfate concentration does not affect the reaction rate for silver leaching that contrast works with different mineralogies [9]. For this reason, more in-depth researches are needed to disclose the behavior of thiosulfate solutions over silver leaching, according to the nature of the minerals and species involved.

For gold leaching, the literature reports that the thiosulfate concentration, joint with some contents of Cu(II) and ammonia, have a detrimental effect on the gold dissolution by the presence of minerals, such as pyrite and other sulfurs [25]. This promotes a self-decomposition of the thiosulfate and, as a consequence, higher consumption of this reactant. In this study, it was found that the thiosulfate concentration had no effect on the reaction rate getting reaction order of $n = -0.09$ (quite similar to that obtained for silver leaching). The latter means that the absence of important amounts of sulfur minerals might avoid the formation of S^0 and S^{-2}, which deposit on the gold surface, reducing the dissolution.

According to the evaluation of the effect of the temperature on the rate of silver leaching, the activation energy found here was $E_a = 3.15$ kJ/mol. This corresponded to a diffusive control, and it validated the model used for the treatment of data [41–43]. In this case, the overall reaction was controlled by oxygen diffusion through the product layer because the chemical reaction of the complexation of Ag is too fast. The above result was consistent with that obtained during the silver leaching contained in a mining waste [38], (even with the absence of Cu(II) where the apparent energy of activation was of $E_a = 1.91$ kJ/mol.

When evaluating the effect of temperature during the gold leaching, it was found that the apparent energy of activation was $E_a = 36.44$ kJ/mol. According to previous studies, this corresponded to a mixed control [41–43]. At low temperatures, the experimental rates were low (0.0011–0.0014 s^{-1}), and diffusive control was dominant since thiosulfate was more stable. Then, at higher temperatures, the instability of thiosulfate could lead to its decomposition, generating S^0 and S^{2-} that might deposit on the gold surface [9]. This avoided a fast chemical reaction, being this step that controlled the process.

Finally, due to the concentration of $\left[S_2O_3^{-2}\right]$ could be affected by the pH (below 4 and above 12), where thiosulfate degradation or decomposition occurred, with the consequent formation of elemental sulfur (especially below a pH 4) [28,38]. Consequently, this work was executed between the valid range pH, where the thiosulfate could be more stable and was not pH-dependent. For the leaching of silver, the order of reaction was n = 5.03, indicating an apparent effect of this variable but only at pH 7. At this condition, the rate of reaction was higher than at pH 8–10. For gold dissolution, the order of reaction was n = 0.94, indicating that the pH did not alter the reaction rate, which indicated the stability of the reactant. This was because of the operational conditions used in this work and the absence of minerals like pyrite that could promote thiosulfate decomposition.

Author Contributions: The conceptualization was done by E.S.-R., J.H.-Á., and E.C.-S.; the methodology was executed by E.R.-C., E.S.-R., and V.R.-L.; validation was done by E.S.-R.; data curation, E.S.-R.; writing—original draft preparation, E.S.-R., E.R.-C., and J.H.-Á.; writing—review and editing, E.S.-R. and R.I.J.; supervision, E.S.-R. and V.R.-L.; funding acquisition, N.T. and R.I.J. All authors have read and agreed to the published version of the manuscript.

Funding: This research received no external funding.

Acknowledgments: Authors want to thank the CONACyT of the Mexican Government for its support given through the Ph.D. scholarship to the student Edmundo R. C. (CVU 430931). Thanks also go to the Autonomous University of the State of Hidalgo, Mexico, especially to the Academic Group of Advanced Materials from the AACTyM-ICBI; University of Antofagasta, Chile; Polytechnic University of Cartagena, Spain, and Northern Catholic University, Chile.

Conflicts of Interest: The authors declare no conflict of interest.

References

1. Hilson, G.; Monhemius, A.J. Alternatives to cyanide in the gold mining industry: What prospects for the future? *J. Clean. Prod.* **2006**, *14*, 1158–1167. [CrossRef]
2. Muir, D.M. A review of the selective leaching of gold from oxidized copper-gold ores with ammonia-cyanide and new insights for plant control and operation. *Miner. Eng.* **2011**, *24*, 576–582. [CrossRef]
3. Jiang, T. *Chemistry of Extractive Metallurgy of Gold*; Hunan Science and Technology Press: Changsha, China, 1998.
4. Hasab, M.G.; Rashchi, F.; Raygan, S. Chloride-hypochlorite leaching and hydrochloric acid washing in multi-stages for extraction of gold from a refractory concentrate. *Hydrometallurgy* **2014**, *142*, 56–59. [CrossRef]
5. Geyne, A.R.; Fries, C. *Geology and Mineral Deposits of the Pachuca—Real Del Monte District, State of Hidalgo, 5-E.*; Consejo de Recursos Naturales no Renovables: Pachuca de Soto, Mexico, 1963; Unpublished work.
6. Hasab, M.G.; Raygan, S.; Rashchi, F. Chloride-hypochlorite leaching of gold from a mechanically activated refractory sulfide concentrate. *Hydrometallurgy* **2013**, *138*, 59–64. [CrossRef]
7. Abbruzzese, C.; Fornari, P.; Massidda, R.; Veglio, F. Thiosulphate leaching for gold hydrometallurgy. *Hydrometallurgy* **1995**, *39*, 265–276. [CrossRef]
8. Xu, B.; Yang, Y.; Li, Q.; Li, G.; Jiang, T. Fluidized roasting-stage leaching of a silver and gold bearing polymetallic sulfide concentrate. *Hydrometallurgy* **2014**, *147*, 79–82. [CrossRef]
9. Bin, X.; Wenhao, K.; Qian, L.; Yongbin, Y.; Tao, J.; Xiaoliang, L. A review of thiosulfate leaching of gold: Focus on thiosulfate consumption and gold recovery from pregnant solution. *Metals* **2017**, *7*, 222.
10. Zheng, S.; Wang, Y.; Chai, L. Research status and prospect of gold leaching in alkaline thiourea solution. *Miner. Eng.* **2006**, *19*, 1301–1306. [CrossRef]
11. Chen, X. Associated Sulfide Minerals in Thiosulfate Leaching of Gold: Problems and Solutions. Ph.D. Thesis, Queen's University, Kingston, ON, Canada, September 2008.
12. Hernández, A.J.; Rivera, L.I.; Patiño, C.F.; Juárez, T.J. Estudio cinético de la lixiviación de plata en el sistema $S_2O^{-2}{}_3$–O_2–Cu^{2+} contenidos en residuos minero-metalúrgicos. *Inf. Tecnol.* **2013**, *24*, 51–58. (In Spanish) [CrossRef]
13. Rivera, L.I. Estudio Cinético de la Precipitación/Lixiviación de Plata en el Sistema O_2–$S_2O^{-2}{}_3$–$S_2O^{-2}{}_4$. Ph.D. Thesis, Universitat de Barcelona, Barcelona, Spain, 2003. (In Spanish)

14. Schmitz, P.A.; Duyvesteyn, S.; Johnson, W.P.; Enloe, L.; McMullen, J. Ammoniacal thiosulfate and sodium cyanide leaching of pre-robbing Goldstrike ore carbonaceous matter. *Hydrometallurgy* **2001**, *60*, 25–40. [CrossRef]
15. Rivera, L.I. Síntesis, Caracterización y Reactividad Alcalina de Compuestos Jarositicos Argentíferos Jarosita de Potasio—Natrojarosita. Bachelor´s Thesis, Universidad Autónoma del Estado de Hidalgo, Pachuca, Mexico, 1994. (In Spanish)
16. Feng, D.; Van Deventer, J.S.J. Thiosulphate leaching of gold in the presence of carboxymethyl cellulose (CMC). *Miner. Eng.* **2011**, *24*, 115–121. [CrossRef]
17. Lampinen, M.; Laari, A.; Turunen, I. Ammoniacal thiosulfate leaching of pressure oxidized sulfide gold concentrate with low reagent consumption. *Hydrometallurgy* **2015**, *151*, 1–9. [CrossRef]
18. Xu, B.; Yang, Y.; Li, Q.; Jiang, T.; Li, G. Stage leaching of a complex polymetallic sulfide concentrate: A focus on the extractions of Ag and Au. *Hydrometallurgy* **2016**, *159*, 87–94. [CrossRef]
19. Jian, W.; Wei, W.; Kaiwei, D.; Yan, F.; Feng, X.W.; Wei, W.; Kaiwei, D.; Yan, F.; Feng, X. Research on leaching of carbonaceous gold ore with copper-ammoniathiosulfate solution. *Miner. Eng.* **2019**, *137*, 232–240.
20. Aylmore, M.G.; Muir, D.M. Thiosulfate leaching of gold—A review. *Miner. Eng.* **2001**, *14*, 135–174. [CrossRef]
21. Grosse, A.C.; Dicinoski, G.W.; Shaw, M.J.; Haddad, P.R. Leaching and recovery of gold using ammoniacal thiosulfate leach liquors (a review). *Hydrometallurgy* **2003**, *69*, 1–21. [CrossRef]
22. Muir, D.M.; Aylmore, M.G. Thiosulphate as an alternative to cyanide for gold processing–issues and impediments. *Miner. Process. Extr. Metall.* **2004**, *113*, 2–12. [CrossRef]
23. Zhang, X.M.; Senanayake, G. A review of ammoniacal thiosulfate leaching of gold: An update useful for further research in Non-cyanide gold lixiviants. *Miner. Process. Extr. Metall. Rev.* **2016**, *37*, 385–411. [CrossRef]
24. Chandra, I.; Jeffrey, M.I. An electrochemical study of the effect of additives and electrolyte on the dissolution of gold in thiosulfate solutions. *Hydrometallurgy* **2004**, *73*, 305–312. [CrossRef]
25. Senanayake, G. The role of ligands and oxidants in thiosulfate leaching of gold. *Gold Bull.* **2005**, *38*, 170–179. [CrossRef]
26. Senanayake, G. Gold leaching by copper (II) in ammoniacal thiosulfate solutions in the presence of additives. Part I: A review of the effect of hard-soft and Lewis acid-base properties and interactions of ions. *Hydrometallurgy* **2012**, *115–116*, 1–20. [CrossRef]
27. Senanayake, G. Gold leaching by copper (II) in ammoniacal thiosulfate solutions in the presence of additives. Part II: Effect of residual Cu (II), pH and redox potentials on reactivity of colloidal gold. *Hydrometallurgy* **2012**, *115–116*, 21–29. [CrossRef]
28. Rivera, I.; Patiño, F.; Roca, A.; Cruells, M. Kinetics of metallic silver leaching in the O_2–thiosulfate system. *Hydrometallurgy* **2015**, *156*, 63–70. [CrossRef]
29. Briones, R.; Lapidus, G.T. The leaching of silver sulfide with the thiosulfate– ammonia–cupric ion system. *Hydrometallurgy* **1998**, *50*, 234–260. [CrossRef]
30. Deutsch, J.L.; Dreisinger, D.B. Silver sulfide leaching with thiosulfate in the presence of additives part I: Copper–ammonia leaching. *Hydrometallurgy* **2013**, *137*, 156–164. [CrossRef]
31. Puente-Siller, D.M.; Fuentes-Aceituno, J.C.; Nava-Alonso, F. A kinetic–thermodynamic study of silver leaching in thiosulfate–copper–ammonia–EDTA solutions. *Hydrometallurgy* **2013**, *134–135*, 124–131. [CrossRef]
32. Puente-Siller, D.M.; Fuentes Aceituno, J.C.; Nava-Alonso, F. Study of thiosulfate leaching of silver sulfide in the presence of EDTA and sodium citrate. Effect of NaOH and NH4 OH. *Hydrometallurgy* **2014**, *149*, 1–11. [CrossRef]
33. Fleming, C.A.; Mcmullen, J.; Thomas, K.G.; Wells, J.A. Recent advances in the development of an alternative to the cyanidation process: Thiosulfate leaching and resin in pulp. *Min. Metall. Explor.* **2003**, *20*, 1–9. [CrossRef]
34. Xu, B.; Yang, Y.; Li, Q.; Yin, W.; Jiang, T.; Li, G. Thiosulfate leaching of Au, Ag and Pd from a high Sn, Pb and Sb bearing decopperized anode slime. *Hydrometallurgy* **2016**, *164*, 278–287. [CrossRef]
35. Wan, R.Y.; LeVier, K.M. Solution chemistry factors for gold thiosulfate heap leaching. *Int. J. Miner. Process.* **2003**, *72*, 311–322. [CrossRef]
36. Melo, H.P.P.; Marcelo, V.H.; Moura, B.A. Leaching of gold and silver from printed circuit board of mobile phones. *Rem Rev. Esc. Minas* **2015**, *68*, 61–68.

37. Cerecedo, S.E.; Rodríguez, L.V.; Hernández, A.J.; Mendoza, A.D.; Reyes, V.M.I.; Moreno, P.E.; Salinas, R.E. Mineralization of rare earths, platinum and gold in a sedimentary deposit, found using an indirect method of exploration. *Asp. Min. Miner. Sci.* **2018**, *1*, 1–9.
38. Salinas, R.E.; Hernández, A.J.; Rivera, L.I.; Cerecedo, S.E.; Reyes, V.M.I.; Correa, C.M.; Rubio, M.D. Leaching of silver contained in mining tailings, using sodium thiosulfate: A kinetic study. *Hydrometallurgy* **2016**, *160*, 6–11. [CrossRef]
39. Jeffrey, M.I.; Breuer, P.L.; Chu, C.K. The importance of controlling oxygen addition during the thiosulfate leaching of gold ores. *Int. J. Miner. Process.* **2003**, *72*, 323–330. [CrossRef]
40. Feng, D.; Van Deventer, J.S.J. The role of oxygen in thiosulphate leaching of gold. *Hydrometallurgy* **2007**, *85*, 193–202. [CrossRef]
41. Sohn, H.Y.; Wadsworth, M.E. (Eds.) *Cinética De Los Procesos De La Metalurgia Extractiva*; Editorial Trillas: Bogota, Colombia, 1986; pp. 141–148. (In Spanish)
42. Huang, H.H. *Unit Processes in Extractive Metallurgy, Montana College of Mineral Science and Technology, Modul 2*; National Science Foundation: Washington, DC, USA, 1989.
43. Ballester, A.; Verdeja, L.F.; Sancho, J. *Metalurgia Extractiva, Fundamentos*; Editorial Síntesis: Madrid, Spain, 2000; Volume 1, pp. 182–189. (In Spanish)
44. Ter-Arakelyan, K. On technological expediency of sodium thiosulphate usage for gold extraction from raw material. *Lzv. V. U. Z. Tsvetn Metall.* **1984**, *5*, 72–76.
45. Yen, W.T.; Aghamirian, M.; Deschenes, G.; Theben, S. Gold extraction from mild refractory ore using ammonium thiosulfate. In Proceedings of the International Symposium on Gold Recovery, Montreal, QC, Canada, 3–6 May 1998.

Article

Leaching Chalcopyrite with an Imidazolium-Based Ionic Liquid and Bromide

Marcelo Rodríguez [1], Luís Ayala [2], Pedro Robles [3,*], Rossana Sepúlveda [4], David Torres [1,5], Francisco Raul Carrillo-Pedroza [6], Ricardo I. Jeldres [7] and Norman Toro [1,5,*]

1 Departamento de Ingeniería Metalúrgica y Minas, Universidad Católica del Norte, Av. Angamos 610, Antofagasta 1270709, Chile; marceloandres.ra@gmail.com (M.R.); david.Torres@sqm.com (D.T.)
2 Faculty of Engineering and Architecture, Universidad Arturo Prat, Almirante Juan José Latorre 2901, Antofagasta 1244260, Chile; luisayala01@unap.cl
3 Escuela de Ingeniería Química, Pontificia Universidad Católica de Valparaíso, Valparaíso 2340000, Chile
4 Departamento de Ingeniería en Metalurgia, Universidad de Atacama, Copiapó 1531772, Chile; rossana.sepulveda@uda.cl
5 Department of Mining, Geological and Cartographic Department, Universidad Politécnica de Cartagena, 30203 Murcia, Spain
6 Facultad de Metalurgia, Universidad Autónoma de Coahuila, Carretera 57 Km 5, Monclova C.P. 25710, Mexico; raul.carrillo@uadec.edu.mx
7 Departamento de Ingeniería Química y Procesos de Minerales, Universidad de Antofagasta, Antofagasta 1270300, Chile; ricardo.jeldres@uantof.cl
* Correspondence: pedro.robles@pucv.cl (P.R.); ntoro@ucn.cl (N.T.); Tel.: +56-552-651-021 (P.R.)

Received: 10 December 2019; Accepted: 24 January 2020; Published: 27 January 2020

Abstract: The unique properties of ionic liquids (ILs) drive the growing number of novel applications in different industries. The main features of ILs are high thermal stability, recyclability, low flash point, and low vapor pressure. This study investigated pure chalcopyrite dissolution in the presence of the ionic liquid 1-butyl-3-methylimidazolium hydrogen sulfate, [BMIm]HSO_4, and a bromide-like complexing agent. The proposed system was compared with acid leaching in sulfate media with the addition of chloride and bromide ions. The results demonstrated that the use of ionic liquid and bromide ions improved the chalcopyrite leaching performance. The best operational conditions were at a temperature of 90 °C, with an ionic liquid concentration of 20% and 100 g/L of bromide.

Keywords: leaching; chalcopyrite; ionic liquid; bromide

1. Introduction

Chalcopyrite ($CuFeS_2$) is the most important source of primary copper sulfides, representing about 70% of global copper reserves. The principal method for its processing is through pyrometallurgical procedures, which is highly efficient. Still, the high operational costs and restricted availability of energy and water make hydrometallurgical methods an attractive opportunity [1–5]. However, the slow dissolution rate in traditional sulfate-acid systems makes this challenging or even impracticable for industrial purposes. Chalcopyrite is the most refractory copper sulfide, so the leaching is even more gradual than other copper sulfides like chalcocite (Cu_2S) and bornite (Cu_5FeS_4) [6,7].

Diverse studies have contemplated further methods for leaching chalcopyrite, including the use of alkaline glycine solution [8,9], black carbon [10], hydrogen peroxide in hydrochloric acid solutions [11], bacteria [12,13], chlorinated media [14], and ionic liquids [15]. Ionic liquids (ILs) are salts formed by an organic cation and an anion which can be organic or inorganic. They are found as liquids at low temperatures, having a wide range of operating temperatures [16]. Among their excellent properties are their high degree of thermal stability, wide electrochemical range, recyclability, low flammability, and low vapor pressure [17]. Additionally, they are adaptable to specific chemical tasks [18–21].

Relating with hydrometallurgy, ILs are considered "green reactive agents", and they have been applied to remove contaminants, extracting metal ions and additives by electrodeposition [19,22–30].

The application of ionic liquids in leaching is one of the most relevant advancements for the treatment of chalcopyrite. These compounds offer an environmentally friendly option since the operations involve lower temperatures, less energy consumption, and reduced acid use. They also have unique attributes as salts that melt at room temperature [22,31]. Hu et al. [31] studied the leaching of chalcopyrite with the addition of hydrogen peroxide as an oxidizing agent, and the ionic liquids 1-hexyl-3-methyl-imidazolium hydrogen sulfate ([hmim][HSO_4]), 1-ethyl-3-methylimidazolium hydrogen sulfate ([emim][HSO_4]), 1-octyl-3-methylimidazolium hydrogen sulfate ([omim][HSO_4]), and 1-butyl-3-methylimidazolium hydrogen sulfate ([bmim][HSO_4]). The authors obtained the highest copper dissolution of 98.3% with 10% (*v/v*) of IL [hmim][HSO_4] in an aqueous solution, with a particle size of under 45 μm, 25% hydrogen peroxide as an oxidant, a temperature of 45 °C, and a leaching time of 120 min. Kuzmina et al. [32] studied chalcopyrite dissolution using the series 1-butylimidazolium hydrogen sulfate ([HC_4im][HSO_4]), 1-butyl-3-methylimidazolium hydrogen sulfate ([bmim][HSO_4]), imidazolium hydrogen sulfate ([HHim][HSO_4]), ethyl ammonium hydrogen sulfate ([N_{0002}][HSO_4]), 1-butyl-3-methylimidazolium dicyanamide ([bmim][$N(CN)_2$]), 1-butylimidazolium nitrate ([HC_4im][NO_3]), and 1-butyl-3-methylimidazolium acetate ([bmim][OAc]), and found that the best IL to leach chalcopyrite was 1-butylimidazolium hydrogen sulfate, although the recovery rate was low without an oxidizing agent. Lastly, the ionic liquid [bmim][HSO_4] has shown the best performance in leaching mineral species, including refractory ores and precious metals [33–35].

Whitehead et al. [35] employed the ionic liquid [bmim][HSO_4] to leach chalcopyrite, resulting in markedly superior performance in the rate of copper dissolution terms. The authors compared the ionic liquid system with the traditional ferric acid medium and obtained similar copper recoveries, but the system employing the ionic liquid was much faster in the initial stages of copper leaching. The system adopted by the authors included a base for comparison using 1 M H_2SO_4 at 70 °C, which resulted in a dissolution of 23.3%. On the other hand, with the use of the ionic liquid from 10% to 100% (*w/w*), the dissolution rates were between 55.7% and 86.6%. Dong et al. [15] evaluated the impact of the concentration of ionic liquid on chalcopyrite solution. The researchers obtained the best copper dissolution of 88% at 70 °C, with an ionic liquid level of 60% and a partial oxygen pressure of 17 kPa.

The literature describes the ionic liquid [bmim][HSO_4] as a chalcopyrite leaching agent with the capacity to produce protons, which occurs according to the following disassociation reactions [15,36]:

$$[\text{bmin}]^{+}[\text{HSO}_4^{-}] \overset{k}{\leftrightarrow} [\text{bmin}]^{+} + \text{H}^{+} + \text{SO}_4^{2-} \tag{1}$$

$$\text{K} = \frac{\left[[\text{bmim}]^{+}\right][\text{H}]^{+}\left[\text{SO}_4^{2-}\right]}{[\text{bmim}]\left[\,\text{HSO}_4^{-}\right]} \tag{2}$$

The main characteristic of the ionic liquid [bmim][HSO_4] is the high decrease in the pH of the leaching solution. With concentrations of [bmim][HSO_4] of 0.02 to 5.10 mol/L, the pH of the solution decreases from 1.7 to −0.61 [15].

Another method to treat refractory ores is through halide systems (fluoride, chloride, bromide, iodide, and astatine), which have environmental advantages over the traditional system since they act as a substitute for cyanide in the leaching of gold and platinum. Leaching ores with halides is characterized by high kinetic rates, complex stability, and the regeneration of leaching agents, among other characteristics [37–44]. The excellent results reported with the addition of halides include the use of iodine in leaching copper and gold [45–47]. The successful use of chlorinated media has been widely reported with concentrates and other copper sulfide products [14,48,49].

Bromide is a possible substitute for cyanide for leaching gold and other precious metals [50–57], which can be complexed by the bromide ion (Br^-) and oxidized by bromine (Br_2). The mechanism proceeds according to the following equations [44,50]:

$$2\,Au + 3Br_2 + 2Br^- \rightarrow 2AuBr_4^- \tag{3}$$

$$Pt + 2Br_2 + 2Br^- \rightarrow PtBr_6^{2-}. \tag{4}$$

$$Pd + Br_2 + 2Br^- \rightarrow PdBr_4^{2-}. \tag{5}$$

There are few publications on copper dissolution with bromide. Still, they are numerous enough to reveal that bromide systems are highly corrosive to metallic copper, being effective in leaching copper sulfides and concentrates. The copper corrosion in the presence of CuBr is an electrochemical process, and the general mechanism proposed is the following [58,59]:

$$Cu + Cu^{2+} + 2Br^- \rightarrow 2CuBr. \tag{6}$$

McDonald et al. [60] described the leaching of chalcopyrite with the addition of copper(II) and bromide/sodium bromide. The central conclusion raised by the authors was that a cupric bromide system is faster than a cupric chloride one. The mechanism proposed by McDonald et al. [60] involves the Cu(I) production as follows:

$$CuFeS_2 + 3Cu^{2+} \frac{NaBr}{H_2O} \rightarrow 4Cu^+ + Fe^{2+} + 2S^0. \tag{7}$$

On the other hand, the mechanism proposed by Han and Meng [61] described the dissolution of chalcopyrite by the following equation, and involve the Cu(II) production (represented by $CuSO_4$):

$$CuFeS_2 + 5/2Br_2 + 3/2H_2SO_4 + 1/6Na_2SO_4 + 2H_2O \leftrightarrow S^0 + 5Br^- + CuSO_4 + 1/3NaFe_3(SO_4)_2(OH)_6 + 5H^+. \tag{8}$$

The reactions shown above represent an ideal bromide system in sulfate media, but in the presence of nitrate ions, bromide reacts to form bromine. The mechanism is shown in the following reaction [61]:

$$5Br^- + 5/3NO_3^- + 20/3H^+ \leftrightarrow 5/2Br_2 + 5/3NO + 10/3H_2O \tag{9}$$

The hydrometallurgical processing of copper sulfides is an interesting alternative to substitute the pyrometallurgical methods. This research displays the experimental results of leaching chalcopyrite with bromide and the ionic liquid 1-butyl-3-methylimidazolium hydrogen sulfate. The approach proposed in this paper represents a promising option, highlighting its effectiveness with refractory species, as well as environmental benefits.

2. Materials and Methods

2.1. Chalcopyrite

The chalcopyrite mineral was provided by the Michilla Mining Company (Antofagasta, Chile). The mineralogical composition was monitored by X-ray diffraction (XRD), using an automatic and computerized X-ray diffractometer (Bruker model Advance D8, Billerica, MA, USA). The results that appear in Figure 1 indicated that it was 99.9% chalcopyrite. The chemical analysis was performed by atomic emission spectrometry via induction-coupled plasma (ICP-AES). The chalcopyrite concentrate sample was digested using aqua regia and hydrofluoric acid. Table 1 shows the chemical composition obtained by ICP-OES; the difference between copper and iron determined the sulfur. The samples were ground in a porcelain mortar to reach sizes between 47 and 38 μm. The procedures were performed in the Applied Geochemistry Laboratory of the Department of Geological Sciences of the Universidad Católica del Norte.

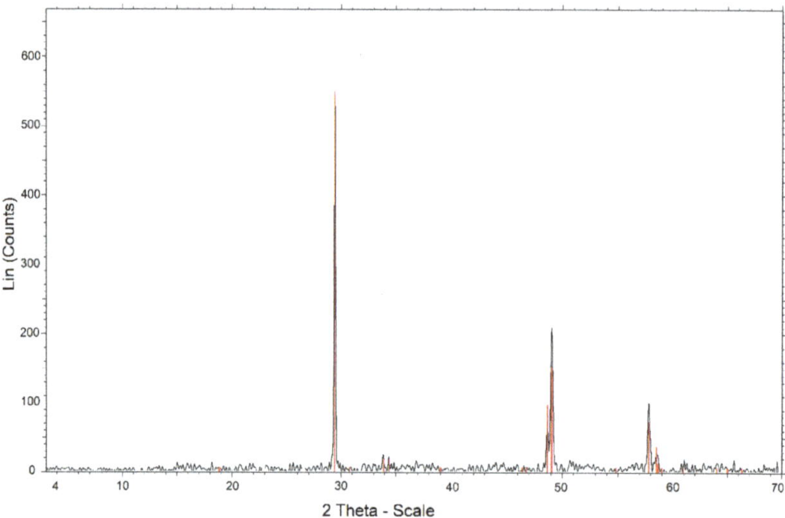

Figure 1. X-ray diffractogram of chalcopyrite ore.

Table 1. Chemical analysis of the sample.

Element	Cu	Fe	S
Content (%)	33.9	30.6	35.5

2.2. Leaching and Leaching Tests

The lixiviant was prepared using the ionic liquid 1-butyl-3-methylimidazolium hydrogen sulfate (94.5–95%, molecular weight of 236.29 g/mol, Sigma Aldrich) and sulfuric acid (95–97%, P.A, Merck, the density of 1.84 kg/L and molecular weight of 98.08 g/mol). Both reagents were dissolved in distilled water. The NaCl and NaBr were of analytical grade, from Merck.

2.3. Leaching Test

Leaching tests were carried out in a 50 mL glass reactor with a 0.01 S/L ratio and an IL concentration of 20% (*v*/*v*) in an aqueous solution. Chalcopyrite ore (200 mg) was kept in agitation and suspension with the use of a five-position magnetic stirrer (IKA ROS, CEP 13087-534, Campinas, Brazil) at 600 rpm. The temperature range tested in the experiments was 30–90 °C controlled by an oil-heated circulator (Julabo). All tests were duplicated, and the measurements (or analyses) were carried out in 5 mL aliquots and diluted using atomic absorption spectrometry (with a variation coefficient of ≤5% and a relative error of 5–10%). The pH and oxidation–reduction potential (ORP) of the leaching solutions were measured with a pH-ORP meter (HANNA HI-4222), by a combination of an ORP electrode cell composed of a platinum working electrode and a saturated Ag/AgCl reference electrode.

3. Results

3.1. Influence of Agitation Velocity

Figure 2 gives the copper extraction from chalcopyrite at different mixing intensities. The optimal extraction rate (31%) was obtained at 800 rpm. Agitating at 200 and 400 rpm, a portion of the particles settled, being unable to remain in the suspension, then the copper dissolution rates were low. The extraction rate was lower at 1000 rpm than at 600 or 800 rpm since part of the mineral was not homogenously agitated and adhered to the reactor wall. A previous study [62] argued that the essential factor is that the mineral particles can keep in suspension, rather than having a high mixing intensity.

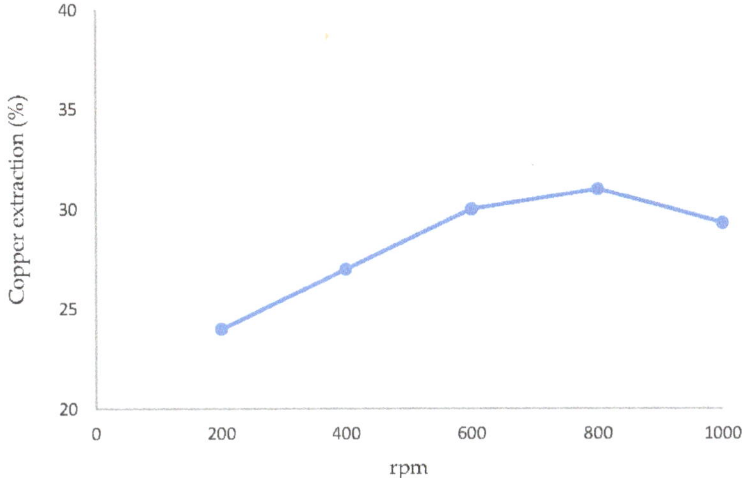

Figure 2. Effect of speed stirring on copper dissolution (temperature 80 °C, 20 g/L of Br^-, 20% (*v*/*v*) [bmim][HSO_4], pulp density 10 g/L, 24 h).

3.2. *Effect on Chloride Concentration*

Figure 3 exhibits the differences in recovery rates over time for three cases: (i) a mixture of sulfuric acid and bromide, (ii) a mixture of sulfuric acid and chloride, and (iii) sulfuric acid alone. The best results were achieved for the mixture of sulfuric acid with chloride. Although no studies have used the same parameters, Encina and Aguayo [63], who worked with galena, concluded that leaching a $CuBr_2$–NaBr system was faster than a copper-chloride system under the same conditions.

The inclusion of IL is compared in Figure 4. Three cases were studied: (i) IL alone, (ii) IL with bromide, and (iii) IL with chloride. The highest recovery rate was 86%, obtained with the mixture between the IL and bromide. The recovery rate was lower with pure ionic liquid than with IL and sulfuric acid (Figure 3). A comparison of Figures 3 and 4 provide the differences between the two leaching agents (H_2SO_4 and IL), with the addition of both halogens (Cl^- and Br^-). Interestingly, the better extraction was achieved with ionic liquids considering bromide as an oxidizing agent (86.0%). The opposite occurred with sulfuric acid, where better results were obtained with chloride (89.2%).

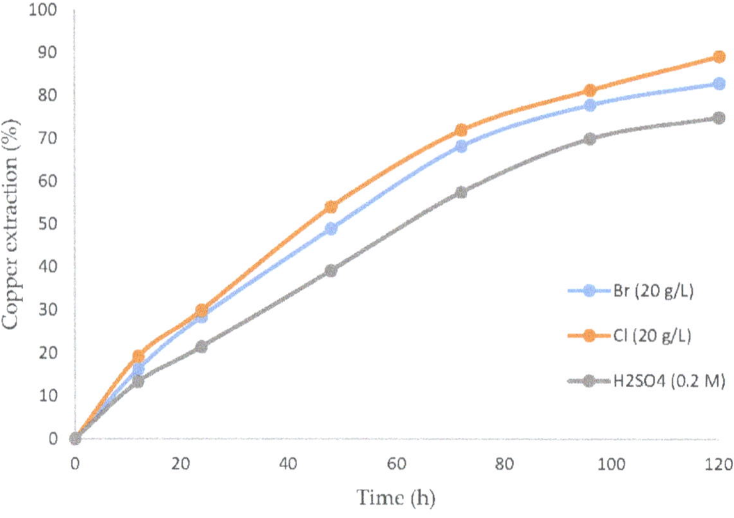

Figure 3. Copper extraction percentage over time (h) comparing: sulfuric acid with bromine, sulfuric acid with chloride, and only sulfuric acid (temperature 80 °C, 20% (*v*/*v*) [bmim][HSO_4], pulp density 10 g/L).

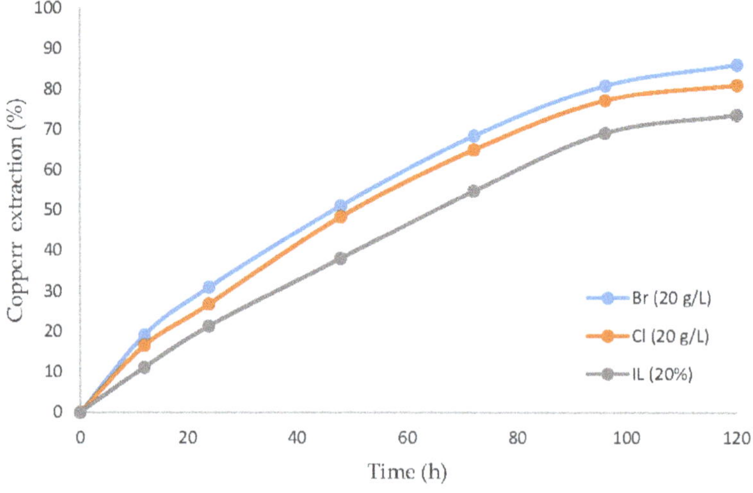

Figure 4. Copper extraction percentage over time (h) comparing: IL with bromine, IL with chloride, and only IL (temperature 80 °C, 20% (*v*/*v*) [bmim][HSO_4], pulp density 10 g/L).

3.3. The Effect of Bromide on the Copper Extraction Rate

Figure 5 presents the results of using a mixture of IL and bromide. As expected, the copper extraction rate was enhanced by increasing the concentration of bromide. The highest copper extraction was 83.6%, obtained with 100 g/L of bromide and a leaching time of 48 h. Although no studies have presented the leaching of chalcopyrite with bromide, it should be noted that there are precedents with similar systems. Our results are similar to those of McDonald et al. [60], who pointed out the benefits of using bromide when dealing with sulfur minerals. Potential and pH levels were respectively in the ranges of 560 to 615 mV and −1.0 to 0.4 in all the tests in this study.

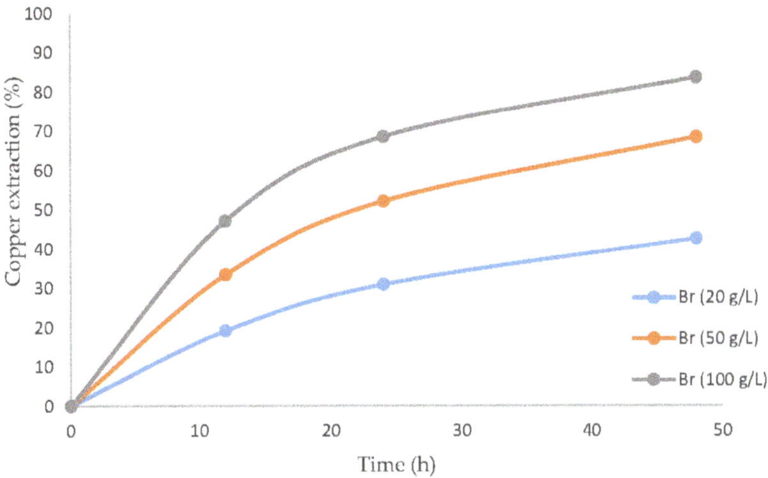

Figure 5. Graph of the copper extraction percentage over time using an ionic liquid and varying concentrations of bromide.

3.4. *Influence of Temperature on Copper Extraction Rates*

Figure 6 shows that copper extraction increased with higher temperatures. This can be explained by the fact that applying heat is the most effective way to provide sufficient activation energy to the molecules to increase the reaction kinetics. The velocity of the chemical reaction approximately doubled with every increase of 10 °C [64], and the highest copper extraction (84.6%) was obtained at 90 °C (maximum temperature in this work). The results were similar to those reported by Aguirre et al. [2], who worked with ionic liquid in chlorinated media.

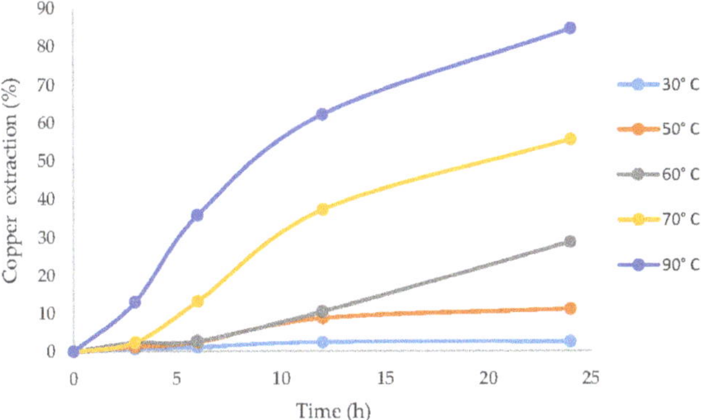

Figure 6. The effect of temperature on copper extraction from chalcopyrite using ionic liquid at 20% (*v/v*) and 100 g/L of Br^-.

The experimental data were analyzed deeply with the recessive nucleus model to control the surface reaction, for mono-sized particles, and explaining Cu extraction from $CuFeS_2$ with the following equation:

$$1 - (1-x)^{\frac{1}{3}} = k_d t, \tag{10}$$

where x is the copper concentration in the solution over time, k_d is the kinetic constant, and t is the reaction time. The plot of the shrinking core model to explain the kinetic reaction mechanism is shown in Figure 7.

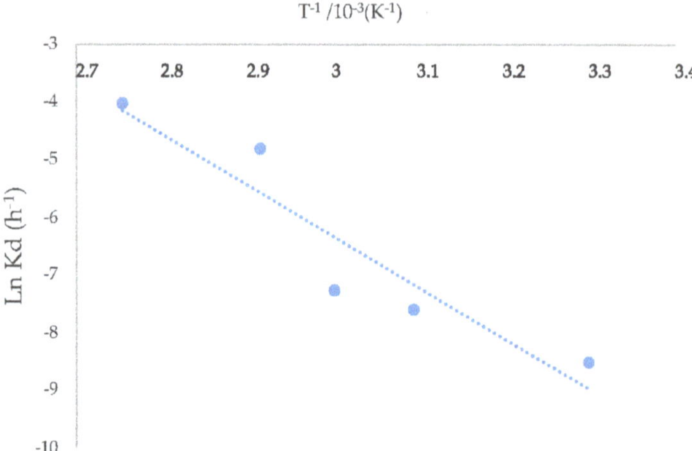

Figure 7. Arrhenius graph for chalcopyrite extraction using ionic liquid at 20% (*v/v*) and 100 g/L of Br^-.

The kinetic constant k_d for each temperature was obtained based on the slopes shown in Figure 7. The kinetics study showed that the shrinking core model was consistent with the experimental data. The Arrhenius graph (Figure 8) was obtained with the apparent kinetic constants k_d. The activation energy in the studied temperature range was 75.73 kJ/mol, similar to reported values for other IL systems [32]. The obtained results corroborate the proposed model for the system under study. On the other hand, the process is controlled by the chemical reaction and not by diffusive processes, this depending on the marked effect of temperature on kinetics.

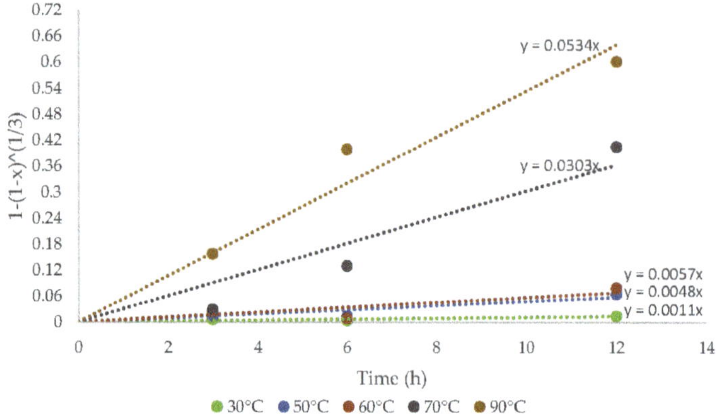

Figure 8. Plot of $1 - (1 - x)^{1/3}$ versus $CuFeS_2$ dissolution time.

Figure 9 shows the solution redox potential values over time for different temperatures. The potential varied between 570 and 600 mV at temperatures of 30 and 50 °C, where a first increase was found, followed by a decrease at the end, which reduced the copper dissolution kinetics. The potential at 60 °C moved between 590 and 610 mV, with copper extraction of 30% after 24 h. No tendency toward mineral passivation was noted at this temperature, in contrast to what observed at 30 and 50 °C. At 70 and 90 °C, the potential values located between 610 and 650 mV and high rates of copper extraction (55% and 85%) were collected within 24 h. Previous studies indicated that the typical potential range for leaching chalcopyrite is between 450 and 750 mV [65,66]. These values were optimized by Velásquez-Yévenes et al. [67], who worked with low chloride concentrations and a controlled potential. The authors stated that the appropriate potential range to leach chalcopyrite is between 550 and 620 mV. However, the potential values obtained in the present study differ from that range, considering that at higher concentrations of chloride, the potential range does not increase, because it is a complexing and non-oxidizing agent [67]. These results concur with those obtained in the present work, in which linear dissolution rates were obtained between 610 and 650 mV.

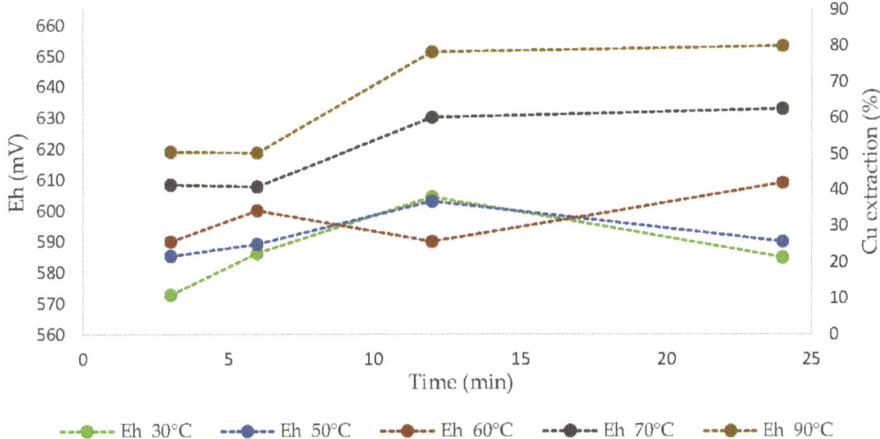

Figure 9. Effect of potential Cu extraction from $CuFeS_2$ using ionic liquid at 20% (*v/v*) and 100 g/L of Br^- at temperatures of 30, 50, 60, 70, and 90 °C.

4. Conclusions

The present study exposes the consequences of leaching chalcopyrite using ionic liquid (1-butyl-3-methyl-imidazolium hydrogen sulfate) and sodium bromide. The assays were conducted to compare ionic liquid and sulfuric acid with the addition of halogens (bromide and chloride). The main findings were:

1. There was no significant improvement in Cu extractions by increasing the agitation rate of the system, and it was only necessary to keep the suspended particles dispersed.
2. When applying ionic liquid dissolved in water, the addition of bromide as an oxidant provided the best performance for the copper dissolution from chalcopyrite, while if sulfuric acid was used, the most suitable outcomes were achieved with chloride as an oxidant.
3. High concentrations of bromide in the system significantly reduced the necessary copper extraction.
4. The best results of this research were obtained at 90 °C, with 20% (*v/v*) of ionic liquid, and 100 g/L of bromide. The collected copper recovery was 85%.
5. High concentrations of bromide (100 g/L) raised the range of electrochemical potential suitable for leaching, with dissolution rates between 610 and 650 mV at 90 °C.

Author Contributions: M.R. and R.S. contributed in research and wrote paper, N.T. and R.I.J. contributed in project administration, L.A., D.T. and P.R. contributed resources, P.R. contributed in review and editing and F.R.C.-P. contributed in data curing. All authors have read and agreed to the published version of the manuscript.

Funding: This research received no external funding.

Acknowledgments: The authors are grateful for the contribution of the Scientific Equipment Unit- MAINI of the Universidad Católica del Norte for facilitating the chemical analysis of the solutions. Pedro Robles thanks the Pontificia Universidad Católica de Valparaíso for the support provided. We also thank Conicyt Fondecyt 11,171,036 and Centro CRHIAM Project Conicyt/Fondap/15130015. Pedro Robles thanks the Pontificia Universidad Católica de Valparaíso for grants Project DI EMERGENTE-039.381/19.

Conflicts of Interest: The authors declare no conflict of interest.

References

1. Bobadilla-Fazzini, R.A.; Pérez, A.; Gautier, V.; Jordan, H.; Parada, P. Primary copper sulfides bioleaching vs. chloride leaching: Advantages and drawbacks. *Hydrometallurgy* **2017**, *168*, 26–31. [CrossRef]
2. Aguirre, C.L.; Toro, N.; Carvajal, N.; Watling, H.; Aguirre, C. Leaching of chalcopyrite (CuFeS2) with an imidazolium-based ionic liquid in the presence of chloride. *Miner. Eng.* **2016**, *99*, 60–66. [CrossRef]
3. Parada, F.; Jeffrey, M.I.; Asselin, E. Hydrometallurgy Leaching kinetics of enargite in alkaline sodium sulphide solutions. *Hydrometallurgy* **2014**, *146*, 48–58. [CrossRef]
4. Choubey, P.K.; Lee, J.; Kim, M.; Kim, H.S. Conversion of chalcopyrite to copper oxide in hypochlorite solution for selective leaching of copper in dilute sulfuric acid solution. *Hydrometallurgy* **2018**, *178*, 224–230. [CrossRef]
5. Baba, A.A.; Ayinla, K.I.; Adekola, F.A.; Ghosh, M.K.; Ayanda, O.S.; Bale, R.B.; Sheik, A.R.; Pradhan, S.R. A Review on Novel Techniques for Chalcopyrite Ore Processing. *Int. J. Min. Eng. Miner. Process.* **2012**, *1*, 1–16. [CrossRef]
6. Watling, H.R. Chalcopyrite hydrometallurgy at atmospheric pressure: 1. Review of acidic sulfate, sulfate-chloride and sulfate-nitrate process options. *Hydrometallurgy* **2013**, *140*, 163–180. [CrossRef]
7. Veloso, T.C.; Peixoto, J.J.M.; Pereira, M.S.; Leao, V.A. Kinetics of chalcopyrite leaching in either ferric sulphate or cupric sulphate media in the presence of NaCl. *Int. J. Miner. Process.* **2016**, *148*, 147–154. [CrossRef]
8. Shin, D.; Ahn, J.; Lee, J. Kinetic study of copper leaching from chalcopyrite concentrate in alkaline glycine solution. *Hydrometallurgy* **2019**, *183*, 71–78. [CrossRef]
9. Eksteen, J.J.; Oraby, E.A.; Tanda, B.C. A conceptual process for copper extraction from chalcopyrite in alkaline glycinate solutions. *Miner. Eng.* **2017**, *108*, 53–66. [CrossRef]
10. Nakazawa, H. Effect of carbon black on chalcopyrite leaching in sulfuric acid media at 50 °C. *Hydrometallurgy* **2018**, *177*, 100–108. [CrossRef]
11. PETROVIĆ, S.J.; BOGDANOVIĆ, G.D.; ANTONIJEVIĆ, M.M. Leaching of chalcopyrite with hydrogen peroxide in hydrochloric acid solution. *Trans. Nonferrous Met. Soc. China* **2018**, *28*, 1444–1455. [CrossRef]
12. Ma, L.; Wang, X.; Liu, X.; Wang, S.; Wang, H. Intensified bioleaching of chalcopyrite by communities with enriched ferrous or sulfur oxidizers. *Bioresour. Technol.* **2018**, *268*, 415–423. [CrossRef] [PubMed]
13. Esmailbagi, M.R.; Schaffie, M.; Kamyabi, A.; Ranjbar, M. Microbial assisted galvanic leaching of chalcopyrite concentrate in continuously stirred bioreactors. *Hydrometallurgy* **2018**, *180*, 139–143. [CrossRef]
14. Velásquez-Yévenes, L.; Torres, D.; Toro, N. Leaching of chalcopyrite ore agglomerated with high chloride concentration and high curing periods. *Hydrometallurgy* **2018**, *181*, 215–220. [CrossRef]
15. Dong, T.; Hua, Y.; Zhang, Q.; Zhou, D. Leaching of chalcopyrite with Brønsted acidic ionic liquid. *Hydrometallurgy* **2009**, *99*, 33–38. [CrossRef]
16. Wei, G.-T.; Yang, Z.; Chen, C.-J. Room temperature ionic liquid as a novel medium for liquid/liquid extraction of metal ions. *Anal. Chim. Acta* **2003**, *488*, 183–192. [CrossRef]
17. Keskin, S.; Kayrak-Talay, D.; Akman, U.; Hortaçsu, Ö. A review of ionic liquids towards supercritical fluid applications. *J. Supercrit. Fluids* **2007**, *43*, 150–180. [CrossRef]
18. Singh, V.; Kaur, S.; Sapehiyia, V.; Singh, J.; Kad, G.L. Microwave accelerated preparation of [bmim][HSO4] ionic liquid: An acid catalyst for improved synthesis of coumarins. *Catal. Commun.* **2005**, *6*, 57–60. [CrossRef]
19. Sepúlveda, R.; Romero, J.; Sánchez, J. Copper removal from aqueous solutions by means of ionic liquids containing a B-diketone and the recovery of metal complexes by supercritical fluid extraction. *J. Chem. Technol. Biotechnol.* **2014**, *89*, 899–908. [CrossRef]

20. Quijada-Maldonado, E.; Romero, J.; Osorio, I. Selective removal of iron(III) from synthetic copper(II) pregnant leach solutions using [bmim][Tf2N] as diluent and TFA as extracting agent. *Hydrometallurgy* **2016**, *159*, 54–59. [CrossRef]
21. Janssen, C.H.C.; Macías-Ruvalcaba, N.A.; Aguilar-Martínez, M.; Kobrak, M.N. Metal extraction to ionic liquids: The relationship between structure, mechanism and application. *Int. Rev. Phys. Chem.* **2015**, *34*, 591–622. [CrossRef]
22. Park, J.; Jung, Y.; Kusumah, P.; Lee, J.; Kwon, K.; Lee, C.K. Application of ionic liquids in hydrometallurgy. *Int. J. Mol. Sci.* **2014**, *15*, 15320–15343. [CrossRef]
23. Liu, Y.; Guo, L.; Zhu, L.; Sun, X.; Chen, J. Removal of Cr(III, VI) by quaternary ammonium and quaternary phosphonium ionic liquids functionalized silica materials. *Chem. Eng. J.* **2010**, *158*, 108–114. [CrossRef]
24. Castillo, J.; Coll, M.T.; Fortuny, A.; Navarro Donoso, P.; Sepúlveda, R.; Sastre, A.M. Cu(II) extraction using quaternary ammonium and quaternary phosphonium based ionic liquid. *Hydrometallurgy* **2014**, *141*, 89–96. [CrossRef]
25. Valdés, H.; Sepúlveda, R.; Romero, J.; Valenzuela, F.; Sánchez, J. Near critical and supercritical fluid extraction of Cu(II) from aqueous solutions using a hollow fiber contactor. *Chem. Eng. Process. Process. Intensif.* **2013**, *65*, 58–67. [CrossRef]
26. Sun, X.; Ji, Y.; Zhang, L.; Chen, J.; Li, D. Separation of cobalt and nickel using inner synergistic extraction from bifunctional ionic liquid extractant (Bif-ILE). *J. Hazard. Mater.* **2010**, *182*, 447–452. [CrossRef]
27. Domańska, U.; Rękawek, A. Extraction of metal ions from aqueous solutions using imidazolium based ionic liquids. *J. Solution Chem.* **2009**, *38*, 739–751. [CrossRef]
28. Sepúlveda, R.; Castillo, J.; Plaza, A.; Sánchez, J.; Torres, A.; Romero, J. Improvement of recovery performance in the solvent extraction of Cu(II) using [bmim][Tf2N] and a B-diketone as extractant and its stripping with supercritical carbon dioxide. *J. Supercrit. Fluids* **2017**, *128*, 26–31. [CrossRef]
29. Zhang, D.; Dong, L.; Li, Y.; Wu, Y.; Ma, Y.; Yang, B. Copper leaching from waste printed circuit boards using typical acidic ionic liquids recovery of e-wastes' surplus value. *Waste Manag.* **2018**, *78*, 191–197. [CrossRef]
30. Huang, J.; Chen, M.; Chen, H.; Chen, S.; Sun, Q. Leaching behavior of copper from waste printed circuit boards with Brønsted acidic ionic liquid. *Waste Manag.* **2014**, *34*, 483–488. [CrossRef]
31. Hu, J.; Tian, G.; Zi, F.; Hu, X. Leaching of chalcopyrite with hydrogen peroxide in 1-hexyl-3-methyl-imidazolium hydrogen sulfate ionic liquid aqueous solution. *Hydrometallurgy* **2017**, *169*, 1–8. [CrossRef]
32. Kuzmina, O.; Symianakis, E.; Godfrey, D.; Albrecht, T.; Welton, T. Ionic liquids for metal extraction from chalcopyrite: Solid, liquid and gas phase studies. *Phys. Chem. Chem. Phys.* **2017**, *19*, 21556–21564. [CrossRef]
33. Rüşen, A.; Topçu, M.A. Investigation of an alternative chemical agent to recover valuable metals from anode slime. *Chem. Pap.* **2018**, *72*, 2879–2891. [CrossRef]
34. Whitehead, J.A.; Lawrance, G.A.; McCluskey, A. "Green" leaching: Recyclable and selective leaching of gold-bearing ore in an ionic liquid. *Green Chem.* **2004**, *6*, 313–315. [CrossRef]
35. Whitehead, J.A.; Zhang, J.; Pereira, N.; McCluskey, A.; Lawrance, G.A. Application of 1-alkyl-3-methyl-imidazolium ionic liquids in the oxidative leaching of sulphidic copper, gold and silver ores. *Hydrometallurgy* **2007**, *88*, 109–120. [CrossRef]
36. Crowhurst, L.; Mawdsley, P.R.; Perez-Arlandis, J.M.; Salter, P.A.; Welton, T. Solvent-solute interactions in ionic liquids. *Phys. Chem. Chem. Phys.* **2003**, *5*, 2790–2794. [CrossRef]
37. Hilson, G.; Monhemius, A.J. Alternatives to cyanide in the gold mining industry: What prospects for the future? *J. Clean. Prod.* **2006**, *14*, 1158–1167. [CrossRef]
38. Lampinen, M.; Seisko, S.; Forsström, O.; Laari, A.; Aromaa, J.; Lundström, M.; Koiranen, T. Mechanism and kinetics of gold leaching by cupric chloride. *Hydrometallurgy* **2017**, *169*, 103–111. [CrossRef]
39. Seisko, S.; Lampinen, M.; Aromaa, J.; Laari, A.; Koiranen, T.; Lundström, M. Kinetics and mechanisms of gold dissolution by ferric chloride leaching. *Miner. Eng.* **2018**, *115*, 131–141. [CrossRef]
40. Ibáñez, T.; Velásquez, L. Lixiviación de la calcopirita en medios clorurados. *Rev. Metal.* **2013**, *49*, 131–144. [CrossRef]
41. Ahtiainen, R.; Lundström, M.; Liipo, J. Preg-robbing verification and prevention in gold chloride-bromide leaching. *Miner. Eng.* **2018**, *128*, 153–159. [CrossRef]
42. Konyratbekova, S.S.; Baikonurova, A.; Akcil, A. Non-cyanide leaching processes in gold hydrometallurgy and iodine-iodide applications: A review. *Miner. Process. Extr. Metall. Rev.* **2015**, *36*, 198–212. [CrossRef]

43. Hojo, M.; Yamamoto, M.; Maeda, T.; Kawano, H.; Okamura, K. Pure gold dissolution in dilute chloric, bromic or iodic acid solution containing abundant halide ions. *J. Mol. Liq.* **2017**, *227*, 295–302. [CrossRef]
44. Mpinga, C.N.; Eksteen, J.J.; Aldrich, C.; Dyer, L. Direct leach approaches to Platinum Group Metal (PGM) ores and concentrates: A review. *Miner. Eng.* **2015**, *78*, 93–113. [CrossRef]
45. Granata, G.; Miura, A.; Liu, W.; Pagnanelli, F.; Tokoro, C. Iodide-assisted leaching of chalcopyrite in acidic ferric sulfate media. *Hydrometallurgy* **2019**, *186*, 244–251. [CrossRef]
46. Baghalha, M. The leaching kinetics of an oxide gold ore with iodide/iodine solutions. *Hydrometallurgy* **2012**, *113–114*, 42–50. [CrossRef]
47. Konyratbekova, S.S.; Baikonurova, A.; Ussoltseva, G.A.; Erust, C.; Akcil, A. Thermodynamic and kinetic of iodine-iodide leaching in gold hydrometallurgy. *Trans. Nonferrous Met. Soc. China* **2015**, *25*, 3774–3783. [CrossRef]
48. Castillo, J.; Sepúlveda, R.; Araya, G.; Guzmán, D.; Toro, N.; Pérez, K.; Rodríguez, M.; Navarra, A. Leaching of white metal in a NaCl-H_2SO_4 system under environmental conditions. *Minerals* **2019**, *9*, 319. [CrossRef]
49. Hidalgo, T.; Kuhar, L.; Beinlich, A.; Putnis, A. Kinetic study of chalcopyrite dissolution with iron(III) chloride in methanesulfonic acid. *Miner. Eng.* **2018**, *125*, 66–74. [CrossRef]
50. Sousa, R.; Futuro, A.; Fiúza, A.; Vila, M.C.; Dinis, M.L. Bromine leaching as an alternative method for gold dissolution. *Miner. Eng.* **2018**, *118*, 16–23. [CrossRef]
51. Yoshimura, A.; Takai, M.; Matsuno, Y. Novel process for recycling gold from secondary sources: Leaching of gold by dimethyl sulfoxide solutions containing copper bromide and precipitation with water. *Hydrometallurgy* **2014**, *149*, 177–182. [CrossRef]
52. Zhang, Y.; Liu, S.; Xie, H.; Zeng, X.; Li, J. Current status on leaching precious metals from waste printed circuit boards. *Procedia Environ. Sci.* **2012**, *16*, 560–568. [CrossRef]
53. Yu, H.; Zi, F.; Hu, X.; Zhong, J.; Nie, Y.; Xiang, P. The copper-ethanediamine-thiosulphate leaching of gold ore containing limonite with cetyltrimethyl ammonium bromide as the synergist. *Hydrometallurgy* **2014**, *150*, 178–183. [CrossRef]
54. van Meersbergen, M.T.; Lorenzen, L.; van Deventer, J.S.J. The electrochemical dissolution of gold in bromine medium. *Miner. Eng.* **1993**, *6*, 1067–1079. [CrossRef]
55. Hojo, M.; Iwasaki, S.; Okamura, K. Pure gold dissolution with hydrogen peroxide as the oxidizer in HBr or HI solution. *J. Mol. Liq.* **2017**, *246*, 372–378. [CrossRef]
56. Dadgar, A. Refractory concentrate gold leaching: Cyanide vs. bromine. *JOM* **1989**, *41*, 37–41. [CrossRef]
57. Encinas-Romero, M.A.; Tiburcio-Munive, G.; Yanez-Montano, M. A Kinetic Study of Gold Leaching in CuBr2-NaBr system. *J. Multidiscip. Eng. Sci. Technol.* **2015**, *2*, 2118–2121.
58. Hibbert, D.B.; Richards, S.; Gonzalves, V. The kinetics and mechanism of the corrosion of copper in acidified copper (II) bromide solution. *Corros. Sci.* **1990**, *30*, 367–376. [CrossRef]
59. Brolo, A.G.; Temperini, M.L.A.; Agostinho, S.M.L. Copper dissolution in bromide medium in the absence and presence of hexamethylenetetramine (HMTA). *Electrochim. Acta* **1998**, *44*, 559–571. [CrossRef]
60. McDonald, G.W.; Darus, H.; Langer, S.H. A cupric bromide process for hydrometallurgical recovery of copper. *Hydrometallurgy* **1990**, *24*, 291–316. [CrossRef]
61. Han, K.N.; Meng, X. Recovery of copper from its sulfides and other sources using halogen reagents and oxidants. *Mining, Metall. Explor.* **2003**, *20*, 160–164. [CrossRef]
62. Toro, N.; Saldaña, M.; Castillo, J.; Higuera, F.; Acosta, R. Leaching of manganese from marine nodules at room temperature with the use of sulfuric acid and the addition of tailings. *Minerals* **2019**, *9*, 289. [CrossRef]
63. Encinas-Romero, M.A.; Aguayo-Salinas, S. Cupric Bromide Process for the Leaching of Galena. In Proceedings of the Second Conference in Hydrometallurgy, Changsha, China, 1993; pp. 227–230.
64. Zabala, M. Comportamiento de la recuperación de cobre en una pila de lixiviación a condiciones ambientales extremas. Master's Thesis, Pontificia Universidad Católica De Valparaíso, Valparaiso, Chile, 30 December 2013.
65. Nicol, M.; Miki, H.; Velásquez-Yévenes, L. The dissolution of chalcopyrite in chloride solutions: Part 3. Mechanisms. *Hydrometallurgy* **2010**, *103*, 86–95. [CrossRef]
66. Nicol, M.; Basson, P. The anodic behaviour of covellite in chloride solutions. *Hydrometallurgy* **2017**, *172*, 60–68. [CrossRef]

67. Yévenes, L.V.; Miki, H.; Nicol, M. The dissolution of chalcopyrite in chloride solutions: Part 2: Effect of various parameters on the rate. *Hydrometallurgy* **2010**, *103*, 80–85. [CrossRef]

Article

Depression of Pyrite in Seawater Flotation by Guar Gum

César I. Castellón [1], Eder C. Piceros [2], Norman Toro [3,4], Pedro Robles [5], Alejandro López-Valdivieso [6] and Ricardo I. Jeldres [1,*]

[1] Departamento de Ingeniería Química y Procesos de Minerales, Facultad de Ingeniería, Universidad de Antofagasta, Av. Angamos 601, Antofagasta 1240000, Chile; castelloniv@gmail.com
[2] Faculty of Engineering and Architecture, Universidad Arturo Prat, PO Box 121, Iquique 1100000, Chile; edpicero@unap.cl
[3] Departamento de Ingeniería Metalúrgica y Minas, Universidad Católica del Norte, Antofagasta 1270709, Chile; ntoro@ucn.cl
[4] Department of Mining, Geological and Cartographic Department, Universidad Politécnica de Cartagena, 30202 Murcia, Spain
[5] Escuela de Ingeniería Química, Pontificia Universidad Católica de Valparaíso, Valparaíso 2340000, Chile; pedro.robles@pucv.cl
[6] Instituto de Metalurgia, Laboratorio de Química de Superficies, Universidad Autónoma de San Luis Potosí, Av. Sierra Leona 550, San Luis Potosi 78210, México; alopez@uaslp.mx
* Correspondence: ricardo.jeldres@uantof.cl; Tel.: +56-552-637-901

Received: 21 January 2020; Accepted: 7 February 2020; Published: 11 February 2020

Abstract: The application of guar gum for pyrite depression in seawater flotation was assessed through microflotation tests, Focused Beam Reflectance Measurements (FBRM), and Particle Vision Measurements (PVM). Potassium amyl xanthate (PAX) and methyl isobutyl carbinol (MIBC) were used as collector and frother, respectively. Chemical species on the pyrite surface were characterized by Fourier-transform infrared spectroscopy (FTIR) spectroscopy. The microflotation tests were performed at pH 8, which is the pH at the copper sulfide processing plants that operate with seawater. Pyrite flotation recovery was correlated with FBRM and PVM characterization to delineate the pyrite depression mechanisms by the guar gum. The high flotation recovery of pyrite with PAX was significantly lowered by guar gum, indicating that this polysaccharide could be used as an effective depressant in flotation with sea water. FTIR analysis showed that PAX and guar gum co-adsorbed on the pyrite surface, but the highly hydrophilic nature of the guar gum embedded the hydrophobicity due to the PAX. FBRM and PVM revealed that the guar gum promoted the formation of flocs whose size depended on the addition of guar gum and PAX. It is proposed that the highest pyrite depression occurred not only because of the hydrophilicity induced by the guar gum, but also due to the formation of large flocs, which could not be transported by the bubbles to the froth phase. Furthermore, it is shown that an overdose of guar gum hindered the depression effect due to redispersion of the flocs.

Keywords: seawater flotation; pyrite depression; guar gum; FBRM

---

1. Introduction

Pyrite is the most abundant iron sulfide mineral, and it is commonly associated as a gangue with ores of base metals such as chalcopyrite, galena, sphalerite, etc. Its presence can be involved in operational terms since it tends to float, even in some instances without the use of collectors [1,2]. Interestingly, it is simple to manage physicochemical changes in its surface, which can lead to significant consequences in its floatability, for example, it has been widely exhibited that the recovery of the pyrite declines by rising the pH [3–5]. This inverse relationship between recovery and pH has been associated with the greater abundance of hydrophilic hydroxides concerning hydrophobic sulfides that found on the pyrite surface. This happens because, at alkaline conditions, ferric hydroxide is generated from the ferrous hydroxide released from inside the pyrite [6–8]. Subsequently, ferric hydroxide that has a hydrophilic nature precipitates on the surface of the pyrite, decreasing its contact angle and consequently lowering its floatability [9,10]. In this way, by regulating the pH with alkalizing agents such as sodium hydroxide, sodium carbonate, or lime, it can make the pyrite no float. An interesting aspect is that lime is more effective compared to sodium hydroxide, which is explained by the participation of calcium ions as, at pH less than 12.5, $Ca(OH)^+$ is the main component of the solution. This hydrophilic element has a high affinity for the surface of pyrite, so it also helps to boost their depression. It is common that in the copper industry lime is used as the sole depressant for pyrite, primarily when flotation is carried out in freshwater [11,12]. Interestingly, the use of seawater is a strategy that is being frequently adopted in sectors that have a shortage of freshwater, highlighting mining plants in countries such as Chile, Australia, Indonesia, etc. [13]. However, when the operations are carried out in seawater, these are unlikely to operate under highly alkaline conditions, mainly for two reasons: (i) the buffer effect of seawater implies that the lime consumption required to reach pH 11 is about ten times higher than in freshwater [14,15]. This excessive addition of lime means a considerable extra cost to the process, in addition to an increase of water hardness. (ii) Calcium and magnesium ions precipitate, which significantly affects the recovery of molybdenite and copper sulfides [16–19]. For these reasons, one of the main recommendations is working at a natural pH (or close to it), separating pyrite through the addition of new depressants reagents. Commercially, cyanide has been used for pyrite depression in several plants; however, due to its toxicity and the associated environmental concerns, further reagents are demanded, including sulfur dioxide, sodium sulfite, or metabisulphite of sodium. Many studies have shown that these reagents suppress xanthate adsorption by reducing the mixed potential to levels below the potential required for the oxidation of xanthate to dixanthogen [20–23]. The dixanthogen is a hydrophobic molecule generated from xanthate. Its presence is one of the leading causes by which pyrite increases its hydrophobicity, but the sulfoxide reagents may favor the formation of hydroxide species on the surface of pyrite; although, in most cases, these reagents do not achieve expected performances, and even environmental problems might arise that in some instances make impossible its implementation. Alternatively, many researchers have attempted to incorporate organic reagents into the processes, which have frequently shown their ability to selectively adsorb on the surface of pyrite, avoiding the collector adsorption and assigning some level of hydrophilicity [24–28]. The structure of these reagents is composed of (i) a hydrocarbon chain; (ii) hydroxyl groups that are distributed through the polymer structure, which are capable of ionizing or forming hydrogen bonds; and (iii) strongly hydrated polar groups such as SO_3^{-2}, COO^-, etc., which are also dispersed throughout the molecule. The biopolymers can be adsorbed to the surface of the pyrite by assigning a higher hydrophilic character to the surface, reducing the chances of adhesion between the particle and the bubble. Mu et al. [29] stated that four mechanisms for the adsorption of biopolymers should be considered: (i) electrochemical attraction, (ii) hydrophobic interaction, (iii) hydrogen bonds, and (iv) chemical interaction. Lopez-Valdivieso et al. [30] pointed that the oxidation state of the pyrite is one of the most relevant aspects for the adsorption of biopolymers like dextrin, wherein the amount of adsorbed molecules was directly correlated to the surface density of ferric hydroxide.

Guar gum, defined as an organic polysaccharide, is a galactomannan. Galactomannans have proven to be more effective depressants than starch, dextrin, and carboxymethyl cellulose (CMC). This has been attributed to the stronger hydrogen bonds formed by the cis-hydroxyl pair of long and linear molecules over a large surface area of particles, causing their agglomeration. Therefore, more effective separation of sulfide ore from gangue minerals has been found in froth flotation using guar gum as a depressant [31,32]. Guar gum applications include hindering the flotation of varied minerals such as talc, potash, chromite [33–36], even promising results have been shown to depress pyrite. However, these studies have been limited to the use of freshwater [37]. In this context, the present research addresses the consequences of using guar gum on pyrite depression in seawater flotation. The assays are carried out at pH 8 to emulate the typical conditions practiced in copper mining plants that use this type of water in their concentration stages. A microscopic analysis seeks to describe the mechanisms involved during the application of this polysaccharide. For this, the properties of pyrite aggregate are directly characterized by the use of the Focused Beam Reflectance Measurement (FBRM) and Particle Vision Measurement (PVM) techniques.

2. Methodology

2.1. Materials

The natural high purity pyrite specimens used in this investigation were acquired from Ward Scientific Establishment Inc. Scanning electron microscopy (SEM), as well as scattered energy x-ray spectroscopy (EDS) are shown in Figure 1 and was used a Zeiss scanning electron microscopy, model EVA MA-10 (CARL ZEISS Ltd., Oberkochen, Germany) and Oxford X-ray detector (OXFORD Instruments, Oxford, UK).

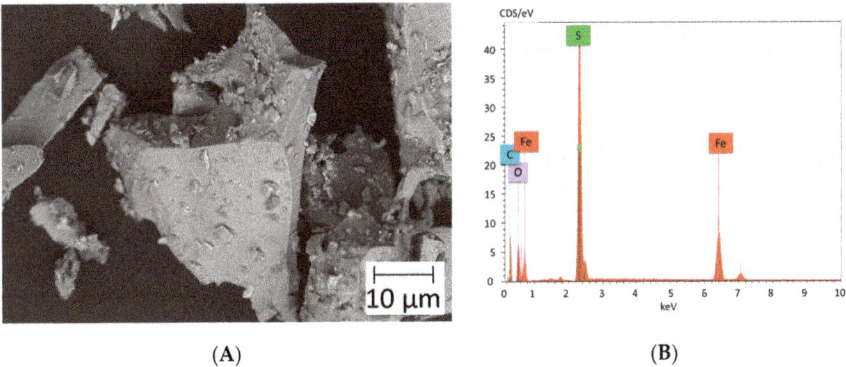

(**A**) (**B**)

Figure 1. (**A**) SEM micrograph of pyrite sample and (**B**) energy x-ray spectroscopy (EDS) spectrum.

Two different sizes of pyrite were prepared, considering a range between 38 and 65 µm (−65 + 38 µm) for micro flotation tests and −38 µm for flocculation kinetics tests. Seawater was obtained 200 m from the coast of San Jorge Bay (Antofagasta, Chile) and it was filtered by UV filter to eliminate bacterial activity. Before being used, seawater passed through a quartz sand filter (50 µm) and a mechanical polyethylene filter (1 µm) to remove insoluble particles. Chemical analysis is presented in Table 1, which were obtained by different analytical techniques (atomic absorption spectrometry (AAS), Argentometric method, and acid–base volumetry).

Table 1. Seawater ion concentration and method of analysis (pH 8).

Ion	Concentration [mg/L]	Method of Analysis
Na^+	9950	Atomic absorption spectrometry
Mg^{2+}	1250	Atomic absorption spectrometry
Ca^{2+}	400	Atomic absorption spectrometry
K^+	380	Atomic absorption spectrometry
Cl^-	19,450	Argentometric method
HCO_3^-	150	Acid–base volumetry

Flotation reagents were potassium amyl xanthate (PAX) as collector and methyl isobutyl carbinol (MIBC) as frother. Commercial xanthates generally have a purity of between 60 and 90%. They usually contain residual alkali hydroxide or metal carbonate that is intentionally added to slow down the decomposition of the product during storage. Because of this, PAX was purified by dissolving it in acetone and recrystallizing from diethyl ether. All aqueous solutions were prepared using distilled water. Guar gum G4129 of high purity was supplied by Sigma-Aldrich (see the chemical structure in Figure 2).

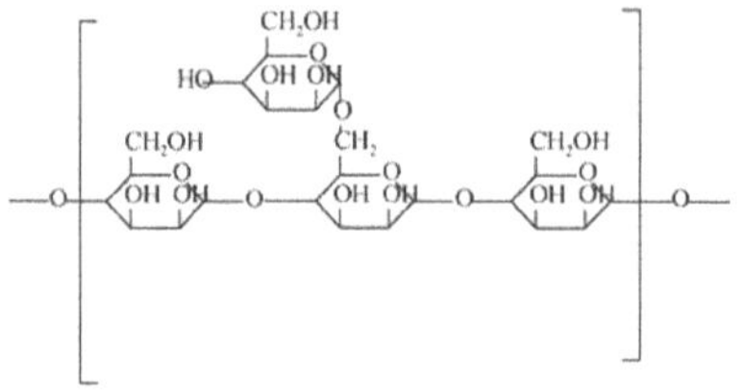

Figure 2. Chemical structure of guar gum.

2.2. *Preparation of Flotation Reagents*

A stock solution of PAX was prepared by adding 750 mg in a 100 mL flask by filling it up with distilled water. From this solution, the required aliquots were taken to achieve the dosages expected in the tests. For the stock solution of MIBC, 0.021 g of the reagent was added to 100 mL of distilled water. From this, 14.3 mL was taken, which corresponded to a concentration of 20 ppm of MIBC. For the stock solution of guar gum (750 mg/L), the powder was stirred in distilled water for 5 h. From this solution, aliquots were taken to reach the corresponding concentration of guar gum.

2.3. *Microflotation Tests*

Microflotation tests were carried out in a 150 mL Partridge–Smith glass cell using air at a flowrate of 100 mL/min. Pyrite–water suspensions were prepared using 2 g of pyrite (considering a range between 38 and 65 μm) and 150 mL aqueous solution. The suspensions were stirred in a beaker with magnetic agitation and conditioned at the required pH for 3 min. Afterwards, PAX was added to the suspension at the desired concentration and contacted with the pyrite for 3 min, then with the frothing agent for another 2 min. Finally, the suspension was transferred to a Partridge–Smith cell and flotation was carried out for 3 min, scraping off the froth every 5 s. The pulp level in the microflotation test was kept constant by adding a background solution prepared at the same chemical reagent concentration as the original aqueous solution. To evaluate the influence of the depressant concentration on the floatability of pyrite, guar gum was added before the collector and contacted with the mineral for 5 min. All flotation tests were performed in triplicate. After the microflotation test, the non-floated and floated products were dried and weighed to calculate the pyrite floatability.

2.4. *Aggregate Characterization*

The effect of reagent conditioning on aggregate characteristics was evaluated; in this case, the mean chord length using the Focused Beam Reflectance Measurement (FBRM) technique. The instrument model was Particle Track G400 with a measurement of every 2 s. Direct images of particle aggregates were obtained with the Particle Vision Measurement (PVM) technique, using the V819 model. Three-hundred milliliters of seawater and 2 grams of pure pyrite were used. After 4 min of mixing, guar gum was added with the desired concentration, and after 5 min, PAX and MIBC were added.

2.5. *Fourier Transform Infrared (FTIR) Spectroscopy*

FTIR studies were performed to provide information on the infrared spectra of pyrite, PAX, guar gum, and the interaction products between the mineral, the collector, and the polysaccharide. A Fourier Infrared Spectrometer (PerkinElmer, Santiago, Chile) that operates in the range of 4000 to 400 cm^{-1} was used for the spectroscopic studies.

2.6. *Surface Tension*

Four sets of surface tension tests were performed at pH 8 and 23 °C: (i) seawater, 20 ppm of MIBC and various guar gum concentrations (0–200 ppm); (ii) seawater and various MIBC levels (0–200 ppm); (iii) seawater and various guar gum concentrations (0–200 ppm); and (iv) seawater, 20 ppm of MIBC, 75 ppm of PAX, and various guar gum concentrations (0–200 ppm). The surface tension was determined by the bubble up technique using a tensiometer Lauda model TD 3 (LAUDA DR. R. WOBSER GMBH & CO, Lauda-Königshofen, Germany).

3. Results and Discussions

3.1. *Microflotations*

Figure 3 presents the recovery of pyrite in seawater as a function of PAX addition in the presence of 20 ppm of MIBC and in the absence of guar gum. The tests were performed at pH 8 to emulate the conditions used in the copper industry that operates with seawater [38]. There is a PAX dosage that maximizes the recovery of pyrite, which was close to 75 ppm. At this PAX addition, the pyrite recovery was 80%. Increasing PAX dosages lead to a little decrease in pyrite recovery.

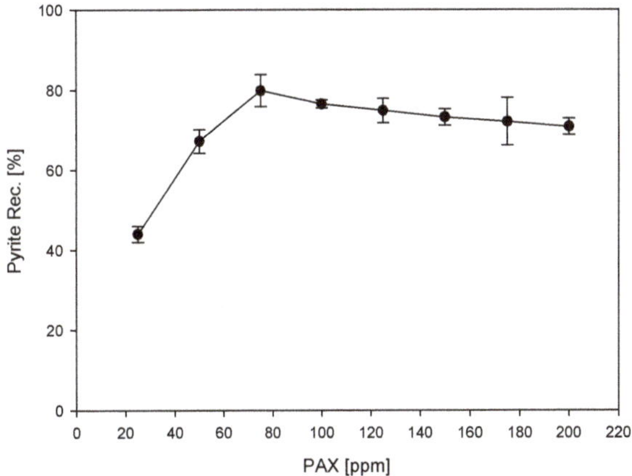

Figure 3. Pyrite recovery as a function of PAX concentration (pH 8, 20 ppm of MIBC).

Dixanthogen formation through the oxidation of the collector ion turns the pyrite surface hydrophobic [39]. This collector oxidation can take place with the reduction of oxygen (Equation (1)). López-Valdivieso et al. [40] proposed that oxidation of the collector ion is coupled with the reduction of surface iron hydroxides (Equation (2)). The pyrite used in this study had iron hydroxides on its surface as will be shown below.

$$ROCS_2^- + \frac{1}{2}O_2 + H_2O \rightarrow (ROCS_2)_2 + 2OH^- \quad (1)$$

$$2ROCS2^- + 2Fe(OH)_3 + 6H^+ = (ROCS2)_2 + 2Fe^{2+} + 6H_2O \quad (2)$$

Figure 4 shows the pyrite flotation recovery as a function of guar gum addition at pH 8 in seawater and in the presence of 75 ppm PAX and 20 ppm MIBC. As the dosage of the polysaccharide increases up to 100 ppm, the recovery of pyrite gradually decreases from 80 to 23%, demonstrating that the polysaccharide is a good pyrite depressant. The OH groups of the guar gum molecules likely adsorbed onto OH sites of ferric hydroxide of the pyrite surface by a similar adsorption mechanism occurring between dextrin and oxidized pyrite [30,41]. The difference between guar gum and dextrin is that two cis-OH groups of the guar gum would link to a ferric hydroxide OH site, whereas, in the case of dextrin, only one OH group would link to a ferric hydroxide site. As the guar gum is a highly hydrophilic big molecule, its adsorption rendered the pyrite surface very hydrophilic. At guar gum dosages greater than 100 ppm, the recovery of pyrite gradually increased reaching 39% at 200 ppm guar gum.

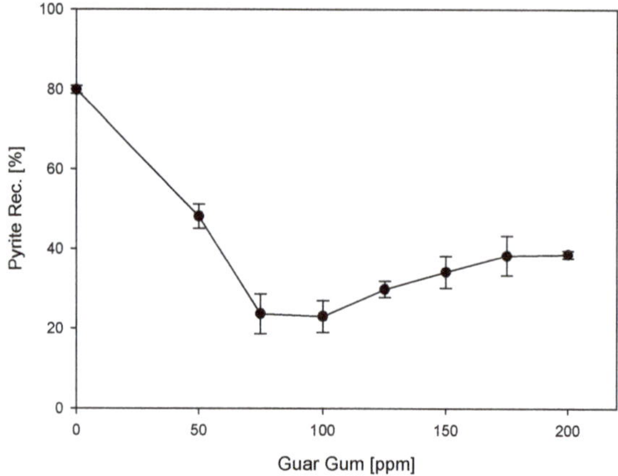

Figure 4. Pyrite recovery as a function of guar gum concentration (pH 8, 75 ppm of PAX, 20 ppm of MIBC).

3.2. Characterization of Aggregates

Figures 5–7 present the characterization of pyrite aggregates throughout their chord length (using the FBRM probe), at various concentrations of guar gum without PAX or MIBC. In a collector and frother free environment, guar gum promoted the growth of aggregates, even at low concentrations. Adding 5 ppm polysaccharide, the aggregates grew from 120 to 160 μm (Figure 5A). At the optimal dose (100 ppm), the maximum chord length was slightly higher than 192 μm (Figure 6A). This value is stabilized at further dosages. At 200 ppm guar gum, the maximum aggregate size was 188 μm, suggesting that the limit zone in which the polysaccharide could act as a flocculating agent was reached. A notable difference in all particle flocculation systems is related to the fragmentation of the floc structure, caused by the hydrodynamic conditions. The FBRM probe provides the particle count classified in different bins according to their size, presenting their size evolution over time. In Figure 5B, it can be seen that after the addition of 5 ppm guar gum, the particle count for flocs smaller than 150 microns radically decreased, with a subtle increase of the flocs having a size between 150 and 300 microns. However, within a few seconds of flocculation, the number of flocs (counts) less than 150 microns increased steadily. When the guar gum dosage was 100 ppm, particles smaller than 150 microns are connected to produce larger stable aggregates, indicating that the aggregates are joined with greater strength. Increasing the dosage of guar gum up to 200 ppm raised the particle count again after a few seconds of flocculation (Figure 7A). The excessive amount of polymer may have saturated the particle surface so the polymer bound a less number of particles. Weak aggregates formed at this dosage.

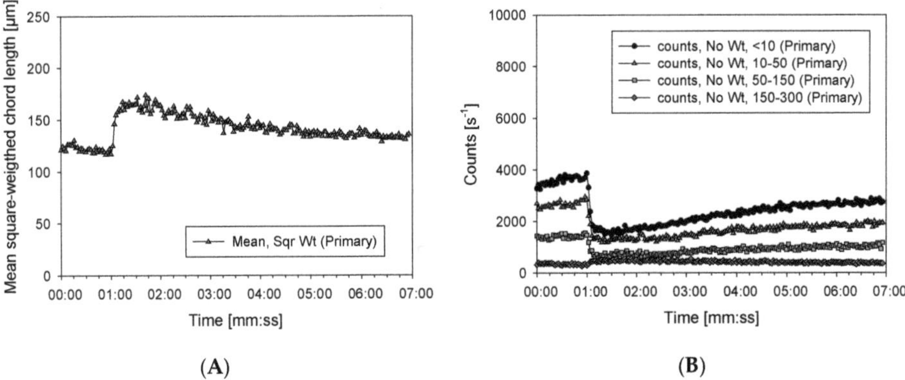

Figure 5. (**A**) Mean square chord length and (**B**) counts evolution of suspension; 5 ppm of guar gum added after 1 min (no frother or collector).

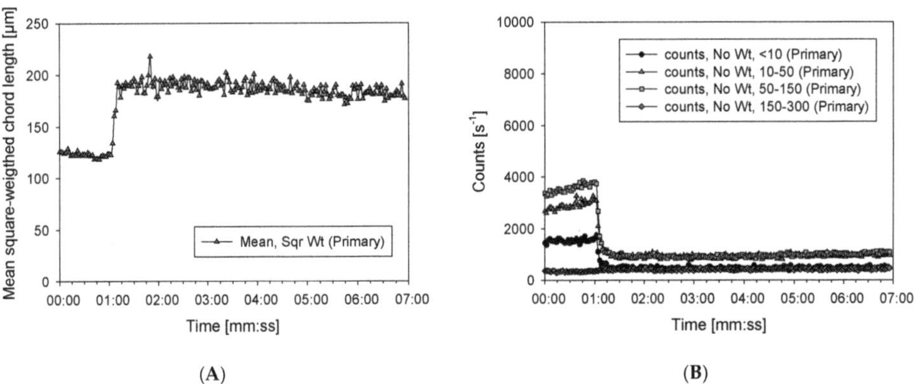

Figure 6. (**A**) Mean square chord length and (**B**) counts evolution of suspension; 100 ppm of guar gum added after 1 min (no frother or collector).

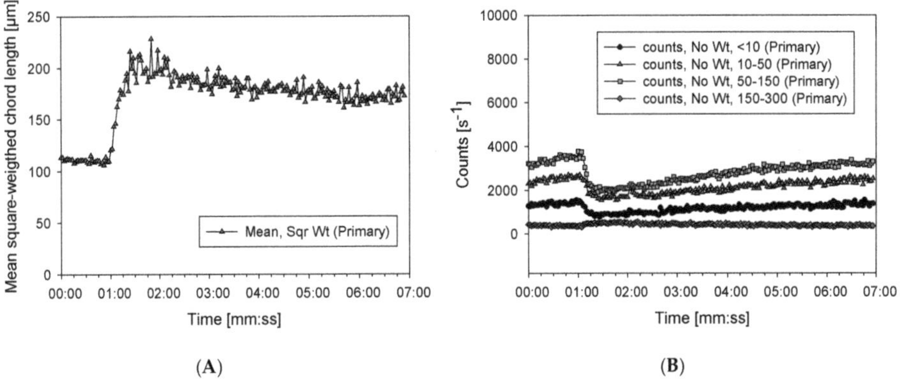

Figure 7. (**A**) Mean square chord length and (**B**) counts evolution of suspension; 200 ppm of guar gum added after 1 min (no frother or collector).

Figures 8–10 exhibit the outcomes for the size of the aggregates at different concentrations of guar gum and in the presence of 75 ppm PAX and 20 ppm MIBC. A notable effect of the collector and the frother is the redispersing of the particles that are bound into the floc, especially at low polymer dosages. As shown in Figure 8A, with 5 ppm guar gum, there was an instantaneous growth of aggregates. However, after 1 min, following the addition of PAX and MIBC, a significant dispersion of particles with rope length less than 150 μm was observed. This seems to indicate that guar gum desorbed from the pyrite surface due to xanthate adsorption. Lopez-Valdivieso et al. [30] show that dextrin desorbed from pyrite surface by xanthates. Dispersion of the particles favored the recovery of pyrite since dispersed particles are more prone to float than those that make up an aggregate. According to Figure 8B, the dispersion due to PAX and MIBC is attenuated by increasing the dosage of the polysaccharide. At the optimum pyrite depression concentration of 100 ppm guar gum, the average decrease in aggregates is gradual (Figure 9A). In addition, particles smaller than 150 μm were released in a substantially lower quantity than at low guar gum dosages (Figure 9B) and not significant presence of particles smaller than 10 μm was quantified. Above 100 ppm guar gum, at a 200 ppm guar gum, particles redispersing increased again (Figure 10B), which coincides with the increase in pyrite recovery (Figure 5).

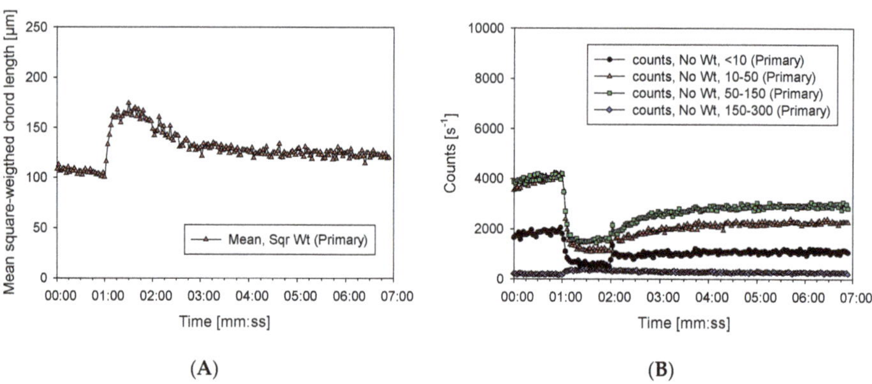

Figure 8. (**A**) Mean square chord length and (**B**) counts evolution of suspension: 5 ppm of guar gum added after 1 min (MIBC 20 ppm, PAX 75 ppm).

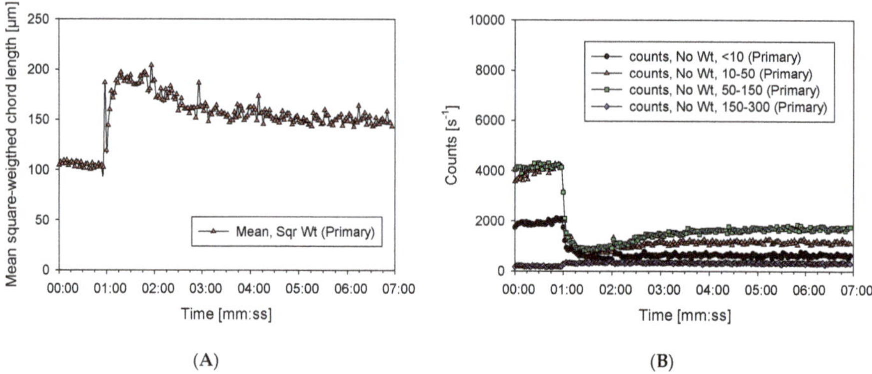

Figure 9. (**A**) Mean square chord length and (**B**) counts evolution of suspension: 100 ppm of guar gum added after 1 min (MIBC 20 ppm, PAX 75 ppm).

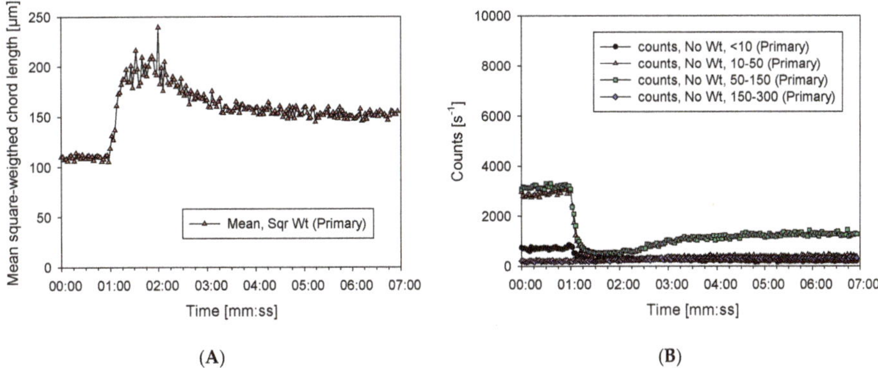

Figure 10. (**A**) Mean square chord length and (**B**) counts evolution of suspension: 200 ppm of guar gum added after 1 min (MIBC 20 ppm, PAX 75 ppm).

Figures 11–13 present images of suspended pyrite particles and their conformation in the absence and presence of guar gum, PAX, and MIBC. In the absence of these reagents (Figure 11), the particles are mostly dispersed and a low number of agglomerates are noted. This was expected, as particles seek to agglomerate in a highly saline environment due to compression of the electrical double layer. The increasing formation of aggregates as the dose of guar gum increases is observed in Figure 12A for 5 ppm and in Figure 13A for 100 ppm. Although the shape of the aggregates is irregular and different from a sphere, it is seen that in both cases the floc structures are compact, and composed of several particles. After the addition of PAX and MIBC, particle redispersion is evident as depicted in Figure 12B (5 ppm), where the aggregates reduce their size but gain in sphericity. At 100 ppm guar gum (Figure 13B), there is no substantial reduction in the extent of the agglomerates. There is an apparent redistribution of the particles changing the shape of the floc structure; however, this is consistent with that reported above for the size of the flocs.

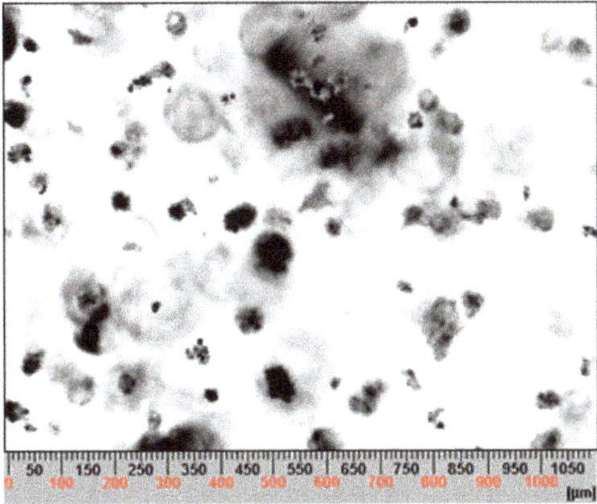

Figure 11. Suspended particles of pyrite in seawater at pH 8 (7 min).

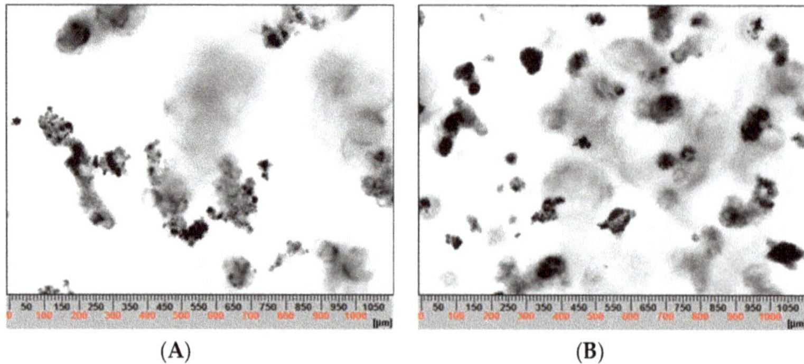

Figure 12. Suspended particles of pyrite in seawater at pH 8 interacting with flotation reagents: (A) 5 ppm of guar gum (7 min) and (B) 5 ppm of guar gum, 75 ppm of PAX, and 20 ppm of MIBC (7 min).

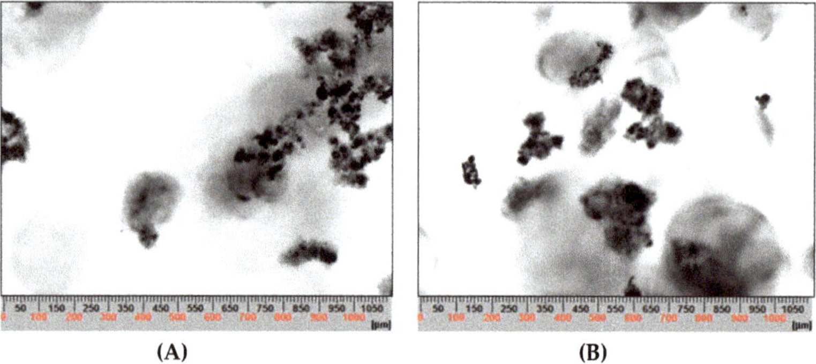

Figure 13. Suspended particles of pyrite in seawater at pH 8 interacting with flotation reagents: (A) 100 ppm of guar gum (7 min) and (B) 100 ppm of guar gum, 75 ppm of PAX, and 20 ppm of MIBC (7 min).

3.3. FTIR Analysis

Figure 14 presents the spectra of guar gum, PAX, pyrite contacted with guar gum, and pyrite contacted with both PAX and guar gum in seawater at pH 8. The bands between 1700 and 3000 cm^{-1} and 1200 and 1500 cm^{-1} are due to the groups CH_2 and CH_3 of the collector alkyl (Figure 14A) [42]. The bands between 3000 and 3700 cm^{-1} are attributed to the hydrogen bonds of hydroxyl groups, due to hydroxides and oxy-hydroxide iron species [42]. The band at 3450 cm^{-1} is due to the stretching vibration of the hydroxyl groups in the guar gum structure. This indicates that hydrogen bonds strongly link to the O–H groups (Figure 14B).

In Figure 14C, the band at 1650 cm^{-1} is characteristic of water molecules indicating sulfate hydration [43]. The intense band at 1650 cm^{-1} suggests a stretch of the glucopyranose ring. This is indicative that the pyrite surface is more hydrated and hydrophilic. The bands between 900 and 1200 cm^{-1} corresponds to the asymmetric oxygen vibration (C-O-C), attributed to the xanthate and may be dixanthogen. The bands between 1000 and 1050 cm^{-1} are due to the vibration C=S [44]. Finally, the bands between 900 and 650 cm^{-1} correspond to oxidized iron species on pyrite such as the oxyhydroxides goethita (α-FeOOH) and limonite (α-$FeOOH{\cdot}nH_2O$)) [43]. Accordingly, it is confirmed that the pyrite surface is oxidized and this oxidation occurs rapidly. The bands in the ranges 1300 to 850 cm^{-1}, belong to sulfate species, such as iron sulfates ($Fe_2(SO_4)_3xH_2O$ between the bands 1000 and 1150 cm^{-1}).

The bands between 500 to 750 cm^{-1} are due to the stretching of the guar gum ring. As noted, they appeared in the guar gum–pyrite spectrum at 750 and 850 cm^{-1} (Figure 14D), indicating Guar Gum on the pyrite surface. In the guar gum–PAX–pyrite spectrum (Figure 14C), they are almost entirely reduced. This is an indication that PAX desorbed some Guar gum and co-adsorption of PAX and guar gum took place on the pyrite.

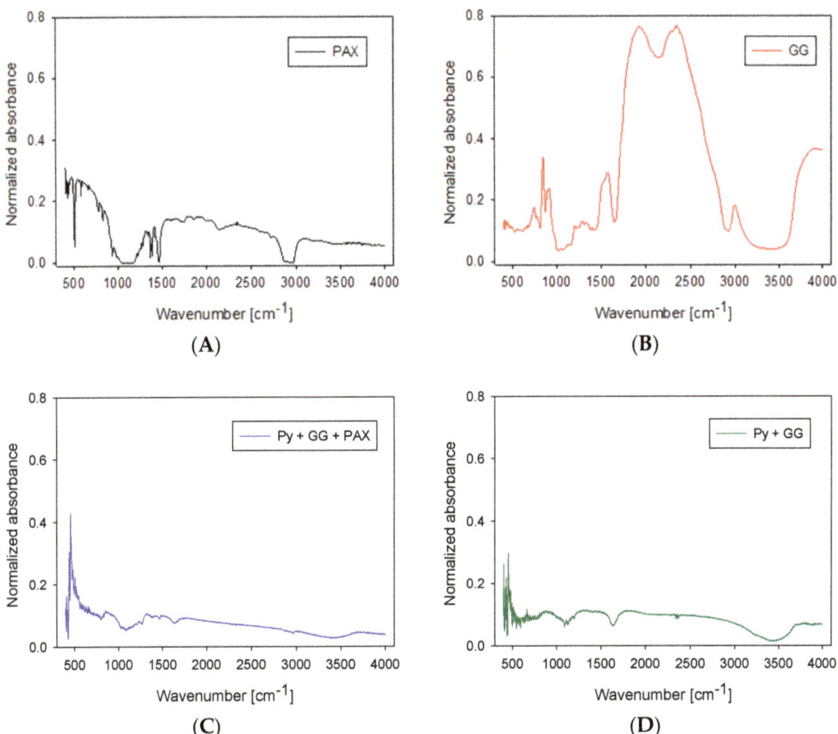

Figure 14. FTIR spectra for (**A**) PAX, (**B**) guar gum (GG), (**C**) pyrite (Py) + GG + PAX, and (**D**) Py + GG (pH 8, 35 ppm of PAX, 20 ppm of guar gum).

3.4. Surface Tension

Figure 15 shows the surface tension of seawater as a function of MIBC and guar gum concentrations. Both reagents affected the surface tension at low concentrations. At 5 ppm, the surface tension dropped by approximately 10%. In this study, a fixed concentration of MIBC (20 ppm) was used while varying the guar gum concentration from 0 to 200 ppm However, a synergistic effect in the surface tension is not noted by adding the two reagents. This result is not in agreement with others reported elsewhere stating that there is a synergistic effect when polymers are mixed with frothers [45,46]. The synergistic effect would cause an overstabilization of the bubbles, which would increase gangue mechanical entrainment lowering the concentrate quality. Accordingly, if guar gum were used in flotation it would not affect the bubbles characteristics so bubble overstabilization would be avoided.

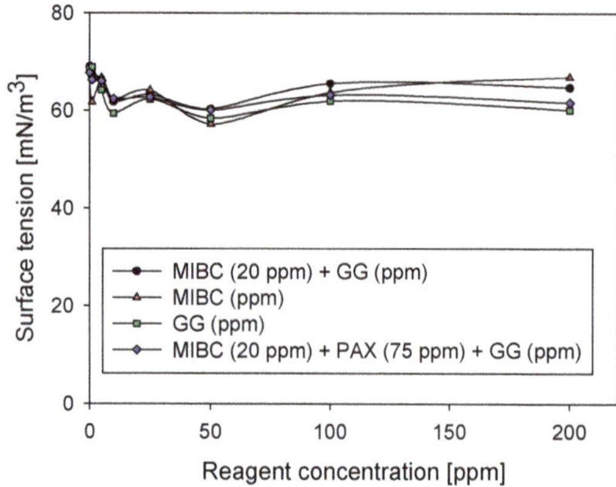

Figure 15. Surface tension as a function of reagents concentrations: (i) varied guar gum (GG) + 20 ppm of MIBC, (ii) varied MIBC, (iii) varied guar gum, and (iv) varied guar gum + 20 ppm of MIBC + 75 ppm of PAX (seawater at pH 8).

4. Discussion

Pyrite is common in cooper and polymetallic ore deposits. It is floatable and shows a high affinity with the flotation collectors that are used to concentrate the valuable metal sulfides. Therefore, flotation of pyrite is desirable as it impacts the quality of the metal concentrate.

Traditionally, pyrite is depressed at highly alkaline conditions. However, this approach cannot be implemented in seawater as precipitation of calcium and magnesium ions affects the performance of valuable minerals. Besides, there is an excessive consumption of lime, due to the buffer effects, that restricts the sustainability of depressing pH at high pH [15]. The current strategy operates at the slurry's natural pH and applies reagents that can selectively depress pyrite [47]. In this context, the organic reagents has shown promising results in the copper, molybdenum, talc, mica, galena, sphalerite, and pyrrhotite flotation [2]. However, their performance in seawater has not been explored yet, where a highly saline environment modifies the electrostatic interactions between surfaces and alters the structure in which water molecules are organized. It brings significant consequences in mineral processing.

Guar gum, used in this study, presented encouraging results, as FTIR studies indicate that this reagent was able to interact with the pyrite, decreasing its recovery from 80% to 23% in the presence of PAX. Although, the use of polysaccharides in general (guar gum, dextrin, starch, etc.) requires that the surface of pyrite should be oxidized at a certain level to achieve the interaction between the OH groups of the guar gum. The more ferric hydroxide on the pyrite surface, the higher is the adsorption of the polysaccharide [30].

The polysaccharide is expected to act by two mechanisms. On the one hand, the coating of the pyrite surface prevents the formation of $ROCS_2$-M bonds, which maintain the connection between the collector and the pyrite surface. Additionally, the in situ analysis of the particles employing the FBRM and PVM techniques revealed their flocculation. The hydrodynamic conditions exert stress that redisperses the particles, especially for low polymer dosages. At the same time, as the concentration of guar gum increases, the redispersing of the flocs decreases, due to the greater strength of the agglomerate structure. As the aggregates of pyrite are massive, even if the bubbles attach to them, the bubbles will not be able to transport them to the flotation froth so they will not float. Although the polymer is a surface active agent, the surface tension results indicate that there is not a synergistic effect with the frother, as has been reported elsewhere [46]. Overdose of guar gum saturate the surface of pyrite, limiting the size of the agglomerates. This leads to an increase in recovery as less massive aggregates formed, which can be carried to the froth phase by bubbles.

5. Conclusions

Guar gum was used to promote pyrite depression in flotations with seawater in the presence of the collector propyl xanthate. The tests were performed at natural pH to emulate the operating conditions of copper concentration plants using seawater. The results were promising, and it was found that the polysaccharide efficiently depressed pyrite in a highly saline environment. The polysaccharide adsorbs on the pyrite surface turning it very hydrophilic on top of the hydrophobicity due to adsorbe collector. The depression of pyrite was accompanied by strong flocculation of particles that generated massive aggregates, which are difficult to be transported by bubbles to the froth phase. However, the collector and frother (PAX and MIBC) redispersed the agglomerates making possible their levitation to the froth by bubbles. This can be reduced by increasing the polymer concentration to minimize the redispersion of the agglomerates. However, an overdose of guar gum leads to a re-stabilization of the agglomerates lowering its depression effect for the pyrite.

Author Contributions: The manuscript was written through the contributions of all authors. R.I.J. designed the research, C.I.C. and E.C.P. performed the experiments, C.I.C. wrote the first draft, R.I.J., P.R., N.T., and A.L.-V. analyzed the results and wrote the manuscript. All authors have read and agreed to the published version of the manuscript.

Funding: This research was funded by Conicyt Fondecyt 11171036 and Centro CRHIAM through Project ANID/FONDAP/15130015.

Acknowledgments: R.I.J. thanks CONICYT Fondecyt 11171036 and Centro de Recursos Hídricos para la Agricultura y la Minería (CRHIAM) through Project ANID/FONDAP/15130015. Pedro Robles thanks the Pontificia Universidad Católica de Valparaíso for the support provided.

Conflicts of Interest: The authors declare no conflicts of interest.

References

1. Wang, X.-H.; Eric Forssberg, K.S. Mechanisms of pyrite flotation with xanthates. *Int. J. Miner. Process.* **1991**, *33*, 275–290. [CrossRef]
2. Mu, Y.; Peng, Y.; Lauten, R.A. The depression of pyrite in selective flotation by different reagent systems—A Literature review. *Miner. Eng.* **2016**, *96–97*, 143–156. [CrossRef]
3. Boulton, A.; Fornasiero, D.; Ralston, J. Depression of iron sulphide flotation in zinc roughers. *Miner. Eng.* **2001**, *14*, 1067–1079. [CrossRef]
4. Jiang, C.L.; Wang, X.H.; Parekh, B.K.; Leonard, J.W. The surface and solution chemistry of pyrite flotation with xanthate in the presence of iron ions. *Colloids Surf. A* **1998**, *136*, 51–62. [CrossRef]
5. Trahar, W.J.; Senior, G.D.; Shannon, L.K. Interactions between sulphide minerals—The collectorless flotation of pyrite. *Int. J. Miner. Process.* **1994**, *40*, 287–321. [CrossRef]
6. Chandra, A.P.; Gerson, A.R. The mechanisms of pyrite oxidation and leaching: A fundamental perspective. *Surf. Sci. Rep.* **2010**, *65*, 293–315. [CrossRef]
7. Karthe, S.; Szargan, R.; Suoninen, E. Oxidation of pyrite surfaces: a photoelectron spectroscopic study. *Appl. Surf. Sci.* **1993**, *72*, 157–170. [CrossRef]
8. Moslemi, H.; Gharabaghi, M. A review on electrochemical behavior of pyrite in the froth flotation process. *J. Ind. Eng. Chem.* **2017**, *47*, 1–18. [CrossRef]
9. Hicyilmaz, C.; Emre Altun, N.; Ekmekci, Z.; Gokagac, G. Quantifying hydrophobicity of pyrite after copper activation and DTPI addition under electrochemically controlled conditions. *Miner. Eng.* **2004**, *17*, 879–890. [CrossRef]
10. Kocabag, D.; Shergold, H.L.; Kelsall, G.H. Natural oleophilicity/hydrophobicity of sulphide minerals, II. Pyrite. *Int. J. Miner. Process.* **1990**, *29*, 211–219. [CrossRef]
11. Zanin, M.; Lambert, H.; du Plessis, C.A. Lime use and functionality in sulphide mineral flotation: A review. *Miner. Eng.* **2019**, *143*, 105922. [CrossRef]
12. Li, Y.; Chen, J.; Kang, D.; Guo, J. Depression of pyrite in alkaline medium and its subsequent activation by copper. *Miner. Eng.* **2012**, *26*, 64–69. [CrossRef]
13. Cisternas, L.A.; Gálvez, E.D. The use of seawater in mining. *Miner. Process. Extr. Metall. Rev.* **2018**, *39*, 18–33. [CrossRef]

14. Jeldres, R.I.; Calisaya, D.; Cisternas, L.A. An improved flotation test method and pyrite depression by an organic reagent during flotation in seawater. *J. South. Afr. Inst. Min. Metall.* **2017**, *117*, 499–504. [CrossRef]
15. Castro, S. Challenges in flotation of Cu-Mo sulfide ores in sea water I. In *Water in Mineral Processing;* SME: Englewood, CO, USA, 2012; pp. 29–40.
16. Jeldres, R.I.; Arancibia-Bravo, M.P.; Reyes, A.; Aguirre, C.E.; Cortes, L.; Cisternas, L.A. The impact of seawater with calcium and magnesium removal for the flotation of copper-molybdenum sulphide ores. *Miner. Eng.* **2017**, *109*, 10–13. [CrossRef]
17. Li, W.; Li, Y.; Xiao, Q.; Wei, Z.; Song, S. The influencing mechanisms of sodium hexametaphosphate on chalcopyrite flotation in the presence of MgCl2 and CaCl2. *Minerals* **2018**, *8*, 150. [CrossRef]
18. Li, Y.; Li, W.; Xiao, Q.; He, N.; Ren, Z.; Lartey, C.; Gerson, A. The influence of common monovalent and divalent chlorides on chalcopyrite flotation. *Minerals* **2017**, *7*, 111. [CrossRef]
19. Suyantara, G.P.W.; Hirajima, T.; Miki, H.; Sasaki, K. Floatability of molybdenite and chalcopyrite in artificial seawater. *Miner. Eng.* **2018**, *115*, 117–130. [CrossRef]
20. Cao, Z.; Chen, X.; Peng, Y. The role of sodium sulfide in the flotation of pyrite depressed in chalcopyrite flotation. *Miner. Eng.* **2018**, *119*, 93–98. [CrossRef]
21. Göktepe, F. Effect of H2O2 and NaSH addition to change the electrochemical potential in flotation of chalcopyrite and pyrite minerals. *Miner. Process. Extr. Metall. Rev.* **2010**, *32*, 24–29. [CrossRef]
22. Hassanzadeh, A.; Hasanzadeh, M. Chalcopyrite and pyrite floatabilities in the presence of sodium sulfide and sodium metabisulfite in a high pyritic copper complex ore. *J. Dispers. Sci. Technol.* **2017**, *38*, 782–788. [CrossRef]
23. Khmeleva, T.N.; Skinner, W.; Beattie, D.A.; Georgiev, T.V. The effect of sulphite on the xanthate-induced flotation of copper-activated pyrite. *Physicochem. Prob. Miner. Process.* **2002**, *36*, 185–195.
24. Kar, B.; Sahoo, H.; Rath, S.S.; Das, B. Investigations on different starches as depressants for iron ore flotation. *Miner. Eng.* **2013**, *49*, 1–6. [CrossRef]
25. Lopez Valdivieso, A.; Sánchez López, A.A.; Song, S.; García Martínez, H.A.; Licón Almada, S. Dextrin as a regulator for the selective flotation of chalcopyrite, galena and pyrite. *Can. Metall. Q.* **2007**, *46*, 301–309. [CrossRef]
26. Wang, Z.; Xie, X.; Xiao, S.; Liu, J. Adsorption behavior of glucose on pyrite surface investigated by TG, FTIR and XRD analyses. *Hydrometallurgy* **2010**, *102*, 87–90. [CrossRef]
27. Sarquís, P.E.; Menéndez-Aguado, J.M.; Mahamud, M.M.; Dzioba, R. Tannins: the organic depressants alternative in selective flotation of sulfides. *J. Clean. Prod.* **2014**, *84*, 723–726. [CrossRef]
28. Rutledge, J.; Anderson, C. Tannins in mineral processing and extractive metallurgy. *Metals* **2015**, *5*, 1520–1542. [CrossRef]
29. Mu, Y.; Peng, Y.; Lauten, R.A. Electrochemistry aspects of pyrite in the presence of potassium amyl xanthate and a lignosulfonate-based biopolymer depressant. *Electrochim. Acta* **2015**, *174*, 133–142. [CrossRef]
30. Lopez-Valdivieso, A.; Sanchez-López, A.A.; Padilla-Ortega, E.; Robledo-Cabrera, A.; Galvez, E.; Cisternas, L.A. Pyrite depression by dextrin in flotation with xanthates. Adsorption and floatability studies. *Physicochem. Prob. Miner. Process.* **2018**, *54*, 1159–1171. [CrossRef]
31. Mathur, N.K. *Industrial Galactomannan Polysaccharides;* CRC Press: Boca Raton, FL, USA, 2016; ISBN 9780429105142.
32. Laskowski, J.S.; Liu, Q.; O'Connor, C.T. Current understanding of the mechanism of polysaccharide adsorption at the mineral/aqueous solution interface. *Int. J. Miner. Process.* **2007**, *84*, 59–68. [CrossRef]
33. Ma, X.; Pawlik, M. Adsorption of guar gum onto quartz from dilute mixed electrolyte solutions. *J. Colloid Interface Sci.* **2006**, *298*, 609–614. [CrossRef] [PubMed]
34. Ma, X.; Pawlik, M. Role of background ions in guar gum adsorption on oxide minerals and kaolinite. *J. Colloid Interface Sci.* **2007**, *313*, 440–448. [CrossRef] [PubMed]
35. Mailula, T.D.; Bradshaw, D.J.; Harris, P.J. The effect of copper sulphate addition on the recovery of chromite in the flotation of UG2 ore. *J. S. Afr. Inst. Min. Metall.* **2003**, *103*.
36. Shortridge, P.; Harris, P.; Bradshaw, D.; Koopal, L. The effect of chemical composition and molecular weight of polysaccharide depressants on the flotation of talc. *Int. J. Miner. Process.* **2000**, *59*, 215–224. [CrossRef]
37. Bicak, O.; Ekmekci, Z.; Bradshaw, D.J.; Harris, P.J. Adsorption of guar gum and CMC on pyrite. *Miner. Eng.* **2007**, *20*, 996–1002. [CrossRef]

38. Jeldres, M.; Piceros, E.C.; Toro, N.; Torres, D.; Robles, P.; Leiva, W.H.; Jeldres, R.I. Copper tailing flocculation in seawater: Relating the yield stress with fractal aggregates at varied mixing conditions. *Metals* **2019**, *9*, 1295. [CrossRef]
39. Hu, Y.; Sun, W.; Wang, D. *Electrochemistry of Flotation of Sulphide Minerals*; Springer: Berlin/Heidelberg, Germany, 2009; ISBN 978-3-540-92178-3.
40. López Valdivieso, A.; Sánchez López, A.A.; Song, S. On the cathodic reaction coupled with the oxidation of xanthates at the pyrite/aqueous solution interface. *Int. J. Miner. Process.* **2005**, *77*, 154–164. [CrossRef]
41. López Valdivieso, A.; Celedón Cervantes, T.; Song, S.; Robledo Cabrera, A.; Laskowski, J. Dextrin as a non-toxic depressant for pyrite in flotation with xanthates as collector. *Miner. Eng.* **2004**, *17*, 1001–1006. [CrossRef]
42. Rath, R.K.; Subramanian, S.; Pradeep, T. Surface chemical studies on pyrite in the presence of polysaccharide-based flotation depressants. *J. Colloid Interface Sci.* **2000**, *229*, 82–91. [CrossRef]
43. Kongolo, M.; Benzaazoua, M.; de Donato, P.; Drouet, B.; Barrès, O. The comparison between amine thioacetate and amyl xanthate collector performances for pyrite flotation and its application to tailings desulphurization. *Miner. Eng.* **2004**, *17*, 505–515. [CrossRef]
44. Leppinen, J.O. FTIR and flotation investigation of the adsorption of ethyl xanthate on activated and non-activated sulfide minerals. *Int. J. Miner. Process.* **1990**, *30*, 245–263. [CrossRef]
45. Tan, S.N.; Pugh, R.J.; Fornasiero, D.; Sedev, R.; Ralston, J. Foaming of polypropylene glycols and glycol/MIBC mixtures. *Miner. Eng.* **2005**, *18*, 179–188. [CrossRef]
46. Wang, Y.; Lauten, R.A.; Peng, Y. The effect of biopolymer dispersants on copper flotation in the presence of kaolinite. *Miner. Eng.* **2016**, *96–97*, 123–129. [CrossRef]
47. Ramos, O.; Castro, S.; Laskowski, J.S. Copper–molybdenum ores flotation in sea water: Floatability and frothability. *Miner. Eng.* **2013**, *53*, 108–112. [CrossRef]

Article

Describing Mining Tailing Flocculation in Seawater by Population Balance Models: Effect of Mixing Intensity

Gonzalo R. Quezada [1], Luís Ayala [2], Williams H. Leiva [3], Norman Toro [4,5,*], Pedro G. Toledo [6], Pedro Robles [7] and Ricardo I. Jeldres [3,*]

1. Water Research Center for Agriculture and Mining (CRHIAM), Concepción 4030000, Chile; gonzaloquezada@udec.cl
2. Faculty of Engineering and Architecture, Universidad Arturo Prat, Almirante Juan José Latorre 2901, Antofagasta 1244260, Chile; luisayala01@unap.cl
3. Departamento de Ingeniería Química y Procesos de Minerales, Facultad de Ingeniería, Universidad de Antofagasta, Av. Angamos 601, Antofagasta 1240000, Chile; wleivajeldres@gmail.com
4. Departamento de Ingeniería Metalúrgica y Minas, Universidad Católica del Norte, Av. Angamos 610, Antofagasta 1270709, Chile
5. Department of Mining, Geological and Cartographic Department, Universidad Politecnica de Cartagena, 30202 Murcia, Spain
6. Department of Chemical Engineering and Laboratory of Surface Analysis (ASIF), Universidad de Concepción, PO Box 160-C, Correo 3, Concepción 4030000, Chile; petoledo@udec.cl
7. Escuela de Ingeniería Química, Pontificia Universidad Católica de Valparaíso, Valparaíso, Chile; pedro.robles@pucv.cl

* Correspondence: ntoro@ucn.cl (N.T.); ricardo.jeldres@uantof.cl (R.I.J.); Tel.: +56-552 651-021 (N.T.); +56-552-637-901 (R.I.J.)

Received: 7 January 2020; Accepted: 29 January 2020; Published: 11 February 2020

Abstract: A population balance model (PBM) is used to describe flocculation of particle tailings in seawater at pH 8 for a range of mixing intensities. The size of the aggregates is represented by the mean chord length, determined by the focused beam reflectance measurement (FBRM) technique. The PBM follows the dynamics of aggregation and breakage processes underlying flocculation and provides a good approximation to the temporal evolution of aggregate size. The structure of the aggregates during flocculation is described by a constant or time-dependent fractal dimension. The results revealed that the compensations between the aggregation and breakage rates lead to a correct representation of the flocculation kinetics of the tailings of particles in seawater and, in addition, that the representation of the flocculation kinetics in optimal conditions is equally good with a constant or variable fractal dimension. The aggregation and breakage functions and their corresponding parameters are sensitive to the choice of the fractal dimension of the aggregates, whether constant or time dependent, however, under optimal conditions, a constant fractal dimension is sufficient. The model is robust and predictive with a few parameters and can be used to find the optimal flocculation conditions at different mixing intensities, and the optimal flocculation time can be used for a cost-effective evaluation of the quality of the flocculant used.

Keywords: clay-based copper tailings; fractal dimension; mixing intensity; population balance model; seawater flocculation

---

1. Introduction

The sustainability of the mining industry faces a challenging scenario for mineral processing, with low grade ore deposits, water scarcity and seawater as an alternative to freshwater. One of the operational priorities is to close the water cycle, a process that can be successful with proper management of the thickening stages. Generally, the concentration stages are carried out at alkaline pH, which means that both particles and flocculant molecules (such as hydrolyzed polyacrylamide) carry anionic charges. Although reagents may have a low affinity with the particles, it is sufficient to form porous aggregates with low density, which facilitate the transport of the thick pulp from the bottom of the thickener to the tailing dams. The fact that industrial water has some ionic charge can be beneficial. There is evidence in the literature that the presence of ions in the liquor is necessary for adequate flocculation, because ions allow the adsorption of polymeric molecules on the charged surfaces [1–3]. There is also evidence that flocculants may exhibit different behaviors when interacting in saline media. For example, studies have indicated that flocculated kaolinite shows decreasing sedimentation rates in environments with high salinity [4,5], while other studies show the opposite [6]. We recently demonstrated that the increase in salinity promotes the adsorption of polyelectrolytes on mineral surfaces but decreases the size of the polymer in solution, resulting in a competitive effect that can benefit or impair particle agglomeration. Thus, varied behaviors are expected [7,8].

There is great interest in studying flocculation from a microscopic level because the mechanisms involved during the aggregation of colloidal particles determine the efficiency of thickeners. Various particle characterization techniques are available, such as laser analysis, dynamic image analysis (DIA), and microscopy [9], however, the focused beam reflectance measurement (FBRM) has significant advantages for flocculation, since it can be used directly in feed slurries, allowing online monitoring of aggregate size without sampling and without dilution [10–14]. Through this technique, the effect of shear rate has been studied in several works [15–18]. In some cases, a maximum aggregate size is reached in a short time after the addition of the flocculant, but then fragmentation of the aggregates and polymer depletion occurs. Interestingly, the systems studied by He et al. [19] have shown that the maximum floc size is reached at low shear rate values and that any further increase in the intensity of the mixture leads to fragmentation.

Several researchers have described the kinetics of growth and fragmentation of aggregates using population balance models (PBM) [20–22] of a wide variety of systems, such as coagulation and flocculation [23], crystallization [24], reduction in mill size [25], and polymerization [26]. A century ago, Smoluchowski [27] was the first to formulate the PBM equations to describe the coalescence growth phenomena. Since then, this procedure has been the basis for describing new systems with mechanisms of aggregation of particles with increasingly richer physics, which has required modifications to the original equations.

Consideration of the fractal dimension as an indicator of the irregular structure of the aggregates has been key [20,28,29]. The fractal dimension has allowed describing the decrease in aggregate size over time due to restructuring mechanisms [30–32]. In all these cases, the fractal dimension increases while flocculation takes place. Recently, Jeldres et al. [33] showed that the fractal dimension of aggregates of flocculated mineral tailings in seawater decreases, while flocculation time and/or mixing intensity increase. These results are important because they define the mechanisms involved during the growth of aggregates, such as fragmentation, collision efficiency, permeability of structures [32,34] and flocculant depletion [21,35].

In this work, PBM is used to describe the flocculation kinetics of quartz-kaolinite tailing particles flocculated in seawater with a high-molecular-weight anionic polymer for a range of mixing intensities. The effect of the shear stress on the flocculation kinetics and on the aggregation and breakage rates of aggregates is analyzed. Optimum mixing conditions are determined to achieve the largest aggregate size. The structure of the aggregates during flocculation is described by a constant and time-dependent fractal dimension. Finally, an attempt is made to determine whether the representation of flocculation kinetics requires a constant average fractal dimension or a time-varying fractal dimension.

2. Model

The PBM equations used in this work were derived from from the work proposed by Spicer and Pratsinis [36] for particle grouping, which combines the discretization approach of Hounslow et al. [37] for aggregations and Kusters et al. [38] for breakage, in which the aggregate size distribution is discretized into a number of bins or channels according to a geometric progression of the volume distribution of the aggregates, that is, $V_{i+1} = 2V_i$. The population balance equation is given by:

$$\begin{aligned}\frac{dN_i}{dt} = \sum_{j=1}^{i-2} 2^{j-(i-1)} Q_{i-1,j} N_{i-1} N_j + \tfrac{1}{2} Q_{i-1,j-1} N_{i-1} N_{i-1} - \sum_{j=1}^{i-1} 2^{j-i} Q_{i,j} N_i N_j - \sum_{j=1}^{i_{max1}} Q_{i,j} N_i N_j \\ - S_i N_i + \sum_{j=1}^{i_{max2}} \Gamma_{i,j} S_j N_j\end{aligned} \tag{1}$$

where N_i is the number concentration of aggregates in bin i. The first two term on the right-hand side of Equation (1) describe the formation of aggregates of size i from smaller aggregates. The following two terms describe the loss of aggregates of size i to greater aggregates. The fifth term on the right-hand side represents the loss of aggregates of size i by fragmentation. The last term represents the gain of aggregates of size i by fragmentation of larger aggregates. The superscript $max1$ and $max2$ stand for the maximum number of intervals used to represent the complete aggregate size spectrum by aggregation and fragmentation, respectively. Functions Q, S and Γ represent aggregation rate, fragmentation rate, and breakage distribution function of aggregates, respectively. All the functions in Equation (1) are empirical in origin and thus parameters involved need to be determined by solving Equation (1) against experimental data. The functions are described next.

2.1. Aggregation Kernel

The function Q is the aggregation kernel defined by the product between the collision frequency (β) and the capture efficiency (α):

$$Q_{i,j} = \beta_{i,j} \alpha_{i,j} \tag{2}$$

Collision frequency β

Fluid flow can penetrate and pass through particle aggregates [39], meaning that the genuine collision frequency is considerably lower than that predicted from rectilinear flow models. To incorporate permeability effects, we use a parameter called 'fluid collection efficiency' η which is the ratio between flow passing through an aggregate and that approaching it (undisturbed). Veerapaeni and Wiesner [40] proposed the function in Equation (3) to account for the collision frequency which includes permeability and fractal dimension, as:

$$\beta_{i,j} = \frac{1}{6}\left(\sqrt{\eta_i} d_i + \sqrt{\eta_j} d_j\right)^3 G \tag{3}$$

where d_i and d_j are the diameters of the aggregates of sizes i and j, respectively, G is the mean value of shear rate, and η is derived from the Brinkman extension to Darcy's law in function of a dimensionless permeability ξ, that is:

$$\eta_i = \frac{9(\xi_i - \tanh\xi_i)}{(38\xi_i + 2\xi_i{}^2 - 3\tanh\xi_i)} \tag{4}$$

where ξ_i is defined as $\xi_i = d_i/2\sqrt{K_i}$ and K_i is the permeability, for which the expression from the Li and Logan [41] is used:

$$K_i = \frac{d_i^2}{72}\left(3 + \frac{3}{1-\phi_i} - \sqrt[3]{\frac{8}{1-\phi_i} - 3}\right) \tag{5}$$

The porosity ϕ is related to fractal dimension (d_f) by the expression of Vainshtein et al. [42]:

$$\phi_i = 1 - C\left(\frac{d_i}{d_0}\right)^{d_f-3} \quad (6)$$

where C is a packing coefficient (generally assumed to be 0.65) and d_0 is the primary particle diameter; d_i and d_0 are related by the expression by Mandelbrot [43]:

$$d_i = d_0\left(\frac{2^{i-1}}{C}\right)^{\frac{1}{d_f}} \quad (7)$$

Collision efficiency α

Several models allow estimating collision efficiency, depending on the type of aggregate and the additive used. In this work, the three-parameter expression of Vajihinejad and Soares [35] is used to capture the depletion of adsorbed high-molecular-weight polymer of the type used here and its rearrangement on the surface of the particles, that is:

$$\alpha_{i,j} = (\alpha_{max} - \alpha_{min})e^{-k_d t} + \alpha_{min} \quad (8)$$

where α_{max} and α_{min} are maximum and minimum collision efficiency, respectively, at steady state conditions and k_d is a decay constant.

2.2. Breakage Kernel

The S function in Equation (1) is the fragmentation rate and the Γ function is the breakage distribution function; these two represent the breakage kernel. The S term is difficult to predict since it has not been theorized; it is common to adjust it to size distribution data. For this work, a two-parameter power law function of the aggregate mass is used [44]:

$$S_i = s_1 G^{s_2} d_i \quad (9)$$

where s_1 and s_2 are the fitted parameters.

For the breakage distribution function, a binary distribution without parameters is used:

$$\Gamma_{i,j} = \begin{cases} \frac{V_j}{V_i} \; for \; j = i+1 \\ 0 \;\; for \; j \neq i+1 \end{cases} \quad (10)$$

2.3. Shear Rate

The shear rate required by the aggregation and breakage kernels is calculated from:

$$G = \left(\frac{\varepsilon \rho_{sus}}{\mu_{sus}}\right)^{\frac{1}{2}} \quad (11)$$

where ε is the average energy dissipation rate given by:

$$\varepsilon = \frac{N_p N^3 D^5}{V} \quad (12)$$

where N_p is the impeller power number which for a plane disk with gentle agitation, as is our case, is 0.6 [45], N is the rotation speed, and D and V are the diameter of the impeller and the working volume of the vessel, respectively. The density of the suspension ρ_{sus} is calculated from:

$$\rho_{sus} = \left(\frac{w}{\rho_s} + \frac{1-w}{\rho_w}\right)^{-1} \tag{13}$$

where w is the mass solid fraction of the solution and ρ_s and ρ_w are the solid and water densities, respectively. Finally, the viscosity of the solution μ_{sus} is measured.

2.4. PBM Solution

In this work, the PBM equation (Equation (1)) is used to describe the flocculation kinetics of synthetic copper tailings in seawater. A solver based on a numerical differentiation formula for stiff ODEs (ode23s) is used in MATLAB environment. The numerical solution requires parameters which are obtained by minimizing the objective function:

$$\mathrm{OF}(\alpha_{max}, \alpha_{min}, k_d, s_1, s_2) = \sum_{t_i}^{t_f}\left(d_{agg,exp} - d_{agg,mod}\right)^2 \tag{14}$$

where $d_{agg,exp}$ is the experimental diameter of the aggregates and $d_{agg,mod}$ is the volume-weighted mean aggregate diameter from the model:

$$d_{mod} = \frac{\sum_{i=1}^{max} N_i d_i^4}{\sum_{i=1}^{max} N_i d_i^3} \tag{15}$$

To solve Equation (14) the MATLAB function *fminsearch* is used; this function uses the Nelder–Mead direct search to find the minimum of an unconstrained multivariable function.

Two criteria validate the model fit and predictions, one is the coefficient of determination (R^2) that measures the closeness of the model values to the experimental values:

$$\mathrm{R}^2 = 1 - \frac{\sum_{i=1}^{max}\left(d_{agg,exp,i} - d_{agg,mod,i}\right)^2}{\sum_{i=1}^{max}\left(d_{agg,exp,i} - d_{agg,\,exp}\right)^2} \tag{16}$$

The other is the goodness of fit:

$$\mathrm{GoF}(\%) = 100\frac{\left\langle d_{agg,\,exp}\right\rangle - \mathrm{std}}{\left\langle d_{agg,\,exp}\right\rangle} \tag{17}$$

where std stands for the standard error calculated from:

$$\mathrm{std} = \left(\frac{1}{n-f}\sum_{t_i}^{t_f}\left(\left\langle d_{agg,\,exp}\right\rangle - \left\langle d_{agg,\,mod}\right\rangle\right)^2\right)^{\frac{1}{2}} \tag{18}$$

where n is the number of data values and f is the number of parameters to be fitted. A GoF of 90% or higher means that the proposed model is able to predict the flocculation kinetics [10]. Finally, conservation of the total volume of particles is verified after every integration to ensure that the simulations are maintaining the particle population.

3. Materials and Methods

3.1. Materials

Quartz particles with a density of 2.65 g/cm^3 acquired from Donde Capo (Chilean Store, Santiago, Chile) were used, which were pulverised and screened with -#325 mesh. The SiO_2 content, estimated by

XRD analysis, was higher than 99 wt%. Kaolin particles were acquired from Ward's Science (Rochester, NY 14586, Rochester, United States), and the XRD analysis indicated 84% kaolinite and 16% halloysite. Seawater from the coast of Antofagasta (Chile) was provided by the Department of Marine Resources from the University of Antofagasta and subsequently filtered by UV for the removal of microorganisms. The ionic composition was determined with varied chemical methods depending on the ion type. By atomic absorption spectrometry: Na^+ 10.5 g/L, Mg^{2+} 1.42 g/L, Ca^{2+} 0.38 g/L, K^+ 0.40 g/L. The composition of Cl^- was determined by argentometric method and was 19.1 g/L. The concentration of HCO^{3-} was determined by acid–base volumetry to be 0.17 mg/L. NaOH was used as a pH modifier; anionic SNF704 and high molecular weight were used as a flocculant.

3.2. Flocculation Kinetics

A suspension of 270 g was prepared, considering a total solids content of 8 wt%, corresponding to a mixture of 80 wt% quartz and 20 wt% kaolin. The slurry was vigorously mixed for 30 min using a 30 mm diameter turbine-type stirrer, placed in the axial position, in a 1 L capacity, 100 mm diameter vessel. The stirrer was placed 20 mm above the vessel bottom. Subsequently, the mixing rate was reduced to 250 rpm, and an established volume of a solution (seawater and polymer) was added, at a proportion fixed by the required polymer dosage. The size distribution was traced using the FBRM technique, which consists of a sensor that measures reflection pulses of a laser beam on suspended particles in the range of 0.25–1000 μm. The main advantage of this technique is that the data is collected in-situ and in real-time; therefore, it does not require sampling or isolation that could contribute to changes in the size and distribution of aggregates by the eventual breakage or agglomeration. The FBRM probe was submerged vertically in the reaction vessel, 10 mm over the stirrer and 20 mm off-axis. The measurement began by tracking the primary particles of the tailings, and after 1 min, the flocculant was added with subsequent growth and destruction of aggregates for 3 min. The flocculation assays where done at pH 8, under ambient temperature and different mixing intensities.

3.3. Viscosity

The calculation of the shear rate requires the viscosity of the slurry. For this, a flocculation process was carried out with the methodology defined in Section 3.2. Subsequently, an aliquot of 56 mL was taken for rheological analysis. An Anton Paar MCR 102 rheometer with the RheoCompass Software was operated in controlled-rate mode, and a sandblasted #CC39 with bob-in-cup geometry was used to reduce wall slip effects. The gap generated between both concentric cylinders was 1.5 mm. Each sample remained for 15 min, ensuring that the rheological properties were obtained in steady state. The range of shear rate was preset between 10–300 s^{-1}.

3.4. Sedimentation

After a pre-set flocculation period (10–80 s), the pulp was gently poured into closed cylinders of 300 cm^3 (35 mm internal diameter) and then slowly inverted twice by hand (the whole cylinder rotation process took 4 s). After 10 min of settling, the supernatant fluid was rescued and stirred to homogenize the suspended solids. Then, a 50 mL aliquot was used for turbidity measurements in a Hanna HI98713 turbidimeter, which performed ten readings in 20 s, delivering the average of the point readings.

3.5. Conditions

Table 1 summarizes parameters and conditions used in this study. With these conditions, shear rate was calculated for all the cases, using Equations (11)–(13).

Table 1. Input parameters and conditions.

i_{max}	30	-
ϕ	0.054	-
c	0.65	-
N_p	0.6	-
D	8.0	cm
V	0.25	L
ρ_s	2600	kg/m^3
ρ_w	1000	kg/m^3
μ_{sus}	0.005	kg/ms
w	0.08	-
d_0	0.0005	cm

4. Results

4.1. Initial Particle Size

Solving the PBM equations requires an initial particle number concentration for each size class and per unit of volume of suspension ($N_{0,i}$), for that, a number distribution from experimental volume distribution is obtained from:

$$N_{0,i} = \phi \frac{v(d_{0,i})}{V_{0,i}} \tag{19}$$

where $v(d_{0,i})$ is the experimental volume fraction of particles with diameter $d_{0,i}$, obtained from Equation (7), and $V_{0,i}$ is the volume of the primary particles following the geometric progression $V_{0,i} = 2^{i-1}V_1$. Figure 1 shows the normalized volume for synthetic tailings in seawater at pH 8. $v(d_{0,i})$ is obtained by interpolating the data in Figure 1 at the required values of $d_{0,i}$. The initial distribution is sensitive to the shear rate, where the tendency is for the particles to aggregate as the intensity of mixing increases. The initial primary particle size is set to 5 μm, which is the mean minimal size with non-zero value in the distribution.

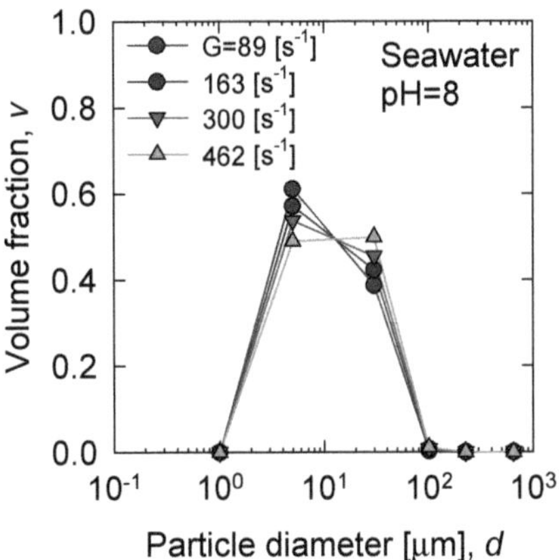

Figure 1. Normalized initial volume distribution of particles of synthetic tailings in seawater at pH 8 for different mixing intensities (shear rate G).

4.2. Fractal Dimension

The fractal dimension defines the structure of the aggregates and appears in several equations describing particle agglomeration. In this study, the fractal dimension (Figure 2) is obtained from combined sedimentation tests with aggregate size measurements using the FBRM probe. The experimental data of hindered settling velocity, mean particle size, and mean aggregate size were used to extract the fractal dimension from the settling velocity model of Heath et al. [11]. In previous work, we found that the fractal dimension of particle tailing aggregates is very close to constant for low values of shear rate [46], in agreement with results from Heath et al. [15,21]. The results of Figure 2 confirm that, for low shear rate, the fractal dimension is a constant independent of flocculation time. However, when the shear rate values are high, the fractal dimension decreases monotonously over time, indicating that fragmentation of aggregates leads to lower-dimension Euclidean structures. In this work, the consequences of using constant fractal dimension in the determination of flocculation kinetics and aggregation and breakage functions for all flocculation conditions are analyzed. The results are compared with those obtained using fractal dimension as a function of time. Note that the constant fractal dimension is defined here as the average between the lowest and highest fractal dimensions for the same system and identical flocculation conditions.

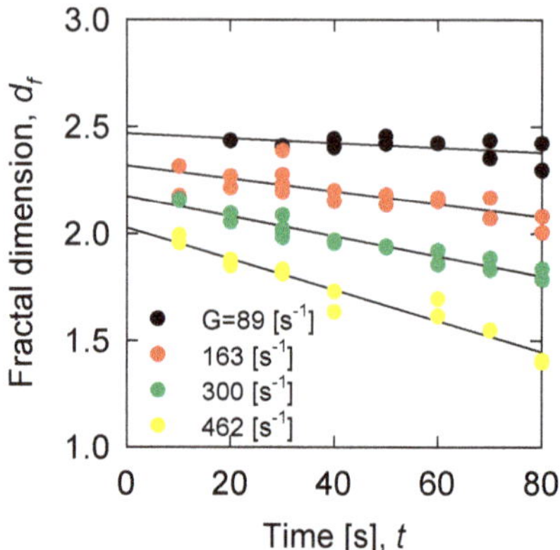

Figure 2. Temporal evolution of the fractal dimension during flocculation of synthetic tailing particles in seawater at pH 8 as a function of shear stress (G).

4.3. Flocculation Kinetics Modelling

The flocculation kinetics of synthetic tailing particles in seawater are shown in Figure 3 as a function of shear stress. While the open circles correspond to experimental data, the solid lines correspond to the best fit with the PBM and the fractal dimension from Figure 2. Low shear requires a long time to give life to small aggregates, while high shear in a short time leads approximately to the same small aggregates; in the first case collisions and captures are unimportant, while in the second case the ruptures are important. There is clearly a mixing intensity that maximizes the size of the aggregates; however, the necessary flocculation time is critical because it defines the utility or lack of utility of the chosen flocculant. In the case of synthetic tailing particles in seawater with the flocculant chosen in this work, the optimum size of the aggregates is ca. 225 μm when the shear rate is 163 s^{-1}. The PBM captures all the complex stages of particle flocculation, that is, initial growth of aggregates by particle bridging and subsequent size reduction as a result of fragmentation.

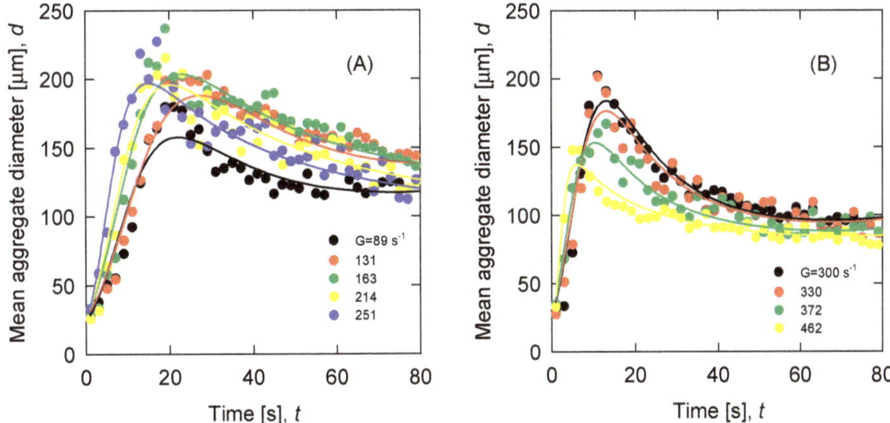

Figure 3. Flocculation kinetics of synthetic tailing particles in seawater at pH 8 as a function of shear stress (G). Open circles correspond to experimental data and solid lines to the best fit with the population balance model (PBM). (**A**): Shear rate range from 89–251 s^{-1}. (**B**): Shear rate range from 300–462 s^{-1}.

Figure 3 shows qualitatively that the model correctly describes the experimental data for all mixing intensities. The quantitative results of GoF and R^2 summarized in Table 2 show that the quality of the PBM is very good in all the cases analyzed and that the effect of representing the fractal dimension of the aggregates as a constant independent of the flocculation time is a correct decision, at least to represent the flocculation kinetics.

Table 2. Quantitative results of GoF and R^2 when the PBM is used with constant fractal dimension (d_f mean) or variable (d_f var) fractal dimension dependent on flocculation time.

Mixing	GoF, %		R^2	
Shear Rate (s^{-1}), Mixing Rate	**d_f var**	**d_f mean**	**d_f var**	**d_f mean**
89 (100 rpm)	90.1	89.2	0.8895	0.8698
131 (130 rpm)	92.8	93.6	0.9449	0.9565
163 (150 rpm)	91.6	91.6	0.9177	0.9174
214 (180 rpm)	90.6	90.2	0.8963	0.8881
251 (200 rpm)	91.6	91.3	0.8956	0.8889
300 (225 rpm)	91.1	90.4	0.9267	0.9151
330 (240 rpm)	89.5	88.2	0.8898	0.8611
372 (260 rpm)	92.5	92.3	0.9055	0.9024
462 (300 rpm)	93.5	94.5	0.9038	0.9319

4.4. Aggregation, Breakage, and Permeability Modelling

The aggregation, rupture, and permeability functions of particle aggregates are the building blocks to describe the flocculation kinetics, in the present case of synthetic tailing particles. Therefore, it is of great interest to evaluate the quality of the PBM representation of these functions when using the fractal dimension of the aggregates as a constant or variable dependent on the flocculation time. Figure 4 summarizes the four results.

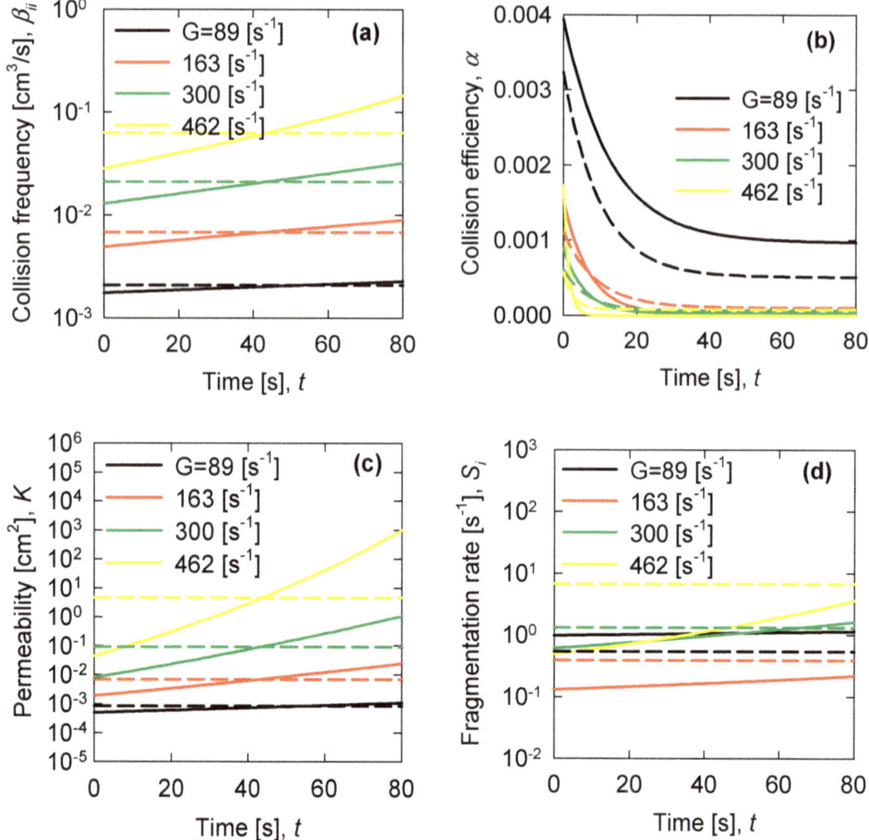

Figure 4. PBM representation of collision frequency (**a**), collision efficiency (**b**), aggregate permeability (**c**) and fragmentation rate (**d**) for the largest aggregates considering constant fractal dimension of the aggregates (dashed lines) and time-dependent fractal dimension (solid lines) for different mixing intensities. Aggregate size range is 200–600 μm (bin $i = 15$).

As expected, the collision frequency increases and the collision efficiency decreases with mixing intensity (Figure 4a,b). The collision frequency remains constant when the fractal dimension is considered constant and grows over time when the fractal dimension is considered dependent on the flocculation time. The collision efficiency shows an exponential decay with flocculation time, at low shear rate the efficiency decreases rapidly when the fractal dimension is used as a function of time, and at high shear rate the collision efficiency is not so different if the fractal dimension is constant or variable. The results suggest that when aggregates are more porous and irregularly structured they are less likely to adhere once they collide.

The permeability of the aggregates increases with the intensity of the mixing and does so more strongly when the shear rate is very high. Permeability is constant when considering a constant fractal dimension and increases with flocculation time when considering a fractal dimension dependent on flocculation time. The product of the collision frequency and the collision efficiency is the aggregation rate; it is expected to be quite different when using constant or variable fractal dimension, however, to represent the optimal mixing (163 s^{-1}) that leads to the larger aggregates (225 μm) the results of Figure 4a,b for collision frequency and collision efficiency, respectively, and the results of Figure 4c for aggregate permeability, suggest that there are no large differences if one or the other fractal dimension is used.

The fragmentation rate is not a monotonous function of the mixing intensity for both constant and variable fractal dimensions (Figure 4d). For low shear rate values, the rate decreases when the mixing intensity increases, and for high shear rate values, the rate increases with the mixing intensity. The fragmentation rate is constant when the constant fractal dimension is used and in general it is very different from the fragmentation rate obtained with a variable fractal dimension. Under optimal mixing conditions (163 s^{-1}), the results of Figure 4c suggest that the fragmentation rate is better represented by the variable fractal dimension, the use of a constant fractal dimension can lead to significant deviations.

It is well known that compensations between aggregation and breakage rates lead to a very good representation of flocculation kinetics; the results in Figure 3 are a good demonstration. However, in addition, this study reveals that under optimal mixing conditions the representation of flocculation kinetics is equally good with a constant or variable fractal dimension.

4.5. Optimized Parameters

The optimization of the PBM model is performed against experimental flocculation data resulting in five parameters: the maximum and minimum collision efficiency (α_{max}, α_{min}), the collision efficiency decay constant (k_d), and two breakage rate kernel parameters (s_1 and s_2). Of great interest is the effect of mixing intensity and the fractal dimension, constant or variable, on these parameters. The three parameters that describe the collision efficiency versus shear rate are presented in Figure 5, for the two conditions of fractal dimension. In general, α_{max} and α_{min} decrease with the shear rate, $\alpha_{min} = 0$ for the whole range of shear rate, except for at the lowest rate, and k_d increases with the shear rate. The effect of the fractal dimension either constant or variable is small except at very high shear rate for α_{max} and k_d. α_{min} is not affected by the choice of fractal dimension. For the optimal mixing conditions (163 s^{-1}), the three parameters of the collision efficiency can be estimated considering a constant fractal dimension. Finally, the two parameters of the breakage rate (s_1, s_2) change erratically with the shear rate but are not affected by the choice of the fractal dimension, whether constant or variable. For the optimal mixing conditions, the breakage rate parameters can be determined assuming a constant fractal dimension.

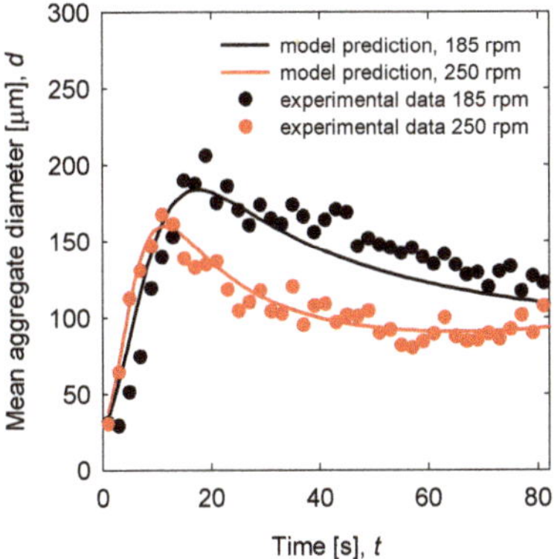

Figure 5. Experimental (symbols) vs. calculated (continuous lines) time evolution of the mean diameter of aggregates of tailing particles as a function of pH in seawater. Constant pH condition at 8.

4.6. Prediction Capability

To test the predictive capacity of PBM with the parameters determined in the preceding sections, new experiments were carried out to determine the kinetics of flocculation of tailing particles in seawater at 185 and 250 rpm. Pulps with the same composition and percentage of solids were considered as in the trials in Section 4.3. The size of primary particles was again 5 μm. Also, the same flocculant and the same operating conditions were used, that is, temperature, flocculant dose, agitation time, and pH. Figure 5 shows the experimental and modeling results. The parameters for PBM were interpolated from Figure 2 for the fractal dimension, Figure 6 for the aggregation kernel and Figure 7 for the breakage rate for each pH. The adjustments are generally good, with GoF of 87.17 for 185 rpm and 92.02 for 250 rpm, which demonstrates the predictive capacity of PBM. Unlike what was found in our previous paper [46] where pH has little influence in kinetics, the shear rate has marked nonlinear dependence on the results of kinetics. Therefore, modeling was possible with good prediction for intervals of 30 s^{-1}.

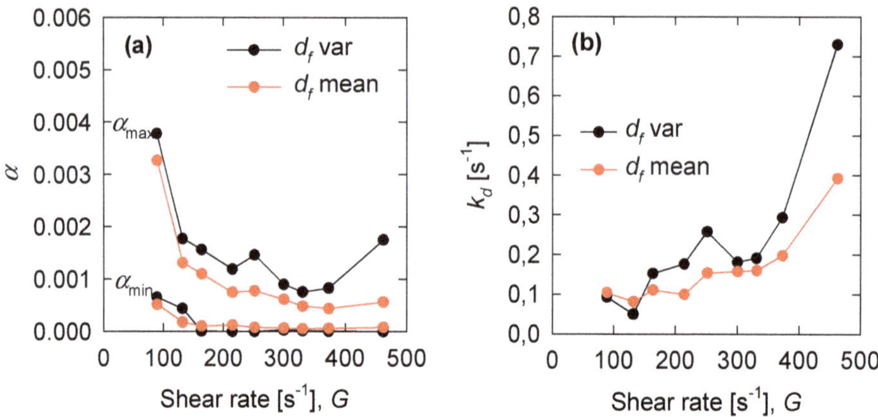

Figure 6. Optimum aggregation parameters vs. shear rate for constant and variable fractal dimension, (**a**) maximum and minimum collision efficiencies and (**b**) collision efficiency decay constant k_d.

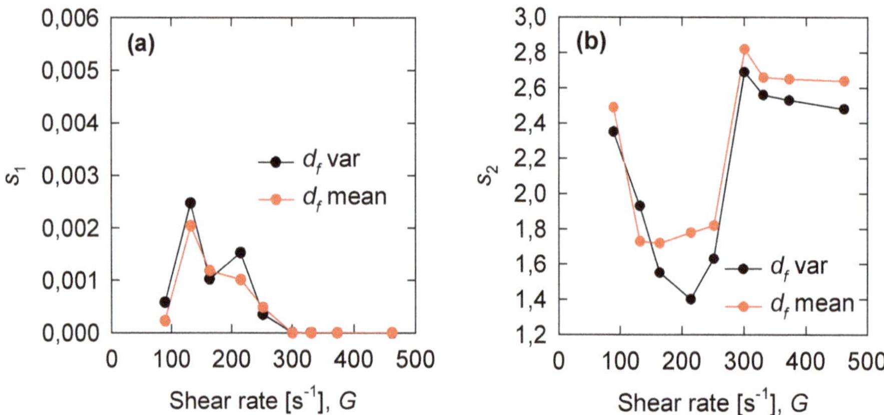

Figure 7. Optimum breakage parameters vs. shear rate for constant and variable fractal dimension, (**a**) s_1 and (**b**) s_2.

5. Conclusions

PBM equations were used to describe the aggregation kinetics of synthetic copper tailings in seawater with a high molecular weight polyacrylamide at pH 8 and a range of mixing intensities. It is common to consider that the fractal dimension of the aggregates is constant throughout the flocculation process, however in this work we found that when the mixing intensity is high, >100 s^{-1}, the fractal dimension of the aggregates decreases monotonously over time, indicating that fragmentation of aggregates leads to lower-dimension Euclidean structures. This study revealed that compensations between aggregation and breakage rates lead to a very good representation of the flocculation kinetics of the particle tailings in seawater, and in addition, that the representation of flocculation kinetics under optimal conditions is equally good with a constant or variable fractal dimension. The mixing intensity of 163 s^{-1} was found to maximize the size of the tailing aggregates, that is, ca. 225 μm at ca. 20 s. The aggregation and breakage functions and corresponding parameters are sensitive to the choice of the fractal dimension of the aggregates, whether constant or time-dependent; however, under optimal conditions, a constant average of the fractal dimension is sufficient. The predictive capacity of the model can be used to find the optimal flocculation conditions based on a few experimental flocculation data at different mixing intensities, and the optimal flocculation time can be used to make decisions regarding the effectiveness of the flocculant used.

Author Contributions: G.R.Q. and W.H.L. contributed in research and wrote paper, N.T., P.R. and R.I.J. contributed in project administration, L.A. contributed resources and P.G.T. contributed in review and validation. All authors have read and agreed to the published version of the manuscript.

Funding: This research was funded by Conicyt Fondecyt 11171036 and Centro CRHIAM Project ANID/FONDAP/15130015.

Acknowledgments: The authors thank Centro CRHIAM Project Conicyt/Fondap/15130015. RIJ thanks CONICYT Fondecyt 11171036. Pedro Robles thanks the Pontificia Universidad Católica de Valparaíso for the support provided.

Conflicts of Interest: The authors declare no conflict of interest.

References

1. Burlamacchi, L.; Ottaviani, M.F.; Ceresa, E.M.; Visca, M. Stability of colloidal TiO2 in the presence of polyelectrolytes and divalent metal ions. *Colloids and Surfaces* **1983**, *7*, 165–182. [CrossRef]
2. Sommerauer, A.; Sussman, D.L.; Stumm, W. The role of complex formation in the flocculation of negatively charged sols with anionic polyelectrolytes. *Kolloid-Zeitschrift Zeitschrift für Polym.* **1968**, *225*, 147–154. [CrossRef]
3. Nabzar, L.; Pefferkorn, E.; Varoqui, R. Polyacrylamide-sodium kaolinite interactions: Flocculation behavior of polymer clay suspensions. *J. Colloid Interface Sci.* **1984**, *102*, 380–388. [CrossRef]
4. Peng, F.F.; Di, P. Effect of multivalent salts-calcium and aluminum on the flocculation of kaolin suspension with anionic polyacrylamide. *J. Colloid Interface Sci.* **1994**, *164*, 229–237. [CrossRef]
5. Wang, S.; Zhang, L.; Yan, B.; Xu, H.; Liu, Q.; Zeng, H. Molecular and surface interactions between polymer flocculant chitosan- g -polyacrylamide and kaolinite particles: Impact of salinity. *J. Phys. Chem. C* **2015**, *119*, 7327–7339. [CrossRef]
6. Ji, Y.; Lu, Q.; Liu, Q.; Zeng, H. Effect of solution salinity on settling of mineral tailings by polymer flocculants. *Colloids Surfaces A Physicochem. Eng. Asp.* **2013**, *430*, 29–38. [CrossRef]
7. Jeldres, R.I.; Piceros, E.C.; Leiva, W.H.; Toledo, P.G.; Herrera, N. Viscoelasticity and yielding properties of flocculated kaolinite sediments in saline water. *Colloids Surfaces A Physicochem. Eng. Asp.* **2017**, *529*, 1009–1015. [CrossRef]
8. Quezada, G.R.; Jeldres, R.I.; Fawell, P.D.; Toledo, P.G. Use of molecular dynamics to study the conformation of an anionic polyelectrolyte in saline medium and its adsorption on a quartz surface. *Miner. Eng.* **2018**, *129*. [CrossRef]
9. Liang, L.; Peng, Y.; Tan, J.; Xie, G. A review of the modern characterization techniques for flocs in mineral processing. *Miner. Eng.* **2015**, *84*, 130–144. [CrossRef]
10. Biggs, C.A.; Lant, P.A. Activated sludge flocculation: On-line determination of floc size and the effect of shear. *Water Res.* **2000**. [CrossRef]
11. Heath, A.R.; Bahri, P.A.; Fawell, P.D.; Farrow, J.B. Polymer flocculation of calcite: Relating the aggregate size to the settling rate. *AIChE J.* **2006**, *52*, 1987–1994. [CrossRef]
12. Gregory, J. Monitoring particle aggregation processes. *Adv. Colloid Interface Sci.* **2009**, *147–148*, 109–123. [CrossRef]

13. Meng, Z.; Hashmi, S.M.; Elimelech, M. Aggregation rate and fractal dimension of fullerene nanoparticles via simultaneous multiangle static and dynamic light scattering measurement. *J. Colloid Interface Sci.* **2013**, *392*, 27–33. [CrossRef]
14. Rong, H.; Gao, B.; Li, J.; Zhang, B.; Sun, S.; Wang, Y.; Yue, Q.; Li, Q. Floc characterization and membrane fouling of polyferric–polymer dual/composite coagulants in coagulation/ultrafiltration hybrid process. *J. Colloid Interface Sci.* **2013**, *412*, 39–45. [CrossRef]
15. Heath, A.R.; Bahri, P.A.; Fawell, P.D.; Farrow, J.B. Polymer flocculation of calcite: Experimental results from turbulent pipe flow. *AIChE J.* **2006**, *52*, 1284–1293. [CrossRef]
16. Owen, A.T.; Fawell, P.D.; Swift, J.D.; Labbett, D.M.; Benn, F.A.; Farrow, J.B. Using turbulent pipe flow to study the factors affecting polymer-bridging flocculation of mineral systems. *Int. J. Miner. Process.* **2008**, *87*, 90–99. [CrossRef]
17. Bubakova, P.; Pivokonsky, M.; Filip, P. Effect of shear rate on aggregate size and structure in the process of aggregation and at steady state. *Powder Technol.* **2013**, *235*, 540–549. [CrossRef]
18. Benn, F.A.; Fawell, P.D.; Halewood, J.; Austin, P.J.; Costine, A.D.; Jones, W.G.; Francis, N.S.; Druett, D.C.; Lester, D. Sedimentation and consolidation of different density aggregates formed by polymer-bridging flocculation. *Chem. Eng. Sci.* **2018**, *184*, 111–125. [CrossRef]
19. He, W.; Nan, J.; Li, H.; Li, S. Characteristic analysis on temporal evolution of floc size and structure in low-shear flow. *Water Res.* **2012**, *46*, 509–520. [CrossRef]
20. Thomas, D.N.; Judd, S.J.; Fawcett, N. Flocculation modelling: a review. *Water Res.* **1999**, *33*, 1579–1592. [CrossRef]
21. Heath, A.R.; Bahri, P.A.; Fawell, P.D.; Farrow, J.B. Polymer flocculation of calcite: Population balance model. *AIChE J.* **2006**, *52*, 1641–1653. [CrossRef]
22. Jeldres, R.I.; Fawell, P.D.; Florio, B.J. Population balance modelling to describe the particle aggregation process: A review. *Powder Technol.* **2018**, *326*, 190–207. [CrossRef]
23. Runkana, V.; Somasundaran, P.; Kapur, P.C. A population balance model for flocculation of colloidal suspensions by polymer bridging. *Chem. Eng. Sci.* **2006**, *61*, 182–191. [CrossRef]
24. Costa, C.B.B.; Maciel, M.R.W.; Filho, R.M. Considerations on the crystallization modeling: Population balance solution. *Comput. Chem. Eng.* **2007**, *31*, 206–218. [CrossRef]
25. Datta, A.; Rajamani, R.K. A direct approach of modeling batch grinding in ball mills using population balance principles and impact energy distribution. *Int. J. Miner. Process.* **2002**, *64*, 181–200. [CrossRef]
26. Kiparissides, C.; Alexopoulos, A.; Roussos, A.; Dompazis, G.; Kotoulas, C. Population balance modeling of particulate polymerization processes. *Ind. Eng. Chem. Res.* **2004**, *43*, 7290–7302. [CrossRef]
27. Von Smoluchowski, M. Versuch einer mathematischen Theorie der Koagulations kinetik kolloider Lösungen. *Zeitschrift fuer Phys. Chemie* **1917**, *129*, 129–168.
28. Flesch, J.C.; Spicer, P.T.; Pratsinis, S.E. Laminar and turbulent shear-induced flocculation of fractal aggregates. *AIChE J.* **1999**, *45*, 1114–1124. [CrossRef]
29. Filippov, A.V.; Zurita, M.; Rosner, D.E. Fractal-like aggregates: Relation between morphology and physical properties. *J. Colloid Interface Sci.* **2000**, *229*, 261–273. [CrossRef]
30. Selomulya, C.; Bushell, G.; Amal, R.; Waite, T.D. Understanding the role of restructuring in flocculation: The application of a population balance model. *Chem. Eng. Sci.* **2003**, *58*, 327–338. [CrossRef]
31. Antunes, E.; Garcia, F.A.P.; Ferreira, P.; Blanco, A.; Negro, C.; Rasteiro, M.G. Modelling PCC flocculation by bridging mechanism using population balances: Effect of polymer characteristics on flocculation. *Chem. Eng. Sci.* **2010**, *65*, 3798–3807. [CrossRef]
32. Jeldres, R.I.; Concha, F.; Toledo, P.G. Population balance modelling of particle flocculation with attention to aggregate restructuring and permeability. *Adv. Colloid Interface Sci.* **2015**, *224*, 62–71. [CrossRef] [PubMed]
33. Jeldres, M.; Piceros, E.C.; Toro, N.; Torres, D.; Robles, P.; Leiva, W.H.; Jeldres, R.I. Copper tailing flocculation in seawater: Relating the yield stress with fractal aggregates at varied mixing conditions. *Metals (Basel).* **2019**, *9*, 1295. [CrossRef]
34. Ahmad, A.L.; Chong, M.F.; Bhatia, S. Population Balance Model (PBM) for flocculation process: Simulation and experimental studies of palm oil mill effluent (POME) pretreatment. *Chem. Eng. J.* **2008**, *140*, 86–100. [CrossRef]
35. Vajihinejad, V.; Soares, J.B.P. Monitoring polymer flocculation in oil sands tailings: A population balance model approach. *Chem. Eng. J.* **2018**, *346*, 447–457. [CrossRef]

36. Spicer, P.T.; Pratsinis, S.E. Shear-induced flocculation: The evolution of floc structure and the shape of the size distribution at steady state. *Water Res.* **1996**, *30*, 1049–1056. [CrossRef]
37. Hounslow, M.J.; Ryall, R.L.; Marshall, V.R. A discretized population balance for nucleation, growth, and aggregation. *AIChE J.* **1988**, *34*, 1821–1832. [CrossRef]
38. Kusters, K.A.; Pratsinis, S.E.; Thoma, S.G.; Smith, D.M. Ultrasonic fragmentation of agglomerate powders. *Chem. Eng. Sci.* **1993**, *48*, 4119–4127. [CrossRef]
39. Thill, A.; Moustier, S.; Aziz, J.; Wiesner, M.R.; Bottero, J.Y. Flocs restructuring during aggregation: Experimental evidence and numerical simulation. *J. Colloid Interface Sci.* **2001**, *243*, 171–182. [CrossRef]
40. Veerapaneni, S.; Wiesner, M.R. Hydrodynamics of fractal aggregates with radially varying permeability. *J. Colloid Interface Sci.* **1996**, *177*, 45–57. [CrossRef]
41. Li, X.Y.; Logan, B.E. Permeability of fractal aggregates. *Water Res.* **2001**. [CrossRef]
42. Vainshtein, P.; Shapiro, M.; Gutfinger, C. Mobility of permeable aggregates: effects of shape and porosity. *J. Aerosol Sci.* **2004**, *35*, 383–404. [CrossRef]
43. Mandelbrot, B.B. Self-affine fractals and fractal dimension. *Phys. Scr.* **1985**, *32*, 257–260. [CrossRef]
44. Pandya, J.D.; Spielman, L.A. Floc breakage in agitated suspensions: Effect of agitation rate. *Chem. Eng. Sci.* **1983**, *38*, 1983–1992. [CrossRef]
45. Pretorius, C.; Wicklein, E.; Rauch-Williams, T.; Samstag, R.; Sigmon, C. How oversized mixers became an industry standard. *Proc. Water Environ. Fed.* **2015**, *11*, 4379–4411. [CrossRef]
46. Quezada, G.R.; Ramos, J.; Jeldres, R.I.; Robles, P.; Toledo, P.G. Analysis of the flocculation process of fine tailings particles in saltwater through a population balance model. *Sep. Purif. Technol.* **2019**, 116319. [CrossRef]

Article

Selective and Mutual Separation of Palladium (II), Platinum (IV), and Rhodium (III) Using Aliphatic Primary Amines

Kazuya Matsumoto *, Yuto Sezaki, Sumito Yamakawa, Yuki Hata and Mitsutoshi Jikei

Department of Materials Science, Graduate School of Engineering Science, Akita University, 1-1 Tegatagakuen-machi, Akita-shi, Akita 010-8502, Japan; sezaki.y@yahoo.ne.jp (Y.S.); smi.kun.12@gmail.com (S.Y.); wimper_h_1114@outlook.jp (Y.H.); mjikei@gipc.akita-u.ac.jp (M.J.)

* Correspondence: kmatsu@gipc.akita-u.ac.jp; Tel.: +81-18-889-2745

Received: 16 February 2020; Accepted: 28 February 2020; Published: 29 February 2020

Abstract: The selective recovery of platinum-group metals (PGMs) remains a huge challenge. Although solvent extraction processes are generally used for PGM separation, the use of organic solvents is problematic because of their toxicity and environmental concerns. Here, we have developed a new PGM recovery method by precipitation from hydrochloric acid (HCl) solutions containing Pd(II), Pt(IV), and Rh(III) using aliphatic primary amines as precipitants. Pt(IV) was precipitated using the amines with alkyl chains longer than hexyl independent of HCl concentration. The precipitation of Pd(II) required longer alkyl amines than octyl, regardless of the HCl concentration. Rh(III) was recovered by precipitation at high HCl concentrations using the amines longer than hexyl. The mutual separation of Pt(IV), Rh(III), and Pd(II), in this order, was successfully achieved by changing the HCl concentrations and alkyl chain lengths of the amines. X-ray photoelectron spectroscopy and thermogravimetric analysis evidently showed that the metal-containing precipitates were ion-pair complexes composed of metal chloro-complex anions and ammonium cations.

Keywords: platinum-group metal; metal precipitation; ion-pair; aliphatic primary amine; hydrochloric acid

1. Introduction

Platinum-group metals (PGMs), particularly Pd, Pt, and Rh, are of crucial importance because of their wide range of applications, such as automobile catalysts and electrical devices [1–5]. Despite the increasing demand for PGMs, their availability remains limited due to their scarcity in Nature and regional maldistribution. Therefore, efficient recovery and separation processes are necessary to recycle PGMs from post-consumer scrap. Typically, solvent extraction is regarded as a practical method to recover and separate PGM ions from metal-containing aqueous solutions. For example, tertiary amines [6,7], organophosphates [8–11], and organosulfides [12,13] have been reported to act as Pd(II) or Pt(IV) extractants. However, solvent extraction processes require organic solvents, which are used as diluents of extractants or extractants themselves, and the use of organic solvents is problematic because of their toxicity and environmental concern. Furthermore, the selective recovery of PGMs remains a massive challenge due to the similarity in their properties.

The recovery mechanisms of PGMs by solvent extraction are generally classified into two types, namely, ligand-metal coordination and ion-pair formation [14]. For the coordination mechanism, metal extraction generally occurs in the order of Pd(II) >> Rh(III) >> Pt(IV); and Rh(III) and Pt(IV) are regarded as kinetically inert [15]. Furthermore, the order of extractability via ion-pair formation for the PGM chloro-complexes formed in an aqueous hydrochloric acid (HCl) solution is reported to be $[MCl_4]^{2-} \cong [MCl_6]^{2-} > [MCl_6]^{3-}$ > aqua species [16]; while the chloro-complex anions of PGMs formed in HCl are reported to be $[PdCl_4]^{2-}$, $[PtCl_6]^{2-}$, and $[RhCl_6]^{3-}$, as well as the chloro-aqua complexes of Rh(III) ($[RhCl_4(H_2O)_2]^-$ and $[RhCl_5(H_2O)]^{2-}$) [14]. Therefore, the extractability of PGMs in ligand-metal coordination and ion-pair formation evidently indicates the difficulty in the selective recovery of Pt(IV) in priority to Pd(II) as well as the Rh(III) recovery in advance of Pd(II) and Pt(IV) by conventional solvent extractions.

Recently, we developed PGM recovery methods based on precipitation using aromatic primary amines as precipitants [17–20]. The selective precipitation of Pt(IV) from HCl solutions containing Pd(II) and Pt(IV) was successfully achieved using 4-hexyloxyaniline at high HCl concentrations [17]. Furthermore, the preferential and selective Rh(III) recovery from the mixture of Pd(II), Pt(IV), and Rh(III) in HCl was successfully accomplished using aromatic primary monoamines [18] or diamines [19,20] as precipitants. The successful precipitation of Rh(III) was achieved based on the formation of unique ion-pair complexes comprising $[RhCl_6]^{3-}$ and anilinium cations of the amines. Although the potential of aromatic primary amines for PGM precipitants has been revealed, studies on the capability of aliphatic primary amines for PGM precipitants, as well as PGM extractants have been limited. Therefore, investigations of the availability of aliphatic primary amines for PGM recycling are valuable in the viewpoint of recycling industry.

Herein, we present a new PGM separation method by precipitation from the mixed solution of Pd(II), Pt(IV), and Rh(III) in HCl using aliphatic primary amines as precipitants. Preferential and selective precipitation of Pt(IV) at low HCl concentrations was achieved using *n*-heptylamine and *n*-octylamine. Moreover, Rh(III) was recovered by precipitation in preference to Pd(II) at high HCl concentrations using *n*-heptylamine and *n*-octylamine. Thereafter, we successfully achieved the mutual separation of Pt(IV), Rh(III), and Pd(II) in this order by changing the HCl concentrations and the alkyl chain lengths of the amines. The mechanism of the selective precipitation of PGMs was studied by the analysis of the metal-containing precipitates.

2. Materials and Methods

2.1. Materials

n-Hexylamine, *n*-heptylamine, *n*-octylamine, *n*-nonylamine, *n*-decylamine, and *n*-dodecylamine were purchased from Tokyo Kasei Kogyo Co., Ltd. (Tokyo, Japan) and used as received. Pd(II) and Pt(IV) standard solutions were purchased from FUJIFILM Wako Pure Chemical Industries, Ltd. (Osaka, Japan) and used as received. Rh(III) standard solution was purchased from Kanto Chemical Co., Inc. (Tokyo, Japan) and used as received.

2.2. Metal Precipitation Experiments from Metal-Containing Solutions

To HCl solutions (1 mL) containing Pd(II), Pt(IV), and Rh(III) (1.0 mM each) were added aliphatic primary amine (0.05 mmol) and the mixtures were shaken vigorously for 30 min at room temperature. After centrifugation (7200× *g*, 10 min), the metal concentration in the supernatant was determined by inductively coupled plasma atomic emission spectroscopy (ICP-AES, Hitachi High-Tech Science Corp., Tokyo, Japan). The used aliphatic primary amines and the HCl concentrations were changed in the metal precipitation experiments.

The metal precipitation experiments from single metal-containing solutions (Pd(II), Pt(IV), or Rh(III): 1.0 mM, 6.0 M HCl) using *n*-nonylamine were performed by the similar procedure described above. The amount of *n*-nonylamine and the shaking time were changed in the metal precipitation experiments.

2.3. Mutual Separation of Pd(II), Pt(IV), and Rh(III)

To a 4 mL of 1.0 M HCl solution containing Pd(II), Pt(IV), and Rh(III) (1.0 mM each) was added *n*-octylamine (0.2 mmol) and the mixture was shaken vigorously for 30 min at room temperature. After centrifugation (7200× *g*, 10 min), the metal concentration in the supernatant was determined by ICP-AES.

Then, an 11 M HCl solution was added to the supernatant to yield a 6.0 M HCl solution containing Pd(II) and Rh(III) (ca. 0.5 mM each). To the resulting solution (4 mL) was added *n*-octylamine (0.05 mmol) and the mixture was shaken vigorously for 30 min at room temperature. After centrifugation (7200× *g*, 10 min), the metal concentration in the supernatant was determined by ICP-AES. Then, to the resulting supernatant (2 mL) was added *n*-decylamine (0.1 mmol) and the mixture was shaken vigorously for 30 min at room temperature. After centrifugation (7200× *g*, 10 min), the metal concentration in the supernatant was determined by ICP-AES.

2.4. Isolation of Metal-Containing Precipitate

To 10 mL of 6.0 M HCl solutions containing Pd(II), Pt(IV) or Rh(III) (2.0 mM) were added *n*-nonylamine (0.2 mmol, for Pd(II)) or *n*-octylamine (0.2 mmol, for Pt(IV) and Rh(III)) and the mixtures were vigorously shaken for 30 min at room temperature. The resulting solid was collected by filtration and washed five times with a 6.0 M HCl solution. The solid was dried at room temperature for 48 h under vacuum. The Rh(III)-containing precipitate using *n*-octylamine from a 2.0 M HCl solution was prepared in the similar manner.

2.5. Measurements

An ICP-AES instrument (SPS5510, Hitachi High-Tech Science Corp., Tokyo, Japan) was used for the metal concentration measurements. X-ray photoelectron spectroscopy (XPS) measurements were performed using an AXIS-ULTRA X-ray photoelectron spectrometer (Kratos Analytical Ltd., Manchester, UK). Element quantification was performed using the relative sensitivity factors supplied with the instrument control software (N 1 s: 0.477, Cl 2 s: 0.493, Pd 3d: 5.356, Pt 4d: 4.637, Rh 3d: 4.822). Thermogravimetric measurements were carried out using a STA7300 (Hitachi High-Tech Science Corp., Tokyo, Japan) at a heating rate of 10 °C/min under air flow (200 mL/min).

3. Results and Discussion

3.1. Metal Precipitation Experiments

The metal recovery experiments from the mixture of Pd(II), Pt(IV), and Rh(III) in HCl were carried out using aliphatic primary amines as metal precipitants. The aliphatic primary amines used herein have linear-chain alkanes (from hexyl to dodecyl). The precipitation percentages of the metals were determined by ICP-AES analysis of the supernatants after centrifugation. Figure 1a–e shows the relationship between the precipitation of metals and HCl concentration. *n*-Hexylamine, which has the shortest alkyl chain in this study, did not work as a precipitant for Pd(II), Pt(IV), and Rh(III) regardless of the HCl concentrations (i.e., no precipitation); however, *n*-heptylamine and *n*-octylamine precipitated most of Pt(IV) (over 80% except for *n*-heptylamine in 1.0 M HCl) in a wide range of HCl concentrations (1.0–8.0 M), whereas Pd(II) precipitation was below 5%. Notably, Rh(III) precipitation occurred using the aliphatic amines except for *n*-hexylamine at HCl concentrations exceeding 1.0 M, and the precipitation percentages increased with the HCl concentrations, resulting in precipitation percentages exceeding 80% at high HCl concentrations (> 4 M HCl). Thus, *n*-heptylamine and *n*-octylamine exclusively precipitated Pt(IV) in a 1.0 M HCl solution, while both Pt(IV) and Rh(III) were co-precipitated at higher HCl concentrations. The use of *n*-nonylamine, *n*-decylamine, and *n*-dodecylamine enabled the efficient precipitation of Pd(II) and Pt(IV) (over 80%) in the studied HCl concentrations. Moreover, the precipitation of Rh(III) increased with the HCl concentrations and exceeded 90% at high HCl concentrations (> 4 M HCl). Thus, *n*-nonylamine, *n*-decylamine, and *n*-dodecylamine can recover Pd(II), Pt(IV), and Rh(III) as precipitates in high yields at high HCl concentrations. These results evidently show that the HCl concentration did not particularly affect the precipitation behavior of Pd(II) and Pt(IV) irrespective of the used amine, whereas Rh(III) precipitation increased with the concentration of HCl. Figure 2a,b shows the effect of the alkyl chain length on metal precipitation in 1.0 M HCl and 6.0 M HCl solutions. The precipitation of Pd(II) required a longer alkyl length than octyl in both 1.0 M HCl and 6.0 M HCl solutions, while the Pt(IV) precipitation occurred using amines with alkyl chains longer than hexyl in both 1.0 M HCl and 6.0 M HCl solutions. The precipitation percentages of Pt(IV) increased in proportion to the length of the alkyl chain and reached approximately 100% at octyl in 1.0 M HCl solutions, while high Pt(IV) precipitation percentages over 80% were achieved using the alkyl amines longer than hexyl in 6.0 M HCl solutions. For the Rh(III) precipitation, the precipitation percentages gradually increased with the length of the alkyl chain and reached 50% by using *n*-dodecylamine in 1.0 M HCl solutions. The percentages of Rh(III) precipitation in 6.0 M HCl solutions exceeded 60% by using the alkyl amines longer than hexyl, although those in the 1.0 M HCl solutions were below 10% except for when *n*-dodecylamine was used. It is noteworthy that a near-quantitative recovery of Rh(III) as precipitates was achieved when the amines had long alkyl chains and the concentration of HCl was high. Furthermore, the metal-containing precipitates obtained in this study were easily collected by filtration.

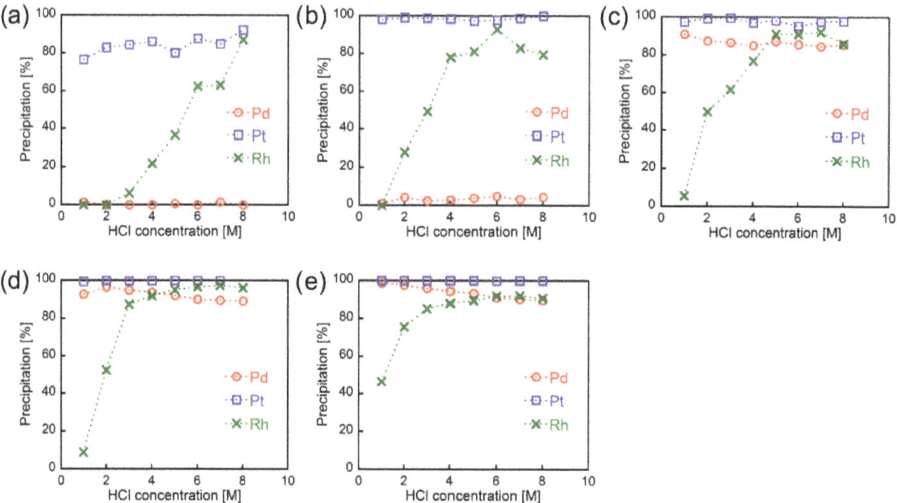

Figure 1. Metal precipitation from HCl solutions containing Pd(II), Pt(IV), and Rh(III) (each at 1 mM) using (**a**) *n*-heptylamine, (**b**) *n*-octylamine, (**c**) *n*-nonylamine, (**d**) *n*-decylamine, and (**e**) *n*-dodecylamine (amine: 50 mM, 30 min of shaking).

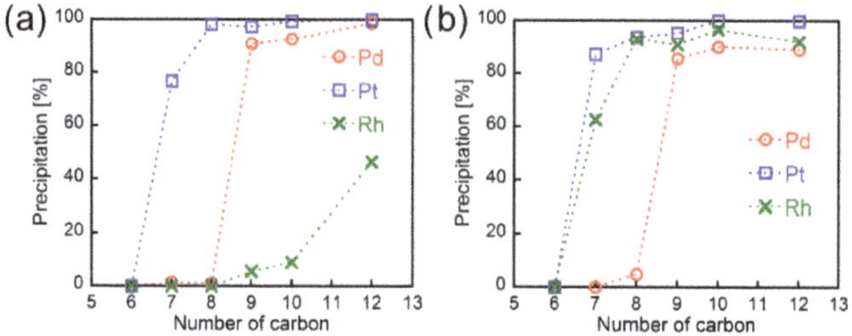

Figure 2. Relationship between metal precipitation and alkyl chain length of the amine in (**a**) 1.0 M HCl solutions and (**b**) 6.0 M HCl solutions (metal: 1 mM each, amine: 50 mM, 30 min of shaking).

3.2. Loading Amount and Shaking Time

The dependence of the metal precipitation percentages on the loading amount of *n*-nonylamine was investigated using 6.0 M HCl solutions containing individual metals (Pd(II), Pt(IV), and Rh(III)). We selected *n*-nonylamine as a typical example because it can precipitate all the three PGMs as shown in Figure 1c. The precipitation percentages of each metal increased with the loading amount of *n*-nonylamine and reached a plateau at [*n*-nonylamine]/[metal] = 10 (mol/mol) when the single metal-containing solutions were used (Figure 3a). There was no significant difference in the precipitation behavior among the three metals.

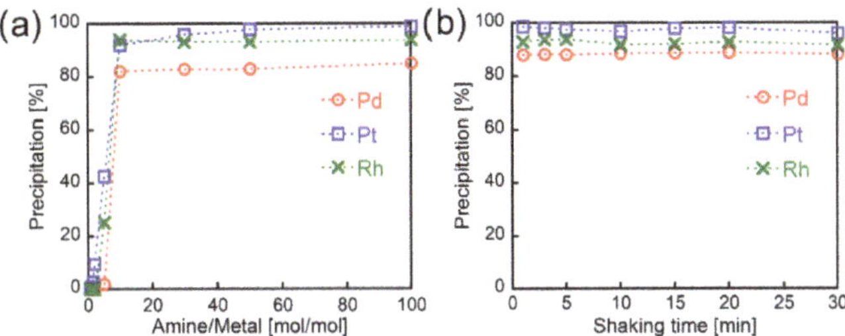

Figure 3. The effect of (**a**) *n*-nonylamine loading (30 min of shaking) and (**b**) shaking time on the metal precipitation (amine/metal = 50 mol/mol) from 6.0 M HCl solutions containing 1 mM individual metals (Pd(II), Pt(IV), and Rh(III)).

The effect of the shaking time on metal precipitation was also examined in the condition of [*n*-nonylamine]/[metals] = 50 (mol/mol) in 6.0 M HCl solutions containing individual metals (Pd(II), Pt(IV), and Rh(III)). Surprisingly, the precipitation percentages of the metals reached a maximum after shaking for 1 min (Figure 3b). This result clearly shows that PGM recovery by precipitation using aliphatic amines can be speedily completed.

3.3. Mutual Separation of Pd(II), Pt(IV), and Rh(III)

Based on the metal precipitation results relating to varying alkyl chain lengths and HCl concentrations, we performed a mutual separation of Pd(II), Pt(IV), and Rh(III) from a metal-mixed HCl solution. In the first precipitation step, *n*-octylamine was used as a precipitant because it can selectively precipitate Pt(IV) from 1.0 M HCl solutions (Figure 1b). As a result, Pt(IV) was recovered using *n*-octylamine from a 1.0 M HCl solution containing Pd(II), Pt(IV), and Rh(III) (1 mM each). Then, the HCl concentration of the resulting solution (Pd, Rh: 1 mM each, 1.0 M HCl) increased by adding an 11 M HCl solution and the 6.0 M HCl solution (Pd, Rh: 0.5 mM each) was obtained. In the second precipitation step, *n*-octylamine was added to the metal-containing 6.0 M HCl solution, and Rh(III) was recovered by precipitation. In the third precipitation step, the remaining Pd(II) was recovered using *n*-decylamine as a precipitant. As shown in Figure 4, each target metal is recovered as a precipitate over 75%, and co-precipitation of other metals is suppressed below 10%. Generally, Pt(IV) recovery in preference to Pd(II) is difficult according to the order of the PGM extractability [15,16]. Furthermore, Rh(III) in HCl cannot be extracted in preference to Pd(II) and Pt(IV) using conventional solvent extraction methods [14–16]. However, we achieved a mutual separation of Pt(IV), Rh(III), and Pd(II) in this order using aliphatic primary amines by changing the HCl concentration and alkyl chain length.

3.4. XPS Analysis of Metal-Containing Precipitates

XPS measurements of the metal-containing precipitates were carried out to analyze elemental composition. The precipitates were prepared by adding *n*-nonylamine (for Pd(II)) or *n*-octylamine (for Pt(IV) and Rh(III)) to the 6.0 M HCl solutions containing individual metals (Pd(II), Pt(IV), and Rh(III)). Prior to the measurements, the precipitates were washed with HCl to remove excess amine. The characteristic XPS peaks assigned to the Pd 3d, Pt 4d, and Rh 3d were observed for the Pd-, Pt-, and Rh-containing precipitates, respectively (Figure 5a–c). Furthermore, the peaks attributed to the N 1s, C 1s, Cl 2s, and Cl 2p were also observed for all the three samples. The N:Cl:metal atomic ratios were calculated from the photoelectron peak areas using the atomic sensitivity factors. The atomic ratios (N:Cl:metal) were calculated to be 2.0:3.9:1.0 (Pd-containing precipitate), 2.0:6.1:1.0 (Pt-containing precipitate), and 6.0:9.1:1.0 (Rh-containing precipitate) (Figure 5a–c). Pd(II) and Pt(IV) exist as chloro-complex anions, $[PdCl_4]^{2-}$ and $[PtCl_6]^{2-}$, in HCl solutions [14]. Since the aliphatic amines used herein form ammonium cations in HCl, the metal-containing precipitates will be the ion-pairs composed of metal chloro-complex anions and ammonium cations. The atomic ratios calculated from the XPS data are consistent with the expected composition of the Pd- and Pt-containing precipitates, as shown in Figure 6a,b. For Rh(III), the predominant species of Rh in 6.0 M HCl is $[RhCl_6]^{3-}$ [21]. Recently, we reported that aromatic primary amines selectively form Rh-containing precipitates comprising anilinium cation/chloride anion/$[RhCl_6]^{3-}$ at a 6:3:1 ratio (N:Cl:Rh = 6:9:1) [18]. The XPS result indicates that aliphatic primary amines also form unique ion-pair complexes composed of ammonium cation/chloride anion/$[RhCl_6]^{3-}$ in a 6:3:1 ratio (Figure 6c).

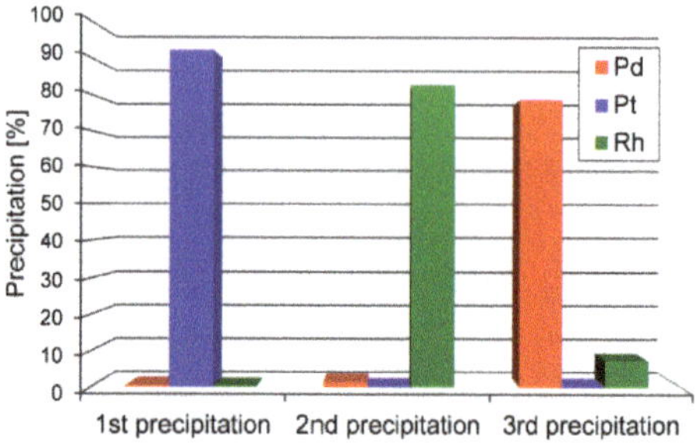

Figure 4. Mutual separation of Pd(II), Pt(IV), and Rh(III) from a 1.0 M HCl solution containing Pd(II), Pt(IV), and Rh(III) (1 mM each).

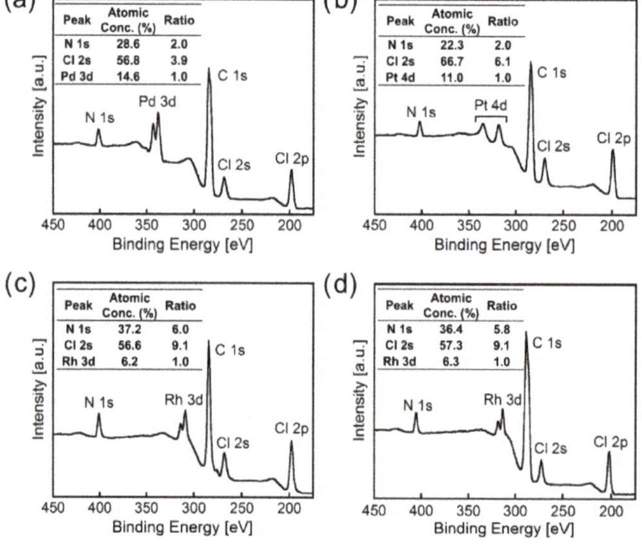

Figure 5. XPS spectra of the precipitates of (**a**) Pd(II) with *n*-nonylamine, (**b**) Pt(IV) with *n*-octylamine, and (c) Rh(III) with *n*-octylamine prepared from 6.0 M HCl solutions, and (**d**) the precipitate of Rh(III) with *n*-octylamine prepared from a 2.0 M HCl solution.

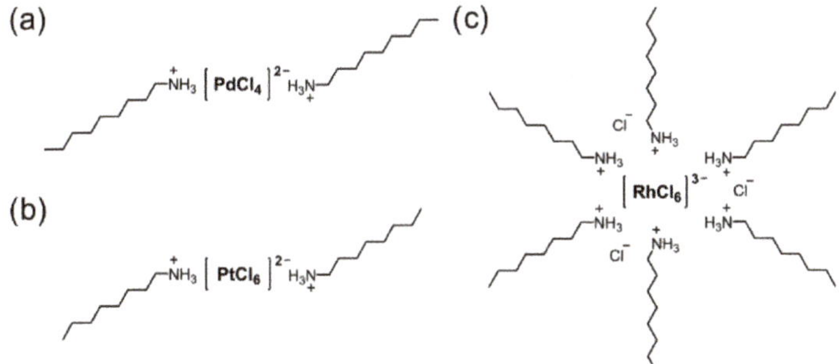

Figure 6. Expected chemical structures of the ion-pair complexes of (**a**) Pd(II) with *n*-nonylamine, (**b**) Pt(IV) with *n*-octylamine, and (**c**) Rh(III) with *n*-octylamine.

3.5. Thermogravimetric Analysis of Metal-Containing Precipitates

The ratios of metals and aliphatic primary amines in the metal-containing precipitates were also examined by thermogravimetric analysis (TGA). The precipitates used for the evaluation were prepared using *n*-nonylamine (for Pd(II)) or *n*-octylamine (for Pt(IV) and Rh(III)) and individual metal-containing 6.0 M HCl solutions. The weight loss observed in the range of 200–500 °C for all the three samples is attributed to the decomposition and volatilization of amines and chlorine (Figure 7a–c). The residual weights of Pd-, Pt-, and Rh-containing precipitates after heating at 750 °C were found to be 22.5%, 28.7%, and 10.3%, respectively. $PdCl_2$, $PtCl_4$, and $RhCl_3$ form Pd oxide (PdO), zerovalent Pt, and Rh oxide (Rh_2O_3) after combustion below 800 °C, respectively [22]. The expected weight fractions after the combustion of ion-pair complexes shown in Figure 6a–c are consistent with the experimental results of the TGA, as summarized in Table 1. The TGA results strongly support that the metal-containing precipitates consist of the ion-pairs depicted in Figure 6a–c.

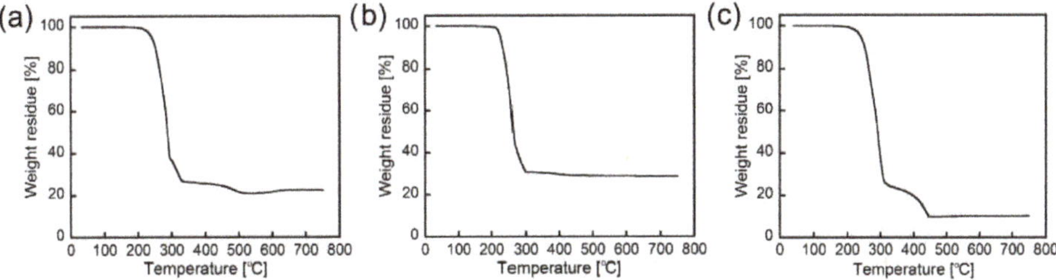

Figure 7. TG curves of the precipitates of (**a**) Pd(II) with *n*-nonylamine, (**b**) Pt(IV) with *n*-octylamine, and (**c**) Rh(III) with *n*-octylamine at a heating rate of 10 °C/min under air flow (200 mL/min).

Table 1. Weight fractions after the combustion of the precipitates.

Metal	Ash at 750 °C [a]	Expected Weight Fraction [b]
Pd(II)	22.8%	22.5%
Pt(IV)	28.7%	29.2%
Rh(III)	10.3%	10.5%

[a] Weight fractions of ash after TGA shown in Figure 7. [b] Expected weight fractions after combustion of ion-pair complexes shown in Figure 6.

3.6. Metal Precipitation Behavior

Figure 2a,b show that the metal precipitation of Pd(II) highly depends on the alkyl chain length of the amines. Since all the amines used herein are primary and have linear alkane chains, there is no difference in the properties of their amine groups, such as electron density and proton affinity. Therefore, the difference among the amines is only their hydrophobicity, i.e., their solubility in HCl. It is assumed that the hydrophobicity of the short alkyl amines (hexyl, heptyl, and octyl) is insufficient to precipitate the ion-pair composed of one Pd chloro-complex anion ($[PdCl_4]^{2-}$) and two ammonium cations, whereas Pt(IV) can be recovered as precipitates using the alkyl amines longer than hexyl. To investigate the difference in the solubility of the metal-containing precipitates, the saturating concentrations were measured for the ion-pairs composed of $[PdCl_4]^{2-}$ or $[PtCl_6]^{2-}$ and *n*-octylammonium cations. The ion-pairs were prepared by evaporating the 1:2 (mol/mol) mixed solution of the metal chloro-complex anions and *n*-octylamine in 1.0 M HCl solutions. The saturating concentrations of the Pd- and Pt-containing ion-pairs in 6.0 M HCl were 5.6 and 0.94, respectively. This result clearly shows that the hydrophilicity of the Pd-containing ion-pairs is higher than that of the Pt-containing ones. Consequently, the amines with alkyl chains longer than octyl are necessary to form Pd-containing precipitates. The difference in the solubility of the Pd- and Pt-containing ion-pairs is derivable from the difference in the number of Cl: the Pd-containing ion-pair possessing only four Cl would be less hydrophobic than the Pt-containing one with six Cl.

The precipitation percentages of Rh(III) increase with the HCl concentrations, although those of Pd(II) and Pt(IV) are quite independent of the HCl concentrations. Rh(III) forms several anion species in HCl, such as $[RhCl_4(H_2O)_2]^-$, $[RhCl_5(H_2O)]^{2-}$, and $[RhCl_6]^{3-}$, and the abundance ratio of $[RhCl_6]^{3-}$ increases with HCl concentrations [18,20,21]. For example, $[RhCl_6]^{3-}$ dominantly exists in 6.0 M HCl ($[RhCl_6]^{3-}$: 91%) [18,20]. From the results of the XPS and TGA of the Rh-containing precipitate using a 6.0 M HCl solution, $[RhCl_6]^{3-}$ is precipitated to form an ion-pair complex shown in Figure 6c. Therefore, we assumed that the preferential precipitation of $[RhCl_6]^{3-}$ occurred independently of the HCl concentrations by the ion-pair formation. To examine the precipitated Rh species, the XPS measurement was carried out for the precipitate formed by adding *n*-octylamine to a Rh-containing 2.0 M HCl solution. From the XPS spectrum of the precipitate composed of *n*-octylamine and Rh, the atomic ratio of N:Cl:Rh was calculated to be 5.8:9.1:1.0, indicating that the Rh-containing precipitate prepared from a 2.0 M HCl solution had the same composition as that from a 6.0 M HCl solution (Figure 5d). Since the distribution of the Rh species in 2.0 M HCl is reported to be 78% of $[RhCl_5(H_2O)]^{2-}$ and 8% of $[RhCl_6]^{3-}$ [18,20], the XPS result evidently shows the selective precipitation of $[RhCl_6]^{3-}$ in 2.0 M HCl using *n*-octylamine. The precipitation percentage of Rh(III) in 2.0 M HCl using *n*-octylamine (28%) was higher than the abundance ratio of $[RhCl_6]^{3-}$ (8%), which would be consequent of an equilibrium shift from $[RhCl_4(H_2O)_2]^-$ and $[RhCl_5(H_2O)]^{2-}$ to $[RhCl_6]^{3-}$. The precipitation behavior of Rh(III) using aliphatic primary amines is similar to that using 4-alkylanilines [18]. On the other hand, only aliphatic primary amines can precipitate Pd(II) and Pt(IV) even at high HCl concentrations: 4-alkylanilines do not form Pd(II)- and Pt(IV)-containing precipitates at high HCl concentrations (> 5 M HCl). Although 4-alkylanilines are advantageous from the viewpoint of Rh(III)-selective precipitation, aliphatic primary amines have the advantages of the capability in the mutual separation of Pd(II), Pt(IV), and Rh(III) as well as in the recovery by precipitation of Pd(II) and Pt(IV) at high HCl concentrations.

4. Conclusions

The recovery of Pd(II), Pt(IV), and Rh(III) by precipitation was achieved using aliphatic primary amines with linear-chain alkanes as precipitants. Pt(IV) was precipitated using the amines with alkyl chains longer than hexyl independent of HCl concentration, while the precipitation of Pd(II) required the longer alkyl amines than octyl, regardless of the HCl concentration. However, Rh(III) was recovered by precipitation at high HCl concentrations using amines longer than hexyl. The mutual separation of Pt(IV), Rh(III), and Pd(II), in this order, was successfully achieved by changing the HCl concentrations and alkyl chain lengths of the amines. XPS and TGA results clearly showed that the metal-containing precipitates are ion-pair complexes composed of metal chloro-complex anions and ammonium cations. It is noteworthy that Rh(III) formed a unique ion-pair complex comprising ammonium cations of aliphatic amines/chloride anions/$[RhCl_6]^{3-}$ = 6:9:1. This PGM precipitation method, using aliphatic primary amines, is a promising application in the PGM recycling.

Author Contributions: Y.S., S.Y., and Y.H. performed the experiments and contributed the data analysis. M.J. participated in the discussion on the results and the preparation of the paper. K.M. designed the research, analyzed the data, and wrote the paper. All authors have read and agreed to the published version of the manuscript.

Funding: This work was supported by the Japan Society for the Promotion of Science (JSPS) KAKENHI (Grant-in-Aid for Young Scientists (B) 17K12835), the Environment Research and Technology Development Fund (3RF-1902) of the Environmental Restoration and Conservation Agency of Japan, and Akita University Support for Fostering Research Project.

Conflicts of Interest: The authors declare no conflict of interest.

References

1. Alonso, E.; Field, F.R.; Kirchain, R.E. Platinum availability for future automotive technologies. *Environ. Sci. Technol.* **2012**, *46*, 12986–12993. [CrossRef] [PubMed]
2. Khaliq, A.; Rhamdhani, M.A.; Brooks, G.; Masood, S. Metal extraction processes for electronic waste and existing industrial routes: A review and Australian perspective. *Resources* **2014**, *3*, 152–179. [CrossRef]
3. Froehlich, P.; Lorenz, T.; Martin, G.; Brett, B.; Bertau, M. Valuable metals—recovery processes, current trends, and recycling strategies. *Angew. Chem. Int. Ed.* **2017**, *56*, 2544–2580. [CrossRef] [PubMed]
4. Kašpar, J.; Fornasiero, P.; Hickey, N. Automotive catalytic converters: Current status and some perspectives. *Catal. Today* **2003**, *77*, 419–449. [CrossRef]
5. Hagelüken, C. Recycling the platinum group metals: A European perspective. *Platinum Met. Rev.* **2012**, *56*, 29–35. [CrossRef]
6. Sun, P.P.; Lee, J.Y.; Lee, M.S. Separation of Pt (IV) and Rh (III) from chloride solution by solvent extraction with amine and neutral extractants. *Mater. Trans.* **2011**, *52*, 2071–2076. [CrossRef]
7. Cieszynska, A.; Wieczorek, D. Extraction and separation of palladium (II), platinum (IV), gold (III) and rhodium (III) using piperidine-based extractants. *Hydrometallurgy* **2018**, *175*, 359–366. [CrossRef]
8. Gupta, B.; Singh, I. Extraction and separation of platinum, palladium and rhodium using Cyanex 923 and their recovery form real samples. *Hydrometallurgy* **2013**, *134–135*, 11–18. [CrossRef]
9. Truong, H.T.; Lee, M.S.; Senanayake, G. Separation of Pt (IV), Rh (III) and Fe (III) in acid chloride leach solutions of glass scraps by solvent extraction with various extractants. *Hydrometallurgy* **2018**, *175*, 232–239. [CrossRef]
10. Firmansyah, M.L.; Kubota, F.; Goto, M. Solvent extraction of Pt (IV), Pd (II), and Rh (III) with the ionic liquid trioctyl(dodecyl) phosphonium chloride. *J. Chem. Technol. Biotechnol.* **2018**, *93*, 1714–1721. [CrossRef]
11. Rzelewska-Piekut, M.; Regel-Rosocka, M. Separation of Pt (IV), Pd (II), Ru (III) and Rh (III) from model chloride solutions by liquid-liquid extraction with phosphonium ionic liquids. *Sep. Purif. Technol.* **2019**, *212*, 791–801. [CrossRef]
12. Narita, H.; Tanaka, M.; Morisaku, K. Palladium extraction with *N,N,N′,N′*-tetra-*n*-octyl-thiodiglycolamide. *Miner. Eng.* **2008**, *21*, 483–488. [CrossRef]
13. Ortet, O.; Paiva, A.P. Development of tertiary thioamide derivatives to recover palladium (II) from simulated complex chloride solutions. *Hydrometallurgy* **2015**, *151*, 33–41. [CrossRef]
14. Bernardis, F.L.; Grant, R.A.; Sherrington, D.C. A review of methods of separation of the platinum-group metals through their chloro-complexes. *React. Funct. Polym.* **2005**, *65*, 205–217. [CrossRef]
15. Rydberg, J.; Cox, M.; Musikas, C.; Choppin, G.R. *Solvent Extraction Principles and Practice*, 2nd ed.; Marcel Dekker, Inc.: New York, NY, USA, 2004.
16. Nikoloski, A.N.; Ang, K.L. Review of the application of ion exchange resins for the recovery of platinum-group metals from hydrochloric acid solutions. *Miner. Process. Extract. Metall. Rev.* **2014**, *35*, 369–389. [CrossRef]
17. Matsumoto, K.; Yamakawa, S.; Jikei, M. Selective recovery of platinum (IV) from palladium (II)-containing solution using 4-(hexyloxy)aniline. *Chem. Lett.* **2017**, *46*, 22–24. [CrossRef]
18. Matsumoto, K.; Yamakawa, S.; Sezaki, Y.; Katagiri, H.; Jikei, M. Preferential precipitation and selective separation of Rh(III) from Pd(II) and Pt(IV) using 4-alkylanilines as precipitants. *ACS Omega* **2019**, *4*, 1868–1873. [CrossRef]

19. Matsumoto, K.; Hata, Y.; Sezaki, Y.; Katagiri, H.; Jikei, M. Highly selective Rh(III) recovery from HCl solutions using aromatic primary diamines via formation of three-dimensional ionic crystals. *ACS Omega* **2019**, *4*, 14613–14620. [CrossRef]
20. Matsumoto, K.; Yamakawa, S.; Haga, K.; Ishibashi, K.; Jikei, M.; Shibayama, A. Selective and preferential separation of rhodium (III) from palladium (II) and platinum (IV) using a *m*-phenylene diamine-containing precipitant. *Sci. Rep.* **2019**, *9*, 12414. [CrossRef]
21. Benguerel, E.; Demopoulos, G.P.; Harris, G.B. Speciation and separation of rhodium (III) from chloride solutions: A critical review. *Hydrometallurgy* **1996**, *40*, 135–152. [CrossRef]
22. Jóźwiak, W.K.; Maniecki, T.P. Influence of atmosphere kind on temperature programmed decomposition of noble metal chlorides. *Thermochim. Acta* **2005**, *435*, 151–161. [CrossRef]

Article

Reducing the Magnesium Content from Seawater to Improve Tailing Flocculation: Description by Population Balance Models

Gonzalo R. Quezada [1], Matías Jeldres [2,3], Norman Toro [4], Pedro Robles [5] and Ricardo I. Jeldres [3,*]

1 Water Research Center for Agriculture and Mining (CRHIAM), Concepción 4030000, Chile; gonzaloquezada@udec.cl

2 Faculty of Engineering and Architecture, Universidad Arturo Prat, Almirante Juan José Latorre 2901, Antofagasta 1244260, Chile; hujeldres@unap.cl

3 Departamento de Ingeniería Química y Procesos de Minerales, Facultad de Ingeniería, Universidad de Antofagasta, Av. Angamos 601, Antofagasta 1240000, Chile

4 Departamento de Ingeniería Metalúrgica y Minas, Universidad Católica del Norte, Antofagasta 1270709, Chile; ntoro@ucn.cl

5 Escuela de Ingeniería Química, Pontificia Universidad Católica de Valparaíso, Valparaíso 2340000, Chile; pedro.robles@pucv.cl

* Correspondence: ricardo.jeldres@uantof.cl; Tel.: +56-552-637-901

Received: 18 January 2020; Accepted: 24 February 2020; Published: 2 March 2020

Abstract: Experimental assays and mathematical models, through population balance models (PBM), were used to characterize the particle aggregation of mining tailings flocculated in seawater. Three systems were considered for preparation of the slurries: i) Seawater at natural pH (pH 7.4), ii) seawater at pH 11, and iii) treated seawater at pH 11. The treated seawater had a reduced magnesium content in order to avoid the formation of solid complexes, which damage the concentration operations. For this, the pH of seawater was raised with lime before being used in the process—generating solid precipitates of magnesium that were removed by vacuum filtration. The mean size of the aggregates were represented by the mean chord length obtained with the Focused beam reflectance measurement (FBRM) technique, and their descriptions, obtained by the PBM, showed an aggregation and a breakage kernel had evolved. The fractal dimension and permeability were included in the model in order to improve the representation of the irregular structure of the aggregates. Then, five parameters were optimized: Three for the aggregation kernel and two for the breakage kernel. The results show that raising the pH from 8 to 11 was severely detrimental to the flocculation performance. Nevertheless, for pH 11, the aggregates slightly exceeded 100 μm, causing undesirable behaviour during the thickening operations. Interestingly, magnesium removal provided a suitable environment to perform the tailings flocculation at alkaline pH, making aggregates with sizes that exceeded 300 μm. Only the fractal dimension changed between pH 8 and treated seawater at pH 11—as reflected in the permeability outcomes. The PBM fitted well with the experimental data, and the parameters showed that the aggregation kernel was dominant at all-polymer dosages. The descriptive capacity of the model might have been utilized as a support in practical decisions regarding the best-operating requirements in the flocculation of copper tailings and water clarification.

Keywords: copper tailings; enhanced flocculation; water recovering; magnesium removal; population balance model; seawater flocculation

1. Introduction

High-salinity resources, such as seawater, are being used in the mining industry in countries such as Australia, Chile, and Indonesia. At first, saltwater was proceeded by desalination plants, primarily through reverse osmosis (RO). Unfortunately, RO has created the new challenge of concentrated brine by-product, which is released into the environment; the effluent has a high temperature and a high demand for electricity, which involves fossil fuel consumption. Therefore, the direct use of seawater without ion removal is an attractive strategy that has recently been implemented in many mining industries. In this matter, engineers and researchers are constantly studying certain operational complexities. Some of these studies focus on the corrosion and incrustation of pipelines, the precipitate complexes, and the buffer effect of seawater. An example is the concentration of Cu/Mo ores, where a high level of pyrite is found, which drastically hinders the efficiency of the flotation process. The depression of pyrite in freshwater lime is used to bring the operation to an alkaline pH condition (>10.5), where the hydrophilic ferric hydroxide is formed. Nevertheless, this condition cannot function in seawater because the formation of solid Ca/Mg complexes strongly decreases the recovery of molybdenite [1,2] and chalcopyrite [3,4]. There is a debate as to the proper explanation of these findings in the sulphide flotation [5–8], but the recovery of water for the sustainability of the mining industry is more important. This is necessary due to the high cost of transporting seawater to mining sites; in Chile, the mining site can be as much as 4000 m above sea level. An essential process in the recovery of water is the thickening stage, where efficiency is linked to the flocculation of the suspended particles, which promotes solid–liquid separation. The flocculants are long water-soluble polyelectrolytes that can bridge multiple particles and form aggregates that are sedimentary because of the effects of gravity. The conventional flocculant used in copper mining is hydrolyzed polyacrylamide (HPAM) with a high molecular weight, which gives high settling rates, low densities, and proper rheological properties of concentrated underflows at relatively low doses and low cost [9,10]. Several efforts have been made in the study of HPAM behaviour in the solution [11,12] and high-saline conditions [13–15]. These studies provide enough proof that bridging flocculation may benefit from a saline medium because ions can act as a bridge between the anionic solid particles and anionic flocculants, commonly known as ion binding. However, the concentrated salt can reduce the size of the reagents by an excess of charge neutralization. These studies were performed on chloride salt solutions; seawater has a complex mix of ions, where Ca and Mg are present, and as we know, at high-alkaline conditions, they precipitate into a hydroxide complex. Recent studies show that kaolinite can adsorb several heavy metal ions [16,17] and also hydroxide complexes [18]; therefore, these precipitates have a strong interaction with solid particles and possibly, the flocculants. These results provide evidence that precipitates can affect the flocculation mechanism and negatively affect the thickener performance. However, this might be avoided by removing magnesium ions prior to incorporating the seawater in the operations.

A good strategy to analyze flocculation performance is by monitoring the aggregate size distribution over time by the focused beam reflectance measurement (FBRM), which has the advantage of being employed directly to the slurry without sampling or the need for dilution [19–22]. In flocculated systems, a maximum floc size is achieved after a short time following flocculant addition; it then follows a decay that emerges from the fragmentation of the aggregates and polymer depletion. Using the FBRM has provided an excellent method to analyze flocculation parameters such as the maximum aggregation size or fragmentation rate [23–26]. In this context, many researchers have described the kinetics of aggregate growth and fragmentation over time by population balance models (PBM) [27–30], which have practical applications in a wide variety of systems [31–35]. The PBM is based on the work by Smoluchowski [27,36], which describes how aggregates evolve over time. Such evolution depends on the mechanisms of the aggregation and the rupture of the equations. To update the PBM to flocculation, it is necessary to improve the physical description of the aggregates by (1) updating the collision efficiency to a time-dependent flocculant depletion [30,35], and (2) to use the fractal dimension as an indicator of the irregular structure of the aggregates [27,36–40]. Recently, we have proven that, for these flocculated systems, it is possible to use a constant fractal dimension when the shear rate is below 200 s^{-1}.

In this work, we use PBM to describe the flocculation kinetics of tailing particles using treated seawater with the magnesium removed in order to improve the water recovery and the performance of the thickening stages. Then, we analyze the implications of precipitates being present for phenomena such as collision frequency, collision efficiency, fragmentation rate, and floc permeability. The outcomes of this work are of particular interest to mining industries that use seawater in concentration stages, and the implementation of this plan might allow for sustainable results without the need to totally remove seawater salts by reverse osmosis.

2. Methodology

2.1. Materials

The seawater (SW) was obtained from the San Jorge Bay in Antofagasta (Chile); to eliminate bacterial activity, the SW was filtered at 1 μm using a UV filter system. This water had a conductivity of 50.4 mS/cm, while its ionic composition was the following: 10.9 g/L Na^+, 1.38 g/L Mg^{2+}, 0.4 g/L Ca^{2+}, 0.39 K^+, 19.6 Cl^-, and 0.15 g/L HCO_3^- [41].

Kaolin was acquired from Ward's Science, and a quantitative X-ray diffraction (XRD) analysis showed that it contained 84 wt% kaolinite ($Al_2Si_2O_5(OH)_4$) and 16 wt% halloysite ($Al_2Si_2O_5(OH)_4{\cdot}2H_2O$) (Figure 1). A D5000 X-ray diffractometer (Siemens S.A., Lac Condes, Chile) was used and the data were processed with Total Pattern Analysis Software (TOPAS) (Siemens S.A., Lac Condes, Chile). Quartz was acquired from a local Chilean store, where the SiO_2 content detected by quantitative XRD was over 99 wt% (see Figure 2). Both quartz and kaolin had a density of 2.6 g/t. A Microtrac S3500 laser diffraction particle size analyzer (Verder Scientific, Newtown, PA, USA) was used. The analysis showed that 10% of the particles were smaller than d10 = 1.8 and 3.8 μm in the kaolin and quartz samples, respectively.

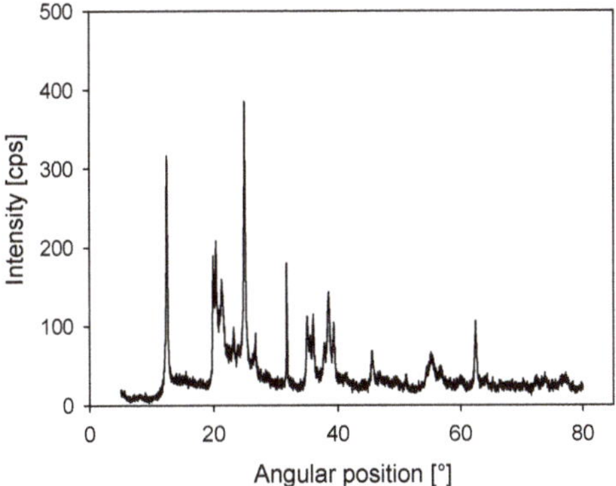

Figure 1. X-ray diffraction (XRD) for kaolin powder.

SNF 704, provided by SNF Chile S.A., was used as an anionic flocculant. The molecular weight was higher than 18 × 106. An initial stock solution was prepared at a concentration of 1 g/L. Then, it was mixed with a low RPM for 24 h before use, stored in a refrigerator, and discarded after two weeks in order to avoid the potential ageing effect. An aliquot of this stock flocculant solution was diluted at 0.1 g/L once a day for use in testing, with unused portions, then discarded. The flocculant dosages were determined in terms of grams of polymer per ton of dry solids (g/ton). The reagents used to modify the pH were lime and sodium hydroxide and were of analytical grade (greater than 98%).

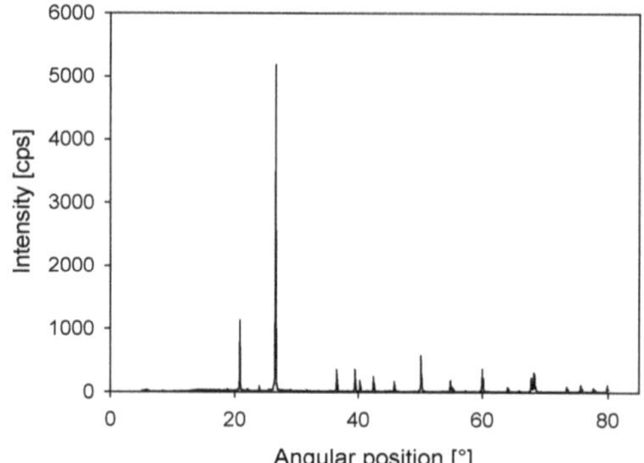

Figure 2. X-ray diffraction (XRD) for quartz powder.

2.2. Magnesium Removal

The removal of magnesium was carried out by increasing the pH of the seawater with 0.06 M of lime in order to form hydroxide magnesium as a white precipitate. In parallel, bicarbonate was reacted with Mg to form magnesium carbonate [41]. After 30 min of intense mixing, the seawater was vacuum filtered, obtaining seawater with an ionic concentration of 10.9 g/L Na^+, 0.01 g/L Mg^{2+}, 2.35 g/L Ca^{2+}, 0.39 K^+, 19.6 Cl^-, and 0.05 g/L HCO_3.

2.3. Flocculant-Suspension

A suspension of 270 g was prepared at 8 wt%, with known masses of the solid phases to give mixtures containing 80 wt% quartz and 20 wt% kaolin. The suspension was vigorously mixed for 30 min using a 30 mm-diameter turbine type stirrer within a 100 mm-diameter vessel with a 1 L capacity. All the experiments were made by placing the stirrer 20 mm above the bottom of the vessel. Subsequently, the mixing rate was controlled at 250 rpm, and the volume of the solution (seawater and polymer) was added in a proportion fixed by the required polymer dosage.

2.4. Batch Settling Tests

These tests were conducted, after 30 s of flocculant-suspension mixing, by gently pouring the slurry into closed, 300 cm^3 cylinders (35 mm internal diameter), and then slowly inverting the cylinder two times by hand (the whole cylinder rotation process took, in all cases, about 4 s). After 10 min of settling, the supernatant fluid was rescued and stirred in order to homogenize the suspended solids. Then, a 50 mL aliquot was used for turbidity measurements in a HANNA HI98713 turbidimeter (Hanna Instruments, Santiago, Chile), which performed ten readings in 20 s, delivering the average at the end of that period.

2.5. Characterization of Aggregates

The FBRM technique was used to record the chord length distribution of the aggregates. The probe was submerged vertically in the reaction vessel, 10 mm over the stirrer and 20 mm off-axis. The FBRM probe featured a laser that was focused through the sapphire window and scanned a circular path at a tangential velocity of 2 m/s. The backscattered light was then received when the laser beam intersected the path of the particle or aggregate. A chord length was determined from the duration of any unusual increase in the backscattered light intensity and laser velocity.

3. Modeling

The PBM equations used in this work are derived from several works [42–44] that discretize the aggregate size into the number i of the bins. Every bin is based on the classical geometric distribution for the aggregate volume $(V_{i+1}) = 2V_i)$. The PBM equation is given by:

$$\begin{aligned}\frac{dN_i}{dt} = & \sum_{j=1}^{i-2} 2^{j-(i-1)} Q_{i-1,j} N_{i-1} N_j + \tfrac{1}{2} Q_{i-1,j-1} N_{i-1} N_{i-1} - \sum_{j=1}^{i-1} 2^{j-i} Q_{i,j} N_i N_j \\ & - \sum_{j=1}^{i_{max1}} Q_{i,j} N_i N_j - S_i N_i + \sum_{j=1}^{i_{max2}} \Gamma_{i,j} S_j N_j\end{aligned} \quad (1)$$

where N_i is the number concentration of the aggregates in bin i, whereas N_1 is the number concentration of primary particles in bin 1. Every term on the right of the equation represents a physical process:

- The first and second terms describe the aggregate formation of size i from smaller aggregates.
- The third and fourth terms describe the aggregation death of size i to higher aggregates.
- The fifth term represents the breakage formation of size i from the rupture of a greater aggregate.
- The sixth term represents the breakage death of size i by creating smaller aggregates.

The superscript max1 is the maximum number of intervals used to represent the complete aggregate size spectrum; max2 corresponds to the largest interval from which the aggregates in the current range are produced.

3.1. Aggregation Kernel

The Q variable is the aggregation kernel, an expression that contains the collision frequency (β) and capture efficiency (α):

$$Q_{i,j} = \beta_{i,j} \alpha_{i,j} \quad (2)$$

Collision Frequency β

It has been shown that fluid flow can penetrate through particle aggregates [45]. This means that the actual collision frequency is considerably lower than that predicted from rectilinear flow models. To add the permeability effects to the collision frequency, we use the parameter called "fluid collection efficiency" η. This measures the ratio between the flow passing through the aggregate and the flow approaching it. Veerapaeni and Wiesner [46] proposed a function to calculate the collision frequency, which includes the permeability and fractal dimension effects:

$$\beta_{i,j} = \frac{1}{6}\left(\sqrt{\eta_i} d_i + \sqrt{\eta_j} d_j\right)^3 G \quad (3)$$

where, d_i and d_j are the diameters of the aggregates sizes of i and j, respectively, G is the shear rate, and η is derived from the Brinkman' extension to Darcy's laws as a function of a dimensionless permeability (ξ) [47]:

$$\eta_i = \frac{9(\xi_i - \tanh\xi_i)}{(3\xi_i + 2\xi_i{}^2 - 3\tanh\xi_i)} \quad (4)$$

where ξ_i is defined as $\xi_i = d_i/2\sqrt{K_i}$ and K_i is the permeability; we use the expression from the work by Li and Logan [47]:

$$K_i = \frac{d_i^2}{72}\left(3 + \frac{3}{1-\phi_i} - \sqrt[3]{\frac{8}{1-\phi_i} - 3}\right) \quad (5)$$

The porosity ϕ is related to the fractal dimension (d_f) using the expression by Vainshtein et al. [48]:

$$\phi_i = 1 - C\left(\frac{d_i}{d_0}\right)^{d_f - 3} \quad (6)$$

where C is a packing coefficient (generally assumed to be 0.65) and d_0 the primary particle diameter. d_i and d_0 are related by the expression proposed by Mandelbrot [49]:

$$d_i = d_0\left(\frac{2^{i-1}}{C}\right)^{\frac{1}{d_f}} \tag{7}$$

Collision Efficiency α

There are several expressions for collision efficiency, which depend on the type of aggregate and the additives. In this case, we use an expression to represent the effect of high-weight polymers; this is the polymer depletion and it rearranges on the adsorbed surface of the particles. This has been implemented in a recent work by Vajihinejad and Soares [35], which showed an exponential decay between two fitted parameters:

$$\alpha_{i,j} = (\alpha_{max} - \alpha_{min})e^{-k_d t} + \alpha_{min} \tag{8}$$

where α_{max} is the maximum collision efficiency, α_{min} is the minimum collision efficiency (at steady-state conditions), and k_d is the collision efficiency decay constant (s^{-1}).

3.2. Breakage Kernel

In regards to Equation (1), the S term is the breakage rate and Γ is the distribution breakage, which represents the breakage kernel. The S term is difficult to predict since there is no theory and is usually fitted to the size distribution data. In this case, we use a power-law function of the aggregate mass, as proposed by Pandya and Spielman [50]:

$$S_i = s_1 G^{s_2} d_i \tag{9}$$

where s_1 and s_2 are the fitted parameters.

To simplify the distribution breakage, we use a binary distribution, where an aggregate breaks into two pieces of equal mass. That is:

$$\Gamma_{i,j} = \begin{cases} \frac{V_j}{V_i} & \text{for } j = i+1 \\ 0 & \text{for } j \neq i+1 \end{cases} \tag{10}$$

3.3. Shear Rate

The shear rate required by the aggregation and breakage kernels is calculated from:

$$G = \left(\frac{\varepsilon \rho_{sus}}{\mu_{sus}}\right)^{\frac{1}{2}} \tag{11}$$

where ε is the average energy dissipation rate:

$$\varepsilon = \frac{N_p N^3 D^5}{V} \tag{12}$$

where N_p is the impeller power number (0.6 in our case for a plane disk with gentle agitation [51]), N is the rotation speed, D and V are, respectively, the diameter of the impeller and the working volume of the vessel. The density of the suspension ρ_{sus} is calculated from:

$$\rho_{sus} = \left(\frac{w}{\rho_s} + \frac{1-w}{\rho_w}\right)^{-1} \tag{13}$$

where w is the solid mass fraction of the solution, ρ_s and ρ_w are, respectively, the solid and water density. Finally, the viscosity of the solution μ_{sus} is measured.

3.4. Solution

To resolve the PBM in Equation (1), we use a solver based on the numerical differentiation formula for stiff ODEs (ode23s) in MATLAB. The optimization was performed with the MATLAB function *fminsearch*, which uses the Nelder–Mead direct search to find the minimum in an unconstrained multivariable function. The objective function (*OF*) used to optimize the parameters is given by:

$$OF(\alpha_{max}, \alpha_{min}, k_d, s_1, s_2) = \sum_{t_i}^{t_f} \left(d_{exp} - d_{mod}\right)^2 \tag{14}$$

where d_{exp} is the experimental diameter of the particles, and d_{mod} is the model diameter obtained by:

$$d_{mod} = \frac{\sum_{i=1}^{max} N_i d_i^4}{\sum_{i=1}^{max} N_i d_i^3} \tag{15}$$

Two criteria validate the model's fit and predictions: One is the coefficient of determination (R^2), which measures the closeness of the model values to the experimental values. That is:

$$R^2 = 1 - \frac{\sum_{i=1}^{max}\left(d_{agg,exp,i} - d_{agg,mod,i}\right)^2}{\sum_{i=1}^{max}\left(d_{agg,exp,i} - d_{agg,exp}\right)^2} \tag{16}$$

and the other is the quality of the fit:

$$GoF(\%) = 100\frac{d_{agg,exp} - \text{std}}{d_{agg,exp}} \tag{17}$$

std stands for the standard error calculated from:

$$\text{std} = \left(\frac{1}{n-f}\sum_{t_i}^{t_f}\left(d_{agg,exp} - d_{agg,mod}\right)^2\right)^{\frac{1}{2}} \tag{18}$$

where n is the number of data values and f is the number of parameters to be fitted. A *GoF* of 90% or higher means that the proposed model can predict the flocculation kinetics. Finally, conservation of the total volume of particles is verified after every integration in order to ensure that the simulations maintain the particle population.

4. Results

4.1. Input Parameters and Distribution

Three cases were studied: Seawater at pH 8 (SW pH 8), seawater at pH 11 (SW pH 11), and treated seawater at pH 11 (T-SW pH 11), with Mg removed. Solving the PBM equations requires a set of input parameters (Table 1). With these parameters, the shear rate can be calculated from Equations (11)–(13) and the rest are used for the PBM equation definitions. A particle number concentration is needed for each size class and for each unit of volume of suspension $N_{0,i}$. For that, a number distribution is obtained from experimental results through $N_{0,i} = \phi v(d_{0,i})/V_{0,i}$, where $v(d_{0,i})$ is the volume fraction of particles with diameter $d_{0,i}$, obtained from Equation (7), ϕ is the solid volume fraction and $V_{0,i}$ is the volume of the primary particles following the geometric progression $V_{0,i} = 2^{i-1}V_1$. To obtain the $v(d_{0,i})$, a volume distribution is needed. Figure 3A shows an experimental volume fraction distribution provided from the FRPM probe for three case studies before the flocculant is added; in this case, all distributions are similar. To calculate the fractal dimension from the settings test, we follow the procedure by Heath et al. [20]. The fractal dimension results are shown in Figure 3B. The fractal dimension is higher when the flocculant dose increases. At pH 11, the fractal dimension decreases abruptly because the solid magnesium precipitates would produce open structures. Nevertheless, the treated seawater overcomes this phenomenon, giving similar structures to those of the pH 8 system.

Table 1. Input parameters and conditions.

i_{max}	30	
ϕ	0.054	
c	0.65	
N_p	0.6	
D	8.0	cm
V	0.25	l
ρ_s	2600	kg/m^3
ρ_w	1000	kg/m^3
μ_{sus}	0.005	kg/(ms)
w	0.08	
d_0	0.0005	cm

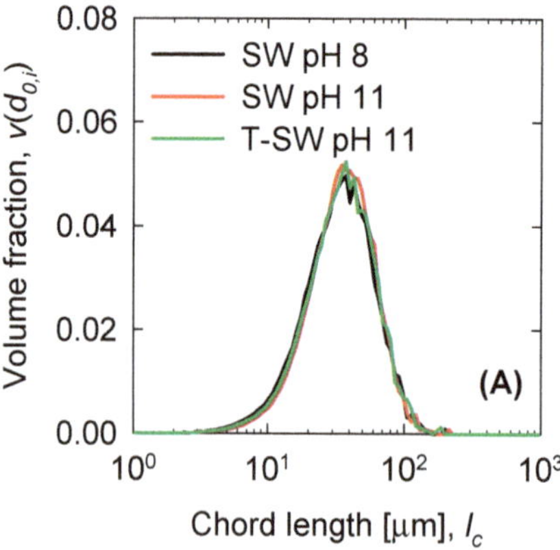

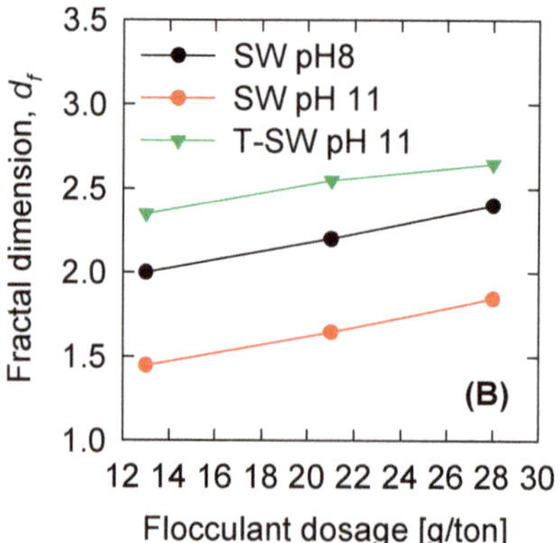

Figure 3. (**A**) Normalized initial volume distribution of particles of synthetic tailings in seawater at a mixing rate of 150 rpm. (**B**) Fractal dimension from the settling experiments.

4.2. Flocculation Kinetics and Modeling

The results of the experimental data are plotted in Figure 4 for the three cases studied in this work. We see that the flocculant doses increase the size of the aggregates, which is similar to other work, as shown in Reference [35]. For SW pH 8, the aggregate increases to a maximum value of 300 μm for the flocculant dose of 28 g/ton. Smaller values are obtained when the flocculant dose decreases. If we increase the pH to 11 (SW pH 11), we see results similar to those in Figure 4B, where the aggregate falls to a smaller value of ~100 μm for all the flocculant doses in this work. This behaviour is primarily related to the formation of hydroxide complexes that hinder aggregate formation. Later, we see the results of Mg removal from the seawater before being added to the slurry at pH 11 (T-SW pH 11). In this case, we have a higher maximum aggregation than when Mg is present as a hydroxide. From this, we can clearly see that Mg hydroxide presence hinders the aggregation process drastically. Then, the PBM modeling in the figures are included in the results as continuous lines. As we see from Table 2, there is a good agreement for the cases where the mean aggregate diameter is higher, but there are limitations associated with the model for the examples that are small in size.

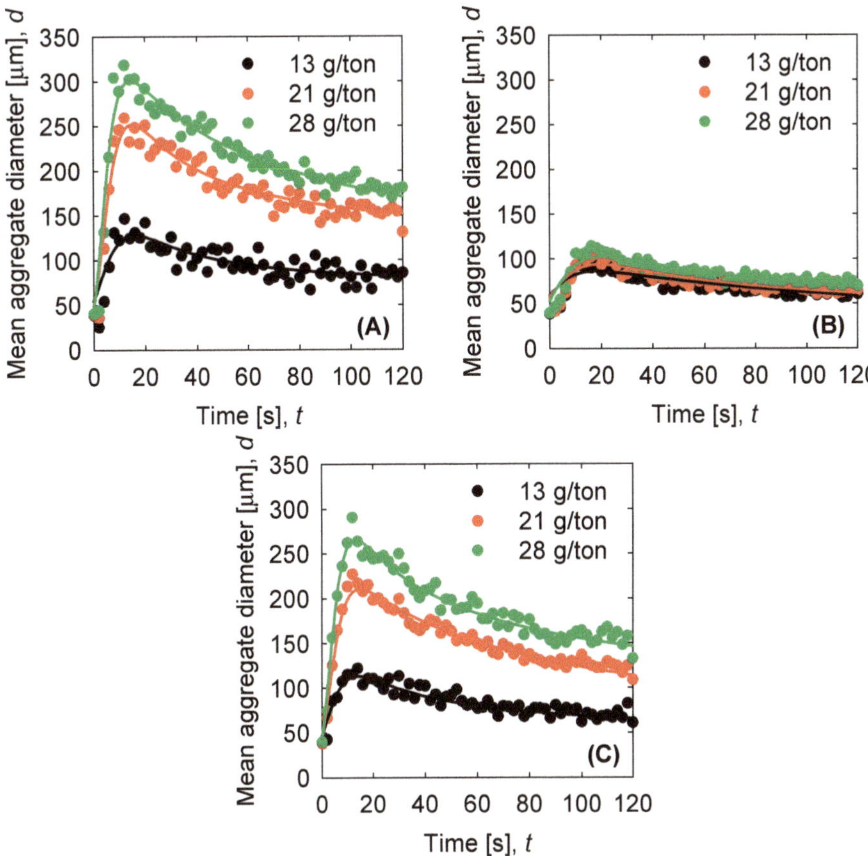

Figure 4. Flocculation kinetics of synthetic tailings in seawater as a function of flocculant doses (mixing rate 150 rpm). Solid circles correspond to experimental data and solid lines to the best fit with the population balance model (PBM). (**A**) Seawater (SW) pH 8. (**B**) SW pH 11. (**C**) Treated (T)-SW pH 11.

Table 2. Quantitative results of *GoF* and R^2 when the PBM model is used.

System	Flocculant Dose, g/ton	*GoF*, %	R^2
SW pH 8	13	86.1	0.675
	21	92.9	0.915
	28	93.8	0.939
SW pH 11	13	89.9	0.629
	21	89.6	0.706
	28	91.7	0.794
T-SW pH 11	13	91.4	0.838
	21	95.3	0.961
	28	94.9	0.956

4.3. Optimized Parameters

The model fitted five parameters: The maximum and minimum collision efficiency (α_{max}, α_{min}), the collision efficiency decay constant (k_d), and two breakage rate kernel parameters (s_1 and s_2). The first three parameters are related to the aggregation kernel and are shown in Figure 5. The α_{max} show a steady increase when the flocculant dose increases. This reflects the activity of the flocculant to aggregate more significant quantities of solids at higher doses. Interestingly, the α_{min} has small values compared to the α_{max}, and all remain below 0.02, meaning that, at steady state (where α_{min} becomes significant), the breakage kernel must also be small. k_d, as we found in previous results, must follow the inverse tendency of the a_{max}, a_{min} because the higher the aggregates, the greater the probability of flocculant depletion. In this case, the fitted parameters also show that behaviour. Finally, if we compare the three trials, we see that the parameters of the SW at pH 8 are similar of the T-SW at pH 11, which implies that the removal of the Mg could keep the aggregation performance similar to that of pH 8. Finally, the two parameters for the breakage kernel are plotted in Figure 6. As discussed above, if the s_1 is small, the a_{min} is also small. This trend must be followed in order to satisfy steady-state conditions in the aggregation process. The s_2 parameters show values between 1 and 2, but as the s_1 parameter is small, its contribution is neglected from the global behavior. From this, we see that high dosage increments contribute to the aggregate kernel rather than the breakage kernel.

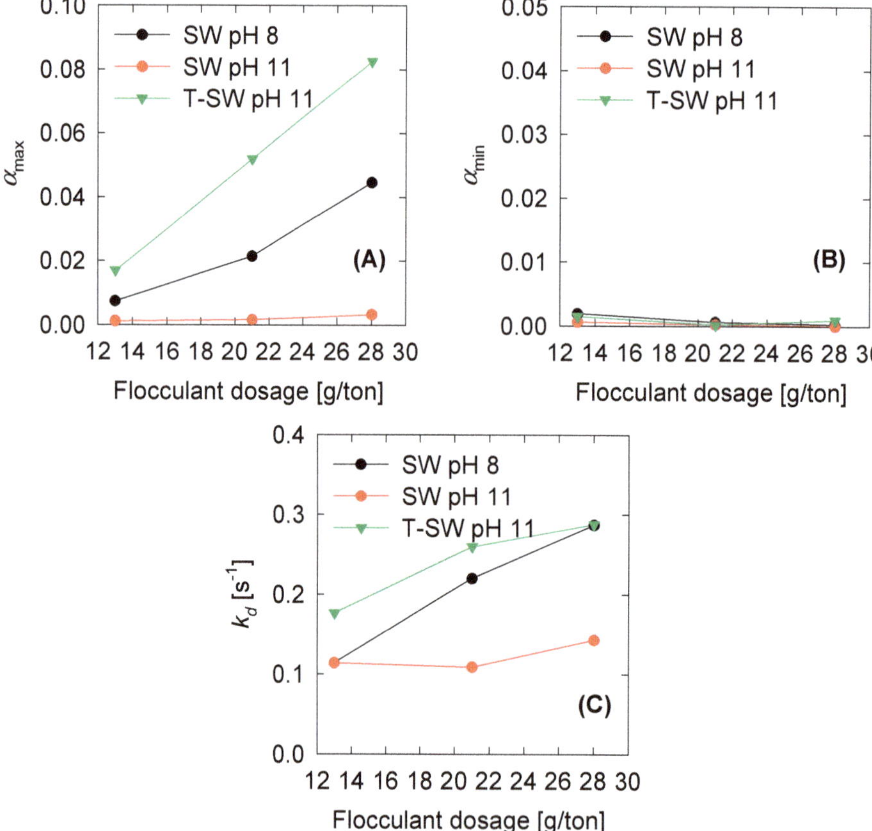

Figure 5. Optimum aggregation parameters vs. shear rate for constant and variable fractal dimension: (**A**) Maximum and (**B**) minimum collision efficiency and (**C**) collision efficiency decay constant k_d.

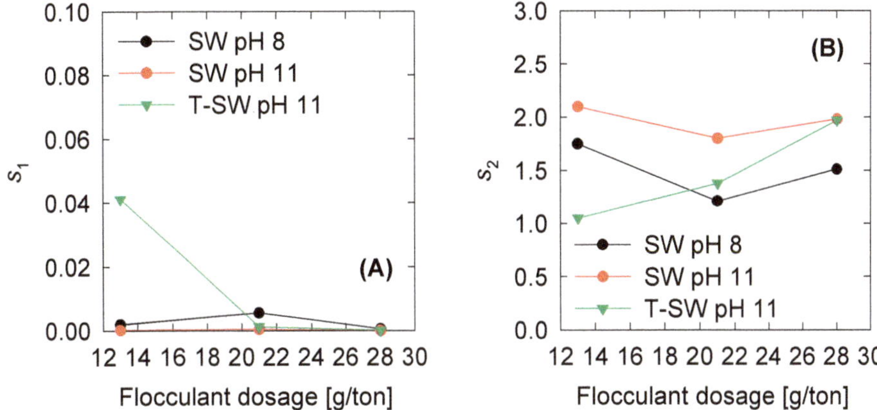

Figure 6. Optimum breakage parameters vs. shear rate for constant and variable fractal dimension: (**A**) S1 and (**B**) collision efficiency decay constant k_d.

4.4. Aggregation, Breakage, and Permeability Modeling

Once we have the optimized parameters, we can compute the physical parameters: Collision efficiency, collision frequency, breakage rate, and permeability with the mean aggregate size from the results in Figure 4. Collision frequencies are shown in Figure 7. We see that high doses of flocculant increase the collision frequency. This can be accounted for by the polymer concentration, which has an extended configuration that increases the likelihood of collisions. For SW pH 11, the small aggregates generated by the appearance of magnesium complexes lead to a low value in the collision frequency. This does not present significant changes with varied flocculant doses. Calculations for collision efficiency are presented in Figure 8 and follow the same trend as the collision frequency, where SW pH 11 has no significant values. In this case, only below 20 s is the parameter relevant to the kinetics, and drops to zero values after 20 s. Breakage rates are shown in Figure 9, which shows that, at SW pH 8, we see that low doses generate higher values because less flocculant cannot maintain the aggregates. We see similar behaviour in SW pH 11 and T-SW pH 11; this is accounted for because of the fractal dimensions. For SW at pH 11, the fractal dimension is low, and the diameter increase coincided with an increased breakage rate. For T-SW at pH 11, the fractal dimension is high, decreasing the aggregate diameter and decreasing the breakage rate, even if the aggregates are big. Finally, permeability results were plotted in Figure 10; we can see that very porous aggregates are found for SW pH 8 and a dose of 21 g/ton. This can be accounted for by the fractal dimension and the large mean size of the aggregates. For SW at pH 11, we see that permeability is lower even if the fractal dimension is small, because the size of the aggregate is small in these cases. Interestingly, the permeability value for T-SW pH 11 is low, as compared to other cases, even if the aggregate sizes are similar to those in SW pH 8. This is because the fractal dimension is high and creates dense aggregates that have little permeability.

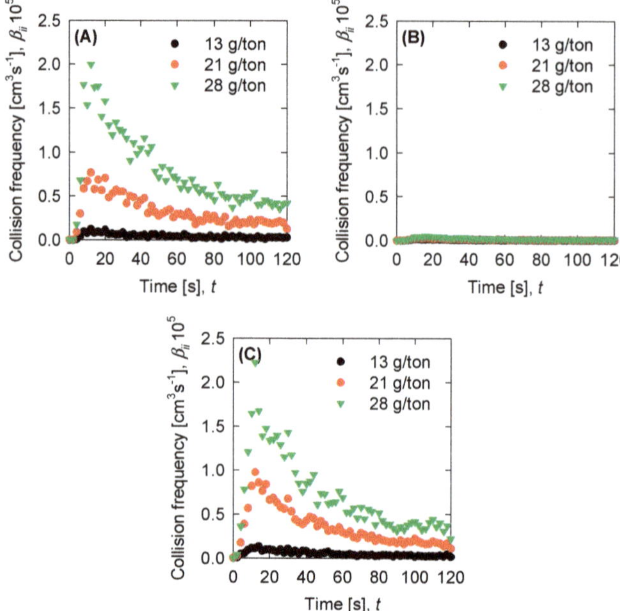

Figure 7. Collision frequency for the experimental data for (**A**) SW pH 8, (**B**) SW pH 11, and (**C**) T-SW pH 11.

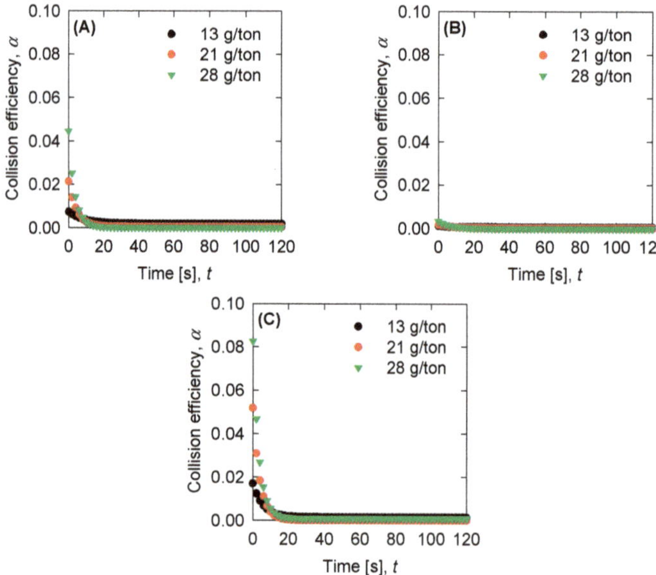

Figure 8. Collision efficiency for the experimental data for (**A**) SW pH 8, (**B**) SW pH 11, and (**C**) T-SW pH 11.

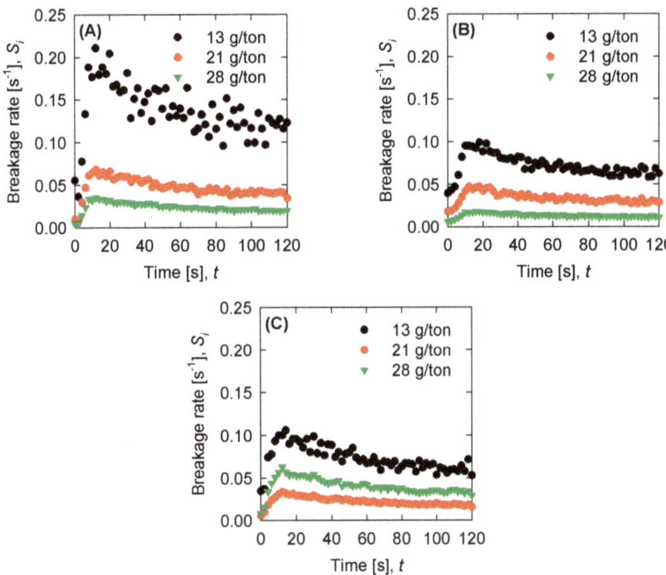

Figure 9. Breakage rate for the experimental data for (**A**) SW pH 8, (**B**) SW pH 11, and (**C**) T-SW pH 11.

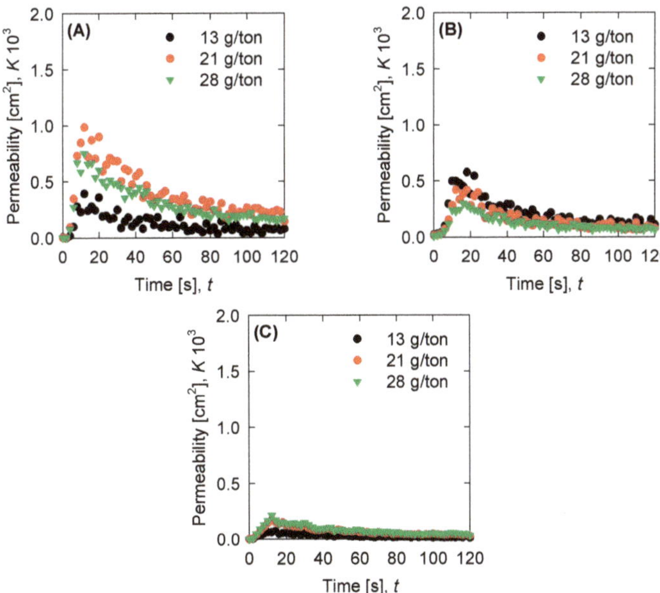

Figure 10. Permeability for the experimental data for (**A**) SW pH 8, (**B**) SW pH 11, and (**C**) T-SW pH 11.

5. Conclusions

Experiments and modeling, through a population balance model (PBM), are used to characterize the particle aggregation of flocculated tailings on untreated and treated seawater (SW) at different flocculant doses. It was found that the Mg hydroxide complex presence could hinder the aggregate kinetics drastically over time; this generates small aggregates between 50–100 microns at pH 11. If the Mg is removed before the flocculation takes place, we can obtain larger aggregates, from 100–200 microns, at the same pH. The results of T-SW pH 11 are in good agreement with the SW pH 8 achievements, which may imply that pH is no longer a restrictive parameter for the aggregation process. The modeling of the aggregation was done with PBM, where fitted parameters can represent the flocculation aggregation kinetics. In this case, the effect of flocculant dose can neglect the impact of the breakage kernel, and the aggregation kernel dominates the flocculation. Additionally, it is found that, without Mg ions, the pH has little effect on both the flocculation kinetics and PBM fitted parameters. Only the fractal dimension shows us the main difference between SW pH 8 and T-SW pH 11, where it is reflected in its permeability. As we can see, experiments and modeling show us that the Mg removal is the main component that hinders particle aggregation, and without its presence, the process can operate similarly at neutral pH and high pH.

Author Contributions: All of the authors contributed to the analysis of the results and writing the paper. All authors have read and agreed to the published version of the manuscript

Funding: This research was funded by ANID Fondecyt 11171036, ANID ACM 170005, and Centro CRHIAM Project ANID/Fondap/15130015.

Acknowledgments: Gonzalo Quezada and Ricardo I. Jeldres thank the Centro CRHIAM Project ANID/Fondap/15130015. Ricardo I. Jeldres thanks ANID Fondecyt 11171036 and ANID ACM 170005. Pedro Robles thanks the Pontificia Universidad Católica de Valparaíso for the support provided.

Conflicts of Interest: The authors declare no conflict of interest.

References

1. Qiu, Z.; Liu, G.; Liu, Q.; Zhong, H. Understanding the roles of high salinity in inhibiting the molybdenite flotation. *Colloids Surfaces A Physicochem. Eng. Asp.* **2016**. [CrossRef]
2. Li, W.; Li, Y.; Wei, Z.; Xiao, Q.; Song, S. Fundamental studies of SHMP in reducing negative effects of divalent ions on molybdenite flotation. *Minerals* **2018**, *8*, 404. [CrossRef]
3. Li, Y.; Li, W.; Xiao, Q.; He, N.; Ren, Z.; Lartey, C.; Gerson, A. The influence of common monovalent and divalent chlorides on chalcopyrite flotation. *Minerals* **2017**, *7*, 111. [CrossRef]
4. Uribe, L.; Gutierrez, L.; Laskowski, J.S.; Castro, S. Role of calcium and magnesium cations in the interactions between kaolinite and chalcopyrite in seawater. *Physicochem. Probl. Miner. Process.* **2017**, *53*, 737–749. [CrossRef]
5. Ramos, O.; Castro, S.; Laskowski, J.S. Copper–molybdenum ores flotation in sea water: Floatability and frothability. *Miner. Eng.* **2013**, *53*, 108–112. [CrossRef]
6. Hirajima, T.; Suyantara, G.P.W.; Ichikawa, O.; Elmahdy, A.M.; Miki, H.; Sasaki, K. Effect of Mg2+ and Ca2+ as divalent seawater cations on the floatability of molybdenite and chalcopyrite. *Miner. Eng.* **2016**, *96–97*, 83–93. [CrossRef]
7. Jeldres, R.I.; Arancibia-Bravo, M.P.; Reyes, A.; Aguirre, C.E.; Cortes, L.; Cisternas, L.A. The impact of seawater with calcium and magnesium removal for the flotation of copper-molybdenum sulphide ores. *Miner. Eng.* **2017**, *109*, 10–13. [CrossRef]
8. Suyantara, G.P.W.; Hirajima, T.; Miki, H.; Sasaki, K. Floatability of molybdenite and chalcopyrite in artificial seawater. *Miner. Eng.* **2018**, *115*, 117–130. [CrossRef]
9. Lee, L.T.; Rahbari, R.; Lecourtier, J.; Chauveteau, G. Adsorption of polyacrylamides on the different faces of kaolinites. *J. Colloid Interface Sci.* **1991**, *147*, 351–357. [CrossRef]
10. Nasser, M.S.; James, A.E. Effect of polyacrylamide polymers on floc size and rheological behaviour of kaolinite suspensions. *Colloids Surfaces A Physicochem. Eng. Asp.* **2007**, *301*, 311–322. [CrossRef]
11. Zhang, L.; Zheng, L.; Pu, J.; Pu, C.; Cui, S. Influence of hydrolyzed polyacrylamide (hpam) molecular weight on the cross-linking reaction of the HPAM/Cr3+ System and transportation of the HPAM/Cr3+ system in microfractures. *Energy Fuels* **2016**. [CrossRef]
12. Abdel-Azeim, S.; Kanj, M.Y. Dynamics, Aggregation, and Interfacial Properties of the Partially Hydrolyzed Polyacrylamide Polymer for Enhanced Oil Recovery Applications: Insights from Molecular Dynamics Simulations. *Energy Fuels* **2018**. [CrossRef]

13. Quezada, G.R.; Jeldres, R.I.; Fawell, P.D.; Toledo, P.G. Use of molecular dynamics to study the conformation of an anionic polyelectrolyte in saline medium and its adsorption on a quartz surface. *Miner. Eng.* **2018**, *129*, 102–105. [CrossRef]
14. Quezada, G.R.; Saavedra, J.H.; Rozas, R.E.; Toledo, P.G. Molecular dynamics simulations of the conformation and diffusion of partially hydrolyzed polyacrylamide in highly saline solutions. *Chem. Eng. Sci.* **2019**. [CrossRef]
15. Jeldres, M.; Piceros, E.C.; Toro, N.; Torres, D.; Robles, P.; Leiva, W.H.; Jeldres, R.I. Copper tailing flocculation in seawater: Relating the yield stress with fractal aggregates at varied mixing conditions. *Metals* **2019**, *9*, 1295. [CrossRef]
16. Jiang, M.Q.; Jin, X.Y.; Lu, X.Q.; Chen, Z.L. Adsorption of Pb(II), Cd(II), Ni(II) and Cu(II) onto natural kaolinite clay. *Desalination* **2010**. [CrossRef]
17. Bhattacharyya, K.G.; Gupta, S.S. Adsorption of a few heavy metals on natural and modified kaolinite and montmorillonite: A review. *Adv. Colloid Interface Sci.* **2008**, *140*, 114–131. [CrossRef]
18. Diamond, S.; Kinter, E.B. Adsorption of calcium hydroxide by montmorillonite and kaolinite. *J. Colloid Interface Sci.* **1966**. [CrossRef]
19. Biggs, C.A.; Lant, P.A. Activated sludge flocculation: On-line determination of floc size and the effect of shear. *Water Res.* **2000**. [CrossRef]
20. Heath, A.R.; Bahri, P.A.; Fawell, P.D.; Farrow, J.B. Polymer flocculation of calcite: Relating the aggregate size to the settling rate. *AIChE J.* **2006**, *52*, 1987–1994. [CrossRef]
21. Gregory, J. Monitoring particle aggregation processes. *Adv. Colloid Interface Sci.* **2009**, *147–148*, 109–123. [CrossRef] [PubMed]
22. Meng, Z.; Hashmi, S.M.; Elimelech, M. Aggregation rate and fractal dimension of fullerene nanoparticles via simultaneous multiangle static and dynamic light scattering measurement. *J. Colloid Interface Sci.* **2013**, *392*, 27–33. [CrossRef] [PubMed]
23. Heath, A.R.; Bahri, P.A.; Fawell, P.D.; Farrow, J.B. Polymer flocculation of calcite: Experimental results from turbulent pipe flow. *AIChE J.* **2006**, *52*, 1284–1293. [CrossRef]
24. Owen, A.T.; Fawell, P.D.; Swift, J.D.; Labbett, D.M.; Benn, F.A.; Farrow, J.B. Using turbulent pipe flow to study the factors affecting polymer-bridging flocculation of mineral systems. *Int. J. Miner. Process.* **2008**, *87*, 90–99. [CrossRef]
25. Bubakova, P.; Pivokonsky, M.; Filip, P. Effect of shear rate on aggregate size and structure in the process of aggregation and at steady state. *Powder Technol.* **2013**, *235*, 540–549. [CrossRef]
26. Benn, F.A.; Fawell, P.D.; Halewood, J.; Austin, P.J.; Costine, A.D.; Jones, W.G.; Francis, N.S.; Druett, D.C.; Lester, D. Sedimentation and consolidation of different density aggregates formed by polymer-bridging flocculation. *Chem. Eng. Sci.* **2018**, *184*, 111–125. [CrossRef]
27. Thomas, D.N.; Judd, S.J.; Fawcett, N. Flocculation modelling: a review. *Water Res.* **1999**, *33*, 1579–1592. [CrossRef]
28. Heath, A.R.; Bahri, P.A.; Fawell, P.D.; Farrow, J.B. Polymer flocculation of calcite: Population balance model. *AIChE J.* **2006**, *52*, 1641–1653. [CrossRef]
29. Jeldres, R.I.; Fawell, P.D.; Florio, B.J. Population balance modelling to describe the particle aggregation process: A review. *Powder Technol.* **2018**, *326*, 190–207. [CrossRef]
30. Quezada, G.R.; Ramos, J.; Jeldres, R.I.; Robles, P.; Toledo, P.G. Analysis of the flocculation process of fine tailings particles in saltwater through a population balance model. *Sep. Purif. Technol.* **2019**, *237*, 116319. [CrossRef]
31. Runkana, V.; Somasundaran, P.; Kapur, P.C. A population balance model for flocculation of colloidal suspensions by polymer bridging. *Chem. Eng. Sci.* **2006**, *61*, 182–191. [CrossRef]
32. Costa, C.B.B.; Maciel, M.R.W.; Filho, R.M. Considerations on the crystallization modeling: Population balance solution. *Comput. Chem. Eng.* **2007**, *31*, 206–218. [CrossRef]
33. Datta, A.; Rajamani, R.K. A direct approach of modeling batch grinding in ball mills using population balance principles and impact energy distribution. *Int. J. Miner. Process.* **2002**, *64*, 181–200. [CrossRef]
34. Kiparissides, C.; Alexopoulos, A.; Roussos, A.; Dompazis, G.; Kotoulas, C. Population balance modeling of particulate polymerization processes. *Ind. Eng. Chem. Res.* **2004**, *43*, 7290–7302. [CrossRef]
35. Vajihinejad, V.; Soares, J.B.P. Monitoring polymer flocculation in oil sands tailings: A population balance model approach. *Chem. Eng. J.* **2018**, *346*, 447–457. [CrossRef]

36. Jeldres, R.I.; Concha, F.; Toledo, P.G. Population balance modelling of particle flocculation with attention to aggregate restructuring and permeability. *Adv. Colloid Interface Sci.* **2015**, *224*, 62–71. [CrossRef]
37. Ahmad, A.L.; Chong, M.F.; Bhatia, S. Population Balance Model (PBM) for flocculation process: Simulation and experimental studies of palm oil mill effluent (POME) pretreatment. *Chem. Eng. J.* **2008**, *140*, 86–100. [CrossRef]
38. Flesch, J.C.; Spicer, P.T.; Pratsinis, S.E. Laminar and turbulent shear-induced flocculation of fractal aggregates. *AIChE J.* **1999**, *45*, 1114–1124. [CrossRef]
39. Selomulya, C.; Bushell, G.; Amal, R.; Waite, T.D. Understanding the role of restructuring in flocculation: The application of a population balance model. *Chem. Eng. Sci.* **2003**, *58*, 327–338. [CrossRef]
40. Antunes, E.; Garcia, F.A.P.; Ferreira, P.; Blanco, A.; Negro, C.; Rasteiro, M.G. Modelling PCC flocculation by bridging mechanism using population balances: Effect of polymer characteristics on flocculation. *Chem. Eng. Sci.* **2010**, *65*, 3798–3807. [CrossRef]
41. Jeldres, M.; Piceros, E.; Robles, P.A.; Toro, N.; Jeldres, R.I. Viscoelasticity of quartz and kaolin slurries in seawater: Importance of magnesium precipitates. *Metals* **2019**, *9*, 1120. [CrossRef]
42. Spicer, P.T.; Pratsinis, S.E. Shear-induced flocculation: The evolution of floc structure and the shape of the size distribution at steady state. *Water Res.* **1996**, *30*, 1049–1056. [CrossRef]
43. Hounslow, M.J.; Ryall, R.L.; Marshall, V.R. A discretized population balance for nucleation, growth, and aggregation. *AIChE J.* **1988**, *34*, 1821–1832. [CrossRef]
44. Kusters, K.A.; Pratsinis, S.E.; Smith, D.M.; Thoma, S.G. Ultrasonic fragmentation of agglomerate powders. *Chem. Eng. Sci.* **1993**, *48*, 4119–4127. [CrossRef]
45. Thill, A.; Moustier, S.; Aziz, J.; Wiesner, M.R.; Bottero, J.Y. Flocs restructuring during aggregation: Experimental evidence and numerical simulation. *J. Colloid Interface Sci.* **2001**, *243*, 171–182. [CrossRef]
46. Veerapaneni, S.; Wiesner, M.R. Hydrodynamics of fractal aggregates with radially varying permeability. *J. Colloid Interface Sci.* **1996**, *177*, 45–57. [CrossRef]
47. Li, X.-Y.; Logan, B.E. Permeability of fractal aggregates. *Water Res.* **2001**, *35*, 3373–3380. [CrossRef]
48. Vainshtein, P.; Shapiro, M.; Gutfinger, C. Mobility of permeable aggregates: effects of shape and porosity. *J. Aerosol Sci.* **2004**, *35*, 383–404. [CrossRef]
49. Mandelbrot, B.B. Self-affine fractals and fractal dimension. *Phys. Scr.* **1985**, *32*, 257–260. [CrossRef]
50. Pandya, J.D.; Spielman, L.A. Floc breakage in agitated suspensions: Effect of agitation rate. *Chem. Eng. Sci.* **1983**, *38*, 1983–1992. [CrossRef]
51. Pretorius, C.; Wicklein, E.; Rauch-Williams, T.; Samstag, R.; Sigmon, C. How oversized mixers became an industry standard. In Proceedings of the 88th Annual Water Environment Federation Technical Exhibition and Conference, WEFTEC 2015, Chicago, IL, USA, 26–30 September 2015.

Article

Hydrometallurgical Treatment of Waste Printed Circuit Boards: Bromine Leaching

Hao Cui [1,*] and Corby Anderson [2]

[1] Nevada Gold Mines, LLC, Elko, NV 89801, USA
[2] Department of Metallurgical & Materials Engineering, Colorado School of Mines, Golden, CO 80401, USA; cganders@mines.edu
* Correspondence: hcui@nevadagoldmines.com; Tel.: +1-775-934-3627

Received: 15 March 2020; Accepted: 30 March 2020; Published: 2 April 2020

Abstract: This paper demonstrates the recovery of valuable metals from shredded Waste Printed Circuit Boards (WPCBs) by bromine leaching. Effects of sodium bromide concentration, bromine concentration, leaching time and inorganic acids were investigated. The most critical factors are sodium concentration and bromine concentration. It was found that more than 95% of copper, silver, lead, gold and nickel could be dissolved simultaneously under the optimal conditions: 50 g/L solid/liquid ratio, 1.17 M NaBr, 0.77 M Br_2, 2 M HCl, 400 RPM agitation speed and 23.5 °C for 10 hours. The study shows that the dissolution of gold from waste printed circuit boards in a Br_2-NaBr system is controlled by film diffusion and chemical reaction.

Keywords: shredded waste printed circuit boards; precious metals; bromine; leaching kinetics

---

1. Introduction

Given its increasing volume and high content of both hazardous and valuable materials contained within, electronic and electric equipment (e-waste) represents an emerging environmental challenge as well as a business opportunity [1]. It is globally recognized that e-waste can be considered as a secondary metallic source to compensate for the excessive demand for mining, enabling a more circular economy for these materials [2].

Printed circuit boards, the key components of electronic devices, are, in their simplest form, a mixture of woven glass reinforced resin and multiple types of metal, and are currently recycled by pyrometallurgy, along with other e-waste products. Table 1 shows the content of selected WPCBs. Waste printed circuit boards can contain up to 60 elements that are classified into two categories: metallic and non-metallic materials [3]. Generally, a printed circuit board is composed of approximately 30–40% metals (such as: copper, iron, nickel, lead, tin, silver, gold, and palladium), 30% organic resin and 30% refractory oxides (mainly glass fibers). The composition of printed circuit boards varies depending on their designs, ages, and applications.

Table 1. Selected material composition of Waste Printed Circuit Boards (WPCBs).

Material	Element	Vasile et al., 2008 [4]	Hino et al., 2009 [5]	Birloaga et al., 2013 [6]	Yang et al., 2009 [7]	Oishi et al., 2007 [8]	Behnamfard et al., 2013 [9]	Average
Organic epoxy resin	C (wt. %)	24.7	18.1	-	-	-	-	21.4
	H (wt. %)	1.38	1.8	-	-	-	-	1.59
	N (wt. %)	0.85	0.32	-	-	-	-	0.59
	Br (wt. %)	4.94	5.07	-	-	-	-	5.01
	Sb (wt. %)	-	0.45	-	-	0.16	0.37	0.33
Inorganic glass fiber		-	37.6	-	-	-	-	37.6
Metallic elements	Cu (wt. %)	13.8	14.6	30.6	25.1	26	19.2	21.5
	Fe (wt. %)	1.97	4.79	15.21	0.66	3.4	1.13	4.53
	Sn (wt. %)	-	5.62	7.36	1.86	4.9	0.69	4.09
	Ni (wt. %)	0.17	1.65	1.58	0.0024	1.5	0.17	0.85
	Zn (wt. %)	-	-	1.86	0.04	2.6	0.84	1.34
	Pb (wt. %)	-	2.96	6.70	0.80	3.0	0.39	2.77
	Au (ppm)	-	205	238	-	-	130	191
	Ag (ppm)	-	450	688	-	630	704	618
	Pd (ppm)	-	220	-	-	-	27	124

As of November, 2017, current payout rates of sorted boards, such as low-grade boards, mid-grade boards, metal multi socket server motherboards, laptop motherboards, hard drive boards, and cell phone boards, ranged from \$0.3 to \$20 per kilogram [10], depending on the various designs, ages, and chemical compositions of the boards. In addition, an estimation of the market value of the metals present in PCBs was conducted to verify if the recovery of metals from the PCBs is cost-effective. For this calculation, an averaged concentration of metals in the printed circuit boards based on data from Table 1 and prices of metals (Cu, Au, Ag, Pd, Sn, Ni, and Zn) on December 1st, 2017 from InvestmentMine [11–18] were used and shown in Table 2. The price of iron was taken from Scrap Metal Pricer on March 21st, 2020 [19].

Table 2. Market value of printed circuit boards.

Metal	Market Price Per Unit (USD)	Unit	Average Composition (%)	Market Value Per Tonne of WPCBs (USD)
Gold	41.5	g	0.0191	7927
Silver	0.54	g	0.0618	334
Palladium	32.5	g	0.0124	4030
Copper	7	kg	21.5	1505
Tin	20	kg	4.09	818
Fe	9.5	kg	4.53	429
Lead	2	kg	2.77	55
Zinc	3	kg	1.34	40
Nickel	11	kg	0.85	94
Total market value per tonne of printed circuit boards (USD)				15,231

This data yields a market value of approximately $ 15,231/ tonne of WPCBs. More than 80% of the overall market value is from precious metals (Au, Ag and Pd), even though the concentration of base metals (Cu, Zn, Ni and Sn) is much greater than the concentration of precious metals present in printed circuit boards.

As our previous review article summarized [20], oxidative acids are extensively used to recover base metals from waste printed circuit boards. Table 3 summarizes the advantages and disadvantages of four selected reagents for base metal extractions. Precious metals are noble in the normal environment and require high oxidation potential during leaching. The most common leaching reagents for precious metal leaching include cyanide, thiourea and thiosulfate because they form stable metal complexes [2]. Table 4 demonstrates a comparison between several common leaching reagents for gold extraction.

Table 3. Comparison of potential leaching reagents for base metal extraction from waste printed circuit boards [2,20–23].

Reagent	Pros	Cons
Sulfuric acid	Highly selective, low reagent cost, well established process for copper ore	At elevated temperature, corrosive
Chloride	Fast kinetics at room temperature, high solubility and activity of base metals, low toxicity	Excessive corrosion, difficul telectrowinning of copper, poor quality of copper
Aqua regia	Fast kinetics, effective	High reagent cost, highly corrosive, low selectivity
Ionic liquids	Thermally stable, environmentally friendly	High cost, excessive dosage

Table 4. Comparison of potential leaching reagents for gold [2,20,24,25].

Reagent	Pros	Cons
Cyanide	Highly effective, low reagent dosage and cost	Difficult to process wastewater, environmental risk, low kinetics
Thiourea	Less toxic, high reaction rate, less interference ions	Poorer stability, high consumption, more expensive than cyanide, downstream metal recovery
Thiosulfate	High selectivity, non-toxic and non-corrosive, fast leaching rate	High consumption of reagent, downstream metal recovery
Halide	High leaching rate, high selectivity, relatively healthy and safe except for bromine	Highly corrosive for chlorine, high consumption for iodine
Aqua regia	Fast kinetics, low reagent dosage	Strongly oxidative and corrosive, difficult to deal with downstream

Using a halide system provides the possibility of direct leaching and recovery of precious metals. Chlorination was extensively introduced to the gold extraction industry in the 1800 s, prior to cyanide leaching, for the treatment of gold sulfide minerals and refractory gold ores. Two other important halide leaching reagents are bromine and iodine. Generally, the major advantage of halide reagents in the gold leaching is their powerful oxidizing ability, leading to a high dissolution rate compared to alkaline cyanide leaching.

Iodine/iodide has been reported as an alternative to cyanidation with air as the oxidant. The stable gold-diiodo and tetraiodo complexes are stable up to pH 14 [26]. Xu et al. reported that the gold extraction reached approximately 95% under the optimum conditions: an iodide concentration of 1.0–1.2%, H_2O_2 concentration of 1–2%, leaching time of 4 h, solid/liquid ratio of 1/10, pH 7 and 25 °C [27]. Batnasan et al. [28] demonstrated an iodine–iodide leaching process to recover valuable metals from waste printed circuit boards. The results indicated that more than 99% of gold was dissolved, while less than 1% of silver and palladium were dissolved at the conditions: iodine/iodide mass ratio of 1/6, pulp density of 10%, agitation speed of 500 rpm, 24 h and 40 °C [28].

Bromine, one of the halide elements, is used in the production of clear brines for the oil drilling industry, bromine-based biocides for water treatment, and brominated flammable retardants [29]. Bromine is the key element in the production of brominated flammable retardants, and manufacturers commonly produce PCBs with flammable retardants to help meet fire safety standards. Thus, it is necessary to monitor and remove bromine/bromide during e-waste recycling processes. It will be beneficial to investigate the bromine chemistry in the extraction of precious metals from e-waste. Unfortunately, studies on bromine leaching of waste printed circuit boards are limited.

This study first characterizes the printed circuit board material to provide a qualitative and quantitative guidance of a methodology to efficiently separate desirable metals from waste printed circuit boards. Based on the material characterization, the investigation of leaching behavior of waste printed circuit boards using bromine to recover valuable metal was conducted, as bromine can provide a high oxidation–reduction potential.

2. Material and Methods

2.1. Material

Waste printed circuit board samples used in this study were provided from Aurubis. Their applications are unknown. All the motherboards were shredded into approximately 2 cm^2 by a pilot scale shredder, and then split into approximately 60 kg by a Jones splitter. A Wiley-Mill shredder (Fellner & Ziegler GmbH, Frankfurt, Germany) shredder with a 1 cm round-hole screen was employed to further grind the material. All of the aluminum heat sinks and lithium ion batteries were removed from the motherboards to prevent being stuck in the Wiley Mill. After mixing and splitting, the 27.8 kg of −1 cm material was then ground again by the Wiley-Mill shredder to −5 mm. The 27.8 kg of −5 mm material was then split to 7 kg samples which were used for subsequent experiments. The size distribution is shown in Figure 1.

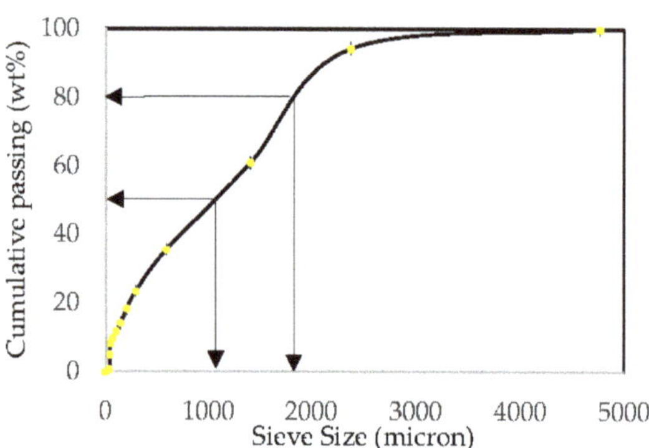

Figure 1. Size distribution of shredded printed circuit boards.

2.2. Chemical Analysis Methods

The elemental analysis of the shredded waste printed circuit boards was conducted by a two-step digestion (Aqua Regia and HF-HCl-HNO_3-H_3BO_3) for Atomic Absorption spectroscopy (AA) (PerkinElmer AAnalyst 400, PerkinElmer, Inc., Waltham, MA, USA) and Inductively Coupled Plasma Mass Spectrometry (ICP-MS) (Thermo Fisher Scientific, MA, USA) analysis.

2.3. Mineral Liberation Analysis

Mineral liberation analysis of the WPCBs was conducted by the Center for Advanced Mineral and Metallurgical Processing (CAMP), at Montana Tech of the University of Montana. The analysis was performed by water/methanol separation to remove the fine low-density particles and hydrophobic material, which mostly consisted of fibrous paper and resin. The residue that has heavier density was treated by a heavy liquid separation using di-iodomethane (d = 3.3 g/mL) to further remove circuit board resins and fibers floated in the heavy liquid. The denser particles remained in the bottom of the separation vessel were then mounted in epoxy blocks for examination using Scanning Electron Microscopy-Mineral Liberation Analysis (SEM-MLA).

2.4. Leaching Tests

A series of bromide leaching experiments were conducted in 250 mL beakers sealed by Parafilm sealing films for a given time period. A schematic diagram of the leaching set-up is given in Figure 2. Since the gold dissolution is known to be exothermic, the 250 mL beaker was placed in a 1 L beaker filled with 300 mL tap water to maintain the desired temperature. A leaching solution of 200 mL with the desired quantity of reagents was used for each experiment. All experiments were carried out using solutions prepared from analytical-grade reagents and distilled water. Unless specified, the solutions contained 10 g of shredded printed circuit boards (particle size-5 mm) and were mixed by a magnetic stir bar at 400 RPM. The oxidation–reduction potential (ORP) and pH were measured by Fisher Scientific^{TM} accumet^{TM} XL 600 m (Fisher Scientific International, Inc., Hampton, USA). Samples of the solution were taken at fixed intervals of time during the leaching experiment and filtered on filter papers (Whatman Grade 42). The filtrate was diluted by 2% HNO_3 (TraceSELECT^{TM} nitric acid, Honeywell International Inc., Charlotte, NC, USA), and then analyzed immediately by atomic absorption spectroscopy in order to determine contents of gold and silver. The residue after leaching was then filtered and rinsed with distilled water. The filtered residue was dried, weighed and treated by a 4-acid digestion (HCl-HNO_3-HF-H_3BO_3). The gold and silver contents were then again determined by AA. ICP-MS was utilized to analyze the rest of elements. The concentrations of bromine and sodium bromide, temperature, and inorganic acids were the variables, which were investigated by parametric study.

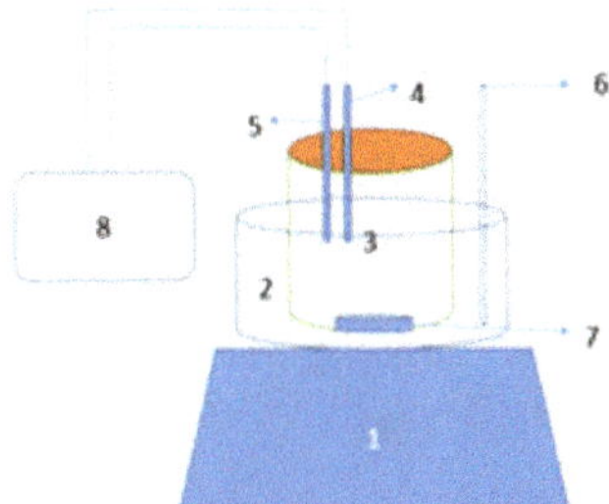

1: Hot plate, 2: Beaker for water bath, 3: Beaker, 4: OPR probe
5: pH probe, 6: Thermometer, 7: Magnetic stir bar, 8: pH meter

Figure 2. Schematic diagram of leaching set-up.

3. Result and Discussions

3.1. Sample Characterization

Table 5 gives a summary of the elemental analysis, where copper is the most abundant metal, followed by iron (6.3%), tin (3.8%) and zinc (3.3%). The most abundant precious metal is silver at 513 ppm. Gold and palladium are also found in the printed circuit boards at 145 and 20 ppm, respectively.

Table 5. Elemental composition of shredded printed circuit boards.

Element	Fe (%)	Ni (%)	Cu (%)	Zn (%)	Pb (%)	Sn (%)	Sb (%)	Pd (ppm)	Ag (ppm)	Au (ppm)
Content	6.3	0.5	24.1	3.3	1.1	3.8	0.3	20	513	145
Standard deviation	0.7	0.06	1.9	0.3	0.05	0.4	0.03	2.6	46	18

The MLA analysis (Table 6) also confirmed that copper was the most abundant metallic element and silver was the primary precious metal found in the printed circuit board. The MLA analysis (Figure 3) also shows that copper is either present as a free copper particle, or associated with other metals, such as silver and gold. Gold is usually found liberated or as gold–nickel alloys (Figures 3 and 4). Moreover, silver either occurs alone or is associated with tin, lead and nickel, as shown in Figures 3 and 5.

Table 6. Selected phase concentrations of the concentrated shredded printed circuit boards.

Phase	Formula	Coarse	Fine	−40 mesh
Copper	Cu	70.4	65.5	69.0
FeMnZnO	$Fe_{3.5}MnZn_{0.5}O_3$	1.13	6.33	6.65
BrCO	$(CH_2)_{40}BrO_6$	0.67	2.15	2.82
SnBi	$Sn_{0.8}Bi_{0.2}$	3.82	2.78	1.53
BaTiO	$BaTiO_3$	0.06	1.67	1.65
PbSn	$Pb_{0.8}Sn_{0.2}$	1.18	1.21	1.21
Carbon_Mix	$C_{10}BaSO_4$	0.11	0.52	0.52
BrSbCO	$(CH_2)_{40}BrO_6Sb_{0.3}$	0.04	0.15	0.17
NdTiO	$NdTiO_3$	1.83	0.14	0.11
Fe60Ni40	$Fe_{0.6}Ni_{0.4}$	1.10	0.58	0.13

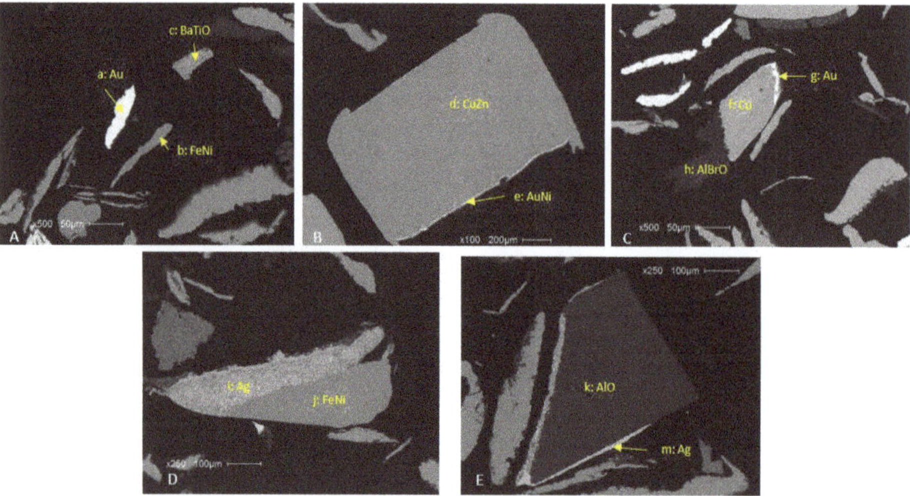

Figure 3. Image of gold and silver associations in shredded printed circuit boards (**A**: isolated gold particles; **B**: gold-nickel alloy coatings; **C**: Au coatings; **D**: Ag with FeNi; **E**: Ag with aluminum oxide).

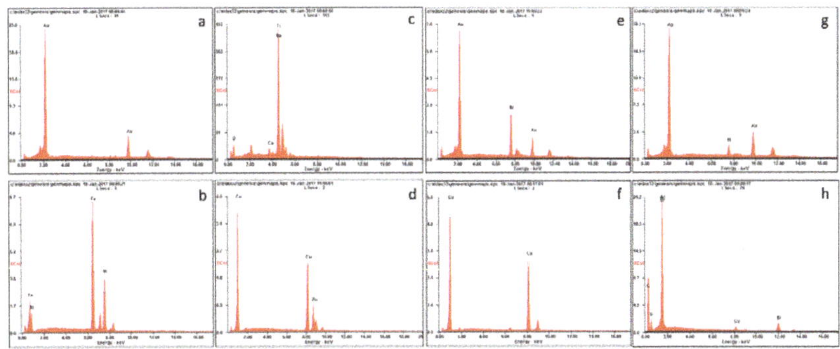

Figure 4. Electron Dispersive Spectroscopy (EDS) spectra corresponding to Figure 3 (The label of spectra is corresponding to the label of compounds in Figure 3A–C).

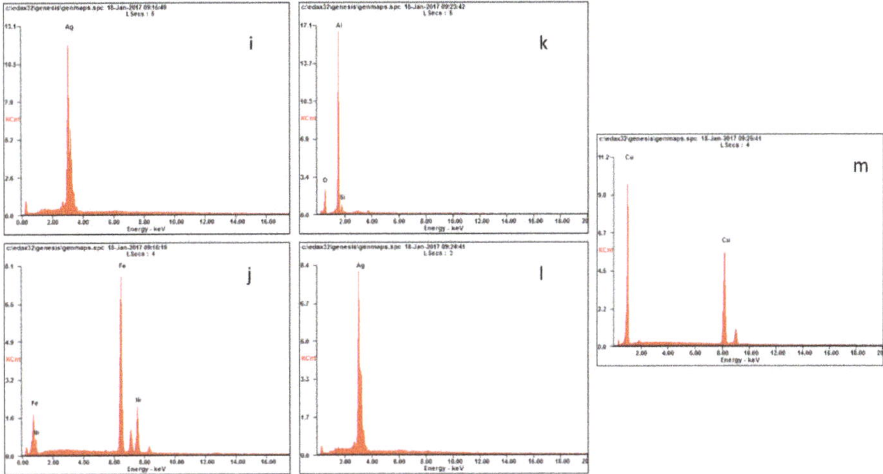

Figure 5. EDS spectra corresponding to Figure 3 (The label of each spectra is corresponding to the same label in Figure 3D,E).

3.2. Bromine Leaching Tests

Methods such as heavy liquid separation, shaking table, flotation, and electrostatic separation proved difficult to extract the metals of interest from shredded waste printed circuit boards with a satisfactory recovery. However, bromine is capable of dissolving gold, silver, palladium and copper. Thus, a direct bromide leaching was utilized to leach the desired metals from printed circuit boards.

Gold dissolves in an aqueous bromine–bromide solution to form both the Au(I) and Au(III) bromide complexes, as follows:

$$Au + 2Br^- = AuBr_2^- + e \quad (1)$$

$$Au + 4Br^- = AuBr_4^- + 3e \quad (2)$$

Figure 6 shows the stability of an Au-Br-H_2O system at room temperature. It is evident that gold is readily formed as $AuBr_4^-$ ions, except for a narrow potential range (approximately 0.7–0.8 V vs. Ag/AgCl) in which the $AuBr_2^-$ ions are dominant. Compared to chlorine (pH 0–7), bromine seems to offer a wider pH range for gold leaching. Eh-pH diagrams for the systems of Ag-Br-H_2O, Pd-Br-H_2O and Cu-Br-H_2O were also constructed for [Br] = 0.775 M and varying concentrations of silver, palladium and copper (Supplementary Materials SM-A1, A2, and A3). These indicate that $AgBr_3^{2-}$ can be found across the entire pH range when the Eh value is below approximately 0.5 V (vs. Ag/AgCl). In the acidic solution, $AgBr_3^{2-}$ exists if the Eh value is less than 1.1 V (vs. Ag/AgCl). Copper can be dissolved as Cu^{2+} and $CuBr^{+}$ in the region in which the Eh value is above 0.3 V (vs. Ag/AgCl) and the pH is less than approximately 6. Moreover, there are four major palladium species that are dissolvable (Pd^{2+}, $PdBr_6^{2-}$, $PdBr_4^{2-}$ and $PdBr^{3-}$). The general Eh-pH region in which palladium is present as soluble species is Eh > 0.3 V (vs. Ag/AgCl) and pH < 12.

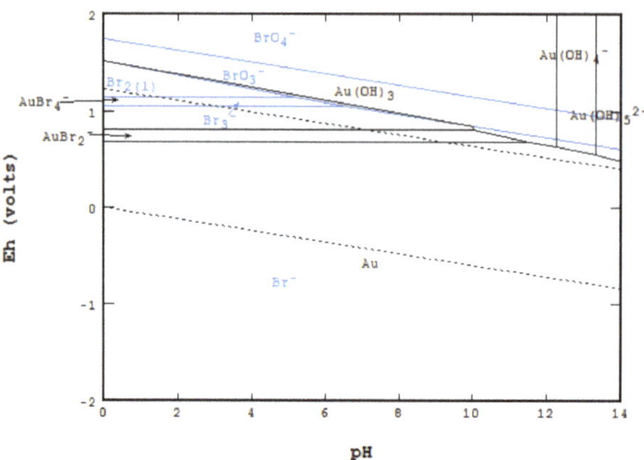

Figure 6. Eh-pH diagram of Br-Au-H_2O system at 25 °C ([Au] = 10–5 M, [Br] = 0.775 M).

3.2.1. Effect of Liquid Bromine

A series of leaching trials were conducted to investigate the effect of initial liquid bromine concentrations on leaching gold and silver from shredded waste printed circuit boards, given that bromine is not only the source of bromide ions, but also acts as an oxidant, offering a high potential (approximately 1.1 V vs. Ag/AgCl).

Figure 7 shows that the dissolution of gold and silver was significantly influenced by bromine concentration. The bromine concentration increase also leads to an increasing gold dissolution after the 10-h leaching. Other than providing a sufficient amount of bromide ions and bromine, the high concentration of bromine also assures the high potential of the solution throughout the leaching trial, further resulting in a high gold extraction.

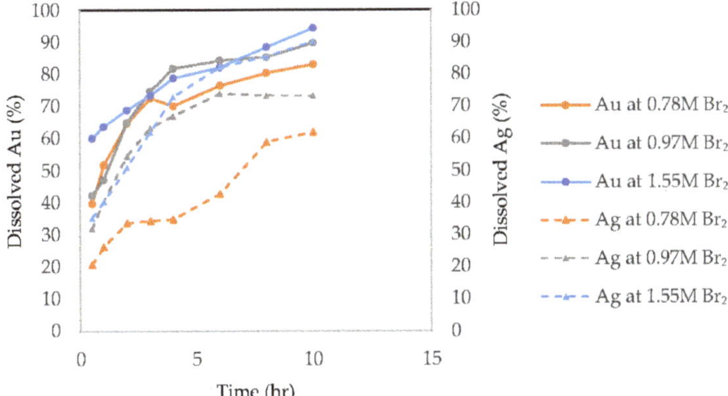

Figure 7. Dissolution of gold and silver at various concentrations of liquid bromine (23.5 °C and natural pH).

In terms of silver, the increasing bromine concentration accelerates the initial silver dissolution rate, until a certain concentration limit (0.97 M Br_2) is reached, beyond which bromine has no further effect. It is because the limiting factor for the initial reaction rate is bromine concentration, when the bromine concentration is lower than 0.97 M; beyond 0.97 M, the reaction rate is likely to be limited by active particle surfaces.

In addition, it is interesting to note that at 0.78 M bromine the dissolved silver content increases quickly in the first two hours of leaching, and then a plateau is observed in the next two hours. Subsequently, the dissolved silver continues to increase, as the leaching proceeds. The plateau could be explained by the coating of silver bromide precipitates. Specifically, silver is thought to be initially dissolved by bromine to become silver–bromide complexes and silver–bromide precipitates in the first two hours. As the leaching proceeds in the next two hours, the $AgBr_{(s)}$ and soluble silver complexes are likely to reach equilibrium. However, after the 4-hour leaching, the free bromide/bromine ions react with $AgBr_{(s)}$ precipitates to become soluble silver–bromide complexes, leading to the further increase of dissolved silver in the solution. At the highest concentration of bromine, bromine is excessive; therefore, silver is always soluble in the solution as silver–bromide complexes, leading to an increasing trend of silver extraction [30].

Regarding other metals, such as copper, zinc, and nickel, more than 90% of copper, zinc and nickel can be dissolved over 10 h (Figure 8), when the concentration of bromine is more than 0.78 M.

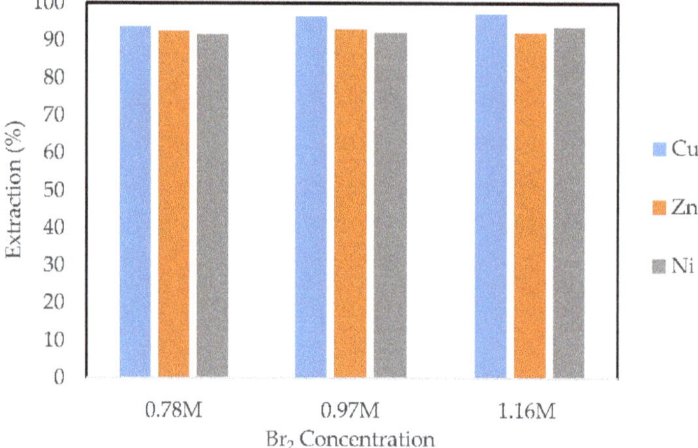

Figure 8. Dissolution of base metals (Cu, Ni and Zn) at various concentrations of bromine (23.5 °C and natural pH).

3.2.2. Effect of Sodium Bromide

As stated in previous studies [26,31,32], both bromine and sodium bromide are responsible for gold dissolution. It is, therefore, of interest to investigate the effect of sodium bromide in the presence of bromine, given that sodium bromide itself cannot provide an ORP that is higher than 0.87 V (vs. SHE).

Taking 0.78 M as the bromine concentration, a series of leaching trials was performed to investigate the effect of sodium bromide at room temperature and natural pH. Figure 9 presents leaching performances at various concentrations of sodium bromide, 1.17 M, 1.75 M and 2.33 M. It is evident that the addition of sodium bromide has a positive effect on the dissolution of precious metals (Au and Ag). The addition of sodium bromide raises the gold dissolution from approximately 83% (no NaBr) to more than 96% (1.75 M NaBr) in 10 h. However, beyond a concentration of 1.75 M, the increase of NaBr has no appreciable effect on gold dissolution for the 10-h leaching.

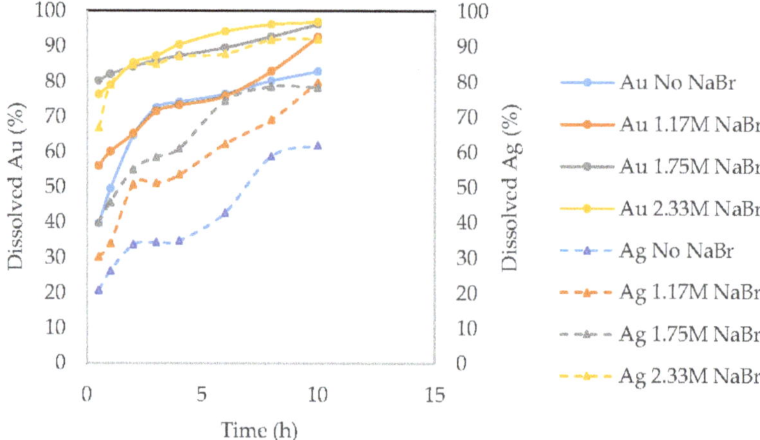

Figure 9. Dissolution of gold and silver at various concentrations of sodium bromide (0.78 M Br_2, 23.5 °C and natural pH).

As shown in Figure 9, it is observed that the initial Ag dissolution rate increases with the NaBr concentration. Meanwhile, a dissolution plateau was also observed when the NaBr concentration is between 0 and 2.33 M. The plateau can be also explained by the formation of AgBr precipitates. Additionally, the initial Au dissolution rate increases significantly as the NaBr concentration increases from 0 to 2.33 M.

Figure 10 indicates that the addition of sodium bromide only has a slightly positive effect on dissolving base metals and palladium, given that the base metals have approached high dissolutions (> 90%). Generally, in order to achieve high extractions for the metals of interest, high consumption of liquid bromine and sodium bromide are likely to be required, which reduces the practical feasibility of this process due to the resulting process economics. Therefore, more efforts should be made to study the effect of other chemical reagents that can either replace or reduce the consumption of bromine and sodium bromide.

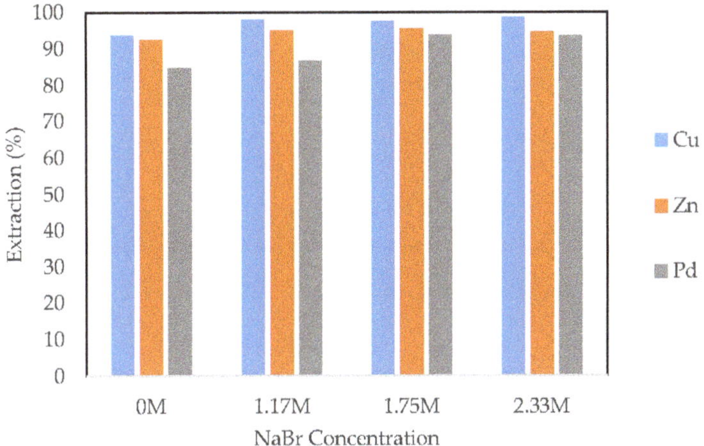

Figure 10. Dissolution of base metals (Cu and Zn) and palladium at various concentrations of sodium bromide (0.78 M Br_2, 23.5 °C and natural pH).

3.2.3. Effect of Selected Acids

Despite the technical feasibility of using bromine and sodium bromide to dissolve valuable metals from shredded waste printed circuit boards, their relatively high prices and the high consumption of sodium bromide during the process limits its industrial feasibility. Therefore, it is important to investigate alternative approaches to either replace sodium bromide or reduce its usage.

In the leaching system, the dissolution of gold and silver requires the presence of bromine, because bromine serves not only as the bromide source, but also offers a high oxidation–reduction potential, which is essential for the reaction to proceed. Meanwhile, sodium bromide provides sufficient bromide ions, along with bromine, for forming soluble gold–bromide/silver–bromide complex ions. With regard to copper, bromine plays a role in maintaining a high potential for copper oxidation. At the same time, as shown in Equation (3), bromine reacts with water to reversibly produce HBr and HBrO. Hydrobromic acid serves as a strong acid to dissolve copper at a high oxidation–reduction potential.

$$Br_2 + H_2O = HBr + HBrO \tag{3}$$

Other base metals are likely to react with hydrobromic acid to form metal ions, and sodium bromide does not appear play a role in their leaching. Therefore, in order to lower the sodium bromide usage, several relatively inexpensive mineral acids (HCl, HNO_3 and H_2SO_4) were investigated to leach shredded waste printed circuit boards.

Given that sulfur is abundantly available [33] and sulfuric acid is relatively cheap, sulfuric acid is widely used for the leaching of a wide variety of metal resources, including oxides, sulfides, silicates, phosphates, and a number of others [34]. Pesic and Sergent (1993) studied the effect of sulfuric acid on gold dissolution and found that the presence of sulfuric acid did not affect the dissolution of pure gold in a bromine–bromide leaching system [32]. Therefore, a study was performed in the presence of two concentrations (0.9 M and 2.7 M) of sulfuric acid under the condition of 0.77 M Br_2, 1.17 M NaBr and 23.5 °C, as well as 400 RPM. Table 7 shows a summary of the experimental data. Notably, the increase of H_2SO_4 concentration significantly decreases the gold dissolution, which could be explained by gold surface passivation. Based on the literature [35], a passive oxide layer may be formed on the surface of gold particles in the presence of sulfuric acid, and the passivation behavior is more pronounced at the high concentration of sulfuric acid.

Table 7. Summary of leaching experiment in the presence of inorganic acids.

Metal	5 v/v% (0.9 M) H_2SO_4 Extraction (%)	15 v/v% (2.7 M) H_2SO_4 Extraction (%)	5 v/v% (2.0 M) HCl Extraction (%)	5 v/v% (1.2 M) HNO_3 Extraction (%)
Ni	95.8	93.4	95.2	94.6
Cu	97.6	96.8	97.9	97.8
Zn	97.2	91.7	92.5	95.8
Sn	98.5	99.2	96.8	97.1
Pd	88.7	91.6	90.0	96.4
Ag	90.0	84.8	96.5	97.2
Au	91.0	71.4	95.6	93.6

To investigate the effect of various mineral acids, a series of 10-h leaching trials was conducted under the condition of 0.77 M Br_2, 1.17 M NaBr and 23.5 °C, as well as 400 RPM. Hydrochloric acid, nitric acid and sulfuric acid were chosen and diluted to a concentration of 5 v/v% (2.0 M HCl, 1.2 M HNO_3, and 0.9 M H_2SO_4).

The experimental results indicated that the presence of the three acids significantly increased the silver extraction, compared to the leaching performance in the absence of mineral acids (shown in Table 7). Sulfuric acid, nitric acid and hydrochloric acid are capable of dissolving more than 95% of base metals, including nickel, zinc, tin and copper. However, sulfuric acid can only dissolve 91% of gold, which is lower than nitric acid and hydrochloric acid. Nitric acid and hydrochloric acid give similar leaching performances; hydrochloric acid, nonetheless, is likely to be superior compared to nitric acid, due to the following reasons:

(1) The cost of nitric acid is higher than that of hydrochloric acid.
(2) Nitric acid is required to be removed in order to successfully complete the metal recovery by solvent extraction and electrowinning.

In order to validate the leaching reproducibility under optimal conditions, more leaching experiments were conducted, and the data is shown in Table 8. The optimal condition is 50 g/L solid/liquid ratio, 1.17 M NaBr, 0.77 M Br_2, 2 M HCl, 400 RPM agitation speed and 23.5 °C for 10 h.

Table 8. Triplicate leaching results under optimal conditions.

Metal	Average Extraction (%)	Standard Deviation
Ni	95.21	1.81
Cu	97.88	0.56
Zn	92.50	5.36
Sn	96.79	0.44
Pb	97.61	0.29
Pd	90.04	5.41
Ag	96.52	0.79
Au	95.59	2.05

To characterize the leaching residue from the triplicate experiments, a SEM-EDS analysis was conducted. Prior to the SEM analysis, the leaching residue was first ground and then rinsed by deionized water and acetone to eliminate the bromine effect and liberate the leaching residue of the waste printed circuit boards. As shown in Figures 11 and 12, it is obvious that there is a gold particle with a particle size of 50 μm. The reason that the gold was not dissolved during the leaching process is that it was not well liberated. Since the residue was further ground before the SEM analysis, the gold was exposed, as shown in Figure 11.

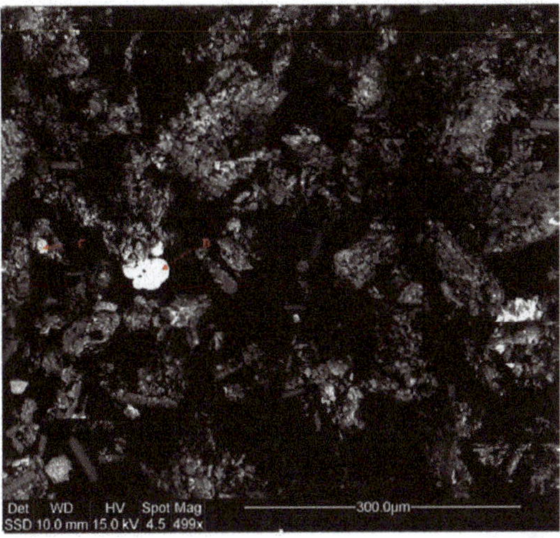

Figure 11. Backscattered Electron (BSE) image of leaching residues.

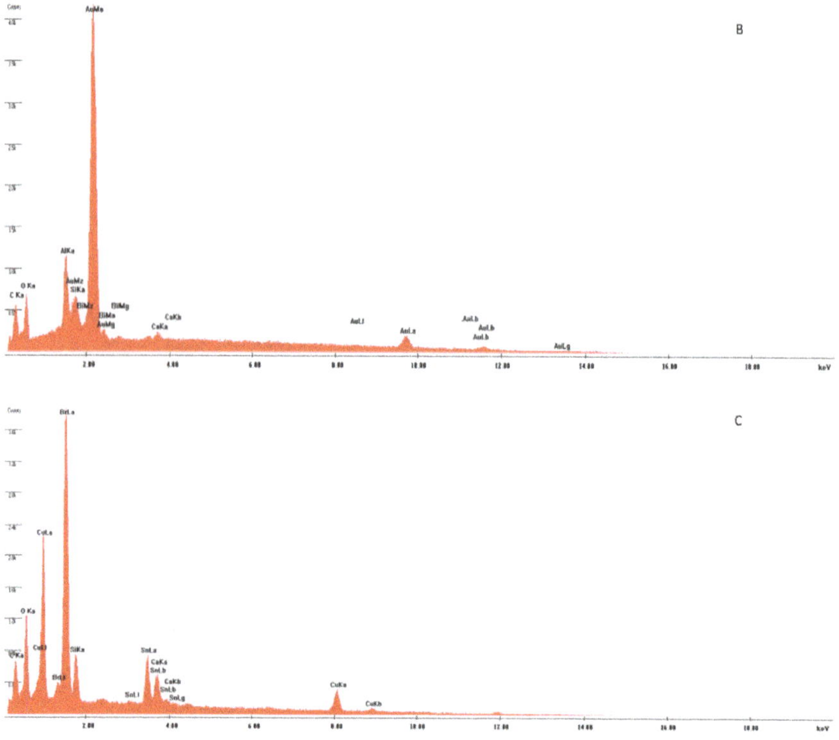

Figure 12. EDS spectra for Point B and C (BSE image).

3.3. Leaching Kinetics

To understand the kinetics of bromide leaching of gold from waste printed circuit boards, it is of interest to first determine a rate expression of the bromine–gold system, which can subsequently be utilized to create a leaching model. To do so, the first step is to determine the variables in the leaching system. In this study, the concentrations of bromine, sodium bromide and copper, and agitation speed, as well as temperature, were the parameters investigated.

In this study, the gold present in waste printed circuit boards is dissolved in a bromine–sodium bromide system to form a gold–bromine complex. The metallic copper in the waste printed circuit boards precipitates the metallic gold from the gold–bromine complex, as shown in Equation (4).

$$Cu + 2AuBr_4^- = Cu^{2+} + 2Au + 8Br^- \tag{4}$$

Therefore, the reaction rate of the gold dissolution can be written in an explicit form shown as Equation (5), and the overall order of the reaction is simply the sum of a, b and c. In this study, given that the concentrations of the reagents are low, it is reasonable to assume that the activity coefficient is equal to 1.

$$r = k \cdot C_{Br2}^{\ a} \cdot C_{NaBr}^{\ b} \cdot C_{Cu}^{\ c} \tag{5}$$

An isolation method was utilized to determine the reaction order for each reactant by isolating each reactant in turn and keeping all other reactants in large excess.

Table 9 illustrates the initial reaction rates of gold dissolution with respect to various concentrations of bromine, copper and sodium bromine at 23.5 °C and 400 RPM. The reaction orders regarding bromine concentration, sodium bromide concentration and copper concentration, respectively, were estimated to be 0.55, 0.16 and −0.41, respectively. The experimental data is displayed in Supplementary Materials (SM-B1, B2, B3 and B4). Therefore, the rate expression can be written as Equation (6).

$$r = k \cdot C_{Br2}{}^{0.55} \cdot C_{NaBr}{}^{0.16} \cdot C_{Cu}{}^{-0.41} \quad (6)$$

Table 9. Gold dissolution rates at various concentrations of Br_2, NaBr and Cu at 23.5 °C and 400 RPM.

Br_2 (mol/L)	NaBr (mol/L)	Cu (mol/L)	Rate (mmol/L·min)
0.388	0	0.156	0.00434
0.776	0	0.156	0.00635
1.94	0	0.156	0.00751
1.94	0	0.261	0.00609
1.94	0.583	0.156	0.01244
1.94	1.166	0.156	0.01386

The boiling point of liquid bromine is around 58.8 °C, so the selected temperatures are 23.5, 34 and 42 °C. The Arrhenius Plot (Supplementary Materials SM-B6) showed that the activation energy was 26.37 KJ/mole (6.3 Kcal/moles). Therefore, by taking 26,365 J/mole for the activity energy, the Equation (6) can be rewritten as follows:

$$r = 0.196 \cdot EXP(-26365/RT) \cdot C_{Br2}{}^{0.55} \cdot C_{NaBr}{}^{0.16} \cdot C_{Cu}{}^{-0.41} \quad (7)$$

The geometry of the shredded printed circuit boards is classified into two categories, spherical particles and flat plates, due to chemical compositions and structures of printed circuit boards. Based on the average thickness of printed circuit boards, minus 149 μm particles were likely to have spherical geometries; whereas the particles larger than 149 μm were considered as being flat. As shown in Figure 1, approximately 15% of the shredded waste printed circuit boards are smaller than 149 μm; while approximately 85% of the boards are larger than 149 μm.

In this study, assuming bromide ions are excessive, silver is completely dissolved as soluble silver bromide complexes. Thus, there is no solid product and the solid A always shrinks, the heterogeneous irreversible reaction can be expressed as follows:

$$aA_{(solid)} + bB_{(aqueous)} = cC_{(aqueous)} \quad (8)$$

There are two assumptions as follows, (1): the concentration of reactant B at the interface of solid A is zero in a diffusion-controlling process; (2): the concentration of reactant B at the interface of solid A is equal to the bulk concentration of B in a chemical-controlling process.

In a quasi-steady state with a constant atmospheric pressure, the kinetic derivation was displayed in Supplementary Materials (SM-C). The fluid film diffusion control and chemical reaction control can be expressed as follows:

Diffusion control for spherical particles

$$t/t_c = 1 - \{0.15(1 - X_{Au})\}^{2/3} \quad (9)$$

Chemical reaction control for spherical particles

$$t/t_c = 1 - \{0.15(1 - X_{Au})\}^{1/3} \quad (10)$$

Diffusion control for flat particles

$$t/t_c = 0.85X_{Au} \quad (11)$$

Chemical reaction control for flat particles

$$t/t_c = 0.85X_{Au} \quad (12)$$

The calculated gold extraction from a selected kinetic test was substituted into Equations (9–12) for each of the possible controlling mechanisms, and the values of t/t_c were plotted as a function of time, as shown in Figure 13.

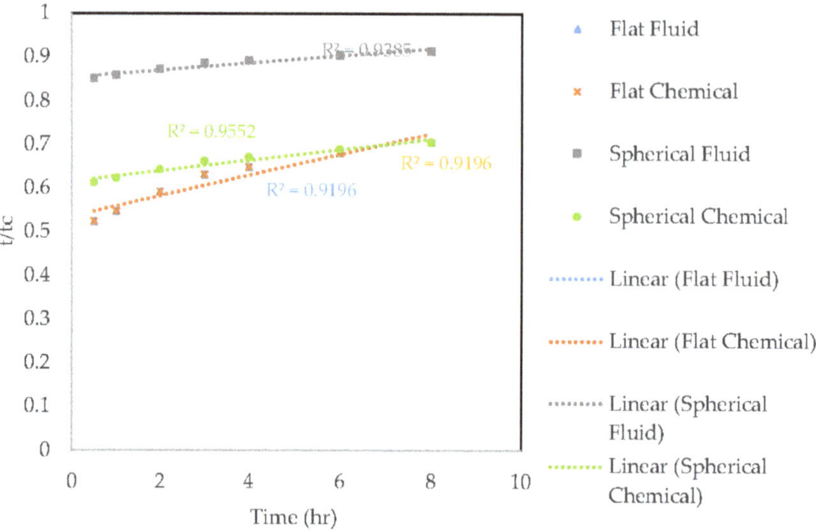

Figure 13. Kinetic calculation for a leaching trial in the presence of 0.776 M Br_2 and 0.166 M NaBr at 23.5 °C and 400 RPM.

R-squared values (Figure 13) for chemical control are relatively higher than fluid control for spherical particles. Moreover, the calculated data shows that the fluid film diffusion controlling is still responsible for the bromide leaching of gold. The mechanism is complex with both a chemical controlling and fluid film diffusion controlling. This could be explained as follows:

1. Waste printed circuit boards are heterogeneous, leading to other metallic elements reacting with bromine and also, possibly reacting with gold ions. Thus, the mechanism of gold dissolution in the bromine system becomes more complex.
2. The diverse presence of gold in the waste printed circuit boards (coating on board surfaces or gold particles in the waste printed circuit boards) also led to the difference of the leaching mechanisms.
3. The metals that are more active than gold may reduce gold from gold–bromine complex ions to gold metal, which may further change the leaching mechanism.
4. There is also a possibility of a mechanism change as the leaching proceeds.
5. In terms of the chemical reaction, three steps may be involved:

    (1) Adsorption of bromine and bromide on the gold surface to form $AuBr_2$ [32],
    (2) Oxidation of $AuBr_2^-$ to produce a stable species, $AuBr_4^-$,
    (3) Copper cementation to reduce $AuBr_4^-$ to Au^0.

Generally, in the current study, gold dissolving in a bromide system is complex, especially in extremely heterogeneous printed circuit boards. By calculating the activation energy (26.37 KJ/mole), the combination of diffusion and chemical controlling are responsible for the gold dissolution.

By assuming that the acceptable R-squared value is 90%, the spherical and flat models demonstrate that the reaction is controlled by chemical reaction and fluid film diffusion. However, additional studies should be performed to eliminate the impurity effect by using a gold disc technique and correlate the existing kinetic model with other factors, such as gold distribution and adsorption.

4. Conclusions

In this study, copper was found to be the most abundant metallic element in waste printed circuit boards, followed by iron and tin. Silver was the primary precious metal present in printed circuit boards at 513 ppm. Silver was liberated, in addition to tin, lead, and nickel. Gold was present at approximately 145 ppm and was usually found liberated or as a gold–nickel alloy.

Given that physical pretreatment was not satisfactory as a result of a significant loss of precious metals, direct bromide leaching was investigated. The thermodynamic calculations illustrated a region where bromine could dissolve gold to form stable tetrabromo gold ions. A series of leaching experiments demonstrate that the concentration of bromine and sodium bromide were the most significant factors influencing bromide leaching. It also appeared that the addition of hydrochloric acid can reduce the usage of sodium bromide. Furthermore, the kinetic results indicate that a combination of chemical and diffusion controlling is responsible for gold dissolution in waste printed circuit boards.

Further investigations will focus on the implementation of a two-step leaching, where the first step will be used to first remove base metals and then, the second step will be to dissolve precious metals by bromine and sodium bromide. Additional studies should be performed in order to eliminate the impurity effect by using a gold disc technique and correlate the existing kinetic with other factors, such as gold distribution and adsorption. Finally, more work needs to be performed to investigate and optimize the metal purification and recovery flowsheet.

Supplementary Materials: The following are available online at http://www.mdpi.com/2075-4701/10/4/462/s1. Figure S1: Eh-pH diagram of Br-Ag-H2O system at 25 °C ([Ag] = 10−4 M, [Br] = 0.775M) (Stabcal). Figure S2: Eh-pH diagram of Br-Pd-H2O system at 25 °C ([Pd] = 10−5 M, [Br] = 0.775M) (Stabcal). Figure S3: Eh-pH diagram of Br-Cu-H2O system at 25 °C ([Cu] = 10−3 M, [Br] = 0.775M) (Stabcal). Figure S4: Effect of bromine concentration on gold dissolution at 23.5 °C and 400 RPM. Figure S5: Reaction orders with bromine during gold dissolution. Figure S6: Effect of sodium bromide on gold dissolution at 23.5 °C and 400 RPM. Figure S7: Effect of copper on gold dissolution at 23.5 °C and 400 RPM. Figure S8: Arrhenius Plot for the gold-bromine leaching system.

Author Contributions: Conceptualization, C.A. and H.C.; methodology, C.A. and H.C.; formal analysis, H.C.; investigation, H.C.; data curation, H.C.; writing—original draft preparation, H.C.; writing—review and editing, C.A.; supervision, C.A. All authors have read and agreed to the published version of the manuscript.

Funding: This research was funded by the Center for Resource, Recovery and Recycling (CR^3).

Conflicts of Interest: The authors declare no conflict of interest.

References

1. Widmer, R.; Oswald-Krapf, H.; Sinha-Khetriwal, D.; Schnellmann., M.; Böni, H. Global perspectives on e-waste. *Environ. Impact Assess. Rev.* **2005**, *25*, 436–458. [CrossRef]
2. Tuncuk, A.; Stazi, V.; Akcil, A.; Yazici, E.Y.; Deveci, H. Aqueous metal recovery techniques from e-scrap: Hydrometallurgy in recycling. *Miner. Eng.* **2012**, *25*, 28–37. [CrossRef]
3. Schluep, M.; Hagelueken, C.; Kuehr, R.; Magalini, F.; Maurer, C.; Meskers, C.; Mueller, E.; Wang, F. *Recycling-from E-Waste to Resources*; Oktoberdruck AG;: Berlin, Germany, 2009; p. 6, United Nationals Environment Programme & United Nations University.
4. Vasile, C.; Brebu, M.A.; Totolin, M.; Yanik, J.; Karayildirim, T.; Darie, H. Feedstock recycling from the printed circuit boards of used computers. *Energy Fuels* **2008**, *22*, 1658–1665. [CrossRef]
5. Hino, T.; Agawa, R.; Moriya, Y.; Nishida, M.; Tsugita, Y.; Araki, T. Techniques to separate metal from waste printed circuit boards from discarded personal computers. *J. Mater. Cycles Waste Manag.* **2009**, *11*, 42–54. [CrossRef]

6. Birloaga, I.; Michelis, I.D.; Ferella, F.; Buzatu, M.; Vegliò, F. Study on the influence of various factors in the hydrometallurgical processing of waste printed circuit boards for copper and gold recovery. *Waste Manag.* **2013**, *33*, 935–941. [CrossRef] [PubMed]
7. Yang, T.; Xu, Z.; Wen, J.; Yang, L. Factors influencing bioleaching copper from waste printed circuit boards by acidithiobacillus ferrooxidants. *Hydrometallurgy* **2009**, *97*, 29–32. [CrossRef]
8. Oishi, T.; Koyama, K.; Alam, S.; Tanaka, M.; Lee., L.C. Recovery of high purity copper cathode from printed circuit boards using ammoniacal sulfate or chloride solutions. *Hydrometallurgy* **2007**, *89*, 82–88. [CrossRef]
9. Behnamfard, A.; Salarirad, M.M.; Veglio, F. Process development for recovery of copper and precious metals from waste printed circuit boards with emphasize on palladium and gold leaching and precipitation. *Waste Manag.* **2013**, *33*, 2354–2363. [CrossRef] [PubMed]
10. Boardsort.com Current Payout Rates-as of 12/15/2017. Available online: http://broadsort.com/payout.php (accessed on 15 December 2017).
11. Gold, InvestmentMine, Commodity and Metal Prices. Available online: http://www.infomine.com/investment/gold/ (accessed on 1 December 2017).
12. Silver, InvestmentMine, Commodity and Metal Prices. Available online: http://www.infomine.com/investment/silver/ (accessed on 1 December 2017).
13. Palladium, InvestmentMine, Commodity and Metal Prices. Available online: http://www.infomine.com/investment/palladium/ (accessed on 1 December 2017).
14. Copper, InvestmentMine, Commodity and Metal Prices. Available online: http://www.infomine.com/investment/copper/ (accessed on 1 December 2017).
15. Tin, InvestmentMine, Commodity and Metal Prices. Available online: http://www.infomine.com/investment/tin/ (accessed on 1 December 2017).
16. Lead, InvestmentMine, Commodity and Metal Prices. Available online: http://www.infomine.com/investment/lead/ (accessed on 1 December 2017).
17. Zinc, InvestmentMine, Commodity and Metal Prices. Available online: http://www.infomine.com/investment/zinc/ (accessed on 1 December 2017).
18. Nickel, InvestmentMine, Commodity and Metal Prices. Available online: http://www.infomine.com/investment/zinc/ (accessed on 1 December 2017).
19. Scrap Iron Prices, SCRAP Metal Pricer. Available online: https://www.scrapmetalpricer.com/ (accessed on 3 December 2017).
20. Cui, H.; Anderson, C. Literature review of hydrometallurgical recycling of printed circuit boards (PCBs). *J. Adv. Chem. Eng.* **2016**, *6*, 142.
21. Yazici, E.Y.; Deveci, H. Ferric sulfate leaching of metals from waste printed circuit boards. *Int. J. Miner. Process.* **2014**, *133*, 39–45. [CrossRef]
22. Yazici, E.Y.; Deveci, H. Cupric chloride leaching (HCl-CuCl2-NaCl) of metals from waste printed circuit boards (WPCBs). *Int. J. Miner. Process.* **2015**, *134*, 89–96. [CrossRef]
23. Zhu, P.; Chen, Y.; Wang, L.Y.; Zhou, M. Treatment of waste printed circuit board by green solvent using ionic liquid. *Waste Manag.* **2012**, *32*, 1914–1918. [CrossRef] [PubMed]
24. Zhang Y. Liu, S.; Xie, H.; Zeng, X.; Li, J. Current status on leaching precious metals from waste printed circuit boards. *Procedia Environ. Sci.* **2012**, *16*, 560–568. [CrossRef]
25. Petter, P.M.H.; Veit, H.M.; Bernardes, A.M. Evaluation of gold and silver leaching from printed circuit board of cellphones. *Waste Manag.* **2014**, *34*, 475–482. [CrossRef] [PubMed]
26. Kelsall, G.H.; Welham, N.J.; Diaz, M.A. Thermodynamics of Cl-H_2O, Br-H_2O, I-H_2O, Au-Cl-H_2O, Au-Br-H_2O and Au-I-H_2O systems at 298K. *J. Electroanal. Chem.* **1993**, *361*, 13–24. [CrossRef]
27. Sahin, M.; Akcil, A.; Erust, C.; Altynbek, S.; Gahan, C.; Tuncuk, A. A potential alternative for precious metal recovery from e-waste: iodine leaching. *Sep. Sci. Technol.* **2015**, *50*, 2587–2595. [CrossRef]
28. Batnasan, A.; Haga, K.; Shibayama, A. Recovery of valuable metals from waste printed circuit boards by using iodine- iodide leaching and precipitation. In *Rare Metal Technology 2018*; Kim., H., Wesstrom, B., Alam, S., Ouchi, T., Azimi, G., Neelameggham, N.R., Wang, S., Guan, X., Eds.; TMS: Pittsburgh, PA, USA, 2018; pp. 131–142.
29. Lyday, P.A. *Bromine, Mineral Yearbook*; U.S. Geological Survey: Reston, VA, USA, 2002.

30. Zhutaeva, G.V.; Shumilova, N.A. Silver. In *Standard Potentials in Aqueous Solution*; Bard, A.J., Parsons, R., Jordan, J., Eds.; International Union of Pure and Applied Chemistry, Marcel Dekker, Inc., CRC Press: Boca Raton, FL, USA, 1985; p. 297.
31. Dadgar, A. Extraction and recovery of gold from concentrates by bromine process. In *Precious Metals '89*; Jha, M.C., Hill, S.D., Eds.; The Minerals, Metals & Materials Society: Pennsylvania, PA, USA, 1988; pp. 227–240.
32. Pesic, B.; Sergent, R.H. Reaction mechanism of gold dissolution with bromine. *Metall. Mater. Trans. B* **1993**, *24*, 419–431. [CrossRef]
33. Apodaca, L.E. *Sulfur, Mineral Commodities*; U.S. Geological Survey: Reston, VA, USA, 2018.
34. Gupta, C.K.; Mukherjee, T.K. Leaching with acid. In *Hydrometallurgy in Extraction Processes*; Gupta, C.K., Mukherjee, T.K., Eds.; CRC Press: Boca Raton, FL, USA, 1990; Volume I, p. 59.
35. Zhang, W.; BAS, A.D.; Ghali, E.; Choi, Y. Passive behavior of gold in sulfuric acid medium. *Trans. Nonferrous Met. Soc. China* **2015**, *25*, 2037–2046. [CrossRef]

Article

Platinum Group Elements Recovery from Used Catalytic Converters by Acidic Fusion and Leaching

Erik Prasetyo [1,*] and Corby Anderson [2]

1 Indonesian Institute of Sciences, Research Unit for Mineral Technology, Jl. Ir. Sutami km. 15 Tanjung Bintang, Lampung Selatan 35361, Indonesia
2 Colorado School of Mines, George Ansell Dept. of Metallurgical and Materials Engineering, Kroll Institute for Extractive Metallurgy, Mining Engineering Department, 1500 Illinois St, Golden, CO 80401, USA; cganders@mines.edu
* Correspondence: erik.prasetyo@lipi.go.id; Tel.: +62-721-350-054

Received: 21 March 2020; Accepted: 1 April 2020; Published: 6 April 2020

Abstract: The recovery of platinum group elements (PGE (platinum group element coating); Pd, Pt, and Rh) from used catalytic converters, using low energy and fewer chemicals, was developed using potassium bisulfate fusion pretreatment, and subsequently leached using hydrochloric acid. In the fusion pre-treatment, potassium bisulfate alone (without the addition of an oxidant) proved to be an effective and selective fusing agent. It altered PGE into a more soluble species and did not react with the cordierite support, based on X-Ray Diffraction (XRD) and metallographic characterization results. The fusion efficacy was due to the transformation of bisulfate into pyrosulfate, which is capable of oxidizing PGE. However, the introduction of potassium through the fusing agent proved to be detrimental, in general, since potassium formed insoluble potassium PGE chloro-complexes during leaching (decreasing the recovery) and required higher HCl concentration and a higher leaching temperature to restore the solubility. Optimization on the fusion and leaching parameter resulted in 106% ± 1.7%, 93.3% ± 0.6%, and 94.3% ± 3.9% recovery for Pd, Pt, and Rh, respectively. These results were achieved at fusion conditions: temperature 550 °C, potassium bisulfate/raw material mass ratio 2.5, and fusion time within 30 min. The leaching conditions were: HCl concentration 5 M, temperature 80 °C, and time within 20 min.

Keywords: platinum group elements; catalytic converters; acidic fusion; acidic leaching; sulfation

---

1. Introduction

Platinum group elements (PGE) have found their use in numerous applications including catalysts, electronic components, jewelry, chemicals, and drugs [1]. Due to their essential role in future modern and green technology, and the fact that the natural resources of these elements exist only in a handful of locations (e.g., South Africa, Russia, and Canada [2]), these elements are categorized as critical elements [3]. These reasons further encourage the effort to explore new research (i.e., secondary resources through recycling processes). Aside from maintaining a steady supply, the processing of secondary resources would reduce the pressure to the environment.

One potential secondary resource for further development is catalytic converters. The majority of PGE (i.e., Pd, Pt, and Rh) would be used as catalysts in vehicles to convert hazardous gases, formed as a result of combustion, into less toxic gases, such as carbon monoxide into carbon dioxide. The content of PGE in catalytic converters varies [4] and could reach 0.5% [5], while the rest would be support, consisting of silicate (e.g., cordierite) and oxide (e.g., alumina, titanium dioxide, and silica).

Several processes had been proposed in order to extract PGE in catalytic converters, which could be grouped into a high temperature process (pyrometallurgy), a low temperature process by leaching (hydrometallurgy), or a combination of the two processes. In pyrometallurgical approaches, the most notable one is metal smelting collection and volatilization [6–11]. In the smelting method, the raw materials containing PGE would be mixed with fluxes, a reductor, and a collector and further heated until reaching the melting point. In this process, PGE would be reduced into metals and incorporated into the collector phase (e.g., Pb, Cu, and Fe). The PGE could be further separated from the collector using an electrochemistry method, while the flux would bind to the gangue phase, such as silicate. The advantages of the smelting method are the simple process of separating PGE from the gangue phase and the high recovery of PGE, with somewhat lower Rh recovery [5]. The other pyrometallurgical approach is volatilization, which transforms the PGE into chloride compounds at a high temperature and recovers as vapor/volatile compounds [12]. In general, the pyrometallurgy approach requires a considerably large energy input and may involve toxic and corrosive gases and compounds (especially chlorine and lead oxide).

In the hydrometallurgy approach, PGE were recovered by a dissolution process using a strong acid (e.g., hydrochloric acid, nitric acid, and sulfuric acid) or a combination of two or more of these acids (e.g., aqua regia), or by the addition of oxidants into these acids (e.g., iodine, bromine, chlorine, and hydrogen peroxide) [13–17], or using a complexant (e.g., cyanide) [18,19]. Aside from the direct leaching of PGE, another approach in hydrometallurgy includes upgrading the PGE in raw materials by leaching out the supporting materials. Mishra (1987) [20] successfully removed the alumina support by exploiting the amphoteric nature of alumina, which could be leached using sulfuric acid. Although hydrometallurgy could effectively deal with low grade raw materials, the consumption of hazardous and highly corrosive chemicals is substantial, due to the relatively stable nature of PGE.

Apart from the pyrometallurgical approach and the hydrometallurgical approach, there is another approach: the fusion method. This basic approach is to transform the insoluble phase into a soluble phase through reaction with an additive (fusing agent) during moderate heat treatment (in general, less than 800 °C), meaning a lower energy input in comparison to pyrometallurgy. The soluble phase produced could be leached in relatively mild conditions (less hazardous chemicals used compared to hydrometallurgy). Generally, the fusion method was applied for PGE upgrading (i.e., matrix decomposition) using alkalies (e.g., sodium hydroxide) [21] or an acidic fusing agent (e.g., potassium bisulfate). Subsequent leaching would remove the matrix and leave the PGE as insoluble residue for further processing stages.

Fusion using bisulfate has attracted attention from several research groups [22–25]. Batista and Afonso [22] intended to upgrade the PGE from a used catalyst by matrix (alumina) decomposition, believing that Pt would not react with bisulfate during fusion and would be accumulated in the residue. The other researchers mentioned, with the same confidence, that bisulfate alone would be ineffective to transform Pt into the soluble phase, so they combined the bisulfate with a strong oxidant (e.g., perchlorates) during fusion to make the transformation applicable. However, none of the studies cited above clearly defined the interaction between fusing agents and PGE or the interaction between fusing agents (i.e., bisulfate and chlorate).

This belief was probably rooted in the fact that PGE (especially Pt) are some of the most inert metals, and Pt is extensively used as a material of apparatus in highly corrosive conditions. For example, the standard procedure of titanium oxide decomposition using potassium pyrosulfate is carried out in a platinum crucible [26]. However, several researchers have documented that Pt could be attacked by a sulfoxide compound at an elevated temperature to form Pt sulfate salt. For example, $Pt_2(HSO_4)_2(SO_4)_2$ was produced by reacting Pt metal and sulfuric acid at 350 °C [27], and $[Pt(S_2O_7)_3]^{2-}$ was produced by reacting Pt metal with oleum (65% SO_3) at 160 °C [28]. In the case of Rh, researchers have successfully produced $Rh_2(SO_4)_3{\cdot}2H_2O$ by reacting Rh metal with sulfuric acid at 465 °C [29].

Our preliminary observation showed that potassium bisulfate alone would effectively alter Pt, Pd, and Rh (Figure 1). Based on observation on the color of the compound produced, the compounds formed were assumed to be PGE-sulfates, which were amenable for further leaching using dilute hydrochloric acid. The hydrochloric acid was chosen in order to stabilize the PGE in an aqueous solution as a chloro-complex. In our studies, the hypotheses we proposed were the transformation of potassium bisulfate into potassium pyrosulfate ($K_2S_2O_7$) during thermal decomposition and, further, the pyrosulfate as a strong oxidant would react to PGE to form PGE-sulfate, according to this reaction:

$$2KHSO_4 \rightarrow K_2S_2O_7 + H_2O \quad (1)$$

$$Pd + 2K_2S_2O_7 \rightarrow PdSO_4 + K_2SO_4 + SO_{2\,(g)} \quad (2)$$

$$Pt + 4K_2S_2O_7 \rightarrow Pt(SO_4)_2 + 4K_2SO_4 + 2SO_{2\,(g)} \quad (3)$$

$$2Rh + 6K_2S_2O_7 \rightarrow Rh_2(SO_4)_3 + 6K_2SO_4 + 3SO_{2\,(g)} \quad (4)$$

Figure 1. PGE salts from left (Pd, Pt, and Rh) produced by reacting PGE metal powder with potassium bisulfate in a muffle furnace at 550 °C for 3 h.

Based on the above hypotheses, the feasibility of recovering PGE from catalytic converters, using potassium bisulfate as the sole fusing agent, followed by hydrochloric acid leaching, would be tested. The parameters investigated included: fusion temperature, mass ratio between raw materials and fusing agents, fusion duration, and leaching parameters (pulp density, HCl concentration, leaching time, and temperature). The characterizations would also be carried out to confirm the fusion and leaching efficacy using X-Ray Diffraction and metallographic observation.

2. Method, Material, and Instrumentation

2.1. Material and Instrumentation

A used catalytic converter sample was obtained from a local scrapyard in Sapporo, Japan. The honeycomb-structured converter was then ball milled and sized with a 270 mesh (53 µm) screen that was used in fusion and leaching studies. Potassium bisulfate, hydrochloric acid, nitric acid, sulfuric acid, and sodium hydroxide were obtained from Merck, Darmstadt, Germany all in analytical grade. Deionized water (MilliQ) was used throughout the experiment.

The leaching experiment was carried out in the Research Unit for Mineral Technology of Indonesian Institute of Sciences, Lampung Selatan, Indonesia. Converter sample total decomposition was carried out using alkali fusion and subsequent HCl-HNO_3-H_2SO_4 digestion, followed by measurement using ICP-OES (Analytik Jena, Plasma Quant 9000 Elite, Jena, Germany) to determine Pd, Pt, and Rh content in the sample after the decomposition (Table 1) or in the liquid phase after the leaching test was completed, in order to calculate the recovery (R, %) according to Equation (5). Characterization on materials both before and after treatment was conducted using XRD (Panalytical, Expert3 Powder, Malvern, UK) in Research Unit for Mineral Technology, Indonesian Institute of Sciences. Metallographic microscopy (Reichert MEF 4M microscope equipped with AxioCam MRc5 camera from Carl-Zeiss, Oberkochen, Germany) was conducted in Dept. of Metallurgical and Materials Engineering, Colorado School of Mines.

$$R = \frac{C_E \times V}{C_o \times m} \times 100\% \quad (5)$$

where:

C_E PGE concentration in liquid phase after fusion and leaching (μg/mL)
C_o PGE content in catalytic converters (μg/g)
m mass of catalytic converters used in leaching (g)
V leaching agent volume (mL)

Table 1. Chemical composition of PGE in used catalytic converters.

Elements	Pd	Pt	Rh
Content (μg/g)	112.0 ± 1.7	1299.4 ± 3.5	316.0 ± 6.6

2.2. Method

Acidic fusion-leaching studies were carried out by a batch method. Typically, 0.5 g of catalytic converter powder was mixed with potassium bisulfate with certain mass ratio in a 30 mL porcelain crucible. The mixture was then introduced into a muffle furnace with a temperature higher than 350 °C. After the fusion was completed, the solid product was put into a conical flask, hydrochloric acid was added, and the mixture was homogenized using an orbital shaker (speed 200 rpm). After leaching was concluded, the supernatant solution was separated by centrifugation and filtration, and the PGE content was determined using ICP-OES. All fusion and leaching data was obtained in duplicates.

3. Results and Discussion

3.1. Characterization Results

Characterizations were conducted on materials before and after treatment, and they were carried out using XRD combined with a metallographic analysis, in order to confirm the efficacy of the fusion and leaching process. The XRD results (Figure 2) on powdered raw materials before and after fusion (fusion temperature 550 °C, $KHSO_4$/raw material mass ratio 2, and fusion time 3 h) show that the cordierite matrix of catalytic converters was partially affected by the fusion process to form potassium alum ($KAl(SO_4)_2 \cdot 12H_2O$). However, based on the metallographic observation of catalytic converters, both before and after fusion, and subsequent leaching using hydrochloric acid 5 M, the cordierite matrix was unaffected, while the PGE coating was clearly washed out (Figure 3). This confirms the efficacy of potassium bisulfate as a fusing agent and its capability as the sole oxidant in transforming PGE into soluble species.

3.2. Acidic Fusion Using $KHSO_4$

To study the effect of fusion on PGE recovery, the leaching parameters were set constant i.e., hydrochloric concentration 5 M, pulp density 20 mL/g, leaching time 6 h, and leaching temperature 30 °C.

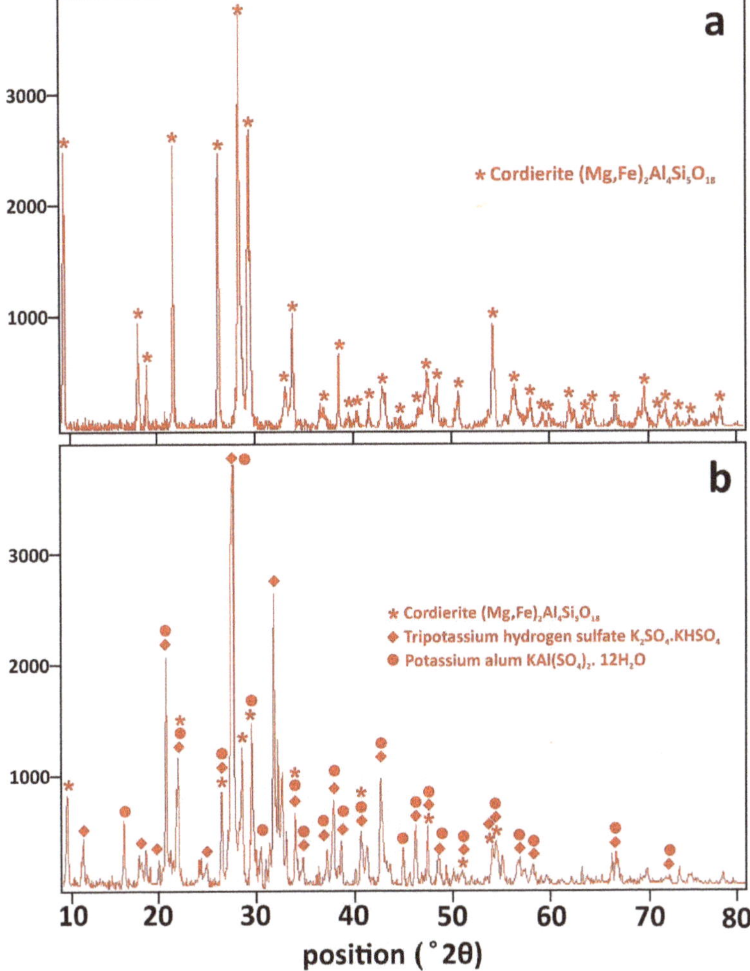

Figure 2. X-Ray powder diffraction of raw materials (**a**) before and (**b**) after fusion with potassium bisulfate.

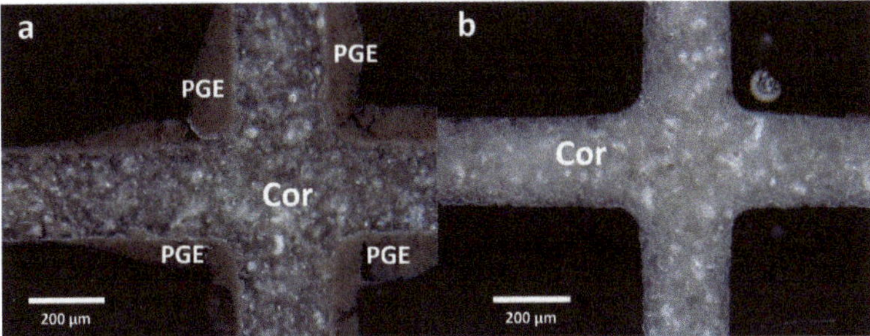

Figure 3. Metallographic observation on catalytic converters mounted on resin (**a**) before and (**b**) after fusion using potassium bisulfate, followed by leaching using hydrochloric acid 5 M. (Cor: cordierite matrix; PGE: platinum group element coating).

3.2.1. Effect of Fusion Temperature

The effect of fusion temperature was studied between 350 and 750 °C, with a constant fusion variable: mass ratio ($KHSO_4$/catalytic converters powder) 2 and fusion duration 3 h. The results were depicted in Figure 4, which showed that the recovery of PGE increased as temperature increased, until it reached the optimum fusion temperature at 550 °C (recovery Pd 92%, Pt 50%, and Rh 78%), and, then, decreased (Pd and Rh) or became relatively constant (Pt). The increasing recovery up to 550 °C, then decreasing recovery, indicates the advantage of thermal decomposition of $KHSO_4$ into $K_2S_2O_7$, which oxidized the PGE present in the raw material. A higher fusion temperature caused further decomposition of potassium pyrosulfate into potassium sulfate (T > 600 °C) [30], which has less oxidative power than pyrosulfate. The XRD data confirmed the transformation of potassium bisulfate into potassium pyrosulfate during fusion up to 650 °C and partial transformation into potassium sulfate at higher temperatures, (Supplementary Materials).

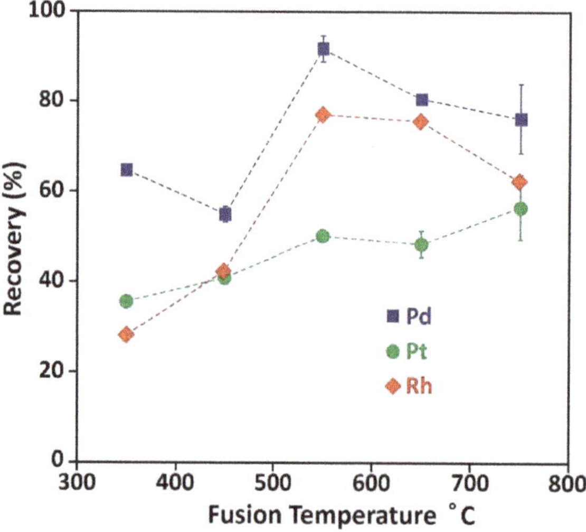

Figure 4. The recovery of PGE related to fusion temperature.

3.2.2. Effect of Mass Ratio

To assess the effect of the amount of fusion agent ($KHSO_4$) added relative to raw materials on the PGE recovery, the fusion was carried out with varied $KHSO_4$/catalytic converters mass ratios from 0.5 to 3. The constant fusion variables were temperature (550 °C) and fusion duration (3 h). The results are depicted in Figure 5. The figure shows the recovery increased as mass ratio increased until an optimum value of 2–2.5 (Pd 93%, Pt 76.5%, and Rh 77.6%), then the recovery decreased as mass ratio increased further. The decreasing recovery at higher mass ratio was probably caused by precipitation of PGE in the form of potassium salt of chloro-complexes (e.g., K_2PtCl_6 [31]) during leaching, due to an excess of potassium ion introduced during the fusion at higher mass ratios (reactions 6–7).

$$Pt(SO_4)_2 + 6HCl \rightarrow H_2PtCl_6 + 2H_2SO_4 \quad (6)$$

$$K_2SO_4 + Pt(SO_4)_2 + 6HCl \rightarrow K_2PtCl_6 \downarrow + 3H_2SO_4 \quad (7)$$

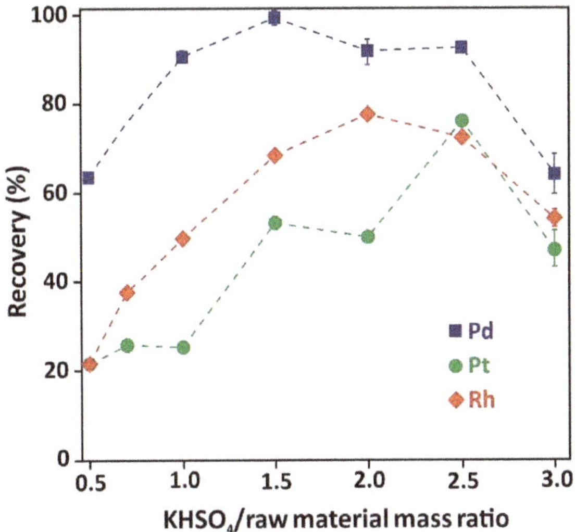

Figure 5. The effect of fusion agent/catalytic converters mass ratios to PGE recovery.

3.2.3. Effect of Fusion Time

To evaluate the effect of fusion time to the PGE recovery, the fusion was conducted at a constant variable $KHSO_4$/raw material mass ratio 2.5 and a fusion temperature of 550 °C, while the fusion time varied between 5 and 240 min. The results shown in Figure 6 demonstrated that the saturation value of PGE recovery could be attained within 30 min of fusion (Pd 98.9%, Pt 63.2%, and Rh 71.3%). Longer fusion time did not significantly increase the recovery.

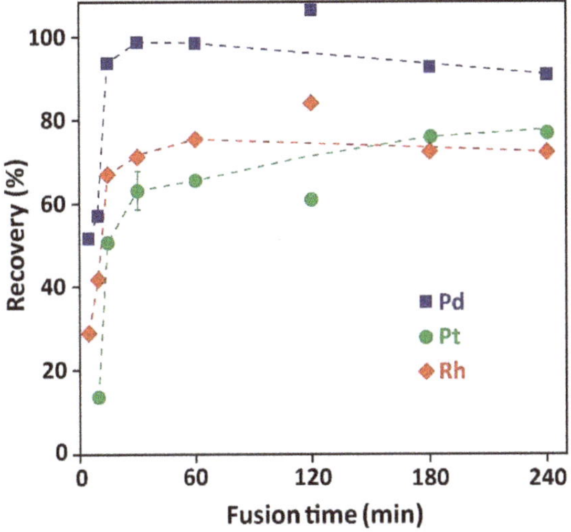

Figure 6. The effect of fusion time on PGE recovery.

3.3. *Acidic Leaching using Hydrochloric Acid after $KHSO_4$ Fusion*

The effect of leaching parameters was studied at constant/optimum fusion parameters (temperature 550 °C, fusion agent/raw material mass ratio 2.5, and fusion time 30 min). Leaching parameters studied included pulp density, HCl concentration, temperature, and time.

3.3.1. Effect of Leaching Pulp Density

The effect of pulp density in the leaching stage was studied in the range of 5 to 25 mL/g. The pulp density (volume/mass) in this case was the ratio of lixiviant added to the mass of solid phase/fusion product. The constant leaching variables were HCl concentration 5 M, temperature 30 °C, and duration 6 h. The results in Figure 7 show the optimum recovery was attained at pulp density 15–20 mL/g.

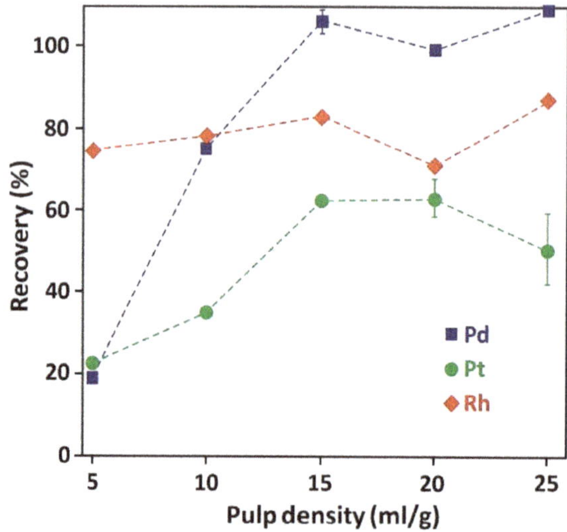

Figure 7. The effect of leaching parameter pulp density on PGE recovery.

3.3.2. Effect of Hydrochloric Acid Concentration

To study the effect of hydrochloric acid as a lixiviant, the leaching temperature, leaching time, and pulp density were fixed constant at 30 °C, 6 h, and 20 mL/g, respectively. Hydrochloric acid concentration as an independent variable was set from 0 to 5 M. The results depicted in Figure 8 show the recovery of Pd and Rh at control conditions (HCl concentration 0/leaching with deionized water) were quite significant (Pd 56.4% and Rh 73.6%), while Pt recovery was only 12%. These results also confirmed the efficacy of potassium bisulfate as the sole fusing agent to transform PGE into a soluble PGE compound. In general, excluding Rh, the PGE recovery increased as the hydrochloric acid concentration increased; this was caused by the stabilization of PGE as chloride complexes in pregnant leach solution. The Rh recovery showed a different trend; the recovery decreased as the hydrochloric acid concentration increased. This was possibly due to the formation of a less soluble Rh chloride complex (K_2RhCl_5) in a higher chloride concentration compared to Rh sulfate [32], Reaction (8). In the case of Pd, the increasing recovery at the higher hydrochloric acid concentration was due to the stabilization of Pd as a chloride complex.

$$Rh_2(SO_4)_3 + 2K_2SO_4 + 10HCl \rightarrow 2K_2RhCl_5 + 5H_2SO_4 \quad (8)$$

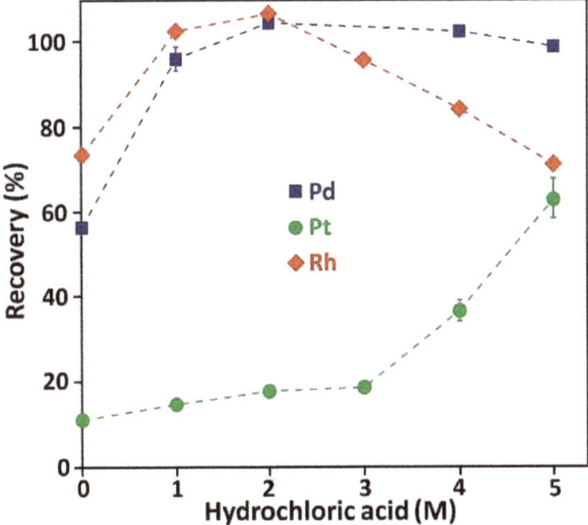

Figure 8. The effect of hydrochloric acid concentration as a lixiviant on the PGE recovery.

Efficient leaching of Pt was only attained at a high concentration of hydrochloric acid. The concentration of potassium seemed to play a major role in the stabilization of the Pt complex in the liquid phase. At lower HCl concentrations, Pt would tend to exist as the potassium complex K_2PtCl_6 (9), which has a very low solubility. As the HCl concentration increased, the Pt complex would be transformed into the highly soluble species H_2PtCl_6 (10), since potassium concentration in the liquid phase would decrease, due to the formation of $KHSO_4$ (11), which has a very low solubility.

$$Pt(SO_4)_2 + K_2SO_4 + 6HCl \rightarrow K_2PtCl_6 \downarrow + 3H_2SO_4 \quad (9)$$

$$K_2PtCl_6 + 2HCl \rightarrow H_2PtCl_6 + 2KCl \quad (10)$$

$$K_2SO_4 + H^+ \rightarrow KHSO_4 \downarrow + K^+ \quad (11)$$

Figure 9 represents the recovery of PGE based on two leaching variables: HCl molarity (1, 2 and 5 M) and leaching temperature (30, 50 and 80 °C), with constant variable pulp density and leaching time (20 mL/g and 1 h, respectively). The temperature clearly increased the recovery of PGE, which was due to increasing solubility of PGE complex in liquid phase during acidic leaching. This was pronounced in the case of Pt.

3.3.3. Effect of Leaching Time

The effect of leaching time (0–60 min) was studied at various temperatures (30, 50, and 80 °C) while hydrochloric acid concentrations and pulp density were set constant at 5 M and 20 mL/g, respectively (Figure 10). It is demonstrated that the leaching reaction occurred fast, and the maximum recovery was attained within 20 min.

Figure 10 also shows that the saturation value of recovery increased, as a result of increasing the leaching temperature. In the case of Pt, the leaching efficiency reached 99.9% ± 2.8% within 5 min at 80 °C, while, for Rh, the recovery increased from 87.2% ± 1.4% (50 °C) to 101.4% ± 6.8% (80 °C). For Pd, the effect of increasing temperature was less significant than it was for Pt and Rh.

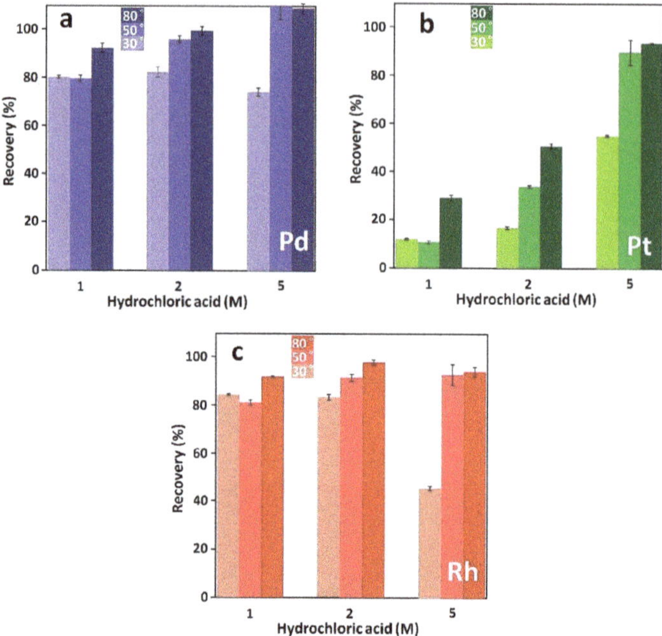

Figure 9. PGE recovery as a function of different hydrochloric acid concentrations and leaching temperatures (**a**) Pd, (**b**) Pt, and (**c**) Rh.

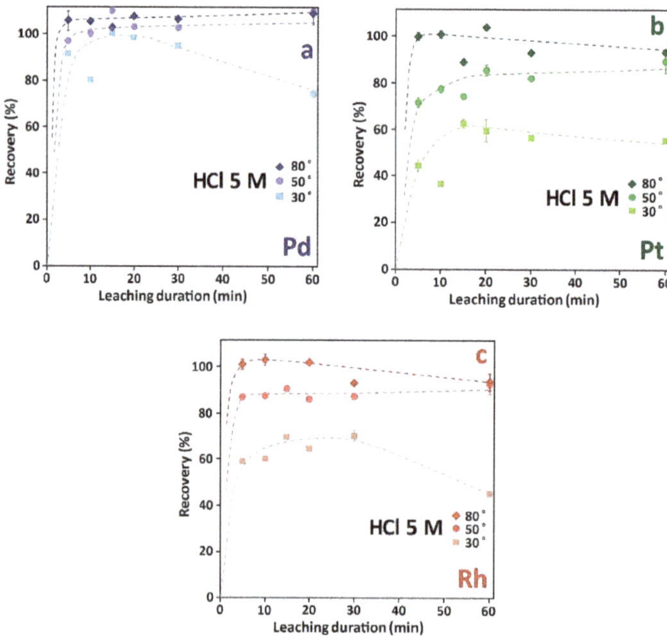

Figure 10. The effect of leaching duration and temperature on PGE recovery (**a**) Pd, (**b**) Pt, and (**c**) Rh at a constant hydrochloric acid concentration.

4. Conclusions

It was demonstrated that potassium bisulfate as the sole fusing agent effectively transformed PGE in catalytic converters into species which were amenable for mild condition leaching processes (HCl < 5 M, T < 80 °C). The advantages of using potassium bisulfate for PGE recovery were lower energy, lower chemical consumption, and selectivity, since potassium bisulfate only reacted with PGE (it did not react with cordierite). This was confirmed by XRD and metallography characterization results.

Generally, all three PGE showed the same trend according to all fusion parameters. The optimum recovery of PGE was achieved at a fusion temperature of 550 °C, a fusion agent-catalytic converters mass ratio of 2.5, and a fusion time within 30 min. In the case of leaching, Pd and Pt showed the same trend, which was that the increase of leaching temperature and HCl concentration would be beneficial. In this case, the recovery of Pd and Pt were 106% ± 1.7% and 93.3% ± 0.6%, respectively. On the other hand, Rh exhibited a different trend, which was that when leaching at room temperature, the increase of HCl concentration led to a recovery decline. However, at higher leaching temperatures, the trend followed the other PGE. At the optimum conditions (temperature 80 °C and HCl 5 M), the Rh recovery reached 94.3% ± 3.9%. A study on the effect of leaching time generally demonstrated that maximum recovery was attained within 20 min.

Supplementary Materials: The following are available online at http://www.mdpi.com/2075-4701/10/4/485/s1, Figure S1: XRD profile obtained from heated KHSO4 powder at 350, 450, 550, 650 and 750 °C.

Author Contributions: E.P. contributed in conceptualization, methodology, funding acquisition, investigation, data analysis and manuscript writing and review. C.A. contributed in conceptualization, funding acquisition, resources and manuscript review. All authors have read and agreed to the published version of the manuscript.

Funding: This research was funded and supported by the Indonesian Institute of Sciences through the "Penelitian Mandiri" scheme FY 2019 and the Fulbright Program through Indonesian Visiting Scholar 2019.

Conflicts of Interest: The authors declare no conflict of interest.

References

1. Mudd, G.M.; Jowitt, S.M.; Werner, T.T. Global platinum group element resources, reserves and mining—A critical assessment. *Sci. Total Environ.* **2018**, *622*, 614–625. [CrossRef] [PubMed]
2. Loferski, P.C.; Ghalayini, Z.T.; Singerling, S.A. *Platinum-Group Metals—2016*; ADVANCE RELEASE; US Geological Survey: Reston, VA, USA, 2016.
3. Dodson, J.R.; Hunt, A.J.; Parker, H.L.; Yang, Y.; Clark, J.H. Elemental sustainability: Towards the total recovery of scarce metals. *Chem. Eng. Process. Process. Intensif.* **2012**, *51*, 69–78. [CrossRef]
4. Saguru, C.; Ndlovu, S.; Moropeng, D. A review of recent studies into hydrometallurgical methods for recovering PGMs from used catalytic converters. *Hydrometallurgy* **2018**, *182*, 44–56. [CrossRef]
5. Peng, Z.; Li, Z.; Lin, X.; Tang, H.; Ye, L.; Ma, Y.; Rao, M.; Zhang, Y.; Li, G.; Jiang, T. Pyrometallurgical recovery of platinum group metals from spent catalysts. *JOM* **2017**, *69*, 1553–1562. [CrossRef]
6. Zhang, L.; Song, Q.; Liu, Y.; Xu, Z. Novel approach for recovery of palladium in spent catalyst from automobile by a capture technology of eutectic copper. *J. Clean. Prod.* **2019**, *239*, 118093. [CrossRef]
7. Benson, M.; Bennett, C.R.; Harry, J.E.; Patel, M.K.; Cross, M. The recovery mechanism of platinum group metals from catalytic converters in spent automotive exhaust systems. *Resour. Conserv. Recycl.* **2000**, *31*, 1–7. [CrossRef]
8. Kolliopoulos, G.; Balomenos, E.; Giannopoulou, I.; Yakoumis, I.; Panias, D. Behavior of platinum group during their pyrometallurgical recovery from spent automotive catalysts. *OALib* **2014**, *1*, 1–9. [CrossRef]
9. Fujima, K.; Morimoto, T.; Itoigawa, H.; Becchaku, D. Development of a process for recycling platinum group metals using molten alkali metal chlorides. *J. Jpn. Inst. Met. Mater.* **2017**, *81*, 168–177. [CrossRef]
10. Yoshimura, A.; Matsuno, Y. A fundamental study of platinum recovery from spent auto catalyst using dry Aqua Regia. *J. Jpn. Inst. Met. Mater.* **2019**, *83*, 23–29. [CrossRef]
11. Ding, Y.; Zheng, H.; Zhang, S.; Liu, B.; Wu, B.; Jian, Z. Highly efficient recovery of platinum, palladium, and rhodium from spent automotive catalysts via iron melting collection. *Resour. Conserv. Recycl.* **2020**, *155*, 104644. [CrossRef]

12. Kim, C.H.; Woo, S.I.; Jeon, S.H. Recovery of platinum-group metals from recycled automotive catalytic converters by carbochlorination. *Ind. Eng. Chem. Res.* **2000**, *39*, 1185–1192. [CrossRef]
13. Marinho, R.S.; da Silva, C.N.; Afonso, J.C.; da Cunha, J.W.S.D. Recovery of platinum, tin and indium from spent catalysts in chloride medium using strong basic anion exchange resins. *J. Hazard. Mater.* **2011**, *192*, 1155–1160. [CrossRef] [PubMed]
14. Harjanto, S.; Cao, Y.; Shibayama, A.; Naitoh, I.; Nanami, T.; Kasahara, K.; Okumura, Y.; Liu, K.; Fujita, T. Leaching of Pt, Pd and Rh from automotive catalyst residue in various chloride based solutions. *Mater. Trans.* **2006**, *47*, 129–135. [CrossRef]
15. Upadhyay, A.K.; Lee, J.c.; Kim, E.y.; Kim, M.s.; Kim, B.S.; Kumar, V. Leaching of platinum group metals (PGMs) from spent automotive catalyst using electro-generated chlorine in HCl solution. *J. Chem. Technol. Biotechnol.* **2013**, *88*, 1991–1999. [CrossRef]
16. Kim, M.S.; Park, S.W.; Lee, J.C.; Choubey, P.K. A novel zero emission concept for electrogenerated chlorine leaching and its application to extraction of platinum group metals from spent automotive catalyst. *Hydrometallurgy* **2016**, *159*, 19–27. [CrossRef]
17. Nogueira, C.A.; Paiva, A.P.; Oliveira, P.C.; Costa, M.C.; da Costa, A.M.R. Oxidative leaching process with cupric ion in hydrochloric acid media for recovery of Pd and Rh from spent catalytic converters. *J. Hazard. Mater.* **2014**, *278*, 82–90. [CrossRef]
18. Chen, J.; Huang, K. A new technique for extraction of platinum group metals by pressure cyanidation. *Hydrometallurgy* **2006**, *82*, 164–171. [CrossRef]
19. Shams, K.; Beiggy, M.R.; Shirazi, A.G. Platinum recovery from a spent industrial dehydrogenation catalyst using cyanide leaching followed by ion exchange. *Appl. Catal. A Gen.* **2004**, *258*, 227–234. [CrossRef]
20. Mishra, R.K. PGM recoveries by atmospheric and autoclave leaching of alumina bead catalyst. *Precious Met.* **1987**, *1987*, 177–195.
21. Yang, M.C.; Sun, E.T.; Zhou, Y.J.; He, X.K. A new process for recovery of Pt and Al from Pt-containing waste catalyst. *Precious Met.* **1996**, *17*, 20–24.
22. Batista, S.G.; Afonso, J.C. Processing of spent platinum-based catalysts via fusion with potassium hydrogenosulfate. *J. Hazard. Mater.* **2010**, *184*, 717–723. [CrossRef] [PubMed]
23. Chen, A.; Wang, S.; Zhang, L.; Peng, J. Optimization of the microwave roasting extraction of palladium and rhodium from spent automobile catalysts using response surface analysis. *Int. J. Miner. Process.* **2015**, *143*, 18–24. [CrossRef]
24. Wang, S.; Chen, A.; Zhang, Z.; Peng, J. Leaching of palladium and rhodium from spent automobile catalysts by microwave roasting. *Environ. Prog. Sustain. Energy* **2014**, *33*, 913–917. [CrossRef]
25. Spooren, J.; Atia, T.A. Combined microwave assisted roasting and leaching to recover platinum group metals from spent automotive catalysts. *Miner. Eng.* **2020**, *146*, 106153. [CrossRef]
26. Gaines, P. Samples Containing Titanium. 2020. Available online: https://www.inorganicventures.com/sample-preparation-guide/samples-containing-titanium (accessed on 9 March 2020).
27. Pley, M.; Wickleder, M.S. $Pt_2(HSO_4)_2(SO_4)_2$, the first binary sulfate of platinum. *J. Inorg. Normal Chem.* **2004**, *630*, 1036–1039.
28. Bruns, J.; Klüner, T.; Wickleder, M.S. Oxidizing elemental platinum with oleum under harsh conditions: The unique tris(disulfato)platinate(IV) $[Pt(S_2O_7)_3]$ 2-anion. *Chem. A Eur. J.* **2014**, *20*, 7222–7227. [CrossRef]
29. Schwarzer, S.; Betke, A.; Logemann, C.; Wickleder, M.S. Oxidizing Rhodium with Sulfuric Acid: The Sulfates $Rh_2(SO_4)_3$ and $Rh_2(SO_4)_3{\cdot}2H_2O$. *Eur. J. Inorg. Chem.* **2017**, *2017*, 752–758. [CrossRef]
30. Hinds, J.I.D. *Inorganic Chemistry: With the Elements of Physical and Theoretical Chemistry*; John Wiley & Sons: New York, NY, USA, 1908.
31. Smith, G.F.; Gring, J.L. The Separation and Determination of the Alkali Metals Using Perchloric Acid. V. Perchloric Acid and Chloroplatinic Acid in the Determination of Small Amounts of Potassium in the Presence of Large Amounts of Sodium. *J. Am. Chem. Soc.* **1933**, *55*, 3957–3961. [CrossRef]
32. Williams, M.L. CRC Handbook of Chemistry and Physics, 76th edition. *Occup. Environ. Med.* **1996**, *53*, 504. [CrossRef]

Correction

Correction: Torres, D. et al. Leaching Manganese Nodules in an Acid Medium and Room Temperature Comparing the Use of Different Fe Reducing Agents. *Metals* 2019, *9*, 1316

David Torres [1], Luís Ayala [2], Manuel Saldaña [2], Manuel Cánovas [3], Ricardo I. Jeldres [4], Steven Nieto [4], Jonathan Castillo [5], Pedro Robles [6] and Norman Toro [2,3,*]

1 Department of Mining, Geological and Cartographic Department, Universidad Politécnica de Cartagena, Paseo Alfonso Xlll N°52, 30203 Cartagena, Spain; david.Torres@sqm.com
2 Faculty of Engineering and Architecture, Universidad Arturo Pratt, Almirante Juan José Latorre 2901, Antofagasta 1244260, Chile; luisayala01@unap.cl (L.A.); manuel.saldana@ucn.cl (M.S.)
3 Departamento de Ingeniería Metalúrgica y Minas, Universidad Católica del Norte, Av. Angamos 610, Antofagasta 1270709, Chile; manuel.canovas@ucn.cl
4 Departamento de Ingeniería Química y Procesos de Minerales, Universidad de Antofagasta, Antofagasta 1240000, Chile; ricardo.jeldres@uantof.cl (R.I.J.); stevennietomejia@gmail.com (S.N.)
5 Departamento de Ingeniería en Metalurgia, Universidad de Atacama, Av. Copayapu 485, Copiapó 1531772, Chile; jonathan.castillo@uda.cl
6 Escuela de Ingeniería Química, Pontificia Universidad Católica de Valparaíso, Valparaíso 2340000, Chile; pedro.robles@pucv.cl
* Correspondence: ntoro@ucn.cl; Tel.: +56-552651021

Received: 8 April 2020; Accepted: 8 April 2020; Published: 14 April 2020

The authors wish to make the following corrections to this paper [1]:

We worked at a temperature of 60 °C for all the tests carried out in which FeS_2 was added as a reducing agent, whereas in the other experiments, other added Fe reducing agents were worked at room temperature (25 °C). This affects the results presented in Table 5 (results when working with FeS2), Figures 4a and 6a. By accident and through carelessness, we did not indicate this important detail in the work methodology. For this reason, we must correct it for the readers, otherwise the reproduction of the results of our experiments will not be possible due to incorrect working parameters. However, we confirm that this error does not affect the conclusions of the manuscript.

We must indicate that it is unlikely that the following series of reactions that were presented in the document could occur at room temperature:

$$3FeS_2 + 4H_2SO_4 = 3FeSO_4 + 4H_2O + 7S \quad (1)$$

$$6FeSO_4 + 4H_2SO_4 = 3Fe_2(SO_4)_3 + 4H_2O + S \quad (2)$$

$$2FeS_2 + 4H_2SO_4 = Fe_2(SO_4)_3 + 4H_2O + 5S \quad (3)$$

$$15MnO_2 + 2FeS_2 + 14H_2SO_4 = Fe_2(SO_4)_3 + 15MnSO_4 + 14H_2O \quad (4)$$

We will update the article and the original version will remain available on the article webpage.

References

1. Torres, D.; Ayala, L.; Saldaña, M.; Cánovas, M.; Jeldres, R.I.; Nieto, S.; Castillo, J.; Robles, P.; Toro, N. Leaching manganese nodules in an acid medium and room temperature comparing the use of different Fe reducing agents. *Metals* **2019**, *9*, 1316. [CrossRef]

Article

An Alternative Process for Leaching Chalcopyrite Concentrate in Nitrate-Acid-Seawater Media with Oxidant Recovery

César I. Castellón [1], Pía C. Hernández [1], Lilian Velásquez-Yévenes [2] and María E. Taboada [1,*]

[1] Departamento de Ingeniería Química y Procesos de Minerales, Facultad de Ingeniería, Universidad de Antofagasta, Av. Angamos 601, Antofagasta 1240000, Chile

[2] Escuela de Ingeniería Civil de Minas, Facultad de Ingeniería, Universidad de Talca, Curicó 3340000, Chile

* Correspondence: mariaelisa.taboada@uantof.cl; Tel.: +56-55-637345

Received: 28 January 2020; Accepted: 14 April 2020; Published: 17 April 2020

Abstract: An alternative copper concentrate leaching process using sodium nitrate and sulfuric acid diluted in seawater followed by gas scrubbing to recover the sodium nitrate has been evaluated. The work involved leaching test carried out under various condition by varying temperature, leaching time, particle size, and concentrations of $NaNO_3$ and H_2SO_4. The amount of copper extracted from the chalcopyrite concentrate leached with seawater, 0.5 M of H_2SO_4 and 0.5 M of $NaNO_3$ increased from 78% at room temperature to 91% at 45 °C in 96 h and 46 h of leaching, respectively. Gas scrubbing with the alkaline solution of NaOH was explored to recover part of the sodium nitrate. The dissolved salts were recovered by evaporation as sodium nitrate and sodium nitrite crystals.

Keywords: chalcopyrite; leaching; nitrate; seawater; gas scrubbing

1. Introduction

Chilean mining is facing a shortage of water resources, depletion of copper oxide ores and those containing secondary sulfide copper. This depletion will leave chalcopyrite ores as the main source of copper for future plants. To assure sustainable production alternative extractive methods must be developed.

Chalcopyrite ($CuFeS_2$) is the most important copper sulfide mineral, representing approximately 70% of the world's known copper reserves [1]. Chalcopyrite is normally associated with pyrite (FeS_2), bornite (Cu_5FeS_4), sphalerite (ZnS), chalcocite (Cu_2S), covellite (CuS), enargite (Cu_3AsS_4) or molybdenite (MoS_2). The extractive metallurgy of chalcopyrite is based largely on a traditional route involving comminution and flotation, smelting and electrorefining, representing around the 85% of the copper production in Chile [2]. Environmental, financial and technical disadvantages of these processes have drawn attention to hydrometallurgy methods as an alternative. The main problem for treating chalcopyrite by a hydrometallurgy route is its refractoriness. This refractoriness is due to the formation of a passivating layer on the surface of chalcopyrite that inhibits the contact of mineral with oxidizing agents, reducing the dissolution rate. It is well known that the dissolution of chalcopyrite is a potential-dependent reaction and many studies have been carried out to elucidate the relationship between the solution potential and this passivating layer [2,3].

Several studies have been carried out in order to increase the dissolution rate of chalcopyrite, including: (i) leaching with strong oxidizing agents [4], (ii) leaching under high pressure and temperature [5], (iii) bacterial leaching [6], (iv) chloride media [2,3,7] and nitrate/nitrite media [8].

There is a considerable body of published information on the dissolution of chalcopyrite in sulfate and chloride media. Most of this literature reports that the dissolution of chalcopyrite is more effective in chloride media due to the greater reactivity of sulfide minerals in this medium. Seawater is then presented as an alternative source of chloride ions [9–13].

Sodium nitrate in an acid medium provides an option for possible leaching of many sulfide minerals, including chalcopyrite, at an acceptable kinetic rate due to the high proton activity created by the presence of sodium nitrate, and a strong oxidizing agent [14]. Sodium nitrate has been used as an oxidizing agent for more than 130 years [15]. However, these applications involve high temperatures and high levels of pressure. The main advantages of using nitrate ions in acidic leaching are: (i) nitrate is a strong oxidizing agent to decompose sulfides (Reaction (1)), (ii) the gases produced (NO_x) can easily be scrubbed, and (iii) scrubbed $NaNO_3$ can be recycled [8,16,17]:

$$NO_3^- + 4H^+ + 3e^- \rightarrow NO_{(g)} + 2H_2O \qquad E^\circ = 0.96\ V \tag{1}$$

Several authors have investigated the possible leaching reactions by considering select elements, ions, solid phases, and gas products [16,18–20]. Sokić et al. [16] proposed the most thermodynamically favorable overall reaction (Reaction (2)) for chalcopyrite dissolution in the $CuFeS_2$-$NaNO_3$-H_2SO_4-H_2O system:

$$C_uF_eS_2 + 4N_aNO_3 + 4H_2SO_4 \rightarrow C_uSO_4 + F_eSO_4 + 2N_{a2}S_{O4} + 4NO_{2(g)} + 2S^\circ + 4H_2O \tag{2}$$

Reaction (2) is accompanied by several secondary reactions shown by Reactions (3) to (8). Oxidative leaching of a chalcopyrite concentrate in an acid medium using nitrate as an oxidizing agent generates elemental sulfur [16,18,21]. The solution creates a strong oxidizing environment, as shown in Reaction (5):

$$FeSO_4 + \frac{1}{3}NaNO_3 + \frac{2}{3}H_2SO_4 \rightarrow \frac{1}{2}Fe_2(SO_4)_3 + \frac{1}{6}Na_2SO_4 + \frac{1}{3}NO + \frac{2}{3}H_2O \tag{3}$$

$$CuFeS_2 + 2Fe_2(SO_4)_3 \rightarrow CuSO_4 + 5FeSO_4 + 2S^\circ \tag{4}$$

$$3NO_{2(g)} + H_2O \rightarrow 2HNO_3 + NO_{(g)} \tag{5}$$

$$NO_{(g)} + 2NaNO_3 + H_2SO_4 \rightarrow 3NO_{2(g)} + Na_2SO_4 + H_2O \tag{6}$$

Sulfur, produced in reactions (2) or/and (4) is then oxidized to sulfate in a sodium nitrate medium, according to reactions (7) and (8). At room temperature, anhydrous sodium sulfate crystallizes [22,23], as shown in Reaction (7):

$$S^\circ + 2NaNO_3 \rightarrow Na_2SO_4 + 2NO_{(g)} \tag{7}$$

$$S^\circ + 3NO_{2(g)} + H_2O \rightarrow H_2SO_4 + 3NO_{(g)} \tag{8}$$

The thermodynamic feasibility of these reactions at atmospheric condition at 298 K and 318 K was estimated from standard Gibbs free energy calculations based on HSC Chemistry software V9 as shown in Table 1 (calculated data are based on the thermodynamic values of ΔG°, ΔS°, and ΔH°, of the chemical elements and their compounds). These calculations show that almost all reactions predict negative Gibbs energy values, which clearly indicates they are all thermodynamically favorable under the given conditions, except for Reaction (5). Reaction (2) is more favorable than Reaction (4) under the same conditions.

According to the reactions shown previously the leaching of chalcopyrite in acidic sodium nitrate solution and at high temperature produces notable amounts of gaseous nitrogen oxide gases or NO_x. These gases are a source of serious environmental problems, therefore, their treatment is mandatory.

Table 1. Standard Gibbs free energies for the $CuFeS_2$-H_2SO_4-$NaNO_3$-H_2O system at 298 K and 318 K (reactions taken Sokić et al. [16] and values corrected using HSC 9.0).

Reaction N°	$\Delta G°_{298\ K}$ (kJ/mol)	$\Delta G°_{318\ K}$ (kJ/mol)
(2)	−349.1	−361.4
(3)	−66.0	−66.6
(4)	−62.9	−63.8
(5)	8.7	14.1
(6)	−16.7	−23.5
(7)	−363.2	−369.3
(8)	−346.5	−345.8

Suchak and Joshi [24] stated that scrubbing NO_x gases is a complex process controlled by mass transfer limitations and involving bulk gas, gas film, interface, liquid film and bulk liquid mechanisms. Other factors that have been reported to affect the overall scrubbing rate and selectivity are temperature, pressure, the composition of NO_x gas, and the partial pressures of oxygen and water in the gas phase [25–30]. In the aerated leaching process, the nitrate ion is reduced to nitrogen oxide gas, $NO_{(g)}$, then the $NO_{(g)}$ is oxidized rapidly (in less than 0.1 s) in the gas phase to nitrogen dioxide, $NO_{2(g)}$. Sodium nitrate ($NaNO_3$) and sodium nitrite ($NaNO_2$) can then be regenerated by bubbling nitrous oxide gas in a dilute solution of sodium hydroxide (NaOH), which can then be crystallized and reused in the leaching stage.

The NO_x gas released from an acidic leaching contains several nitrogen oxides, mainly NO, NO_2, N_2O_3 and N_2O_4. The most important reactions are showed from reactions (9)–(11):

$$2NO_{(g)} + O_{2(g)} \rightarrow 2NO_{2(g)} \tag{9}$$

$$2NO_{2(g)} \rightarrow N_2O_{4(g)} \tag{10}$$

$$NO_{(g)} + NO_{2(g)} \rightarrow N_2O_{3(g)} \tag{11}$$

When NO_x is in contact with an alkaline solution of NaOH, it forms nitrite and nitrate ions. The liquid phase reactions are presented from reaction (12) to (14). The nitrogen oxides (NO_2, N_2O_3, and N_2O_4) are scrubbed in the liquid film phase [31] but the selectivity is poor which leads to the formation of a mixture of nitrite and nitrate ions [29]. According to Suchak et al. [29] chemical reactions (12) to (14) show that $NaNO_2$ is more selective than $NaNO_3$ and $NO_{2(g)}$ captured in the aqueous solution is generally produced through disproportionation reactions:

$$2NO_{2(g)} + 2NaOH_{(aq)} \rightarrow NaNO_{3(aq)} + NaNO_{2(aq)} + H_2O_{(aq)} \tag{12}$$

$$N_2O_{4(g)} + 2NaOH_{(aq)} \rightarrow NaNO_{3(aq)} + NaNO_{2(aq)} + H_2O_{(aq)} \tag{13}$$

$$N_2O_{3(g)} + 2NaOH_{(aq)} \rightarrow 2NaNO_{2(aq)} + H_2O_{(aq)} \tag{14}$$

The oxidation of NO, the formation in the gas phase of N_2O_3 and N_2O_4, and the scrubbing of NO_2, N_2O_3 and N_2O_4 in the liquid phase are all exothermic reactions. Thus, their equilibria shift to the right as temperature decreases. The gas dispersion rate has a significant effect on kinetics, and smaller bubbles provide a better contact with the alkaline scrubbing solution. Table 2 shows several reactions and related enthalpies calculated using HSC Chemistry Software V.9.0 (Outokumpu Research Oy, Helsinki, Finland).

The most effective methods to scrub NO_x are using urea + nitric acid [32], using ammonium bisulfate [33] and using sodium hydroxide as the scrubbing solution.

This paper proposes a feasible process to leach chalcopyrite concentrate with a mixture of sodium nitrate and diluted sulfuric acid dissolved in seawater as a source of chloride ions at atmospheric pressure and moderate temperature (≤45 °C). Due to its low cost the process includes a scrubbing stage using NaOH to capture NO_x gases that could enhance the formation of $NaNO_x$ salt (or a mixture of $NaNO_3$ and $NaNO_2$), the $NaNO_x$ (oxidants) can be recovered by crystallization and reused again as the leaching process oxidants.

Table 2. Thermodynamic data for gas and liquid phase reactions (reactions taken Suchak et al. [15] and values corrected using HSC V.9.0).

Reaction N°	$\Delta H°_{298\ K}$ (kJ/mol)	$\Delta G°_{298\ K}$ (kJ/mol)
(9)	−114.5	−70.9
(10)	−56.9	−4.5
(11)	−40.5	1.9
(12)	−204.6	−171.0
(13)	−147.7	−166.6
(14)	−118.6	−129.3

2. Materials and Methods

2.1. Chalcopyrite Concentrate and Reagents

The chalcopyrite concentrate sample was supplied by a Chilean mining company. Two particle sizes were used with P_{80} of 29.8 and 60.66 μm, respectively. The particle sizes distributions were determined using a Microtrac model S3500 laser diffraction Particle Size Analyser (PSA) (Verder Scientific, Newtown, PA, USA). The chemical composition was determined by digestion which was then analyzed using inductively coupled plasma atomic emission spectroscopy (ICP-AES, ICPE-9000, Shimadzu, Tokyo, Japan). The mineralogical characterization was determined by quantitative evaluation of the minerals by scanning electron microscopy (QEMSCAN) and morphology of the samples were determined by Scanning Electron Microscopy (SEM), equipped with a microscope-coupled X-ray dispersive energy analyzer, SEM-EDX (model JSM 6360 LV, JEOL Ltd., Tokyo, Japan). Table 3 shows the chemical analysis and mineralogical composition of the concentrate sample. Chemical analysis showed that the concentrate included mostly Cu, Fe and S. The XRD pattern (Siemens D5000 X-ray diffractometer, Bruker, Billerica, MA, USA) shows that the sample was mainly composed of chalcopyrite and pyrite, with small amounts of covellite and chalcanthite. Based on this data, the sample is composed of 63% of copper sulfides, following by 24.3% of other sulphides.

Table 3. Chemical and mineralogical analysis of the chalcopyrite concentrate.

Chemical Analysis		Mineralogical Analysis		
Element	wt.%	Minerals	Formula	wt.%
Si	2.3	Quartz	SiO_2	2.2
Fe	32.7	Pyrite	FeS_2	23.3
Al	1.0	Chalcopyrite	$CuFeS_2$	61.5
Mg	0.2	Covellite	CuS	1.5
Ca	0.3	Molybdenite	MoS_2	0.4
Cu	25.0	Dolomite	$CaMg(CO_3)_2$	1.4
Zn	0.5	Boehmite	AlOOH	2.0
Ti	0.5	Chalcanthite	$CuSO_4{\cdot}5H_2O$	1.2
Mo	0.2	Albite	$NaAlSi_3O_8$	2.2
K	0.4	Muscovite	$KAl_2(Si_3Al)O_{10}(OH)_2$	1.9
C	0.2	Biotite	$K(Mg, Fe^{2+})_3(Si_3Al)O_{10}(OH)_2$	0.7
Na	0.3	Sphalerite	$(Zn_x, Fe_{1-x})S$	0.6
S	36.3	Gypsum	$CaSO_4{\cdot}2H_2O$	0.8
		Clinochlore	$(Mg, Fe^{2+})_5Al(Si_3Al)O_{10}(OH)_8$	0.2

Sodium nitrate (99.5% absolute, Merck, Darmstadt, Germany) and sulfuric acid (95–97%, Merck, Darmstadt, Germany) were used in the leaching tests. The scrubbing solution was prepared with NaOH (99.0% absolute, Merck, Darmstadt, Germany) and distilled water.

Seawater was used as source of water and chloride ions in all experiments. It was collected from the coast in San Jorge Bay (Antofagasta, Chile). The sample of seawater was passed through a quartz sand filter (50 μm) and a mechanical polyethylene filter (1 μm) to remove insoluble particulate matter. Table 4 shows the composition of the seawater, which was obtained by different analytical techniques (argentometric method, atomic absorption spectrometry-AAS and volumetric analysis).

Table 4. Major composition of seawater from San Jorge Bay, Chile.

Ion	Concentration [mg/L]	Method of Analysis
Na^+	9950	Atomic absorption spectrometry
Mg^{2+}	1250	Atomic absorption spectrometry
Ca^{2+}	400	Atomic absorption spectrometry
K^+	380	Atomic absorption spectrometry
Cl^-	19,450	Argentometric method
HCO_3^-	150	Acid-base volumetry

2.2. Experimental Procedure

The experimental test work was carried out in three stages. The first stage was the leaching of chalcopyrite concentrate. A Taguchi L6(2^3) orthogonal array design experimental setup was employed in Series I to clarify the effects of sulfuric acid and nitrate concentration and a Taguchi L4(2^2) orthogonal array in Series II for temperature and particle size on copper extraction. Leaching experiments were performed in 2 L jacketed glass reactors. Each reactor was loaded with 1 L of leaching solution (0.1, 0.5 and 1.0 M of sulfuric acid 0.1 and 0.5 M of sodium nitrate and seawater as the source of water and chloride ions). Once the solution reached the desired temperature (room and 45 °C), 50 g of solid sample (P_{80} of 29.8 and 60.66 μm) was added to the reactor.

The pulp was stirred to a homogenous mix with a propeller at a rotation speed of 450 rpm. A 10-mL aliquot of the leached solution was withdrawn periodically during the test and analyzed for Cu and Fe using the atomic absorption spectrometry (AAS method, model 2380, Perkin Elmer, Wellesley, MA, USA). Redox potential (ORP) and pH were measured throughout the test with a portable meter (model HI991003, Hanna, St. Louis, MO, USA). All experiments were conducted in duplicate. The solid residues were carefully filtered, washed with distilled water, dried at 60 °C and samples were taken for mineralogical characterization and particle size determination.

The second stage was the alkaline gas scrubbing of NO_x produced in the leaching stage at 45 °C (Test 9, see Table 5). At this temperature, enough NO_x gases were produced. Scrubbing was conducted in duplicate using a 1.0 M sodium hydroxide solution. The initial pH of the scrubbing solution was 13. The gas was transported to the scrubber by pumping air to the reactor through a tube. The scrubber column was 25 cm high, the volume of the absorbing solution was 250 mL, and air was supplied at the rate of 3.0 L/min. The flue gases were dispersed into the solution using a fritted bubble disperser. The bubble size of the dispersed gas was generally >1 mm. All scrubbing tests were carried out at 2 °C. Figure 1 shows the schematic diagram of leaching and scrubbing tests. The amounts of $NaNO_2$-$NaNO_3$ and NaOH salt formation produced from the scrubbing solution were obtained by chemical analysis and ionic balance using atomic absorption spectrometry (AAS), volumetric analysis and colorimetric assays.

Last stage of the work was the recovery of nitrate-nitrite salts by evaporation. The scrubbing solution was dried at 100 °C to determine the amount of salts that crystallized with the evaporating solvent. Thermal stability was studied from crystallized solids. Nitrite-nitrate salts were characterized by differential scanning calorimetry using a Mettler Toledo TGA/DSC 1 StarSystem, (NETZSCH, Bavaria, Germany). A 10 mg sample was used in the TGA/DSC determinations. This equipment can measure the heat capacity of a sample to determine the melting temperature and quantify the degree of crystallinity and characterization of the material.

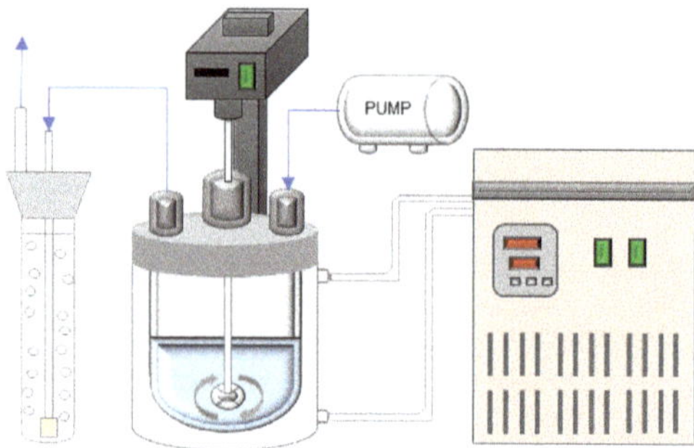

Figure 1. Schematic diagram of leaching-scrubbing system.

3. Results and Discussion

Table 5 summarize the leaching test results of Series I and Series II. The tests are discussed in the following sections. The tests (1 to 10) were carried out in duplicate.

Table 5. Serie I (Leach time 94 h, and Serie II (Leach time 46 h). Leaching test results in seawater.

Series	Test N°	Particle Size (μm)	Temperature (°C)	[H_2SO_4] (M)	[$NaNO_3$] (M)	ORP Range (mV vs. Ag/AgCl)	Cu Ext. (wt.%)	Cu Ext. (wt.%) (Duplicate)	Cu Ext. (wt.%) Average
I	1	−60.66	Room	0.1	0.1	380–463	11.5	11.4	11.5
	2	−60.66	Room	0.1	0.5	390–760	27.6	27.0	27.3
	3	−60.66	Room	0.5	0.1	410–793	46.1	47.7	46.9
	4	−60.66	Room	0.5	0.5	734–826	76.0	77.8	76.9
	5	−60.66	Room	1.0	0.1	690–791	54.9	56.5	55.7
	6	−60.66	Room	0.5	0	435–660	41.2	39.2	40.2
II	7	−29.80	45	0.5	0.5	742–801	87.5	89.5	88.5
	8	−29.80	Room	0.5	0.5	615–750	66.7	66.1	66.0
	9	−60.66	45	0.5	0.5	743–764	91.1	90.6	90.8
	10	−60.66	Room	0.5	0.5	751–787	71.2	73.9	72.5

3.1. *Series I. Leaching Test Results at Room Temperature*

3.1.1. Variation of Redox Potential (ORP)

Figure 2 shows the variations of the oxidation redox potential (ORP) according to leaching time. It is clear that redox potential increases when the dissolution of copper increases (see Tests 2 and 3, at 56 and 26 h, respectively).

The results of tests 2 and 3 indicate that ion oxidation increases with higher levels of aciditsiy in the solution due to NO_x formation as an oxidant. NO_x gas did not form immediately, but was observed with extending leaching time. Once the NO_x is formed, this gas tends to oxidize rapidly due to the action of the injected oxygen, turning into a brown gas, which is characteristic of $NO_{2(g)}$. The potential remained above 700 mV vs. Ag/AgCl, facilitating the leaching of chalcopyrite.

The redox potential of Test 1 with 0.1 M H_2SO_4–0.1 M $NaNO_3$ solution remained low, between 400 and 450 mV vs. Ag/AgCl, indicating a low level of chalcopyrite oxidation under these conditions. When 1.0 and 0.5 M of acid and 0.5 M of $NaNO_3$ were added to the leaching solution the redox potential remained above 700 mV vs. Ag/AgCl throughout tests 4 and 5 enhancing the dissolution of copper.

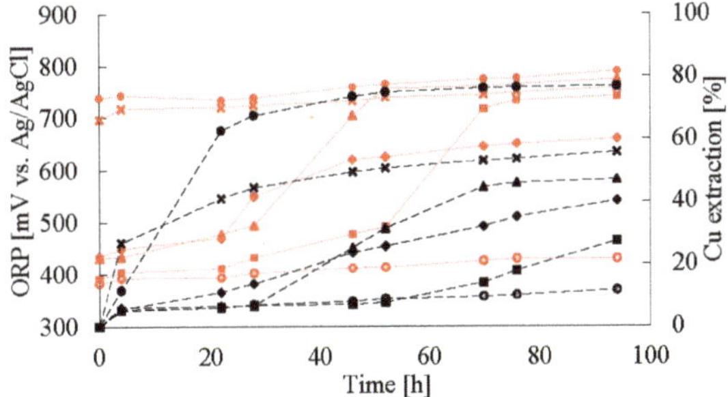

Figure 2. Dissolution of copper (%) from chalcopyrite concentrate and averaged ORP (mV vs. Ag/AgCl) values over time (hours) for Series I at different concentrations of H_2SO_4 and $NaNO_3$. (○) Test 1 (H_2SO_4 = 0.1 M and $NaNO_3$ = 0.1 M); (■) Test 2 (H_2SO_4 = 0.1 M and $NaNO_3$ = 0.5 M); (▲) Test 3 (H_2SO_4 = 0.5 M and $NaNO_3$ = 0.1 M); (•) Test 4 (H_2SO_4 = 0.5 M and $NaNO_3$ = 0.5 M); (X) Test 5 (H_2SO_4 = 1.0 M and $NaNO_3$ = 0.1 M); (◆) Test 6 (H_2SO_4 = 0.5 M and $NaNO_3$ = 0 M). (Test conditions: 94 h, P80 = 60.66 μm, 450 rpm and room temperature). The copper recoveries are showing with black dotted line in axis secondary and averaged ORP values are showing with red dotted line.

3.1.2. Effect of the Nitrate Concentration

The influence of nitrate ion concentrations on copper extraction was investigated at room temperature. Figure 2 shows the effect of the addition of initial $NaNO_3$ concentration. Copper extraction in solutions with 0.1 M of H_2SO_4 increased from 11.5% to 27.3% when the concentration of $NaNO_3$ increased from 0.1 M (Test 1) to 0.5 M (Test 2). The dissolution of copper leached in a solution with 0.5 M of H_2SO_4 increased from around 47 to 77% when the concentration of $NaNO_3$ increased from 0.1 M (Test 3) to 0.5 M (Test 4). This result agree with other found in literature [4] and shows that nitrates with a standard potential of 0.96 V are a powerful oxidants.

3.1.3. Effect of Sulfuric Acid Concentration

Figure 3a shows the results of leaching copper concentrate at different concentration of acid and sodium nitrate. The dissolution of copper clearly is affected by the initial addition of both sulfuric acid and sodium nitrate concentrations to the leaching solutions. Copper extraction clearly increased with higher acid concentration (Tests 1, 3 and 5). Without enough H^+ ions in the system (Test 1), copper extraction was negligible (10–11.5%). The experiments also showed little or no effect of sodium nitrate on copper recovery under the conditions studied (Test 1).

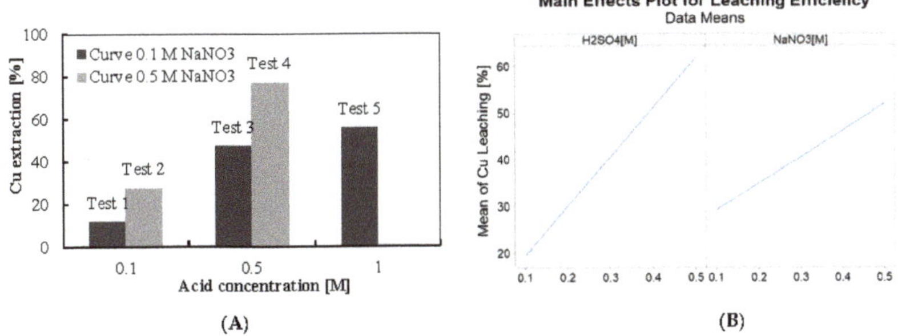

Figure 3. (**A**) Effect of the acid concentration on the extend of the copper concentrate at room temperature and (**B**) Effect of concentrations of H_2SO_4 and $NaNO_3$ on the leaching efficiency of copper concentrate.

The copper extraction rate in Test 5, with a high acid and low sodium nitrate content, was 55.7%. Although the low concentration of the oxidant the potential reminds high. This could be due to the formation of a new oxidant in the NO_x system (perhaps Fe^{3+} according with Figure 7), but less powerful than $NaNO_3$.

As shown in Figure 3A, the best copper extraction rate (76.9%) was obtained in Test 4, with a solution containing sulfuric acid and sodium nitrate both at a concentration of 0.5 M. These highlights, presented in Figure 3B, shown the importance of the role of the concentration of H^+ ions compared to that of the sodium nitrate (Tests 1–3 and Tests 2–4). However, increasing amount of nitrate ion concentration enhances the copper extraction rate for all acid concentration used in the leaching solution. These results are agreed with results from Hernández et al. [11].

Table 6 shows the effect of parameters (H_2SO_4 and $NaNO_3$) using the analysis of variance (ANOVA) module of the Minitab 19 software (version 19, Pennsylvania State University, State College, PA, USA). According to Table 6, H_2SO_4 had the largest contribution (65.45%) for copper extraction compared to $NaNO_3$ with 32.6% approximately.

Table 6. Analysis of variance (ANOVA) for Cu extraction.

Parameters	Degree Freedom	Sum of Squares	Contribution	Media of Squares	Fisher Ratio
H_2SO_4	2	1683.50	65.45%	1137.21	22.56
$NaNO_3$	2	838.34	32.59%	419.17	8.32
Error		50.41	1.96%	50.41	
Total		2572.25	100.00%		

The data of Tests 3–5, showed in Figure 2, also showed that dissolution rates slowed down in the final stage of the leaching. This is thought to be due to a formation of a layer of sulfur on the chalcopyrite particles surface (cf. Figure 4a,b). The morphology of the layer shows irregular grains with porosity that perhaps allows the solution to flow into the particles of chalcopyrite. This suggests that dissolution reaction is initially controlled by a surface mechanism, and later by diffusion reaction mechanism [16]. The extent of sulfur formation was influenced by reagent concentrations and the leaching conditions.

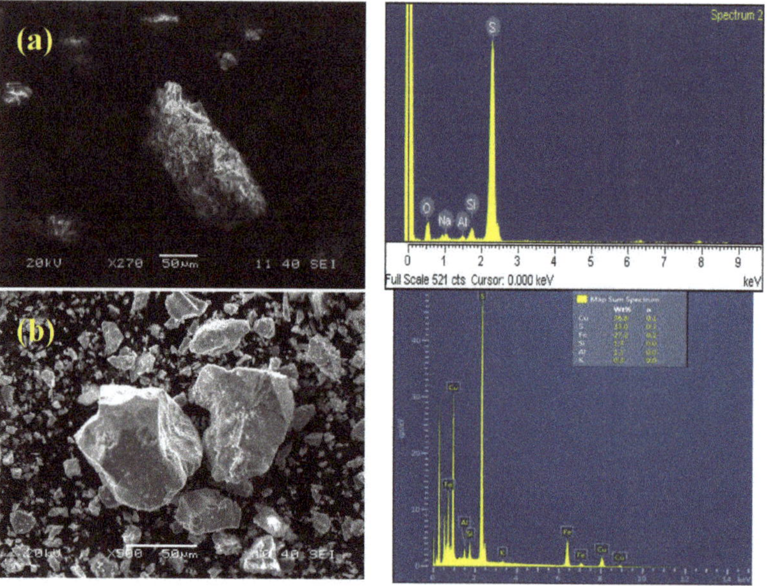

Figure 4. SEM micrographs and EDS spectrums of (**a**) residue (Test 4) and (**b**) Initial chalcopyrite concentrate sample.

3.2. Series II. The Effect of Temperature and Particle Size on the Leaching Rate

The effects of increase temperature and particle size on the dissolution of copper Tests 7 to 10 of Series II (Table 5). It is not surprising that the leaching rate increases as the temperature increases from room temperature (tests 8 and 10) to 45 °C (Tests 7 and 9) seen in Figure 5. Extraction was relatively rapid at both temperatures during the first 24 h (80% and 60% of copper extraction at 45 °C and room temperature respectively), but then slowed down, which was related to the formation of a thin sulfur layer on the chalcopyrite surface grains as shows in Figure 4a (see Section 3.2.1).

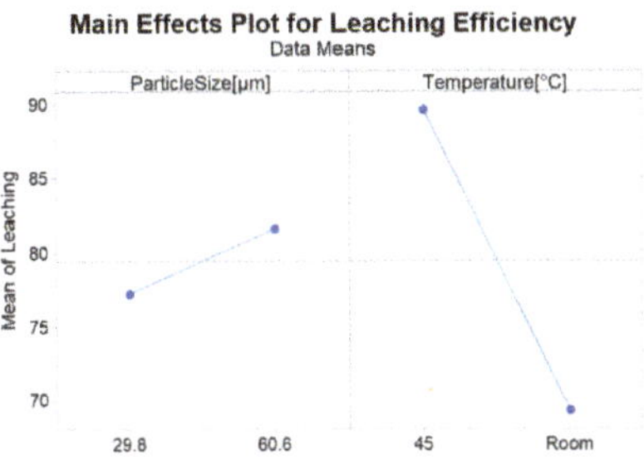

Figure 5. The effect of particle size and temperature on leaching efficiency of copper concentrate.

It seems that particle sizes between 29.8 and 60.66 µm have a negligible effect on the leaching of chalcopyrite under all conditions studied, according to Table 5. All ORP values (Series II) were over 700 mV vs. Ag/AgCl. This was due to the presence of relatively high concentrations (0.5 M) of $NaNO_3$ and H_2SO_4. It also shows that the ORP of finer particle size ore takes longer to reach 700 mV when leaching is carried out at room temperature (Test 8).

3.2.1. Characterization of the Residues

The solid leaching residue at 45 °C, with a P_{80} of 60.66 µm (Test 9) was characterized by XRD (Table 7) and SEM-EDS analyses. Results are shown in Figure 6.

Table 7. Mineral composition of the residue (Test 9).

Mineral	Formula	(wt.%)
Chalcopyrite	$CuFeS_2$	5.9
Quartz	SiO_2	11.5
Sulfur	S	46.3
Chlorite	$(Mg, Fe)_6(Si, Al)_4 O_{10}(OH)_8$	1.5
Albite	$NaAlSi_3O_8$	7.7
Orthoclase	$KAlSi_3O_8$	7.9
Gypsum	$CaSO_4{\cdot}2H_2O$	3.1
Goethite	FeO(OH)	2.2
Biotite	$KMg_3(Si_3Al)O_{10}(OH)_2$	11.8
Langbeinite	$K_2Mg_2(SO_4)_3$	2.1

Table 7 shows the mineral phases in the leaching residue identified by XRD, indicating the presence mainly of elemental sulfur (46%) and gangue minerals (46%), with 5.9% of unleached chalcopyrite. The presence of large quantities of sulfur confirms the expectation that elemental sulfur forms during the leaching. The presence of sulfur is also confirmed by SEM and EDS spectra, as shown in Figure 6. This residue was selected to analyze the products formed due to the highest extraction of copper. The spectrum and the micrograph corroborated the presence of a sulfur layer on the chalcopyrite surface. The grains of sulfur have intergranular porosity. This porosity decreases with the leaching progress. The data also shows that other major sulfide minerals were completely dissolved in the first

48 h. It is noted also the absence of pyrite, sphalerite and molybdenite in the residue (see Table 7), as compared to the feed material (chalcopyrite concentrate) shown in Table 4.

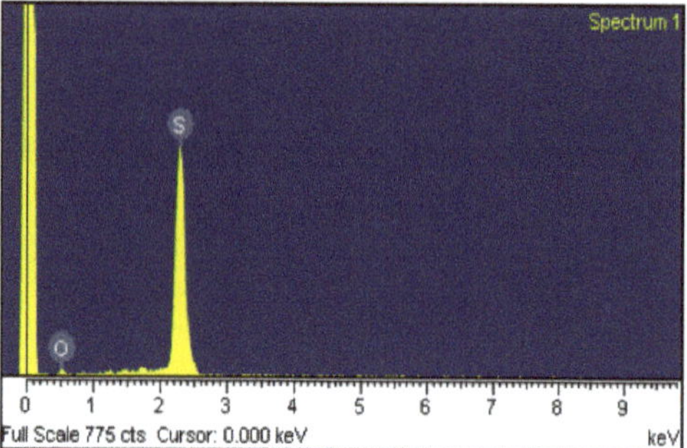

Figure 6. SEM micrograph and EDS spectrum of residue (Test 9).

In order to confirm that pyrite could leach under the studied conditions, a pure sample of pyrite (P_{80} = 60.66 μm) was leached with 0.5 M of H_2SO_4 and 0.5 M of $NaNO_3$, the result showed that the entire pyrite sample was leached. According to thermodynamic calculations NO^{3-} ions and the acidity allow Fe^{3+} to remain in the solution and therefore acts as an oxidant. At higher ORP (600–800 mV vs. Ag/AgCl) according to the Eh-pH diagram (Figure 7) the Fe^{2+} then could re-oxidize to Fe^{3+}.

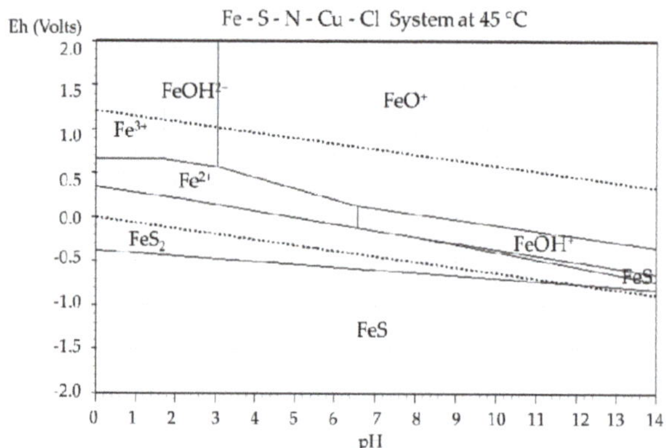

Figure 7. Diagram Eh-pH of Fe-S-N-Cu-Cl at 45 °C, during the interaction of chalcopyrite concentrate (pyrite included) (HSC Chemistry 9.0 version).

3.3. Flue Gas Scrubbing Tests

Gas scrubbing was tested with a solution of sodium hydroxide. Table 8 shows the results of the analysis of the scrubbing solution before the crystallization stage.

Table 8. Chemical analysis of nitrogen oxide scrubbing tests.

Scrubbing System (1.0 M NaOH)	Chemical Analysis		
	Na^+ (mg/L)	OH^- (mg/L)	NO_2 (mg/L)
Test A	21,200	2581	23,758
Test B	23,450	2374	23,820

The amounts of $NaNO_2$ and $NaNO_3$ produced and of absorbed gas ($NO_{2(g)}$), as shown in Table 8, were estimated with an ionic balance. The results are shown in Figures 8 and 9. The data shows a relationship between the scrubbing solution concentration and the percentage of crystallized salts. The alkaline concentration of the wash solution (1.0 M NaOH) allowed $NaNO_3$ and $NaNO_2$ salts to crystallize with a lower quantity of mass.

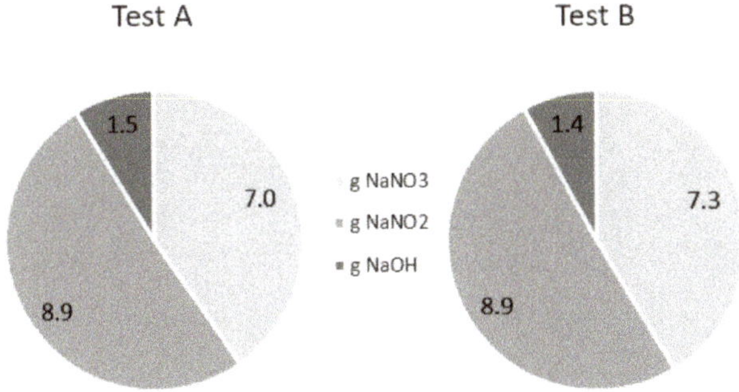

Figure 8. Amount of $NaNO_2$ - $NaNO_3$ and NaOH salt formation (obtained from chemical analysis and ionic balance).

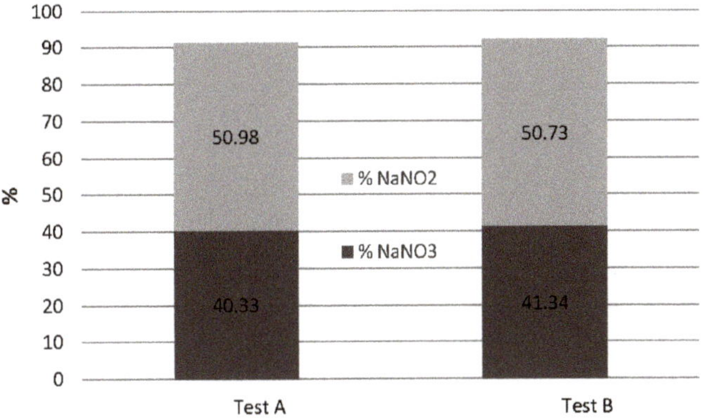

Figure 9. Percentages of $NaNO_2$ - $NaNO_3$ and NaOH salt formed.

Differential scanning calorimetry (DSC) analyses were performed on the scrubbing systems (Tests A and B). Figures 10 and 11 shows the results of Tests A and Test B, in which the peaks occur at 228.48 °C and 186.65 °C, respectively. These values are different from those reported by Janz et al. [34] and Janz and Tomkins [35] for pure $NaNO_2$ (281 °C) and for pure $NaNO_3$ (307 °C). The difference between the crystallized products in this study and those in the literature are related to the combination of $NaNO_2$ and $NaNO_3$ in the samples of this study, showing the presence of a eutectic salt as a result of absorption and crystallization by evaporation. Raman spectroscopy measurements and differential scanning calorimetry data on solidified mixtures of different compositions have provided support for a simple eutectic diagram with a solid at 230 °C ranging from 0.25 to 0.80 in mole fraction of sodium nitrate [36].

The formation of sodium and nitrogen salts is also shown (qualitatively) by SEM-EDS patterns in sample from Test A (Figure 12). The particles crystallize, as shown by EDS analysis, in small agglomerates of nitrogen salts combined with sodium. NO_x gas scrubbing is a relatively complex operation that requires optimizing several parameters like the NO to NO_2 ratio, initial concentrations, bubble size, temperature and the type of alkaline solution, as these parameters influence overall scrubbing efficiency.

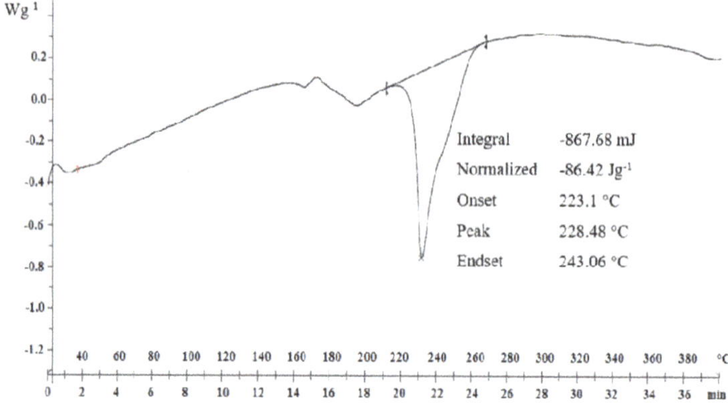

Figure 10. DSC curve for products crystallized from NO_x gases in Test A.

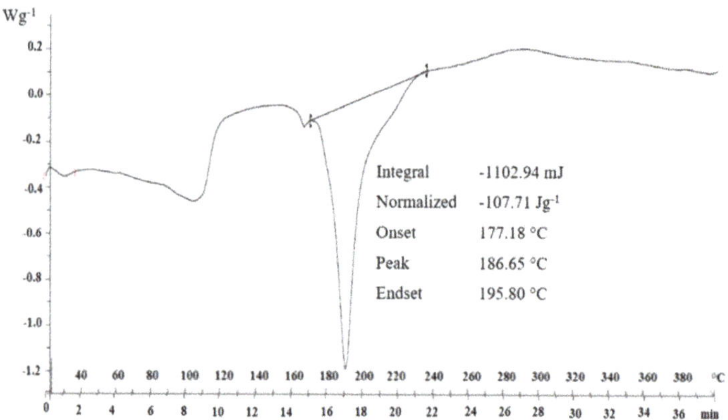

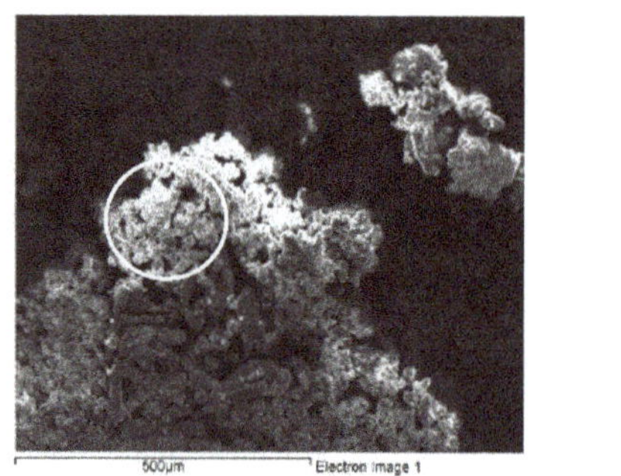

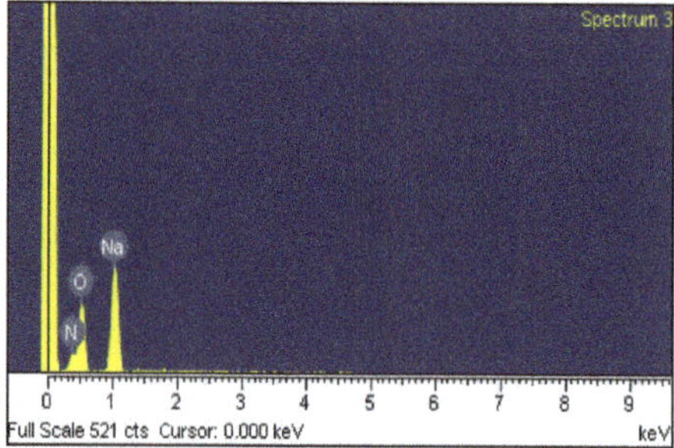

Figure 11. DSC curve for products crystallized from NO_x gases in Test B.

Figure 12. SEM micrograph and EDS spectrum of solids crystallized from Test A.

3.4. Proposed Process for Leaching Chalcopyrite and Recovering Oxidants

This paper proposes a process for leach chalcopyrite using nitrate-acid-seawater media, with subsequent recovery and recycling of oxidant salts. The recovery of sodium nitrate and sodium nitrite from nitrous gases is also shown in the process flowsheet in Figure 13.

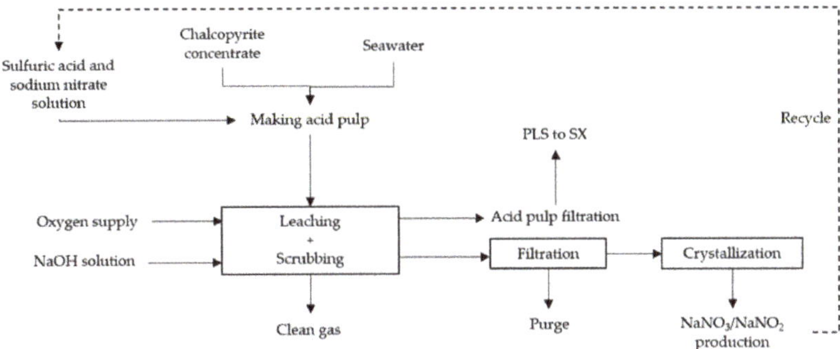

Figure 13. Proposed process for chalcopyrite leaching.

The process includes leaching, absorption and crystallization. Leaching should be carried out in closed reactors with controlled agitation to ensure continuous solid-liquid contact. The pregnant leaching solution (PLS) will continue with solvent extraction (SX) and electrowinning (EW) stages. It should be noted that there are reagents that have been created and are capable of extracting copper from a matrix with a high concentration of nitrate and chloride ions [37–42]. The second nitrous gas absorption stage recovers contaminating gases (NO_X) formed by dissolving the concentrate that would otherwise escape into the environment. The last stage involves recovering salts for reusing, which can be carried out by crystallization, or by returning the absorption solution in the case of an acid or neutral absorbent.

4. Conclusions

Leaching of a chalcopyrite concentrate using sulfuric acid and sodium nitrate in seawater was studied. The results of this study suggest that sodium nitrate in the presence of seawater is an effective oxidation agent that makes it possible to replace other lixiviants and recovery copper faster.

The highest copper extraction of 91% was obtained when the chalcopyrite concentrate with a P_{80} of 60.66 μm was leached with 0.5 M H_2SO_4 and 0.5 M $NaNO_3$ at 45 °C for a period of 46 h.

This study shows that the rate of leaching of the finely ground ore concentrate is relatively fast for the first 24 h but then slows down. This is related to the formation of a passive layer of elemental sulfur on the chalcopyrite grains that was confirmed by XRD and SEM/EDS analysis. The analysis also showed that pyrite, sphalerite and molybdenite were completely dissolved within the first 48 h.

Leaching of chalcopyrite concentrate with sulfuric acid and sodium nitrate generated NO and NO_2 gases. Scrubbing parts of these gases in a sodium hydroxide solution recovered sodium nitrate and sodium nitrite. The amount of sodium nitrite recovered tended to exceed the amount of sodium nitrate. The recovery of sodium nitrate can be beneficial to an industrial-scale leaching operation processing copper concentrate.

Leaching technologies with the absorption stage described above are technically feasible and environmentally friendly. Future studies should consider improving the environmental effects of leaching and making the process more economical. These might be the recyclability of recrystallized salts and the amenability of the PLS to solvent extraction for Cu recovery.

Author Contributions: C.I.C. contributed in research and wrote paper, P.C.H. contributed in project. L.V.-Y. and M.E.T. contributed in review and editing. All authors have read and agreed to the published version of the manuscript. All of the authors contributed to analyzing the results and writing the paper.

Funding: This research was funded by ANID-PFCHA/National Doctoral Program/2017-21171313, FONDECYT Project 11170179 and CODELCO Piensa Mineria scholarship.

Acknowledgments: César I. Castellón thanks ANID-PFCHA/National Doctoral Program/2017-21171313, Piensa Mineria scholarship from CODELCO and María E. Taboada thanks the Universidad de Antofagasta for the support provided.

Conflicts of Interest: The authors declare no conflict of interest.

References

1. Li, Y.; Kawashima, N.; Li, J.; Chandra, A.; Gerson, A.R. A review of the structure, and fundamental mechanisms and kinetics of the leaching of chalcopyrite. *Adv. Colloid Interface Sci.* **2013**, *197*, 1–32. [CrossRef]
2. Velásquez-Yévenes, L.; Nicol, M.; Miki, H. The dissolution of chalcopyrite in chloride solutions: Part 1. The effect of solution potential. *Hydrometallurgy* **2010**, *103*, 108–113. [CrossRef]
3. Nicol, M.; Miki, H.; Velásquez-Yévenes, L. The dissolution of chalcopyrite in chloride solutions: Part 3. Mechanisms. *Hydrometallurgy* **2010**, *103*, 86–95. [CrossRef]
4. Watling, H. Chalcopyrite hydrometallurgy at atmospheric pressure: 1. Review of acidic sulfate, sulfate–chloride and sulfate–nitrate process options. *Hydrometallurgy* **2013**, *140*, 163–180. [CrossRef]
5. McDonald, R.; Muir, D. Pressure oxidation leaching of chalcopyrite. Part I. Comparison of high and low temperature reaction kinetics and products. *Hydrometallurgy* **2007**, *86*, 191–205. [CrossRef]
6. Romero, R.; Mazuelos, A.; Palencia, I.; Carranza, F. Copper recovery from chalcopyrite concentrates by the BRISA process. *Hydrometallurgy* **2003**, *70*, 205–215. [CrossRef]
7. Watling, H. Chalcopyrite hydrometallurgy at atmospheric pressure: 2. Review of acidic chloride process options. *Hydrometallurgy* **2014**, *146*, 96–110. [CrossRef]
8. Gok, O.; Anderson, C.G. Dissolution of low-grade chalcopyrite concentrate in acidified nitrite electrolyte. *Hydrometallurgy* **2013**, *134–135*, 40–46. [CrossRef]
9. Cisternas, L. *Uso de Agua de mar en la mineria. Fundamentos y Aplicaciones*; RIL Editores: Santiago, Chile, 2014.
10. Torres, C.; Taboada, M.; Graber, T.; Herreros, O.; Ghorbani, Y.; Watling, H. The effect of seawater based media on copper dissolution from low-grade copper ore. *Miner. Eng.* **2015**, *71*, 139–145. [CrossRef]
11. Hernández, P.C.; Taboada, M.E.; Herreros, O.O.; Graber, T.A.; Ghorbani, Y. Leaching of Chalcopyrite in Acidified Nitrate Using Seawater-Based Media. *Minerals* **2018**, *8*, 238. [CrossRef]
12. Hernández, P.; Taboada, M.; Herreros, O.; Torres, C.; Ghorbani, Y. Chalcopyrite dissolution using seawater-based acidic media in the presence of oxidants. *Hydrometallurgy* **2015**, *157*, 325–332. [CrossRef]
13. Velásquez-Yévenes, L.; Quezada-Reyes, V. Influence of seawater and discard brine on the dissolution of copper ore and copper concentrate. *Hydrometallurgy* **2018**, *180*, 88–95. [CrossRef]
14. Hernández, P.C.; Dupont, J.; Herreros, O.O.; Jimenez, Y.P.; Torres, C.M. Accelerating copper leaching from sulfide ores in acid-nitrate-chloride media using agglomeration and curing as pretreatment. *Minerals* **2019**, *9*, 250. [CrossRef]
15. Stetefeldt, G.A. Process of Treating Sulphides. U.S. Patent 287,737, 30 October 1883.
16. Sokić, M.D.; Marković, B.; Živković, D. Kinetics of chalcopyrite leaching by sodium nitrate in sulphuric acid. *Hydrometallurgy* **2009**, *95*, 273–279. [CrossRef]
17. Pacovic, N.V. *Hydrometalurgy SRIF*; Bor, Serbia, 1980; Chapter 3.
18. Vračar, R.Ž.; Vučković, N.; Kamberović, Ž. Leaching of copper (I) sulphide by sulphuric acid solution with addition of sodium nitrate. *Hydrometallurgy* **2003**, *70*, 143–151. [CrossRef]
19. Sokić, M.D.; Matković, V.L.; Marković, B.R.; Štrbac, N.D.; Živković, D.T. Passivation of chalcopyrite during the leaching with sulphuric acid solution in presence of sodium nitrate. *Hem. Ind.* **2010**, *64*, 343–350.
20. Sokić, M.; Radosavljević, S.; Marković, B.; Matković, V.; Štrbac, N.; Kamberović, Ž.; Živković, D. Influence of chalcopyrite structure on their leaching by sodium nitrate in sulphuric acid. *Metall. Mater. Eng.* **2014**, *20*, 53–60. [CrossRef]
21. Bredenhann, R.; Van Vuuren, C. The leaching behaviour of a nickel concentrate in an oxidative sulphuric acid solution. *Miner. Eng.* **1999**, *12*, 687–692. [CrossRef]

22. Linnow, K.; Steiger, M.; Lemster, C.; De Clercq, H.; Jovanović, M. In situ Raman observation of the crystallization in $NaNO_3$–Na_2 SO_4–H_2O solution droplets. *Environ. Earth Sci.* **2013**, *69*, 1609–1620. [CrossRef]
23. Rodriguez-Navarro, C.; Doehne, E.; Sebastian, E. How does sodium sulfate crystallize? Implications for the decay and testing of building materials. *Cem. Concr. Res.* **2000**, *30*, 1527–1534. [CrossRef]
24. Suchak, N.; Joshi, J. Simulation and optimization of NOx absorption system in nitric acid manufacture. *AIChE J.* **1994**, *40*, 944–956. [CrossRef]
25. Kameoka, Y.; Pigford, R.L. Absorption of nitrogen dioxide into water, sulfuric acid, sodium hydroxide, and alkaline sodium sulfite aqueous solutions. *Ind. Eng. Chem. Fundam.* **1977**, *16*, 163–169. [CrossRef]
26. Aoki, M.; Tanaka, H.; Komiyama, H.; Inoue, H. Simultaneous absorption of NO and NO_2 into alkaline solutions. *J. Chem. Eng. Jpn.* **1982**, *15*, 362–367. [CrossRef]
27. Carta, G. Scrubbing of nitrogen oxides with nitric acid solutions. *Chem. Eng. Commun.* **1986**, *42*, 157–170. [CrossRef]
28. Carta, G. Role of nitrous acid in the absorption of nitrogen oxides in alkaline solutions. *Ind. Eng. Chem. Fundam.* **1984**, *23*, 260–264. [CrossRef]
29. Suchak, N.J.; Jethani, K.; Joshi, J.B. Absorption of nitrogen oxides in alkaline solutions: Selective manufacture of sodium nitrite. *Ind. Eng. Chem. Res.* **1990**, *29*, 1492–1502. [CrossRef]
30. Xiao, Z.; Li, D.; Wang, F.; Sun, Z.; Lin, Z. Simultaneous Removal of NO and SO_2 with A New Recycling Micro-nano Bubble Oxidation-absorption Process Based on HA-Na. *Sep. Purif. Technol.* **2020**, *242*, 116788. [CrossRef]
31. Joshi, J.; Mahajani, V.; Juvekar, V. Invited review absorption of NOx gases. *Chem. Eng. Commun.* **1985**, *33*, 1–92. [CrossRef]
32. Duncan, J.L.; McLarnon, C.R.; Alix, F.R. NOx, Hg, and SO2 Removal Using Ammonia. U.S. Patent 6,936,231, 30 August 2005.
33. Hignett, T.P. *Fertilizer Manual*; Springer Science & Business Media: Berlin/Heidelberg, Germany, 2013; Volume 15.
34. Janz, G.J.; Ward, A.T.; Reeves, R.D. *Molten Salt Data. Electrical Conductance, Density, and Viscosity*; DTIC Document: Los Angeles, CA, USA, 1964.
35. Janz, G.; Tomkins, R. Molten Salts: Volume 5, Part 2. Additional Single and Multi-Component Salt Systems. Electrical Conductance, Density, Viscosity and Surface Tension Data. *J. Phys. Chem. Ref. Data* **1983**, *12*, 591–815. [CrossRef]
36. Berg, R.W.; Kerridge, D.H.; Larsen, P.H. $NaNO_{2+}$ $NaNO_3$ phase diagram: New data from DSC and Raman spectroscopy. *J. Chem. Eng. Data* **2006**, *51*, 34–39. [CrossRef]
37. Yáñez, H.S.A.; Soderstrom, M.; Bednarski, T. Nitration in copper SX? Cytec Acorga provides a new reagent. In *Hydrocopper 2009*; Gecamin Ltd.: Santiago, Chile, 2009.
38. Zambra, R.Q.A.; Rivera, G.; Castro, O. Use of NR® reagents in presence of nitrate ion in SX: A revision of the present moment. In *Hydroprocess 2013*; Gecamin Ltd.: Santiago, Chile, 2013.
39. Hamzah, B.; Jalaluddin, N.; Wahab, A.W.; Upe, A. Copper (II) extraction from nitric acid solution with 1-phenyl-3-methyl-4-benzoyl-5-pyrazolone as a cation carrier by liquid membrane emulsion. *J. Chem.* **2010**, *7*, 239–245. [CrossRef]
40. Virnig, M.J.; Mattison, P.L.; Hein, H.C. Processes for the Recovery of Copper from Aqueous Solutions Containing Nitrate Ions. U.S. Patent 6,702,872, 22 July 2004.
41. Yáñez, H.A.; Ardiles, L.; Reyes, F.; Joly, P.; Mejias, J. Washing Stages in the Control of Chloride Transfer from Leach Solutions to the Electrolyte in Copper Solvent Extraction Process. In *Hydroprocess 2018*; Gecamin Ltd.: Santiago, Chile, 2018.
42. Yáñez, H.A.; Ardiles, L.; Del Río, C. High Chloride in PLS and their Impact on Copper Solvent Extraction. In *Hydroprocess 2017*; Gecamin Digital Publications: Santiago, Chile, 2017.

Article

Recovery of Gold from the Refractory Gold Concentrate Using Microwave Assisted Leaching

Kanghee Cho [1], Hyunsoo Kim [2], Eunji Myung [2], Oyunbileg Purev [2], Nagchoul Choi [1,*] and Cheonyoung Park [2,*]

[1] Research Institute of Agriculture and Life Sciences, Seoul National University, Seoul 08826, Korea; kanghee1226@hanmail.net
[2] Department of Energy and Resource Engineering, Chosun University, Gwangju 61452, Korea; star8538@naver.com (H.K.); ej6865@naver.com (E.M.); oyunbileg@chosun.kr (O.P.)
* Correspondence: nagchoul@empas.com (N.C.); cybpark@chosun.ac.kr (C.P.); Tel.: +82-62-230-7878 (N.C.)

Received: 23 March 2020; Accepted: 27 April 2020; Published: 28 April 2020

Abstract: Microwave technology has been confirmed to be suitable for use in a wide range of mineral leaching processes. Compared to conventional leaching, microwave-assisted leaching has significant advantages. It is a proven process, because of its short processing time and reduced energy. The purpose of this study was to enhance the gold content in a refractory gold concentrate using microwave-assisted leaching. The leaching efficiencies of metal ions (As, Cu, Zn, Fe, and Pb) and recovery of gold from refractory gold concentrate were investigated via nitric acid leaching followed by microwave treatment. As the acid concentration increased, metal ion leaching increased. In the refractory gold concentrate leaching experiments, nitric acid leaching at high temperatures could limit the decomposition of sulfide minerals, because of the passive layer in the refractory gold concentrate. Microwave-assisted leaching experiments for gold recovery were conducted for the refractory gold concentrate. More extreme reaction conditions (nitric acid concentration > 1.0 M) facilitated the decomposition of passivation species derived from metal ion dissolution and the liberation of gangue minerals on the sulfide surface. The recovery rate of gold in the leach residue was improved with microwave-assisted leaching, with a gold recovery of ~132.55 g/t after 20 min of the leaching experiment (2.0 M nitric acid), according to fire assays.

Keywords: recovery; gold; refractory; nitric acid; microwave

1. Introduction

Gold has excellent physical and chemical properties, and is one of the most important noble metals. The current rapid decline in high-grade gold ores and readily available low-grade gold ores has made the mineral processing industry increasingly reliant on complex and refractory gold ores [1,2]. Consequently, mineral processing challenges related to the complexities of ore mineralogy and the process parameters, such as the impacts of associated minerals, are important research questions [3]. Mineralogical studies aim to characterize complex sulfides and show the interrelations of the target minerals in the refractory minerals. In addition, these studies usually analyze gold-bearing sulfide minerals for follow-up processes and efficient gold recovery. Gold can be found in complex sulfide minerals, not only due to the presence of invisible gold, but also due to the existence of solid solutions [4]. Gold is commonly associated with sulfide minerals, particularly pyrite and arsenopyrite [5]. A refractory gold ore typically contains different sulfide minerals with various gold concentrations [6]. Gold is highly encapsulated in the sulfide matrix in refractory gold ores. Pretreatment is an important process to recover gold from sulfide minerals. Gold-bearing ores ordinarily contain complex or refractory ores of sulfide minerals that interfere with gold recovery efficiency. It is necessary to remove sulfide minerals prior to leaching, and the most appropriate pretreatment is flotation and roasting. However, most of the sulfide minerals in a refractory gold concentrate react to form a harmful gas during roasting and the iron oxide produced during roasting encapsulates invisible or fine gold [7]. Pyrite is the most abundant sulfide mineral in refractory gold ore and its oxidation is thermodynamically favorable in the environment [8].

The surface oxidation of pyrite is a particularly important step for leaching valuable metals from sulfide minerals. The presence of invisible gold has been established in pyrite, which is a common type of gold-bearing sulfide mineral that is mostly found in association with other sulfide minerals. The gold present in sulfide minerals can be divided into visible gold and invisible gold. While visible gold can be observed with an optical microscope, invisible gold is very difficult to observe with these microscopes. The size of invisible gold may be on the order of nanometers, its presence is indicated by significant refractory imposed by the sulfide minerals (e.g., arsenopyrite, pyrite) [9]. The hindrance of gold recovery from complex sulfide minerals has been attributed to the formation of a passive layer on the mineral surface. Recent studies [10–14] have been conducted on the passive layer effect in sulfide minerals and the role played by sulfur in leaching solutions. In particular, the leaching of galena (PbS) has some problems, due to the low solubility of galena without the presence of an oxidant, the formation of precipitation, and the disposal of a large amount of iron that dissolves along with the lead. Therefore, leaching of the galena has been investigated by many researchers, who have proposed different nitric acid based leaching systems, including HNO_3 [10], H_2O_2-HNO_3 [13] and Fe(III)-HNO_3 [14]. The kinetics of lead dissolution from galena are slow, due to the formation of the passivation layer on the surface of sulfide by oxidation of sulfide, to form elemental sulfur under oxidative conditions. It is known that a partial dissolution or decomposition of sulfide minerals by the leaching solution lowers the leaching rate by forming a passive layer [11].

Unfortunately, the contributions of reactive sulfur species and the mechanisms of interaction with the leach residue surface are unclear. Therefore, studying the effects of relational minerals on gold is important. The processing of intractable substances presents challenges related to their complexity or refractory minerals. Increasing concerns regarding environmental protection has triggered efforts to identify alternatives that are environmentally friendly. Due to environmental concerns over cyanidation in hydrometallurgy, considerable attention has been dedicated to alternative non-cyanide solutions for refractory gold mineral leaching. Nitric acid has been recognized as one of the most promising reagents to pre-treatment process, and reduce cyanide consumption by refractory gold concentrates. The nitric acid leaching process is advantageous in that it can produce highly oxidizing conditions and is therefore an effective leaching agent for most sulfides [10]. However, a disadvantage of the nitric acid leaching process is the oxidation of sulfide, both to elemental sulfur and sulfate, in many cases in almost equal parts. This results in an increase in reagent consumption and the necessity of handling sulfate [10,13].

The use of microwave-assisted leaching processes has several advantages, including reduced energy consumption and elevated reproducibility [15,16]. These characteristics are the main drivers of metal ions in complex sulfide minerals. For example, selective leaching was successfully applied to lead smelting residues. Kim et al. (2017) reported that, when microwave assisted extraction (MAE) and autoclave leaching were performed to solubilize other valuable metals (Cu, Ni, Zn) from the matte, MAE has higher oxidation power than autoclave leaching [11]. This can explain the higher conversion of Fe sulfides to Fe oxides compared to autoclave leaching, as well as the higher leaching efficiency of Cu, Ni and Zn from their sulfides. Moreover, MAE is a simple process that can save energy and processing costs. Microwave heating has also been applied by Choi et al. [17] to perform pre-treatment, followed by a thiourea leaching step for gold extraction from gold concentrate. Due to the many advantages of microwaves, they have been widely used in mineral pretreatment. The main purpose of this work is to investigate the leaching behavior of the metals As, Cu, Fe, Zn, and Pb and passive layer decomposition in the refractory gold concentrate associated with nitric acid leaching under various conditions, using a microwave system. Furthermore, to increase the recovery of gold, nitric acid was used during microwave-assisted leaching.

2. Materials and Methods

2.1. Refractory Gold ore and Concentrates

The refractory gold ore and concentrate were obtained from a gold mine in Haenam, Korea. Refractory gold ores, including sulfide minerals such as pyrite and chalcopyrite, were used to investigate the influence of mineralogical characteristics on nitric acid leaching. A polished section of the ore mineral was prepared and studied microscopically under reflected light, to identify mineralogical properties. Polished sections were prepared by placing refractory gold ore in an epoxy resin which, after curing, was polished to ensure flatness. The textures of pyrite in the gold ore were investigated using nitric acid etching. The etching method involves the application of a few drops of 65% nitric acid on the polished mounts.

The refractory gold concentrate was obtained through a flotation process. The mineral composition of the refractory gold concentrate was analyzed using XRD. The chemical characterization of the surface species was performed using X-ray photoelectron spectroscopy (XPS). Gold is an element with a severe nugget effect [7,18], which may cause errors in the analysis based on the sampling process. To minimize this, the whole sample was sufficiently mixed and the sample was prepared using the cone and quartering method. To determine the chemical composition of the refractory gold concentrate, the sample was digested with aqua regia. The solution chemistry was analyzed using ICP-OES. The content of gold was analyzed using a fire assay [19,20].

2.2. Leaching Experiment

2.2.1. Microwave-Assisted Leaching: Effect of Nitric Acid Concentration and Temperature

The samples were sieved through a < 170 mesh screen. For the leaching experiment of the refractory gold concentrate, a microwave system (2.45 GHz, MARS 6, CEM Corporation, Matthews, NC, USA) was used. The microwave system was equipped with a digital temperature control sensor, which allowed the temperature to be accurately measured in real-time. The leaching experiments were conducted at different temperatures (40, 80, and 120 °C) and nitric acid concentrations (0.1, 0.5, and 1.0 M). Each leaching experiment consisted of several heating steps to reach a set temperature. A 100 mL disposable Teflon vessel containing 20 mL of HNO_3 at various concentrations and 1.0 g of the refractory gold concentrate was heated in the microwave. At the end of the reaction, the microwave system was switched off and the reactor was allowed to cool to room temperature. The oxidation and reduction potentials of the leaching solution were measured using an ORP (Orion 3-Star, Thermo Fisher Scientific, Santa Clara, CL, USA) meter. The leaching solution was filtered through a syringe filter. The weight of the leach residue was measured after leaching. The solution was analyzed by ICP-OES. Each experiment was performed in duplicate. The amount of metal ions leached was calculated using the following equation:

$$\mathrm{L} = \frac{M_1}{M_0} \times 100 \tag{1}$$

where L is the leaching percentage of metal ions, M_0 is the metal ion content of the sample before leaching, and M_1 is the metal ion content of the sample after leaching.

The rate of metal ion leaching was determined using the following equation [17]:

$$\mathrm{E} = E_I\left(1 - e^{-kt}\right) \tag{2}$$

where E (%) is the metal ion concentration in the leaching solution at time t, E_I (%) is the maximum concentration of metal ions, and K (min^{-1}) is the leaching rate constant.

2.2.2. Microwave-Assisted Leaching: Recovery of Gold

Microwave-assisted leaching was performed by heating a 500 mL Pyrex glass containing 200 mL of nitric acid (1.0, 3.0, and 5.0 M) and 20 g of the refractory gold concentrate. Figure 1 shows the schematic of the laboratory-scale microwave-assisted leaching developed in this study. The flask was placed in the center of the microwave oven and heated. The processing time was set to 15 min for all experiments. After the reaction period was over, the contents were cooled to ambient temperature and removed for solid-liquid separation. After each leaching experiment, the leach residue was filtered through filter paper and the content of gold in the leach residue was analyzed by fire assay. The weight of the leach residue was measured after leaching.

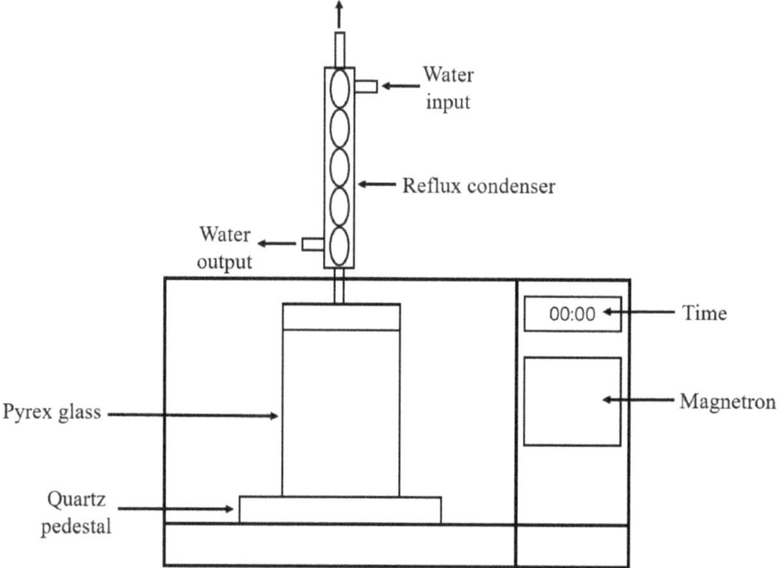

Figure 1. Schematic diagram of microwave-assisted leaching system.

2.3. *Characteristics of the Residue*

The leach residues obtained were analyzed by XPS and XRD to determine the surface arsenic, sulfur, and iron species and the mineralogical composition, respectively. Morphological studies were carried out on the leach residue using SEM-EDS (Scanning Electron Microscopy-Energy-dispersive X-ray spectroscopy) (S4800, Hitachi, Matsuda, Japan).

2.4. *Analysis Method*

The particle specific surface area of the leach residue was determined by laser diffraction (Mastersizer 2000, Malvern Panalytical Ltd., Malvern, UK). After the leaching experiment, the leaching solution was analyzed using inductively coupled plasma optical emission spectrometry (ICP-OES; Optima Model 5300 DV, Perkin Elmer, Norwalk, CT, USA), to determine the amount of metal leaching. The chemical characterization of the surface species was performed by XPS (K-Alpha+, Thermo Fisher Scientific, Waltham, MA, USA) to determine As, Fe, and S elements in the measured sample. The refractory gold concentrate and leach residue were subjected to XRD (X'Pert Pro MRD, PANalytical, Almelo, Netherlands) analysis. Cu-Kα X-rays were used and 2θ of 10°–70° was analyzed with an acceleration voltage of 40 kV, a current of 30 mA, and a scanning speed of 2°/min.

3. Results

3.1. Characteristics of Refractory Gold Ore and Concentrate

The refractory gold mineral sample was observed using an optical microscope. As described in Figure 2, the internal textures of the pyrite samples were investigated using nitric acid etching. Minerals containing arsenic are deeply etched by nitric acid; therefore, this procedure can greatly assist in highlighting pyrite growth zones [18]. However, arsenopyrite was not found. Optical microscopy images of pyrite with core and rim zones were mainly targeted for this study. The main features observed were pyrite and chalcopyrite; native gold and electrum were not found. The general crystals were angular and irregularly shaped, with a grain size distribution. Some typical textures, such as pores and cracks, were found in the pyrite. Chalcopyrite crystals contained in the pyrite were also observed.

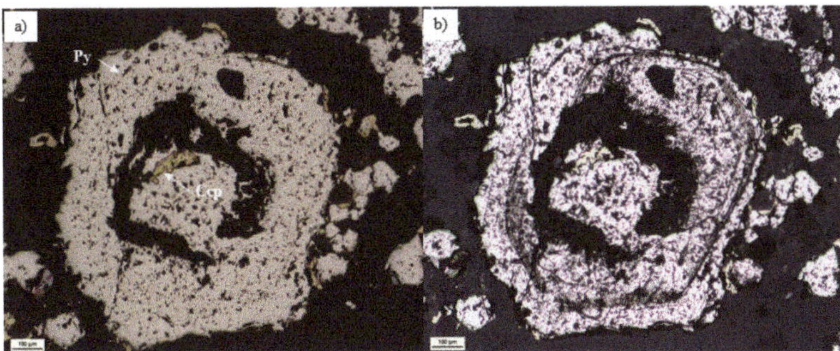

Figure 2. Microphotographs of gold ore samples. Reflected light microscopy of the etching of pyrite grains (**a**) before and (**b**) after etching in HNO_3 solution. Ccp is chalcopyrite and Py is pyrite.

Figure 3 presents the SEM-EDS analysis for the refractory gold concentrate. Idiomorphic cubic pyrite was observed. Generally, the Fe:S ratio is approximately 1:2 in pyrite; however, this phase has an S/Fe ratio of > 3, identical to that of pyrite. EDS analysis of the surface of pyrite showed the presence of arsenic, iron, sulfur, and gold, suggesting the formation of sulfur-containing arsenic and iron compounds. XRD analysis of the refractory gold concentrate revealed that consists of pyrite and quartz. The main chemical composition of the concentrate is listed in Table 1.

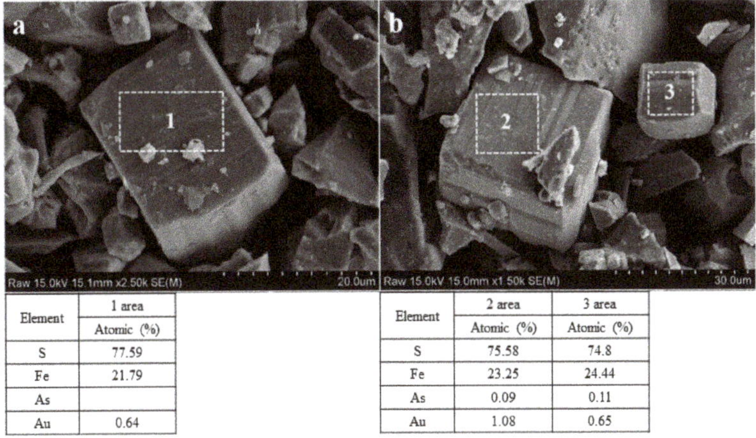

Element	1 area
	Atomic (%)
S	77.59
Fe	21.79
As	
Au	0.64

Element	2 area	3 area
	Atomic (%)	Atomic (%)
S	75.58	74.8
Fe	23.25	24.44
As	0.09	0.11
Au	1.08	0.65

Figure 3. SEM-EDS (Scanning Electron Microscopy-Energy-dispersive X-ray spectroscopy) of pyrite in the refractory gold concentrate.

Table 1. Chemical composition of the complex sulfide concentrate.

Cu (wt%)	Pb (wt%)	Zn (wt%)	Fe (wt%)	As (g/t)	Au (g/t)
0.19	0.15	0.63	43.86	0.10	94.37

3.2. Leaching Experiments

3.2.1. Effect of Nitric Acid Concentration

As shown in Figure 4, As, Cu, Fe, and Zn show increasing leaching efficiency with increasing nitric acid concentration from 0.1 to 1.0 M. Au was not detected under most of the experimental conditions. The increase in the concentration of nitric acid should accelerate the reaction rate of the refractory gold concentrate. As this effect is evident only at the highest nitric acid concentrations, it suggests that oxidation is the most likely factor. Based on the experimental results, the leaching parameters of the leaching experiment conditions estimated using Equation (2) are presented in Table S1. Initially, the metal ion leaching rate was slow, but steadily increased with increased nitric acid concentration. The sample mass decreased from 19.0% at 0.1 M, 27.0% at 0.5 M and 51.5% at 1.0 M. Nitric acid concentration was shown to be effective for the decomposition of the dominant sulfide minerals, due to the oxidation of insoluble sulfides to water-soluble sulfate phases during leaching.

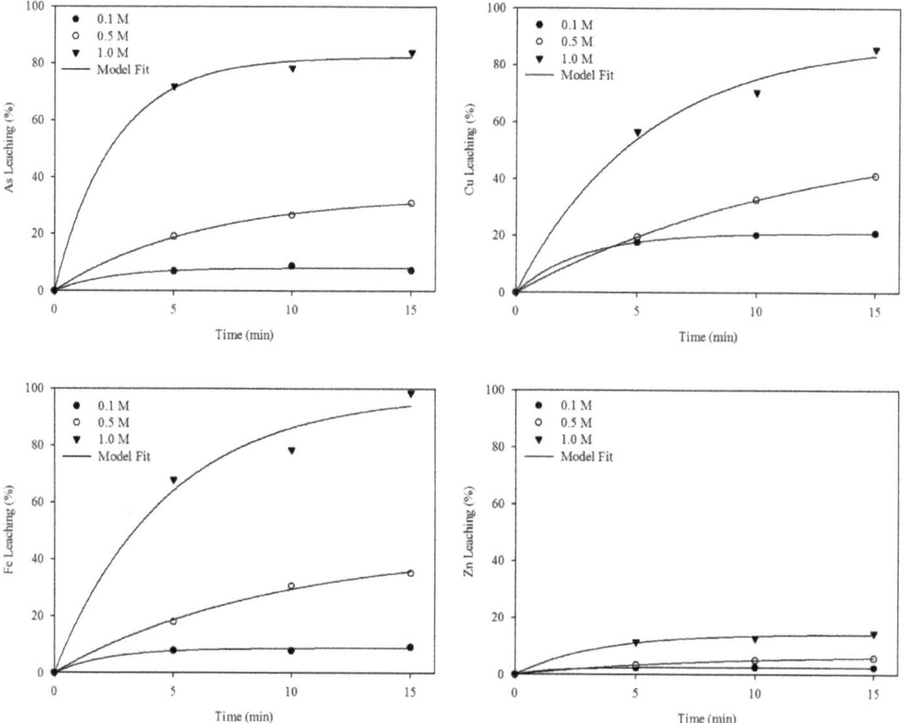

Figure 4. Effect of HNO_3 concentration on the leaching efficiencies of As, Cu, Zn, and Fe from the refractory gold concentrate. The leaching conditions were a reaction time of 15 min, reaction temperature of 80 °C, and HNO_3 concentration of 0.1, 0.5, or 1.0 M. According to Equation (2) in Table 2.

Table 2. Refractory gold concentrate and leach residue surface atomic composition with different treatments. The leaching conditions were a reaction temperature of 80 °C reaction time of 15 min, and HNO_3 concentration of 0.1 and 0.5 M.

Experiment	S2p (%)	Fe2p (%)	S/Fe ratio (%)	As3d (%)
Raw	44.52	39.64	1.12	6.39
Residue 0.1 M	77.79	17.93	4.34	0.11
Residue 0.5 M	78.50	18.69	4.20	-

3.2.2. Effect of Temperature

The results are presented in Figure 5. The As leaching increased from 84.0% to 92.5%, as the temperature increased from 80 to 120 °C. Initially, the As leaching rate was slow, but steadily increased with increased temperature. However, Cu leaching decreased from 85.8% at 80 °C to 75.8% at 120 °C and Zn leaching decreased from 14.6% at 80 °C to 10.5% at 120 °C. The Fe leach ability was 98.7% at 80 °C and slightly decreased to 94.6% at 120 °C. As the temperature increased from 80 to 120 °C, the Cu leaching rate constant increased from 0.18 to 0.36 min^{-1} and the Fe leaching rate constant increased from 0.21 to 0.66 min^{-1}. Based on the experimental results, the leaching parameters of the leaching experiment conditions estimated using Equation (2) are presented in Table S2. Due to the effect of temperature, it is necessary for sulfide to be oxidized into sulfate, with a minimized formation of elemental sulfur that impedes the subsequent recovery of gold.

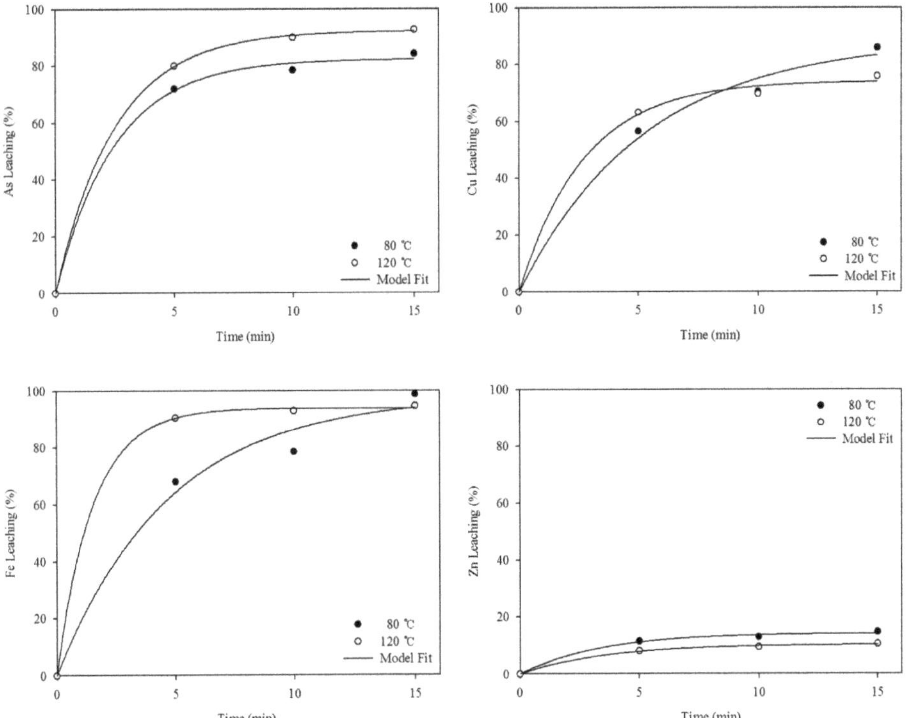

Figure 5. Effect of temperature on the leaching efficiencies of As, Cu, Zn, and Fe from the refractory gold concentrate. The leaching conditions were a reaction temperature between 80 and 120 °C, HNO_3 concentration of 1.0 M, and reaction time of 15 min. According to Equation (2) in Table S2.

Figure 6 shows the effect of temperature on the Pb leach ability from the refractory gold concentrate at nitric acid 1.0 M. Pb leaching was increased for the first 5 min; the rate decreased after that, and leaching efficiency decreased from 57.8% at 40 °C, 15.3% at 80 °C, and was not detected at 120 °C. The ORP (mV) was reduced by leaching, particularly at the end of the leaching time. The ORP slowly decreased to 686 mV at 40 °C, 638 mV at 80 °C, and 623 mV at 120 °C (Figure 6b). These results indicate that lead complexes, such as $PbNO^{3+}$, $Pb(NO_3)_2$, $Pb(NO_3)_3^-$ and $Pb(NO_3)_4^{2-}$, are formed by oxidation at high-temperatures [10]. The respective reactions may involve the formation of a passive layer of elemental sulfur, which is then leached, and passivation by lead sulfate or basic sulfate [10]. The lower Pb leaching at > 80 °C is likely due to passivation by increased elemental sulfur formation. This shows that it is possible to obtain lead in solution, although lead may not precipitate at lower potentials. XRD analysis of the leach residues mainly show untreated pyrite, with a small quantity of quartz and sulfur. The reaction produced elemental sulfur (Figure 7) and there is a clear indication that surface passivation prevented further leaching. Therefore, the decrease in Pb leaching efficiency may be due to the consumption of oxidants and acids.

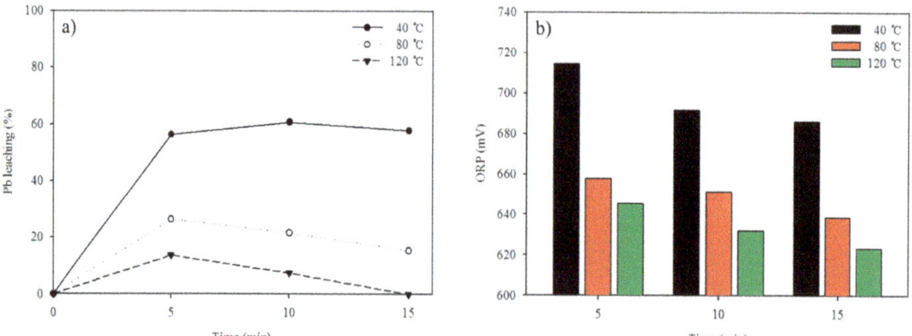

Figure 6. Effect of temperature on the leaching efficiencies of Pb (**a**) and ORP (mV) (**b**) from the refractory gold concentrate. The leaching conditions were an HNO_3 concentration of 1.0 M, reaction time of 15 min, and reaction temperatures of 40, 80, and 120 °C.

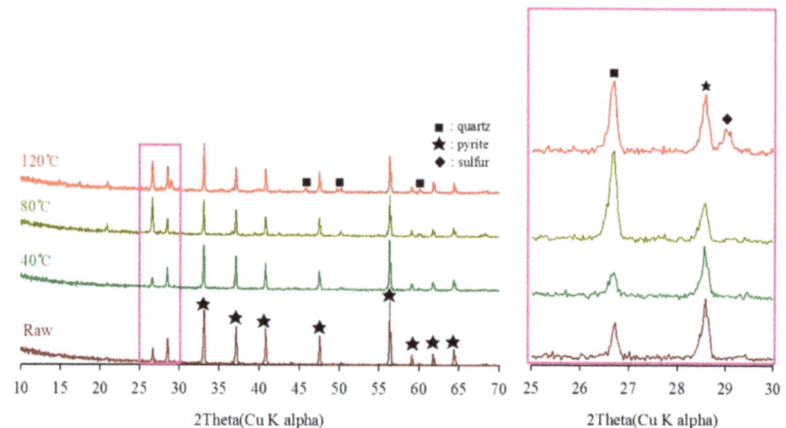

Figure 7. XRD patterns of raw and leach residue from nitric acid leaching at 40, 80, and 120 °C. The leaching conditions were an HNO_3 concentration of 1.0 M and a reaction time of 15 min.

As the temperature increased, Pb leaching decreased, which mainly occurred due to the formation of a passive layer. The leaching efficiencies of Fe in the leaching experiment were not significant

(Figure 8). The Fe leaching rate was 90.5% at 40 °C, 93.1% at 80 °C, and 94.61% at 120 °C after 15 min. Sulfide oxidation [21] is strongly affected by temperature (> 100 °C) and a temperature increase has a negative effect on Pb leaching. It is noteworthy that the Pb leaching efficiency was lower than the Fe leaching efficiency. Therefore, it is difficult to simultaneously leach Pb and other base metals, such as As, Cu, and Zn, from sulfide minerals in nitric acid at elevated temperatures. In addition, Pb shows a different leaching behavior compared with the other metal ions, due to the leaching of elemental sulfur to sulfate that occurs after the complete oxidation of sulfide to sulfur [13].

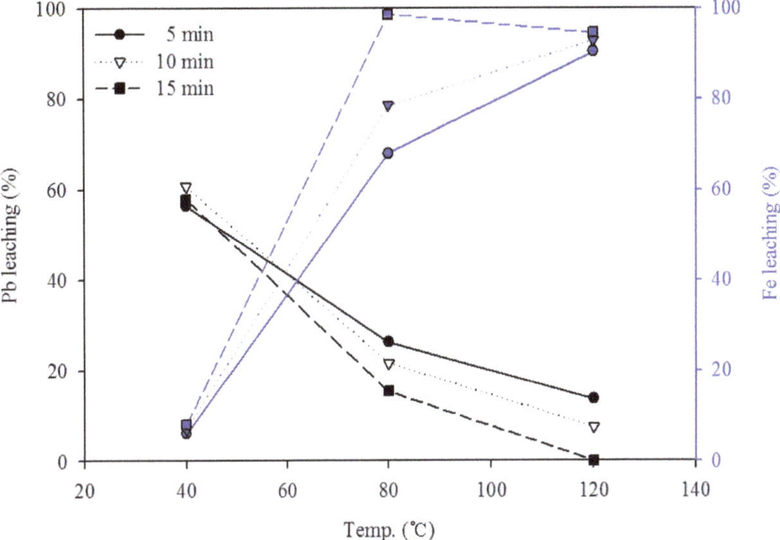

Figure 8. Effect of temperature on the leaching efficiencies of Pb and Fe from the refractory gold concentrate. The leaching conditions were an HNO_3 concentration of 1.0 M and a reaction temperature of 40, 80 and 120 °C.

3.3. Characterization of the Passive Layer in the Refractory Gold Minerals

XPS analysis was conducted to detect the dissolution changes of metal ions (S, Fe, and As) on the surface of minerals. Table 2 shows the changes in the chemical environment of S, Fe, and As, before and after leaching. These results indicate a change in the surface of the metal ions during nitric acid leaching. The Fe atom percent was 39.64% in the raw sample and 17.93% at 0.1 M, 18.69% at 0.5 M in the nitric acid leaching residue, indicating that the superficial pyrite in the minerals did not significantly dissolve during the reaction time of 10 min. Some Fe existed in the crystal lattice of the sulfide minerals in the refractory gold concentrate, due to the superficial pyrite on the minerals dissolving. The XPS analysis indicates that the pyrite in the refractory gold concentrate did not significantly dissolve during the nitric acid leaching at < 0.5 M. The S/Fe ratio determines the surface of sulfide mineral decomposition and affects the efficiency of passive layer removal [22]. S/Fe ratio in nitric acid leaching (reaction time of 10 min) reaches 4.34 at 0.1 M and 4.20 at 0.5 M. This indicates that a relatively lower proportion of iron was consumed. However, the atomic percentages of arsenic on the surface of the raw concentrate decreased from 6.39% to 0.11% at 0.1 M and was not detected at 0.5 M. This indicates that a relatively large proportion of arsenic was consumed.

According to the XPS reference [23], the Fe2p peak with a binding energy of 708.17 eV and the S 2p peak with a binding energy of 160.93 eV belong to pyrite. In the nitric acid leaching residue from leaching at < 0.5 M, the Fe2p peak with a binding energy of 708.54 eV and the S2p peak with a binding energy of 161.02 eV were assigned to pyrite, both having small changes compared with those in the raw sample. The S2p spectrum demonstrates that S exists in multiple oxidation states, including monosulfide (S^{2-}), disulfide ($S_2{}^{2-}$), polysulfide ($S_n{}^{2-}$), elemental sulfur (S^0), thiosulfate ($S_2O_3{}^{2-}$), and sulfate ($SO_4{}^{2-}$). Raw sample disulfide species of sulfide minerals are oxidized to sulfate via several intermediate steps [23,24]. The leaching of sulfide is complex, involving the dissolution of various sulfur species. For the leaching of sulfide minerals, the preferential release of metal ions such as Fe, Cu, Zn, etc., into the leaching solution results in the passive layer of surface S^{2-}, accounting for the formation of $S_2O_3{}^{2-}$ and $SO_4{}^{2-}$. This was possibly due to the S-S bonding being weaker than the Fe-S bonding, meaning that the S-S bonds were more easily broken in the lattice pyrite. The passivation layer on the residue surface mainly consisted of iron and sulfur species, as seen in Figures 9 and 10.

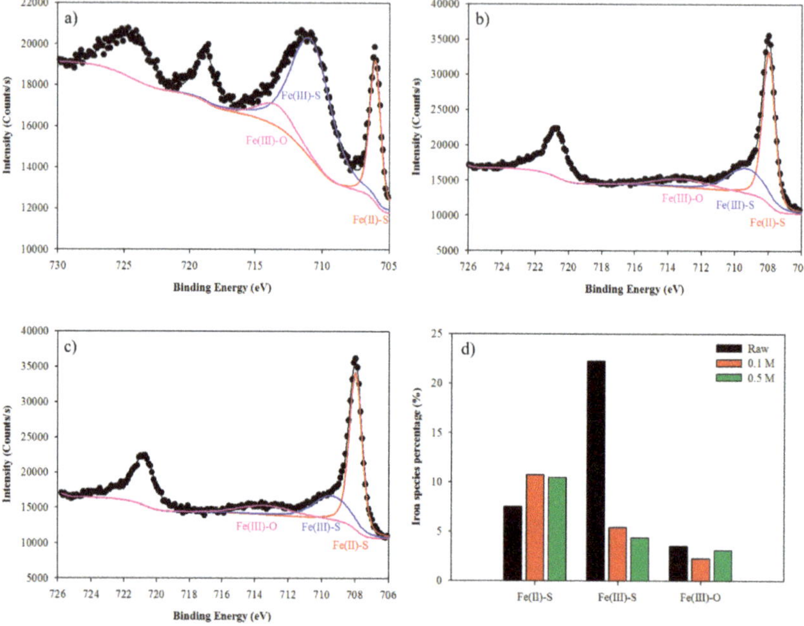

Figure 9. XPS spectra of Fe2p for (**a**) raw concentrate, (**b**) leach residue (0.1 M), (**c**) leach residue (0.5 M), and (**d**) distribution of the surface Fe species. The leaching conditions were 80 °C and 15 min.

Under different conditions, refractory gold concentrate leaching results in the formation of a passive layer on its surface, consisting of sulfur- and iron-rich layers. XRD analyses were conducted to confirm the changes in refractory sulfide minerals during leaching; the results of these analyses are presented in Figure 11. The intensities of the pyrite peak (111) decreased, while those of the gangue peak increased in the leach residues. Sulfur was not found in the leach residues, which means that the sulfur in the sulfides was transformed into thiosulfate or sulfate, as opposed to elemental sulfur. The (111) plane shows a higher oxidation rate than the (100) and (110) planes [24]. This is likely due to the S-S bond in the pyrite, which is weaker than that in Fe-S. This indicates that the fraction of dissolved metal ions increased as the concentration of nitric acid increased. The SEM image (Figure 12) shows the morphology of the residue after leaching for 10 min, which is not noticeably different from that of the refractory gold concentrate. However, the surface of the leach residue became uneven; it is obvious that multiple holes appeared on the surface of the leach residue particles. It can be concluded that pyrite in the sulfide minerals is not decomposed by nitric acid at < 0.5 M (Figure 11). Therefore, we investigated the increase in nitric acid concentration to enhance the recovery of gold.

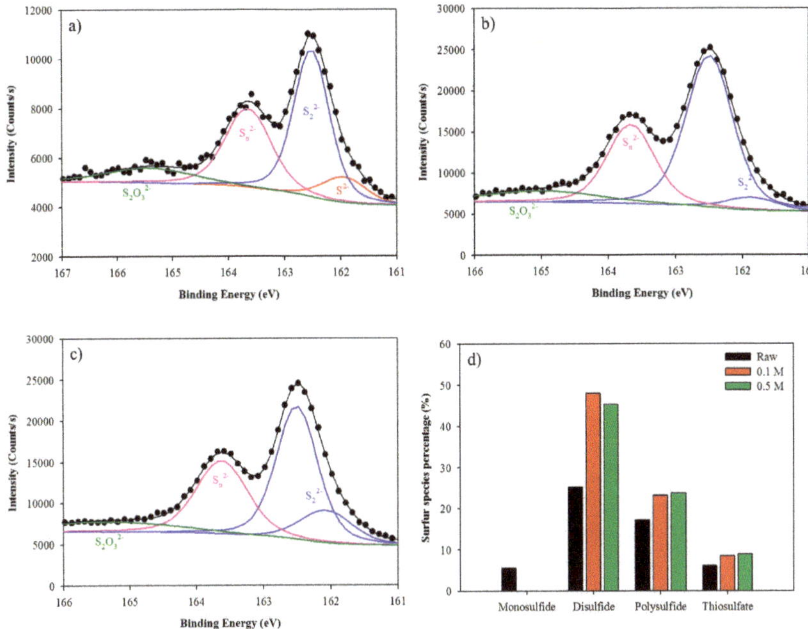

Figure 10. XPS spectra of S2p for (**a**) raw concentrate, (**b**) leach residue (0.1 M), (**c**) leach residue (0.5 M), and (**d**) distribution of the surface S species. The leaching conditions were 80 °C and 15 min.

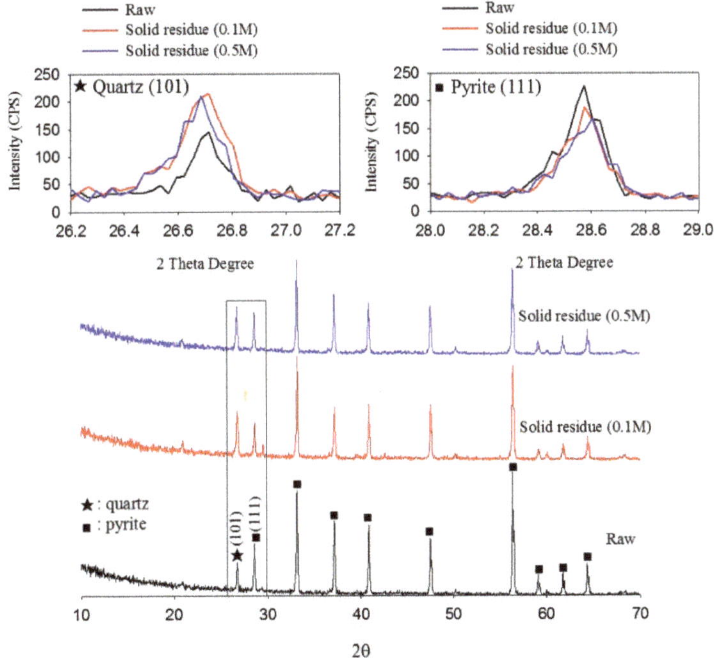

Figure 11. XRD patterns of raw and leach residue from nitric acid leaching at 0.1 and 0.5 M. The leaching conditions were 80 °C and 15 min.

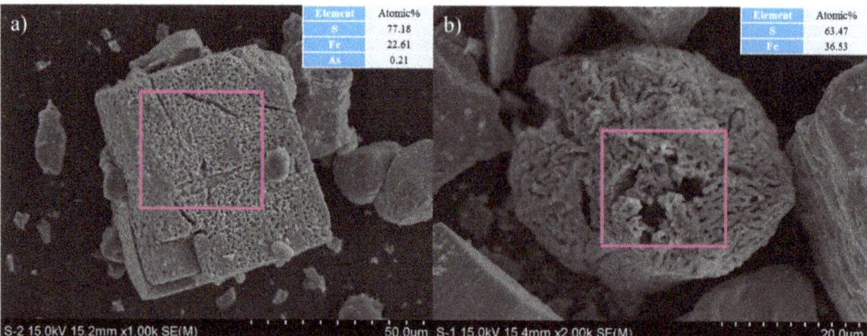

Figure 12. SEM images of leach residues. The leaching conditions were 80 °C, 15 min, and HNO_3 concentration of (**a**) 0.1 M and (**b**) 0.5 M.

3.4. Recovery of Gold by Microwave-Assisted Leaching

The gold content of the leach residue was analyzed using a fire assay. The effect of nitric acid concentration on gold recovery was studied, with a leaching time of 15 min (Figure 13). The original sample showed a gold content of 94.37 g/t according to the fire assay. After the leaching process with 3.0 M nitric acid for 10 min, a gold content of 126.39 g/t was obtained. Compared to the untreated refractory gold concentrate, the higher gold recovery confirms that a large portion of the gold was refractory in nature, with the gold occurring as either solid solution components in sulfide minerals or encapsulated in the sulfide minerals. The weight of the leach residue decreased when the nitric acid concentration increased. The sample mass decreased by a maximum of 69.60% after leaching at 5.0 M, indicating that the sulfide minerals were decomposed or dissolved by microwave-assisted leaching.

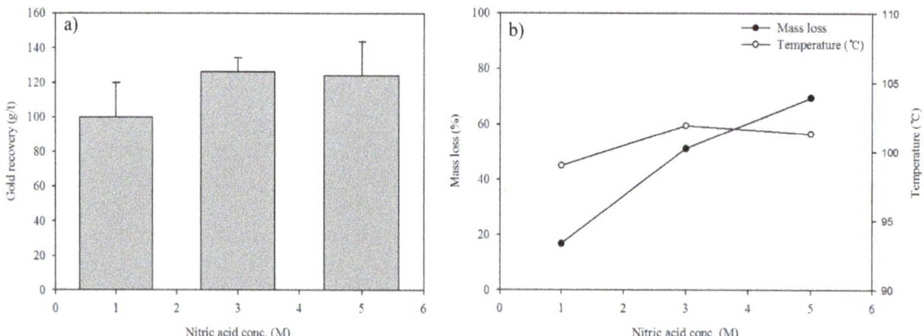

Figure 13. Effect of HNO_3 concentration on the (**a**) recovery of gold and (**b**) mass loss and reaction temperature from the refractory gold concentrate. The leaching conditions were a reaction time of 15 min and an HNO_3 concentration of 1.0, 3.0 and 5.0 M.

Microwave-assisted leaching experiments for gold recovery were conducted for the refractory gold concentrate for different reaction times. The results showed that the reaction time enhanced the recovery of gold. The gold recovery was approximately 132.55 g/t after 20 min of leaching with 2.0 M nitric acid, whereas the mass loss in leach residue similarly decreased with reaction time (Figure 14). The sample mass decreased by a maximum of 40.5% after 20 min. The microwave-assisted leaching experiment resulted in the loss of the constituents in the refractory concentrate.

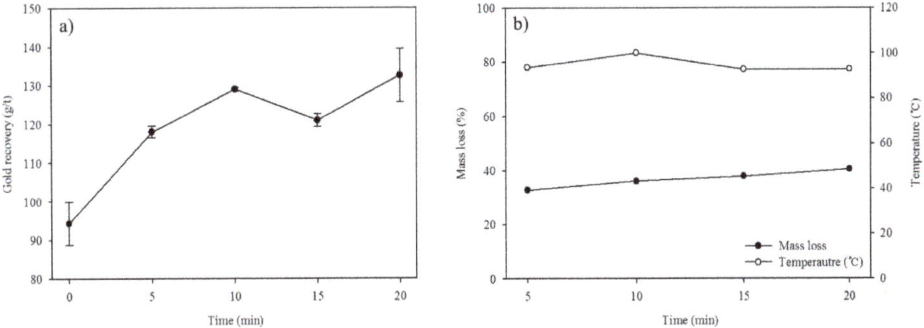

Figure 14. Effect of leaching time on (**a**) recovery of gold and (**b**) mass loss and reaction temperature from the refractory gold concentrate. The leaching conditions were an HNO_3 concentration of 2.0 M.

Leach residues were collected by performing microwave-assisted leaching experiments for each nitric acid concentration. When XRD analysis was conducted on these leach residues, quartz, pyrite, and sulfur were detected (Figure 15). While pyrite was detected with 3 M nitric acid, it disappeared with 5 M nitric acid. This means that the intensities of the pyrite peak decreased, while those of the gangue peak increased in the leach residues, thus indicating the dissolution of pyrite during microwave-assisted leaching. However, sulfur appeared in the leach residue at 3.0 M and 5.0 M nitric acid, even though it was not detected in the concentrate. This appears to be due to the reactions of the pyrite included in the refractory gold concentrate with nitric acid, as shown in reactions (3) and (4) [25].

$$FeS_2 + 8HNO_3 = Fe(NO_3)_3 + 2H_2SO_4 + 2H_2O + 5NO \quad (3)$$

$$2FeS_2 + 8HNO_3 = Fe_2(SO_4)_3 + S^0 + 8NO + 4H_2O \quad (4)$$

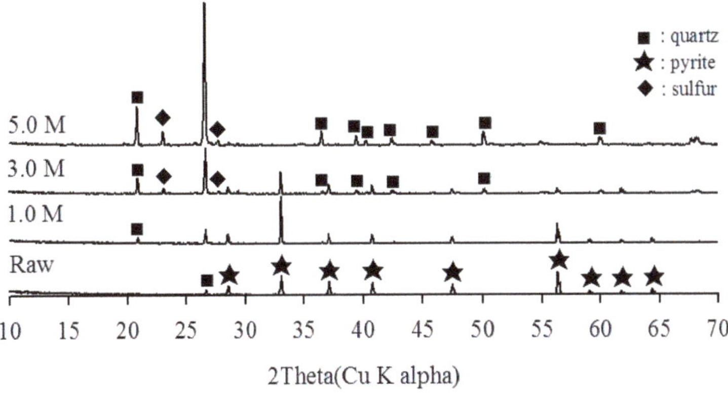

Figure 15. XRD patterns of raw and leach residue from nitric acid leaching at 1.0, 3.0 and 5.0 M. The leaching time was 15 min.

From the redox reactions above, it can be inferred that sulfur was generated from the leaching decomposition. However, most studies [13,19] have reported that S^0 is responsible for the blocking of the surface. The transformation of S species, such as S^0, on the surface is the most important intermediate during the leaching of sulfide minerals, and it is considered to be the main component hindering the dissolution of sulfide minerals. The results also revealed that a larger proportion of sulfur was transformed and the generated hydroxyl precipitated (Figure 16), which could be confirmed by SEM-EDS. More extreme reaction conditions, such as the increase in nitric acid concentration from 1.0 M to 5.0 M, facilitated the decomposition of passivation species derived from metal ion dissolution and the liberation of gangue minerals from the sulfide surface. After leaching, SEM-EDS analysis was conducted on the leach residues (Figure 16). Based on the EDS analysis, the particles contain sulfur. Leaching increased with time, due to the oxidation of insoluble sulfides to soluble sulfate phases. A larger proportion of sulfur was transformed from the pyrite lattice in the refractory sulfide, thus leaving many vacant areas and microstructures, which can effectively liberate encapsulated gold and improve the recovery of gold.

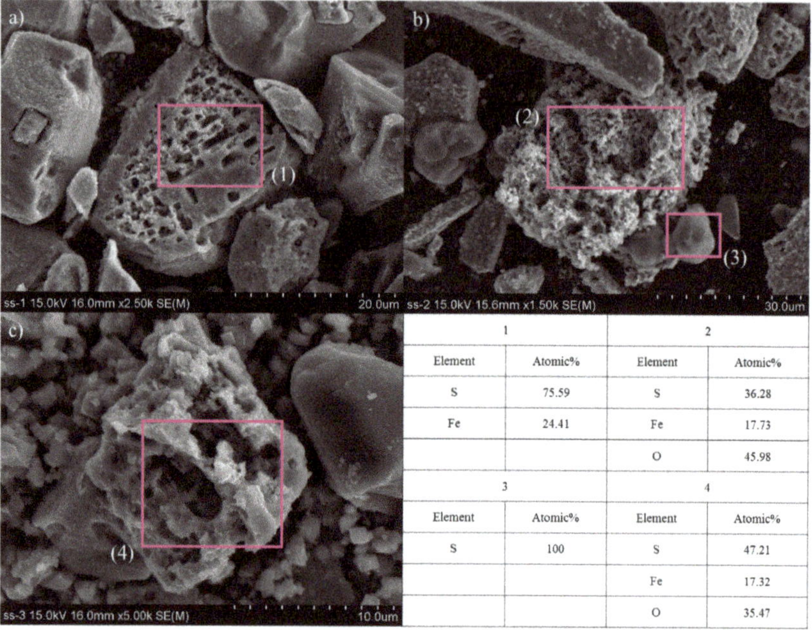

1		2	
Element	Atomic%	Element	Atomic%
S	75.59	S	36.28
Fe	24.41	Fe	17.73
		O	45.98
3		4	
Element	Atomic%	Element	Atomic%
S	100	S	47.21
		Fe	17.32
		O	35.47

Figure 16. SEM images of the leach residues from nitric acid leaching at (a) 1.0 M, (b) 3.0 M and (c) 5.0 M. The leaching time was 15 min.

4. Conclusions

The leaching efficiencies of metal ions (As, Cu, Zn, Fe, and Pb) and the recovery of gold from the refractory gold concentrate were investigated using a microwave system. As the acid concentration increased, the metal ion leaching increased, but the leach residues for the low nitric acid concentration (> 0.5 M) consisted mainly of untreated pyrite. The S/Fe ratio determines the surface of the leach residue, which affects the efficiency of passive layer removal. S/Fe ratio in nitric acid leaching with a reaction time of 10 min reached 4.34 at 0.1 M and 4.20 at 0.5 M. It was found that a relatively lower proportion of iron was consumed, and the pyrite in the refractory gold concentrate did not significantly dissolve during nitric acid leaching at > 0.5 M.

In the refractory gold concentrate leaching experiments, nitric acid leaching at high-temperature could limit the decomposition of sulfide minerals, due to the passive layer in the refractory gold concentrate. During the leaching process, As, Cu, Fe, and Zn, leaching increased with time, due to the oxidation of insoluble sulfides to soluble sulfate phases. Initially, the As leaching rate was slow, but steadily increased with increased nitric acid concentration. For the refractory gold concentrate, the As, Cu, Fe, and Zn leaching increased from 7.28% to 84.0% (As), 20.8% to 85.8% (Cu), 9.19% to 98.7% (Fe), and 2.46% to 14.9% (Zn), upon increasing the HNO_3 concentration from 0.1M to 1.0M (80 °C,

S/L ration 5, leaching time 15 min). As the temperature increased from 80 to 120 °C, the Cu leaching rate constant increased from 0.18 to 0.36 min^{-1} and the Fe leaching rate constant increased from 0.21 to 0.66 min^{-1}. However, Pb leaching decreased at > 80 °C, due to complex lead and passivation, by increased elemental sulfur formation from the high-temperature oxidation.

Microwave-assisted leaching experiments for gold recovery were conducted for the refractory gold concentrate. More extreme reaction conditions, such as the increase in nitric acid concentration from 1.0 to 5.0 M, facilitated the decomposition of passivation species derived from metal ion dissolution and the liberation of gangue minerals from the sulfide surface. From the comparison between the XRD patterns of the refractory gold concentrate and the leach residues after leaching with different nitric acid concentrations, it can be concluded that pyrite in the sulfide minerals can be destroyed. The SEM-EDS analyses of the leach residue showed that dissolution and decomposition of pyrite in the complex sulfide concentrate leave many vacant areas and microstructures, which can effectively liberate encapsulated gold and improve the recovery of gold.

The recovery rate of gold in the leach residue was improved with microwave-assisted leaching and the gold recovery was about 132.55 g/t after 20 min of the leaching experiment (nitric acid at 2.0 M), according to fire assays. The effect of the increase in nitric acid concentration was consistent with increased exposure to reactive sulfide minerals, which could effectively liberate, encapsulate and improve the gold recovery rate.

The current rapid decline in high-grade gold ores has made the mineral processing industry increasingly reliant on complex and refractory gold ores. At present, several approaches have been employed for mineral processing, including metallurgy and hydro-metallurgy. The leaching process is a method used in hydrometallurgy, which is used to leach base and precious metals from source materials. Particularly, percolation leaching methods, such as Heap leaching, dump leaching, bioleaching and in situ leaching, have been very effective in extracting metals from low grade ores, which could not otherwise be economically extracted. However, the challenge of this facility has been the handling of waste and control of environmental pollution caused by toxic leakages from heaps [26]. In addition, the formation of complex or refractory substances has limited its application. The main reason for this is that the complex or refractory ores require more processing and the various approaches used for gold recovery are challenging to perform in aqueous solutions, owing to surface passivation.

Gold-bearing ores contain pyrite, chalcopyrite, arsenopyrite, and galena, which interfere with gold recovery and lower its efficiency. Therefore, studying the effects of relational minerals on gold is important. Among the alternative processes, the use of microwave additives followed by leaching is important for the recovery of precious metals from complex or refractory minerals. To increase the recovery of gold, the microwave-assisted leaching process may be used as a process to address the formation of complex or refractory sulfide minerals.

Supplementary Materials: The following are available online at http://www.mdpi.com/2075-4701/10/5/571/s1, Table S1: The summary of leaching parameters for As, Cu, Zn, and Fe leaching using Equation (2). The leaching conditions were a reaction time of 15 min, reaction temperature of 80 °C, and HNO3 concentration of 0.1, 0.5, or 1.0 M. Table S2: The summary of leaching parameter for As, Cu, Zn, and Fe leaching, using Equation (2). The leaching conditions were a reaction temperature between 80 and 120 °C, HNO3 concentration of 1.0 M, and reaction time of 15 min.

Author Contributions: H.K.: Experiment, Investigation; E.M.: Formal analysis, Visualization; O.P.: Formal analysis; N.C.: Funding acquisition, Conceptualization; C.P.: Project administration, Conceptualization; K.C.: Writing-original draft, Writing-review & editing. All authors have read and agreed to the published version of the manuscript.

Funding: This study was supported by the Korea Ministry of Environment (MOE), South Korea, as an Advanced Industrial Technology Development Project (No. 2016000140010).

Conflicts of Interest: The authors declare no conflicts of interest.

References

1. Adams, M.D. *Gold Ore Processing: Project Development and Operations*, 2nd ed.; Elsevier: Cambridge, MA, USA, 2016; pp. 57–94.
2. Senanayake, G. A review of effects of silver, lead, sulfide and carbonaceous matter on gold cyanidation and mechanistic interpretation. *Hydrometallurgy* **2008**, *90*, 46–73. [CrossRef]
3. Feng, D.; Van Deventer, J. Effect of thiosulphate salts on ammoniacal thiosulphate leaching of gold. *Hydrometallurgy* **2010**, *105*, 120–126. [CrossRef]
4. Hough, R.; Noble, R.; Reich, M. Natural gold nanoparticles. *Ore Geol. Rev.* **2011**, *42*, 55–61. [CrossRef]
5. Asamoah, R.K.; Skinner, W.; Addai-Mensah, J. Alkaline cyanide leaching of refractory gold flotation concentrates and bio-oxidised products: The effect of process variables. *Hydrometallurgy* **2018**, *179*, 79–93. [CrossRef]
6. Cho, K.H.; Lee, J.J.; Park, C.Y. Liberation of Gold Using Microwave-Nitric Acid Leaching and Separation-Recovery of Native Gold by Hydro-Separation. *Minerals* **2020**, *10*, 327. [CrossRef]
7. Kusebauch, C.; Oelze, M.; Gleeson, S.A. Partitioning of arsenic between hydrothermal fluid and pyrite during experimental siderite replacement. *Chem. Geol.* **2018**, *500*, 136–147. [CrossRef]
8. Li, J.; Dabrowski, B.; Miller, J.; Acar, S.; Dietrich, M.; LeVier, K.; Wan, R. The influence of pyrite pre-oxidation on gold recovery by cyanidation. *Miner. Eng.* **2006**, *19*, 883–895. [CrossRef]
9. Zia, Y.; Mohammadnejad, S.; Abdollahy, M. Gold passivation by sulfur species: A molecular picture. *Miner. Eng.* **2019**, *134*, 215–221. [CrossRef]
10. Zárate-Gutiérrez, R.; Lapidus, G.; Morales, R. Aqueous oxidation of galena and pyrite with nitric acid at moderate temperatures. *Hydrometallurgy* **2012**, *115*, 57–63. [CrossRef]
11. Kim, E.; Horckmans, L.; Spooren, J.; Vrancken, K.; Quaghebeur, M.; Broos, K. Selective leaching of Pb, Cu, Ni and Zn from secondary lead smelting residues. *Hydrometallurgy* **2017**, *169*, 372–381. [CrossRef]
12. Han, B.; Altansukh, B.; Haga, K.; Takasaki, Y.; Shibayama, A. Leaching and Kinetic Study on Pressure Oxidation of Chalcopyrite in H_2SO_4 Solution and the Effect of Pyrite on Chalcopyrite Leaching. *J. Sustain. Metall.* **2017**, *3*, 528–542. [CrossRef]
13. Aydoğan, S.; Erdemoğlu, M.; Uçar, G.; Aras, A. Kinetics of galena dissolution in nitric acid solutions with hydrogen peroxide. *Hydrometallurgy* **2007**, *88*, 52–57. [CrossRef]
14. Kholmogorov, A.G.; Pashkov, G.L.; Mikhlina, E.V.; Shashina, L.V.; Zhizhaev, A.M. Activation of hydrometallurgical treatment of PbS in nitric solutions. *Chem. Sustain. Dev.* **2003**, *11*, 879–881.
15. Al-Harahsheh, M.; Kingman, S. Microwave-assisted leaching—A review. *Hydrometallurgy* **2004**, *73*, 189–203. [CrossRef]
16. Zhang, X.; Liu, F.; Xue, X.; Jiang, T. Effects of microwave and conventional blank roasting on oxidation behavior, microstructure and surface morphology of vanadium slag with high chromium content. *J. Ailloy. Compd.* **2016**, *686*, 356–365. [CrossRef]
17. Choi, N.-C.; Kim, B.-J.; Cho, K.; Lee, S.; Park, C.-Y. Microwave pretreatment for thiourea leaching for gold concentrate. *Metals* **2017**, *7*, 404. [CrossRef]
18. Cromie, P.; Makoundi, C.; Zaw, K.; Cooke, D.R.; White, N.; Ryan, C. Geochemistry of Au-bearing pyrite from the Sepon Mineral District, Laos DPR, Southeast Asia: Implications for ore genesis. *J. Asian Earth Sci.* **2018**, *164*, 194–218. [CrossRef]
19. Santos-Munguía, P.; Nava-Alonso, F.; Rodríguez-Chávez, V.; Alonso-González, O. Hidden gold in fire assay of gold telluride ores. *Miner. Eng.* **2019**, *141*, 105844. [CrossRef]
20. Choi, N.-C.; Kim, B.-J.; Cho, K.-H.; You, D.-S.; Park, C.-Y. Enhancement of gold recovery during lead fire assay by salt-roasting. *Geosystem Eng.* **2014**, *17*, 226–234. [CrossRef]
21. Kim, E.; Roosen, J.; Horckmans, L.; Spooren, J.; Broos, K.; Binnemans, K.; Vrancken, K.C.; Quaghebeur, M. Process development for hydrometallurgical recovery of valuable metals from sulfide-rich residue generated in a secondary lead smelter. *Hydrometallurgy* **2017**, *169*, 589–598. [CrossRef]
22. Gan, M.; Gu, C.; Ding, J.; Zhu, J.; Liu, X.; Qiu, G. Hexavalent chromium remediation based on the synergistic effect between chemoautotrophic bacteria and sulfide minerals. *Ecotoxicol. Environ. Saf.* **2019**, *173*, 118–130. [CrossRef]

23. Zhao, L.; Chen, Y.; Liu, Y.; Luo, C.; Wu, D. Enhanced degradation of chloramphenicol at alkaline conditions by S (-II) assisted heterogeneous Fenton-like reactions using pyrite. *Chemosphere* **2017**, *188*, 557–566. [CrossRef] [PubMed]
24. Feng, J.; Tian, H.; Huang, Y.; Ding, Z.; Yin, Z. Directional Oxidation of Pyrite in Acid Solution. *Minerals* **2019**, *9*, 7. [CrossRef]
25. Rogozhnikov, D.A.; Shoppert, A.A.; Dizer, O.A.; Karimov, K.A.; Rusalev, R.E. Leaching Kinetics of Sulfides from Refractory Gold Concentrates by Nitric Acid. *Metals* **2019**, *9*, 465. [CrossRef]
26. Ilankoon, I.M.S.K.; Tang, Y.; Ghorbani, Y.; Northey, S.; Yellishetty, M.; Deng, X.; McBride, D. The current state and future directions of percolation leaching in the Chinese mining industry: Challenges and opportunities. *Miner. Eng.* **2018**, *125*, 206–222. [CrossRef]

Article

Acid and Acid-Alkali Treatment Methods of Al-Chloride Solution Obtained by the Leaching of Coal Fly Ash to Produce Sandy Grade Alumina

Dmitry Valeev [1,2,*], **Andrei Shoppert** [3], **Alexandra Mikhailova** [4] **and Alex Kondratiev** [2]

1 Laboratory of New Metallurgical Processes, A.A. Baikov Institute of Metallurgy and Materials Science, Russian Academy of Sciences, 49, Leninsky Prospect, 119334 Moscow, Russia
2 Scientific Research Centre "Thermochemistry of Materials", National University of Science & Technology "MISIS", 4, Leninsky Prospect, 119049 Moscow, Russia; al.v.kondratiev@gmail.com
3 Department of Non-ferrous Metals Metallurgy, Ural Federal University named after the first President of Russia B.N. Yeltsin, Yekaterinburg 620002, Russia; andreyshop@list.ru
4 Laboratory of Crystal Structure Studies, A.A. Baikov Institute of Metallurgy and Materials Science, Russian Academy of Sciences, 49, Leninsky Prospect, 119334 Moscow, Russia; sasham1@mail.ru
* Correspondence: dmvaleev@yandex.ru; Tel.: +7-905-781-6667

Received: 13 April 2020; Accepted: 28 April 2020; Published: 29 April 2020

Abstract: Sandy grade alumina is a valuable intermediate material that is mainly produced by the Bayer process and used for manufacturing primary metallic aluminum. Coal fly ash is generated in coal-fired power plants as a by-product of coal combustion that consists of submicron ash particles and is considered to be a potentially hazardous technogenic waste. The present paper demonstrates that the Al-chloride solution obtained by leaching coal fly ash can be further processed to obtain sandy grade alumina, which is essentially suitable for metallic aluminum production. The novel process developed in the present study involves the production of amorphous alumina via the calcination of aluminium chloride hexahydrate obtained by salting-out from acid Al-Cl liquor. Following this, alkaline treatment with further Al_2O_3 dissolution and recrystallization as $Al(OH)_3$ particles is applied, and a final calcination step is employed to obtain sandy grade alumina with minimum impurities. The process does not require high-pressure equipment and reutilizes the alkaline liquor and gibbsite particles from the Bayer process, which allows the sandy grade alumina production costs to be to significantly reduced. The present article also discusses the main technological parameters of the acid treatment and the amounts of major impurities in the sandy grade alumina obtained by the different (acid and acid-alkali) methods.

Keywords: coal fly ash; sandy grade alumina; $AlCl_3{\cdot}6H_2O$; crystallization; salting-out method; gibbsite; precipitation

1. Introduction

Sandy grade alumina (SGA) is the aluminum oxide that is used for metallic aluminum production and is mainly produced by the Bayer alkaline method. In this method, bauxite is used as a raw material, which may differ in quality and composition. For bauxite with a high silica content, the sintering process is used [1,2]. A decrease in the bauxite quality and orientation of the main aluminum producers (Russia and China) with regards to their own resources has led to investigations of alternative raw materials for SGA production [3]. In fact, it can be obtained from both natural (kaolin clay [4,5] or pyrophyllite ore [6]) and technogenic (red mud [7], coal fly ash (CFA) [8], and secondary aluminum dross [9,10]) materials. The CFA waste generated from coal-fired power plants is one of the most promising raw materials for SGA production [11], since it does not require preliminary grinding, as in the case of using kaolin clay, or the removal of alkali, as in the case of using red mud [12]. The CFA volume produced annually varies in each country, depending on the share of coal in the electricity generation. In Russia, 25 million tons are generated annually, while only 8% was utilized in 2019. The Russian Government plans to increase this quantity to 50% by 2035 [13].

Since CFA has a low silicon modulus (the mass ratio of Al_2O_3 to SiO_2), in the range of 0.45–0.55, acid treatment (leaching) is the main approach employed to extract aluminum from CFA. Sulfuric acid (H_2SO_4) [14], ammonium bisulfate ($(NH_4)_2SO_4$) [15], potassium bisulfate ($KHSO_4$) [16], and hydrochloric acid (HCl) [17] are normally used for aluminum leaching. The use of HCl is more promising, since the chloride liquor can be used as a coagulant, polyaluminum chloride (PAC), for water treatment after leaching [18]. In addition to aluminum, valuable metals can be extracted from the liquor, including gallium (Ga) [19], scandium (Sc) [20], and rare earth elements (REE) [21], which can effectively be recovered as separate concentrates.

Various methods have been used to precipitate aluminum from acidic solutions and subsequently obtain alumina: salting-out by the addition of salts, alkali precipitation, isothermal evaporation [22], and direct spray pyrolysis of an acid solution at T = 1000 °C [23]. In the salting-out method, sodium chloride (NaCl) [24], calcium chloride ($CaCl_2$) [25], ammonium chloride (NH_4Cl), magnesium chloride ($MgCl_2$), or strontium chloride ($SrCl_2$) can be used [26]. The use of alkaline reagents, such as ammonia hydrate (NH_4OH) or sodium hydroxide (NaOH), leads to the formation of gibbsite ($Al(OH)_3$), which is accompanied, however, by an irreversible loss of hydrochloric acid [27].

The salting-out of aluminum chloride hexahydrate (ACH: $AlCl_3{\cdot}6H_2O$) performed by flushing gaseous HCl through the solution leads to a decrease in the ACH solubility and its precipitation [28]. This method is the most promising one, since it allows the reuse of hydrochloric acid and permits ACH crystals to be obtained with a minimum amount of impurities. The alkali leaching of alumina after ACH calcination is another way of providing a low impurity content.

In preceding studies [29], we have shown that coal fly ash can be utilized by HCl autoclave leaching to produce an Al-chloride solution, which can readily be used as a water coagulant. In the present study, we propose a novel method that allows sandy grade alumina suitable for the Bayer process to be produced from Al-chloride liquor, which was obtained after leaching coal fly ash. The method, for which the flowsheet is given in Figure 1, consists of two stages: first, conventional acid treatment (salting-out), and then, combined acid-alkali treatment, with intermediate calcination stages. Contrary to the conventional route (shown by red arrows), this method (shown by green arrows) provides SGA with minimum impurities, which can directly be utilized to produce primary metallic aluminum. The present article also discusses the main technological parameters of the salting-out process and the behavior of the main impurities, and compares the SGA obtained by the conventional acid and the novel acid/alkaline treatment methods.

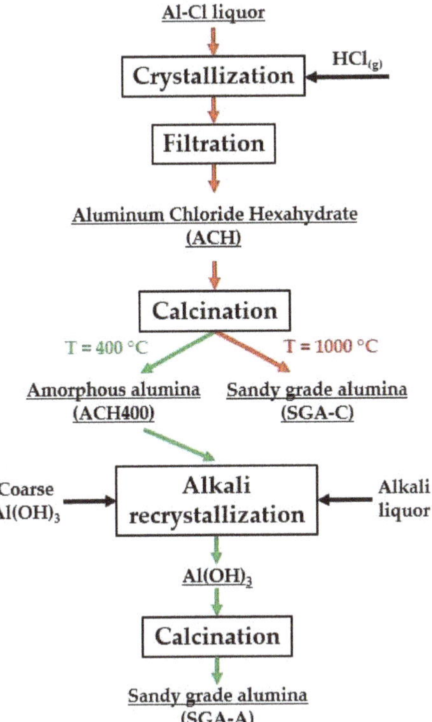

Figure 1. A flowsheet of sandy grade alumina (SGA) production from the Al-chloride liquor by the conventional acid and novel acid-alkali treatment methods (red arrows show the traditional acid method used for SGA production, and green arrows show the new acid-alkali method).

2. Materials and Methods

2.1. Materials

Raw CFA was collected from the Omsk TPP-4 ash residue storage (Google Earth coordinates: 55.129231, 73.167497). Magnetite concentrate and unburned carbon were separated from CFA by wet magnetic separation and by foam flotation with kerosene as a collector, respectively [30]. The non-magnetic fraction, after the enrichment process, was subjected to high-pressure HCl leaching to maximize the aluminum extraction from the liquor (T = 210 °C, C_{HCl} = 30%, τ = 3 h, and the solid liquid ratio = 1:5) [29]. The chemical composition of the liquor is presented in Table 1.

Table 1. The chemical composition (g/l), measured by inductively coupled plasma optical emission spectrometry (ICP-OES), of the aluminum chloride liquor obtained by high-pressure HCl leaching of coal fly ash (CFA).

Al	Ca	Fe	Ti	Na	K	Si	Sc	Cl_{total}
22.3	3.8	2.5	0.3	0.2	0.1	4.5×10^{-3}	2.2×10^{-3}	335

2.2. Analytical Methods

The metal concentrations in the liquor and solid samples were measured by inductively coupled plasma optical emission spectrometry (ICP-OES) using an atomic absorption spectrometer AA-240FS

(Varian, Belrose, Australia). Solid samples were dissolved beforehand by distilled water (ACH) or hydrochloric acid (alumina).

An X-ray diffraction analysis (XRD) of samples was carried out using an Ultima IV diffractometer (Rigaku, Tokyo, Japan) with Cu-Kα; the radiation was in a range of 2θ angles of 10–100° with a 0.02° step. The quantitative and qualitative structure of samples were investigated using the PDXL (Rigaku) program.

The change of mass and temperature of phase transformations of the ACH sample were determined via differential scanning calorimetry (DSC) and thermogravimetric analysis (TGA), using an STA 409 Luxx synchronous thermal analyzer (Netzsch, Selb, Germany) combined with a QMS 403 Aeolos quadrupole mass spectrometer with a capillary connection (Netzsch, Selb, Germany). The heating rate was 5 °C/min in the temperature range of 100–1200 °C.

The morphology of the samples was examined by scanning electron microscopy (SEM) using EVO LS 10 (Carl Zeiss, Oberkochen, Germany) and Vega 3 (Tescan, Brno, Czech Republic) microscopes.

The particle size distribution and specific surface area of the samples were determined by the laser diffraction method (LD) using an Analysette 22 NanoTec analyzer (Fritsch, Idar-Oberstein Germany) and by the Brunauer–Emmett–Teller method (BET) using a NOVA 1200e apparatus (Quantachrome Instruments, Hook, UK).

2.3. Experiments

The salting-out of ACH from the aluminum chloride liquor was carried out by the flushing of gaseous HCl mixed with Ar at a rate of 1.5 g/min (Figure 2) through a Drexel flask with 10 mL of the liquor, which was thermostated at temperatures of 0, 25, 50, 75, and 95 °C for 10 min. The ACH crystals obtained were separated from the liquor in a Schott funnel. Polypropylene, which was stable in hydrochloric acid, was used as a filter cloth, and acetone was used as a washing liquid. After washing, the ACH crystals obtained were dried in a drying oven at 105 °C for 2 h, and the impurities in the crystals were then analyzed.

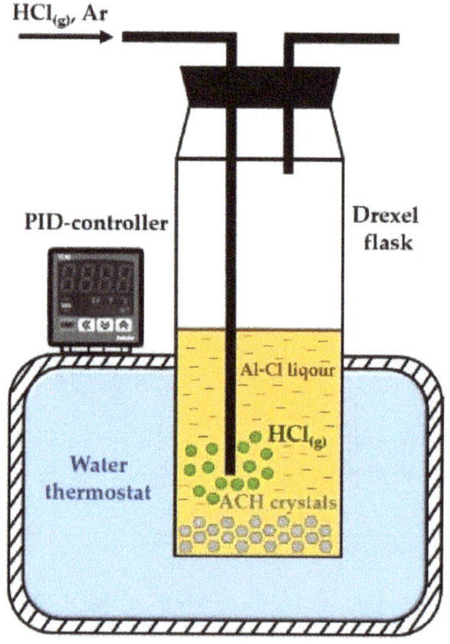

Figure 2. Schematic of the salting-out experimental set-up.

Amorphous Al_2O_3 (ACH400) was obtained by the calcination of ACH crystals in an electric tube furnace (SNOL, Moscow, Russia) in a corundum boat. The heating time to T = 400 °C was 15 min, the weight of the ACH sample was 15 g, and the calcination time was 1 h. Exhaust gases (HCl) were passed through a Drexel flask with distilled water to regenerate HCl.

The leaching of ACH400 was carried out in a thermostatically controlled stainless-steel reactor equipped with an overhead stirrer. The alkaline aluminate liquor (500 mL) with a concentration of Na_2O_k = 150 g/L and α_k = 3.4 was loaded into the reactor. After reaching 60 °C, 200 g/L of a 2:1 mixture of aluminum hydroxide from the Ural Aluminum Smelter (coarse $Al(OH)_3$ in Figure 1) and ACH400, respectively, was added to the reactor. Then, the pulp was maintained for 36 h with constant-speed stirring of 200 rpm. The precipitate obtained was separated from the liquor by filtration, washed with hot water, and dried at 105 °C for 8 h.

The conventional SGA obtained from ACH crystals (SGA-C) and the novel SGA obtained from $Al(OH)_3$ (SGA-A) were then annealed in a muffle furnace HTC 03/18/3N/PE (Nabertherm, Lilienthal, Germany) in open corundum crucibles. The heating time to T = 1000 °C was 1 h, the weight of the sample was 5 g, and the calcination time was 1 h.

2.4. Equations

The ACH crystal size distribution was determined by an analysis of the SEM images. It was impossible to use LD analysis, since the crystals dissolve in water. The Sturges equation [31] was employed to find the number of intervals of the particle size distribution histogram (n):

$$n = 1 + 3.22 \times \mathrm{Log}\,(N), \tag{1}$$

where N is the number of measured ACH particles.

The data obtained were analyzed by the Origin Pro 8 Software, and 1120, 790, and 813 ACH particles were determined for each ACH sample (0, 50, and 95 °C, respectively).

The ACH recovery level (R) from the aluminum chloride liquor was calculated by the following equation:

$$R = (Al_1/Al_2) \times 100\%, \tag{2}$$

where Al_1 and Al_2 represent the aluminum content (g/l) in the aluminum chloride liquor after and before salting-out experiments.

3. Results

3.1. ACH Precipitation

The temperature of the Al chloride liquor and the HCl content are the two most important parameters of the salting-out process, which affect both the precipitation of the main impurities (Fe, Ca, Na, and K) and the yield of the ACH crystals [32,33]. It was found that the temperature of the liquor does not significantly affect the ACH solubility. At 125 °C, this value was 4.012 mol/kg, and at 0 °C, it was 3.283 mol/kg [34]. The HCl concentration imposes a much greater effect: with water liquor (0 mol/kg HCl), the ACH solubility was 31.20%, while with 10 mol/kg HCl, this value decreased to 0.13%. The solubility of iron chloride hexahydrate was practically unchanged and stayed at a ~50% level [35,36]. The effect of temperature on the ACH yield is illustrated in Figure 3. Decreasing the temperature from 95 to 0 °C enhances the ACH yield from 58% to 99%, respectively.

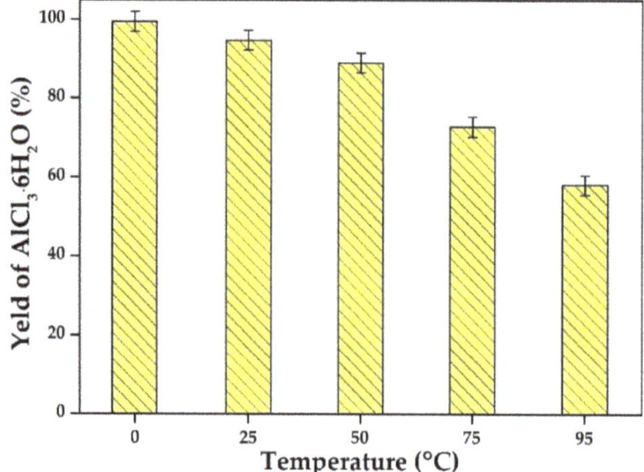

Figure 3. The effect of the aluminum chloride liquor temperature on the distilled water (ACH) yield in the salting-out process.

The dependence obtained correlates with the ACH crystal size distribution shown in Figure 4. At 95 °C, the number of small crystals ≤100 μm is significantly lower than that at 50 and 0 °C. The average crystal size for 0, 50, and 95 °C is 85, 127, and 135 μm, respectively. A large number of small crystals at 0 °C can simultaneously work as precursors for the formation of new ones.

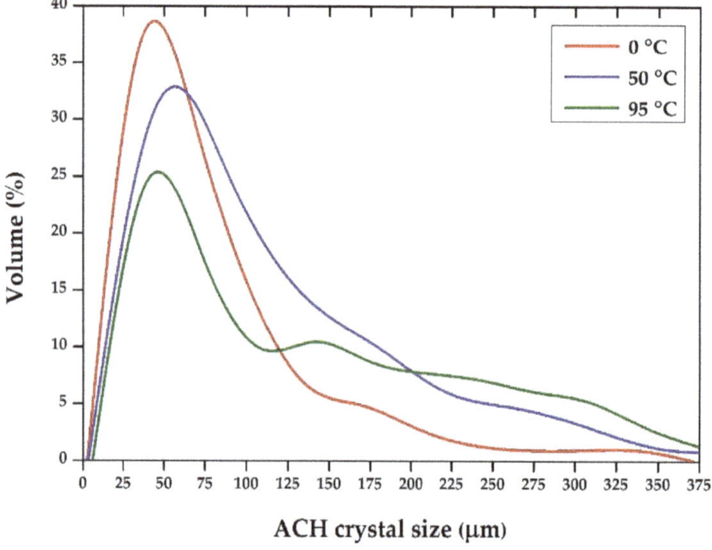

Figure 4. The crystal size distribution of ACH, obtained at 0, 50, and 95 °C.

Figure 5a–c provide the SEM images of ACH crystals obtained at different temperatures. It can be seen that the temperature greatly affects the ACH crystal size, which can be associated with a change in the supersaturation and density of the liquor caused by an increasing temperature [22,34,37]. An increase of the density, apparently, leads to a decrease in the supersaturation of the liquor at T > 50 °C and results in a lower degree of the ACH yield. With high supersaturation, the rate of nucleation often exceeds the crystal growth rate [38,39], which accounts for the formation of smaller crystals at 0 °C (Figure 5a). However, at the surface of large crystals (>100 μm), the offset of secondary nucleation and crystallization can be observed (see Figure 5b, where yellow arrows indicate secondary crystals). This is most clearly seen in the sample obtained at 95 °C, where the sizes of the secondary crystals exceed 100 μm (Figure 5c). The secondary nucleation of gibbsite was also studied by Li et al. [40].

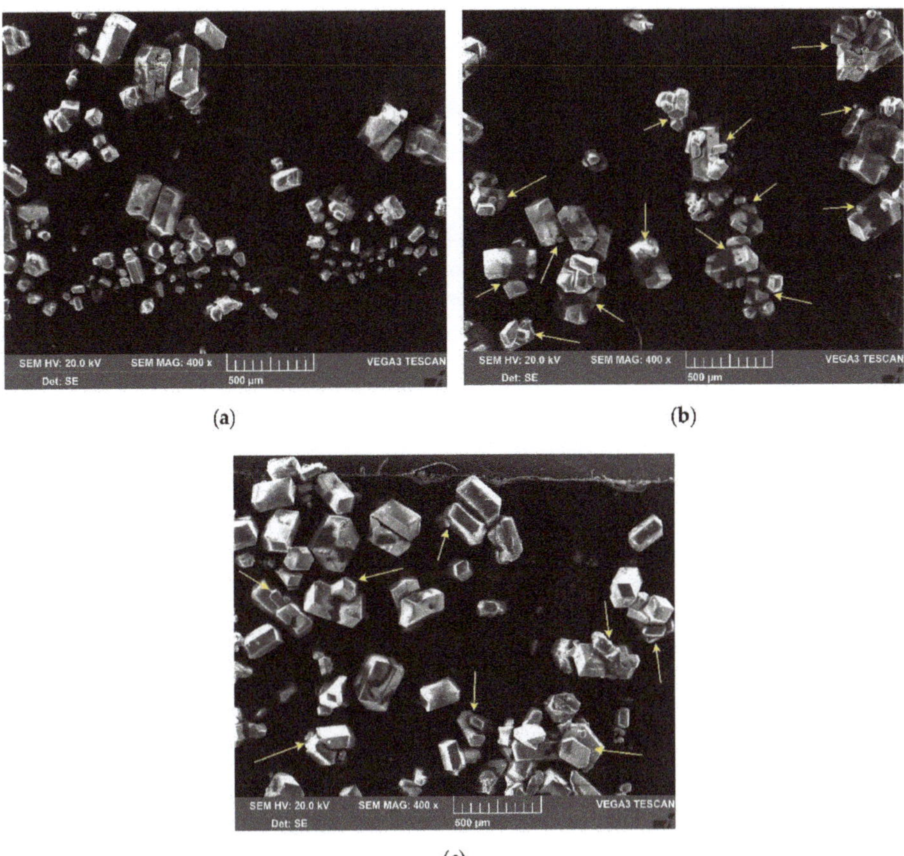

Figure 5. The SEM images of aluminum chloride hexahydrate (ACH) crystals obtained at (**a**) 0 °C, (**b**) 50 °C, and (**c**) 95 °C (yellow arrows indicate secondary nucleation and crystallization on the surface of primary crystals).

The chemical composition of the ACH crystals obtained at 0–95 °C is presented in Table 2. It can be seen that the total content of impurities in the crystals stays within a range of 0.05–0.1 wt.%. It can also be concluded from the results presented in Table 2 that a decrease of temperature of the aluminum chloride solution reduces the solubility of K, Na, and Si impurities, which was also confirmed by Gao et al. for $CaCl_2$ and KCl [41].

Table 2. The chemical composition, measured by ISP-OES, of impurities in ACH obtained at 0–95 °C (wt.%).

Salting-out Temperature	Cr	Fe	K	Mn	Na	P	Si	Ti	V	Zn
0 °C	0.0038	0.01	0.0073	0.0005	0.039	0.021	0.011	0.0075	0.00037	0.0001
25 °C	0.0020	0.0027	0.0083	0.00016	0.0075	0.017	0.013	0.0060	0.00036	0.0001
50 °C	0.0004	0.0035	0.0085	0.00007	0.0075	0.013	0.021	0.0027	0.00032	0.0001
75 °C	0.0004	0.0041	0.0091	0.0001	0.0075	0.012	0.021	0.0022	0.00027	0.0002
95 °C	0.0003	0.0041	0.0110	0.00005	0.0090	0.011	0.025	0.0012	0.000021	0.0003

Despite the fact that a purer ACH can be obtained at 95 °C, the results of the ACH yield and crystal size advocate the use of a low temperature (e.g., 0 °C) as optimal for the salting-out process.

3.2. ACH Calcination for SGA Production

To study ACH decomposition by calcination, a large number of ACH crystals were produced in an additional experiment using 200 mL of the liquor, T = 0 °C, and a duration of 1 h. Figure 6a shows the phase composition of the crystals obtained. It can be seen that the impurities did not form individual phases and were incorporated into the crystalline phase. Figure 6b shows the ACH crystal size distribution. The average ACH crystal size was 156.57 µm. Increasing the salting-out time leads to a significant increase in the crystal size. It can be seen from Figure 6c–d that surfaces of large particles >200 µm became nucleation sites for smaller ones. Some crystals were greater than 800 µm in size (Figure 6d).

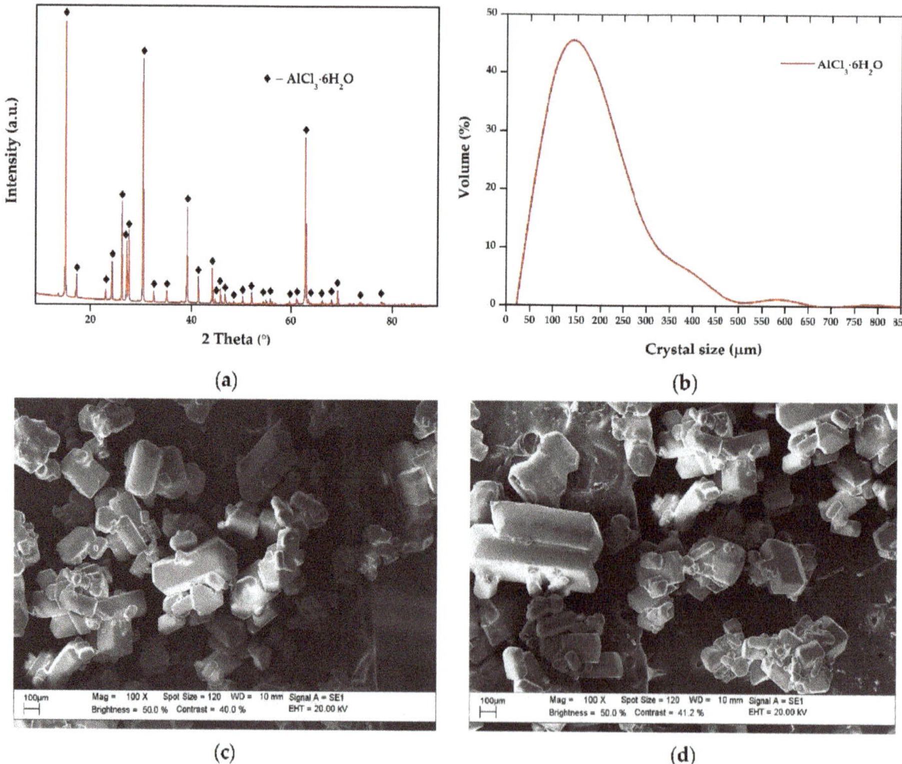

Figure 6. ACH crystals obtained at T_{liqour} = 0 °C, V_{liqour} = 200 mL, and duration 1 h: (**a**) XRD analysis; (**b**) distribution of the crystal size; (**c**) and (**d**) SEM images.

The TGA/DSC analysis was carried out in air atmosphere to determine the mass loss and main thermal effects present upon the heating of ACH. There are two endothermic peaks and one exothermic peak shown in Figure 7. The first peak (before 100 °C) indicates the evaporation of physical water with a loss of ~1 wt.%. The second peak at 180 °C indicates ACH decomposition, which is accompanied by a loss of more than 78 wt.% from the total mass. Ivanov et al. [42] showed that this peak corresponds to the decomposition of ACH to Al_2O_3, H_2O, and HCl by the following reaction:

$$2AlCl_3{\cdot}6H_2O_{(s)} = Al_2O_{3(s)} + 6HCl_{(g)}\uparrow + 9H_2O_{(g)}\uparrow. \quad (3)$$

A small exothermic peak at 850 °C indicates the phase transformation of amorphous alumina into γ-Al_2O_3; a small peak at 950 °C can indicate the transformation of γ-Al_2O_3 into α-Al_2O_3 [43]. The process of ACH decomposition could only be completed above 250 °C, when the ACH phase became undetectable by XRD, while the Al-product could be transformed into amorphous alumina until the temperature reached ~850 °C, when the γ-Al_2O_3 phase appeared.

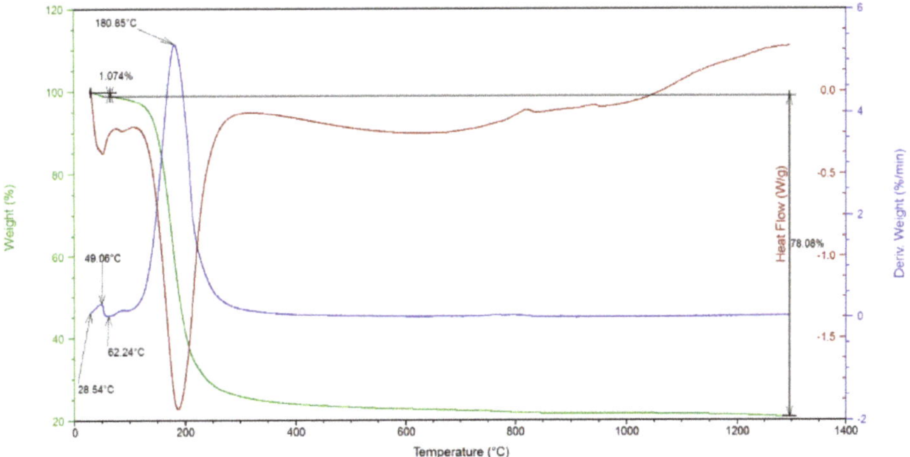

Figure 7. TGA/DSC curves of the $AlCl_3{\cdot}6H_2O$ sample at a heating rate of 5 °C/min from 28 to 1300 °C (initial mass of the sample: 43.374 mg; atmosphere: air; the green line shows the weight loss, the blue line shows the derivative of the weight change rate, and the red line shows the heat flow).

The direct production of sandy grade alumina (SGA-C) from ACH crystals was carried out by calcination at 1000 °C for 1 h. Using these technological parameters, it is possible to obtain SGA with the smallest chlorine content (less of 0.2 wt.%) [44] and a high specific surface (>60 m^2/g). A further increase of the temperature will result in a high content of α-Al_2O_3 and a decrease in the specific surface to less than 60 m^2/g [45].

3.3. ACH Calcination to Produce Amorphous ACH400

Further study will focus on producing amorphous ACH400 and obtaining sandy grade alumina from it. The idea is to remove chlorine and water from ACH, but to avoid the crystallization of alumina, which can occur at temperatures higher than 800 °C. It can be seen from Section 3.2 that the decomposition of ACH crystals into aluminum oxide via reaction (3) starts at temperatures of around 100–150 °C. However, the decomposition of ACH usually occurs in two steps: first, it decomposes into an aluminum oxychloride compound at temperatures up to 350 °C, which can be presented by the following reaction:

$$2AlCl_3{\cdot}6H_2O_{(s)} = Al_2(OH)_4Cl_2{\cdot}H_2O_{(s)} + 4HCl_{(g)}\uparrow + 7H_2O_{(g)}\uparrow, \quad (4)$$

and second, the aluminum oxychloride decomposes into Al_2O_3 at higher temperatures. However, if the heating rate is high enough, reactions (3) and (4) occur almost simultaneously (see, for example, [42]), and this results in one broad peak from 100 to 300 °C, as can be seen in Figure 7. In the present study, it was found that in order to maximize the removal of chlorine ions and thereby ACH400 formation, it is necessary to anneal ACH at temperatures above ~400 °C.

Taking into account the above, ACH400 was obtained by the calcination of ACH crystals at T = 400 °C for 1 h. This was further used to obtain sandy grade alumina (SGA-A) by a novel alkali treatment method described in Section 3.5.

3.4. Comparison of SGA-C and ACH400

This section provides a comparison of the sandy grade alumina (SGA-C) produced by high-temperature calcination and amorphous alumina (ACH400) produced by the low-temperature calcination of ACH crystals. The samples of Al_2O_3 obtained have different phase compositions, particle sizes, and forms (Figure 8). Figure 8a presents the results of XRD analysis of powders. It can be seen that the sample calcined at 400 °C is fully amorphous, while the sample calcined at 1000 °C contains crystalline phases, which are stable and metastable polymorphs of Al_2O_3. The main phase is found to be metastable γ-Al_2O_3, while the most stable form, corundum (α-Al_2O_3), does not exceed 10 wt.%. After low-temperature ACH calcination with the subsequent removal of H_2O and HCl, the particle size of the initial ACH crystals increases. As a result, ACH400 has an average particle size of 95.19 µm, while that of SGA-C calcined at 1000 °C is 43.67 µm (Figure 8b). Amorphous alumina powder retains an initial hexagonal shape of the particles similar to that of crystalline SGA-C (Figure 8c,d). The ACH400 particle surface partly remains smooth and is partly covered with long pores that are ~5 µm in length, which were formed when water and HCl were evaporated (Figure 8e). The SGA-C powder becomes brittle, and the entire surface of the SGA-C particles contains micro- and nanopores (Figure 8f).

The BET analysis of samples showed that the flat surface of the initial ACH crystals provides a low specific surface area of about 0.7 m^2/g. During ACH decomposition, pores appeared on the particle surface, which led to an increase of the specific surface area for ACH400 to 36.9 m^2/g. Further calcination at 1000 °C resulted in the entire surface of the particles consisting of pores, and the specific surface area for SGA-C increased to 71.2 m^2/g. Due to such a high specific surface area of the SGA-C obtained by the acid treatment, its dissolution rate in cryolite (Na_3AlF_6) was faster than the SGA obtained by the Bayer alkaline method [46].

The chemical composition of the initial ACH, ACH400, and SGA-C samples is shown in Table 3. It can be seen that the amount of impurities increases and the chlorine content decreases significantly when increasing the calcination temperature.

Table 3. The chemical composition, measured by ISP-OES, of impurities in ACH, amorphous alumina, and sandy grade alumina (wt.%).

Samples	Cr	Fe	K	Mn	Na	P	Si	Ti	V	Zn	Cl
ACH	0.0041	0.015	0.0081	0.0006	0.043	0.022	0.012	0.0094	0.00044	0.0002	-
ACH400	0.073	0.027	0.015	0.001	0.074	0.038	0.022	0.016	0.0008	0.0004	5.27
SGA-C	0.076	0.029	0.016	0.001	0.081	0.041	0.023	0.018	0.0009	0.0004	0.15

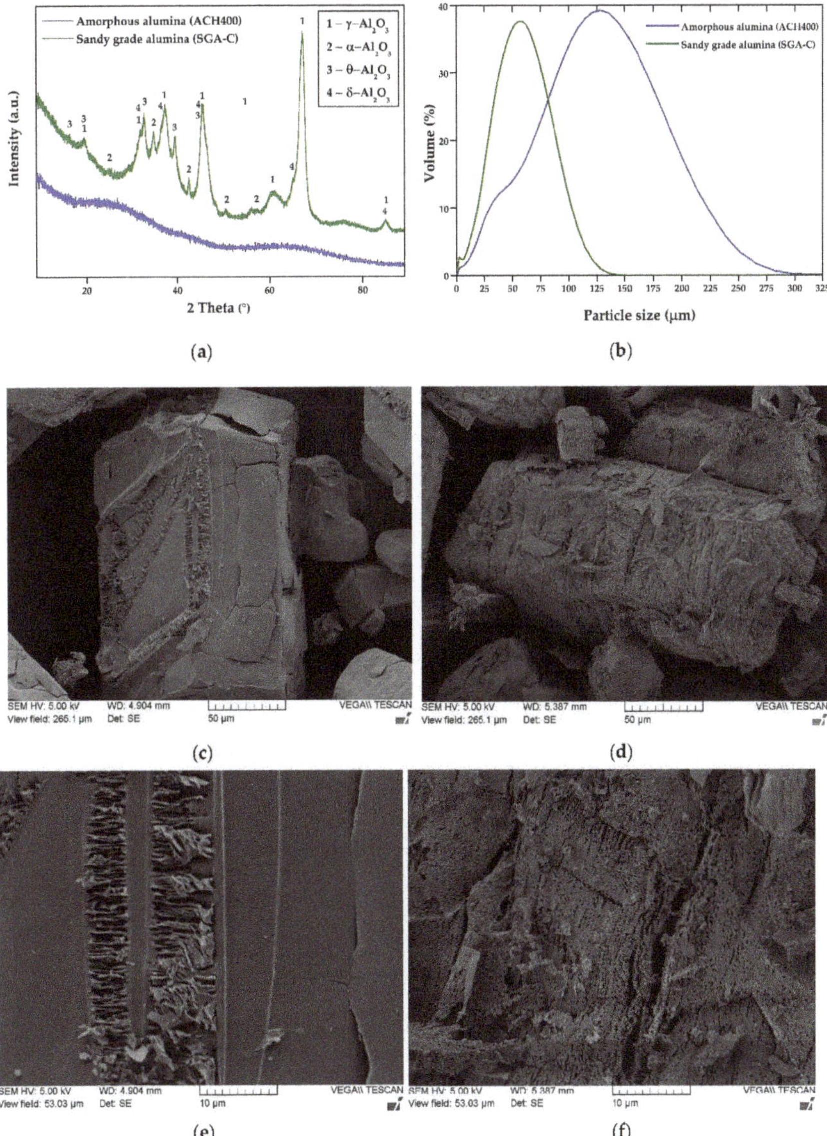

Figure 8. Data of ACH400 (T = 400 °C; τ = 1 h) and SGA-C (T = 1000 °C; τ = 1 h) obtained from ACH calcination: (**a**) XRD analysis; (**b**) distribution of the particle size; (**c**) and (**d**) SEM images of the particle shape for ACH400 and SGA-C powders, respectively; (**e**) and (**f**) SEM images of the particle surface for ACH400 and SGA-C powders, respectively.

3.5. SGA Production from Amorphous Alumina

The SGA-C obtained by the direct acid treatment has a high specific surface area; however, the particle shape and the content of impurities do not meet the technical standards accepted at RUSAL plants [44]. It is possible to reduce the impurities content, if recrystallization or solvent extraction methods are utilized [47].

Another approach involves alkaline treatment of the Al_2O_3 by the Bayer process [48]. In this case, Al_2O_3 from the acid treatment stage is first leached in an alkaline aluminate solution to obtain a liquor supersaturated by alumina. For this process, high-pressure leaching equipment should be used. It is also necessary to obtain the alkaline aluminate liquor with a caustic module (the molar ratio of Na_2O to Al_2O_3 in the solution) below 1.5 [49]. Then, the supersaturated alkaline aluminate liquor can enter a precipitation stage, where gibbsite ($Al(OH)_3$) precipitates out in the presence of a large number of seed crystals (to achieve that, the ratio of Al_2O_3 in the liquor to that in the seed crystals must be greater than 4). To obtain coarse particles of $Al(OH)_3$ (only 10% of particles should have a size below 45 μm) and a precipitation rate of more than 50% required for SGA production, the duration of the whole process should be 48–72 h [50]. Therefore, in order to produce SGA that suits the required standards, the use of high-pressure equipment and large decomposers, as well as additional energy and raw materials, is required, which significantly increases the cost of the final product.

It is also possible to apply the above alkaline recrystallization process at lower material/energy costs using an enlargement method [51] based on the difference in solubility of coarse particles of gibbsite or boehmite (AlOOH) and amorphous alumina (or highly dispersed hydroxide) [52,53]. Previous studies on the leaching of amorphous alumina in hot alkaline liquor have shown that, in this case, the process can occur at atmospheric pressure [54,55] and no high-pressure leaching equipment is necessary.

In this paper, we have proposed, for the first time, the use of the intermediate product of the acid treatment stage, the amorphous ACH400, in the alkaline recrystallization process. In the present research, the amorphous alumina obtained after the acid treatment stage was subjected to further alkaline treatment and subsequent recrystallization by the mother liquor and coarse $Al(OH)_3$ particles, respectively, formed during a stage of the Bayer process on the Ural Aluminum Plant (UAP). As a result, the amorphous alumina dissolved, increasing the supersaturation of the aluminate solution, which in turn led to the precipitation of aluminum hydroxide on the coarse $Al(OH)_3$ particles. Finally, the sandy grade alumina could be obtained after filtration and calcination. The proposed process flowsheet is shown in Figure 9.

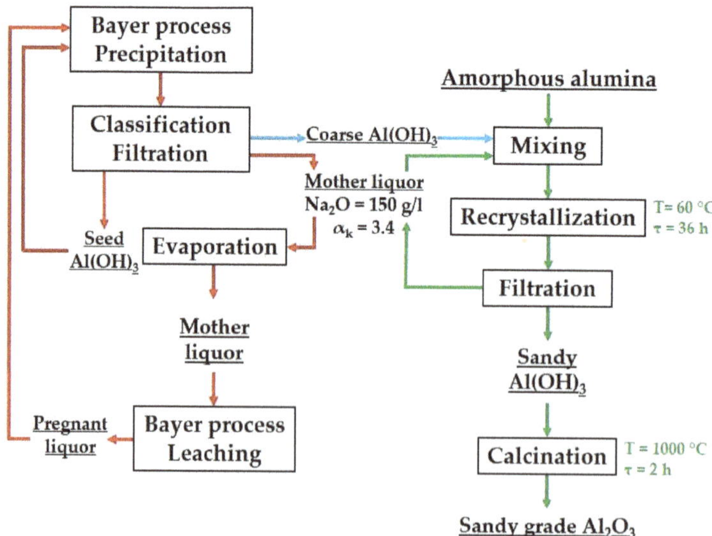

Figure 9. Flowsheet of SGA production from amorphous alumina obtained by low-temperature ACH calcination.

The results of the proposed treatment are discussed in the rest of the section. The particle size distribution of the initial $Al(OH)_3$ particles produced on the UAP and of $Al(OH)_3$ particles after the recrystallization stage is shown in Figure 10. The initial powder contains about 25% of a fraction of fewer than 50 μm, which prevents the production of SGA from this powder on the UAP. To increase the size of the coarse $Al(OH)_3$ particles, ACH400 powder was used. When mixed with an alkaline liquor, due to the high specific surface area (36.9 m^2/g for ACH400 and 0.1 m^2/g for coarse $Al(OH)_3$), it will be dissolved first, and then precipitated on the $Al(OH)_3$ surface. This process can be described by the following reactions:

$$Al_2O_{3(s)} + NaOH_{(l)} = NaAl(OH)_{4(l)}, \quad (5)$$

$$NaAl(OH)_{4(l)} + Al(OH)_{3(s)} = 2Al(OH)_{3(s)} + NaOH_{(l)}. \quad (6)$$

According to reaction (5), amorphous alumina is dissolved in the liquor to form hydrated sodium aluminate, thereby increasing the supersaturation of the liquor relative to the coarse $Al(OH)_3$ particles [52]. Next, dissolved alumina in the form of hydrated sodium aluminate precipitates on the surface of $Al(OH)_3$ via reaction (6). Li et al. showed that $Al(OH)_3$ agglomeration also proceeds during the late stages of the precipitation process [56]. It can be seen (Figure 10) that the particle size distribution after recrystallization shifted towards a greater average size and a higher volume fraction of coarse particles.

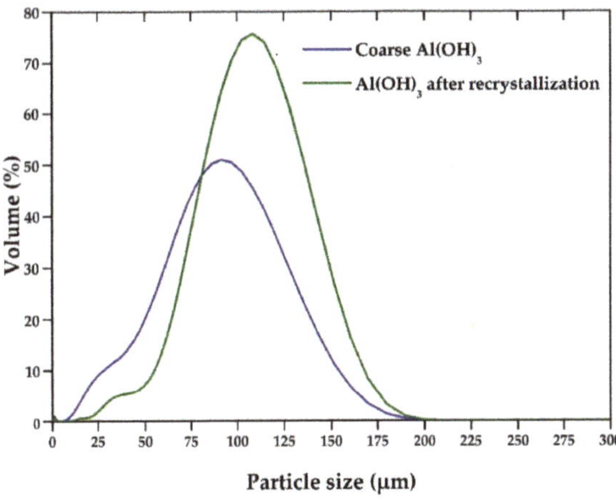

Figure 10. Distribution of particle size for coarse $Al(OH)_3$ and $Al(OH)_3$ after recrystallization.

Figure 11a–d shows a comparison of the morphology of the coarse $Al(OH)_3$ particles and $Al(OH)_3$ particles after recrystallization. Initially, the coarse $Al(OH)_3$ had a large number of small particles, for which the size was less than 20 μm (Figure 11a). The $Al(OH)_3$ sample after recrystallization mainly consisted of large particles with a size of 100–125 μm (Figure 11c); a large number of small particles were not found, as was the case before (Figure 11a). An analysis of the particle shape of the coarse $Al(OH)_3$ revealed voids and surface inhomogeneity (Figure 11b), whereas, after recrystallization, the $Al(OH)_3$ particles had an almost uniform structure of a dense pack of small particles, consisting of larger agglomerates (Figure 11d).

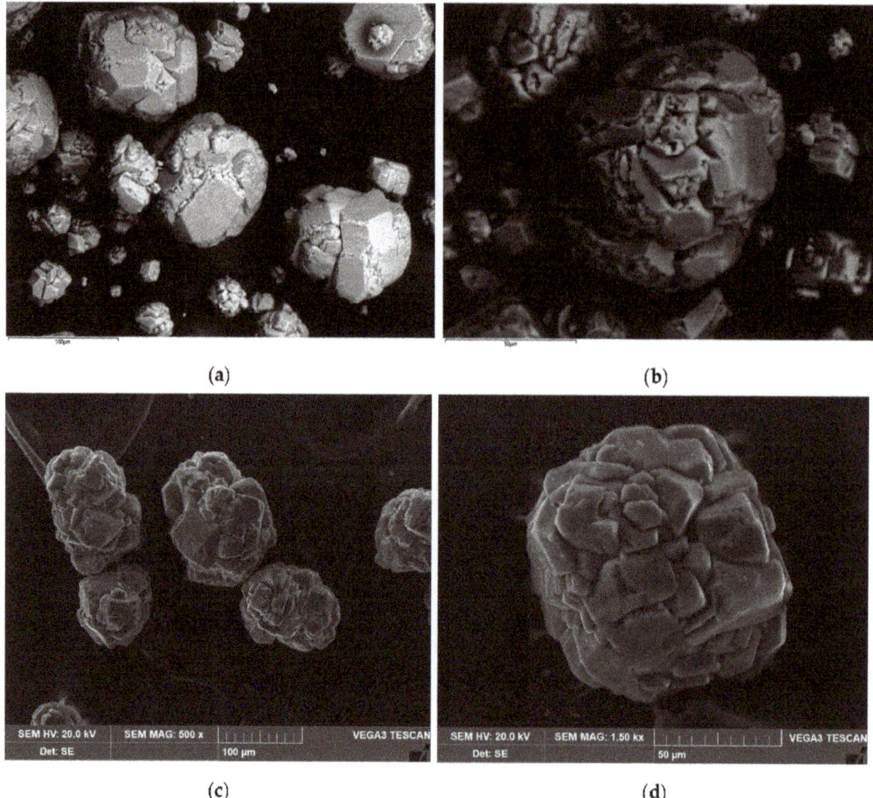

Figure 11. SEM images of (**a**,**b**) coarse $Al(OH)_3$ and (**c**,**d**) $Al(OH)_3$ after recrystallization.

The proposed process proceeds at atmospheric pressure and the liquor temperature of 60 °C. The Al_2O_3 precipitation degree after 36 h was only 4.5%. The XRD analysis shows that the powder obtained only consists of the gibbsite phase (Figure 12). The low Al_2O_3 precipitation degree and the absence of amorphous alumina in the recrystallized sample (Figure 11c) confirm that the growth of $Al(OH)_3$ occurs due to the leaching and recrystallization of ACH400 powder.

After filtration, the mother liquor was reused to leach a new portion of amorphous alumina. For SGA production, it is necessary to calcinate $Al(OH)_3$ at T = 1000 °C and τ = 1 h. The removal of external moisture and crystalline water in $Al(OH)_3$ occurs according to the following reaction:

$$2Al(OH)_{3(s)} = Al_2O_{3(s)} + 3H_2O_{(g)}\uparrow. \quad (7)$$

It can be seen from Figure 10 that the number of particles smaller than 45 μm becomes lower than 5%. Since calcination in this research was carried out in a muffle furnace, the attrition index (AI) was not studied. However, Wind et al. showed that decreasing the fraction of particles greater than 45 μm increases the AI in the calcination furnace [57]. Therefore, if the $Al(OH)_3$ after recrystallization contains 5% of particles smaller than 45 μm, the AI is 5%. This means that, even after abrasion, the obtained SGA-A must comply with the SGA requirements.

The XRD pattern of the SGA-C obtained after $Al(OH)_3$ calcination is shown in Figure 12. SGA-C consists of two phases: 95% γ-Al_2O_3 and 5% α-Al_2O_3.

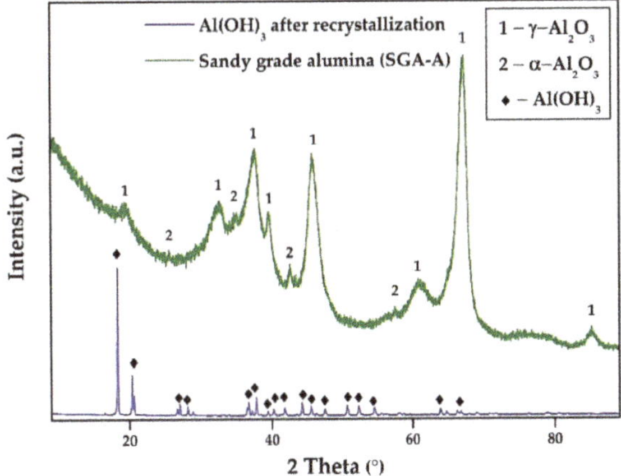

Figure 12. XRD analysis of $Al(OH)_3$ after the recrystallization process and SGA-A (T = 1000 °C, τ = 1 h).

The chemical composition of SGA-C is presented in Table 4. Since ACH400 was completely dissolved in the alkali liquor, the maximum dilution of impurities also occurred. Therefore, the chlorine content in the SGA-A was quite low. Senyuta et al. [48] showed that, in a real metallurgical process, chlorine can be accumulated in circulating liquors at the amount of 40–90 g/L, which may lead to an increase of the chlorine content in the SGA to about 0.013 wt.%.

Table 4. The chemical composition, measured by ISP-OES, of impurities in SGA-C (wt.%).

Cr_2O_3	Fe_2O_3	K_2O	MnO_2	Na_2O	P_2O_5	SiO_2	TiO_2	V_2O_5	ZnO	Cl
0.0005	0.01	0.008	0.0001	0.31	0.0009	0.003	0.002	0.0004	0.0001	0.004

Using the proposed novel process, it is possible to reduce the cost of SGA production, which is necessary for modern electrolysis processes. Acid/alkali treatment enables the leaching stage of the amorphous alumina at high pressure to be eliminated; the long precipitation time to be significantly shortened; and the sandy grade alumina, which is appropriate with regards to the SGA standards, to be produced.

4. Conclusions

In this work, a novel method has been developed for sandy grade alumina production from the Al-chloride liquor obtained by the HCl leaching of coal fly ash. In this method, the amorphous alumina obtained after traditional acid treatment of the Al liquor is subjected to alkaline treatment by the mother liquor from a Ural Aluminum Plant and to subsequent precipitation using the coarse $Al(OH)_3$ particles from UAP, with further calcination to obtain SGA suitable for metallic aluminum production. The following points concerning the technological parameters of various stages of the novel process can be made:

1. The temperature of the salting-out process affects the ACH crystal size and impurities content. At T = 0 and 95 °C, the total amount of impurities is 0.1 and 0.06 wt.%, and the average particle size is 85 and 135 μm, respectively;
2. ACH calcination at T = 400 °C produces amorphous alumina with the specific surface of 36.9 m^2/g and content of chlorine ions of 5.27 wt.%;
3. ACH calcination at T = 1000 °C allows SGA production with the specific surface area of 71.2 m^2/g and chlorine ion content of 0.15 wt.%;

4. Alkaline recrystallization of amorphous alumina obtained after the calcination of ACH at 400 °C at T = 60 °C and Na_2O = 150 g/L, allows the average particle size of coarse $Al(OH)_3$ to be increased from 76 to 99 microns;
5. The calcination of recrystallized $Al(OH)_3$ at T = 1000 °C and maintenance for 1 h allows SGA with a content of chlorine ions of 0.004 wt.% to be obtained.

Author Contributions: Conceptualization, D.V., A.S., and A.K.; formal analysis, A.M.; funding acquisition, D.V.; investigation, D.V. and A.S.; project administration, D.V.; resources, D.V.; writing—original draft, D.V., A.S., and A.K.; writing—review and editing, D.V., A.S., and A.K. All authors have read and agreed to the published version of the manuscript.

Funding: This work was financially supported by the Russian Science Foundation Project No. 18-79-00305.

Acknowledgments: The authors would like to appreciate the assistance provided by The Center for Collective Use Testing Analytical Center of the JSC "Scientific-research institute of chemical technology" and that personally given to Natalya Ognevskaya for the chemical analysis of solid and liquid samples.

Conflicts of Interest: The authors declare no conflicts of interest.

References

1. Anawar, H.M.; Strezov, V.; Adyel, T.M.; Ahmed, G. Sustainable and Economically Profitable Reuse of Bauxite Mining Waste with Life Cycle Assessment. In *Sustainable and Economic Waste Management Resource Recovery Techniques*; Anawar, H.M., Strezov, V., Abhilash, Eds.; CRC Press: Boca Raton, MA, USA, 2019; p. 328. ISBN 9780429279072.
2. Smith, P. The processing of high silica bauxites - Review of existing and potential processes. *Hydrometallurgy* **2009**, *98*, 162–176. [CrossRef]
3. Senyuta, A.; Panov, A.; Suss, A.; Layner, Y. Innovative Technology for Alumina Production from Low-Grade Raw Materials. In *Light Metals 2013*; Springer: Cham, Switzerland, 2013; pp. 203–208. ISBN 978-3-319-65136-1.
4. Chen, J.; Li, X.; Cai, W.; Shi, Y.; Hui, X.; Cai, Z.; Jin, W.; Fan, J. High-efficiency extraction of aluminum from low-grade kaolin via a novel low-temperature activation method for the preparation of poly-aluminum-ferric-sulfate coagulant. *J. Clean. Prod.* **2020**, *257*, 120399. [CrossRef]
5. Brichkin, V.N.; Kurtenkov, R.V.; Eldeeb, A.B.; Bormotov, I.S. State and development options for the raw material base of aluminum in non-bauxite regions. *Obogashchenie Rud* **2019**, *2019*, 31–37. [CrossRef]
6. Barry, T.S.; Uysal, T.; Birinci, M.; Erdemoğlu, M. Thermal and Mechanical Activation in Acid Leaching Processes of Non-bauxite Ores Available for Alumina Production—A Review. *Mining, Metall. Explor.* **2019**, *36*, 557–569. [CrossRef]
7. Zinoveev, D.V.; Grudinskii, P.I.; Dyubanov, V.G.; Kovalenko, L.V.; Leont'ev, L.I. Global recycling experience of red mud—A review. Part i: Pyrometallurgical methods. *Izv. Ferr. Metall.* **2018**, *61*, 843–858. [CrossRef]
8. Gao, Y.; Liang, K.; Gou, Y.; Wei, S.; Shen, W.; Cheng, F. Aluminum extraction technologies from high aluminum fly ash. *Rev. Chem. Eng.* **2020**. [CrossRef]
9. Mahinroosta, M.; Allahverdi, A. Enhanced alumina recovery from secondary aluminum dross for high purity nanostructured γ-alumina powder production: Kinetic study. *J. Environ. Manage.* **2018**, *212*, 278–291. [CrossRef]
10. Yang, Q.; Li, Q.; Zhang, G.; Shi, Q.; Feng, H. Investigation of leaching kinetics of aluminum extraction from secondary aluminum dross with use of hydrochloric acid. *Hydrometallurgy* **2019**, *187*, 158–167. [CrossRef]
11. Ding, J.; Ma, S.; Shen, S.; Xie, Z.; Zheng, S.; Zhang, Y. Research and industrialization progress of recovering alumina from fly ash: A concise review. *Waste Manag.* **2017**, *60*, 375–387. [CrossRef]
12. Wang, Y.; Zhang, T.; Lyu, G.; Guo, F.; Zhang, W.; Zhang, Y. Recovery of alkali and alumina from bauxite residue (red mud) and complete reuse of the treated residue. *J. Clean. Prod.* **2018**, *188*, 456–465. [CrossRef]
13. Project Energy Strategy of the Russian Federation for the Period Until 2035. Available online: https://minenergo.gov.ru/node/1920 (accessed on 18 December 2019).
14. Shi, Y.; Jiang, K.-X.; Zhang, T.-A. A cleaner electrolysis process to recover alumina from synthetic sulfuric acid leachate of coal fly ash. *Hydrometallurgy* **2020**, *191*, 105196. [CrossRef]
15. Wu, Y.; Yang, X.; Li, L.; Wang, Y.; Li, M. Kinetics of extracting alumina by leaching coal fly ash with ammonium hydrogen sulfate solution. *Chem. Pap.* **2019**, *73*, 2289–2295. [CrossRef]

16. Guo, C.; Zou, J.; Ma, S.; Yang, J.; Wang, K. Alumina extraction from coal fly ash via low-temperature potassium bisulfate calcination. *Minerals* **2019**, *9*, 585. [CrossRef]
17. Ma, Z.; Zhang, S.; Zhang, H.; Cheng, F. Novel extraction of valuable metals from circulating fluidized bed-derived high-alumina fly ash by acid–alkali–based alternate method. *J. Clean. Prod.* **2019**, *230*, 302–313. [CrossRef]
18. Zhang, Y.; Li, M.; Liu, D.; Hou, X.; Zou, J.; Ma, X.; Shang, F.; Wang, Z. Aluminum and iron leaching from power plant coal fly ash for preparation of polymeric aluminum ferric chloride. *Environ. Technol. (United Kingd.)* **2019**, *40*, 1568–1575. [CrossRef]
19. Huang, J.; Wang, Y.; Zhou, G.; Gu, Y. Investigation on the Effect of Roasting and Leaching Parameters on Recovery of Gallium from Solid Waste Coal Fly Ash. *Metals (Basel)* **2019**, *9*, 1251. [CrossRef]
20. Li, G.; Ye, Q.; Deng, B.; Luo, J.; Rao, M.; Peng, Z.; Jiang, T. Extraction of scandium from scandium-rich material derived from bauxite ore residues. *Hydrometallurgy* **2018**, *176*, 62–68. [CrossRef]
21. Pan, J.; Nie, T.; Vaziri Hassas, B.; Rezaee, M.; Wen, Z.; Zhou, C. Recovery of rare earth elements from coal fly ash by integrated physical separation and acid leaching. *Chemosphere* **2020**, *248*, 126112. [CrossRef]
22. Wu, L.; Zhao, J.; Xue, F.; Cheng, H.; Cheng, F. Phase equilibrium andseparation of $AlCl_3$-$FeCl_3$-HCl-H_2Osystem [$AlCl_3$-$FeCl_3$-HCl-H_2O 体系的相平衡及相分离]. *Guocheng Gongcheng Xuebao/The Chin. J. Process Eng.* **2020**, *20*, 318–323.
23. Long, W.; Ting-An, Z.; Guozhi, L.; Weiguang, Z.; Yan, L.; Zhihe, D.; Liping, N. A new green process to produce activated alumina by spray pyrolysis. *Green Process. Synth.* **2018**, *7*, 464–469. [CrossRef]
24. Cheng, H.; Wu, L.; Zhang, J.; Lv, H.; Guo, Y.; Cheng, F. Experimental investigation on the direct crystallization of high-purity $AlCl_3{\cdot}6H_2O$ from the $AlCl_3$-NaCl-H_2O(-HCl-C_2H_5OH) system. *Hydrometallurgy* **2019**, *185*, 238–243. [CrossRef]
25. Yu, X.; Liu, M.; Zheng, Q.; Chen, S.; Zou, F.; Zeng, Y. Measurement and Correlation of Phase Equilibria of Ammonium, Calcium, Aluminum, and Chloride in Aqueous Solution at 298.15 K. *J. Chem. Eng. Data* **2019**, *64*, 3514–3520. [CrossRef]
26. Yu, X.; Zheng, Q.; Wang, L.; Liu, M.; Cheng, X.; Zeng, Y. Solid-liquid phase equilibrium determination and correlation of ternary systems $NH_4Cl+AlCl_3+H_2O$, $MgCl_2+AlCl_3+H_2O$ and $SrCl_2+AlCl_3+H_2O$ at 298 K. *Fluid Phase Equilib.* **2020**, *507*, 112426. [CrossRef]
27. Mahinroosta, M.; Allahverdi, A. A promising green process for synthesis of high purity activated-alumina nanopowder from secondary aluminum dross. *J. Clean. Prod.* **2018**, *179*, 93–102. [CrossRef]
28. Guo, Y.; Yang, X.; Cui, H.; Cheng, F.; Yang, F. Crystallization behavior of $AlCl_3{\cdot}6H_2O$ in hydrochloric system. *Huagong Xuebao/CIESC J.* **2014**, *65*, 3960–3967.
29. Valeev, D.; Kunilova, I.; Alpatov, A.; Mikhailova, A.; Goldberg, M.; Kondratiev, A. Complex utilisation of ekibastuz brown coal fly ash: Iron & carbon separation and aluminum extraction. *J. Clean. Prod.* **2019**, *218*, 192–201.
30. Valeev, D.; Kunilova, I.; Alpatov, A.; Varnavskaya, A.; Ju, D. Magnetite and carbon extraction from coal fly ash using magnetic separation and flotation methods. *Minerals* **2019**, *9*, 320. [CrossRef]
31. Sturges, H.A. The Choice of a Class Interval. *J. Am. Stat. Assoc.* **1926**, *21*, 65–66. [CrossRef]
32. Miles, G.L. Some Studies in the System $AlCl_3$-$FeCl_3$-KCl-NaCl-HCl-H_2O at 25, 30 an4 35°. *J. Am. Chem. Soc.* **1947**, *69*, 1716–1719. [CrossRef]
33. Guo, Y.; Lv, H.; Yang, X.; Cheng, F. $AlCl_3{\cdot}6H_2O$ recovery from the acid leaching liquor of coal gangue by using concentrated hydrochloric inpouring. *Sep. Purif. Technol.* **2015**, *151*, 117–183. [CrossRef]
34. Christov, C.; Dickson, A.G.; Moller, N. Thermodynamic modeling of aqueous aluminum chemistry and solid-liquid equilibria to high solution concentration and temperature. I. the acidic H-Al-Na-K-Cl-H_2O system from 0 to 100∈°C. *J. Solution Chem.* **2007**, *36*, 1495–1523. [CrossRef]
35. Cheng, H.; Wu, L.; Cao, L.; Zhao, J.; Xue, F.; Cheng, F. Phase Diagram of $AlCl_3$-$FeCl_3$-H_2O(-HCl) Salt Water System at 298.15 K and Its Application in the Crystallization of $AlCl_3{\cdot}6H_2O$. *J. Chem. Eng. Data* **2019**, *64*, 5089–5094. [CrossRef]
36. Pak, V.I.; Kirov, S.S.; Mamzurina, O.I.; Nalivayko, A.Y. Understanding the regularities of aluminum chloride hexahydrate crystallization from hydrochloric acid solutions resultant from leaching of Russian kaolin clays. Part 1. process kinetics. *Tsvetnye Met.* **2020**, *2020*, 47–53. [CrossRef]
37. Golubev, V.O.; Chistyakov, D.G.; Brichkin, V.N.; Postika, M.F. Population balance of aluminate solution decomposition: Physical modelling and model setup. *Tsvetnye Met.* **2019**, *2019*, 75–81. [CrossRef]

38. Li, Y.; Zhang, Y.; Yang, C.; Zhang, Y. Precipitating sandy aluminium hydroxide from sodium aluminate solution by the neutralization of sodium bicarbonate. *Hydrometallurgy* **2009**, *98*, 52–57. [CrossRef]
39. Elduayen-Echave, B.; Lizarralde, I.; Larraona, G.S.; Ayesa, E.; Grau, P. A New Mass-Based Discretized Population Balance Model for Precipitation Processes: Application to Struvite Precipitation. *Water Res.* **2019**, *155*, 26–41. [CrossRef]
40. Li, J.; Addai-Mensah, J.; Thilagam, A.; Gerson, A.R. Growth mechanisms and kinetics of gibbsite crystallization: Experimental and quantum chemical study. *Cryst. Growth Des.* **2012**, *12*, 3096–3103. [CrossRef]
41. Gao, W.; Li, Z.; Asselin, E. Solubility of $AlCl_3{\cdot}6H_2O$ in the Fe(II) + Mg + Ca + K + Cl + H_2O system and its salting-out crystallization with $FeCl_2$. *Ind. Eng. Chem. Res.* **2013**, *52*, 14282–14290. [CrossRef]
42. Ivanov, V.V.; Kirik, S.D.; Shubin, A.A.; Blokhina, I.A.; Denisov, V.M.; Irtugo, L.A. Thermolysis of acidic aluminum chloride solution and its products. *Ceram. Int.* **2013**, *39*, 3843–3848. [CrossRef]
43. Zhang, N.; Yang, Y.; Wang, Z.; Shi, Z.; Gao, B.; Hu, X.; Tao, W.; Liu, F.; Yu, J. Study on the thermal decomposition of aluminium chloride hexahydrate. *Can. Metall. Q.* **2018**, *57*, 235–244. [CrossRef]
44. Suss, A.; Senyuta, A.; Kravchenya, M.; Smirnov, A.; Panov, A. The quality of alumina produced by the hydrochloric acid process and potential for improvement. In Proceedings of the The International Committee for Study of Bauxite, Alumina & Aluminium (ICSOBA), Dubai, UAE, 29 November 2015; Volume 44, pp. 1–8.
45. Zhao, L. Calcination of aluminum chloride hexahydrate (ach) for alumina production: Implications for alumina extraction from aluminum rich fly ash (ARFA). *Arch. Metall. Mater.* **2018**, *63*, 235–240.
46. Pak, V.I.; Kirov, S.S.; Nalivaiko, A.Y.; Ozherelkov, D.Y.; Gromov, A.A. Obtaining alumina from kaolin clay via aluminum chloride. *Materials (Basel).* **2019**, *12*, 3938. [CrossRef] [PubMed]
47. Cui, L.; Cheng, F.; Zhou, J. Preparation of high purity $AlCl_3{\cdot}6H_2O$ crystals from coal mining waste based on iron(III) removal using undiluted ionic liquids. *Sep. Purif. Technol.* **2016**, *167*, 45–54. [CrossRef]
48. Senyuta, A.; Panov, A.; Milshin, O.; Slobodyanyuk, E.; Smirnov, A. *Method for Producing Metallurgical Alumina (Variants)*; Rospatent: Moscow, Russia, 2016; p. 20.
49. Panias, D.; Asimidis, P.; Paspaliaris, I. Solubility of boehmite in concentrated sodium hydroxide solutions: Model development and assessment. *Hydrometallurgy* **2001**, *59*, 15–29. [CrossRef]
50. Çelikel, B.; Demir, G.K.; Kayacl, M.; Baygul, M.; Suarez, C.E. Precipitation area upgrade at ETI aluminum. In *Light Metals*; Suarez, C.E., Ed.; Springer: Cham, Switzerland, 2012; pp. 129–133.
51. Liu, G.; Li, Z.; Qi, T.; Li, X.; Zhou, Q.; Peng, Z. Two-Stage Process for Precipitating Coarse Boehmite from Sodium Aluminate Solution. *JOM* **2017**, *69*, 1888–1893. [CrossRef]
52. Li, X.-B.; Yan, L.; Zhao, D.-F.; Zhou, Q.-S.; Liu, G.-H.; Peng, Z.-H.; Yang, S.-S.; Qi, T.-G. Relationship between $Al(OH)_3$ solubility and particle size in synthetic Bayer liquors. *Trans. Nonferrous Met. Soc. China (Engl. Ed.)* **2013**, *23*, 1472–1479. [CrossRef]
53. Li, H.; Addai-Mensah, J.; Thomas, J.C.; Gerson, A.R. The influence of Al(III) supersaturation and NaOH concentration on the rate of crystallization of $Al(OH)_3$ precursor particles from sodium aluminate solutions. *J. Colloid Interface Sci.* **2005**, *286*, 511–519. [CrossRef]
54. Alex, T.C.; Kumar, R.; Roy, S.K.; Mehrotra, S.P. Mechanical Activation of Al-oxyhydroxide Minerals—A Review. *Miner. Process. Extr. Metall. Rev.* **2016**, *37*, 1–26. [CrossRef]
55. Alex, T.C.; Kumar, R.; Roy, S.K.; Mehrotra, S.P. Towards ambient pressure leaching of boehmite through mechanical activation. *Hydrometallurgy* **2014**, *144–145*, 99–106. [CrossRef]
56. Li, X.-B.; Feng, G.-T.; Zhou, Q.-S.; Peng, Z.-H.; Liu, G.-H. Phenomena in late period of seeded precipitation of sodium aluminate solution. *Trans. Nonferrous Met. Soc. China (Engl. Ed.)* **2006**, *16*, 947–950. [CrossRef]
57. Wind, S.; Raahauge, B.E. Experience with commissioning new generation gas suspension calciner. *Miner. Met. Mater. Ser.* **2016**, 155–162.

Article

Monosodium Glutamate as Selective Lixiviant for Alkaline Leaching of Zinc and Copper from Electric Arc Furnace Dust

Erik Prasetyo [1,*], Corby Anderson [2], Fajar Nurjaman [1], Muhammad Al Muttaqii [1], Anton Sapto Handoko [1], Fathan Bahfie [1] and Fika Rofiek Mufakhir [1]

1 Research Unit for Mineral Technology, Indonesian Institute of Sciences, Jl. Ir. Sutami km. 15 Tanjung Bintang, Lampung Selatan 35361, Indonesia; nurjaman_80@yahoo.com (F.N.); almuttaqiimuhammad@gmail.com (M.A.M.); e_electrical@yahoo.com (A.S.H.); fathanbahfie@gmail.com (F.B.); fika.cupiw@gmail.com (F.R.M.)

2 George Ansell Department of Metallurgical and Materials Engineering, Kroll Institute for Extractive Metallurgy, Mining Engineering Department, Colorado School of Mines, 1500 Illinois St, Golden, CO 80401, USA; cganders@mines.edu

* Correspondence: erik.prasetyo@lipi.go.id; Tel.: +62-721-350-054

Received: 27 April 2020; Accepted: 12 May 2020; Published: 15 May 2020

Abstract: The efficacy of monosodium glutamate (MSG) as a lixiviant for the selective and sustainable leaching of zinc and copper from electric arc furnace dust was tested. Batch leaching studies and XRD, XRF and SEM-EDS characterization confirmed the high leaching efficiency of zinc (reaching 99%) and copper (reaching 86%) leaving behind Fe, Al, Ca and Mg in the leaching residue. The separation factor (concentration ratio in pregnant leach solution) between zinc vs. other elements, and copper vs. other elements in the optimum condition could reach 11,700 and 250 times, respectively. The optimum conditions for the leaching scheme were pH 9, MSG concentration 1 M and pulp density 50 g/L. Kinetic studies (leaching time and temperature) revealed that the saturation value of leaching efficiency was attained within 2 h for zinc and 4 h for copper. Modeling of the kinetic experimental data indicated that the role of temperature on the leaching process was minor. The study also demonstrated the possibility of MSG recycling from pregnant leach solutions by precipitation as glutamic acid (>90% recovery).

Keywords: electric arc furnace dust; monosodium glutamate; leaching; zinc; copper

1. Introduction

The demand for base metals such as zinc (Zn) and copper (Cu) has increased in recent years. Natural (primary) resources of both metals are dominated by sulfide ores, which unfortunately have declined both in quantity and grade causing discrepancies between supply and demand [1]. This in turn encourages the exploration of new resources, which not only includes primary resources but also secondary ones through recycling. Aside from helping address the resource depletion problem, the exploitation of secondary resources also mitigates environmental and resource sustainability problems. One of the secondary resources of base metal, which holds potential for further processing, is electric arc furnace (EAF) dust. The dust is a waste of steel making and is classified as hazardous [2]. The amount of dust produced during steel making is significant, where 11–20 kg of dust is generated for each ton of steel produced [3].

In addition to exploring new resources, extraction is also progressing toward green technology, which prioritizes the principles of efficiency, sustainability, safety and environmental impact. The EAF dust not only contains Zn and Cu as target elements but also others, e.g., lead (Pb), chromium (Cr), iron (Fe), aluminum (Al), calcium (Ca) and magnesium (Mg). As Pb and Cr (VI) represent pollutants, their release into the environment during the extraction of Zn and Cu must be minimized and as such, a comprehensive study is required to formulate applicable technology to maximize benefits from EAF dust and minimize the impact of pollutants (heavy metals) to the environment.

One of the most widely applied methods to recover Zn and Cu from EAF dust is hydrometallurgy, especially in the case where Zn in EAF dust exists as zinc oxide. Other common Zn phases, i.e., zinc ferrite requires pre-treatment (e.g. fusion) [4,5]. The hydrometallurgical approach for Zn and Cu recovery involves a leaching process using chemical lixiviants to dissolve these metals in aqueous solution. Dissolved Zn and Cu in the pregnant leach solution can then be further separated and purified using solvent extraction [6,7], ion exchange [8] or electrochemical methods [9,10]. Several leaching schemes that use various lixiviants have been previously proposed in order to recover Zn and/or Cu from EAF dust at ambient and elevated temperatures, respectively. Halli et al., 2017 [11] had screened potential lixiviants including nitric acid, citric acid, sodium hydroxide and ethanol for metal recovery from EAF dust. Proposed lixiviants by other researchers include sulfuric acid [12–14], hydrochloric acid [15], oxalic acid [16], citric acid [17], sodium hydroxide [18–20], sodium carbonate/sodium bicarbonate [21], ammonia [22], imminodiacetic acid (IDA) [23] and nitrilotriacetic acid (NTA) [24].

In general, the leaching schemes proposed in the literature can be classified into four categories:

1. Acidic leaching using strong acid, e.g., sulfuric acid, hydrochloric acid.
2. Acidic leaching using organic lixiviants, e.g., oxalic acid, citric acid.
3. Alkaline leaching using bases, e.g., sodium hydroxide, sodium carbonate.
4. Alkaline leaching using organic lixiviants, e.g., nitrilotriacetic acid (NTA).

The first and second schemes are advantageous in terms of recovery efficiency but both lack selectivity. Other elements, especially Fe and Ca which would cause fouling in further separation and purification steps, and Pb or Cr which are toxic, are also extracted [25]. The third scheme possesses an advantage in terms of selectivity but the extraction efficiency is low, requiring very low pulp density ratio, high lixiviant concentration and high leaching temperature to attain satisfying recovery. In addition, the caustic and corrosive nature of lixiviant requires special handling and equipment during operation and also poses a risk to the environment. The same is true of schemes using strong acid as lixiviant to certain extent, especially leaching using high concentration (concentrated) mineral acid, e.g., Halli et al., 2017 [11].

Considering these factors, the fourth scheme presented is preferable, due to its relatively selective recovery of Zn and Cu, higher leaching efficiency, safer handling in general compared to lixiviants in the other schemes and possibility to recover the lixiviant such as NTA. Although the leaching of base metal in this scheme is considered as selective, previous studies by Yang et al. (2016) [24] revealed relatively high extraction of iron, while the recovery of other elements such as Ca, Mg and Al during leaching was yet to be addressed, where these elements play an important role in further separation and purification steps. The lixiviants in the fourth category rely on their ability to form complexes with metal ions. Aminocarboxylic acid ligands, e.g., IDA, NTA or ethylenediaminetetraacetic acid (EDTA) possess relatively strong yet homogenous binding capacity to the metal ions, not only target elements such as Cu and Zn, but also matrix elements i.e. Fe, Al, Mg and Ca.

In order to partition the matrix elements into the solid phase and the valuable elements into the aqueous phase, it is suggested to use a ligand which complexes preferentially to Cu and Zn and shows weaker affinity to Ca, Mg, Fe and Al. Lixiviants such as amino acids broadly meet these criteria. Glycine has been extensively studied in copper [26] and gold [27] alkaline leaching, but to the best of our knowledge, no study on the efficacy of amino acid on Zn recovery has been carried out so far. Aside from glycine, another amino acid with potential to be developed as lixiviant is glutamic acid. In this research an amino acid sodium salt, i.e., monosodium glutamate (MSG) is proposed as a novel lixiviant. The advantages of MSG as a lixiviant include its wide availability and low price, and low risk to environment (biodegradable). MSG in alkaline conditions could serve as a powerful ligand to bind several transition metals, e.g., Cu, which in turn increase the extraction efficiency. The complex formation of these metals with MSG have been reported and exploited in chemical analysis [28,29] and our preliminary research also confirmed its efficacy to extract Zn [30]. Another advantage of using MSG is its reusability, since MSG can be recovered from the pregnant leach solution as glutamic acid precipitate by acidification of pregnant leach solution (PLS). Considering these possible advantages of using MSG, a study to confirm its efficacy in Zn and Cu extraction from EAF dust is warranted. The effect of leaching parameters on the leaching efficiency such as pH, lixiviant concentration, pulp density and kinetics, including selectivity toward other elements are investigated herein.

2. Species Distribution Modeling for Leaching Efficiency and Selectivity Prediction

The leaching efficiency and selectivity are hypothetically controlled by the distribution of species in the aqueous phase. Factors controlling this distribution include pH and concentration of metal ions and ligands/lixiviants involved in the system. The effect of pH on the recovery and selectivity could be predicted using a species distribution diagram, which was constructed using Visual MINTEQ 3.1 (KTH Royal Institute of Technology, Sweden). [31]. Figure 1 shows the species distribution of Zn and Cu in glutamate-H_2O system as a function of pH.

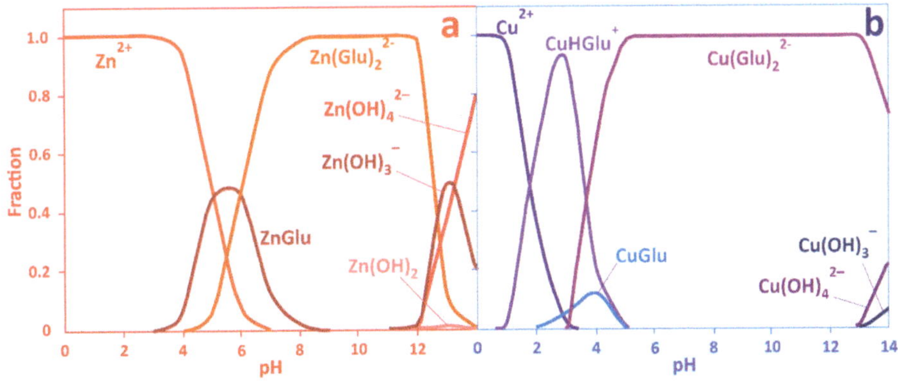

Figure 1. Species distribution of Zn (**a**) and Cu (**b**) in glutamate-H2O system as function of pH. Glutamate concentration 1 M, Zn^{2+} 300 mM and Cu^{2+} 8 mM.

Based on Figure 1, in acidic conditions the species are dominated by free Zn^{2+} and Cu^{2+} ions. Glutamate as chelate effectively binds Zn and Cu from weakly acidic to alkaline conditions, and prevents the precipitation of hydroxides at strong alkaline pH. The figure also shows that Zn starts to hydrolyze at pH 12, while Cu at pH 13. The program was also used to model the species distribution according to pH for other elements (Supplementary file: Figure S1), i.e., Fe, Mg, Ca and Al. In general Fe and Al start to hydrolyze at weakly acidic pH (±4), while Mg and Ca tend to hydrolyze at medium alkaline conditions (pH 12).

Aside from species distribution, leaching selectivity can be predicted using the complex formation constant (K_f) (Table 1). The constant was calculated for each complex, which might exist in the solution phase. Table 1 also lists the formation constants of metal ions with EDTA, NTA and glycinate as comparisons to assess the selectivity of organic compounds as lixiviants. If a lixiviant is expected to selectively bind one metal ion (α) and not another metal ion (β), then the value of $K_{f\alpha}/K_{f\beta}$ has to be higher than 10^6 ($\log K_{f\alpha} - log\ K_{f\beta} \geq 6$) [32]. The value of complex formation constants for glutamate, which were calculated based on the model generated by Visual Minteq in Table 1, indicates the possibility of separation of Zn and Cu from Mg and Ca, and also Zn from Cu using glutamate, and demonstrates better selectivity over the other chelators such as EDTA, NTA and glycine (amino acid).

The weaker complex formation constants for Fe-glutamate compared to Fe-EDTA and Fe-NTA indicate the possibility of separation of Zn and Cu from Fe through Fe hydrolysis in alkaline condition.

Table 1. Log complex formation constant of metal ions with three organic compounds as chelator/lixiviants.

Ions	log K_f EDTA [33]	log K_f NTA [34]	log K_f Glycinate [35]	log K_f Glutamate (This Study)
Zn^{2+}	16.4	10.7	5.5	8.9
Cu^{2+}	18.4	12.7	8.3	14.9
Fe^{3+}	24.2	24.3	10.3	11.8
Mg^{2+}	8.7	8.2	3.5	1.8
Ca^{2+}	10.6	7.0	1.4	1.1

3. Experimental

3.1. Materials and Instrumentation

EAF dust samples were kindly supplied by local smelters around Jakarta and Banten Province, Indonesia. A sieving test showed that more than 90% of dust passed through a 270 mesh (53 μm), and this size fraction was used in further leaching tests. All leaching experiment and characterization studies were performed in the Research Unit for Mineral Technology—Indonesian Institute of Sciences, Bandar Lampung, Indonesia, or otherwise stated.

XRF characterization (Panalytical X'Pert 3 Powder with Omnian Standard) was carried out to determine the major chemical components in EAF dust (Table 2), and shows the most dominant constituents to be zinc and aluminum. The mineralogical phases were determined using X-ray powder diffraction (Panalytical, Expert3 Powder). Quantitative determination of metal contents in EAF dust (Table 3) and pregnant leach solution after leaching for recovery and separation factor calculation were carried out using atomic absorption spectrophotometry (AAS, Shimadzu AA7000, Japan) and Inductively Coupled Plasma-Optical Emission Spectrometry (ICP-OES, Analytik Jena, Plasma Quant 9000 Elite, Germany). SEM-EDS characterization to investigate the change of morphology and elemental distribution on the surface of EAF dust grain before and after leaching was performed using a SEM (Hitachi SUM 3500, Japan) at the Research Unit for Natural Product Technology—Indonesian Institute of Sciences, Yogyakarta, Indonesia. The advanced mineral identification and characterization system (AMICS) analysis to determine and map the mineral phases was carried out by Eagle Engineering, Butte, Montana, USA.

Chemicals such as sodium hydroxide, sulfuric acid, nitric acid, hydrochloric acid, and standard solutions were obtained from Merck, Darmstadt all in analytical grade, while monosodium glutamate was produced by PT Ajinomoto Indonesia (99% purity) and used as received. The solution pH was adjusted using dilute sulfuric acid or sodium hydroxide and monitored using a pH meter (Oakton 45, Vernon Hills, IL, USA), while deionized water (MilliQ) was used throughout the experiment.

Table 2. Major chemical composition of electric arc furnace (EAF) dust.

Component	MgO	Al_2O_3	SiO_2	SO_3	K_2O	CaO	TiO_2	MnO	Fe_2O_3	NiO	CuO	ZnO	Total
wt.%	0.59	36.0	2.22	0.34	0.12	2.23	0.18	0.87	2.72	0.60	1.29	51.8	99.2

Table 3. Chemical composition of important base metals in EAF dust (after aqua regia digestion and AAS/ICP-OES determination).

Element	Cr	Cu	Fe	Ni	Pb	Zn
wt.%	0.016 ± 0.002	0.898 ± 0.021	1.432 ± 0.029	0.522 ± 0.008	0.075 ± 0.016	38.652 ± 0.428

3.2. Leaching Procedure

Leaching studies were carried out using a batch method. In general, 1 g of EAF dust was mixed with 20 mL lixiviant in 250 mL sealed flask. The mixture was homogenized using an orbital shaker at 200 rpm. After the leaching was completed, the supernatant solution was separated using centrifugation and filtration (Whatman 42). The metal concentration was determined using AAS or ICP-OES. Metal recovery (R) was calculated using Equation (1). To evaluate the selectivity of the leaching scheme, a separation factor (SF) between Zn or Cu and other metals was evaluated using Equation (2). All leaching data were obtained in duplicates.

$$R = \frac{C_E \times V}{C_o \times m} \times 100\% \quad (1)$$

where:

C_E Zn or Cu concentration in supernatant solution (mg/L)
C_o Zn or Cu content in EAF dust (mg/g)
m mass of EAF dust used in leaching (g)
V leaching agent volume (L)

$$SF = \frac{C_Z}{C_M} \quad (2)$$

where:

C_Z, Zn or Cu concentration in pregnant leach solution (mg/L).
C_M, other metal concentration in pregnant leach solution (mg/L).

4. Results and Discussion

4.1. Characterization

EAF dust characterization using XRD showed the mineral phase composition was dominated by zinc oxide (ZnO). XRD of the dust residue obtained after leaching (leaching condition pH 9, MSG concentration 1 M, pulp density 50 g/L and 12-hour homogenization) revealed the phases were dominated by spinel (Mg-Al oxide), calcium aluminum oxide and zinc oxide phases (Figure 2). These elements such as Al, Mg, Ca and Fe were found to be relatively retained in the solid phase during the leaching process. The XRD characterization results are supported by AMICS analysis results (Supplementary file: Tables S1 and S2 and Figure S2), which revealed the major Zn phase to be ZnO and gahnite ($ZnAl_2O_4$) in original material.

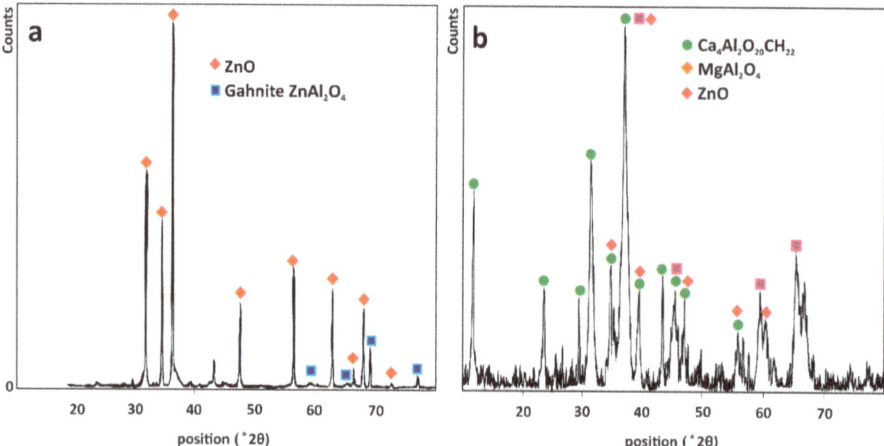

Figure 2. Phases present in EAF dust (**a**) before and (**b**) after leaching.

The change in elemental distribution of EAF dust before and after leaching was also explored by SEM-EDS characterization (Figure 3). Surface morphology of EAF dust changed from porous and grainy to relatively smooth. Intensity decrease of Zn and Cu peaks on the EDS profile after leaching confirmed the efficacy of MSG as lixiviant in EAF dust leaching. Qualitative mapping on the surface reveals that elements such as Al, Fe and Ca were relatively retained and even enriched after leaching (Table 4), indicating the selective nature of the leaching using MSG.

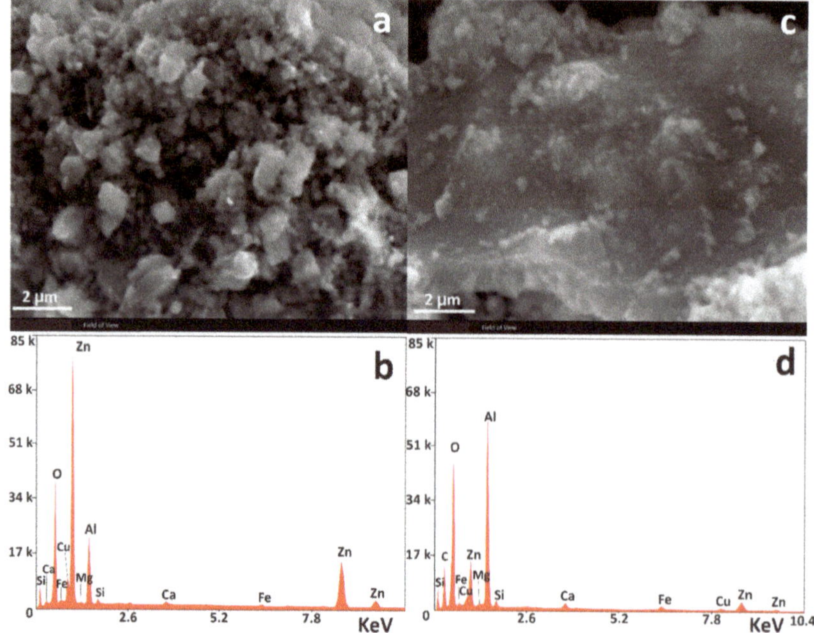

Figure 3. Morphology (**a**) before and (**b**) after leaching, including EDS intensities of elements based on mapping on the respective field (**c**) before and (**d**) after leaching.

Table 4. Qualitative mapping of elements by EDS on grain surface before and after leaching.

Elements	O K	Mg K	Al K	Si K	Ca K	Fe K	Zn K
Before leaching (wt %)	24.08	1.62	15.19	1.06	0.57	0.37	56.95
After leaching (wt %)	35.53	0.57	20.42	0.90	1.00	1.30	6.99

4.2. Effect of pH

The effect of pH on the Zn and Cu leaching efficiency and selective recovery was studied between 6 and 11. The constant variables included MSG concentration 1 M, pulp density 50 g/L and an agitation time of 12 h at room temperature. pH ranges between 6–11 were chosen according to the modeling in Section 2 in order to suppress the solubility of other elements (Mg, Ca, Al and Fe). The leaching efficiency and leaching selectivity are shown in Figures 4 and 5, respectively.

Based on Figure 4, the optimum pH to recover Zn and Cu are 9. Low recovery at lower pH was due to the weaker or repulsive interaction between glutamate and Zn or Cu since the glutamate species was dominated as a protonated species, e.g., H_3Glu^+, $HGlu^-$. Lower recovery at pH higher than 9 was probably due to the hydrolysis of Zn and Cu, which was lower compared to the modeling results using Visual Minteq (pH > 12 for Zn and pH > 13 for Cu). In the case of other elements, all but Al extraction decreased as pH became alkaline. Increasing recovery of Al in alkaline conditions (0.5% at pH 11) was due to the amphoteric characteristic of Al, which is soluble in excess of alkali as $Al(OH)_4^-$.

Increasing Al recovery at higher pH and decreasing Zn or Cu contributed to the sharp decrease of the selectivity factor of Zn or Cu toward Al (Figure 5). The figure also shows the optimum pH to separate Zn and Cu are 9 and 10, respectively. The highest separation factor is obtained toward iron, which could reach an order of 10^4 (with Zn) and 400 (with Cu). Lower separation factors were obtained between Zn or Cu and Mg and Ca (reaching an order of 700).

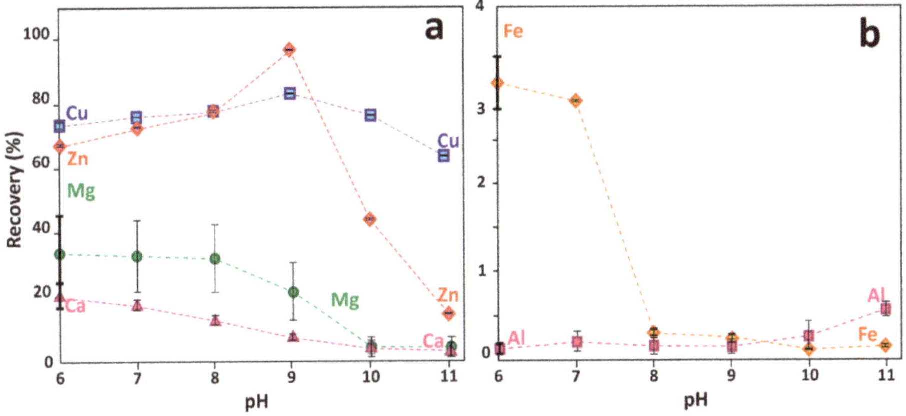

Figure 4. The recovery of (**a**) Zn, Cu, Mg and Ca and (**b**) Fe and Al as function of pH.

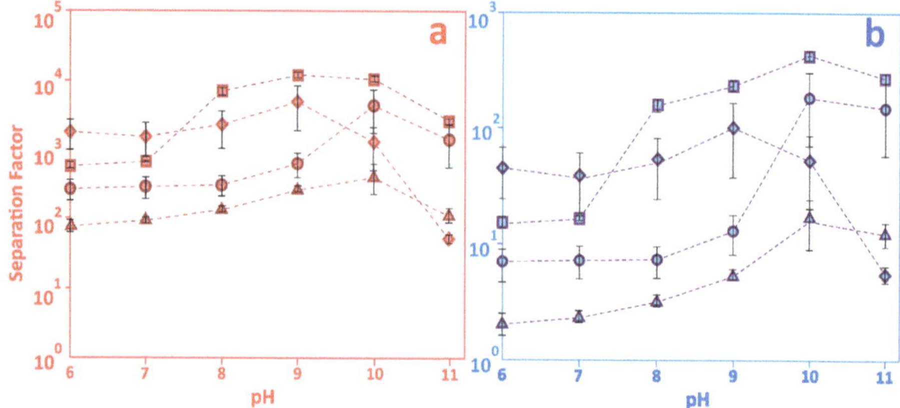

Figure 5. Separation factor of (**a**) Zn: [Zn]/[Fe] (squares), [Zn]/[Al] (diamonds), [Zn]/[Mg] (circles), [Zn]/[Ca] (triangles) and (**b**) Cu: [Cu]/[Fe] (squares), [Cu]/[Al] (diamonds), [Cu]/[Mg] (circles), [Cu]/[Ca] (triangles) in pregnant leach solution as function of pH.

4.3. *Effect of MSG Concentration*

The effects of MSG concentration to metal recovery and Zn and Cu selectivity over other elements were studied between 0.1 M and 2 M, with constant variables pH 9, pulp density 50 g/L and 12 h leaching time at room temperature. The results in Figure 6 show that MSG concentration has positive effects on metal recovery. Based on modeling using Visual Minteq (Supplementary file: Figure S3), the minimum MSG concentration to completely solubilize Zn is 0.7 M, and at lower MSG concentration, Zn speciation is dominated by hydroxide precipitate. In the case of Cu, the model shows that MSG concentration has little effect on speciation, which contradicts the results in Figure 6. This is probably due to competition with Zn, which has a 40 times larger concentration. The leaching process at higher MSG concentration increased the recovery of other elements, e.g., Fe, due to increasing soluble species ($FeGlu^+$ complex) (Supplementary file: Figure S3). The results in Figure 6 show the optimum recovery of Zn and Cu are attained at 1.2 and 0.7 M, respectively, which at higher MSG concentration did not significantly increase the recovery of both metals. At higher MSG concentration, the recovery of other metals, especially Mg and Ca became substantial, decreasing the selectivity (Figure 7). The optimum MSG concentration based on data in Figure 6 is well supported by results in Figure 7.

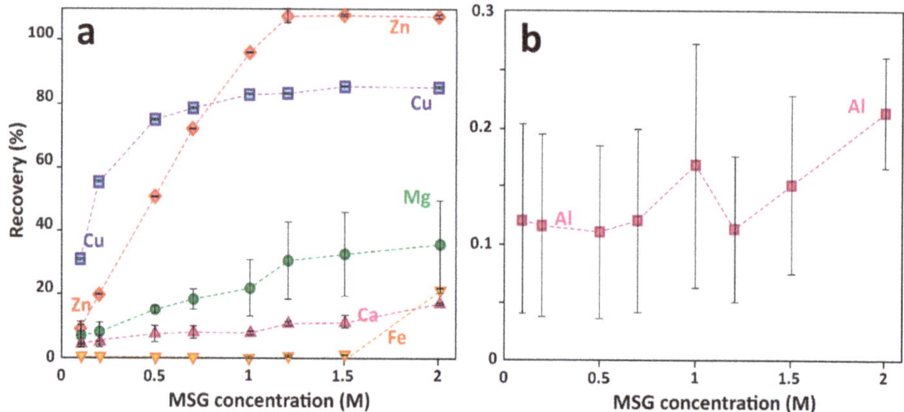

Figure 6. The recovery of (**a**) Zn, Cu, Mg, Ca and Fe and (**b**) Al as function of MSG concentration.

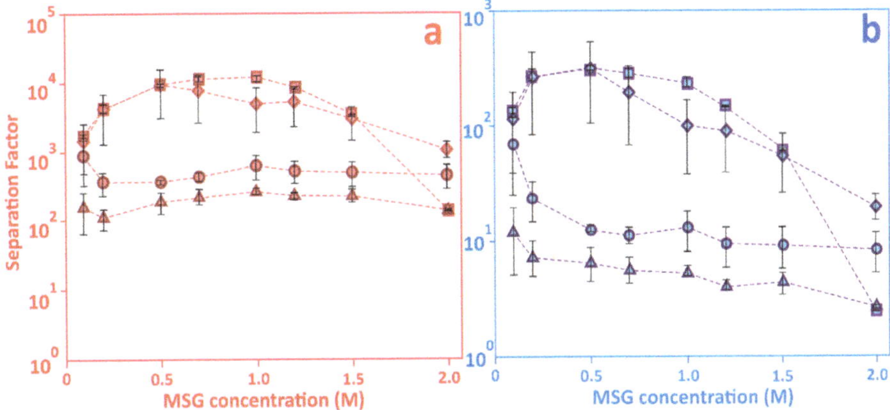

Figure 7. Separation factor of (**a**) Zn: [Zn]/[Fe] (squares), [Zn]/[Al] (diamonds), [Zn]/[Mg] (circles), [Zn]/[Ca] (triangles) and (**b**) Cu: [Cu]/[Fe] (squares), [Cu]/[Al] (diamonds), [Cu]/[Mg] (circles), [Cu]/[Ca] (triangles) in pregnant leach solution as function of MSG concentration.

4.4. Effect of Pulp Density

To evaluate the effect of pulp density (ratio between lixiviant and EAF dust) on the recovery and selectivity, leaching was carried out at constant variables pH 9, MSG concentration 1 M, 12 h leach at room temperature, while pulp density was varied between 33 and 200 g/L. The results in Figure 8 show the recovery decreased as pulp density increased. Maximum pulp density to obtain optimum recovery of Zn and Cu were 50 and 100 g/L, respectively. In the case of Mg and Ca the optimum pulp density was 66.67 g/L, while in the case of Fe and Al, the recovery was very low (less than 1 percent).

The plot of separation factor value as a function of pulp density (Figure 9) reveals that pulp density has a negative effect on the separation factor. This was probably caused by the increase of solubility of elements especially Mg, Ca, Fe and Al at larger volumes in alkaline conditions. This also indicates the leaching of EAF dust using MSG as lixiviant in alkaline conditions should be performed using continuous methods, e.g., column leaching or heap leaching instead of a batch method, since in the continuous method the volume could be minimized to optimize the selectivity.

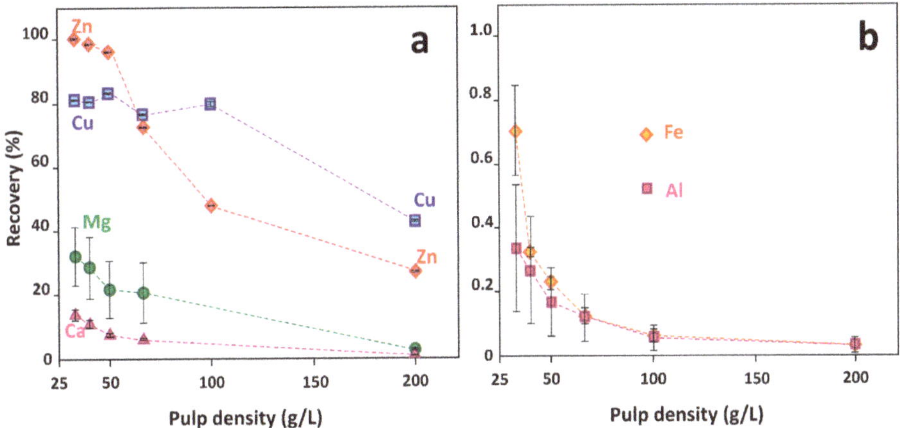

Figure 8. The recovery of (**a**) Zn, Cu, Mg and Ca and (**b**) Fe and Al as function of pulp density.

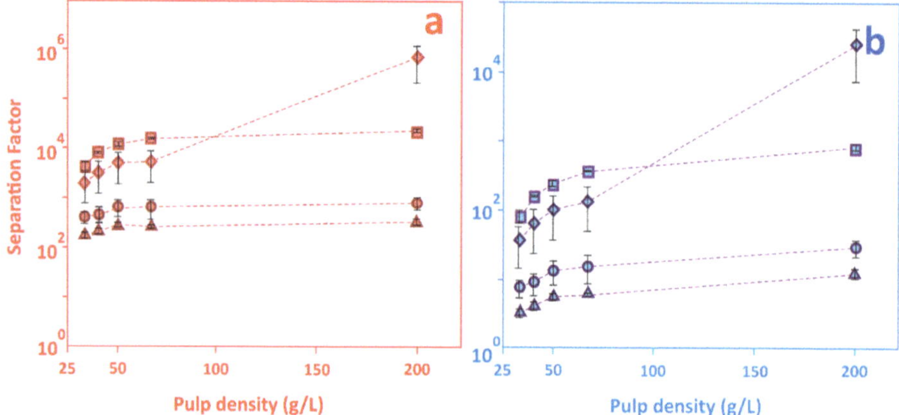

Figure 9. Separation factor of (**a**) Zn: [Zn]/[Fe] (squares), [Zn]/[Al] (diamonds), [Zn]/[Mg] (circles), [Zn]/[Ca] (triangles) and (**b**) Cu: [Cu]/[Fe] (squares), [Cu]/[Al] (diamonds), [Cu]/[Mg] (circles), [Cu]/[Ca] (triangles) in pregnant leach solution as function of pulp density.

4.5. Kinetic Studies (Effect of Leaching Time and Temperature)

Kinetic studies were carried out at a time range up to 12 h at three different temperatures (30, 55 and 80 °C), with constant variables pH 9, MSG concentration 1 M and pulp density 50 g/L. The results in Figure 10 show that maximum leaching efficiency was attained within 120 min for Zn and 4 h for Cu at 30 °C. At higher leaching temperature, recovery reached saturation in a shorter period. Figure 10b shows the recovery of Cu is delayed about 1 h. Based on Eh monitoring, in the early stage of leaching the aqueous phase was reductive, which inhibited the oxidation and complexation of Cu by lixiviant. As leaching progressed, the aqueous phase became oxidative, favoring the oxidation and complexation of Cu.

To describe the kinetic process of leaching, three kinetic models were tested to model experimental data: shrinking core model (chemical reaction control), shrinking particle model (film diffusion) and interface transfer and diffusion by Dickinson and Heal (1999) [36]. The fitting results (R^2, coefficient of correlation) of each model are listed in Table 5.

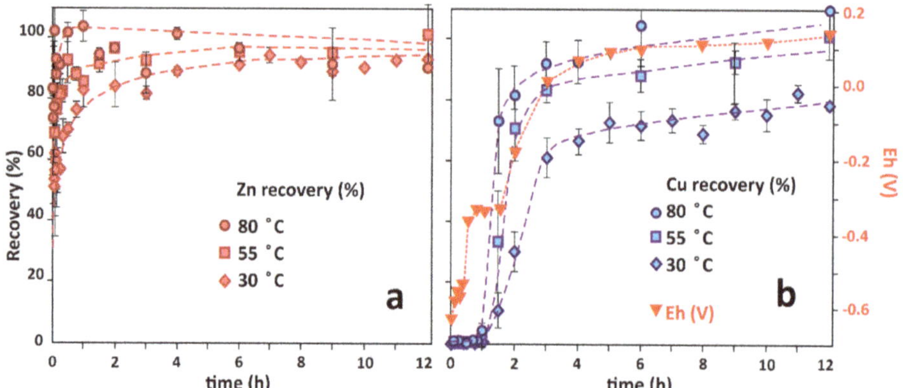

Figure 10. The effect of leaching time and temperature on recovery of (**a**) Zn and (**b**) Cu. In b, the change of Eh in aqueous phase during leaching was obtained at 30 °C (single measurement).

Table 5. Correlation coefficient for experimental data fitting using three kinetic models (SCM, shrinking core model; SPM, shrinking particle model).

Element	Model	Equation	Coefficient of Correlation, R^2		
			30 °C	55 °C	80 °C
Zn	SCM (chemical reaction control)	$kt = 1 - (1 - R)^{1/3}$	0.768	0.715	0.710
	SPM (film diffusion)	$kt = 1 - (1 - R)^{2/3}$	0.717	0.635	0.622
	Interface transfer and diffusion	$kt = 1/3\ ln(1 - x) - [1 - (1 - x)^{-1/3}]$	0.886	0.916	0.895
Cu	SCM (chemical reaction control)	$kt = 1 - (1 - R)^{1/3}$	0.880	0.847	0.912
	SPM (film diffusion)	$kt = 1 - (1 - R)^{2/3}$	0.860	0.802	0.885
	Interface transfer and diffusion	$kt = 1/3\ ln(1 - x) - [1 - (1 - x) - {}^{1/3}]$	0.930	0.969	0.931

k, apparent rate constant (min^{-1}); t, time (min); R, metal recovery.

Based on the coefficient of the correlation value in Table 5, the interface transfer and diffusion model is the best model to describe the experimental data. Although the shrinking core model and shrinking particle model are more popular models to explain kinetic data, both models are generally based on the assumption of constant lixiviant concentration during leaching [37]. In the leaching process, the concentration of glutamate would decrease due to complex formation. In this study, the model developed by Dickinson and Heal (1999) [36] was adequately applied to analyze the dissolution of Zn and Cu in EAF dust using glutamate.

The value of the apparent rate constant (k) obtained for each temperature using the interface transfer and diffusion model (Table 6) was used to determine the activation energy (E_a, kJ/mol) using the Arrhenius Equation (3), where A, T, and R are the frequency factor, temperature, and gas constant, respectively. Relatively low energy activation for both Zn and Cu indicates that the effect of temperature on the leaching process is minor.

$$k = Ae^{-\frac{E_a}{RT}} \text{ or } lnk = lnA - \frac{E_a}{RT} \tag{3}$$

Table 6. Apparent rate constant (k) attained from fitting of experimental data using interface transfer and diffusion model, which was used to obtain the activation energy (E_a) using linear regression, including the coefficient of correlation (R^2).

Element	k (min^{-1})			E_a (kJ/mol)	R^2
	30 °C	55 °C	80 °C		
Zn	7.3×10^{-4}	1.9×10^{-3}	5.4×10^{-3}	35.5	0.995
Cu	3.0×10^{-4}	1.0×10^{-3}	2.2×10^{-3}	35.6	0.994

4.6. Monosodium Glutamate Recovery

Due to the significant amount of MSG used in leaching and to sustain the leaching scheme, the regeneration of glutamate is required. Glutamate could be recovered from the pregnant leach solution as precipitate by acidifying the solution to the pH between 2–4.5. In this pH range glutamate is precipitated as neutral species glutamic acid, H_2Glu (Supplementary file: Figure S4). The addition of more acid (pH less than 1) caused the formation of soluble cationic species H_3Glu^+. The experiment performed to recover glutamate from the pregnant leach solution showed that the optimum pH was 3, which in this pH more than 90% of the glutamate was recovered as glutamic acid (Figure 11). The recovery efficiency of glutamate was determined gravimetrically.

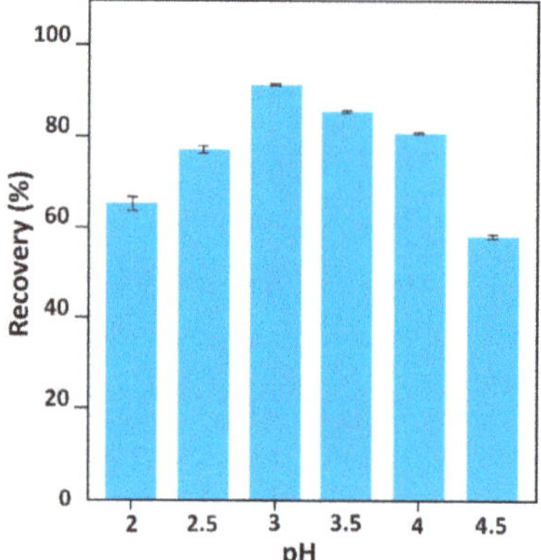

Figure 11. Monosodium glutamate (MSG) recovery efficiency as function of pH.

5. Conclusions

Monosodium glutamate effectively and selectively recovered Zn and Cu from EAF dust, based on batch leaching studies according to leaching efficiency and separation factor values. The optimum conditions for the leaching scheme are pH 9, lixiviant concentration 1 M and pulp density 50 g/L. Studies on the effect of pulp density on the metal recovery and separation factor showed that pulp density correlated negatively to the metal recovery and positively to the separation factor of Zn and Cu to the other elements. This indicates the leaching is better performed using a continuous method, considering the selective recovery of Zn and Cu from other elements. Kinetic studies showed the leaching efficiency reached a saturation value in less than 2 and 4 h for Zn and Cu, respectively. The activation energy obtained from the experimental data modeling revealed the effect of temperature on the leaching process was minor. Further the use of MSG as lixiviant offers a sustainable leaching scheme since MSG is recoverable from the pregnant leach solution and reusable for the next leaching cycle.

Supplementary Materials: The following are available online at http://www.mdpi.com/2075-4701/10/5/644/s1, Figure S1: Species distribution of (a) Al, (b) Fe, (c) Mg and (d) Ca in glutamate-H_2O system as function of pH. Glutamate concentration 1 M, Al^{2+} 350 mM, Fe^{3+} 10 mM, Mg^{2+} 10 mM and Ca^{2+} 20 mM, Table S1: AMICS Mineralogy for Zinc EAF Dust Sample, Figure S2: Identified AMICS Minerals for Zinc EAF Dust Sample, Table S2, AMICS Color Scheme, Figure S3: Species distribution of (a) Zn and (b) Fe in glutamate-H_2O system as function of glutamate concentration at pH 9. Zn^{2+} 300 mM and Fe^{3+} 10 mM, Figure S4: Species distribution of glutamate in glutamate-H_2O system as function of pH. Glutamate concentration 1 M.

Author Contributions: Conceptualization, E.P.; data curation, E.P.; formal analysis, E.P. and C.A.; funding acquisition, E.P. and C.A.; investigation, E.P. and C.A.; methodology, E.P.; project administration, F.R.M., E.P.; resources, E.P., C.A., F.N., M.A.M., A.S.H., F.R.M. and F.B.; validation, E.P.; visualization, M.A.M. and F.B.; writing–original draft, E.P.; writing–review & editing, E.P. and C.A. All authors have read and agreed to the published version of the manuscript.

Funding: This research was funded and supported by Indonesian Institute of Sciences FY 2019, Fulbright Program through Visiting Scholar Scheme 2019–2020, Indonesian Ministry of Research and Technology Insinas Program FY 2020, PRN (National Priority Research) of Indonesia FY 2020.

Acknowledgments: We would like to express our gratitude to the two reviewers for improving this manuscript.

Conflicts of Interest: The authors declare no conflict of interest.

References

1. Ejtemaei, M.; Gharabaghi, M.; Irannajad, M. A review of zinc oxide mineral beneficiation using flotation method. *Adv. Colloid Interface Sci.* **2014**, *206*, 68–78. [CrossRef] [PubMed]
2. Martins, F.M.; dos Neto, J.M.R.; da Cunha, C.J. Mineral phases of weathered and recent electric arc furnace dust. *J. Hazard. Mater.* **2008**, *154*, 417–425. [CrossRef]

3. Suetens, T.; Klaasen, B.; Van Acker, K.; Blanpain, B. Comparison of electric arc furnace dust treatment technologies using exergy efficiency. *J. Clean. Prod.* **2014**, *65*, 152–167. [CrossRef]
4. Youcai, Z.; Stanforth, R. Extraction of zinc from zinc ferrites by fusion with caustic soda. *Miner. Eng.* **2000**, *13*, 1417–1421. [CrossRef]
5. Wang, C.; Guo, Y.F.; Wang, S.; Chen, F.; Tan, Y.J.; Zheng, F.Q.; Yang, L.Z. Characteristics of the reduction behavior of zinc ferrite and ammonia leaching after roasting. *Int. J. Miner. Metall. Mater.* **2020**, *13*, 1417–1421. [CrossRef]
6. Lupi, C.; Pilone, D. Effectiveness of saponified D2EHPA in Zn(II) selective extraction from concentrated sulphuric solutions. *Miner. Eng.* **2020**, *150*, 106278. [CrossRef]
7. Zhu, Z.; Cheng, C.Y. A Study on zinc recovery from leach solutions using ionquest 801 and its mixture with D2EHPA. *Miner. Eng.* **2012**, *39*, 117–123. [CrossRef]
8. Zhou, K.; Wu, Y.; Zhang, X.; Peng, C.; Cheng, Y.; Chen, W. Removal of Zn(II) from manganese-zinc chloride waste liquor using ion-exchange with D201 resin. *Hydrometallurgy* **2019**, *190*, 105171. [CrossRef]
9. Rudnik, E. Recovery of zinc from zinc ash by leaching in sulphuric acid and electrowinning. *Hydrometallurgy* **2019**, *188*, 256–263. [CrossRef]
10. Youcai, Z.; Chenglong, Z.; Youcai, Z.; Chenglong, Z. Electrowinning of Zinc and Lead from Alkaline Solutions. In *Pollution Control and Resource Reuse for Alkaline Hydrometallurgy of Amphoteric Metal Hazardous Wastes*; Springer: Berlin/Heidelberg, Germany, 2017. [CrossRef]
11. Halli, P.; Hamuyuni, J.; Revitzer, H.; Lundström, M. Selection of leaching media for metal dissolution from electric arc furnace dust. *J. Clean. Prod.* **2017**, *164*, 265–276. [CrossRef]
12. Kukurugya, F.; Vindt, T.; Havlík, T. Behavior of Zinc, Iron and Calcium from Electric Arc Furnace (EAF) dust in hydrometallurgical processing in sulfuric acid solutions: Thermodynamic and kinetic aspects. *Hydrometallurgy* **2015**, *154*, 20–32. [CrossRef]
13. Montenegro, V.; Agatzini-Leonardou, S.; Oustadakis, P.; Tsakiridis, P. Hydrometallurgical treatment of EAF dust by direct sulphuric acid leaching at atmospheric pressure. *Waste Biomass Valoriz.* **2016**, *7*, 1531–1548. [CrossRef]
14. Rudnik, E. Investigation of industrial waste materials for hydrometallurgical recovery of Zinc. *Miner. Eng.* **2019**, *139*. [CrossRef]
15. Teo, Y.Y.; Lee, H.S.; Low, Y.C.; Choong, S.W.; Low, K.O. Hydrometallurgical extraction of Zinc and Iron from Electric Arc Furnace Dust (EAFD) Using Hydrochloric Acid. *J. Phys. Sci.* **2018**, *29*, 49–54. [CrossRef]
16. Kusumaningrum, R.; Fitroturokhmah, A.; Sinaga, G.S.T.; Wismogroho, A.S.; Widayatno, W.B.; Diguna, L.J.; Amal, M.I. Study: leaching of zinc dust from electric arc furnace waste using oxalic acid. In *IOP Conference Series: Materials Science and Engineering*; IOP Publishing: Bristol, UK, 2019. [CrossRef]
17. Halli, P.; Hamuyuni, J.; Leikola, M.; Lundström, M. Developing a sustainable solution for recycling electric arc furnace dust via organic acid leaching. *Miner. Eng.* **2018**, *124*, 1–9. [CrossRef]
18. Dutra, A.J.B.; Paiva, P.R.P.; Tavares, L.M. Alkaline leaching of zinc from electric arc furnace steel dust. *Miner. Eng.* **2006**, *19*, 478–485. [CrossRef]
19. Palimakaą, P.; Pietrzyk, S.; Stępień, M.; Ciećko, K.; Nejman, I. Zinc Recovery from steelmaking dust by hydrometallurgical methods. *Metals (Basel)* **2018**, *8*, 547. [CrossRef]
20. Zhang, D.; Ling, H.; Yang, T.; Liu, W.; Chen, L. Selective Leaching of Zinc from Electric Arc Furnace dust by a hydrothermal reduction method in a sodium hydroxide system. *J. Clean. Prod.* **2019**, *224*, 536–544. [CrossRef]
21. Al-Makhadmeh, L.A.; Batiha, M.A.; Al-Harahsheh, M.S.; Altarawneh, I.S.; Rawadieh, S.E. The effectiveness of Zn leaching from EAFD using caustic soda. *Water. Air. Soil Pollut.* **2018**, *229*, 33. [CrossRef]
22. Ma, A.; Zheng, X.; Shi, S.; He, H.; Rao, Y.; Luo, G.; Lu, F. Study on recovery of Zinc from metallurgical solid waste residue by ammoniacal leaching. In *Minerals, Metals and Materials Series*; Springer: Berlin/Heidelberg Germany, 2019. [CrossRef]
23. Zhang, D.; Zhang, X.; Yang, T.; Rao, S.; Hu, W.; Liu, W.; Chen, L. Selective leaching of Zinc from blast furnace dust with mono-ligand and mixed-ligand complex leaching systems. *Hydrometallurgy* **2017**, *169*, 219–228. [CrossRef]
24. Yang, T.; Rao, S.; Zhang, D.; Wen, J.; Liu, W.; Chen, L.; Zhang, X. Leaching of low grade zinc oxide ores in nitrilotriacetic acid solutions. *Hydrometallurgy* **2016**, *161*, 107–111. [CrossRef]

25. Herrero, D.; Arias, P.L.; Güemez, B.; Barrio, V.L.; Cambra, J.F.; Requies, J. Hydrometallurgical process development for the production of a zinc sulphate liquor suitable for electrowinning. *Miner. Eng.* **2010**, *23*, 511–517. [CrossRef]
26. Eksteen, J.J.; Oraby, E.A.; Tanda, B.C. A conceptual process for copper extraction from chalcopyrite in alkaline glycinate solutions. *Miner. Eng.* **2017**, *108*, 53–66. [CrossRef]
27. Oraby, E.A.; Eksteen, J.J.; Karrech, A.; Attar, M. Gold extraction from paleochannel ores using an aerated alkaline glycine lixiviant for consideration in heap and in-situ leaching applications. *Miner. Eng.* **2019**, *138*, 112–118. [CrossRef]
28. Prasetyo, E. Monosodium Glutamate For Simple Photometric Iron Analysis. In *IOP Conference Series: Materials Science and Engineering*; IOP Publishing: Bristol, UK, 2018. [CrossRef]
29. Prasetyo, E. Simple method of copper analysis using monosodium glutamate and its application in ore analysis. *Mineralogia* **2012**, *43*, 137–146. [CrossRef]
30. Prasetyo, E.; Bahfie, F.; Al Muttaqii, M.; Handoko, A.S.; Nurjaman, F. Zinc extraction from electric arc furnace dust using amino acid leaching. In *AIP Conference Proceeding*; AIP Publishing: College Park, MA, USA, 2020.
31. Gustafsson, J.P. Visual MINTEQ Ver. 3.1. Available online: https://vminteq.lwr.kth.se/visual-minteq-ver-3-1/ (accessed on 4 August 2019).
32. Leonard, M. Vogel's textbook of quantitative chemical analysis. *Endeavour* **1990**, *14*, 100. [CrossRef]
33. Anderegg, G. Critical survey of stability constants of EDTA complexes. In *Critical Survey of Stability Constants of EDTA Complexes*; Pergamon: Oxford, UK, 1977. [CrossRef]
34. Anderegg, G. Critical survey of stability constants of Nta complexes. *Pure Appl. Chem.* **2013**, *54*, 2693–2758. [CrossRef]
35. Kiss, T.; Sovago, I.; Gergely, A. Critical survey of stability constants of complexes of glycine. *Pure Appl. Chem.* **2007**, *63*, 597–638. [CrossRef]
36. Dickinson, C.F.; Heal, G.R. Solid-liquid diffusion controlled rate equations. *Thermochim. Acta* **1999**, *340*, 89–103. [CrossRef]
37. Rao, S.; Yang, T.; Zhang, D.; Liu, W.F.; Chen, L.; Hao, Z.; Xiao, Q.; Wen, J.F. Leaching of low grade zinc oxide ores in nh4cl-nh3 solutions with nitrilotriacetic acid as complexing agents. *Hydrometallurgy* **2015**, *158*, 101–106. [CrossRef]

Article

Effect of Preliminary Alkali Desilication on Ammonia Pressure Leaching of Low-Grade Copper–Silver Concentrate

Kirill Karimov [1], Andrei Shoppert [1,*], Denis Rogozhnikov [1], Evgeniy Kuzas [1], Semen Zakhar'yan [2] and Stanislav Naboichenko [1]

1 Department of Non-ferrous Metals Metallurgy, Ural Federal University, 620002 Yekaterinburg, Russia; kirill_karimov07@mail.ru (K.K.); darogozhnikov@yandex.ru (D.R.); e.kuzas@ya.ru (E.K.); ssnaboichenko@mail.ru (S.N.)

2 TOO Kazgidromed, Karaganda 100009, Kazakhstan; szakharyan@yandex.kz

* Correspondence: andreyshop@list.ru; Tel.: +7-9220243963

Received: 19 May 2020; Accepted: 15 June 2020; Published: 18 June 2020

Abstract: Ammonia leaching is a promising method for processing low-grade copper ores, especially those containing large amounts of oxidized copper. In this paper, we study the effect of Si-containing minerals on the kinetics of Cu and Ag leaching from low-grade copper concentrates. The results of experiments on the pressure leaching of the initial copper concentrate in an ammonium/ammonium-carbonate solution with oxygen as an oxidizing agent are in good agreement with the shrinking core model in the intra-diffusion mode: in this case, the activation energies were 53.50 kJ/mol for Cu and 90.35 kJ/mol for Ag. Energy-dispersive X-ray spectroscopy analysis (EDX) analysis showed that reagent diffusion to Cu-bearing minerals can be limited by aluminosilicate minerals of the gangue. The recovery rate for copper and silver increases significantly after a preliminary alkaline desilication of the concentrate, and the new shrinking core model is the most adequate, showing that the process is limited by diffusion through the product layer and interfacial diffusion. The activation energy of the process increases to 86.76 kJ/mol for Cu and 92.15 kJ/mol for Ag. Using the time-to-a-given-fraction method, it has been shown that a high activation energy is required in the later stages of the process, when the most resistant sulfide minerals of copper and silver apparently remain.

Keywords: ammonia leaching; high-pressure leaching; copper extraction; silver extraction; desilication; kinetics

---

1. Introduction

The depletion of reserves of high-quality raw materials of non-ferrous metals is an increasingly significant problem around the globe today. As a result, low-grade ores and concentrates are involved in processing [1,2], and enterprises have to sacrifice recovery and quality to maintain the required levels of productivity.

The deterioration of technological parameters and indicators is due to a decrease in the content of valuable components and an increase in harmful impurities, along with the fact that non-ferrous and noble metals often exist in minerals as fine impregnations in gangue with mutual intergrowth and the presence of nanodispersed, colloidal grains, etc. [3,4].

Traditional methods of processing and the beneficiation of such raw materials are characterized by significant losses, since such finely disseminated components cannot be separated and recovered to an acceptable degree to obtain monometallic concentrates and intermediates, which leads to significant losses of valuable metals and the stockpiling of various industrial products and waste.

At the same time, the use of such off-balance materials in processing can increase the term of exhaustion of natural resources by several times, which is greatly important for sustainable economic development and environmental safety. Therefore, science and industry are in a constant search for new effective ways of processing such troublesome raw materials.

Currently, the world has developed a large number of diverse technologies for processing such complicated raw materials. The most common technologies involve the ultrafine grinding of raw materials [5–7]; various versions of hydrochemical oxidation with different oxidizing agents, including microorganisms [8–11], nitric acid and its decomposition products [12–15]; processes at elevated temperatures and pressures [16,17]; and combined technologies with various types of preliminary roasting, which are less and less popular due to environmental problems and low technical and economic indicators [18,19].

One of the most widespread technologies in the industry involves sulfuric acid and neutral pressure leaching for the quantitative oxidation of sulfides containing valuable metals (pyrite, arsenopyrite, chalcopyrite, etc.), followed by the cyanidation of the obtained cakes to recover noble metals and non-ferrous metals from solutions [20]. The recovery of silver by cyanidation from sulfuric acid autoclave leachates is difficult due to intergrowths with quartzite and the formation of argentojarosite, inert with respect to the cyanide ion [21]. Therefore, leaching in ammonia media can be used to avoid the formation of argentojarosite and to increase the degree of silver recovery directly at the pressure-leaching stage [22].

Ammonia has been widely used as an effective leaching agent in many hydrometallurgical processes over many years. Ammonia can be used to leach base metals (Cu, Ni, Co and Zn), as well as precious metals (Ag and Au), owing to the formation of soluble and very strong ammonia complexes. Various ammonium salts [23] can act as complexing agents. For example, for sulfide copper minerals and silver, the formation of complexes can be represented as follows (Equations (1) and (2)):

$$CuS + 2NH_3 + 2O_2 \rightarrow Cu(NH_3)_2{}^{2+} + SO_4{}^{2-}, \quad (1)$$

$$Cu(NH_3)_2{}^{2+} + 2NH_3 \rightarrow Cu(NH_3)_4{}^{2+}, \quad (2)$$

$$Ag_2O + 4NH_3 + H_2O \rightarrow 2[Ag(NH_3)_2]^+ + 2OH^-, \quad (3)$$

$$2Ag + 2NH_3 + 0.5O_2 + H_2O \rightarrow 2[Ag(NH_3)_2]^+ + 2OH^-. \quad (4)$$

Ammonia is also considered an attractive leaching agent because of its low toxicity, low cost and ease of regeneration by evaporation from alkaline solutions [24]. Selectivity is an important advantage of the ammonia medium in hydrometallurgy, meaning that the desired metals can be dissolved and the undesired iron segregated in the same unit.

For many years, Alexander Mining (has plants using the "AmmLeach" technology in the Democratic Republic of the Congo and Australia) led the implementation and development of ammonia leaching for the recovery of copper from sulfide, oxidized ores and tailings, as well as the recovery of cobalt [25]. According to Alexander Mining, the key advantages of this technology include the ability to leach in atmospheric conditions, the low consumption of reagents to process acid-intensive materials, the ability to use standard equipment, high selectivity, and a low environmental impact.

The first industrial application of an ammonia leaching process related to the hydrometallurgical recovery of copper from oxidized ores was put into operation in 1916 [26]. Since that time, a large number of studies have been carried out to study the kinetics of leaching various copper ores and copper-containing wastes in ammonia media at atmospheric [27–31] and high pressure [32,33]. For example, Liu et al. [30] studied the leaching of calcareous bornite ore in a solution containing ammonia and ammonium persulfate. The study showed that copper recovery rises from 82.7 to 90.3% as the temperature increases from 303.15 to 333.15 K. It was also revealed that copper recovery reaches 77.8% and 88.1% at ammonia concentrations of 1.0 mol/dm^3 and 2.5 mol/dm^3, respectively, and that an increase in the concentration of ammonium persulfate from 0.5 mol/dm^3 to 2.0 mol/dm^3 effects an increase in copper recovery from 80.9% to 90.3%. Baba et al. [34] investigated the kinetics of leaching with ammonium sulfate solutions of complex covellite ore mined in Nigeria. It was revealed that the process was controlled by the diffusion kinetics, with an activation energy of 37.37 kJ/mol. In this case, copper recovery was 86.2% under optimal conditions.

Ammonia leaching with oxygen as an oxidizing agent was used commercially in 1954 to extract copper, nickel and cobalt from sulfide ores and concentrates by Sherritt Gordon [23] in Fort Saskatchewan. In the 1970s, Anaconda developed the Arbiter ammonia leaching process [24]. In one study [35], it was shown that for a concentrate containing 25% copper in the form of chalcopyrite in a solution of ammonium hydroxide and ammonium carbonate with the use of oxygen as an oxidizing agent under optimal conditions (HH_4OH concentration of 5 mol/dm^3; $(NH_4)_2CO_3$, 0.3 mol/dm^3; L/S, 20 to 1; stirring speed, 1000 rpm; temperature, 333 K; and oxygen flow rate, 1 dm^3/min), only 70% of copper could be recovered into solution. A kinetic analysis of this process shows that the leaching rate was controlled by diffusion through the product layer, and the activation energy was 25 kJ/mol.

Ammonia leaching can also be used to extract precious metals. The ammonia–cyanide process was used commercially by Hunt to process tails from Comstock Lode in Nevada and copper–iron gold ores in Dale, California [36]. The industrial application of the Hunt process started in 2014, when ammonia leaching was introduced at the Gadabay plant [26].

Few studies, however, have focused on the kinetics of simultaneous copper and silver recovery by ammonia leaching from low-grade concentrates or ores under pressure [22,24,37]. Therefore, the aim of this work is to investigate the kinetics and mechanisms of the ammonia pressure leaching of low-grade sulfide raw materials containing copper and silver, in order to study the effects of Si-containing minerals on the kinetics of Cu and Ag leaching.

2. Materials and Methods

2.1. Materials

All the reagents used in this study were of analytical purity. A 25% ammonia solution was provided by JSC Azot (Kemerovo, Russia), and caustic soda was provided by JSC Bashkir Soda Company (Sterlitamak, Russia).

To achieve the required concentrations of solutions, ammonia and caustic alkali solutions were diluted with distilled water obtained on the apparatus manufactured by GFL GmbH (Burgwedel, Germany).

The original high-siliceous copper concentrate was provided by an ore processing plant at a Zhezkazgan mine, Kazakhstan [38]. The concentrate was obtained by a two-stage flotation enrichment (rougher and scavenger) of the initial low-grade ore containing 0.53% of copper and 12.53 ppm of Ag. The yield of the concentrate was 7.09%, while the copper recovery 93.7%. The average particle size of the concentrate was 30 μm; 90% of the particles were smaller than 61 μm.

Table 1 shows the chemical composition of the obtained concentrate. Figure 1 shows an X-ray diffraction pattern of this concentrate. As can be seen from Table 1 and Figure 1, the initial concentrate contains a large amount of aluminosilicates and quartz, while the content of sulfur and copper does not exceed 5%. It is obvious that the iron is partly present in oxidized form (chlorites).

Table 1. Chemical composition of initial low-grade copper–silver concentrate, wt. %.

Cu	Fe	S	SiO_2	Al_2O_3	Na_2O	Ag	Other
7.0	8.2	5.7	55.1	10.5	2.2	140 ppm	11.3

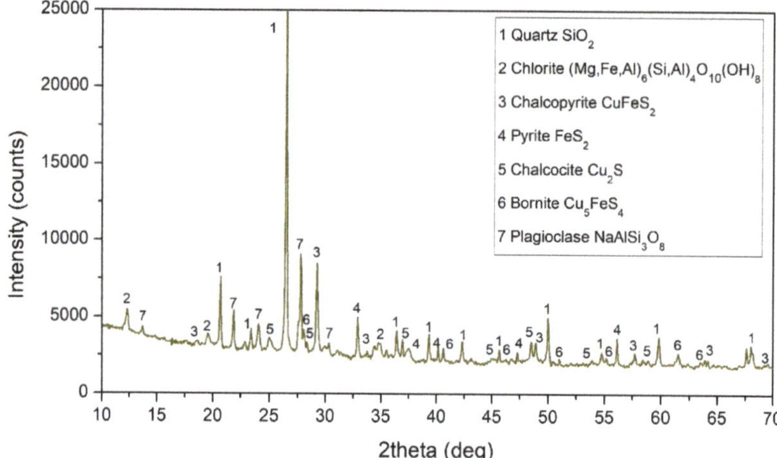

Figure 1. XRD pattern of initial low-grade copper–silver concentrate.

Table 2 shows the distribution of copper-containing minerals in the concentrate. The amount of oxidized copper was determined by leaching the initial concentrate in 10% sulfuric acid solution for 2 h at 95 °C without an oxidizing agent. The distribution of primary and secondary copper sulfides was determined by the x-ray analysis of the concentrate after the complete removal of aluminosilicates [39]. It can be seen that the largest part of the copper is represented by primary and secondary sulfides, but oxidized minerals are also present.

Table 2. Copper distribution in initial low-grade copper–silver concentrate, by minerals.

Copper Form.	Content, wt. %
Primary sulfides	14.7
Secondary sulfides	66.5
Oxidized form	18.8
Total	100

2.2. Analytical Methods

The chemical analysis of the original material and the resulting solid products of the studied processes was performed using an Axios MAX X-ray fluorescence spectrometer (XRF) (Malvern Panalytical Ltd., Almelo, The Netherlands). The phase analysis was performed on an XRD 7000 Maxima diffractometer (Shimadzu Corp., Tokyo, Japan).

Scanning electron microscopy (SEM) was performed using a JSM-6390LV microscope (JEOL Ltd., Tokyo, Japan) equipped with a module for energy-dispersive X-ray spectroscopy analysis (EDX).

Samples from each experiment were taken at the selected time intervals, and the obtained solutions were analyzed using inductively coupled plasma mass spectrometry (ICP-MS—NexION 300S quadrupole mass spectrometer, PerkinElmer Inc., Waltham, MA, USA).

2.3. Experiments

Laboratory experiments for NaOH and NH_4OH leaching were carried out using a 0.6 dm^3 autoclave reactor (Parr Instrument, Moline, IL, USA), equipped with sample collection vessel. The reactor was thermostated. The materials were stirred using an overhead mixer at 700 rpm, which ensured a uniform density of the pulp and eliminated diffusion limitations. A portion of the raw material weighing 100 g was added to a prepared solution of NaOH (300 g/dm^3) or NH_4OH and

$(NH_4)_2SO_4$ (with solution concentrations of 0.5 mol/dm^3 $(NH_4)_2SO_4$ and 7 mol/dm^3 NH_3); the L/S ratio in all the experiments was 5 to 1. Then, the resulting pulp was heated in an autoclave with constant stirring. After the required temperature was reached, the oxygen supply and time countdown began. At the predetermined time intervals, a portion of the leaching pulp was taken with help of excessive pressure in reactor and cooled in a sealed vessel to atmospheric temperature. The obtained pulp was filtered in a Buchner funnel (ECROSKHIM Co., Ltd., St. Petersburg, Russia); the solutions were sent for ICP-MS analysis. At the end of experiment, the leaching cake was washed with distilled water, dried at 100 °C to a constant weight, weighed and sent for XRF analysis. All the experiments were performed twice, and the mean values are presented here.

3. Results and Discussion

3.1. Leaching Kinetics of the Initial Concentrate

The pressure leaching of sulfide copper concentrate in an ammonia environment is possible only in the presence of an additional oxidizing agent, necessary for the conversion of copper and silver into oxidized forms. However, even with a strong oxidizing agent, the efficiency of copper and silver recovery will strongly depend on the process temperature. In order to determine the limiting stage and to better understand the mechanisms of the reactions, we studied the kinetics of the ammonia leaching of a sample of the original concentrate at various temperatures and otherwise-equal conditions. The oxygen pressure in reactor was 0.6 MPa, and the total pressure was 0.9 MPa at 120 °C and 1.8 MPa at 160 °C. The obtained leaching kinetic curves at various temperatures ranging from 120 to 160 °C are shown in Figure 2.

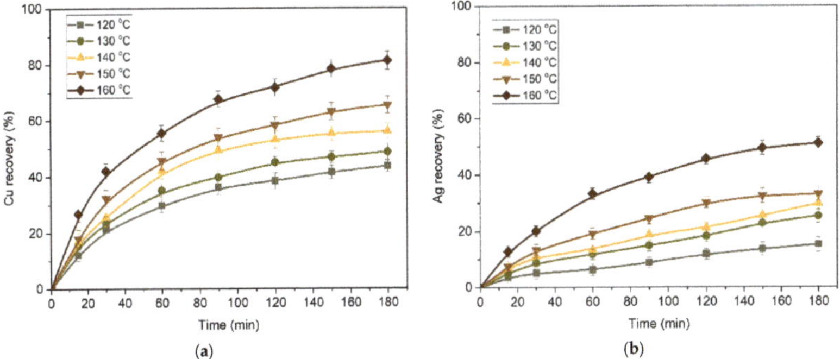

Figure 2. Effect of temperature and ammonia leaching duration on the degree of Cu (**a**) and Ag (**b**) recovery into solution.

Figure 2a shows that as temperature the increased from 120 to 160 °C, the degree of copper recovery into solution increased significantly. After 3 h of leaching at 120 °C, the degree of copper recovery into the solution is 36.0% and reaches 81.3% at 160 °C within the same amount of time. The degree of silver recovery also increases (Figure 2b). However, unlike copper, silver can be effectively recovered into solution only at temperatures above 160 °C. After 3 h of leaching at 160 °C, the recovery of silver in the solution was 51.1%, while at below 130 °C, it did not exceed 15–16%. According to published data [28], oxidized copper leaches quite efficiently into an ammonia solution, even at atmospheric pressure. The oxidation rate for copper sulfides in an ammonia solution in autoclave conditions is also quite high, even at temperatures below 140 °C [22,35], which means that the process under study has certain limitations.

As shown in Figure 3, the corresponding experimental data for the leaching of silver and copper from the initial concentrate were linearized using various shrinking core models (SCM) [40–42] (Table 3), and the most adequate one proved to be the internal diffusion model, $1–2/3\alpha - (1 - \alpha)^{2/3} = Kt$, where α

is the degree of component recovery into the solution, K is the reaction rate constant (min^{-1}) and t is the duration (min). This result indicates that the leaching of copper and silver is limited by the rate of the supply of oxygen or a leaching reagent to the reaction surface through a layer of the product or unreacted substance.

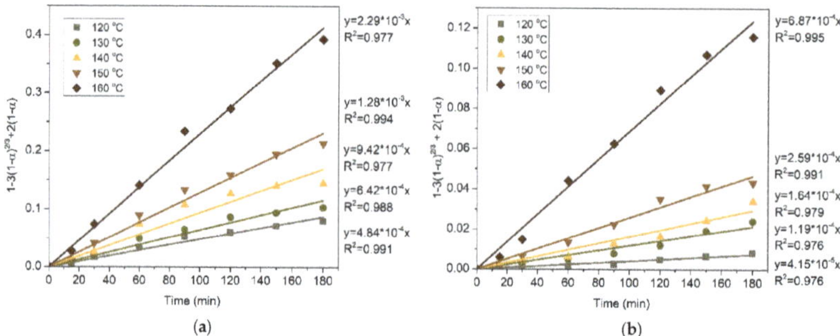

Figure 3. Dependence of 1–2/3α − $(1 - \alpha)^{2/3}$ on time for Cu (**a**) and Ag (**b**) recovery into solution.

Table 3. SCM equation fitting for Cu recovery from the initial low-grade concentrate.

#	Limiting step	Equation	R^2				
			120 °C	130 °C	140 °C	150 °C	160 °C
1	Diffusion through the product layer (sp)	$1 - 3(1 - \alpha)^{2/3} + 2(1 - \alpha)$	0.991	0.988	0.977	0.994	0.977
2	Diffusion through the liquid film (sp)	α	0.667	0.662	0.626	0.625	0.610
3	Surface chemical reactions (sp)	$1 - (1 - \alpha)^{1/3}$	0.750	0.729	0.805	0.737	0.845
4	New shrinking core model	$1/3\ln(1 - \alpha) + [(1 - \alpha)^{-1/3} - 1]$	0.989	0.993	0.960	0.989	0.968
5	Hybrid (1st equation + 3rd equation)	$(1 - 3(1 - \alpha)^{2/3} + 2(1 - \alpha)) + (1 - (1 - \alpha)^{1/3})$	0.855	0.846	0.841	0.846	0.947

sp—spherical particles; α—the degree of Cu recovery into the solution.

Figure 4 shows a plot of lnk versus the inverse of temperature, 1/T (K^{-1}). In accordance with the Arrhenius equation, the slope of the straight line in Figure 4 was used to determine the apparent activation energy of the leaching of copper and silver as 53.50 kJ/mol and 90.35 kJ/mol, respectively. It follows that the silver leaching reaction requires higher activation energies than copper leaching. This is probably due to the fact that the copper in the initial material is represented by various minerals, including oxides, some easily soluble at lower temperatures. By contrast, silver and a certain part of sulfide copper are more refractory. It is also possible that the copper and silver minerals are finely impregnated in the aluminosilicate rock, resulting in diffusion limitations. As shown before [43], a high value of the activation energy is not always representative of a kinetically controlled reaction.

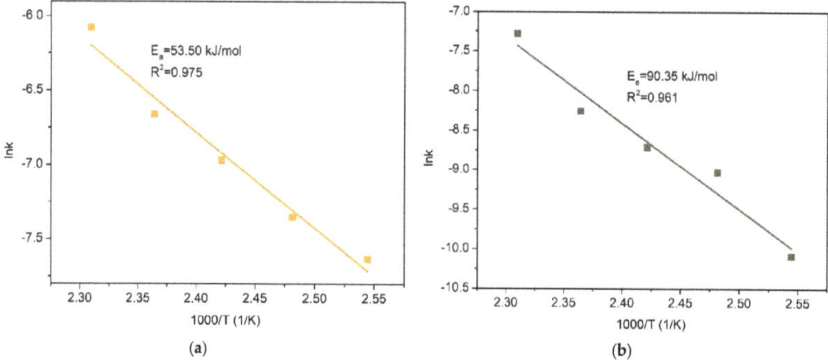

Figure 4. Dependence of lnk-1000/T for the recovery of copper (**a**) and silver (**b**) into solution.

To reveal the effect of aluminosilicates on the mechanism of leaching, electron micrograph and EDX analyses of the initial concentrate particles (Figure 5) were performed. According to Figure 5, there are two types of grain with different morphologies in the initial concentrate: grains with a smooth surface with ultrafine particles scattered on them and grains with a rough surface. The result of the EDX analysis (Table 4) shows that the first type of grain is most likely a sulfide mineral. The EDX analysis also indicates that sulfides are bound to aluminosilicates by mutual germination, since silicon is distributed, although unevenly, over the entire surface, including at accumulations of copper and silver. At the same time, silver is inextricably linked with copper accumulation areas.

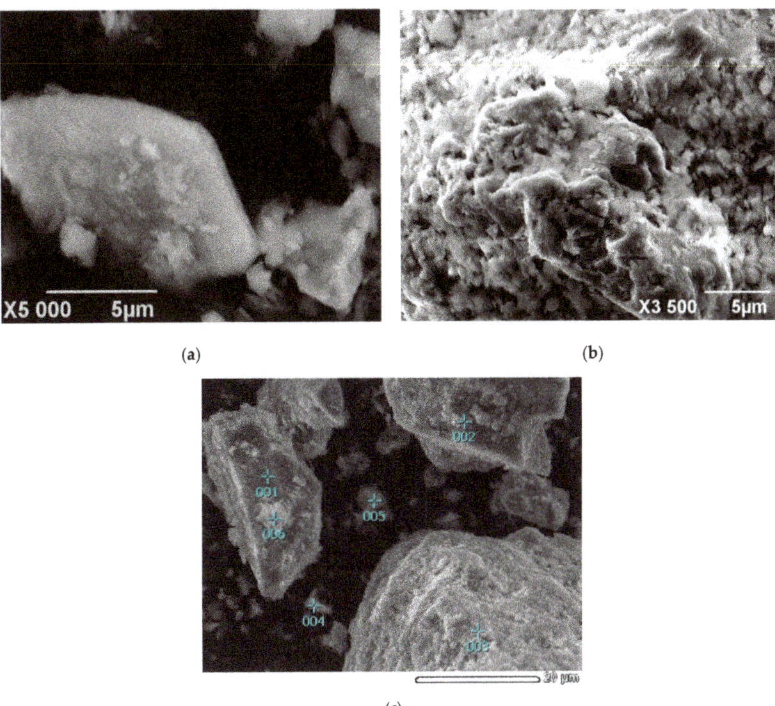

(a) (b)

(c)

Figure 5. SEM images of typical grains of the initial concentrate (**a**, **b**) and SEM image with the position of an EDX spectrum (**c**).

Table 4. The elemental compositions (energy-dispersive X-ray spectroscopy analysis (EDX) spectrum) of the initial concentrate, wt. %.

Element	Al	Si	S	Fe	Cu	Ag	O	Total
Figure 5c. Point 001	2.78	21.07	2.83	18.39	18.60	0.15	36.16	100
Figure 5c. Point 002	3.10	32.29	0.29	11.12	7.24	0.00	45.96	100
Figure 5c. Point 003	3.58	40.46	0.00	2.92	1.89	0.00	51.15	100
Figure 5c. Point 004	2.91	10.43	0.62	27.41	26.33	0.50	31.80	100
Figure 5c. Point 005	1.49	5.61	0.29	53.19	7.29	0.00	32.13	100
Figure 5c. Point 006	4.40	14.31	5.74	12.16	34.85	0.83	27.71	100

Thus, we can assume the inhibitory effect of aluminosilicates, as their layer on the particle surface prevents the diffusion of oxygen and ammonia. Aluminosilicates can also adsorb non-ferrous metals from solution [44,45]. Therefore, the direct pressure ammonia leaching of the concentrate without pre-treatment does not allow copper and silver recovery from a low-grade concentrate into the solution at temperatures below 160 °C, which is explained by the diffusion limitations created by a layer of aluminosilicates on the surface of sulfides.

In order to remove the passivating effect of aluminosilicates, one can use physical (mechanical activation, fusion) and chemical methods (preliminary alkaline desilication). Therefore, to further study the effect of aluminosilicates on copper and silver recovery into an ammonia solution, we studied the process kinetics after extracting silica from the original material by pressure alkaline leaching.

3.2. *Effect of Preliminary Desilication on Ammonia Leaching of Low-Grade Copper–Silver Concentrate*

To exclude the possible effects of silica-containing minerals, we investigated the preliminary removal of silicon from the concentrate by pressure alkaline leaching under the following conditions: NaOH = 300 g/dm^3, L/S = 5:1, t = 200 °C, total pressure = 1.5 MPa, and duration = 2 h [39]. The method allowed the recovery of 63% of the silicon into the solution, while the cake had the following composition (Table 5):

Table 5. Chemical composition of desilicated copper concentrate, wt. %.

Cu	Fe	S	SiO_2	Al_2O_3	Na_2O	Ag	Other
10.7	12.6	8.7	30.5	15.1	8.8	210 ppm	13.6

3.2.1. Kinetics of Copper and Silver Leaching from a Concentrate after Preliminary Desilication

The effect of temperature and the duration of leaching on copper and silver recovery from a desilicated concentrate is shown in Figure 6.

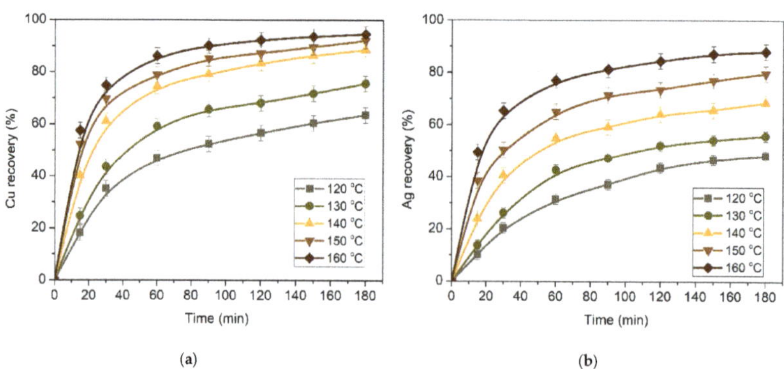

Figure 6. Dependence of the recovery of copper (**a**) and silver (**b**) into the solution on the temperature and duration of leaching of desilicated concentrate.

Figure 6 shows that, as in the case of the initial concentrate, the temperature has a significant effect on the speed of the process; after 60 min of leaching, copper recovery increases from 63% to 94%. A significant influence of the duration in the range 0–60 min is observed at all studied temperatures; after 60 min, the rate of Cu and Ag recovery in the solution decreases significantly, which is most likely due to diffusion limitations. Nevertheless, the rate of Cu and Ag recovery into solution from a desilicated concentrate is much higher, especially clearly seen in case of silver and apparently associated with a decrease in the passivating or sorption effect of aluminosilicates after desilication and with an increase in the surface area of sulfide particles that are in contact with solvent.

As shown in Figure 7, the corresponding experimental data on Cu and Ag leaching from a desilicated concentrate were also linearized using various shrinking core models. In this case, the most adequate was a new model proposed by Dickinson and Heal [40], $1/3\ln(1-\alpha+((1-\alpha)^{-1/3}-1)=kt$, where α is the degree of a component's recovery into solution, k is the reaction rate constant (min^{-1}), and t is the duration (min). This result indicates that the leaching of copper and silver is limited by the rate of the supply of oxygen or leaching agent to the reaction surface through the interfacial transfer and diffusion across the product layer. Thus, we can assume that in this case, the process mode changes: interfacial transfer is added to the internal diffusion.

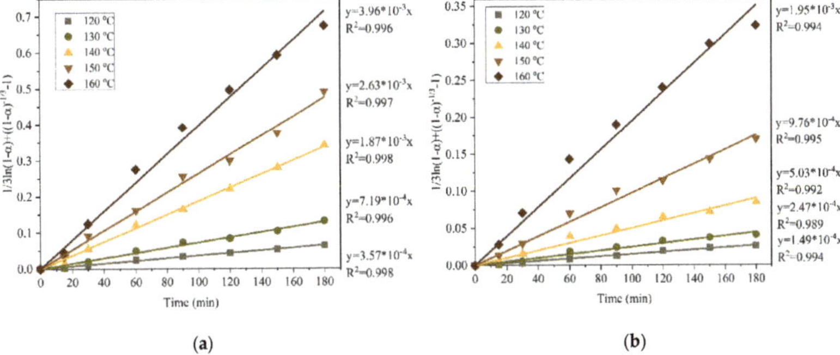

Figure 7. Dependence of $1/3\ln(1-\alpha+((1-\alpha)^{-1/3}-1)$ on time for Cu (**a**) and Ag (**b**) recovery into solution.

Based on the slope of the straight line in Figure 8, the apparent activation energy of the leaching is 86.76 kJ/mol for Cu and 92.15 kJ/mol for Ag, respectively. This indicates an increase in the effect of temperature on the leaching of copper.

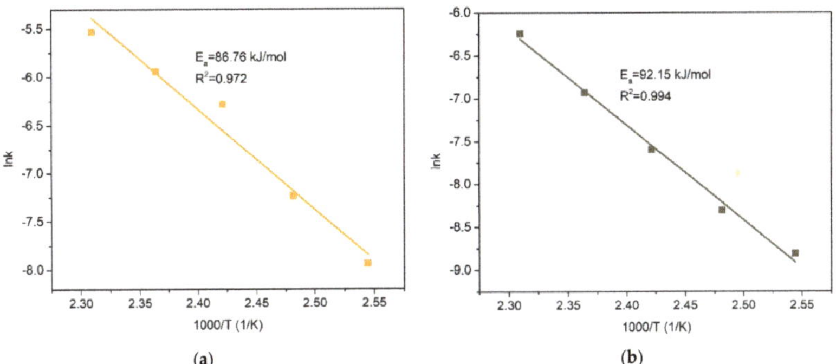

Figure 8. Dependence of lnk−1000/T for the recovery of copper (**a**) and silver (**b**) into solution from a desilicated concentrate.

The obtained high value of the activation energy along with the high degree of description of the kinetic curves by the new SCM equation, which is applicable for the processes restricted by the supply of reagents through the diffusion layer, indicates a complex mechanism of the autoclave leaching of copper and silver in an ammonia environment.

The SCM can only be used to determine the average values of the activation energy, which may lead to the omission of some reaction features. The time-to-a-given-fraction method [46] was used to calculate the apparent activation energy at different points in the leaching process. The time required to achieve a certain degree of leaching and the apparent activation energy, E_a, are related according to Equation (5).

$$\ln t_x = \text{const} - \ln A + E_a/RT, \tag{5}$$

The slope of the graph plotted with $\ln t_x$ and 1000/T coordinates allows one to calculate the apparent activation energy.

The calculation of activation energy by the graphical method was performed as shown in Figure 9.

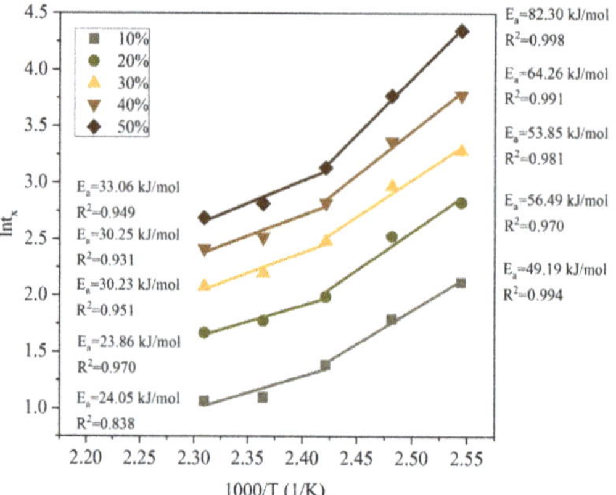

Figure 9. Dependence of lnt_x on 1000/T for various degrees of recovery of copper.

According to Figure 9, the calculated values of the apparent activation energy according to the time-to-a-given-fraction method change from 39.77 to 60.95 kJ/mol as the degree of copper recovery into the solution increases from 10% to 50%. These values are lower than the activation values obtained earlier in the SCM model. The low activation energies in the initial period of the ammonia pressure leaching of copper from desilicated concentrate is explained by the increased dissolution rates for less refractory oxidized compounds and secondary copper sulfides. Additionally, there seems to be a change in the slope of the curves at 140 °C (2.42 1/K), suggesting a change in the rate-determining mechanism at higher temperatures. It can be assumed that the effect of kinetic difficulties at low temperatures may be higher. Chalcocite is dissolved according to the following reactions (Equations (6) and (7)):

$$2Cu_2S + 4NH_4OH + 2(NH_4)_2SO_4 + O_2 = 2Cu(NH_3)_4SO_4 + 2CuS + 6H_2O \quad (6)$$

$$CuS + 4NH_4OH + 2O_2 = Cu(NH_3)_4SO_4 + 4H_2O, \quad (7)$$

The dissolution of covellite and other refractory sulfides (chalcopyrite, bornite) at subsequent stages of leaching occurs at lower rates due to lower solubility, which explains the increase in activation energy [47].

The efficiency of copper and silver recovery will strongly depend on the oxygen pressure [24]. The effect of oxygen pressure in the range of 0.2 to 0.8 MPa on copper recovery at a leaching temperature of 140 °C is shown in Figure 10.

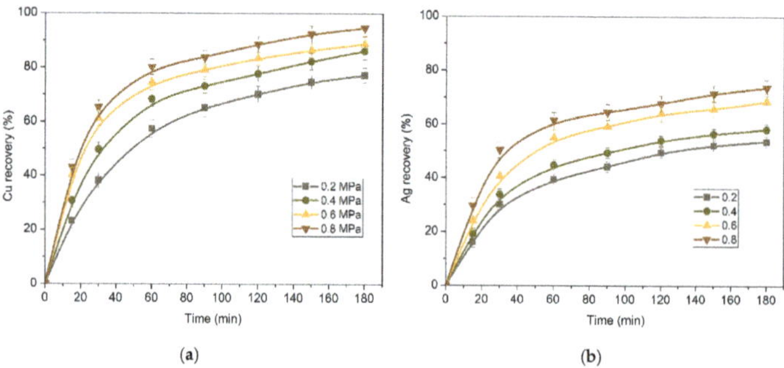

Figure 10. Dependence of Cu (**a**) and Ag (**b**) recovery on oxygen pressure and duration (temperature: 140 °C).

The results in Figure 10 show the positive effect of an increase in oxygen pressure on the recovery of Cu and Ag into solution. As the oxygen pressure increases from 0.2 to 0.8 MPa, copper recovery increases from 77.2% to 93.6% over 180 min of leaching, and silver recovery increases from 53.6% to 73.7%. An increase in the process duration for the entire studied range also positively affects the transition of Cu and Ag into solution. The final concentration of copper in the solution at an oxygen pressure of 0.8 MPa was 20.3 g/dm^3, and the concentration of Ag was 0.4 mg/dm^3. This solution can be used for the extraction of valuable metals. The final concentration of iron in the solution in all the experiments was 0.6–1 mg/dm^3.

Using the slopes of the straight lines obtained by substituting data for Cu and Ag recovery at oxygen pressures of 0.2–0.8 MPa and a temperature of 140 °C into the new SCM (Figure 11a,b), we plotted a graph with the $\ln k_c$-$\ln P_{02}$ coordinates to determine the partial order with respect to oxygen pressure (Figure 11c,d) [47]. As a result, the empirical orders with respect to oxygen pressure were 0.959 for Cu and 0.932 for Ag, both close to 1, which may also indicate external diffusion limitations [48].

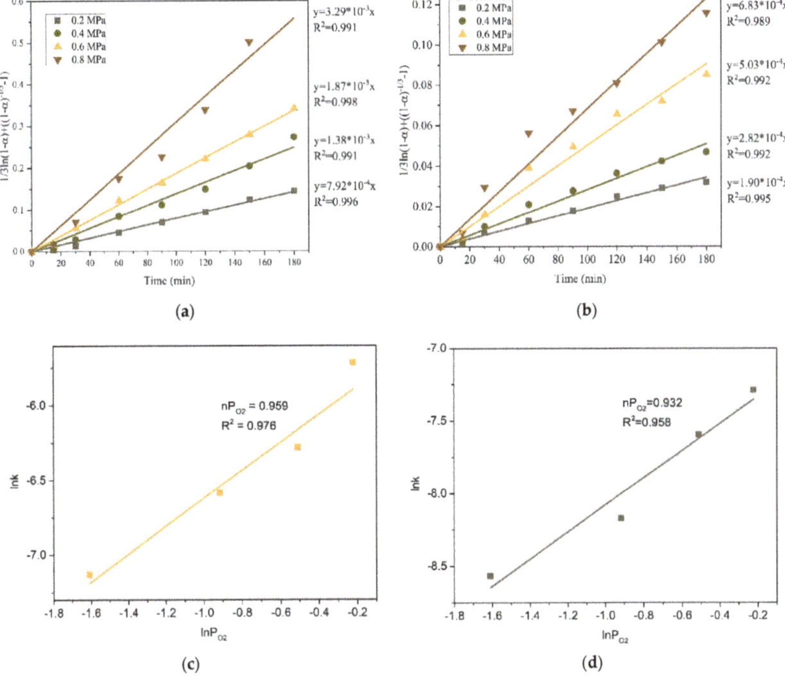

Figure 11. Dependence of $1/3\ln(1-\alpha)+((1-\alpha)^{-1/3}-1)$ on oxygen pressure for Cu (**a**) and Ag (**b**) recovery into solution, and dependence of $\ln k_c$ on $\ln P_{02}$ to determine the oxygen order for Cu (**c**) and Ag (**d**).

3.2.2. Characteristics of Solid Residue

To study the effect of preliminary desilication on the mechanism of leaching, electron micrographs and maps of the surface (SEM-EDX) of solid residue particles obtained at oxygen pressures of 0.8 MPa and a temperature of 140 °C after 3 h of leaching were examined. The SEM image and map-scans for the desilicated concentrate are presented in Figure 12.

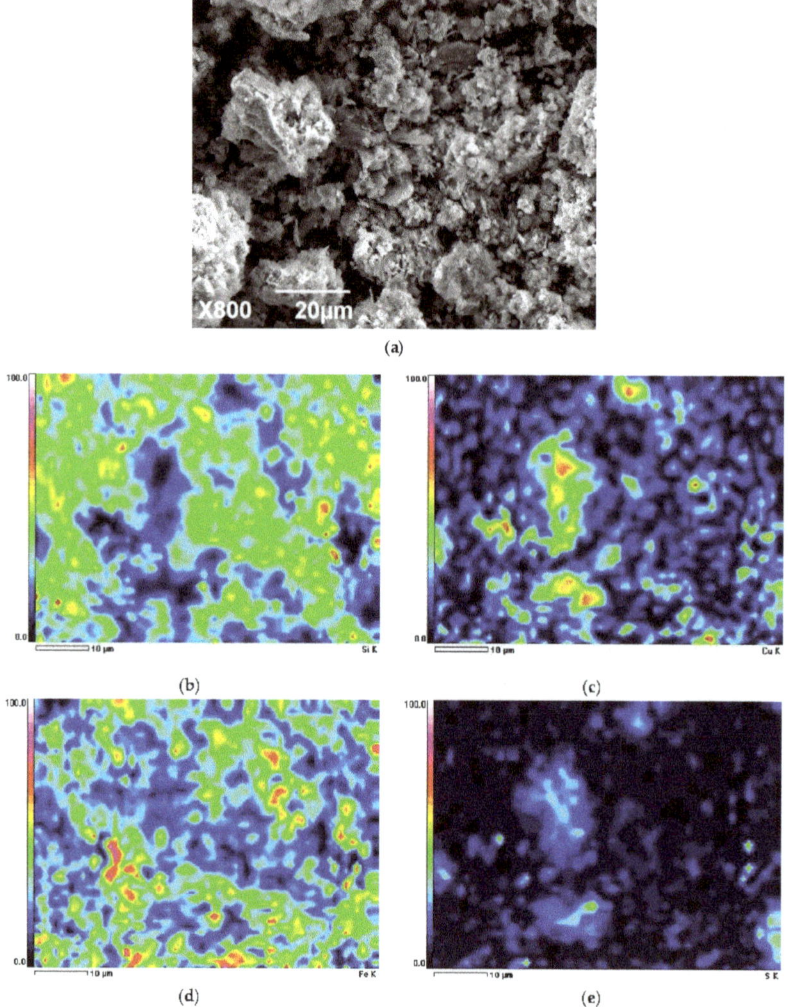

Figure 12. SEM image of desilicated concentrate (**a**), and map-scans (**b**–**e**) for Si, Cu, Fe and S, respectively.

In contrast to the initial material, the cake after pressure desilication contains much less silica (Table 5), although it is also distributed over the entire surface, and has a heterogeneous surface with many cavities and voids (Figure 12a,b), which facilitates the access of reagents to sulfide minerals. The presence of pores is explained by the dissolution of Si-bearing minerals in an alkaline solution. Additionally, spherical particles are formed on the surface, which is associated with the precipitation of sodium hydroaluminosilicate [49]. However, sodium hydroaluminosilicate can also inhibit the process [50] but apparently to a lesser extent due to its high porosity. The existence of amorphous inclusions in the cake is confirmed by the presence of noise in the XRD pattern of the cake after desilication (Figure 13). In addition, the XRD pattern in Figure 13 shows that the main phases after desilication are cancrinite, quartz and pyrite, as well as a small amount of copper minerals.

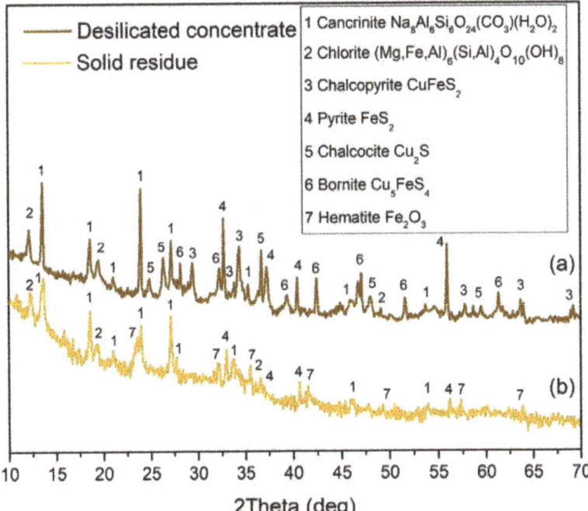

Figure 13. XRD pattern of desilicated concentrate (**a**), and the product of its leaching in ammonia solution (**b**).

The cakes after desilication have a relatively uniform distribution of the above elements, especially Si, which is explained by the formation of sodium hydroaluminosilicates on the surface of copper and iron minerals (Figure 12b).

The cakes after the pressure ammonia leaching of desilicated concentrate were also analyzed by scanning electron microscopy to assess morphological changes. The SEM and map-scanning images obtained for the ammonia pressure leaching cake of the desilicated concentrate are presented in Figure 14.

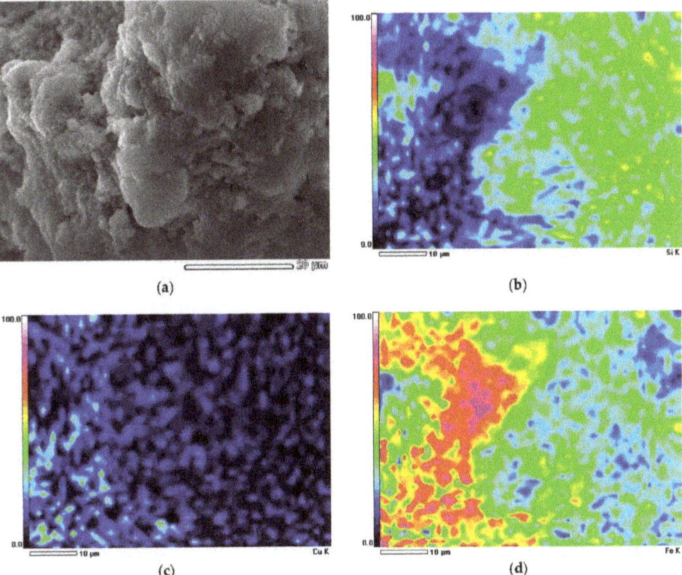

Figure 14. SEM image of ammonia leaching residue (**a**), and map-scans (**b–d**) for Si, Cu and Ag, respectively.

According to the SEM images, the residue after ammonia pressure leaching is represented by particles larger than 30 μm with a rough surface (Figure 14a). The diffraction pattern of this cake (Figure 13b) shows that cancrinite and hematite are the main minerals in the material.

The copper content decreases significantly after ammonia pressure leaching (Figure 14c). It follows that preliminary alkaline desilication makes it possible to reduce the diffusion limitations caused by aluminosilicates and quartz (into which sulfide minerals are apparently finely impregnated) due to the formation of a porous surface during leaching. A schematic diagram of the process is presented in Figure 15.

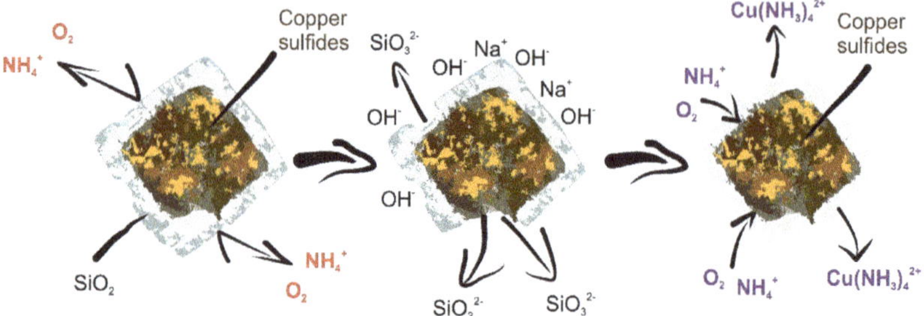

Figure 15. Schematic diagram of the effect of pre-desilication on recovery of Cu and Ag during ammonia leaching.

Thus, a new environmentally friendly method of processing such a low-grade copper–silver concentrate is proposed, eliminating the need for ultra-fine grinding, which leads to the formation of a large amount of fine and toxic waste. Silica that is converted to an alkaline solution, in turn, can be used to obtain additional valuable products. Ammonia can also be easily regenerated by distillation [31].

4. Conclusions

The effect of aluminosilicate minerals and their preliminary alkaline leaching on the ammonia pressure leaching of low-grade copper concentrates has been investigated. Using new methods for studying the kinetics of the processes, it has been shown that the limiting stage of the process changes over the course of the leaching. The following conclusions have been reached:

1. In addition to sulfide minerals of copper, the initial concentrate contains a large amount of oxidized copper and aluminosilicate compounds that coat valuable components, preventing their leaching into the solution.
2. The results of the experiments on the pressure leaching of the initial copper concentrate in an ammonium/ammonium-sulphate solution in the presence of oxygen are in good agreement with the shrinking core model in the intra-diffusion mode; the activation energies for copper and silver in this case were 53.50 kJ/mol and 90.35 kJ/mol, respectively.
3. After the preliminary alkaline desilication of the concentrate, the recovery rate for Cu and Ag increases significantly, and the new shrinking core model proves to be the most adequate, indicating that the process is limited by diffusion through the product layer and interfacial diffusion. The partial order with respect to oxygen pressure is close to 1, which may also indicate external diffusion limitations. The activation energy of the process increases to 86.76 kJ/mol for Cu and 92.15 kJ/mol for Ag.
4. Using the time-to-a-given-fraction method, it was shown that a high activation energy is necessary in the later stages of the process at low temperatures, when the most refractory sulfide minerals of Cu and Ag appear to be unleached.

Author Contributions: Conceptualization, S.N.; methodology, K.K. and A.S.; validation, S.Z. and E.K.; formal analysis, K.K. and D.R.; investigation, A.S. and E.K.; resources, S.Z.; data curation, D.R.; writing—original draft preparation, K.K.; writing—review and editing, A.S. and S.N.; visualization, A.S. and E.K.; supervision, D.R.; project administration, S.Z.; funding acquisition, S.N. All authors have read and agreed to the published version of the manuscript.

Funding: This work was financially supported by the Russian Science Foundation Project No. 18-19-00186. The SEM-EDX analyses were funded by State Assignment, grant number 0836-2020-0020.

Acknowledgments: Technicians at the Ural Branch of Russian Academy of Sciences are acknowledged for their assistance with the XRD, XRF, SEM, EDS and ICP-MS analyses.

Conflicts of Interest: The authors declare no conflict of interest.

References

1. Peters, E. Hydrometallurgical process innovation. *Hydrometallurgy* **1992**, *29*, 431–459. [CrossRef]
2. Sinclair, L.; Thompson, J. In situ leaching of copper: Challenges and future prospects. *Hydrometallurgy* **2015**, *157*, 306–324. [CrossRef]
3. Ignatkina, V.A.; Bocharov, V.A.; Makavetskas, A.R.; Kayumov, A.A.; Aksenova, D.D.; Khachatryan, L.S.; Fishchenko, Y.Y. Rational Processing of Refractory Copper-Bearing Ores. *Russ. J. NonFerr. Metals* **2018**, *59*, 364–373. [CrossRef]
4. Singer, D.A. Future copper resources. *Ore Geol. Rev.* **2017**, *86*, 271–279. [CrossRef]
5. Palaniandy, S. Impact of mechanochemical effect on chalcopyrite leaching. *Int. J. Miner. Process.* **2015**, *136*, 56–65. [CrossRef]
6. Wang, G.; Yang, H.; Liu, Y.; Tong, L.; Auwalu, A. Study on the mechanical activation of malachite and the leaching of complex copper ore in the Luanshya mining area, Zambia. *Int. J. Miner. Metall. Mater.* **2020**, *27*, 292–300. [CrossRef]
7. Ye, X.; Gredelj, S.; Skinner, W.; Grano, S.R. Regrinding sulphide minerals—Breakage mechanisms in milling and their influence on surface properties and flotation behaviour. *Powder Technol.* **2010**, *203*, 133–147. [CrossRef]
8. Dong, Y.; Lin, H.; Xu, X.; Zhou, S. Bioleaching of different copper sulfides by Acidithiobacillus ferrooxidans and its adsorption on minerals. *Hydrometallurgy* **2013**, *140*, 42–47. [CrossRef]
9. Halinen, A.-K.; Rahunen, N.; Kaksonen, A.H.; Puhakka, J.A. Heap bioleaching of a complex sulfide ore. *Hydrometallurgy* **2009**, *98*, 92–100. [CrossRef]
10. Petersen, J. Heap leaching as a key technology for recovery of values from low-grade ores—A brief overview. *Hydrometallurgy* **2016**, *165*, 206–212. [CrossRef]
11. Watling, H.R. The bioleaching of sulphide minerals with emphasis on copper sulphides—A review. *Hydrometallurgy* **2006**, *84*, 81–108. [CrossRef]
12. Anderson, C.G.; Harrison, K.D.; Krys, L.E. Theoretical considerations of sodium nitrite oxidation and fine grinding in refractory precious-metal concentrate pressure leaching. *Min. Metall. Explor.* **1996**, *13*, 4–11. [CrossRef]
13. Rogozhnikov, D.A.; Mamyachenkov, S.V.; Anisimova, O.S. Nitric Acid Leaching of Copper-Zinc Sulfide Middlings. *Metallurgist* **2016**, *60*, 229–233. [CrossRef]
14. Rogozhnikov, D.A.; Kolmachikhin, B.V. Polymetallic Ore Concentration Middlings Nitric Acid Leaching Kinetics. *SSP* **2017**, *265*, 1065–1070. [CrossRef]
15. Xie, Y.; Xu, Y.; Yan, L.; Yang, R. Recovery of nickel, copper and cobalt from low-grade Ni–Cu sulfide tailings. *Hydrometallurgy* **2005**, *80*, 54–58. [CrossRef]
16. Ukraintsev, I.V.; Petrov, G.V.; Ivanov, B.S.; Boduen, A.Y. Autoclave conditioning of a low-grade sulphide copper concentrate. *Tsm* **2016**, *10*, 43–48. [CrossRef]
17. McDonald, R.G.; Muir, D.M. Pressure oxidation leaching of chalcopyrite. Part I. Comparison of high and low temperature reaction kinetics and products. *Hydrometallurgy* **2007**, *86*, 191–205. [CrossRef]
18. Cui, F.; Mu, W.; Wang, S.; Xin, H.; Xu, Q.; Zhai, Y.; Luo, S. Sodium sulfate activation mechanism on co-sulfating roasting to nickel-copper sulfide concentrate in metal extractions, microtopography and kinetics. *Miner. Eng.* **2018**, *123*, 104–116. [CrossRef]
19. Medvedev, A.S.; So, T.; Ptitsyn, A.M. Combined processing technology of the Udokan sulfide copper concentrate. *Russ. J. NonFerr. Met.* **2012**, *53*, 125–128. [CrossRef]

20. Sun, X.; Chen, B.; Yang, X.; Liu, Y. Technological conditions and kinetics of leaching copper from complex copper oxide ore. *J. Cent. South Univ. Technol.* **2009**, *16*, 936–941. [CrossRef]
21. Dutrizac, J.E.; Jambor, J.L. Jarosites and Their Application in Hydrometallurgy. *Rev. Mineral. Geochem.* **2000**, *40*, 405–452. [CrossRef]
22. Boduen, A.Y.; Petrov, G.V.; Serebryakov, M.A. Investigation of ammonia autoclave leaching of silver and rhenium containing ill-conditioned copper concentrate. *Tsm* **2016**, *10*, 23–28. [CrossRef]
23. Forward, F.A.; Mackiw, V.N. Chemistry of the Ammonia Pressure Process for Leaching Ni, Cu, and Co from Sherritt Gordon Sulphide Concentrates. *JOM* **1955**, *7*, 457–463. [CrossRef]
24. Meng, X.; Han, K.N. The Principles and Applications of Ammonia Leaching of Metals—A Review. *Mineral Process. Extr. Metall. Rev.* **1996**, *16*, 23–61. [CrossRef]
25. MetaLeach. Available online: http://www.metaleach.com/ (accessed on 23 April 2020).
26. Hedjazi, F.; Monhemius, A.J. Industrial application of ammonia-assisted cyanide leaching for copper-gold ores. *Miner. Eng.* **2018**, *126*, 123–129. [CrossRef]
27. Baba, A.A.; Ghosh, M.K.; Pradhan, S.R.; Rao, D.S.; Baral, A.; Adekola, F.A. Characterization and kinetic study on ammonia leaching of complex copper ore. *Trans. NonFerr. Met. Soc. China* **2014**, *24*, 1587–1595. [CrossRef]
28. Bingöl, D.; Canbazoğlu, M.; Aydoğan, S. Dissolution kinetics of malachite in ammonia/ammonium carbonate leaching. *Hydrometallurgy* **2005**, *76*, 55–62. [CrossRef]
29. Liu, Z.; Yin, Z.; Hu, H.; Chen, Q. Dissolution kinetics of malachite in ammonia/ammonium sulphate solution. *J. Cent. South Univ. Technol.* **2012**, *19*, 903–910. [CrossRef]
30. Liu, Z.X.; Yin, Z.L.; Xiong, S.F.; Chen, Y.G.; Chen, Q.Y. Leaching and kinetic modeling of calcareous bornite in ammonia ammonium sulfate solution with sodium persulfate. *Hydrometallurgy* **2014**, *144–145*, 86–90. [CrossRef]
31. Radmehr, V.; Koleini, S.M.J.; Khalesi, M.R.; Tavakoli Mohammadi, M.R. Ammonia Leaching: A New Approach of Copper Industry in Hydrometallurgical Processes. *J. Inst. Eng. India Ser. D* **2013**, *94*, 95–104. [CrossRef]
32. Turan, M.D.; Arslanoğlu, H.; Altundoğan, H.S. Optimization of the leaching conditions of chalcopyrite concentrate using ammonium persulfate in an autoclave system. *J. Taiwan Inst. Chem. Eng.* **2015**, *50*, 49–55. [CrossRef]
33. Turan, M.D.; Altundoğan, H.S. Leaching of copper from chalcopyrite concentrate by using ammonium persulphate in an autoclave: Determination of most suitable impeller type by using response surface methodology. *J. Cent. South Univ.* **2013**, *20*, 622–628. [CrossRef]
34. Baba, A.A.; Balogun, A.F.; Olaoluwa, D.T.; Bale, R.B.; Adekola, F.A.; Alabi, A.G.F. Leaching kinetics of a Nigerian complex covellite ore by the ammonia-ammonium sulfate solution. *Korean J. Chem. Eng.* **2017**, *34*, 1133–1140. [CrossRef]
35. Nabizadeh, A.; Aghazadeh, V. Dissolution study of chalcopyrite concentrate in oxidative ammonia/ammonium carbonate solutions at moderate temperature and ambient pressure. *Hydrometallurgy* **2015**, *152*, 61–68. [CrossRef]
36. Muir, D.M. A review of the selective leaching of gold from oxidised copper–gold ores with ammonia–cyanide and new insights for plant control and operation. *Miner. Eng.* **2011**, *24*, 576–582. [CrossRef]
37. Boduen, A.Y.; Fokina, S.B.; Petrov, G.V.; Andreev, Y.V. Ammonia autoclave technology for the processing of low-grade concentrates generated in flotation concentration of cupriferous sandstones. *Obogashchenie Rud* **2019**, *2*, 33–38. [CrossRef]
38. Baranov, V.F. The use of foreign experience in the development of a reconstruction option for the Zhezkazgan concentration complex. *Obogashchenie Rud* **2020**, *1*, 54–59. [CrossRef]
39. Shoppert, A.A.; Loginova, I.V.; Pismak, V.N. Investigating of a Low-Grade Copper Concentrate Desilication by Alkali Pressure Leaching. *MSF* **2019**, *946*, 608–614. [CrossRef]
40. Dickinson, C.F.; Heal, G.R. Solid–liquid diffusion controlled rate equations. *Thermochim. Acta* **1999**, *340–341*, 89–103. [CrossRef]
41. Levenspiel, O. *Chemical Reaction Engineering*, 3rd ed.; Wiley: New York, NY, USA, 1999; ISBN 978-0-471-25424-9.
42. Rogozhnikov, D.A.; Shoppert, A.A.; Dizer, O.A.; Karimov, K.A.; Rusalev, R.E. Leaching Kinetics of Sulfides from Refractory Gold Concentrates by Nitric Acid. *Metals* **2019**, *9*, 465. [CrossRef]

43. Baba, A.A.; Adekola, F.A. Hydrometallurgical processing of a Nigerian sphalerite in hydrochloric acid: Characterization and dissolution kinetics. *Hydrometallurgy* **2010**, *101*, 69–75. [CrossRef]
44. Karimov, K.A.; Rogozhnikov, D.A.; Naboichenko, S.S.; Karimova, L.M.; Zakhar'yan, S.V. Autoclave Ammonia Leaching of Silver from Low-Grade Copper Concentrates. *Metallurgist* **2018**, *62*, 783–789. [CrossRef]
45. Naboichenko, S.S. *Avtoklavnaja Gidrometallurgija Cvetnych Metallov*; GOU UGTU-UPI: Ekaterinburg, Russia, 2002; ISBN 978-5-321-00065-6.
46. Hidalgo, T.; Kuhar, L.; Beinlich, A.; Putnis, A. Kinetics and mineralogical analysis of copper dissolution from a bornite/chalcopyrite composite sample in ferric-chloride and methanesulfonic-acid solutions. *Hydrometallurgy* **2019**, *188*, 140–156. [CrossRef]
47. Tanda, B.C.; Eksteen, J.J.; Oraby, E.A. Kinetics of chalcocite leaching in oxygenated alkaline glycine solutions. *Hydrometallurgy* **2018**, *178*, 264–273. [CrossRef]
48. Lorenzo-Tallafigo, J.; Iglesias-Gonzalez, N.; Romero, R.; Mazuelos, A.; Carranza, F. Ferric leaching of the sphalerite contained in a bulk concentrate: Kinetic study. *Miner. Eng.* **2018**, *125*, 50–59. [CrossRef]
49. Smith, P. The processing of high silica bauxites—Review of existing and potential processes. *Hydrometallurgy* **2009**, *98*, 162–176. [CrossRef]
50. Wang, Y.; Li, X.; Zhou, Q.; Qi, T.; Liu, G.; Peng, Z.; Zhou, K. Effects of Si-bearing minerals on the conversion of hematite into magnetite during reductive Bayer digestion. *Hydrometallurgy* **2019**, *189*, 105126. [CrossRef]

Article

Mechanism and Kinetics of Malachite Dissolution in an NH_4OH System

Alvaro Aracena [1,*], Javiera Pino [1] and Oscar Jerez [2]

1 Escuela de Ingeniería Química, Pontificia Universidad Católica de Valparaíso, Avenida Brasil 2162, Valparaíso 2362854, Chile; javiera.p.rocco@gmail.com
2 Instituto de Geología Económica Aplicada (GEA), Universidad de Concepción, Casilla 160-C, Concepción 4070386, Chile; ojerez@udec.cl
* Correspondence: alvaro.aracena@pucv.cl; Tel.: +56-322-372-614

Received: 22 May 2020; Accepted: 15 June 2020; Published: 24 June 2020

Abstract: Copper oxide minerals composed of carbonates consume high quantities of leaching reagent. The present research proposes an alternative procedure for malachite leaching ($Cu_2CO_3(OH)_2$) through the use of only compound, ammonium hydroxide (NH_4OH). Preliminary studies were also carried out for the dissolution of malachite in an acid system. The variables evaluated were solution pH, stirring rate, temperature, NH_4OH concentration, particle size, solid/liquid ratio and different ammonium reagents. The experiments were carried out in a stirred batch system with controlled temperatures and stirring rates. For the acid dissolution system, sulfuric acid consumption reached excessive values (986 kg H_2SO_4/ton of malachite), invalidating the dissolution in these common systems. On the other hand, for the ammoniacal system, there was no acid consumption and the results show that copper recovery was very high, reaching values of 84.1% for a concentration of 0.2 mol/dm^3 of NH_4OH and an experiment time of 7200 s. The theoretical/thermodynamic calculations indicate that the solution pH was a significant factor in maintaining the copper soluble as $Cu(NH_3)_4{}^{2+}$. This was validated by the experimental results and solid analysis by X-ray diffraction (XRD), from which the reaction mechanisms were obtained. A heterogeneous kinetic model was obtained from the diffusion model in a porous layer for particles that begin the reaction as nonporous but which become porous during the reaction as the original solid splits and cracks to form a highly porous structure. The reaction order for the NH_4OH concentration was 3.2 and was inversely proportional to the square of the initial radius of the particle. The activation energy was calculated at 36.1 kJ/mol in the temperature range of 278 to 313 K.

Keywords: malachite; carbonate; leaching; ammonium hydroxide; heterogeneous model

---

1. Introduction

1.1. *Consumption of Sulfuric Acid Due to Impurities*

Copper oxide compounds are often treated using hydrometallurgy, specifically through the use of chemical leaching with acidic leaching (dissolution) solutions composed mainly of diluted sulfuric acid (H_2SO_4). However, when copper oxides contain a large quantity of carbonates ($CaCO_3$, $MgCO_3$) or hydroxides ($Al(OH)_3$, $Ca(OH)_2$), acid consumption increases enormously, to a level that makes metallurgical treatment economically inviable [1,2]. This consumption is mainly because carbonates as well as hydroxides are quicker to react with sulfuric acid than the copper oxides, because they are very soluble in acids [3,4], while in some cases the copper compounds can contain considerable amounts of carbonate and hydroxide in their crystalline system. This is the case with basic copper carbonates, such as azurite ($Cu_3(CO_3)_2(OH)_2$) and malachite ($Cu_2CO_3(OH)_2$). This leads to the complication of compounds that excessively consume leaching reagent (H_2SO_4), making the leaching process inefficient and hindering copper recovery. There have been different works that have shown excessive consumption of acid with malachite; Bingöl [5] worked with a malachite mineral and obtained high copper recoveries (90%), along with other impurities, generating a consumption of 450 kg of acid per ton of ore. The same author tried to analyze the dissolution kinetics of malachite ore, but unfortunately he could not use a heterogeneous kinetic model (because the chemical reaction was very fast), obtaining a complete dissolution in a very short time caused by the present impurities that excessively consumed sulfuric acid (a group of minerals including pyroxene, quartz, goethite and magnetite, among others). Instead, Nicol [6] worked on the dissolution kinetics of malachite with H_2SO_4 (0.033 to 0.15 mol/dm^3), finding that the kinetics was governed by the chemical reaction on the surface. This could be achieved because he worked with malachite without impurities (i.e., acid consumers). In his work [6], he did not find the consumption of H_2SO_4 per ton of malachite.

An alternative option for the treatment of copper oxide minerals containing carbonates is to use leaching in an alkaline system, i.e., in an ammonium system. The main objective is to decrease acid consumption so that the process becomes more economically viable. In addition, the use of an ammoniacal system promotes dissolution selectivity as well as the reduction of corrosive attacks. Aracena [7,8] treated copper oxide minerals in an ammonium hydroxide system (NH_4OH). The experimental work was conducted using a stirred system with controlled temperature. The oxidized copper compounds such as tenorite (CuO) and cuprite (Cu_2O) were of high purity. The results obtained showed that copper can be extracted from tenorite (particle size of 5 μm; 0.45 mol/L (mol/dm^3) NH_4OH; pH = 10.5; temperature of 298 K; time of 300 min) and cuprite (particle size of 5 μm; 0.10 mol/L (mol/dm^3) NH_4OH; pH = 10.5; temperature of 318 K; time of 240 min), up to recovery values of 98% and 82%, respectively. The reaction mechanisms established in each study were the following:

$$2CuO + NH_4OH + 3NH_4^+ \rightarrow 2Cu^{2+} + 4NH_3 + 2H_2O + OH^- \quad (1)$$

$$2Cu_2O + 8NH_4OH + O_2 + 8NH_3 \rightarrow 4Cu(NH_3)_4^{2+} + 4H_2O + 8OH^- \quad (2)$$

The kinetic model representing tenorite and cuprite leaching was a chemical reaction on the surface. The activation energies calculated for tenorite and cuprite were 59.0 and 44.36 kJ/mol, respectively.

1.2. Leaching of $Cu_2CO_3(OH)_2$ Through the Use of Ammoniacal Systems

Several studies have been carried out with malachite leaching using an ammoniacal system. Ekmeyapar [9] studied malachite leaching using ammonium nitrate solutions (NH_4NO_3) and varying the working conditions of NH_4NO_3 concentration, particle size, stirring rate and temperature. The tests were carried out in batches, and the most significant results showed that, at a temperature close to 70 °C (343 K), copper recovery of 98% was obtained after 75 min (4500 s). The ammonium nitrate concentration was 4.0 mol/L (4.0 mol/dm^3). The reaction mechanism proposed by the researchers is given by the following expression:

$$Cu_2CO_3(OH)_2 + 4NH_4NO_3 \rightarrow Cu^{2+} + 4NH_3 + 4NO_3^- + CO_2 + 3H_2O \quad (3)$$

It was concluded that the reaction of the malachite with the ammonium nitrate followed a mixed-kinetics control, comprising two sequential mechanisms: for a temperature range of 30 to 50 °C (303 to 323 K), the leaching rate was controlled by the chemical reaction, giving E_a = 95.10 kJ/mol; for temperatures of 50 to 70 °C (323 to 343 K), the dissolution rate was controlled by diffusion in the porous layer, as shown in the activation energy, which was 29.50 kJ/mol for this temperate range.

Bingöl [10] showed the effects of variables such as the ratio between the two reagents, pH, temperature, stirring rate, solid/liquid ratio (S/L), particle size and leaching time, on the leaching of malachite in ammonium hydroxide and ammonium carbonate solutions. The experiments were conducted in a batch system. The results showed that the optimal ratio for leaching malachite was using a mixed solution of 5.0 M (5.0 mol/dm^3) NH_4OH and 0.3 M (0.3 mol/dm^3) $(NH_4)_2CO_3$, with a leaching time of 120 min (7200 s), temperature of 25 °C (298 K), stirring rate of 300 rpm and particle size below 450 μm, obtaining copper recovery of 98%. Finally, the researchers indicate that the dissolution kinetics of the malachite in ammonium hydroxide with ammonium carbonate was controlled by transfer on the interface and diffusion in the porous layer, obtaining an activation energy of 15 kJ/mol.

Künkül [11] performed experiments with a malachite mineral in a magnetically stirred reactor, analyzing variables such as particle size, ammonium concentration, solid/liquid ratio and temperature. Künkül found that by increasing the concentration of ammonium and the temperature while decreasing the solid/liquid ratio and particle size, a high-copper solution was obtained. The most effective parameter was the particle size (−125 μm). The porous layer diffusion model (produced by the SiO_2 that was around the malachite) represented the dissolution, with an activation energy of 22.338 kJ/mol being found.

Arzutug [12] used NH_3-saturated water to leach malachite mineral. The leaching experiments were carried out in a 250 mL (0.25 dm^3) glass reactor equipped with gas inlet and outlet tubes. The results showed that malachite reached its maximum dissolution (approximately 96%) for a particle size between +75–90 μm, temperature of 45 °C (318 K), ammonia concentration of 7.68 mol/dm^3, solid/liquid ratio of 2/100 g/mL (2/0.1 g/dm^3) and stirring speed of 400 rpm. Arzutug [12] found that malachite leaching was well represented by the second-order pseudokinetics with an activation energy value of 85.16 kJ/mol.

Studies related to the recovery of copper and iron from malachite minerals have been conducted using sulfuric acid (H_2SO_4) and hydrogen peroxide (H_2O_2) as leaching agents [13]. This last reagent was used as an oxidant for iron (Fe^{2+} passed to Fe^{3+}). The experiments were developed inside a 400 mL (0.4 dm^3) Pyrex beaker in a temperature-controlled shaking bath. The results showed that recoveries of copper and iron close to 99% and 36% were achieved, respectively, using a temperature of 80 °C (353 K), 1.6 M H_2SO_4 (1.6 mol/dm^3), 700 rpm and 10 g (1×10^{-2} kg) of solids. In addition, in [13], copper was precipitated as copper sulfate pentahydrate through the use of ethanol, methanol and sulfuric acid (99%, 98% and 73% precipitation, respectively).

Other studies have worked with different indirect analysis methods to obtain the best malachite dissolution parameters, such as by using the Taguchi method (as in Kurşuncu et al. [14]). Other studies have considered leaching malachite in organic solutions such as 5-SSA (5-sulfosalicylic acid) [15].

The studies mentioned above do not show the intrinsic influence of NH_4OH on leaching of malachite (only when mixed with other reagents), that is, the effectiveness and the reactions that occur are not clear. Therefore, they fail to obtain the reaction mechanisms of the reagent or the model representing the reaction or its kinetic parameters. In addition, one of the studies had to be carried out with a high reaction temperature (343 K) to obtain maximum malachite dissolution. The present study aims to obtain the malachite dissolution mechanisms in an ammonium system such as NH_4OH by analyzing the variables of solution pH, stirring rate, temperature, ammonium hydroxide concentration, solid/liquid ratio and different ammonium reagents. The heterogeneous kinetic model representing the leaching of $Cu_2CO_3(OH)_2$ will also be obtained along with its kinetic parameters, such as activation energy and reaction order with regard to NH_4OH.

2. Mechanism of Malachite Dissolution

The dissolution of malachite is given by the chemical process, that is, copper carbonate has a covalent bond and is insoluble to water but soluble in the presence of certain ions in the solution. Künkül [11] also indicated that malachite follows a simple dissolution process, where the oxide/reduction reaction (electron exchange) is not involved. Since it has a low value of the dissociation constant [16], this copper carbonate dissolves according to the reaction (4).

$$Cu_2CO_3(OH)_2 \rightarrow 2Cu^{2+} + CO_3^{2-} + 2(OH)^- \qquad k_{sp} = 4.0 \times 10^{-29} \tag{4}$$

In conditions of high basicity, cupric ions (Cu^{2+}) can precipitate due to the formation of copper hydroxide species ($Cu(OH)_2$). However, in the presence of ammonia ions, the solubility of copper species is very high [7,8,17]. In order to corroborate this complex reaction, a speciation diagram of the Cu-NH_3 system was constructed at a concentration of 0.005 mol/dm^3 copper and a temperature of 343 K (Figure 1). It can be seen that as the concentration of NH_3 increases, the amount of copper complexes increases. Thus, copper ammine ($(Cu(NH_3)^{2+}$, log k = 3.71), copper bi-ammine ($(Cu(NH_3)_2{}^{2+}$, log k = 3.07), copper tri-ammine ($(Cu(NH_3)_3{}^{2+}$, log k = 2.54) and copper tetra-ammine ($(Cu(NH_3)_4{}^{2+}$, log k = 1.79) are obtained. Copper tetra-ammine is stable from an ammonia concentration of 0.01 mol/dm^3 (log{NH_3} = −2) and is completely stable for an NH_3 concentration of 0.10 mol/dm^3 (log{NH_3} = −1). The other species of copper tetra-ammine (ammine, bi and tri) are thermodynamically unstable [17].

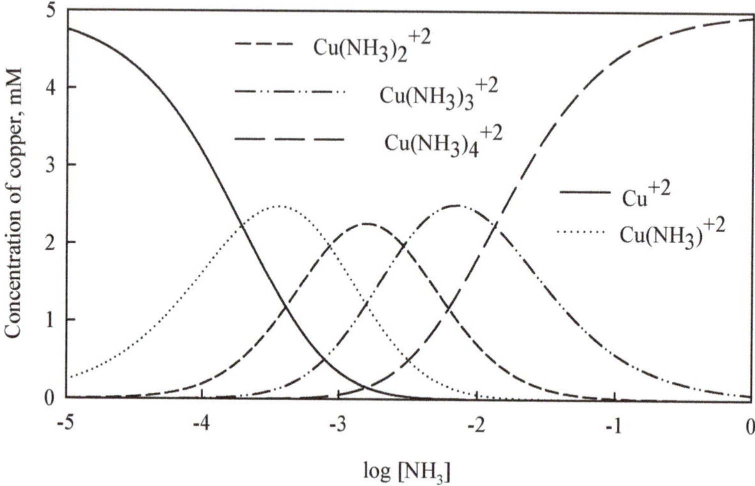

Figure 1. Formation of copper and ammonia complexes based on the concentration of NH_3.

Ammonium hydroxide, when in contact with water, dissociates (reaction (5)) to form several ionic species, including the hydronium ion (H_3O^+). This species has been the main oxidant of several copper oxides [7]. In this case, as the malachite proceeds by a chemical process, H_3O^+ would have no chance of reacting with anything besides the ions generated in reaction (4); therefore, it hydrolyzes to produce water. Thus, the overall dissociation reaction of the ammonium hydroxide can be given by the reaction 6.

$$NH_4OH + H_2O \rightarrow NH_3 + H_3O^+ + OH^- \quad (5)$$

$$NH_4OH \rightarrow NH_3 + H_2O \quad (6)$$

Therefore, copper in an ammonium system and base system is stable in the copper tetra-ammine form (shown in Figure 1), as represented by Equation (7).

$$2Cu^{2+} + 8NH_3 \rightarrow 2Cu(NH_3)_4{}^{2+} \quad (7)$$

Thus, the general equation that represents the malachite leaching with ammonium hydroxide is given by Equation (8).

$$Cu_2CO_3(OH)_2 + 8NH_4OH \rightarrow 2Cu(NH_3)_4{}^{2+} + CO_3{}^{2-} + 2(OH)^- + 8H_2O \quad (8)$$

Given how carbonates balance with water [18] Equation (8) can be expressed as:

$$Cu_2CO_3(OH)_2 + 8NH_4OH \rightarrow 2Cu(NH_3)_4{}^{2+} + HCO_3{}^{-} + 3(OH)^{-} + 7H_2O \quad (9)$$

Therefore, malachite leaching with ammonium hydroxide alone can be carried out without problems, considering the pH values of the solution.

3. Materials and Methods

3.1. Malachite Samples

Malachite samples (Sigma Aldrich, Santiago, Chile) were obtained from Sigma Aldrich in the form of very fine powders (less than 5 μm) with a purity of 99.5%. Pelletizing was used for the experiments with different particle sizes. The particles were agglomerated by controlled pressure, creating spheres measuring 12, 24 and 36 μm.

3.2. Acid Tests

The experimental development for the acid tests was carried out in a 1 dm^3 capacity reactor. A mechanical stirrer with Teflon rod was used. A condensate system added to the reactor served to minimize evaporation. A thermocouple was used to record the temperature of the solution. The amount of malachite used was 1.0×10^{-3} kg. The volume of sulfuric acid leaching solution was 0.4 dm^3. According to the studied pH value, concentrated sulfuric acid (98% purity) was added. The working temperature was 294 K. After 5400 s of each experiment, the liquid samples were filtered and sent for analysis by atomic absorption spectroscopy (AAS) with a Hitachi Z-8100 Zeeman kit (Hitachi High-Technologies Corporation, Tokyo, Japan).

3.3. Ammoniacal Leaching

A batch experiment system was used. The details of the reactor were established in a prior study [17]. Briefly, the equipment comprised a heating blanket, water cooled condenser to minimize evaporation losses, thermocouple and mechanical stirrer. ginstruments came from Hess (Santiago, Chile). The glass reactor had a total volume of 2.0 dm^3.

The reactor was subsequently loaded with 1.0 dm^3 of ammonium hydroxide leaching solution. In some cases, other ammonium reagents were added: ammonium sulfate ($(NH_4)_2SO_4$) (Arquimed, Santiago, Chile) with 99.9% purity from Arquimed, ammonium fluoride (NH_4F) with 99.6% purity from AnalaR (AnalaR NORMAPUR®, Santiago, Chile) and ammonium nitrate (NH_4NO_3) with 98.0% purity from Vimaroni (Vimaroni, Santiago, Chile). Depending on the experiment, the leaching solution was heated or cooled. A mass of 1.0×10^{-3} kg of malachite sample was then added to the reactor. The reaction was initiated, and liquid samples were extracted at regular time intervals for subsequent analysis (AAS). At the end of each experiment, the solution was filtered and the residue was then washed and dried for X-ray diffraction (XRD) analysis using a Bruker diffractometer (Bruker Scientific LLC, Billerica, MA, USA) model D4 Endeavor, operated with Cu radiation and Ni Kβ radiation filter.

4. Results and Discussion

4.1. Sulfuric Acid System Analysis

Experiments related to obtaining the specific consumption of H_2SO_4 were developed from malachite, without the interference of impurities (carbonates). The results of the leaching tests of malachite with sulfuric acid are found in Figure 2. This figure shows the consumption of sulfuric acid per dry ton of malachite used, along with copper recovery depending on the pH of the solution. The pH values were 1.0, 2.0, 2.5, 3.0 and 4.0.

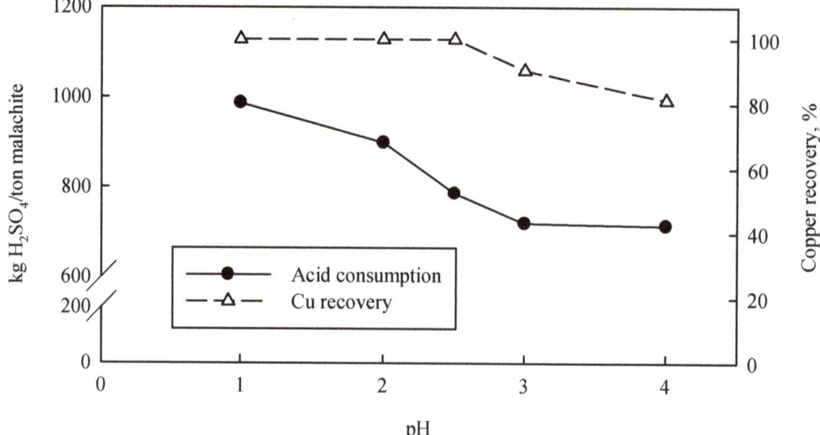

Figure 2. Sulfuric acid consumption (kg H_2SO_4/ton malachite) and copper recovery (%) depending on the pH of the solution.

It can be seen in this figure that the acid consumption is very high (excessive), reaching values of 986 kg H_2SO_4/ton of malachite used for a pH of 1.0. Copper recovery reached up to 99.9%. As the pH increased (became less acidic), the consumption of H_2SO_4 and the recovery of copper decreased, reaching values of 713 kg H_2SO_4/ton malachite and 81.2% copper, respectively. For efficient acid leaching to be carried out in terms of acid consumption, such consumptions generally must reach values of 50 kg H_2SO_4/ton of mineral [1]. With these results shown in Figure 2, acid leaching from malachite could not be carried out. This high consumption may be due to the reaction of malachite with sulfuric acid (reaction (10)) [6].

$$Cu_2CO_3(OH)_2 + 4H^+ = 2Cu^{2+} + CO_2 + 3H_2O \quad (10)$$

Thus, the malachite leaching process using sulfuric acid at room temperature would not be advisable to perform. Therefore, it becomes attractive to be able to develop the leaching of malachite in a basic system, since low sulfuric acid consumption and increased copper recovery are promoted. The leaching solution chosen was ammonium hydroxide.

4.2. Zone of Malachite Dissolution in Ammoniacal System

In order to study the zone of malachite dissolution in an ammonium system, tests were conducted at different pH values (6.0, 10.5 and 13.0), an ammonium hydroxide concentration of 0.1 mol/dm^3, a temperature of 298 K and stirring at 500 rpm. The solid/liquid ratio used was 1/1000, and the results are shown in Figure 3 as copper recovery as a function of solution pH.

It can be seen that at a pH of 10.5, the copper recovery reached a value of 63.9%. This level of recovery was mainly due to the formation of copper tetra-ammine, as shown in Figure 1. At a pH of 6.0 or 13.0, the copper recovery levels reached only 1.9%. This behavior may be due to copper precipitation in the form of CuO, based on the thermodynamic study shown in the Cu-NH_3-H_2O stability diagram (Figure 1). In order to corroborate the malachite dissolution mechanisms for different pH values, the solid samples obtained after 7200 s in tests carried out at pH levels of 10.5 and 13.0 were sent for XRD analysis. The results are shown in Figure 4.

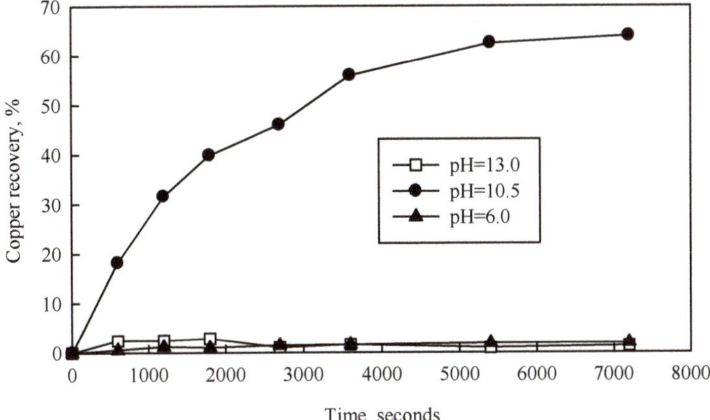

Figure 3. The effect of the solution pH on the dissolution of $Cu_2CO_3(OH)_2$. Working conditions: NH_4OH = 0.1 mol/dm^3, temperature = 298 K, particle size = 5 µm, stirring rate = 500 rpm and S/L ratio = 1:1000.

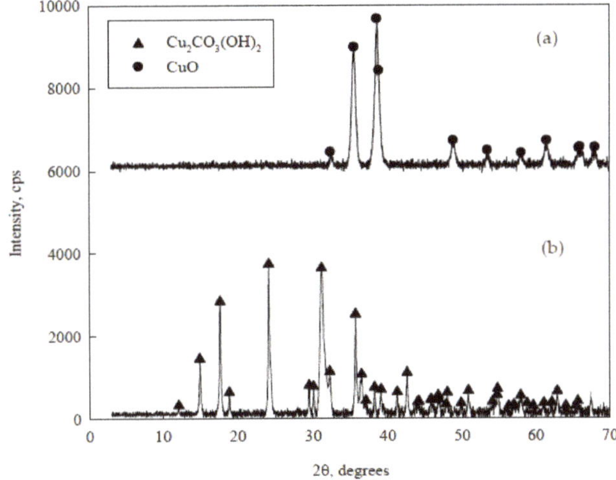

Figure 4. XRD analysis of solid samples: (**a**) samples obtained at a pH of 10.5; (**b**) samples obtained at a pH of 13.0.

As can be seen in Figure 4a, the peaks indicate malachite without the presence of any other compound associated with copper or ammonia. Therefore, the dissolution mechanism of $Cu_2CO_3(OH)_2$ should be that shown in Equation (8). In Figure 4b, no malachite is seen, though there are important peaks for copper oxide, such as tenorite. To corroborate this formation of copper oxide, a predominance diagram was built for the Cu-NH_3-H_2O system for three different temperatures (278, 298 and 313 K). The thermodynamic data were taken from the database of the HSC Chemistry 6.0 program [19]. The copper and ammonium concentrations were 0.0043 and 0.1 mol/dm^3, respectively. The potential values used were above 0.4 V. The diagram is presented in Figure 5. It can be seen that the cupric ion is stable at pH less than 4.8. However, it becomes stable again in the pH range from 9.2 to 12.5. This regained stability is due to complexing with the ammonium ion, generating copper tetra-ammine ($Cu(NH_3)_4^{2+}$). This stability is seen for all potential values. The other copper complexes (other ammines) are not considered due to their thermodynamic instability. In addition, by increasing the temperature, the range of stability of the copper tetra-ammine moves to more acidic pH values, from a pH of 9.8 (278 K) to 9.0 (313 K). Outside these pH ranges, the copper oxidizes and precipitates as cuprite. This happens at all temperatures.

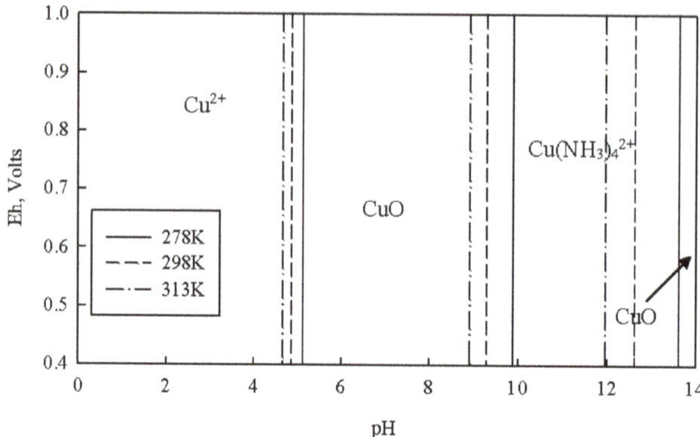

Figure 5. Diagram of $Cu\text{-}NH_3\text{-}H_2O$ medium stability at a copper concentration of 0.0043 mol/dm^3 and ammonia concentration of 0.1 mol/dm^3. The solid line corresponds to the equilibrium arising at 278 K, the dotted line is equilibrium at 298 K and the dashed line is equilibrium at 313 K.

Therefore, and according to what is seen in Figure 3, at the more basic pH (13.0), copper recovery was low due to the formation of this oxide (CuO), as was posited thermodynamically previously (Figure 5). The copper must be present in solution as copper tetra-ammine, but in highly base conditions it is precipitated to form tenorite, leading to the low level of recovery of the metal of interest (cupric ions).

On the other hand, the curves shown in Figure 1 show that there would be only one form of chemical reaction of malachite with NH_4OH, which could be represented by a single heterogeneous kinetic model (as will be seen later). Oudenne and Olson [20] studied the kinetics of leaching from malachite in an ammonium carbonate solution, where they found that there were two reaction stages: stage I, where a 10% reaction was obtained (quickly) but then became slow, and then stage II where 90% reaction is obtained (total malachite dissolution). Oudenne pointed out that in stage I the reaction became very slow due to a blockage in the surface generated by an intermediate compound formed in the reaction, $Cu(OH)_2$. This compound can be dissolved by the intervention of the hydronium ion. In our case, the generation of the intermediate compound was not evident; therefore, the dissolution of the malachite was always carried out by the chemical process (reaction (9)).

Based on the results observed for the effect of pH, the subsequent experiments were all carried out at a pH of 10.5.

4.3. Evaluation of Stirring Rate

The stirring of the leaching solution was assessed in the range of 200 to 600 rpm, including a test without stirring (0 rpm). The tests were carried out in a solution of 0.1 mol/dm^3 NH_4OH, at 298 K, with a solid/liquid ratio of 1/1000. The results are shown in Figure 6.

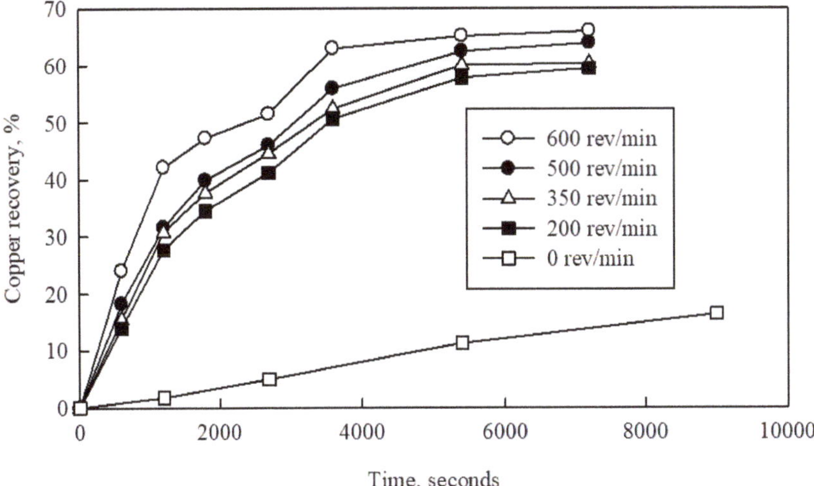

Figure 6. The effect of stirring rate on the malachite dissolution. Working conditions: NH_4OH = 0.1 mol/dm^3, temperature = 298 K, pH = 10.5, particle size = 5 µm and S/L ratio = 1:1000.

The figure shows that copper recovery increases as the stirring rate increases. Thus, for a stirring rate of 200 rpm, recovery of 64.3% was obtained after 7200 s; when increasing the stirring to 500 rpm, the Cu recovery reached a value of 71.0% for the same experiment time. For higher stirring rates, the copper recovery remained similar. This is due to the phenomenon of mass transference no longer playing a significant role at higher stirring rates. Therefore, all subsequent experiments were conducted at a stirring rate of 500 rpm to ensure that the stirring rate was not affected by mass transfer.

It should be noted that the recovery rate increased with time during the experiment conducted without stirring (0 rpm), reaching a maximum copper recovery of 16.4% after 9000 s.

4.4. Temperature Analysis

The effect of temperature on the dissolution rate of $Cu_2CO_3(OH)_2$ was assessed. The range of temperatures tested was 278 to 313 K. The working conditions used were 0.1 mol/dm^3 NH_4OH and a solid/liquid ratio of 1/1000. As can be seen in Figure 7, there was a significant effect on the early dissolution times, with this effect decreasing after 3600 s (except for the curve generated at 313 K). Maximum dissolution reached a value close to 72% for a temperature of 313 K. It can also be seen that at the lower temperature (278 K), which is close to the freezing point of water (273 K), significant copper recovery was also obtained (56.1%) after 7200 s.

The differences may be due to the changes in the kinetic constants involved in the malachite dissolution processes.

It should also be considered that ammonia volatilizes slowly in equilibrium with ammonia in solution (ammonia dissociation constant = 1.77×10^{-5} at 298 K) increasing with temperature, thus decreasing its concentration in the solution as the dissolution time elapses. This dissociation of Cu-NH_3 will follow the series set forth by the diagram in Figure 1, beginning with $Cu(NH_3)_4^{2+}$ until reaching the form of Cu^{2+}, as the concentration of NH_3 decreases (in solution). Then, the cupric ion under conditions of high alkalinity would precipitate as an oxide, thus decreasing the concentration of copper in the solution.

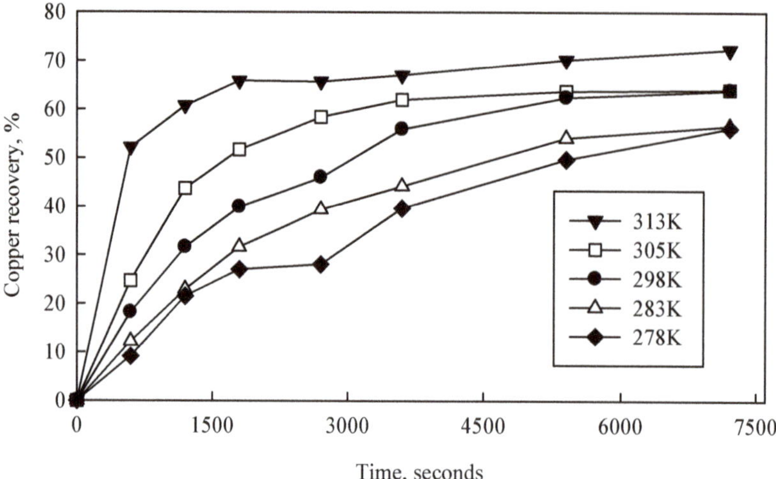

Figure 7. Analysis of the effect of temperature on the rate of malachite dissolution. Working conditions: NH_4OH = 0.1 mol/dm^3, pH = 10.5, particle size = 5 μm, stirring rate = 500 rpm and S/L ratio = 1:1000.

4.5. Effect of NH_4OH Concentration

The study of the ammonium medium was carried out at 298 K with a solid/liquid ratio of 1/1000. The concentration values ranged from 0.01 to 0.2 mol/dm^3 NH_4OH. The results are shown in Figure 8. Based on Equation (9) and using stoichiometry, the minimum required NH_4OH concentration to extract copper from malachite was found to be 0.036 mol/dm^3. It can be seen in Figure 8 that no copper was recovered from the malachite when using a concentration of 0.01 mol/dm^3. However, when increasing the concentration to 0.05 mol/dm^3 (equal to or greater than the stoichiometric level), copper recovery reached a value of 19.3% after 7200 s. For the maximum concentration of ammonium hydroxide (0.2 mol/dm^3), copper recovery obtained reached a value of 84.1% for the same experiment time.

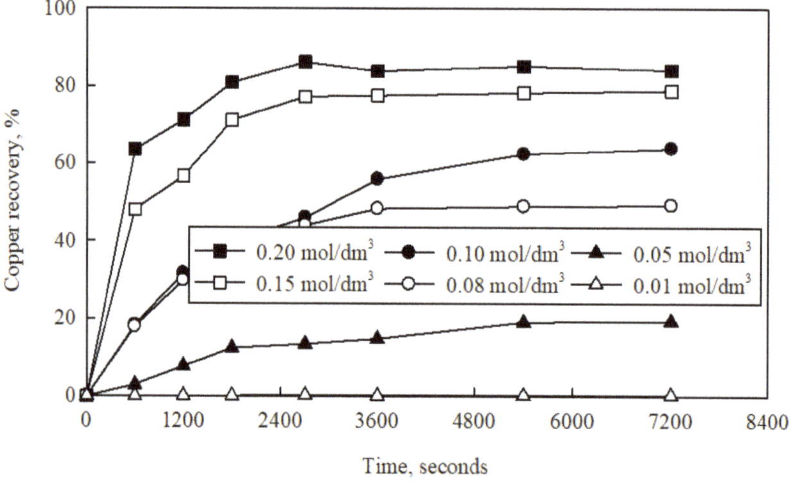

Figure 8. The effect of ammonium hydroxide concentration. Working conditions: temperature = 298 K, particle size = 5 μm, stirring rate = 500 rpm, pH = 10.5 and S/L ratio = 1:1000.

4.6. Evaluation of Particle Size

The effect of the particle size of the malachite on its leaching rate was also evaluated. Four tests were carried out with different particle sizes: 5, 12, 24 and 36 μm. The temperature and NH_4OH concentration were set at 298 K and 0.1 mol/dm^3, respectively. Figure 9 shows the copper recovery as a function of time for the different particle sizes.

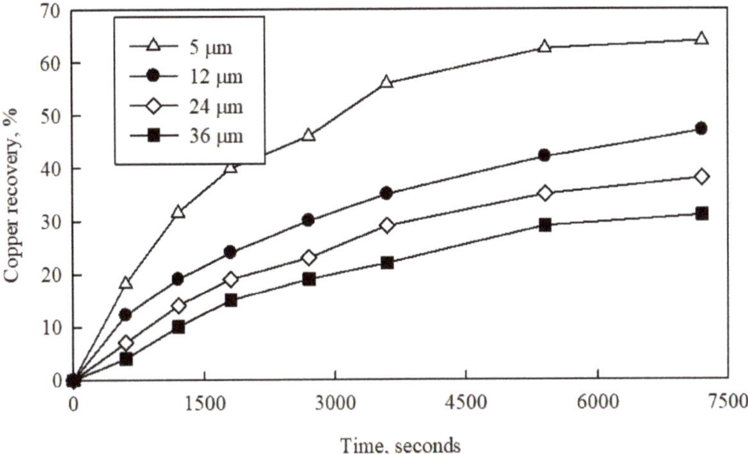

Figure 9. The effect of average particle size on copper recovery. Working conditions: NH_4OH = 0.1 mol/dm^3, temperature = 298 K, pH = 10.5, stirring rate = 500 rpm and S/L ratio = 1:1000.

It can be seen in the figure that there is increased copper recovery as the particle size decreases. Thus, for a particle size of 36 μm (and a time of 7200 s), copper recovery of 31.0% was obtained; at a particle size close to 7 times smaller, copper recovery increased almost 2-fold. This is mainly because the smaller particle size increases the area of the reaction interface between the $Cu_2CO_3(OH)_2$ and ammonia molecules.

4.7. Analysis of the Solid/Liquid Ratio

In order to evaluate the solid/liquid ratio, tests were carried out lasting 3600 s. The working conditions were a temperature of 298 K and NH_4OH concentration of 0.1 mol/dm^3. Different ammonium solution volumes were used, ranging from 0.1 to 0.8 dm^3, maintaining a constant mass of malachite of 1×10^{-3} kg. Figure 10 summarizes the copper extraction results as a function of the solid/liquid ratio.

The main objective was to obtain the maximum possible level of copper recovery with the lowest solution volume. This was achieved using a ratio of 0.6 dm^3/kg, which reported a copper recovery of 72.0% due to a more efficient reaction medium between the diffusion of NH_3 and $Cu_2CO_3(OH)_2$. It can also be seen that Cu recovery reached only 7.0% when using the lowest ratio (0.1 dm^3/kg).

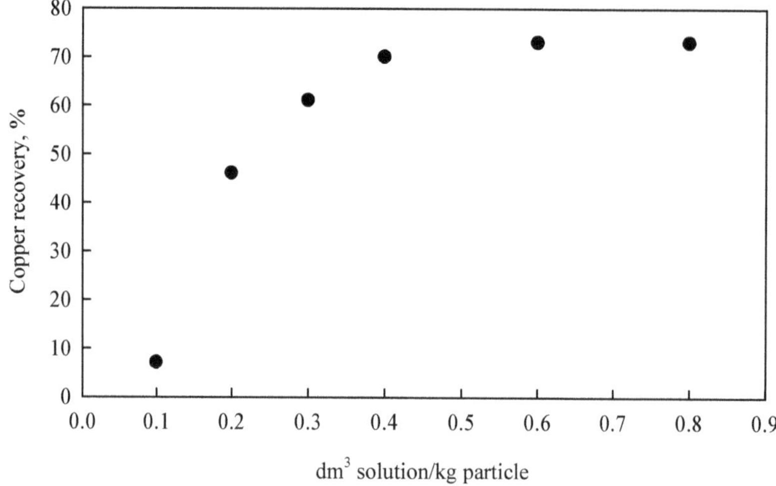

Figure 10. Evaluation of the effect of the solid/liquid ratio on copper recovery. Working conditions: NH_4OH = 0.1 mol/dm^3, temperature = 298 K, particle size = 5 µm, pH = 10.5, stirring rate = 500 rpm and experiment time = 3600 s.

4.8. Effect of Different Ammonium Reagents

Figure 11 shows the results for copper recovery as a function of time for different ammonium reagents: $(NH_4)_2SO_4$, NH_4F, NH_4NO_3 and NH_4OH. There is a clear positive effect on the dissolution of $Cu_2CO_3(OH)_2$ for the four reagents used, obtaining values close to 24.2% after 7200 s for the ammonium nitrate. However, when using ammonium fluoride and ammonium sulfate, the copper recovery values reached only 9.2% and 80.9%, respectively, for the same experiment time. For these two reagents (NH_4F and $(NH_4)_2SO_4$), the copper recovery becomes extremely slow after 20 min.

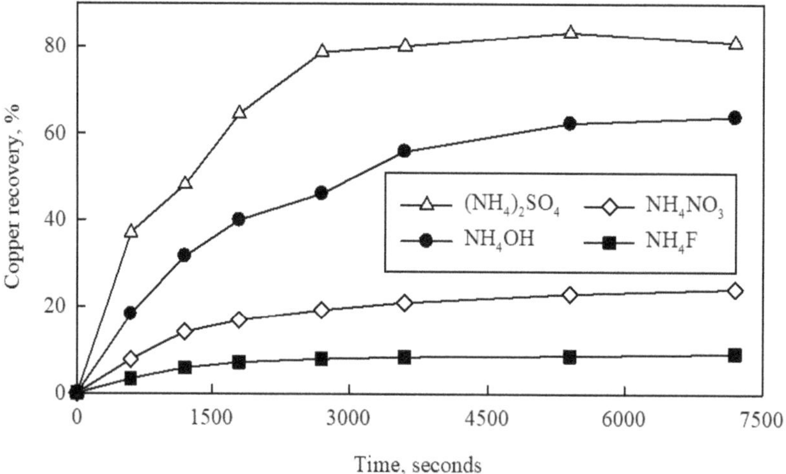

Figure 11. Evaluation of the influence of different ammonium reagents on the malachite leaching rate. Working conditions: NH_4OH = 0.1 mol/dm^3, temperature = 298 K, particle size = 5 µm, pH = 10.5, stirring rate = 500 rpm and S/L ratio = 1:1000.

It should be noted that the pH values remained constant at 10.5. Therefore, due to the lack of prior studies of oxide leaching using ammonium reagents, the use of other ammonium reagents requires additional research to find the maximum possible dissolution of malachite.

The copper generated in the dissolution of malachite can be recovered by a process of solvent extraction (SX) with electrowinning (EW). Some research studies have used SX to recover copper and ammonia using sterically hindered β-diketone [21], to recover copper using LIX 54 [22] or with the use of liquid membranes using LIX 84I [23]. Therefore, copper concentrated by SX can be obtained through electrowinning as metallic copper (with a cathode of high purity) or precipitated as copper sulfate.

4.9. Dissolution Kinetics

According to Figure 7, the effect of temperature on the rate of malachite dissolution was not significant. This suggests that the malachite dissolution is governed by a process of diffusion in a porous layer due to the dissolution of the particle as it cracks. This model represents particles that begin to react as nonporous but become porous during the reaction, i.e., the original solid cracks and splits to form a porous structure resembling a granular material, with each grain reacting through a decreasing core mechanism (shrinking core model). Therefore, the reaction rate follows a shrinking core model in which diffusion is controlled by the porous layer with an initial radius for a constant reagent concentration, expressed as in the following equation [24]:

$$1-\frac{2}{3}\alpha-(1-\alpha)^{\frac{2}{3}} = k_{app}\,t \tag{11}$$

In this equation, the converted fraction, α, represents the conversion of the malachite at time t. The apparent reaction rate constant is represented by the following expression:

$$k_{app} = k_o\frac{b\,[NH_4OH]^n}{r_o^2}e^{-E_a/RT} \tag{12}$$

In this expression, $[NH_4OH]$ and n are the concentration and order to the reaction with regard to the ammonium hydroxide concentration, respectively; k_o is the intrinsic reaction rate constant; b is the stoichiometric constant given by Equation (9), which relates the molarity between the ammonium hydroxide and the malachite; and r_o is the initial radius of the particles.

Using the experimental data shown in Figure 7, a graph was built to represent the diffusion model through the porous layer as a function of time for the temperature range of 278 to 313 K. Figure 12 indicates the results; it can be seen that there is a good linear fit of the experimental data, with the regression coefficients (R^2) being close to 0.96 for the entire temperature range. These high values of R^2 validate the Kinetic Equation (11). The apparent reaction rate constants (the gradient of each straight line) are presented in Table 1.

Table 1. Value of each apparent kinetics constant for the five temperatures studied.

T [°C (K)]	1000/T (1/K)	k_{app}, 1/s
5 (278)	3.5971	6.3×10^{-6}
10 (283)	3.5336	7.9×10^{-6}
25 (298)	3.3557	12.0×10^{-6}
32 (305)	3.2787	18.8×10^{-6}
40 (313)	3.1949	44.7×10^{-6}

The experimental data in Figure 8 for the NH_4OH concentration range of 0.08 to 0.2 mol/dm^3 were also used to build a graph of the diffusion model in the porous layer as a function of time (Figure 13). It can be seen in this figure that a good fit is obtained for the experimental data for all the straight lines generated ($R^2 > 0.92$).

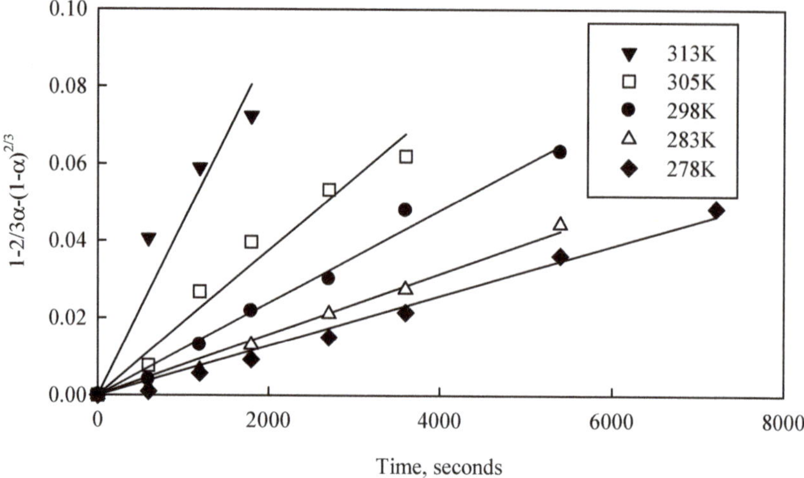

Figure 12. Analysis of malachite leaching kinetics as a function of temperature. The working conditions are the same as those in Figure 7.

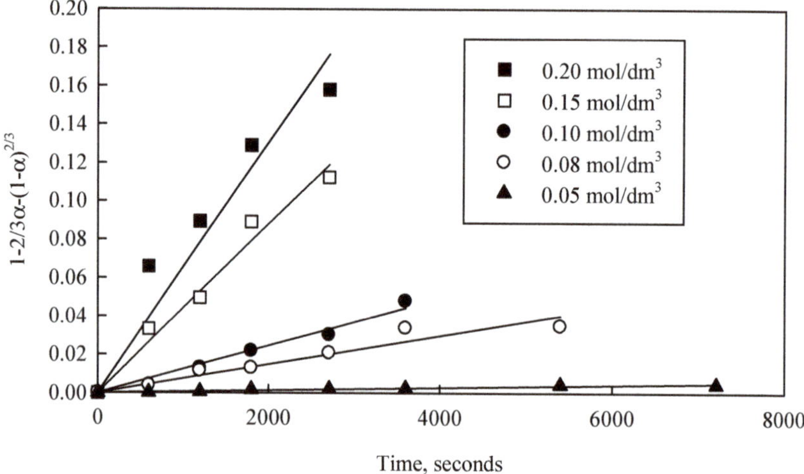

Figure 13. Evaluation of the kinetic model for different ammonium hydroxide concentrations. Working conditions are the same as those in Figure 8.

The values of k_{app} were used to build a graph of $\ln(k_{app})$ as a function of $\ln([NH_4OH])$, as shown in Figure 14. This figure shows a good linear fit, with R^2 values reaching 0.95. The gradient of the straight line corresponds to the value of the reaction order (n) for the specific ammonium hydroxide concentration. Therefore, the reaction order calculated for the malachite dissolution is 3.2.

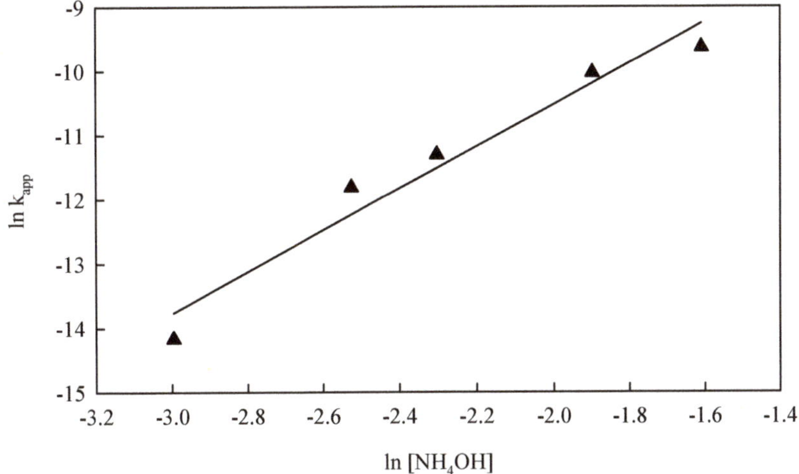

Figure 14. Reaction order calculated with regard to the concentration of NH_4OH.

For a kinetic model that is controlled by diffusion through a porous layer, the apparent constant values should vary linearly with the inverse square of the initial particle radius, as seen in Equation (12). In order to verify this, the particle size data (Figure 9) were entered into Equation (12), generating the graph shown in Figure 15.

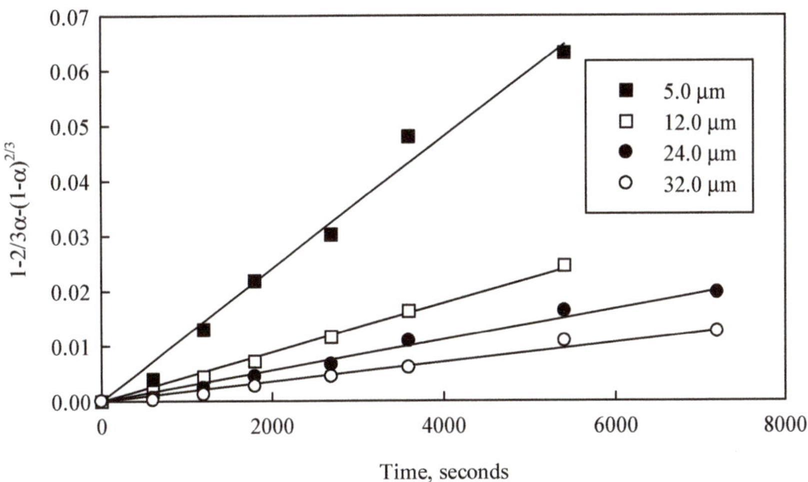

Figure 15. Malachite leaching kinetics for different average particle sizes. Working conditions are the same as those in Figure 7.

It can be seen in Figure 15 that a good correlation was obtained (R^2 close to 0.98), validating Equation (11) for the diffusion model in a porous layer generated postleaching. The values of k_{app} obtained from Figure 15 were graphed in Figure 16 as a function of the inverse square of the initial radius, as shown in Equation (12). The linear dependence of the data shown in Figure 16 ($R^2 > 0.99$) therefore validates the kinetic model used in the present study.

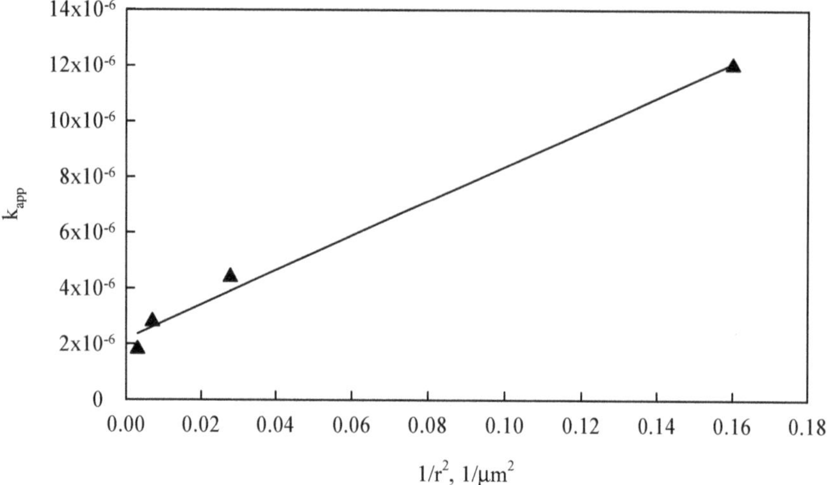

Figure 16. Evaluation of the inverse of the average particle size.

In order to calculate the activation energy (E_a), the apparent reaction rate constants obtained in Table 1 were used along with the ammonium hydroxide concentration and reaction order values of 0.1 mol/dm^3 and 3.2, respectively. The value of b was 1/8, according to the stoichiometry in Equation (9). These values were substituted into Equation (12). Table 2 shows the results of the intrinsic reaction rate constant values as a function of the temperature range used in the study.

Table 2. Intrinsic reaction rate constants for the malachite leaching in NH_4OH.

T, K	k_o, 1/s μm^2 1/(mol/dm^3)$^{3.2}$
278	49.92×10^{-2}
283	62.60×10^{-2}
298	95.09×10^{-2}
305	148.98×10^{-2}
313	354.22×10^{-2}

An Arrhenius plot was then built using the values of k_o for the temperature range in study. Figure 17 shows a good linear fit ($R^2 = 0.90$) for the temperature dependence with regard to the kinetics constants. The activation energy was calculated as 36.1 kJ/mol for the temperature range of 278 to 313 K. This value is typical for a diffusion model through a porous layer. Therefore, the kinetic equation representing the malachite leaching in an ammonium system (NH_4OH) is that shown in expression (13):

$$1-\frac{2}{3}\alpha-(1-\alpha)^{\frac{2}{3}} = 2.85\times 10^{6}\frac{\frac{1}{8}\,[NH_4OH]^{3.2}}{r_o^2}e^{-36.1/RT}\,t \tag{13}$$

where R is the gas constant and is equal to 8.314 J/mol/K, [NH_4OH] is in mol/dm^3, r_o is in μm, t is in seconds, T is in Kelvin and k_o equals 2.85×10^6 1/s μm^2 1/(mol/dm^3)$^{3.2}$.

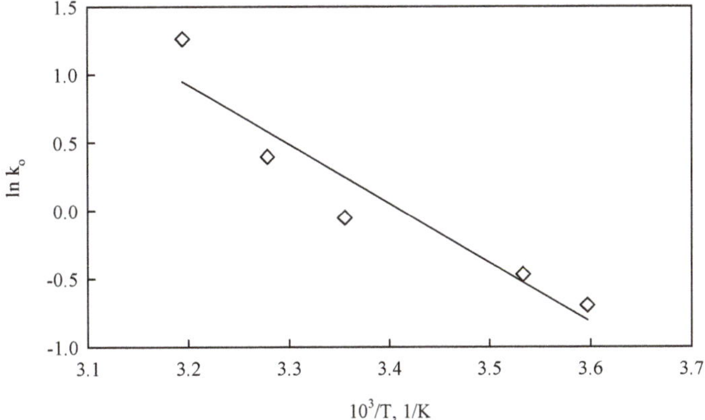

Figure 17. Arrhenius plot for the temperature range of 278 to 313 K.

5. Conclusions

This research aimed to obtain the reaction mechanism and analyze the kinetics of malachite leaching with the use of ammonium hydroxide at different temperatures. The innovative use of this leaching solution was done mainly because the principle component of malachite is carbonate, which consumes large amounts of sulfuric acid (the most commonly used leaching compound). For our study, the excessive value of acid consumption by malachite was 986 kg H_2SO_4/ton of malachite. The use of NH_4OH avoids the need to use H_2SO_4, leading to useful metal (copper) extraction.

The results obtained are promising, showing copper recovery above 82% (ammoniacal system). Increasing the temperature and ammonium hydroxide concentration led to increased copper recovery, while decreasing the particle size also caused an increase in the recovery rate. The pH of the solution was also a significant factor in the malachite leaching process.

Malachite dissolution is governed by a process of diffusion in a porous layer due to the dissolution of the particle as it cracks, i.e., the original solid cracks and splits to form a porous structure resembling a granular material, with each grain reacting through a decreasing core mechanism.

Author Contributions: Conceptualization, A.A.; formal analysis, O.J.; funding acquisition, A.A.; investigation, A.A., O.J.; methodology, J.P.; validation, J.P., A.A., O.J.; writing—original draft, A.A., O.J. authors have read and agreed to the published version of the manuscript.

Funding: The APC was financed by the Office of the Vicerrectoría de Investigación y Estudios Avanzados of the Pontificia Universidad Católica de Valparaíso.

Conflicts of Interest: The authors declare no conflict of interest.

References

1. Domic, E. *Hidrometalurgia, Fundamentos, Procesos Y Aplicaciones (in Spanish)*; Santiago de Chile: Santiago, Chile, 2001; ISBN: 9562910830, 9789562910835.
2. YouCai, L.; Wei, Y.; Jian-Gang, F.; Li-Feng, L.; Dong, Q. Leaching kinetics of copper flotation tailings in aqueous ammonia/ammonium carbonate solution. *Can. J. Chem. Eng.* **2013**, *91*, 770–775. [CrossRef]
3. Pokrovsky, O.S.; Golubev, S.V.; Schott, J. Dissolution kinetics of calcite, dolomite and magnesite at 25 °C and 0 to 50 atm pCO2. *Chem. Geol.* **2005**, *217*, 239–255. [CrossRef]
4. Vignes, A. *Extrative Metallurgy 2: Metallurgical Reaction Processes*; Wiley & Sons: Hoboken, NJ, USA, 2013.
5. Bingöl, D.; Canbazoğlu, M. Dissolution kinetics of malachite in sulphuric acid. *Hydrometallurgy* **2004**, *72*, 159–162. [CrossRef]
6. Nicol, M. The kinetics of the dissolution of malachite in acid solutions. *Hydrometallurgy* **2018**, *177*, 214–217. [CrossRef]

7. Aracena, A.; Vivar, Y.; Jerez, O.; Vásquez, D. Kinetics of dissolution of tenorite in ammonium media. *Miner. Process. Extr. Metall. Rev.* **2015**, *36*, 317–323. [CrossRef]
8. Aracena, A.; Pérez, F.; Carvajal, D. Leaching of cuprite through NH_4OH in basic systems. *Trans. Nonferrous Met. Soc. China* **2018**, *28*, 2545–2552. [CrossRef]
9. Ekmeyapar, A.; Aktas, E.; Künkül, A.; Demirkiran, N. Investigation of leaching kinetics of copper from malachite ore in ammonium nitrate solutions. *Metall. Mater. Trans. B* **2012**, *43*, 764–772. [CrossRef]
10. Bingöl, D.; Canbazoğlu, M.; Aydoğan, S. Dissolution kinetics of malachite in ammonia/ammonium carbonate leaching. *Hydrometallurgy* **2005**, *76*, 55–62. [CrossRef]
11. Künkül, A.; Muhtar, M.; Yapici, S.; Demirbağ, A. Leaching kinetics of malachite in ammonia solutions. *Int. J. Miner. Process.* **1994**, *41*, 167–182. [CrossRef]
12. Arzutug, M.E.; Kocakerim, M.M.; Copur, M. Leaching of malaqhite ore in NH3-saturated wáter. *Ind. Eng. Chem. Res.* **2004**, *43*, 4118–4123. [CrossRef]
13. Kokes, H.; Morcali, M.H.; Acma, E. Dissolution of copper and iron from malachite ore and precipitation of copper sulfate pentahydrate by chemical process. *Eng. Sci. Technol. Int. J.* **2014**, *17*, 39–44. [CrossRef]
14. Kurşuncu, B.; Yaraş, A.; Arslanoglu, H. Application of multi criteria decision making methods to leaching process of copper from malachite ore. *Sigma J. Eng. Nat. Sci.* **2018**, *36*, 783–794.
15. Deng, J.; Wen, S.; Yin, Q.; Wu, D.; Sun, Q. Leaching of malachite using 5-sulfosalicylic acid. *J. Taiwan Inst. Chem. Eng.* **2017**, *71*, 20–27. [CrossRef]
16. Mortiner, C.E. *Chemistry; A Conceptual Approach*, 4rd ed.; Van Nostrand: New York, NY, USA, 1979; pp. 796–798.
17. Aracena, A.; Fernández, F.; Jerez, O.; Jaques, A. Converter slag leaching in ammonia medium/column system with subsequent crystallisation with NaSH. *Hydrometallurgy* **2019**, *188*, 31–37. [CrossRef]
18. Tanda, B.C.; Eksteen, J.J.; Oraby, E.A. An investigation into the leaching behaviour of copper oxide minerals in aqueous alkaline glycine solutions. *Hydrometallurgy* **2017**, *167*, 153–162. [CrossRef]
19. *HSC Chemistry, version 6.0*; OutoKumpu Research Py: Pori, Finland, 1999.
20. Oudenne, P.D.; Olson, F.A. Leaching kinetics of malachite in ammonium carbonate solutions. *Metall. Mater. Trans. B* **1983**, *14*, 33–40. [CrossRef]
21. Hu, H.; Liu, C.; Han, X.; Liang, Q.; Chen, Q. Solvent extraction of copper and ammonia from amoniacal solutions using sterically hindered b-diketone. *Trans. Nonferrous Met. Soc. China* **2010**, *20*, 2026–2031. [CrossRef]
22. Alguacil, F.J.; Alonso, M. Recovery of copper from amoniacal ammonium sulfate médium by LIX 54. *J. Chem. Technol. Biotechnol.* **1999**, *74*, 1171–1175. [CrossRef]
23. Sengupta, B.; Bhakhar, M.S.; Sengupta, R. Extraction of copper from amoniacal solutions into emulsion liquid membranes using LIX 84 I®. *Hydrometallurgy* **2007**, *89*, 311–318. [CrossRef]
24. Sohn, H.Y.; Wadsworth, M.E. *Rate Processes of Extractive Metallurgy*; Plenum Press: New York, NY, USA, 1979.

 *metals*

Review

Hydrometallurgical Recovery of Rare Earth Elements from NdFeB Permanent Magnet Scrap: A Review

Yuanbo Zhang [1,*], Foquan Gu [1,*], Zijian Su [1], Shuo Liu [1], Corby Anderson [2] and Tao Jiang [1]

1 School of Minerals Processing and Bioengineering, Central South University, Changsha 410083, China; szjcsu@163.com (Z.S.); lsus91@163.com (S.L.); jiangtao@csu.edu.cn (T.J.)

2 Department of Metallurgical and Materials Engineering, Colorado School of Mines, Golden, CO 80401, USA; cganders@mines.edu

* Correspondence: sintering@csu.edu.cn (Y.Z.); gufoquan@csu.edu.cn (F.G.)

Received: 30 May 2020; Accepted: 15 June 2020; Published: 24 June 2020

Abstract: NdFeB permanent magnet scrap is regarded as an important secondary resource which contains rare earth elements (REEs) such as Nd, Pr and Dy. Recovering these valuable REEs from the NdFeB permanent magnet scrap not only increases economic potential, but it also helps to reduce problems relating to disposal and the environment. Hydrometallurgical routes are considered to be the primary choice for recovering the REEs because of higher REEs recovery and its application to all types of magnet compositions. In this paper, the authors firstly reviewed the chemical and physical properties of NdFeB permanent magnet scrap, and then carried out an in-depth discussion on a variety of hydrometallurgical processes for recovering REEs from the NdFeB permanent magnet scrap. The methods mainly included selective leaching or complete leaching processes followed by precipitation, solvent extraction or ionic liquids extraction processes. Particular attention is devoted to the specific technical challenge that emerges in the hydrometallurgical recovery of REEs from NdFeB permanent magnet scrap and to the corresponding potential measures for improving REEs recovery by promoting the processing efficiency. This summarized review will be useful for researchers who are developing processes for recovering REEs from NdFeB permanent magnet scrap.

Keywords: rare earth elements; NdFeB permanent magnet; hydrometallurgical; recovery

1. Introduction

NdFeB magnets are considered as the strongest permanent magnets with the highest energy product BH_{max} (200–440 kJ/m^3) of all permanent magnets [1]. They are widely used in wind turbines, hybrid electric vehicles, hydro-electric turbine generators, etc. [2]. Figure 1 shows the proportion of the different applications around the global NdFeB market.

Depending on the application field of NdFeB magnets, they have different life cycles and weight. The life cycles of NdFeB magnets range from 2–3 years for consumer electronics to 20–30 years in wind turbines. Meanwhile, the weight of NdFeB magnets ranges from less than 1 g for small electronics to 1–2 t for wind turbines [4]. The phase of NdFeB magnets based on (Nd, Pr, Dy)-Fe-B and other trace elements, with REEs contents of 27–32 wt.%, Fe content of 67–73 wt.%, B content of about 1 wt.% [5,6], and other minor metals, are determined by the applications of NdFeB magnets. Obviously, recovering REEs from NdFeB magnet scrap has good economic benefits.

Recyclable materials of NdFeB magnets scrap mainly include: (1) swarf originating from magnet manufacturing; (2) large magnets in wind turbines; (3) small magnets in End-of-Life consumer products. About 20–30 wt.% scrap is generated during the NdFeB magnets cutting and grinding processes, and 95% of those scraps can be recycled [7,8]. The NdFeB magnets used in wind turbines, hybrid vehicles and electric vehicles are easy to directly recycle or re-use [9]. However, it remains a social and technological challenge to collect and recover the magnets from the End-of-Life of small consumer electronics. In view of the potential economic and environmental benefits of utilizing the NdFeB magnet scrap, it is significant to develop appropriate methods for treating the NdFeB magnet scrap. At present, recycling of NdFeB magnet scrap mainly focus on the recovery of REEs. The disclosed methods include direct re-use in current form/shape [10–15], reprocessing of alloys to magnets after hydrogen decrepitation [12,14–18], pyrometallurgical methods [19–28], gas-phase extraction [29,30], and hydrometallurgical methods [31–37]. In Table 1, an overview of different methods for NdFeB magnet scrap recycling is given.

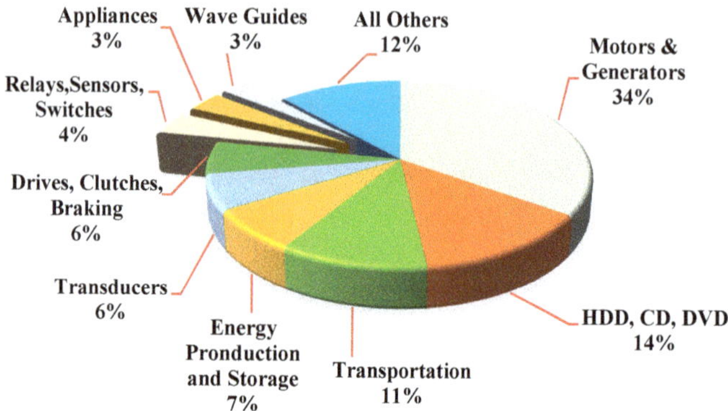

Figure 1. The proportion of different applications around the global NdFeB magnet market, data from [3].

Table 1. Overview of the advantages and disadvantages of different methods for NdFeB magnet scrap recycling, Reproduced with permission from [38]; published by Elsevier, 2013.

Method	Advantages	Disadvantages
Direct re-use in current form/shape	• Most economical way of recycling (low energy input, no consumption of chemicals) • No waste generated	• Only for large, easily accessible magnets (wind turbines, large electric motors and generators in hybrid and electric vehicles) • Not available in large quantities in scrap today
Reprocessing of alloys to magnets after hydrogen decrepitation	• Less energy input required than for hydrometallurgical and pyrometallurgical routes • No waste generated • Especially suited for hard disk drives (little compositional change over the years)	• Not applicable to mixed scrap feed, which contains magnets with large compositional variations • Not applicable to oxidized magnets
Pyrometallurgical methods	• Generally applicable to all types of magnet compositions • No generation of waste water • Fewer processing steps than hydrometallurgical methods • Direct melting allows master alloys to be obtained • Liquid metal extraction allows REEs to be obtained in metallic state	• Larger energy input required • Direct smelting and liquid metal extraction cannot be applied to oxidized magnets • Electroslag refining and the glass slag method generate large amounts of solid waste

Table 1. *Cont.*

Method	Advantages	Disadvantages
Gas-phase extraction	• Generally applicable to all types of magnet compositions • Applicable to non-oxidized and oxidized alloys • No generation of waste water	• Consumption of large amounts of chlorine gas • Aluminum chloride is very corrosive
Hydrometallurgical methods	• Generally applicable to all types of magnet compositions • Applicable to non-oxidized and oxidized alloys • Same processing steps as those for extraction of rare earths from primary ores	• Many process steps required before obtaining new magnets • Consumption of large amounts of chemicals • Generation of large amounts of waste water

Recycling of NdFeB magnet scrap has been researched broadly and various methods have been carried out. Among these methods, the hydrometallurgical methods seem to be the most prominent for recovering REEs from NdFeB magnet scrap because hydrometallurgical methods can be used to treat all types of magnets. More importantly, hydrometallurgical methods can be well connected with the existing REEs production industry. The purpose of the present paper is to provide an overview of REEs recovering from NdFeB magnet scrap by hydrometallurgical processes. The recycling potential of NdFeB permanent magnet scraps is firstly clarified, and then the chemical and physical characteristics of NdFeB permanent magnet scrap are provided. Lastly, hydrometallurgical methods for recovering REEs from NdFeB permanent magnet scrap are reviewed. This paper aims to offer a useful guideline for sustainable recovering REEs from NdFeB permanent magnet scrap.

2. Recycling Potential of NdFeB Permanent Magnet Scrap

According to the United States Geological Survey (USGS), the global annual consumption of rare earth oxides is about 120,000 t [39]. The market share of global consumption of rare earth oxides is shown in Figure 2; approximately 26,400 t of rare earth oxides are used in the permanent magnet market, accounting for a large proportion among all rare earth oxides consumption.

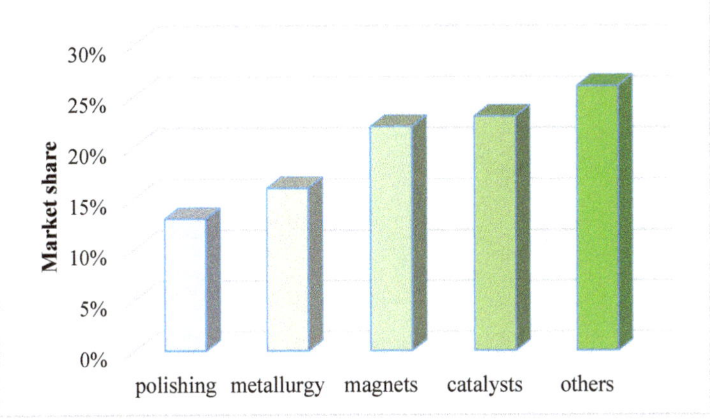

Figure 2. Market share of global consumption of rare earth oxides (based on data from the United States Geological Survey (USGS) [39]).

Alonso et al. [40] estimated the growths of market share of global REEs consumption according to the applications. As seen from Figure 3, the fraction of REEs demand in the magnets increased continuously until 2035. The magnets will become the most in-demand materials of REEs, and the fraction of REEs demand will be close to 50%, because of the rapidly developing technologies for clean energy and transportation (e.g., electric vehicles and wind turbines).

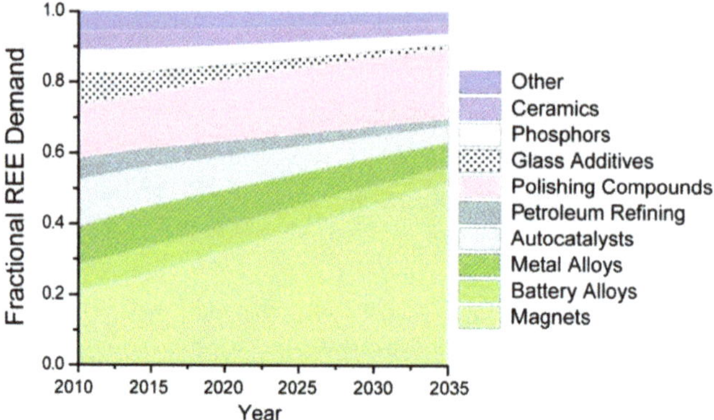

Figure 3. Predicted growths of market share of global rare earth elements (REEs) consumption [40]; published by American Chemical Society, 2012.

Schulze et al. [41] calculated the NdFeB demand for magnets used in different application groups and net availability of secondary NdFeB supply from End-of-Life (EOL) magnets from 2020 to 2030 based on low and high NdFeB demand scenario. As shown in Figure 4, the demand of NdFeB and the net availability of secondary NdFeB supply from EOL magnets both increase gradually. The demand of NdFeBis about 240 kt and 633 kt in 2030 for the low and high NdFeB demand scenario, respectively. Meanwhile, the net availability of secondary NdFeB supply from EOL magnets is about 27 kt and 54 kt in 2030 for the low and high NdFeB demand scenario, respectively. Obviously, the recovery of NdFeB magnets possesses huge potential value.

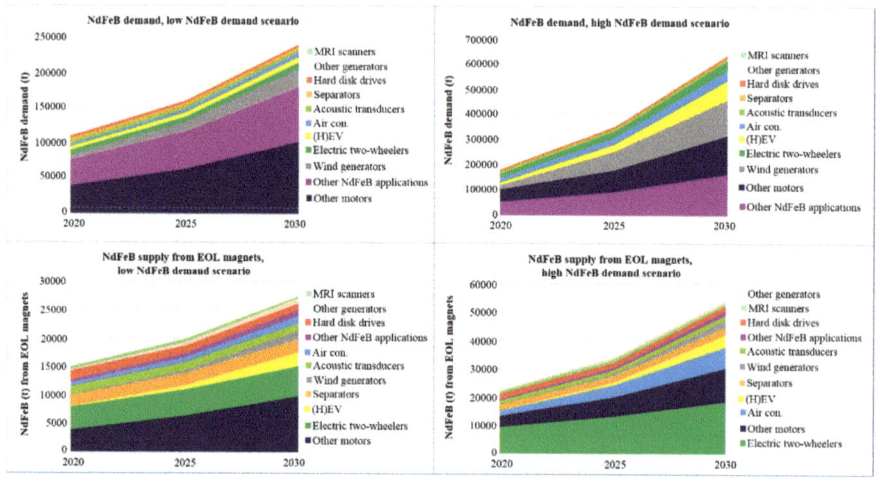

Figure 4. NdFeB demand and net supply from End-of-Life (EOL) magnets (losses during collection and disassembly have been subtracted) [41]; published by Elsevier, 2016.

The historical data (1983–2007) of NdFeB permanent magnets from China, Japan, the United States and Europe were used to estimate the global stock of REEs in NdFeB permanent magnets. The results showed that about 62.6 kt Nd, 15.7 kt Pr, 15.7 kt Dy and 3.1 kt Tb were stocked in NdFeB permanent magnets from 1983 to 2007. If these stocks are effectively recovered, they can serve as a valuable supplement to the geological stocks because they are about four times of the 2007 annual extraction of the individual elements [42]. Guyonnet et al. [43] presented an analysis of flows and stocks of some REEs along the value chain in Europe. The analysis indicated that in 2010, about 580 t Nd and 70 t Dy were wasted. The Sankey diagrams illustrate the serious imbalance of flows of REEs in NdFeB magnets along the value chain, and the Europe mainly depends on the import of finished products. They also indicated that the recirculation flow of Nd in NdFeB magnets is expected to be 170–230 t in 2020.

Schulze and Buchert [44] quantitatively analyzed the global recycling potential of EOL magnets from different application groups and industrial waste by dynamic material flow analysis. The modelled scenario shows that 18–22% of the global Nd and Pr and 20–23% of Dy and Tb used in NdFeB magnet production can be provided by EOL magnets and industrial scrap in 2020, 2025 and 2030. In another study [45], the authors estimated annual waste flows of Nd and Dy from three common permanent magnets. The results indicate that for some time to come, compared to the rapidly growing global REEs demand, the waste stream generated by permanent magnets will remain small, and the global recycling potential for the next few decades is also limited. Due to the small amount of waste, recycling at an economically advantageous scale is impossible. However, in the long term, waste stream generated by permanent magnets will increase dramatically and meet a large portion of the total demand for REEs. Therefore, those authors highlight that the NdFeB magnet is one of the most important secondary resources for the recovery of Nd and Dy. It is also suggested that decision makers should develop recycling technology through pilot projects, which should take about five to ten years to establish recycling practices [41].

There is no denying that recycling is a key technology for metal recycling from various sources. However, at present, recycling of REEs is within 1%, but the current situation demands that we improve the recycling of REEs from REEs-bearing wastes [38]. The NdFeB permanent magnet is an important REEs demanded material, and the NdFeB permanent magnet scrap has huge potential for REEs recovery.

3. Chemical and Physical Characteristics of NdFeB Permanent Magnet Scrap

The chemical and physical properties of NdFeB permanent magnet scrap are the starting point for choosing a suitable recycling process. These characteristics mainly include chemical composition, phase composition, and microstructural morphology.

3.1. Chemical Composition

Table 2 lists typical chemical compositions of NdFeB permanent magnet scraps. In fact, according to the compositions of REEs, NdFeB permanent magnet scraps can be classified into three categories, namely low REEs scraps (the contents of REEs < 20 wt.%), medium REEs scraps (the contents of REEs about 20–30 wt.%), and high REEs scraps (the contents of REEs > 30 wt.%). All types of NdFeB permanent magnet scraps contain Nd, sometimes, Tb, Dy, and Gd are also added to replace some fractions of Nd to increase the operating temperature and intrinsic coercivity of NdFeB permanent magnet [46,47]. Pr, La, and Sm are generally added to replace Nd at a lower production cost [48]. The addition of Co can improve the Curie temperature of the magnet [49]. The addition of Al, Cu, Nb, and Ni is proposed to replace some Fe, to increase the coercivity of NdFeB permanent magnet [50].

Table 2. Typical chemical compositions of NdFeB magnet scrap (wt.%).

Typical	Nd	Fe	B	Pr	Dy	Co	Sm	La	Nb	Gd	Tb	Cu	Al	Ni	Refs.
Low REE sscraps	10.70	79.20	5.75	2.68	0.43	0.79	-	-	-	-	-	0.11	0.19	-	[16]
	14.00	78.00	6.00	-	0.60	-	-	-	0.40	-	-	-	0.70	-	[19]
Medium REEs scraps	26.10	63.50	0.73	0.68	2.68	2.99	-	-	-	0.02	-	-	0.35	-	[51]
	25.38	61.09	1.00	2.62	1.08	1.42	-	-	-	-	-	-	0.95	2.03	[52]
	28.00	68.00	1.00	-	1.00	-	-	-	-	-	-	-		1.00	[53]
	23.70	66.10	0.91	0.12	2.42	3.34	-	-	-	-	-	-		3.00	[13]
	24.43	64.07	0.97	-	-	1.67	-	-	0.37	-	-	0.15	0.20	-	[54]
	21.00	70.60	1.04	-	6.30	0.57	-	-	-	-	-	0.15	-	-	[8]
High REEs scraps	19.40	66.30	0.96	6.43	5.21	-	0.77	-	-	-	-	-	-	0.87	[31]
	30.73	61.60	0.96	4.39	-	-	-	1.58	-	-	-	-	0.83	-	[25]
	18.80	63.90	1.02	5.98	5.93	0.42	-	-	-	1.51	-	-	1.04		[55]
	25.95	58.16	1.00	0.34	4.21	4.22	-	-	-	-	-	-	0.34	0.02	[34]
	25.95	58.16	1.00	0.07	4.21	4.22	-	-	0.83	-	-	-	0.34	0.02	[56]
	22.57	67.15	0.98	7.10	0.79	0.74	-	-	-	-	0.41	-	-	-	[14]

3.2. Phase Composition

The main phase of the NdFeB magnet scrap is $Nd_2Fe_{14}B$ [54,57], which accounts for 96–98%. Herbst et al. indicated the unit cell structure of $Nd_2Fe_{14}B$. The space group is $P4_2/mnm$, and there are four $Nd_2Fe_{14}B$ units (68 atoms) per unit cell. All the Nd and B atoms, but only four of the 56 Fe atoms, reside in the z = 0 and z = 0.5 planes. Between these, the other Fe atoms form puckered, yet fully connected, hexagonal nets [58]. Apart from $Nd_2Fe_{14}B$, NdO, Nd_2O_3, and the minor $NdFe_4B_4$ may appear in the grain boundary [51,55].

3.3. Microstructural Morphology

The coercivity of the NdFeB magnets is closely related to the interface microstructure between main phase ($Nd_2Fe_{14}B$) and the grain boundary phase (Nd-rich) [59]. Figure 5 shows the SEM images of the NdFeB magnet. Grain sizes of the NdFeB magnet are very small (Figure 5 left), and small agglomerates of the Nd-rich phase are also observed. The field emission gun scanning electron microscope image (Figure 5 right) shows a uniform continuous coating of $Nd_2Fe_{14}B$ grains, where the thickness of the Nd-rich grain boundary phase is a few nanometers, separating the individual grains. Scanning electron microscopy-energy dispersive X-ray spectroscopy (SEM-EDS) analysis was carried out to identify the compositional variation in the NdFeB magnets, and the results are listed in Figure 6. The results indicate that the Nd is concentrated in the grain boundaries instead of within the grains, and the Fe is concentrated within the grains.

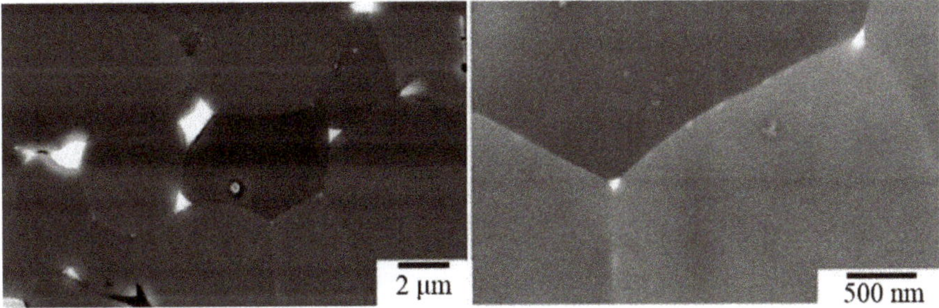

Figure 5. SEM images of NdFeB permanent magnets [59]; published by Elsevier, 2004.

EDS results		
	Spot 1	Spot 2
Nd	87.9	26.1
Fe	2	73.9
Dy	5.4	-
O	4.7	-

Figure 6. SEM-EDS analysis of NdFeB permanent magnets [54]; published by Elsevier, 2014.

The distribution of Nd, Dy, C, Al, Si, Fe, Ce, and Pr in the NdFeB magnet scrap was analyzed using energy dispersive X-ray spectroscopy (EDX) mapping. As shown in Figure 7, the Fe is most abundant in the matrix, while Nd and Pr are located in the grains and concentrated on the grain boundaries. Dy and Ce are nearly evenly distributed over the surface [60]. According to Önal et al., the area dominated by Fe indicates the $Nd_2Fe_{14}B$ phase and the area highlighted by Nd and O represent the grain boundary phase. An area with a high B concentration represents the presence of $Nd_1Fe_4B_4$ phase. They also concluded that all target metals are distributed over the entire microstructure of the magnet and the structure needs to be completely destroyed to recover valuable metals from the NdFeB magnet [61].

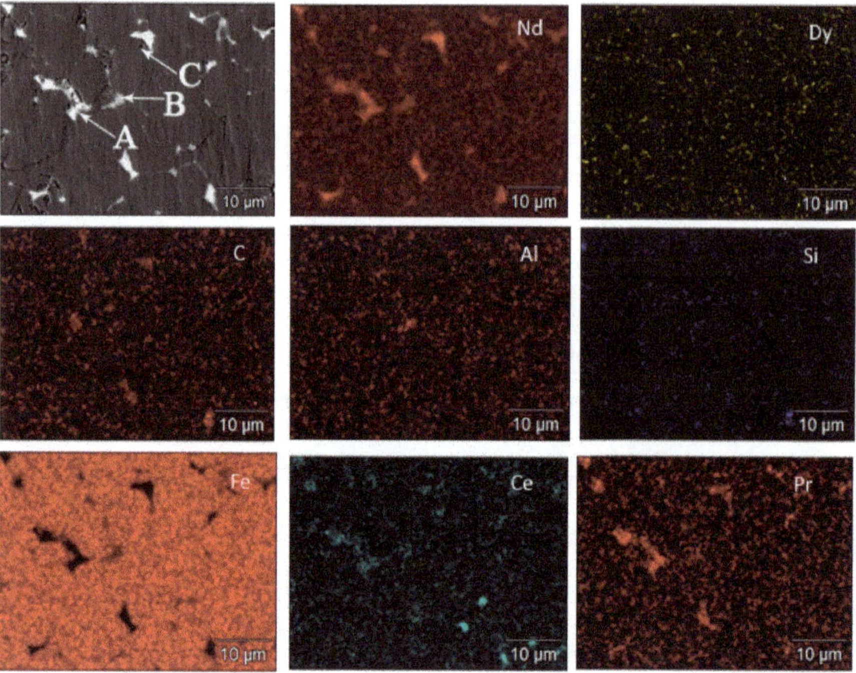

Figure 7. High-resolution EDX mapping of the NdFeB magnet scrap [60]; published by Elsevier, 2020.

4. Hydrometallurgical Processes for Recovering REEs from the NdFeB Magnet Scrap

For the treatment of NdFeB permanent magnet scrap, many hydrometallurgical processes have been developed or are under development. A brief schematic diagram of these hydrometallurgical processes is shown in Figure 8. The principle processes employed during hydrometallurgical treatment of REEs resources mainly include leaching, and REEs separation process [62–64]. In the present study, leaching technologies used for NdFeB permanent magnet scrap include selective leaching process and complete leaching process, which are followed by REEs separation technologies consisting of precipitation process, solvent extraction process and ionic liquids extraction process.

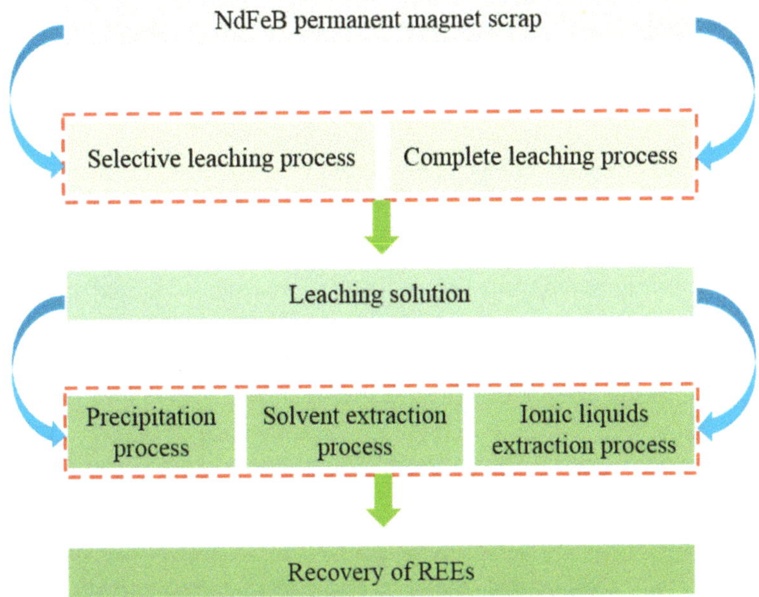

Figure 8. Schematic diagram of hydrometallurgical processes.

4.1. *Leaching Technologies Used for NdFeB Permanent Magnet Scraps*

Leaching is always the first step to dissolve the REEs in the magnet scraps [65,66]. According to the complexity level, different dissolution methods are used. Dissolution of the magnet scrap can be performed in two different ways: (1) selective leaching of NdFeB permanent magnet scrap depending on the solubility of metal at different conditions, (2) complete leaching of the NdFeB permanent magnet scrap.

4.1.1. Selective Leaching Process

The selective leaching process depends on the solubility between REEs and Fe. In order to improve the selectivity and efficiency, many pretreatment processes (e.g., roasting) have been carried out.

Thermal oxidation is often used to improve the selectivity before leaching [63,64]. During the thermal oxidation process, the relatively easily soluble Nd_2O_3 and insoluble Fe_2O_3 were formed to hinder the leaching of Fe in acidic solution. It was reported that, after oxidative roasting at 900 °C for 360 min, the roasted sample was leached by 0.02 mol/L HCl at 180 °C for 120 min, and the recovery of REEs and iron were 99% and 5%, respectively [67,68]. Similar research [59] reported that, under conditions of oxidative roasting at 900 °C for 480 min and subsequent dissolution with 37 wt.% HCl at 80 °C for 900 min, the leaching percentages of Nd and Dy were both above 90%. According to Kumari et al., the roasting of the magnet enhances the selectivity as well as the leaching efficiency. 98% of the REEs were selectively leached and iron oxide was left in the leaching residue under the conditions of roasting temperature of 850 °C, roasting time of 360 min, HCl concentration of 0.5 mol/L, leaching temperature of 95 °C, and leaching time of 300 min. At leaching temperature of 75–95 °C, the leaching of REEs follows the mixed controlled kinetic model with Ea of 30.1 kJ/mol [69]. An identical leaching process with HNO_3 as leaching agent was carried out and very similar results were obtained. Leaching with HCl and HNO_3 in the presence of a concentrated chloride or nitrate matrix (3.75 mol/L of $CaCl_2$ or 7.5 mol/L of NH_4NO_3) was also tested to prove that the leachate could be used directly into a cheap solvent extraction systems to further extract and purify the REEs [70,71]. However, during the oxidation roasting process, the inevitable formation of neodymium-iron mixed oxide ($NdFeO_3$) will hinder the leaching of Nd. To avoid the formation of $NdFeO_3$, Martina et al. [72], roasting the NdFeB scrap in Ar atmosphere at $PO_2 \leqq 10–20$ atm with 5 wt.% C at 1400 °C for 120 min. The results showed that the roasting sample included a metallic Fe phase and B-Dy-Nd phase. After the recovery of metallic iron by mechanical treatment, the REEs in the NdFeB scrap can be completely dissolved in the water-containing ionic liquid [Hbet][Tf_2N] in 20 min.

Rabatho et al. described a process for recovering Nd and Dy from a NdFeB manufacturing process via selective leaching process. The leaching agents were 1 mol/L HNO_3+0.3 mol/L H_2O_2. The leaching of Nd and Dy was up to 98% and 81%, respectively, and the leaching of Fe was below 15% [73]. Another research reported that the pretreatment of the corrosion process increased the selectivity between Nd and Fe at room temperature, and nearly 100% of Nd was recovered from NdFeB magnet scrap [74].

An alkaline treatment of NdFeB magnet at various NaOH concentration was carried out. When the equivalents of NaOH was 10, the leaching of Nd and Dy was 91.6% and 94.6%, respectively, and the leaching of Fe of 24.2%, resulting in the highest selective leaching efficiency [60]. Using a sample prepared under the grinding and alkaline roasting treatments, 94.2%, 93.1%, 1.0% of Nd, Dy, Fe can be selectively leached at 90 °C in 1 mol/cm^3 acetic acid solution with 1% pulp density [75].

Itoh et al. proposed a new recovery process for REEs by selective chlorination roasting of the NdFeB magnet and leaching of the roasted sample in distilled water, the leaching of REEs reached 87% when chlorinating at 300 °C for 3 h, with a nearly negligible content of Fe in the solution [30]. Önal et al. developed a sulfation selective roasting followed by water leaching process. A suitable selective roasting and water leaching treatment showed that 95–100% Nd, Dy, Pr, Gd, Tb, and Eu were leached, while Fe remained in the leaching residue to form a marketable Fe_2O_3-based by-product [60]. This process offers a simple and controllable processing alternative that is completely compatible with the existing REEs production process [76]. It also faces some disadvantages; one disadvantage was that the sulfation roasting required relatively high temperature (750–800 °C). The other disadvantage was related to the low solubility limit of REE sulfate, which may have negative impact on process capability. More seriously, these solubility limitations can also cause problems in subsequent downstream processes, where organic/inorganic solvents are introduced into the leachate at higher temperatures (e.g., 60–80 °C) [34]. In order to avoid those disadvantages, they replaced sulfuric acid with nitric acid. Similar to sulfation roasting, the REEs nitrates are expected to remain soluble, and the solubility limit of the REEs nitrates is higher than that of its sulfate counterparts [55].

Venkatesan et al. reported an electrochemical route to selectively leach REEs from the NdFeB magnet scrap. At first, part of the magnet scrap was leached by HCl. A portion of the leachate was collected with the undissolved magnet scrap on the anolyte side of the two-chamber reactor (Figure 9), which was separated by an anion exchange membrane, and the catholyte was composed of NaCl solution. The Fe(II) in the leachate was oxidized and precipitated as $Fe(OH)_3$, and more than 95% of the REEs were dissolved in the solution [77]. In addition, when the NdFeB magnet scrap was completely leached with HCl, the Fe(II) in the leachate can be selectively oxidized to Fe(III) [78]. In another research, a route for recovering REEs from NdFeB magnet scrap based on electrochemical leaching was also verified (Figure 10). NdFeB magnet scrap was taken as an anode along with an inert anode in an electrochemical reactor (NH_4Cl was used as the electrolyte) to ensure the elements in the magnet scrap into the respective hydroxides, then leaching with HCl. The leaching of REEs and Co exceed 97% with Fe was left in the leaching residue [79].

The selective leaching process is considered to be a relatively commercial process for recovering REEs from NdFeB permanent magnet scrap in spite of several disadvantages [80–83]. The major shortcoming of this process is insufficient separation efficiency, and it is hard to avoid the unwanted elements going into a solution.

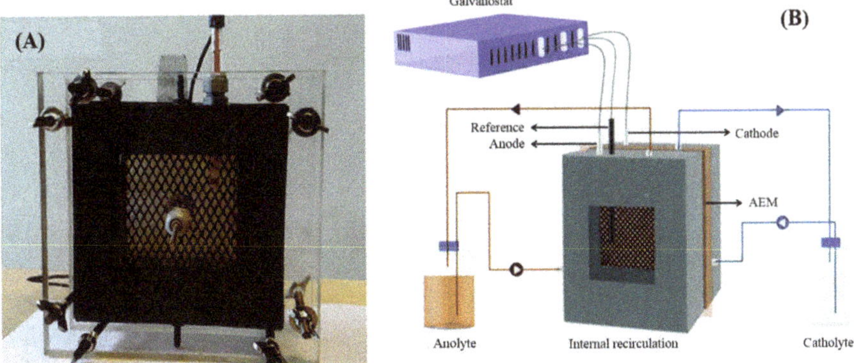

Figure 9. (**A**) Photo and (**B**) schematic illustration of the membrane electrochemical reactor [77]; published by Royal Society of Chemistry, 2018.

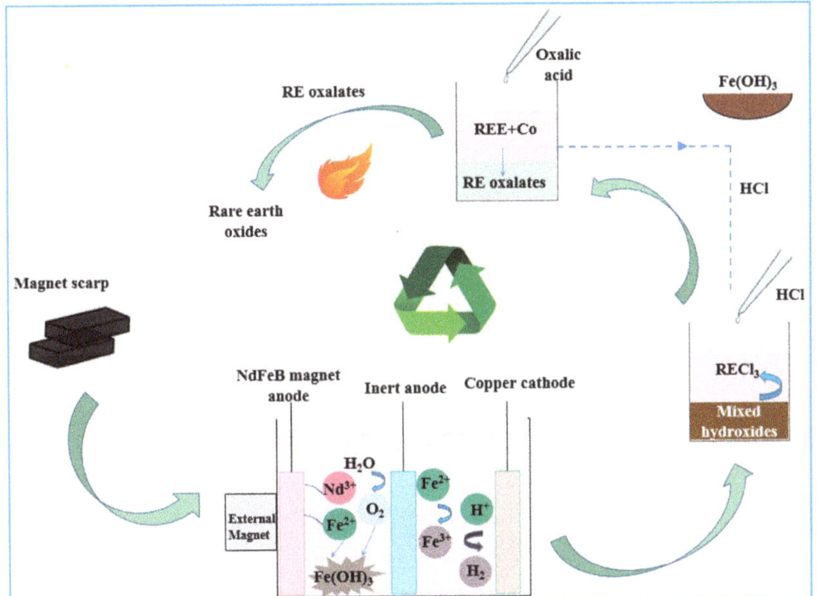

Figure 10. Process flowsheet for the proposed recycling scheme [79]; published by American Chemical Society, 2018.

4.1.2. Complete Leaching Process

The aim of complete leaching process is to dissolve the magnet completely. According to Itakura et al., a commercially available Ni-coated NdFeB sintered magnet was leached by 3 mol/L HCl and 0.2 mol/L $H_2C_2O_4$ at 110 °C for 6 h [53]. The NdFeB magnet went into a solution and Nd was formed a precipitate of neodymium oxalate and more than 99% of Nd was recovered. By addition of $Ca(OH)_2$, the B can be recovered from the highly acidic waste water by formation of $CaB_2O_5{\cdot}H_2O$ [84,85]. Abrahami et al. indicated the feasibility of directly leaching of NdFeB magnet scrap by H_2SO_4, although the composition of the scrap was complex, the combination of different steps still achieved a high recovery with a relatively pure rare earth double salt product (98.4%). Many impurities (mainly iron) were also dissolved in sulfuric acid, and these impurities can be removed after precipitation of

REEs [86]. A decomposition and leaching process of NdFeB permanent magnet scrap by oxidation roasting and sulfuric acid leaching were examined by Yoon et al. [87]. Under the conditions of roasting temperature of 500 °C for sintered scrap and of 700 °C for bonded scrap followed by H_2SO_4 (2 mol/L) leaching at 50 °C for 120 min, the leaching of Nd was over 99.4%, but 95.7% of Fe was also dissolved into the solution, which was similar to the results obtained by Layman and Palmer [88]. HCl and HNO_3 could also be used to completely leach magnet scrap [89]. The feasibility of organic acids for leaching REEs was verified by Gergoric et al. the results showed that after leaching with 1 mol/L acetic acid or citric acid at 25 °C for 24 h, the leaching of REEs exceeded 95% [90].

Electrochemical leaching was investigated to leach Fe and REEs from NdFeB magnet scrap with H_2SO_4 and $H_2C_2O_4$ by Makarova et al. [60]. The scanning Kelvin probe force microscopy results showed that the Local Volta potential difference between $Nd_2Fe_{14}B$ ϕ-phase and the Nd-rich anodic phases exceeded 500 mV, which indicated preferential selective leaching of the Nd-rich phase. A 3D printed Ti basket (Figure 11) was used to leach metals from the magnet scrap, the experimental results found that the acid concentration and current density strongly affected the leaching of REEs, and the addition of oxalic acid reduced the energy consumption and improved the recovery of REEs. The mechanism of dissolution was shown in Figure 12, fast and preferential leaching occurs in the less noble Nd-rich phases located around the Fe-rich $Nd_2Fe_{14}B$ grains (ϕ-phase), then whole grains of the ϕ-phase eventually falls off from the surface.

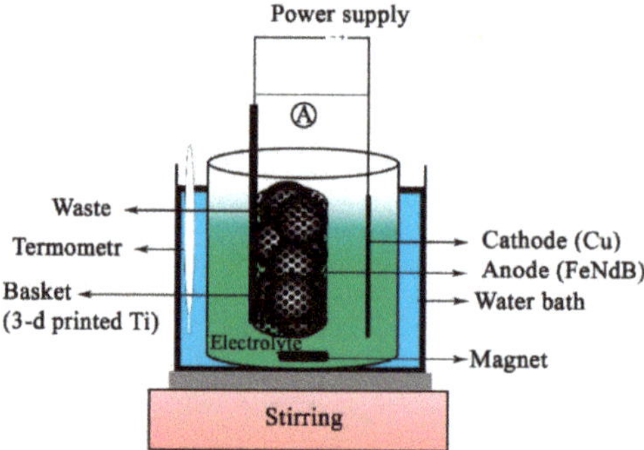

Figure 11. Schematic picture of electro-leaching in a Ti basket [60]; published by Elsevier, 2020.

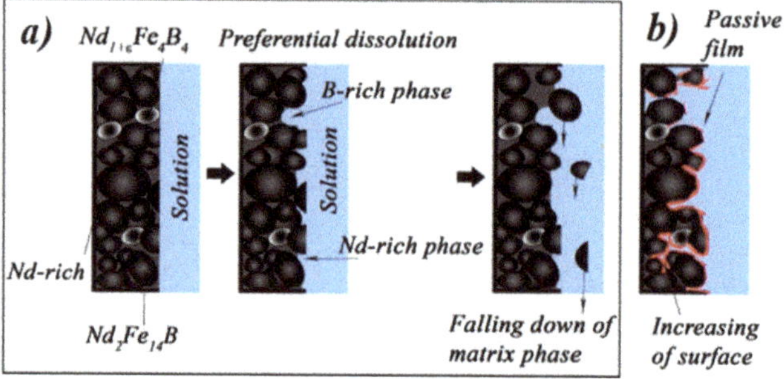

Figure 12. Schematic illustration of leaching process on the surface of NdFeB magnet scrap in (**a**) H_2SO_4, (**b**) mixture of H_2SO_4 and $H_2C_2O_4$ [60]; published by Elsevier, 2020.

Auerbach et al. [82] suggested recovering REEs from NdFeB magnet scrap by means of bioleaching with various bacteria. The *Acidithiobacillus ferrooxidans* and *Leptospirillum ferrooxidans* was confirmed to have highest leaching efficiencies with the leaching of Dy, Nd, and Pr of 86%, 91%, and 100%, respectively. However, due to the non-selective leaching of bacteria, further separation and purification processes should be carried out. Precipitation with concentrated $H_2C_2O_4$ and a two-step extraction process with the ionic liquid Cyphos IL 101 and subsequent treatment with DEHPA are considered to be the most effective methods. Extraction rates up to 100% with a purity of 98% were achieved.

The complete leaching process features a simple and easy operation, but the main shortcomings of this process include relatively high consumption of leaching agent and subsequent difficulties in removing impurities.

4.2. REEs Separation Technologies Used for NdFeB Permanent Magnet Scrap

Leaching liquor obtained through various leaching processes should be treated by different separation techniques such as precipitation, solvent extraction, and ionic liquids extraction to selectively separate REEs from the liquor [83,90]. Note that precipitation and solvent extraction can also use a combination of these. A brief comparison of these separation technologies is shown in Table 3.

Table 3. Comparison of different REEs separation technologies for NdFeB permanent magnet scraps.

Methods.	Advantages	Disadvantages
Precipitation	Low cost The process is simple	Low recovery Product is impurity Hard to obtain single REE
Solvent extraction	High recovery Can obtain high purity single REE	High cost The process is complicate Generation of large amount of waste
Ionic liquids extraction	High recovery Can obtain high purity single REE Efficient Environmentally friendly	High cost It is difficult to prepare ionic liquids system

4.2.1. Precipitation Process

The precipitation methods could be classified into two categories, namely selective precipitation and co-precipitation. According to the leaching process, the principle flowsheets of the precipitation processes are shown in Figure 13.

The resulting leaching solution is treated by a precipitation method with pH changes to separate Nd. Nd can usually be precipitated by direct methods such as fluoride method with HF and oxalate method with $H_2C_2O_4$ [83,91]. It has been found that strongly acidic solvents are suitable for the dissolution of NdFeB magnet scrap. However, these solvents cannot produce precipitation of the Nd compound. Therefore, some precipitating agents for Nd is needed. NaCl, ethanol and $H_2C_2O_4$ were used as precipitating agents by Itakura et al. It was indicated that these precipitating agents led to the formation of insoluble Nd compounds. However, Fe was contained in the precipitate when H_2SO_4/NaCl or H_2SO_4/ethanol were used as precipitating agents. But in a mixed aqueous solvent of 3.0 mol/L HCl and 0.2 mol/L $H_2C_2O_4$ via hydrothermal treatment at 110 °C for 360 min. About 99% of Nd contained in the magnet scrap was recovered as $Nd_2(C_2O_4)_3 \cdot xH_2O$ with a purity of 99.8% [56]. Mechano-chemical treatment with a mixed aqueous solution of HCl and $(COOH)_2$ has been proved to be an efficient method that can selectively recover REEs as oxalate from NdFeB magnet scrap without external heating. The optimal concentrations of HCl and $(COOH)_2$ were found to be 0.2 mol/L and 0.25 mol/L, respectively. The recovery and purity of REEs were 95.3% and 95.0%, respectively [92]. Over 99% of Nd can be recovered from a Fe-Nd solution by using H_3PO_4 and ascorbic acid via selective precipitation process [31]. Nd also can be recovered as Mn_2O_3-type Nd_2O_3, with the recovery of Nd of 97% via oxalic acid precipitation from NdFeB magnet scrap [74].

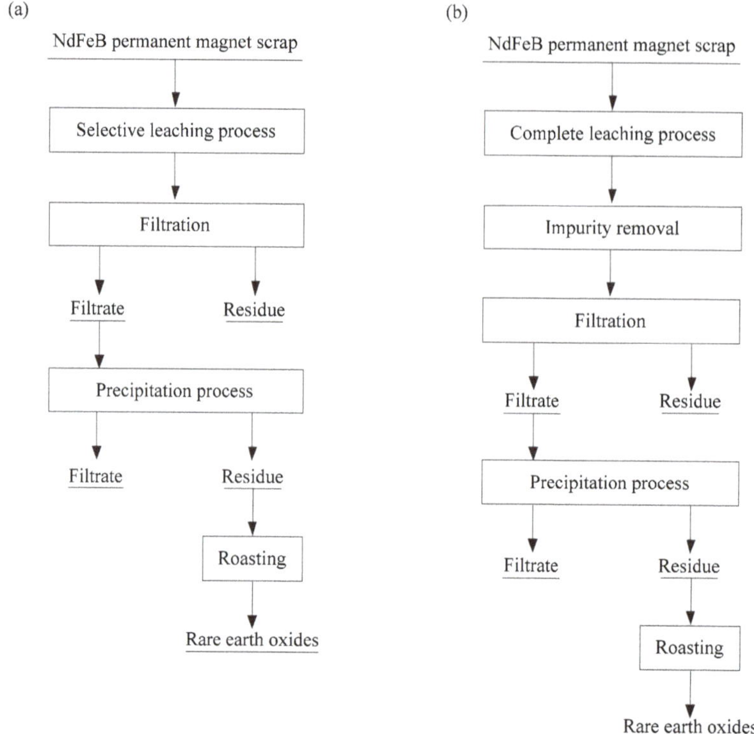

Figure 13. Principle flow of precipitation processes. (**a**) selective precipitation (**b**) co-precipitation.

Rabatho et al. tried to recover REEs from NdFeB magnetic waste sludge via selective precipitation process. Fe impurity in solution was first removed as $Fe(OH)_3$ with losses of Nd and Dy of 22.50% and 23.65%, respectively, by addition of NaOH solution to control the solution pH of 3, then Nd and Dy were precipitated by addition of $H_2C_2O_4$. 91.5% of Nd and 81.8% of Dy were recovered from the solution. After roasting the Nd and Dy containing precipitate at 800 °C, an Nd_2O_3 product with purity of 68% could be obtained, and the final recovery of Nd and Dy of 69.7% and 51% achieved, respectively [73]. Similarly, Kikuchi et al. researched precipitation of Fe^{3+} from NdFeB magnet scrap HNO_3 leaching solution by adjusting pH to 4.3 with the addition of NaOH [93]. As shown in Figure 14, the REEs were selectively dissolved from a crushed and roasted NdFeB magnet by acid, then purified by solvent extraction and precipitated as pure REE oxalate salt [56].

Although these methods can effectively separate Nd from other metals, due to the difficulty of filtering NdF_3 (because of the addition of HF), the production cost of Nd oxalate is high and non-selective, so it is not preferred. Double salt precipitation ($Nd_2(SO_4)_3$, $Na_2SO_4{\cdot}6H_2O$) may be an option for the precipitation of Nd with NaOH, and the REEs can be separated from Fe inexpensively without filtration problems. Nd was successfully separated from the optimized H_2SO_4 leaching solution and Nd was separated in the form of heavy salt precipitation with a heavy salt content of 75.41% [89]. Although the composition of NdFeB magnet scrap is complex, the combination of different steps makes it possible to obtain a relatively pure rare earth double salt product (98.4%) with a high recovery by using the precipitation method [82,94].

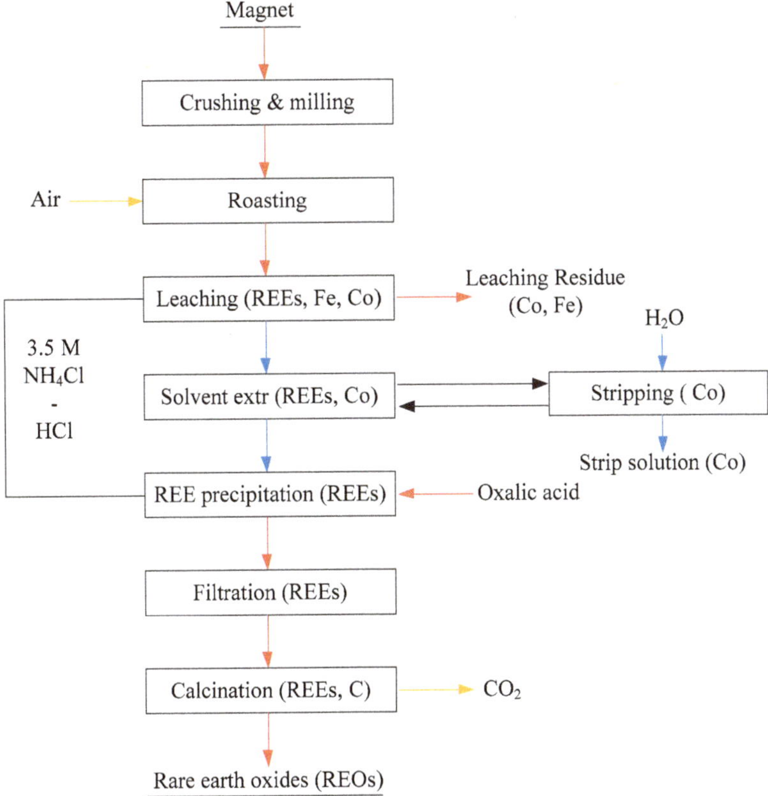

Figure 14. Flow chart of the recycling scheme. Yellow arrows: gas steams, blue arrows: aqueous streams, black arrows: ionic liquid steam, and red arrows: solid streams [56]; published by Royal Society of Chemistry, 2014.

Compared with the selective precipitation method, the co-precipitation method can simultaneously recover the valuable elements in the NdFeB magnet leaching solution [37,95,96]. Using this method, a composite powder containing REEs, Fe and Co can be obtained. The powder can be directly used as raw material for the preparation of all types of NdFeB magnetic powder with controlling to add desirable elements. Based on the thermodynamic simulation and calculation of MATLAB in the $H_2C_2O_4$-NH_3 co-precipitation system, an effective co-precipitation route (see Figure 15) was carried out to obtain a composite powder containing more than 99.4% of valuable elements, such as Fe_2O_3, $Fe_2O_3{\cdot}Nd_2O_3$, and Pr_2CoO_4. These valuable elements can be directly used as raw materials for the preparation of recycled NdFeB magnetic powder [37].

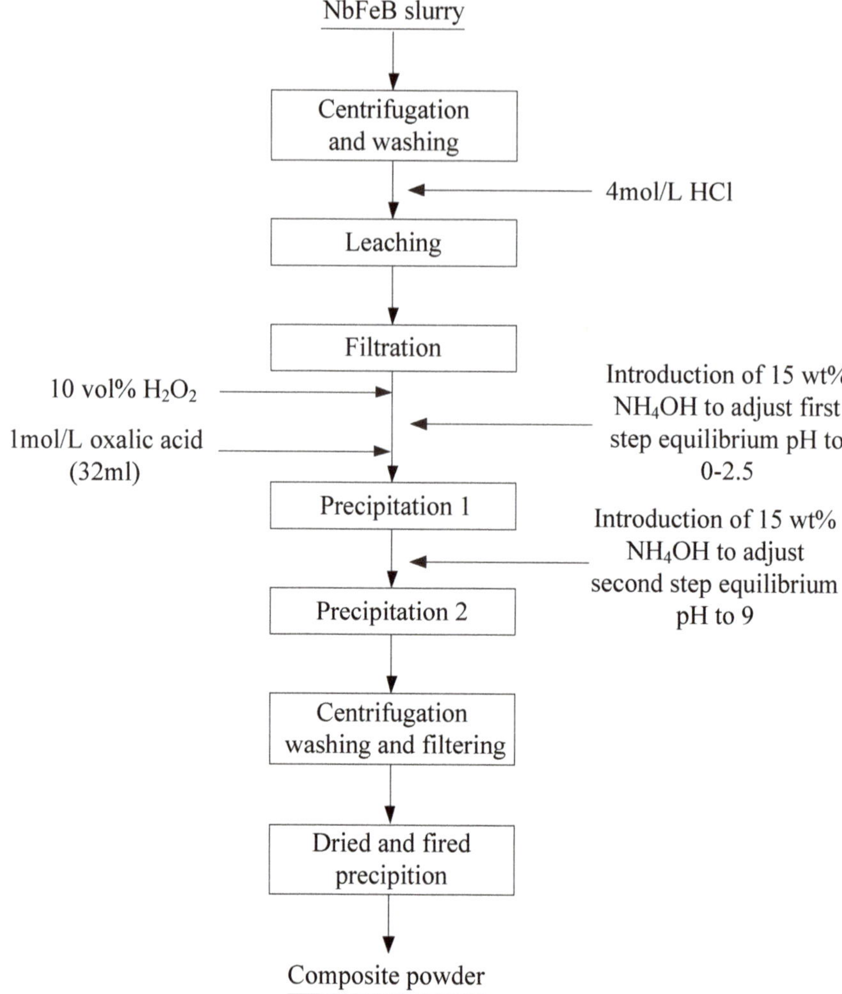

Figure 15. Scheme for the synthesis of a composite powder from NdFeB magnet scrap leaching solution using a co-precipitation method [37]; published by Elsevier, 2014.

4.2.2. Solvent Extraction Process

Because of the chemical similarity of REEs, separation of each other from different sources is one of the greatest challenges in the recycling of REEs [97,98]. Among various separation techniques, solvent extraction can be an effective method for separating and extracting individual metal or obtaining mixed solutions and compounds. Solvent extraction is a process for the selective extraction of a target component from an aqueous solution with one or more water immiscible organic reagents. To develop feasible and eco-friendly processes, extensive studies have been carried out for the extraction of various REEs by solvent extraction process using different organic extractants. Corresponding salient features of extractants are presented in Table 4. According to the solvent extraction mechanism, the solvent extraction process could be classified into four categories, including acidic solvent extraction (cationic solvent extraction), ion-pair solvent extraction (anion solvent extraction), neutral solvent extraction, and synergetic solvent extraction [99]. The properties of their extraction and separation, the formation of complex, separation factor, etc., have been highlighted as given below.

Table 4. Different organic solvents used for the extraction of REEs.

Reagents Class	Extractants	Chemical Name	Structure
Acidic extractant	D2EHPA	Di-2-ethylhexyl phosphoric acid	$(CH_3\text{-}(CH_2)_3\text{-}CH(C_2H_5)\text{-}CH_2O)_2P(=O)OH$
	EHEHPA	2-ethylhexylphosphonic acid mono-2-ethylhexyl ester	$CH_3\text{-}(CH_2)_3\text{-}CH(C_2H_5)\text{-}CH_2O\text{-}P(=O)(OH)\text{-}CH_2\text{-}CH(C_2H_5)\text{-}(CH_2)_3\text{-}CH_3$
	PC88A	2-Ethylhexyl phosphonic acid mono 2-ethylhexyl ester	$CH_3\text{-}(CH_2)_3\text{-}CH(C_2H_5)\text{-}CH_2O\text{-}P(=O)(OH)\text{-}CH_2\text{-}CH(C_2H_5)\text{-}(CH_2)_3\text{-}CH_3$
	Cyanex 302	Bis(2,4,4-trimethylpentyl) monothiophosphinic acid	S, P, OH
	Cyanex 272	Di-2,4,4,-trimethylpentyl phosphinic acid	$(CH_3\text{-}C(CH_3)_2\text{-}CH_2\text{-}CH(CH_3)\text{-}CH_2)_2P(=O)OH$
Anion extractant	Aliquat 336	Tri-octyl methylammonium chloride	$(R_1, R_2{=}C_6)$ H_3C, H_3C, N^+, H_3C, CH_3, Cl^-
Neutral extractant	TBP	Tri-n-butyl phosphate	$(CH_3\text{-}(CH_2)_3O)_2P(=O)O(CH_2)_3CH_3$
	TODGA	Tetraoctyldigylcol amide	$(C_8H_{17})_2N\text{-}C(=O)\text{-}CH_2\text{-}O\text{-}CH_2\text{-}C(=O)\text{-}N(C_8H_{17})_2$

Acidic Solvent Extraction

Acidic solvent extraction uses weak organic acid as the extractants. Acidic extractants are used to extract and separate REEs because they form cationic species in aqueous solution. The general extraction mechanism shows as the following equation [99]:

$$M^{n+} + nHA = MA_{n(org)} + nH^{+}{}_{(aq)} \quad (1)$$

Various acidic organophosphorous extractants have been used in REEs extraction processes. D2EHPA, di-(2-ethylhexyl) phosphoric acid (HDEHP), PC88A and EHEHPA are the most widely used solvents [100,101]. HCl, H_2SO_4 and HNO_3 were utilized as extraction media for the extraction of Nd, Eu and Tm using D2EHPA as an extractant. The results indicated that the equilibrium constants increased as the order: Nd < Eu < Tm [102,103]. Proximate researches confirmed that the extraction results of lanthanides with D2EHPA from HNO_3 were poorer than those from HCl and H_2SO_4 solutions [97]. The selectivity sequence for extracting REEs from 0.5 mol/L HCl solution with 0.75 mol/L D2EHPA in toluene was Lu > Yb > Tm > Tb > Eu > Pm > Pr > Ce > La, and the average separation factor of two adjacent REEs was 2.5 [104]. Although the REEs extraction efficiency of D2EHPA is very high, however, the difficulty of stripping the loaded extractant limits its utilization, especially for the extraction of heavy REEs. Lately, PC88A has attracted considerable attention to replace D2EHPA in the separation of REEs because of its higher separation factor for REEs (Table 5) [105]. Mohammadi et al. compared the separation of Nd, Dy and Y by D2EHPA and EPEHPA, the calculated separation factors at equilibrium pH = 1 indicated that D2EHPA was the most effective for separating Nd from Y and Dy, and the EHEHPA showed the highest separation factor for Y and Dy [106].

Table 5. Separation factors for extraction of rare earths by DEHPA and PC88A.

Rare Earths Pair	DEHPA	PC88A
Ce/La	2.98	6.83
Pr/Ce	2.05	2.03
Nd/Pr	1.38	1.55
Sm/Nd	6.58	10.60
Eu/Sm	1.90	2.30
Gd/Eu	1.43	1.50
Tb/Gd	0.93	5.80
Dy/Tb	2.40	2.82
Ho/Dy	1.90	2.00
Er/Ho	2.25	2.73
Er/Y	1.37	1.43
Tm/Er	2.90	3.34
Yb/Tm	3.09	3.56
Lu/Yb	1.86	1.78

The PC88A in kerosene exists in the form of dimer, and the extraction of Nd with PC88A can be expressed as the Equation (2) [107]. According to Lee et al., the extraction reaction of Nd with PC88A in chloride solution was identified by the graphical method as shown in Equation (3). They also observed that the distribution coefficients of Nd increased linearly with the equilibrium pH range of 0.62–1.01 [108].

$$Nd^{3+} + \frac{3+x}{2}H_2A_{2,org} = NdA_3xHA_{org} + 3H^{+} \quad (2)$$

$$Nd^{3+} + 1.5H_2A_{2,org} = NdA_3HA_{org} + 3H_{aq}{}^{+} \quad (3)$$

where H_2A_2 and org represents the PC88A dimer and the organic phase, respectively.

The acidic extractants saponified with NaOH was researched to overcome the adversely affects caused by acid liberated during the acidic extraction process [106,109–111]. Moreover, partial saponification was preferred to avoid gel formation and the solution of the saponified extractant in the aqueous phase. Compared with Nd distribution coefficients of PC88A, the use of 40% saponified PC88A significantly improved the extraction of Nd [108]. A comparative study between Cyanex 302, PC88A and Cyanex 272, NaCyanex 302, NaPC88A and NaCyanex 272 for extracting Nd from a chloride solution has been conducted by Padhan et al. [33]. It was found that Cyanex 272 had the highest extraction rate and Cyanex 302 showed the lowest extraction rate. It is reported that the synthesized Primene 81R·Cyanex 572 ionic liquids can overcome the shortcomings of Cyanex 272 and Cyanex 572 of the sensitivity to pH value. 99.99% of Nd can be extracted from Nd/Tb/Dy containing aqueous by two stages counter-current extraction process with 0.30 mol/L Primene 81R·Cyanex 572 ionic liquid, and without pH adjusting [112]. Because saponified Cyanex 302 was an effective commercial reagent for extracting Nd, Padhan et al. [111] used it as the extractant to separate Nd and Dy. A maximum separation factor (D_{Dy}/D_{Nd}) of 53.65 was observed at pH = 1.2. Extraction of Dy was 98% with co-extraction of Nd only 7.22% after two stages of counter current extraction in 0.125 mol/L NaCyanex 302 at A:O = 1:1. Then, 99.79% of Nd can be recovered with 0.2 mol/L NaCyanex 302 in two counter current extraction stages at A:O = 1:1.

Ion-Pair Solvent Extraction

Ion-pair solvent extraction is effective in the presence of strong anionic ligands because of ion-pair solvent extraction metal ions as anionic complexes. Ion-pair solvent extract mainly nitrogen and oxygen containing organic compounds, such as tri-alkyl methylamine (Primene JMT) and tri-octylmethylammonium nitrate (Aliquat 336).

According to early work, separation factors for adjacent REEs with primary or tertiary amines were higher in sulfate media than in chloride media, so that sulfate media was more promising for ion-pair solvent extraction [113,114]. El-Yamani and Shabana indicated that the extraction of La from sulfate solutions with Primene JMT was extracted according to the following Reactions (4) and (5) [115]. The extraction reaction of quaternary ammonium salts could be also simply represented as Reaction (6) [116,117] where RNH_2 denotes the Primene JMT in the organic phase, Ln denotes the rare earth ion and $R_4N^+NO_3{}^-$ the quaternary ammonium nitrate salt.

$$2\overline{RNH_2} + H_2SO_4 = \overline{(RNH_3)_2SO_4} \quad (4)$$

$$2La(SO_4)_2{}^{3-} + 3\overline{(RNH_3)_2SO_4} = 2\overline{(RNH_3)_3La(SO_4)_3} + 3SO_4{}^{2-} \quad (5)$$

$$Ln^{3+} + 3NO_3{}^- + x\overline{(R_4N^+NO_3{}^-{}_n} = \overline{Ln(NO_3)_3\, xnR_4N^+NO_3{}^-} \quad (6)$$

Amines is mainly used for the separation of Pr, Nd, Y and heavy lanthanoids [117]. Lu et al. [118] obtained a high purity (>99%) of Nd with recovery of 95% from didymium nitrate solution by Aliquat 336 in a 45-stage tube-type mixer-settler. Another study described a solvent-extraction process for the recovery of magnet-grade Nd_2O_3 from a light rare earth nitrate liquor used a 0.50 mol/L Aliquat 336 nitrate in Shellsol AB in 8 extraction and 6 scrubbing stages [119].

Neutral Solvent Extraction

Various neutral extractants have been used for REEs separations, and it has been confirmed that TODGA is a promising extractant [120] and could be used for the extraction of REEs from the NdFeB magnet scrap leaching solution. TODGA forms a strong tridentate complex with metal ions, and, compared with other ions in aqueous solution, it has previously exhibited particularly good extraction performance for lanthanides and action elements in terms of the selectivity [121].

The effect of the diluent on the extraction and the selectivity of the TODGA was researched by Gergoric et al. [52]. The efficiency of the diluents decreases in the following order: hexane < cyclohexanone < Solvent 70 < toluene < 1-octanol. With the exception of cyclohexanone, the distribution ratio of extractable substances decreases with the polarity of the diluent [122]. It was also found that, in all diluents, the distribution ratio of REEs increased with TODGA concentration increased. A supported liquid membrane processes for extracting REEs from NdFeB magnets with TODGA or Cyanex 923 as extractants were evaluated to define the distribution coefficient and selectivity of Nd and Dy, respectively. It was found that TODGA has superior selectivity in REEs recovery than Cyanex 923. REEs solution in HNO_3 showed higher distribution coefficients and selectivity than in HCl. Lower molar concentration of HNO_3 in the strip resulted in higher recovery of Nd [123].

Synergetic Solvent Extraction

The phenomenon that the distribution coefficient of some extracted substances is greater than the sum of the partition coefficients when two or more extractants are used alone under the same conditions is called the synergistic effects. Many types of synergistic solvent extraction systems for extracting and separating REEs have been reported, including mixtures of acidic extractants, mixtures of neutral extractants, and combinations thereof [124–127].

In the past ten years, the use of different mixtures of acidic organophosphorous extractants to improve the extraction efficiency and selectivity of REEs has attracted attention. Extraction of REEs with a mixture of D2EHPA and EHEHPA was reported to be a promising method, which could not only decrease the acidity required for stripping the loaded D2EHPA but also increase the extraction efficiency of EHEHPA [128,129]. Higher selectivity and extraction efficiency of REEs can be achieved by using mixtures of D2EHPA and EHEHPA. The extraction reaction of REE with cationic extractants D2EHPA and EHEHPA, represented as Reaction (7) [130,131].

$$REE^{3+} + mH_2A_2 = REEA_3(HA)_{2m-3} + 3H^+ \quad (7)$$

where H_2A_2 represents the dimeric form of the extractant.

It was suggested that the improvement of the extraction capacity of the mixed system may be attributed to the breakage of the dimers of D2EHPA and EHEHPA [132,133]. As the HCl concentration increases, with the exception of the cationic exchange reaction in Reaction (7), a solvating reaction has been proposed, as shown in Reaction (8) [131].

$$REE^{3+} + 3Cl^- + mH_2A_2 = REECl_3(HA)_{2m}, \quad (8)$$

Extraction of Nd, Dy, and Y from HCl solution by using D2EHPA, EHEHPA and their mixtures were investigated by Mohammadi et al. [105], the extraction order in general was found to be Y > Dy > Nd. The extraction efficiency of Y, Dy, and Nd increased with the extractant concentration increased and the acidity decreased. Mixture of EHEHPA and D2EHPA improves the extraction of Nd when extractants concentration of 0.15 mol/L and equilibrium pH = 1. Separation factors calculated for an equilibrium pH of 1 show that at low concentration of extractant (0.06 mol/L and 0.09 mol/L) the mixture of EHEHPA and D2EHPA beneficial to separate Dy and Y. The REE distribution ratios and EHEHPA/D2EHPA ratios of the two extractants and their mixtures at different extractant concentrations indicate that the stoichiometrically different complexes of Dy, Nd and Y depend on the concentration of REEs and D2EHPA concentration.

The synergistic extraction of Nd from HNO_3 medium using the mixture of Cyanex 272 and Cyanex 921/Cyanex 923 (B) has been studied by Panda et al. [134]. The extraction of Nd from 0.001 mol/L HNO_3 using 0.6 mol/L Cyanex 272 in kerosene was 95.5%. It was found that the calculated synergy coefficient for the extraction of Nd using a mixture of 0.1 mol/L Cyanex 272 and Cyanex 923 was higher than that for the extraction of a mixture of 0.1 mol/L Cyanex 272 and Cyanex 921 in 0.001 mol/L HNO_3 solution.

To increase the extraction of Pr and Nd by Cyanex 272, the mixture of Cyanex 272 with Alamine 336, TOA or TEHA have been employed from chloride solution [135]. Among those mixture systems, the Cyanex 272 and Alamine 336 mixture performed the highest synergism enhancement factors for Pr (14.2) and Nd (10.1). The extraction reaction of Pr and Nd with the mixture of Cyanex 272 and Alamine 336 progresses gradually. At first, the REEs were extracted by Cyanex 272 with a cationic exchange mechanism (Equation (9)). Secondly, during the extraction process, the protons released by Cyanex 272 were simultaneously extracted into organic matter through Alamine336. The overall extraction reaction of Pr and Nd with the mixture of Cyanex 272 and Alamine 336 could be written as Equation (10).

$$RE^{3+} + Cl^- + 2H_2A_2 = REClA_2{\cdot}2HA + 2H^+ \ (RE = Pr, Nd) \tag{9}$$

$$RE^{3+} + (x+1)Cl^- + 2H_2A_2 + xNR_3 = RECl(HA_2)_2 + xR_3NHCl + (2-x)H^+ (RE = Pr, Nd) \tag{10}$$

where the value of x ($0 \leq x \leq 2$) is dependent on the concentration ratio of Alamine 336 to the chloride ion.

Hence, it can be concluded that the chloride ion concentration in the aqueous phase is significantly affecting the extraction of Nd and Pr in the mixture of Cyanex 272 and Alamine 336.

A similar experiment was performed on the extraction of Nd with a mixture of Cyanex 302 (HB) and Alamine 308 by Kumar et al. The mixture showed a significant synergistic effect with a synergistic factor of 44.1. Raction for extracting Nd with this mixture is shows in Equation (11) [136].

$$Nd^{3+} + Cl^- + (HB)_2 + 2R_3N = NdClB_2{\cdot}2R_3N + 2H^+ \tag{11}$$

Because of the extractants used in synergistic systems are generally consist of a mixture of acidic and neutral extractant, the acids released during the extraction of REEs with these acidic extractants can adversely affect extraction. Thus, a present work was directed to study the extraction behavior of La and Nd from HNO_3 solution using a mixture of TOPO and TRPO neutral extractants in kerosene to keep the best conditions to recover and separate REEs. Extraction of Y from HNO_3 medium with a mixture of neutral organophosphorus reagents gave synergistic behavior and forming a neutral complex of the form of $\overline{M(NO_3)_3(TOPO)(TRPO)}$, and the reaction in case of synergism could be written as Equation (12) [137]. The extraction order for the REEs studied was Nd > Y > La.

$$M^{3+} + 3NO_3^{\ -} + \overline{TOPO} + \overline{TRPO} = \overline{M(NO_3)_3(TOPO)(TRPO)} \tag{12}$$

4.2.3. Ionic Liquids Extraction Process

Ionic liquids (ILs) have the ability to solvate many compounds [138]. Meanwhile, ILs was considered to be environment friendly solvents with the potential to replace traditional organic solvent [139]. Recently, emphasis has been placed on such solvents in cleaner production processes [140,141].

Non-Functional Ionic Liquids

ILs made with imidazolium salts, which are denoted as $[C_n mim]$, were the first potential ILs to be used for extracting REEs. $[C_8mim][PF_6]$ was first tested for separating Ce (IV) from HNO_3 solutions containing La (III) and Th (IV) [142]. It was found that the distribution ratios of Ce (IV) and Th (IV) using pure $[C_8mim][PF_6]$ as the extract phase was similar to that using HDEHP or DEHPA in n-heptane, which indicated that $[C_8mim][PF_6]$ is possible to act as both extractant and diluent.

In traditional organic solvents, metal ions are always extracted with a neutral extractant together with their counter anions, which is called the neutral mechanism. However, the cation exchange mechanism, an anion exchange mechanism, or neutral mechanism can appear in the IL system.

A general equation of the cation exchange mechanism can be written as in Equation (13) or demonstrated as in Figure 16a, where L is the extractant, M is the REE, and C is the cation of IL.

$$M^{m+} + mL + mC_{IL}{}^{+} = ML_{m,IL}{}^{m+} + mC^{+} \quad (13)$$

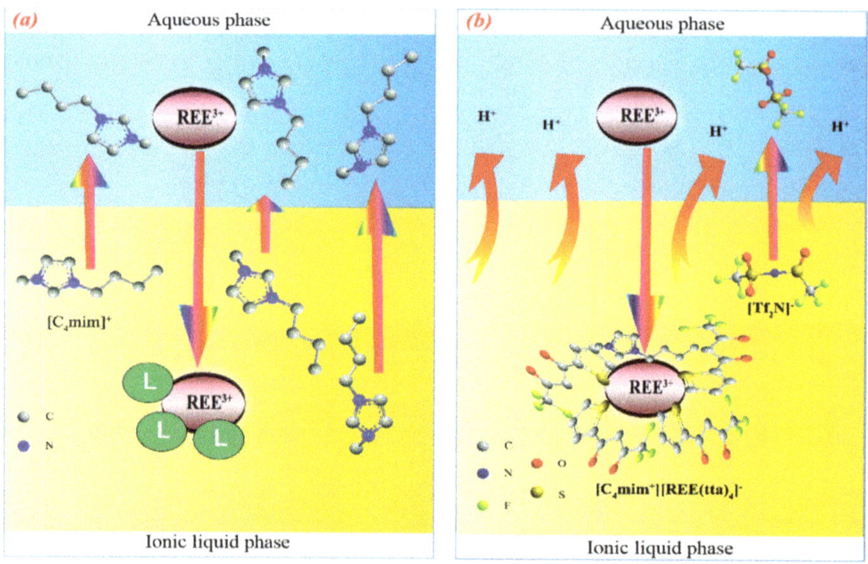

Figure 16. (**a**) Cation exchange in an ionic liquids (IL) system, (**b**) Anion exchange in IL system [143]; published by Royal Society of Chemistry, 2017.

The cation exchange mechanism largely depends on the hydrophobic character of ILs, Nd^{3+} extraction with Cyanex 923 in five ILs with the same anion but different cations showed that the extraction efficiency of Nd^{3+} in ILs with a small hydrophilic cation such as $[C_4mim][Tf_2N]$ and $[N_{1444}][Tf_2N]$ was higher than that in ILs with a hydrophobic cation such as $[C_{10}mim][Tf_2N]$, $[P_{66614}][Tf_2N]$, or $[N_{1888}][Tf_2N]$ [144]. In other words, hydrophilic cations promote cation exchange, while hydrophobic cations inhibit cation exchange.

Anion exchange was observed in the biphasic aqueous/IL system by Jensen et al. $[C_4mim][Tf_2N]$ functioned via liquid anion exchange mechanisms accelerates the formation of $REE(tta)_4{}^{-}$ in the IL phase, while $REE(tta)_3$ were generally formed in organic solvents or cationic complexes, $REE(tta)_2{}^{+}$, as observed in previously reported IL systems. $[Tf_2N]^{-}$ was transferred to the aqueous phase to keep the charge neutrality, and $[C_4mim^{+}][REE(tta)_4{}^{-}]$ be part of the IL phase without greatly changing the general structure of the IL [145]. The full equilibrium could be described as in Equation (14), and the details of an anion exchange mechanism in an IL, as shown in Figure 16b.

$$REE^{3+} + 4Htta + [C_4mim][Tf_2N]_{IL} = [C_4mim+][REE(tta)_4{}^{-}]_{IL} + 4H^{+} + [Tf_2N]^{-} \quad (14)$$

Part of the ionic liquid components in the anion exchange system will still be transferred to the water phase, which will contaminate the water phase and consume IL. Therefore, the neutral mechanism opposite to the conventional organic solvent mechanism can also be used in the IL system. According to Kubota et al. [146] the mechanism in $[C_nmim][Tf_2N]$/DODGAA is the same as that in n-dodecane, that is, protons exchange in the reaction between DODGAA and REEs. The extraction equation of REE^{3+} ions with DODGAA (HL) is given in Equation (15) and Figure 17.

$$M^{3+} + 3HL_{IL} = ML_{3,\,IL} + 3H^{+} \tag{15}$$

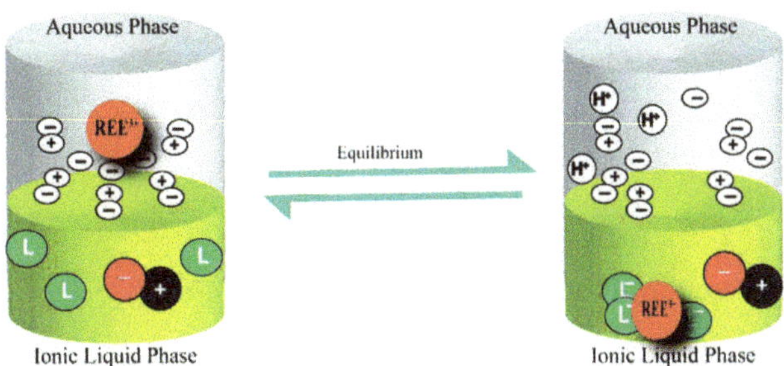

Figure 17. Neutral mechanism in non-functional IL-based extraction system [143]; published by Royal Society of Chemistry, 2017.

The same group used mainly $[C_8mim][Tf_2N]$ with DODGAA as the extractant for extraction of a series of REEs, the results showed that all the REEs can be extracted at low pH from a H_2SO_4 solution. The ILs were confirmed to be applicable for the separation of REEs from a variety of resources [147]. Finally, they used $[C_4mim][Tf_2N]$/DODGAA to recover REEs from waste fluorescent lamps, the $[C_4mim][Tf_2N]$/DODGAA showed a high affinity for REEs in liquid-liquid extraction [148]. $[C_4mim][Tf_2N]$/DODGAA showed to be a promising system for recovering REEs from fluorescent lamps or NdFeB magnets containing large amounts of Zn and Fe.

Functional Ionic Liquids

Since one of the most important characteristics of IL is to improve its chemical and physical properties by combining appropriate anion/cation pairs, functional groups can be introduced into anionic or cationic compositions. IL with functional groups is called functional IL, and due to its unique chemical and physical properties, it has attracted much attention in the metal recycling process [149–152].

The ionic liquids $[C_6mim]$[DEHP], $[C_6mpyr]$[DEHP] and $[N_{4444}]$[DEHP] with bis(2-ethylhexyl)phosphate anions were prepared and the extraction behavior of Nd with the ILs extraction in HNO_3 medium were studied. Compared with the quaternary ammonium analog $[N_{4444}]$[DEHP], except the extraction properties of Nd in the ionic liquids $[C_6mim]$[DEHP] and $[C_6mpyr]$[DEHP] were significantly different, the extraction efficiency of Nd can exceed 99%, and the extraction process can be selected by the ionic liquid cation [148]. In another research, because the functional phosphate moiety of DEHPA, Nd^{3+} was extracted by $[C_6mim][Tf_2N]$/DEHPA and $[C_6mpyr][Tf_2N]$/DEHPA systems, the Equation (16) can be proposed for the extraction mechanism of Nd by DEHPA in $[C_6mim][NTf_2]$ and $[C_6mpyr][NTf_2]$ [153]. These non-functional ILs only acted as dilutants and were not included in the extraction mechanism, as shown in Figure 18. When the functional IL $[C_6mim]$[DEHP] was used to extract Nd^{3+}, both the anion and cation were involved in the extraction so that neither cation nor anion was lost to the aqueous phase.

$$Nd^{3+} + \overline{3DEHPA} = \overline{Nd(DEHP)_3} + 3H^{+} \tag{16}$$

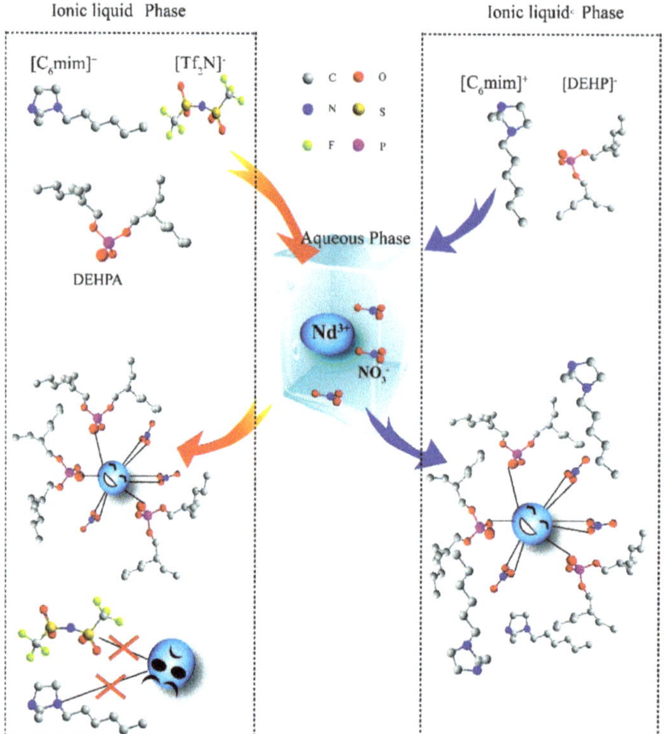

Figure 18. Different Nd^{3+} extraction behaviors in $[C_6mpyr][Tf_2N]$/DEHPA and $[C_6mim][DEHP]$ systems [143]; published by Royal Society of Chemistry, 2017.

The use of these type of functionalized ionic liquids for separating REEs indicated that the light REEs La and Ce are poorly extracted, however, Nd and Yb are strongly extracted by the $[C_6mim][DEHP]/[C_6mim][NTf_2]$ and $[C_6mpyr][DEHP]/[C_6mpyr][NTf_2]$ ionic liquid extraction systems. A separation factor of 340 is obtained for Nd over La in the $[C_6mim][DEHP]$. Pr is weakly extracted by the $[C_6mpyr][DEHP]/[C_6mpyr][NTf_2]$ system, but rather strongly by the $[C_6mim][DEHP]/[C_6mim][NTf_2]$ system. Overall, these DEHPA-based ILs were suitable for recovering Nd from NdFeB magnet scrap.

A new recycling process for NdFeB magnets was carried out by Dupont and Binnemans, based on the carboxyl-functionalized ionic liquid: betainium bis(trifluoromethylsulfonyl)imide, $[Hbet][Tf_2N]$, a combined leaching/extraction step was proposed, and the detailed recycling process as shown in Figure 19 [36]. First, roasting of NdFeB magnets convert all elements in the magnets to their respective oxides, because these oxides are more easily dissolved in $[Hbet][Tf_2N]$-H_2O systems. The leaching of the roasted NdFeB was tested using a 1:1 wt/wt $[Hbet][Tf_2N]$-H_2O at 80 °C to form a homogeneous phase. After that, the solution was cooled to 25 °C, the mixture could be separated into a valuable REEs/Co-rich aqueous phase and an iron-rich ionic liquid phase. The two separated phases were stripped with $H_2C_2O_4$ to remove the Fe and produce a REEs/Co oxalate. Then, aqueous ammonia was added to separate REEs and Co. The obtained REEs oxalate with a purity higher than 99.9% and can be calcined to obtain Nd_2O_3 and Dy_2O_3 mixture, which is the precursors of NdFeB magnets. The stripping step can automatically regenerate the ionic liquid, and the ionic liquid recovery was also considered. Therefore, the proposed closed-loop system only generates little waste and offers selectivity, which makes this a promising green method for recovering NdFeB magnets [149].

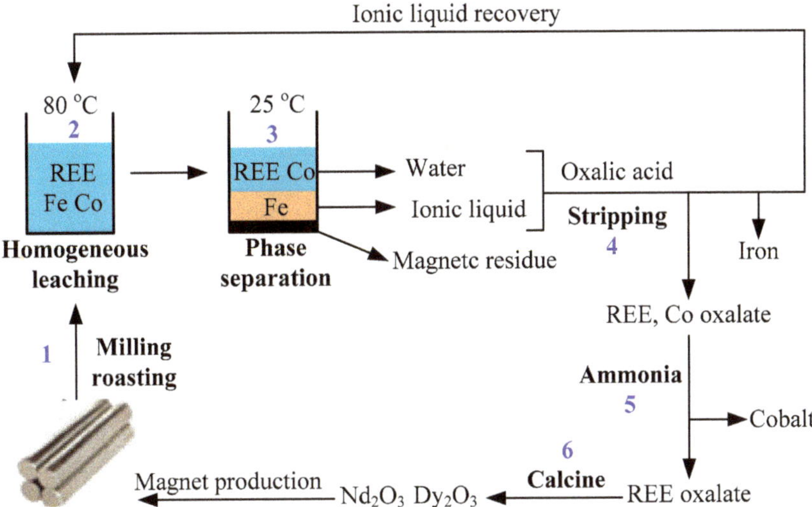

Figure 19. Overview of the proposed recycling process for roasted NdFeB magnets [36]; published by Royal Society of Chemistry, 2015.

Bi-Functional Ionic Liquids

Ionic liquid extractants prepared from ammoniumand phosphonium are considered to be bi-functional IL extractants (Bif-ILs) because the cations and anions of ILs are involved in the extraction [150].

Rout et al. [148] synthesized two ionic liquids derived from Aliquat 336: trioctylmethylammonium bis(2-ethylhexyl)phosphate, [A336][DEHP], and trioctylmethylammonium bis(2-ethylhexyl)diglycolamate, [A336][DGA]. These ionic liquids were applied to the separation of Eu from Am. The extraction of Eu in these Bif-ILs showed a strong dependence on the properties of molecular diluent used. In their further researcher [151], [A336][DGA] combined with [A336][NO_3] to extract Nd from La ions, and the Nd extraction behavior of [A336][DGA] in [A336][NO_3] was compared with that of [A336][DGA] in the Cl-containing ionic liquid diluent [A336][Cl]. The results shown that extraction of Nd can reach nearly 100%, and the nitrate media was found to be more suitable for extracting Nd than in chloride media. The extraction mechanism in a system with the ionic liquid diluent [A336][NO_3] is significant different from that of extraction systems with molecular diluents. A rational mechanism for extracting Nd by [A336][DGA] in [A336][NO_3] at pH 2–5 with 0.1 M salting-out agent in the feed phase is shown as Equation (17).

$$Nd^{3+} + 3NO_3^{-} + \overline{[A336][DGA]} = \overline{Nd(NO_3)_3 \bullet [A336][DGA]} \tag{17}$$

A comparative study was carried out between Bi-ILs [A336][CA-12] (tricaprylmethylammonium secoctylphenoxy acetic acid)/[A336][CA-100] (tricaprylmethylammonium secnonylphenoxy acetic acid), organic carboxylic acids CA-12/CA-100 and neutral organophosphorus extractants TBP/P350 for extracting REEs in HNO_3 medium [148]. The efficiency of different extractants for metal extraction follows the order: [A336][CA-12]/[A336][CA-100] > CA-12/CA-100 > TBP/P350. Extraction and separation for REEs from chloride medium used [A336][CA-12] and [A336][CA-100] as extractants have also researched by Wang et al. [152] The results showed that at the same conditions, the extraction capacity of [A336][CA-12] and [A336][CA-100] was higher than that of CA-12, CA-100, TBP, and P350. [A336][CA-12] and [A336][CA-100] in the present of NaCl, can efficiently extraction of REEs at low acidity, avoiding the harm that the conventional acidity and neutral extractants produced in the extraction.

The inner synergistic effect of Bi-ILs extractants using [A336][P204] as an extractant for solvent extraction of Eu was reported. The distribution coefficients of Eu in [A336][P204], [A336][P507], [A336][CA-12], [A336][CA-100], [A336][Cyanex272] and corresponding mixtures of their precursors were carried out to explore whether there are similar inner synergistic effects in some other Bi-ILs [153]. As shown in Table 6, obviously, the distribution coefficients of the Bi-ILs are all higher than their mixed precursors, showing good inner synergistic effect of these Bi-ILs [154]. HDEHP and HEH [EHP] have been developed into 6 types acid-base coupling bifunctionalized ionic liquids (ABC-BILs) extractants. As with the mixture of HDEHP and HEH[EHP], the combined [DEHP]2 type ABC-BILs and [EHEHP]2 type ABC-BILs revealed synergistic extraction effects for REEs in 7 different ABC-BILs combinations. The synergy coefficients of REEs also confirmed the synergistic extraction effects from combined ABC-BILs [155].

Table 6. Comparison of the distribution ratios obtained using mixed extractants and Bi-ILs in hydrochloric acid and nitric acid media [154]; published by Elsevier, 2010.

Extractant	Hydrochloric Acid Media		Nitric Acid Media	
	D	D_{IL}/D_{mix}	D	D_{IL}/D_{mix}
A336 + P204	0.071	27.86	0.912	23.79
[A336][P204]	1.97	-	21.7	-
A336 + P507	0.192	13.28	0.269	70.63
[A336][P507]	2.55	-	19	-
A336 + CA-12	0.169	3.04	0.067	19.79
[A336][CA-12]	0.513	-	1.33	-
A336 + CA-100	0.054	8.28	0.055	15.57
[A336][CA-100]	0.445	-	0.855	-
A336 + Cyanex272	0.928	2.55	0.912	3.44
[A336][Cyanex272]	2.37	-	3.14	-

It was indicated that the IL consisting the mixture of Aliquat 336 and Cyanex 272 showed good selectivity and extractability for REEs, therefore, a detailed study has been researched to extract Nd and Pr using Aliquat 336 based ionic liquid from the NdBFe magnet scrap leach liquor. A comparative study showed that the extraction efficiency of Bi-ILs trioctylmethylammoniumbis (2,4,4-trimethylpentyl)phosphate (R_4NCy) and trioctylmethylammonium di(2-ethylhexyl)phosphate (R_4ND) was higher than the conventional extractants Aliquat 336, Cyanex 272 and D2EHPA under the same conditions. The extraction efficiency of different extractants for Nd and Pr is: R_4NCy > R_4ND > Cyanex 272 > D2EHPA > Aliquat 336, with a maximum extraction of Nd and Pr of 98.97%, 99.02%, respectively.

5. Conclusions

NdFeB permanent magnet scrap is an important secondary resource that contains a number of valuable REEs. Recovery of REEs from this scrap via appropriate methods such as hydrometallurgical processes has both remarkable economic and environmental benefits. The authors reviewed the chemical, physical characteristics of NdFeB permanent magnet scrap and the main hydrometallurgical processes for recovering REEs from the magnet scrap, from the leaching process to the separation process. A variety of leaching technologies have been developed for REEs recovery depending upon their mineralogy, REEs occurrence and engineering feasibility. Both selective leaching and complete leaching are interactively used for leaching REEs from the magnet scrap, among which complete leaching treatment was found to be acceptable from industrial point of view. Although electrochemical selectively leaching seems to be a promising method, more systematic research is still needed. The REEs separation process mainly include precipitation process, solvent extraction process and ionic liquids extraction process. The precipitation process is a simple and easy separation process for REEs by different precipitating reagents; however, the problem of incomplete separation needs to be addressed. The leaching liquor generated is put to solvent extraction studies using different process such as acidic solvent extraction, ion-pair solvent extraction, neutral solvent extraction and synergetic solvent extraction. D2EHPA and PC88A have been considered feasible for recovering REEs from the leaching liquors. The physical properties of ionic liquids make them potentially valuable replacements for traditional organic solvents used in liquid-liquid separation processes. However, the extraction mechanisms of ionic liquids extraction are still indefinite, and the high price of most types of ionic liquids hinders its widespread use. Future research focus should be on the understanding of extraction mechanisms, which is very important for the better design of extraction systems from laboratory curiosities to industrial processes. Technically, for further optimization of treatment of the magnet scrap, hydrometallurgical recovery of REEs from the magnet scrap has to confront the challenges of many process steps required before obtaining REEs, consumption of large amount of chemicals, generation of large amount waste water and effluents. To resolve these issues, it is necessary to use high selective leaching agents, recycle the leaching agents and establish a closed-loop system. This review is expected to serve as a useful guideline for promoting treatment of NdFeB permanent magnet scrap by a hydrometallurgical processes.

Author Contributions: Conceptualization: Y.Z.; Investigation: F.G.; Data Curation: Z.S.; Formal analysis: S.L.; Writing—original draft preparation: F.G.; Writing—review and editing: Y.Z. and C.A.; Supervision: T.J. All authors have read and agreed to the published version of the manuscript.

Funding: This research was funded by the National Natural Science Foundation of China under Grant U1960114.

Conflicts of Interest: The authors declare no conflict of interest.

References

1. Jiles, D. *Introduction to Magnetism and Magnetic Materials*, 2nd ed.; Chapman & Hall: New York, NY, USA, 1998.
2. Coey, J.M.D. Permanent magnet applications. *J. Magn. Magn. Mater.* **2002**, *248*, 441–456. [CrossRef]
3. Shaw, S.; Constantinides, S. Permanent magnets: The demand for rare earths. In Proceedings of the 8th International Rare Earths Conference, Hong Kong, China, 13–15 November 2012.
4. Yang, Y.; Walton, A.; Sheridan, R.; Güth, K.; Gauß, R.; Gutfleisch, O.; Buchert, M.; Steenari, B.-M.; Van Gerven, T.; Jones, P.T.; et al. REE Recovery from end-of-life NdFeB permanent magnet scrap: A critical review. *J. Sustain. Metall.* **2016**, *3*, 122–149. [CrossRef]
5. Sagawa, M.; Fujimura, S.; Yamamoto, H.; Matsuura, Y.; Hiraga, K. Permanent magnet materials based on the rare earth-iron-boron tetragonal compounds. *IEEE Trans. Magn.* **1984**, *20*, 1584–1589. [CrossRef]
6. Peiró, L.T.; Méndez, G.V.; Ayres, R.U. Material flow analysis of scarce metals: Sources, functions, end-uses and aspects for future supply. *Environ. Sci. Technol.* **2013**, *47*, 2939–2947. [CrossRef] [PubMed]
7. Japan Oil Gas and Metals National Corporation. *Mineral Resources Material Flow*; Japan Oil Gas and Metals National Corporation: Tokyo, Japan, 2008. Available online: mric.jogmec.go.jp/public/report/2012-12/2012120122_REs.pdf (accessed on 29 May 2020).
8. Mochizuki, Y.; Tsubouchi, N.; Sugawara, K. Selective recycling of rare earth elements from Dy containing NdFeB magnets by chlorination. *ACS Sustain. Chem. Eng.* **2013**, *1*, 655–662. [CrossRef]
9. Zakotnik, M.; Tudor, C.O.; Peiró, L.T.; Afiuny, P.; Skomski, R.; Hatch, G.P. Analysis of energy usage in Nd–Fe–B magnet to magnet recycling. *Environ. Technol. Inno.* **2016**, *5*, 117–126. [CrossRef]
10. Lee, K.; Yoo, K.; Yoon, H.-S.; Kim, C.J.; Chung, K.W. Demagnetization followed by remagnetization of waste NdFeB magnet for reuse. *Geosyst. Eng.* **2013**, *16*, 286–288. [CrossRef]
11. Sheridan, R.S.; Williams, A.J.; Harris, I.R.; Walton, A. Improved HDDR processing route for production of anisotropic powder from sintered NdFeB type magnets. *J. Magn. Magn. Mater.* **2014**, *350*, 114–118. [CrossRef]

12. Gutfleisch, O.; Güth, K.; Woodcock, T.G.; Schultz, L. Recycling used Nd-Fe-B sintered magnets via a hydrogen-based route to produce anisotropic, resin bonded magnets. *Adv. Energy Mater.* **2013**, *3*, 151–155. [CrossRef]
13. Itoh, M.; Masuda, M.; Suzuki, S.; Machida, K.-I. Recycling of rare earth sintered magnets as isotropic bonded magnets by melt-spinning. *J. Alloy. Compd.* **2004**, *374*, 393–396. [CrossRef]
14. Li, C.; Liu, W.Q.; Yue, M.; Liu, Y.Q.; Zhang, D.T.; Zuo, T.Y. Waste Nd-Fe-B sintered magnet recycling by doping with rare earth rich alloys. *IEEE Trans. Magn.* **2014**, *50*, 2015403. [CrossRef]
15. Hogberg, S.; Holboll, J.; Mijatovic, N.; Jensen, B.B.; Bendixen, F.B. Direct reuse of rare earth permanent magnets-coating integrity. *IEEE Trans. Magn.* **2017**, *53*, 1–9. [CrossRef]
16. Xia, M.; Abrahamsen, A.B.; Bahl, C.R.H.; Veluri, B.; Søegaard, A.I.; Bøjsøe, P. Hydrogen decrepitation press-less process recycling of ndfeb sintered magnets. *J. Magn. Magn. Mater.* **2017**, *441*, 55–61. [CrossRef]
17. Sheridan, R.S.; Sillitoe, R.; Zakotnik, M. Anisotropic powder from sintered NdFeB magnets by the HDDR processing route. *J. Magn. Magn. Mater.* **2012**, *324*, 63–67. [CrossRef]
18. Zakotnik, M.; Harris, I.R.; Williams, A.J. Possible methods of recycling NdFeB-type sintered magnets using the HD/degassing process. *J. Alloy. Compd.* **2008**, *450*, 525–531. [CrossRef]
19. Bian, Y.Y.; Guo, S.; Tang, K.; Jiang, L.; Lu, C.; Lu, X.; Ding, W. Recovery of rare earth elements from permanent magnet scraps by pyrometallurgical process. *Rare Met.* **2015**. [CrossRef]
20. Hua, Z.; Wang, L.; Wang, J.; Xiao, Y.; Yang, Y.; Zhao, Z.; Liu, M. Extraction of rare earth elements from NdFeB scrap by AlF_3-NaF melts. *Mater. Sci. Tech.* **2014**, *31*, 1007–1010. [CrossRef]
21. Hua, Z.; Wang, J.; Wang, L.; Zhao, Z.; Li, X.; Xiao, Y.; Yang, Y. Selective extraction of rare earth elements from NdFeB scrap by molten chlorides. *ACS Sustain. Chem. Eng.* **2014**, *2*, 2536–2543. [CrossRef]
22. Saito, T.; Sato, H.; Ozawa, S.; Yu, J.; Motegi, T. The extraction of Nd from waste Nd-Fe-B alloys by the glass slag method. *J. Alloy. Compd.* **2003**, *353*, 189–193. [CrossRef]
23. Takeda, O.; Okabe, T.H.; Umetsu, Y. Phase equilibrium of the system Ag-Fe-Nd, and Nd extraction from magnet scraps using molten silver. *J. Alloy. Compd.* **2004**, *379*, 305–313. [CrossRef]
24. Moore, M.; Gebert, A.; Stoica, M. A route for recycling Nd from Nd-Fe-B magnets using Cu melts. *J. Alloy. Compd.* **2015**, *647*, 997–1006. [CrossRef]
25. Bian, Y.; Guo, S.; Jiang, L.; Liu, J.; Tang, K.; Ding, W. Recycling of rare earth elements from NdFeB magnet by VIM-HMS method. *ACS Sustain. Chem. Eng.* **2016**, *4*, 810–818. [CrossRef]
26. Bian, Y.; Tang, K.; Gabriella, R. A thermodynamic assessment of the Nd-C system. *Calphad* **2015**, *51*, 206–210. [CrossRef]
27. Miura, K.; Itoh, M.; Machida, K.I. Extraction and Recycling Characteristics of Fe element from Nd-Fe-B sintered magnet powder scrap by carbonylation. *J. Alloy. Compd.* **2008**, *466*, 228–232. [CrossRef]
28. Firdaus, M.; Rhamdhani, M.A.; Durandet, Y.; Rankin, W.J.; McGregor, K. Review of high-temperature recycling of rare earth (Nd/Dy) from magnet waste. *J. Sustain. Metall.* **2016**, *2*, 276–295. [CrossRef]
29. Uda, T. Recycling of rare earths from magnet sludge by $FeCl_2$. *Mater. Trans.* **2002**, *43*, 55–62. [CrossRef]
30. Itoh, M.; Miura, K.; Machida, K.I. Novel rare earth recycling process on Nd-Fe-B magnet scrap by selective chlorination using NH_4Cl. *J. Alloy. Compd.* **2009**, *477*, 484–487. [CrossRef]
31. Yamada, E.; Murakami, H.; Nishihama, S.; Yoshizuka, K. Separation process of dysprosium and neodymium from waste neodymium magnet. *Sep. Purif. Technol.* **2018**, *192*, 62–68. [CrossRef]
32. Padhan, E.; Sarangi, K. Recycling of Nd and Pr from NdFeB magnet leachates with bi-functional ionic liquids based on Aliquat 336 and Cyanex 272. *Hydrometallurgy* **2017**, *167*, 134–140. [CrossRef]
33. Padhan, E.; Nayak, A.K.; Sarangi, K. Recovery of neodymium and dysprosium from NdFeB magnet swarf. *Hydrometallurgy* **2017**, *174*, 210–215. [CrossRef]
34. Riaño, S.; Binnemans, K. Extraction and separation of neodymium and dysprosium from used NdFeB magnets: An application of ionic liquids in solvent extraction towards the recycling of magnets. *Green Chem.* **2015**, *17*, 2931–2942. [CrossRef]
35. Kim, D.; Powell, L.; Delmau, L.H.; Peterson, E.S.; Herchenroeder, J.; Bhave, R.R. Selective extraction of rare earth elements from permanent magnet scraps with membrane solvent extraction. *Environ. Sci. Technol.* **2015**, *49*, 9452–9459. [CrossRef] [PubMed]
36. Dupont, D.; Binnemans, K. Recycling of rare earths from NdFeB magnets using a combined leaching/extraction system based on the acidity and thermomorphism of the ionic liquid [Hbet][Tf_2N]. *Green Chem.* **2015**, *17*, 2150–2163. [CrossRef]

37. Lai, W.; Liu, M.; Li, C.; Suo, H.L.; Yue, M. Recycling of a composite powder from NdFeB slurry by co-precipitation. *Hydrometallurgy* **2014**, *150*, 27–33. [CrossRef]
38. Binnemans, K.; Jones, P.T.; Blanpain, B.; Van Gerven, T.; Yang, Y.; Walton, A.; Buchert, M. Recycling of rare earths: A critical review. *J. Clean. Prod.* **2013**, *51*, 1–22. [CrossRef]
39. USGS (2010–2016) Minerals Yearbook, Rare Earths. Available online: http://minerals.usgs.gov/minerals/pubs/commodity/rare_earths/index.html#myb (accessed on 29 May 2020).
40. Alonso, E.; Sherman, A.M.; Wallington, T.J.; Everson, M.P.; Field, R.; Roth, R.; Kirchain, R.E. Evaluating rare earth element availability: A case with revolutionary demand from clean technologies. *Environ. Sci. Technol.* **2012**, *46*, 3406–3414. [CrossRef] [PubMed]
41. Schulze, R.; Buchert, M. Estimates of global REE recycling potentials from NdFeB magnet material. *Resour. Conserv. Recy.* **2016**, *113*, 12–27. [CrossRef]
42. Du, X.; Graedel, T.E. Global rare earth in-use stocks in NdFeB permanent magnets. *J. Ind. Ecol.* **2011**, *15*, 836–843. [CrossRef]
43. Guyonnet, D.; Planchon, M.; Rollat, A.; Escalon, V.; Tuduri, J.; Charles, N.; Vaxelaire, S.; Dubois, D.; Fargier, H. Material flow analysis applied to rare earth elements in Europe. *J. Clean. Prod.* **2015**, *107*, 215–228. [CrossRef]
44. Schüler, D.; Buchert, M.; Liu, R.; Dittrich, S.; Merz, C. *Study on Rare Earths and Their Recycling (Öko-Institut eV, Darmstadt)*; The Greens/EFA Group in the European Parliament: Brussels, Belgium, 2011.
45. Rademaker, J.H.; Kleijn, R.; Yang, Y. Recycling as a strategy against rare earth element criticality: A systemic evaluation of the potential yield of NdFeB magnet recycling. *Environ. Sci. Technol.* **2013**, *47*, 10129–10136. [CrossRef]
46. Seo, Y.; Morimoto, S. Comparison of dysprosium security strategies in Japan for 2010–2030. *Resour. Policy* **2014**, *39*, 15–20. [CrossRef]
47. Gutfleisch, O.; Willard, M.A.; Bruck, E.; Chen, C.; Sankar, S.G.; Liu, J. Magnetic materials and devices for the 21st century: Stronger, lighter, and more energy efficient. *Adv. Mater.* **2011**, *23*, 821–842. [CrossRef] [PubMed]
48. *Critical Materials Strategy*; U.S. Department of Energy, Advanced Research Projects Agency-Energy: Washington, DC, USA, 2010.
49. Brown, D.; Ma, B.; Chen, Z. Developments in the processing and properties of NdFeb-type permanent magnets. *J. Magn. Magn. Mater.* **2002**, *248*, 432–440. [CrossRef]
50. Zhou, S.; Dong, Q.; Gao, X. *Sintered NdFeB Rare Earth Permanent Magnetic Materials and Technology*; Metallurgical Industry Press: Beijing, China, 2011.
51. Önal, M.A.R.; Borra, C.R.; Guo, M.X.; Blanpain, B.; Gerven, T.V. Hydrometallurgical recycling of NdFeB magnets: Complete leaching, iron removal and electrolysis. *J. Rare Earth.* **2017**, *35*, 574–584. [CrossRef]
52. Gergoric, M.; Ekberg, C.; Foreman, M.R.S.J.; Steenari, B.-M.; Retegan, T. Characterization and leaching of neodymium magnet waste and solvent extraction of the rare-earth elements using TODGA. *J. Sustain. Metall.* **2017**, *3*, 638–645. [CrossRef]
53. Itakura, T.; Sasai, R.; Itoh, H. Resource recycling from Nd-Fe-B sintered magnet by hydrothermal treatment. *J. Alloy. Compd.* **2006**, *408–412*, 1382–1385. [CrossRef]
54. Chae, H.J.; Kim, Y.D.; Kim, B.S.; Kim, J.G.; Kim, T.S. Experimental investigation of diffusion behavior between molten Mg and Nd-Fe-B magnets. *J. Alloy. Compd.* **2014**, *586*, S143–S149. [CrossRef]
55. Önal, M.A.R.; Aktan, E.; Borra, C.R.; Blanpain, B.; Gerven, T.V.; Guo, M.X. Recycling of NdFeB magnets using nitration, calcination and water leaching for REE recycling. *Hydrometallurgy* **2017**, *167*, 115–123. [CrossRef]
56. Hoogerstraete, T.V.; Blanpain, B.; Gerven, T.V.; Binnemans, K. From NdFeB magnets towards the rare-earth oxides: A recycling process consuming only oxalic acid. *RSC Adv.* **2014**, *4*, 64099–64111. [CrossRef]
57. Herbst, J.F.; Croat, J.J.; Pinkerton, F.E. Relationships between crystal structure and magnetic properties in $Nd_2Fe_{14}B$. *Phys. Rev. B* **1984**, *29*, 4176. [CrossRef]
58. Chung, K.W.; Kim, C.J.; Yoon, H.S. Novel extraction process of rare earth elements from NdFeB powders via alkaline treatment. *Arch. Metall. Mater.* **2015**, *60*, 1301–1305. [CrossRef]
59. Khlopkov, K.; Gutfleisch, O.; Eckert, D.; Hinz, D.; Wallb, B.; Rodewald, W.; Müller, K.-H.; Schultz, L. Local texture in Nd-Fe-B sintered magnets with maximised energy density. *J. Alloy. Compd.* **2004**, *365*, 259–265. [CrossRef]
60. Makarova, I.; Soboleva, E.; Osipenko, M.; Kurilo, I.; Laatikainen, M.; Repo, E. Electrochemical leaching of rare-earth elements from spent NdFeB magnets. *Hydrometallurgy* **2020**, *192*, 105264. [CrossRef]

61. Önal, M.A.R.; Borra, C.R.; Guo, M.; Blanpain, B.; Gerven, T.V. Recycling of NdFeB magnets using sulfation, selective roasting, and water leaching. *J. Sustain. Metall.* **2015**, *1*, 199–215. [CrossRef]
62. Jha, M.K.; Kumari, A.; Panda, R.; Kumar, J.R.; Yoo, K.; Lee, J.Y. Review on hydrometallurgical recovery of rare earth metals. *Hydrometallurgical* **2016**, *165*, 2–26. [CrossRef]
63. Jyothi, R.K.; Thenepalli, T.; Ahn, J.W.; Parhi, P.K.; Chung, K.W.; Lee, J.-Y. Review of rare earth elements recovery from secondary resources for clean energy technologies: Grand opportunities to create wealth from waste by saving energy. *J. Clean. Prod.* **2020**, *267*, 122048. [CrossRef]
64. Kumari, A.; Panda, R.; Jha, M.K.; Kumar, J.R.; Lee, J.Y. Process development to recover rare earth metals from monazite mineral: A review. *Miner. Eng.* **2015**, *79*, 102–115. [CrossRef]
65. Lee, J.C.; Kim, W.B.; Jeong, J.; Yoon, I.J. Extraction of neodymium from Nd-Fe-B magnet scraps by sulfuric Acid. *J. Korean Inst. Met. Mater.* **1998**, *36*, 967.
66. Niinae, M.; Yamaugchi, K.; Nakahiro, Y. Study on recycling of rare earth magnet scrap. *J. Mining Mater. Pro. Ins.* **1994**, *110*, 337.
67. Koyama, K.; Kitajima, A.; Tanaka, M. Selective leaching of rare-earth elements from an Nd-Fe-B magnet. *Kidorui (Rare Earths)* **2009**, *54*, 36–37.
68. Koyama, K.; Tanaka, M. *The Latest Technology Trend and Resource Strategy of Rare Earths*; Machida, K., Ed.; CMC Press: Tokyo, Japan, 2011; pp. 127–131.
69. Kumari, A.; Sinha, M.K.; Pramanik, S.; Sahu, S.K. Recovery of rare earths from spent NdFeB magnets of wind turbine: Leaching and kinetic aspects. *Waste Manag.* **2018**, *75*, 486–498. [CrossRef]
70. Hoogerstraete, T.V.; Wellens, S.; Verachtert, K.; Binnemans, K. Removal of transition metals from rare earths by solvent extraction with an undiluted phosphonium ionic liquid: Separations relevant to rare-earth magnet recycling. *Green Chem.* **2013**, *15*, 919–927. [CrossRef]
71. Hoogerstraete, T.V.; Binnemans, K. Highly efficient separation of rare earths from nickel and cobalt by solvent extraction with the ionic liquid trihexyl (tetradecyl) phosphonium nitrate: A process relevant to the recycling of rare earths from permanent magnets and nickel metal hydride batteries. *Green Chem.* **2014**, *16*, 1594–1606.
72. Martina, O.; Amy, V.B.; Bart, B.; Koen, B. Selective roasting of Nd-Fe-B permanent magnets as a pretreatment step for intensifed leaching with an ionic liquid. *J. Sustain. Metall.* **2020**, *6*, 91–102.
73. Rabatho, J.P.; Tongamp, W.; Takasaki, Y.; Haga, K.; Shibayama, A. Recycling of Nd and Dy from rare earth magnetic waste sludge by hydrometallurgical process. *J. Mater. Cycles Waste* **2012**, *15*, 171–178. [CrossRef]
74. Kataoka, Y.; Ono, T.; Tsubota, M.; Kitagawa, J. Improved room-temperature-selectivity between Nd and Fe in Nd recovery from Nd-Fe-B magnet. *AIP Adv.* **2015**, *5*, 117212. [CrossRef]
75. Yoon, H.S.; Kim, C.J.; Chung, K.W.; Jeon, S.; Park, I.; Yoo, K.; Jha, M.K. The effect of grinding and roasting conditions on the selective leaching of Nd and Dy from NdFeB magnet scraps. *Metals* **2015**, *5*, 1306–1314. [CrossRef]
76. Xie, F.; Zhang, T.A.; Dreisinger, D.; Doyle, F. A critical review on solvent extraction of rare earths from aqueous solutions. *Miner. Eng.* **2014**, *56*, 10–28. [CrossRef]
77. Venkatesan, P.; Vander Hoogerstraete, T.; Hennebel, T.; Binnemans, K.; Sietsma, J.; Yang, Y. Selective electrochemical extraction of REEs from NdFeB magnet waste at room temperature. *Green Chem.* **2018**, *20*, 1065–1073. [CrossRef]
78. Venkatesan, P.; Sun, Z.H.I.; Sietsma, J.; Yang, Y. An environmentally friendly electro-oxidative approach to recover valuable elements from NdFeB magnet waste. *Sep. Purif. Technol.* **2018**, *191*, 384–391. [CrossRef]
79. Venkatesan, P.; Vander Hoogerstraete, T.; Binnemans, K.; Sun, Z.; Sietsma, J.; Yang, Y. Selective extraction of rare-earth elements from NdFeB magnets by a room-temperature electrolysis pretreatment step. *ACS Sustain. Chem. Eng.* **2018**, *6*, 9375–9382. [CrossRef]
80. Yoon, H.-S.; Kim, C.-J.; Chung, K.W.; Kim, S.-D.; Kumar, J.R. Process development for recovery of dysprosium from permanent magnet scraps leach liquor by hydrometallurgical techniques. *Can. Metall. Q.* **2015**, *54*, 318–327. [CrossRef]
81. Yoon, H.-S.; Kim, C.-J.; Chung, K.W.; Kim, S.-D.; Kumar, J.R. Recovery process development for the rare earths from permanent magnets scraps leach liquors. *J. Brazil. Chem. Soc.* **2015**, *26*, 1143–1151. [CrossRef]
82. Lyman, J.W.; Palmer, G.R. Recycling of rare earths and iron from NdFeB magnet scrap. High Temp. *Mater. Proc.* **1993**, *11*, 175–187.
83. Lee, C.-H.; Chen, Y.J.; Liao, C.H.; Popuri, S.R.; TSAI, S.L.; Hung, C.E. Selective leaching process for neodymium recycling from scrap Nd-Fe-B magnet. *Metall. Mater. Trans. A* **2013**, *44*, 5825–5833. [CrossRef]

84. Itakura, T.; Sasai, R.; Itoh, H. A novel recovery method for treating wastewater containing fluoride and fluoroboric acid. *Bull. Chem. Soc. Jpn.* **2006**, *79*, 1303–1307. [CrossRef]
85. Itakura, T.; Sasai, R.; Itoh, H. In situ solid/liquid separation effect for high-yield recycling of boron and fluorine from aqueous media containing borate or fluoroborate ions. *Bull. Chem. Soc. Jpn.* **2007**, *80*, 2014–2018. [CrossRef]
86. Auerbach, R.; Bokelmann, K.; Stauber, R.; Gutfleisch, O.; Schnell, S.; Ratering, S. Recycling of rare earth metals out of end of life magnets by bioleaching with various bacteria as an example of an intelligent recycling strategy. *Miner. Eng.* **2019**, *134*, 104–117. [CrossRef]
87. Yoon, H.S.; Kim, C.J.; Lee, J.Y.; Kim, S.D.; Lee, J.C. Separation of neodymium from NdFeB permanent magnet scrap. *J. Korean Inst. Resour. Recycl.* **2003**, *12*, 57–63.
88. Yoon, H.-S.; Kim, C.-J.; Chung, K.-W.; Kim, S.-D.; Lee, J.-Y.; Kumar, J.R. Solvent extraction, separation and recovery of the dysprosium (Dy) and neodymium (Nd) from aqueous solutions: Waste recycling strategies for permanent magnet processing. *Hydrometallurgy* **2016**, *165*, 27–43. [CrossRef]
89. Reddy, B.R.; Kumar, J.R. Rare earths extraction, separation and recovery from phosphoric acid media. *Solvent Extr. Ion Exc.* **2016**, *34*, 226–240. [CrossRef]
90. Gergoric, M.; Ravaux, C.; Steenari, B.-M.; Espegren, F.; Retegan, T. Leaching and Recovery of Rare-Earth Elements from Neodymium Magnet Waste Using Organic Acids. *Metals* **2018**, *8*, 721. [CrossRef]
91. Lee, J.Y.; Jha, A.K.; Kumari, A.; Kumar, J.R.; Jha, M.K.; Kumar, V. Neodymium recovery by precipitation from synthetic leach liquor of concentrated rare earth mineral. *J. Metall. Mater. Sci.* **2011**, *53*, 349–354.
92. Sasai, R.; Shimamura, N. Technique for recovering rare-earth metals from spent sintered Nd-Fe-B magnets without external heating. *J. Asian Ceram. Soc.* **2018**, *4*, 155–158. [CrossRef]
93. Kikuchi, Y.; Matsumiya, M.; Kawakami, S. Extraction of rare earth ions from Nd-Fe-B magnet wastes with TBP in Tricaprylmethylammonium nitrate. *Solvent Extr. Dev.* **2014**, *21*, 137–145. [CrossRef]
94. Saito, T.; Sato, H.; Motegi, T. Recovery of rare earths from sludges containing rare-earth elements. *J. Alloy. Compd.* **2006**, *425*, 145–147. [CrossRef]
95. Cai, J.; Dai, J.; Zhou, X. Thermodynamic analysis on preparation of zinc doped Co_2-Y planar hexagonal ferrite powder by chemical co-precipitation method. *Chin. J. Inorg. Chem.* **2008**, *24*, 1943–1948.
96. Su, J.; Su, Y.; Lai, Z. Thermodynamic analysis of preparation of multiple carbonate of Ni, Co and Mn by co-precipitation method. *J. Chin. Ceram. Soc.* **2006**, *34*, 695–698.
97. Kołodyńska, D.; Hubicki, Z. Investigation of sorption and separation of lanthanides on the ion exchangers of various types. In *Ion Exchange Technologies*; Kilislioglu, A., Ed.; InTech: London, UK, 2012; ISBN 978-953-51-0836-8.
98. Huang, L.H. *Chemical Beneficiation*; Metallurgical Industry Press: Beijing, China, 2012.
99. Deshpande, S.M.; Mishra, S.L.; Gajankush, R.B.; Thakur, N.V.; Koppiker, K.S. Recovery of high purity Y_2O_3 by solvent extraction route using organo-phosphorus extractants. Miner. *Process. Extr. Metall. Rev. Int. J.* **1992**, *10*, 267–273. [CrossRef]
100. Yin, S.; Wu, W.; Zhang, B.; Zhang, F.; Luo, Y.; Li, S.; Bian, X. Study on separation technology of Pr and Nd in D2EHPA-HCl-LA coordination extraction system. *J. Rare Earths* **2012**, *28*, 111–115. [CrossRef]
101. Lee, M.-S.; Lee, J.Y.; Kim, J.S.; Lee, G.S. Solvent extraction of neodymium ions from hydrochloric acid solution using PC88A and saponified PC88A. *Sep. Purif. Technol.* **2005**, *46*, 72–78. [CrossRef]
102. El-Kot, A.M. Solvent extraction of neodymium, europium and thulium by di-(2-ethylhexyl) phosphoric acid. *J. Radioanal. Nucl. Chem.* **1993**, *170*, 207–214. [CrossRef]
103. Onoda, H.; Nakamura, R. Recycling of neodymium from an iron–neodymium solution using phosphoric acid. *J. Environ. Chem. Eng.* **2014**, *2*, 1186–1190. [CrossRef]
104. Preston, J.S.; Preez, A.C. The separation of europium from a middle rare earth concentrate by combined chemical reduction, precipitation and solvent-extraction methods. *J. Chem. Technol. Biot.* **1996**, *65*, 93–101. [CrossRef]
105. Maharana, L.N.; Nair, V.R. Production of value added rare earths from monazite by solvent extraction. In *Light Metals*; Kvande, H., Ed.; TMS 2005: Warrendale, PA, USA, 2005; pp. 1163–1166.
106. Mohammadi, M.; Forsberg, K.; Kloo, L.; Cruz, J.M.D.L.; Rasmuson, A. Separation of ND(III), DY(III) and Y(III) by solvent extraction using D2EHPA and EHEHPA. *Hydrometallurgy* **2015**, *156*, 215–224. [CrossRef]

107. Sarangi, K.; Reddy, B.R.; Das, R.P. Extraction studies of cobalt(II) and nickel(II) from chloride solutions using Na-Cyanex 272: Separation of Co(II)/Ni(II) by the sodium salts of D2EHPA, PC88A and Cyanex 272 and their mixtures. *Hydrometallurgy* **1999**, *52*, 253–265. [CrossRef]
108. Thakur, N.V.; Jayawant, D.V.; Iyer, N.S.; Koppiker, K.S. Separation of neodymium from lighter rare earths using alkyl phosphonic acid, PC88A. *Hydrometallurgy* **1993**, *34*, 99–108. [CrossRef]
109. Devi, N.B.; Nathsarma, K.C.; Chakravortty, V. Separation and recycling of cobalt(II) and nickel(II) from sulphate solutions using sodium salts of D2EHPA, PC88A and Cyanex 272. *Hydrometallurgy* **1999**, *49*, 47–61. [CrossRef]
110. Devi, N.B.; Nathsarma, K.C.; Chakravortty, V. Separation of divalent manganese and cobalt ions from sulphate solutions using sodium salts of D2EHPA, PC88A and Cyanex 272. *Hydrometallurgy* **2000**, *54*, 117–131. [CrossRef]
111. Padhan, E.; Sarangi, K. Solvent extraction of Nd using organo-phosphorus extractants from chloride media. *Chem. Technol. Biotechnol.* **2015**, *90*, 1869–1875. [CrossRef]
112. Pavóna, S.; Fortunya, A.; Collb, M.T.; Sastrec, A.M. Neodymium recovery from NdFeB magnet wastes using Primene 81R·Cyanex 572 IL by solvent extraction. *J. Environ. Manag.* **2018**, *222*, 359–367. [CrossRef] [PubMed]
113. Rice, A.C.; Stone, C.A. *Amines in Liquid–Liquid Extraction of Rare Earth Elements*; US Department of the Interior, Bureau of Mines: Washington, DC, USA, 1961; p. 5923.
114. Hsu, K.H.; Huang, C.H.; King, T.C.; Li, P.K. Separation of praseodymium and neodymium in high purity (99.9%) by counter-current exchange extraction and its mechanism. In Proceedings of the International Solvent Extraction Conference'80 (ISEC'80), Liege, Belgium, 6–12 September 1980; Society of Chemical Industry: London, UK; Volume 2, pp. 80–82.
115. El-Yamani, I.S.; Shabana, E.L. Solvent extraction of lanthanum(III) from sulfuric acid solutions by Primene JMT. *J. Less-Common Metals* **1985**, *105*, 255–261. [CrossRef]
116. Huang, C.H.; Jin, T.Z.; Li, B.G.; Li, J.R.; Xu, G.X. Studies on extraction mechanism of the rare earths with quaternary ammonium salts. In *Proceedings of the International Solvent Extraction Conference—ISEC'86*; Dechema: Frankfurt, Germany, 1986; pp. 215–221.
117. Cerna, M.; Volaufova, E.; Rod, V. Extraction of light rare earth elements by amines at high inorganic nitrate concentration. *Hydrometallurgy* **1992**, *28*, 339–352. [CrossRef]
118. Lu, D.; Horing, J.S.; HOH, Y.C. The separation of neodymium by quaternary amine form didymium nitrate solution. *J. Less-Common Metals* **1989**, *149*, 219–224. [CrossRef]
119. Preston, J.S. The recycling of rare earth oxides from a phosphoric acid byproduct, Part 4: The preparation of magnet-grade neodymium oxide from the light rare earth fraction. *Hydrometallurgy* **1996**, *42*, 151–167. [CrossRef]
120. Mincher, B. Radiation chemistry in the reprocessing and recycling of spent nuclear fuels. In *Reprocessing and Recycling of Spent Nuclear Fuel*; Taylor, R., Ed.; Elsevier: Cambridge, UK, 2015; pp. 191–213.
121. Tachimori, S.; Sasaki, Y.; Suzuki, S. Modification of TODGA-n-dodecane solvent with a monoamide for high loading of lanthanides(III) and actinides(III). *Solvent Extr. Ion. Exch.* **2002**, *20*, 687–699. [CrossRef]
122. Marcus, Y. Principles of Solubility and Solutions. In *Solvent Extraction Principles and Practise*, 2nd ed.; Rydberg, J., Cox, M., Musicas, C., Choppin, G.R., Eds.; Marcel Dekker: New York, NY, USA, 2004; pp. 27–81.
123. Kim, D. A supported liquid membrane system for the selective recycling of rare earth elements from neodymium-based permanent magnets. *Sep. Sci. Technol.* **2017**, *51*, 1716–1726. [CrossRef]
124. Wang, X.; Li, W.; Meng, S.; Li, D. The extraction of rare earths using mixtures of acidic phosphorus based reagents or their thio-analogues. *J. Chem. Technol. Biot.* **2006**, *81*, 761–766. [CrossRef]
125. Santhi, P.B.; Reddy, M.L.P.; Ramamohan, T.R.; Damodaran, A.D. Liquid-liquid extraction of yttrium (III) with mixtures of organophosphorous extractants: Theoretical analysis of extraction behavior. *Hydrometallurgy* **1991**, *27*, 169–177. [CrossRef]
126. Tian, M.; Song, N.; Wang, D.; Quan, X.; Jia, Q.; Liao, W.; Lin, L. Applications of the binary mixture of sec-octylphenoxyacetic acid and 8-hydroxyquinoline to the extraction of rare earth elements. *Hydrometallurgy* **2012**, *111–112*, 109–113. [CrossRef]
127. Tian, M.; Jia, Q.; Liao, W. Studies on synergistic solvent extraction of rare earth elements from nitrate medium by mixtures of 8-hydroxyquinoline with Cyanex 301 or Cyanex 302. *J. Rare Earth.* **2013**, *31*, 604–608. [CrossRef]

128. Huang, X.; Li, J.; Zhang, Y.; Long, Z.; Wang, C.; Xue, X. Synergistic extraction of Nd^{3+} and Sm^{3+} with the mixtures of P204 and P507 in acidic sulfate solutions. *Chin. J. Nonferrous Met.* **2008**, *18*, 366–371.
129. Luo, X.; Huang, X.; Zhu, Z.; Long, Z.; Liu, Y. Synergistic extraction of cerium from sulfuric acid medium using mixture of 2-ethylhexyl phosphonic acid mono 2-ethylhexyl ester and Di-(2-ethyl hexyl) phosphoric acid as extractant. *J. Rare Earth.* **2009**, *27*, 119–122. [CrossRef]
130. Baes, C.F.J. The extraction of metallic species by dialkylphosphoric acids. *J. Inorg. Nucl. Chem.* **1962**, *24*, 707–720. [CrossRef]
131. Sato, T. Liquid–liquid extraction of rare-earth elements from aqueous acid solutions by acid organophosphorous compounds. *Hydrometallurgy* **1989**, *22*, 121–140. [CrossRef]
132. Zhang, C.; Wang, L.; Huang, X.; Dong, D.; Long, Z.; Zhang, Y. Yttrium extraction from chloride solution with a synergistic system of 2-ethylhexyl phosphonic acid mono-(2-ethylhexyl) ester and bis (2, 4, 4-trimethylpentyl) phosphinic acid. *Hydrometallurgy* **2014**, *147–148*, 7–12. [CrossRef]
133. Zhang, F.; Wu, W.; Bian, X.; Zeng, W. Synergistic extraction and separation of lanthanum (III) and cerium (III) using a mixture of 2-ethylhexylphosphonic mono-2-ethylhexyl ester and di-2-ethylhexyl phosphoric acid in the presence of two complexing agents containing lactic acid and citric acid. *Hydrometallurgy* **2014**, *149*, 238–243. [CrossRef]
134. Panda, N.; Devi, N.B.; Mishra, S. Extraction of neodymium(III) using binary mixture of Cyanex 272 and Cyanex 921/Cyanex 923 in kerosene. *J. Radioanal. Nucl. Chem.* **2013**, *296*, 1205–1211. [CrossRef]
135. Liu, Y.; Jeon, H.S.; Lee, M.S. Solvent extraction of Pr and Nd from chloride solution by the mixtures of Cyanex 272 and amine extractants. *Hydrometallurgy* **2014**, *150*, 61–67. [CrossRef]
136. Kumar, B.N.; Reddy, B.R.; Kantam, M.L.; Kumar, J.R.; Lee, J.Y. Synergistic solvent extraction of neodymium(iii) from chloride solutions using a mixture of triisooctylamine and bis(2,4,4-Trimethylpentyl) monothiophosphinic acid. *Sep. Sci. Technol.* **2014**, *49*, 130–136. [CrossRef]
137. El-Nadi, Y.A. Lanthanum and neodymium from Egyptian monazite: Synergistic extractive separation using organophosphorus reagents. *Hydrometallurgy* **2012**, *119–120*, 23–29. [CrossRef]
138. Huddleston, J.G.; Willauer, H.D.; Swatloski, R.P.; Visser, A.E.; Rogers, R.D. Room temperature ionic liquids as novel media for 'clean' liquid–liquid extraction. *Chem. Commun.* **1998**, *16*, 1765–1766. [CrossRef]
139. Nakashima, F.; Maruyama, T.; Goto, M. Feasibility of ionic liquids as alternative separation media for industrial solvent extraction processes. *Ind. Eng. Chem. Res.* **2005**, *44*, 4368–4372. [CrossRef]
140. Billard, I.; Ouadi, A.; Gaillard, C. Liquid-liquid extraction of actinides, lanthanides, and fission products by use of ionic liquids: From discovery to understanding. *Anal. Bioanal. Chem.* **2011**, *400*, 1555–1566. [CrossRef]
141. Wei, G.; Yang, Z.; Chen, C. Room temperature ionic liquid as a novel medium for liquid/liquid extraction of metal ions. *Anal. Chim. Acta* **2003**, *488*, 183–192. [CrossRef]
142. Zuo, Y.; Liu, Y.; Chen, J.; Li, D.Q. The separation of cerium(IV) from nitric acid solutions containing thorium(IV) and lanthanides(III) using pure $[C_8mim]PF_6$ as extracting phase. *Ind. Eng. Chem. Res.* **2008**, *47*, 2349–2355. [CrossRef]
143. Wang, K. Recovery of rare earth elements with ionic liquids. *Green Chem.* **2017**, *19*, 4469–4493. [CrossRef]
144. Rout, A.; Binnemans, K. Influence of the ionic liquid cation on the solvent extraction of trivalent rare-earth ions by mixtures of Cyanex 923 and ionic liquids. *Dalton Trans.* **2015**, *44*, 1379–1387. [CrossRef]
145. Jensen, M.P.; Neuefeind, J.; Beitz, J.V.; Skanthakumar, S.; Soderholm, L. Mechanisms of metal ion transfer into room-temperature ionic liquids: The role of anion exchange. *J. Am. Chem. Soc.* **2003**, *125*, 15466–15473. [CrossRef]
146. Fukikoa, K.; Yousuke, S.; Yuzo, B.; Yusuke, K.; Kojiro, S.; Noiho, K.; Masahiro, G. Application of ionic liquids to extraction separation of rare earth metals with an effective diglycol amic acid extractant. *J. Chem. Eng. Jpn.* **2011**, *44*, 307–312.
147. Yang, H.; Wang, W.; Cui, H.; Chen, J. Extraction mechanism of rare earths with bifuncional ionic liquids (Bif-ILs) [A336][CA-12]/[A336][CA-100] in nitrate medium. *Chin. J. Anal. Chem.* **2011**, *39*, 1561–1566. [CrossRef]
148. Rout, A.; Kotlarska, J.; Dehaen, W.; Binnemans, K. Liquid–liquid extraction of neodymium(III) by dialkylphosphate ionic liquids from acidic medium: The importance of the ionic liquid cation. *Phys. Chem. Chem. Phys.* **2013**, *15*, 16533–16541. [CrossRef]
149. Qi, G.; Hino, M.; Yazawa, A. Experimental study on the reduction-diffusion process to produce Fe-Nd, Fe-Sm, Co-Nd and Co-Sm alloys. *Mater. Trans. JIM* **1990**, *31*, 463–470. [CrossRef]

150. Liu, Y.; Chen, J.; Li, D. Application and perspective of ionic liquids on rare earths green separation. *Sep. Sci. Technol.* **2012**, *47*, 223–232. [CrossRef]
151. Rout, A.; Binnemans, K. Solvent extraction of neodymium(III) by functionalized ionic liquid trioctylmethylammonium dioctyl diglycolamate in fluorine-free ionic liquid diluent. *Ind. Eng. Chem. Res.* **2014**, *53*, 6500–6508. [CrossRef]
152. Wang, W.; Yang, H.; Cui, H.; Zhang, D.; Liu, Y.; Chen, J. Application of bifunctional ionic liquid extractants [A336][CA-12] and [A336][CA-100] to the lanthanum extraction and separation from rare earths in the chloride medium. *Ind. Eng. Chem. Res.* **2011**, *50*, 7534–7541. [CrossRef]
153. Rout, A.; Karmakar, S.; Venkatesan, K.A.; Srinivasan, T.G.; Vasudeva, P.R. Room temperature ionic liquid diluent for the mutual separation of europium(III) from americium(III). *Sep. Purif. Technol.* **2011**, *81*, 109–115. [CrossRef]
154. Sun, X.; Ji, Y.; Hu, F.; He, B.; Chen, J.; Li, D. The inner synergistic effect of bifunctional ionic liquid extractant for solvent extraction. *Talanta* **2010**, *81*, 1877–1883. [CrossRef] [PubMed]
155. Sun, X.; Waters, K.E. The adjustable synergistic effects between acid-base coupling bifunctional ionic liquid extractants for rare earth separation. *AICHE J.* **2014**, *60*, 3859–3868. [CrossRef]

Article

Application of Multistage Concentration (MSC) Electrodialysis to Concentrate Lithium from Lithium-Containing Waste Solution

Yeonchul Cho [1], Kihun Kim [1], Jaewoo Ahn [1,*] and Jaeheon Lee [2]

1 Department of Advanced Materials Sci. & Eng., Daejin University, 1007 Hoguk-ro, Pocheon-si, Gyeonggi-do 11159, Korea; bhyc1642@naver.com (Y.C.); gihun2019@gmail.com (K.K.)
2 Department of Mining and Geological Engineering, University of Arizona, Tucson, AZ 85721, USA; jaeheon@email.arizona.edu
* Correspondence: jwahn@daejin.ac.kr; Tel.: +82-31-539-1982

Received: 26 May 2020; Accepted: 25 June 2020; Published: 28 June 2020

Abstract: In order to manufacture lithium carbonate to be used as a raw material for a secondary lithium battery, lithium sulfate solution is used as a precursor, and the concentration of lithium is required to be 10 g/L or more. Electrodialysis (ED) was used as a method of concentrating lithium in a low-concentration lithium sulfate solution, and multistage concentration (MSC) electrodialysis was used to increase the concentration ratio (%). When MSC was performed using a raw material solution containing a large amount of sodium sulfate, the process lead time was increased by 60 min. And the concentration ratio (%) of lithium decreased as the number of concentration stages increased. In order to remove sodium sulfate, methanol was added to the raw material solution to precipitate sodium sulfate, and when it was added in a volume ratio of 0.4, lithium was not lost. Using a solution in which sodium sulfate was partially removed, fourth-stage concentration ED was performed to obtain a lithium sulfate solution with a lithium concentration of 10 g/L.

Keywords: electrodialysis (ED); lithium; lithium-ion battery; lithium sulfate; multistage concentration (MSC)

1. Introduction

1.1. General Introduction

Metal resources currently in use can be divided into mine resources and renewable resources. Urban mining means that metal resources, which are industrial raw materials, are widely distributed in the form of products or waste, which means that they exist quantitatively on a mine scale. These are renewable resources. The amounts of waste catalyst, waste lithium batteries, and electronic waste are steadily increasing, and among them, the amount of waste lithium batteries is greatly increased. As the demand for lithium, a raw material that is a key component of electric vehicle batteries, is increasing, technology for recovering it from waste resources is required [1,2]. The market demand for electrolytes that make up lithium-ion batteries is expected to have an annual average growth rate of 42% from 2019 to 2025 [3]. Lithium is the main material of the fourth industrial revolution. Lithium-based batteries are rechargeable chemical batteries with excellent technical performance and high working voltage and energy density [4]. In addition, the charging and discharging life cycle is long, and they can be implemented with light weights, so the demand in the electric vehicle market has greatly increased [5]. In the Republic of Korea, because the amount of lithium is very small, the country is dependent on imports, so if the supply is insufficient compared to global demand, the domestic industry will be affected. So there is a need for a method to maintain an internally stable lithium supply and demand.

The amount of lithium in saltwater is about twice as large as that in ore, and many studies have been conducted on recovering lithium from brine [6–10]. In order to recover lithium from brine, a purification process is required due to the large amount of impurities. The concentration of lithium is also low, so the energy consumption required is high and the process takes a long time [6–12]. Lithium carbonate is mainly used in industry, and in order to prepare lithium carbonate from a lithium sulfate solution, the concentration of lithium ions in the lithium sulfate solution must be at least 10 g/L [13]. In general, the recovery of lithium from brine can be done by commercialized processes with natural evaporation and evaporative concentration, but this has the disadvantage of low energy efficiency and recovery. Lithium recovery by solvent extraction is effective at a lithium concentration of 1.0 g/L or more, and there are increased process cost and environmental pollution due to the use of organic extractant [11,14,15]. Electrodialysis (ED) is a process used to concentrate the desired component or separate it from impurities by selectively passing ions through an ion exchange membrane in an electric field. It is environmentally friendly because almost no byproducts are generated during the process and no additional chemical treatment is required to reuse the reagents afterward. Research using ED to recover lithium has been reported, and ion exchange membranes used for ED have been developed. Most of the previous studies use seawater or brine as a raw material and contain about 200 ppb lithium and impurities (potassium, magnesium, and sodium) [6–9].

Therefore, in this study, seawater or brine was not used as a raw material, but electrodialysis was used to concentrate lithium in the raffinate after the hydrometallurgy of cobalt and nickel. A basic study was conducted to concentrate high concentration lithium from the waste solution containing about 3000 ppm lithium.

1.2. Theoretical Background of Electrodialysis

In electrodialysis, when an electric field is applied, cations move to the cathode and anions move to the anode by an electric potential gradient, and ions are selectively passed through an ion exchange membrane that has a charge, thereby selectively concentrating a target component or separating impurities. At this time, the cation and the anion exchange membranes form a pair, and they can be effectively concentrated by being installed in several pairs. Figure 1 shows the ion exchange principle and solution flow in a general electrodialysis apparatus. The raw material solution is introduced into the solution chamber, and ions move to be divided into a diluted or a concentrated solution. When a membrane and laminar flow boundary layer is formed, an increase in current density can rapidly decrease ions in the boundary layer, generating concentration polarization. When the concentration polarization increases, the electrical resistance increases and the required power increases rapidly, which can cause decomposition of water. The pH change of the solution generated by the electrolysis of water can damage the ion exchange membrane or cause scaling. In addition, a spacer, which is an insulating material, is placed between the membranes to prevent short circuit between them. This spacer can form turbulence that reduces polarization due to concentration gradient. As a main variable in research and at industrial sites, there is a water migration phenomenon from the dilution chamber to the concentration chamber. If the water transfer rate is high, the ion concentration ratio can be reduced, so the effect on water movement cannot be ignored. Although osmotic pressure due to a concentration difference is hardly generated while an electric field is applied to the ion exchange membrane, metal ions and hydrated water molecules can move together by electrolytic osmosis to increase the water transfer [16,17].

Generally, high volume ratio concentration (HVRC) and multistage concentration (MSC) are the methods of concentrating target components using ED. HVRC is a method of concentrating a target component by using the volume difference between a concentrate and a diluent. In this study, two or more stages of ED were performed through a continuous process, and the MSC method was used for the concentrated solution at high concentration. Usually, the optimal conditions in the first stage are confirmed by the HVRC method and concentrated to a high concentration by the MSC method [17]. In a previous study by the authors, the conditions for lithium concentration in waste liquid containing lithium were established by HVRC, and in this study, a high concentration of lithium was also achieved by MSC.

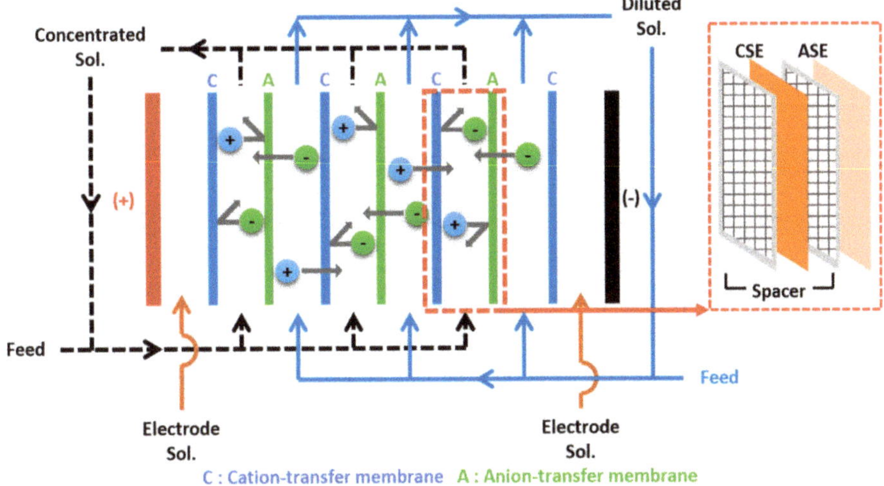

Figure 1. Schematic diagram of the electrodialysis process for the ion transfer.

2. Materials and Methods

2.1. Materials and Equipment

Similar to the composition of lithium-containing waste solution, a simulated solution was prepared using reagent-grade sulfate salt, and the composition is shown in Table 1. It contained 3.3 g/L lithium and a large amount of sodium as the remaining solution after cobalt and nickel were recovered from a waste secondary lithium battery. The simulated solutions were prepared using lithium sulfate ($Li_2SO_4{\cdot}H_2O$, 99%, JUNSEI) and sodium sulfate (Na_2SO_4, 99%, DAEJUNG). The electrodialysis equipment used to concentrate lithium was a CJT-055 electrodialyzer (Chang Jo Tech. Co., Seoul, Korea). A photograph of the equipment and the main specifications of the equipment are shown in Figure 2 and Table 2, respectively. Each chamber can be injected with 2.0 L of solution, and the power supply can operate within the DC current range of 0–30 volts and 0–5 amperes. In order to study the ion exchange membrane and the equipment, the upper limit of the current was set to 3 ampere. Titanium electrodes coated with platinum were used for the anode and cathode. The cartridge consisted of 10 pairs of NEOSEPTA (Ion exchange membrane, ASTOM Co., Tokyo, Japan) cation-exchange (CEM) and anion-exchange (AEM) membranes and an additional CEM spaced between the ion exchange membranes. The effective area of the ion exchange membranes in the cartridge is 55 cm^2, and the main details of the membranes used in the study are shown in Table 3. Membrane stability is ensured below 40 °C, and the pH can be in a wide range from 0 to 14.

Table 1. Composition of Li-containing solution used in study.

Element	Li	Na	SO_4
Concentration (g/L)	3.3	60.0	148.0

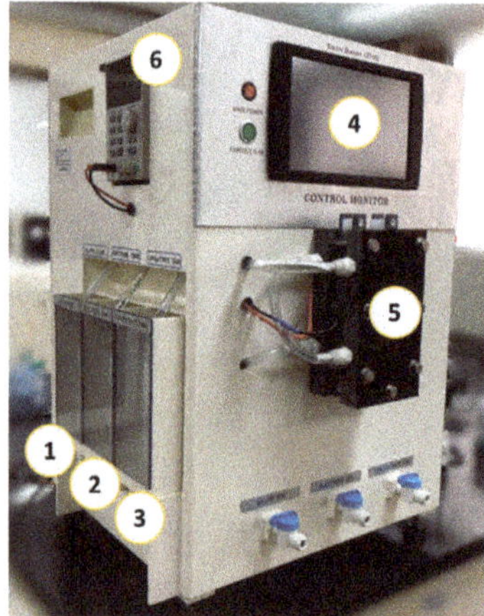

Figure 2. Photograph of the electrodialysis equipment.

Table 2. Specifications of Electro Dialyzer.

CJT-055		
Chamber	Dilute chamber	2.0 L (PVC)
	Electrode chamber	2.0 L (PVC)
	Concentrate chamber	2.0 L (PVC)
Effective membrane area		55 cm^2
Power supply (Rectifier)		DC, 30 V, 5 A
Primary power source		1 Ø, 220 V, 50 Hz/60 Hz
Cathode/Anode		Pt coated Ti (Pt/Ti)

Table 3. Characteristics of ion exchange membrane (NEOSEPTA).

NEOSEPTA	CEM (Cation Exchange Membrane)	AEM (Anion Exchange Membrane)
Type	Strong acid (Na type)	Strong base (Cl type)
Electric resistance *	1.8 ($\Omega \cdot cm^2$)	2.6 ($\Omega \cdot cm^2$)
Thickness	0.16 (mm)	0.15 (mm)
Characteristic	high mechanical strength	
Burst strength**	≥0.35 (MPa)	
Temperature	≤40 (°C)	
pH	0–14	

* Electric resistance: Measured on alternative current after equilibration with a 0.5 N NaCl solution at 25 °C; ** Burst Strength: Mullen Bursting strength.

2.2. Methods

The raw material solution was added to the dilution and concentration chambers at a constant volume ratio. Sodium sulfate (Na_2SO_4), used as the electrode solution, was prepared at a constant concentration, then 500 mL was added to the electrode chamber. The solution was sufficiently circulated for 5 min to remove residual air in the ion exchange cartridge and tube, then voltage was applied. It was applied as constant voltage of 10 volts, and the electrical conductivity of the solution in the dilution and concentration chambers over time was measured through the system's equipment. The experiment was terminated when the electrical conductivity of the solution in the dilution chamber became 1.00 mS/cm, where S (siemens) is the unit of conductivity and the reciprocal of the ohm (Ω) of electrical resistance, and 1 S = 1 A/V = 1 /Ω. At this time, there was a difference in the end time of the experiment depending on the volume ratio of the input solution and the initial concentration of ions. After the end of the experiment, samples were collected from the concentration and dilution chambers and analyzed by ICP-AES (Optima-4300 DV, Perkin Elmer Inc., Waltham, MA, US). The average energy consumption, average flux, concentration ratio, and water migration were respectively calculated by Equations (1)–(4), and the current was measured at intervals of 1 min through the device system to calculate the average current value, which was used in Equations (1) and (2).

The average energy consumption (E, kWh/m^3) can be obtained by Equation (1) [18], where U(V) is applied voltage, I(A) is applied current, and V(L) is the total volume of raw material solution (concentration chamber + dilution chamber input).

$$E = \int_0^t \frac{UI}{V} dt \quad (1)$$

The average flux (J, mol/m^{2}h) can be obtained by Equation (2), where m (g) is the mass of Li concentrated in the concentrated chamber, N is the number of pairs of alternating cation and anion exchange membranes, A (m^2) is the effective area of the exchange membrane, M is the molecular weight of lithium (g/mol), and t is the dialysis time (h).

$$J = \frac{m}{N \cdot A \cdot M \cdot t} \quad (2)$$

The concentration ratio (C_R, %) can be obtained by Equation (3), where C_f is the final concentration and C_i is the initial concentration of solution in the concentration chamber.

$$C_R\,(\%) = \frac{C_f}{C_i} \times 100 \quad (3)$$

The water migration (W_m,%) can be obtained by Equation (4), where V_D^f is the final volume of solution and V_D^i is the initial volume of solution in the dilution chamber.

$$W_m(\%) = \frac{V_D^i - V_D^f}{V_D^i} \times 100 \quad (4)$$

3. Results

3.1. *Effects of Electrode Solution Concentration*

As a result of electrodialysis by HVRC in the authors' previous study [19], the effects of the applied voltage and the volume ratio of raw material solution entering the initial concentration and dilution chamber were examined. Prior to MSC electrodialysis, the concentration of sodium sulfate to be used as the electrode solution was adjusted to 0.1–1.0 M to examine the effect of the solution concentration. Then 500 mL of raw material solution was added to the dilution and concentration chambers, and 500 mL of electrode solution was added to the electrode solution chamber. The experiment was conducted by applying a constant voltage of 10 V. Figure 3 shows the electrical conductivity of the diluent and concentrate solutions over time. As the concentration of the electrode solution increased, less time was required for the process to decrease the electrical conductivity of the diluent to 1.00 mS/cm. When 0.1 M Na_2SO_4 was used, it took 3 h 30 min, with 0.7 M Na_2SO_4 it was 2 h 5 min, and with 1.0 M Na_2SO_4 it was 1 h 55 min. With 0.7 M and 1.0 M Na_2SO_4 compared to 0.1 M Na_2SO_4, the process lead time was reduced by more than 1 h. Table 4 shows the ion concentration in the concentrate solution after the end of the experiment. The same experiment was conducted three times, and the average value of each ion concentration was calculated. As the concentration of the solution increased, the concentration of Li increased. With 0.7 M Na_2SO_4, lithium was concentrated to 4.53 g/L, and with 1.0 M Na_2SO_4, the concentration was slightly reduced to 4.19 g/L. Figure 4 shows energy consumption and average flux. As shown in Figure 4a, there was no significant difference in energy consumption from 41 to 43 kWh/m^2 depending on the concentration of electrode solution. As shown in Figure 4b, as the concentration of electrode solution increased, the average flux increased, and it can be seen that the ion migration rate per unit time is high. Therefore, it was effective to use a concentration of 0.7 M Na_2SO_4 or higher in the ED process.

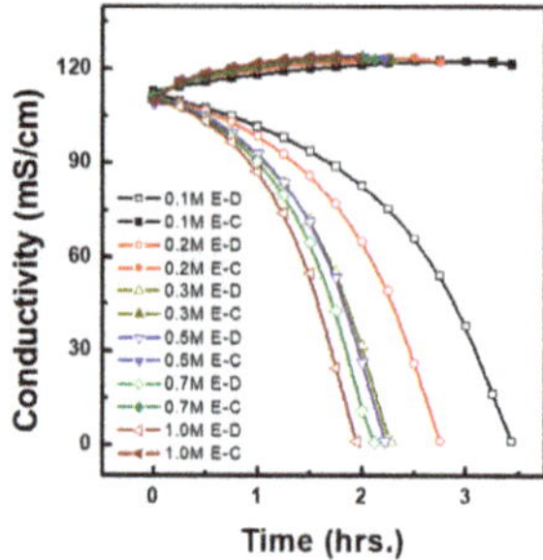

Figure 3. Effect of electrode solution concentration (M) on the electrical conductivity (D: Dilute solution, C: Concentrate solution, 10 V, D:C [1:1]).

Table 4. Analytical results of the concentrated solution after electrodialysis (ED) process (10V, D:C [1:1]).

Concentration of Electrode Solution	Concentration (g/L)			Process Lead Time (min)
	Li	Na	SO_4	
Initial solution	3.30	60.00	148.20	-
0.1 M Na_2SO_4	3.38	57.86	167.65	205
0.2 M Na_2SO_4	3.38	53.66	158.89	165
0.3 M Na_2SO_4	3.69	59.32	174.99	135
0.5 M Na_2SO_4	3.68	56.99	169.99	135
0.7 M Na_2SO_4	4.53	66.45	201.52	130
1.0 M Na_2SO_4	4.19	64.95	193.67	120

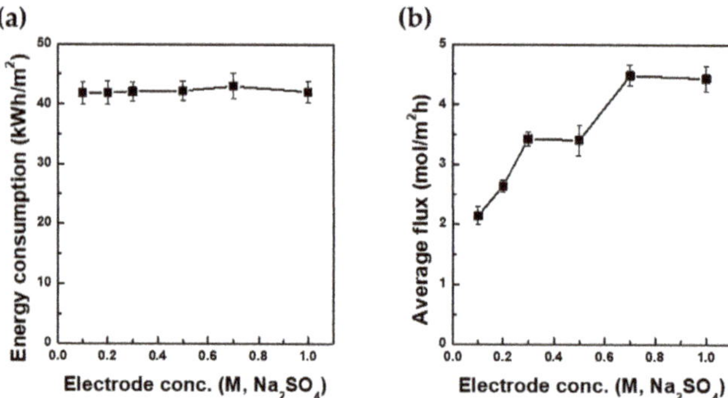

Figure 4. Effect of electrode solution concentration (M) by (**a**) energy consumption and (**b**) average flux (10 V, D:C [1:1]).

3.2. Concentration of Lithium by MSC Electrodialysis

As a result of electrodialysis by HVRC in the authors' previous study [19], the highest Li concentration was obtained with applied voltage 10 V, volume ratio (diluent:concentrate) = 3:1, and 0.7 M Na_2SO_4 electrode solution. This was selected as an optimal condition and applied to the MSC, and 900 mL was added to the dilution chamber and 300 mL to the concentration chamber to start the experiment. The ED process after the first stage was used as a feed solution in the second stage, and the third stage was supplied with feed solution in the same way. In order to confirm the continuous process, the raw material solution entering the second and third stages was not prepared and not used as simulation solution. The first stage was repeated 10 times, the second stage five times, and the third stage three times. The concentrate solution from each stage was collected in a bottle and used as a raw material solution in the next stage. The samples were taken from bottles to analyze the concentration of ions. The concentration of Li in each stage was compared, as shown in Figure 5. The concentration of lithium increased with the number of stages, and after the third stage, the lithium concentration was more than twice the initial concentration, at 6.95 g/L. Table 5 shows the results of ICP analysis for Li, Na, and SO_4 according to the stage, lithium concentration, and process lead time. Three components increased in concentration as the number of stages increased. In the case of lithium, the concentration increased from the initial concentration or from the previous stage, but the concentration ratio (%) tended to decrease compared to the previous stage. As the ions concentration in the raw material solution increases, the amount of water hydrated and moved during the ED process also increases. Therefore, as the number of stage increases, the volume of the concentrated solution increases after the process. As shown in Table 6, the volume changes according to the water migration acted to reduce the C_R (%) of lithium. The ion concentration in the raw material solution input at each stage increased, so the process lead time increased. The process lead time for ED in the first stage was 210 min, and in the third stage it increased significantly to 320 min.

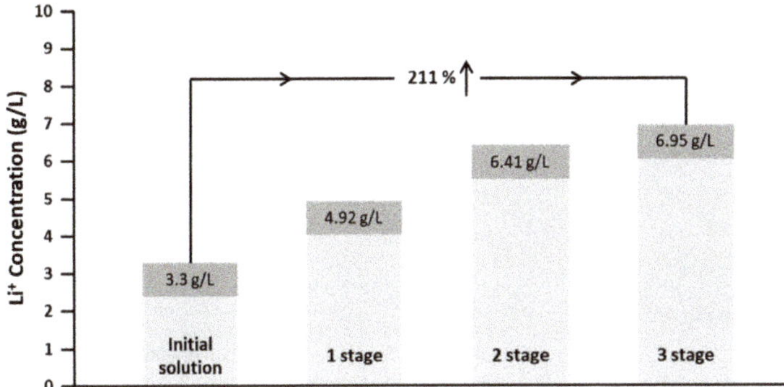

Figure 5. Effect of number of concentration stage on the Lithium concentration (10 V, 0.7 M Na_2SO_4, D:C [3:1]).

Table 5. Analytical results of the concentrated solution after multistage concentration (MSC) process (10 V, 0.7 M Na_2SO_4, D:C [3:1]).

Stage	Concentration (g/L)			Concentration (%) Compared to Previous Stage Sol.	Concentration (%) Compared to Initial Sol.	Process Lead Time (min)
	Li	Na	SO_4			
Initial sol.	3.30	60.00	148.20	-	-	-
1st	4.92	42.05	147.88	149.09%	149.09%	210
2nd	6.41	58.30	213.87	130.28%	194.24%	270
3rd	6.95	63.58	224.84	108.42%	210.61%	320

Table 6. Volume change of solution in chamber after MSC process (10 V, 0.7 M Na_2SO_4, D:C [3:1]).

Stage	Volume Change of		Water Migration (%)
	Concentrated Sol.	Diluted Sol.	
1st	300 mL → 670 mL	900 mL → 530 mL	41.11%
2nd	300 mL → 870 mL	900 mL → 330 mL	63.33%
3rd	300 mL → 1050 mL	900 mL → 150 mL	83.33%

However, as shown in Table 5, the concentrations of sodium and sulfate in the first stage of ED were lower than those of the initial raw material solution. Repeating the first stage and storing the concentrated solution, white columnar crystals occurred, as shown in Figure 6a. These crystals were dried and crushed and recovered as a powder, and then analyzed by XRD, and were found to be Na_2SO_4 (thenardite, syn) (Figure 6b). Seen in the form of crystals recovered before drying, it is considered to be $Na_2SO_4{\cdot}10H_2O$. After the first stage, the crystals were separated into a solid–liquid, and after separation, the solution was used in a second stage ED process. When the concentration of sodium sulfate in the initial raw material solution is high, crystals may be generated inside the ion exchange cartridge during ED. These crystals can cause problems such as damage to the ion exchange membrane and reduced current efficiency, so it is necessary to remove sodium sulfate from the raw material solution before the ED process.

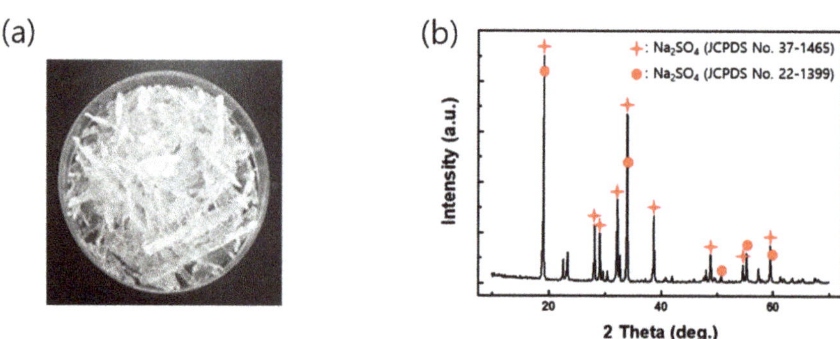

Figure 6. Photograph and result of XRD analysis of precipitate from 1st stage ED process by (**a**) precipitate and (**b**) XRD analysis of precipitate.

3.3. Sodium Sulfate Removal Process

In the previous experiment, when the concentration of Na and SO_4 in the raw material solution was high, a problem occurred: $Na_2SO_4{\cdot}10H_2O$ crystals were generated during ED. Methanol was used as a precipitant to remove sodium sulfate in solution. Since sodium sulfate has low solubility in methanol, sodium sulfate may be precipitated when methanol is added to the sodium sulfate solution [20,21]. Methanol was added to the solution based on the composition of Table 1 in a constant volume ratio, followed by stirring for 30 min to induce a reaction. After the end of the reaction, filtration was performed to separate solids. Methanol was added in a volume ratio of 0.1 to 1.0 relative to the volume of the raw material solution, and as the volume of added methanol increased, the removal rate (%) of Li, Na, and SO_4 increased, as shown in Figure 7. When the volume ratio of added methanol was 0.6 or more, the loss rate of lithium was 14% or more. It is preferable to proceed with the Na_2SO_4 removal process at a volume ratio of 0.4 or less of methanol with a lithium loss (%) of less than 0.3%. When the volume ratio of added methanol was 0.4, lithium loss was about 2%, Na removal was 65%, and SO_4 removal was 51%. Figure 8a shows the powder generated in the Na_2SO_4 removal process, and in Figure 8b the results of XRD analysis show that it was sodium sulfate. It is considered that the recovered powder can be used to prepare the electrode solution for ED process without purification.

In addition, methanol is likely to be recovered through fractional distillation after sodium sulfate filtration. The recovered sodium sulfate and methanol can be reused to reduce the overall process cost.

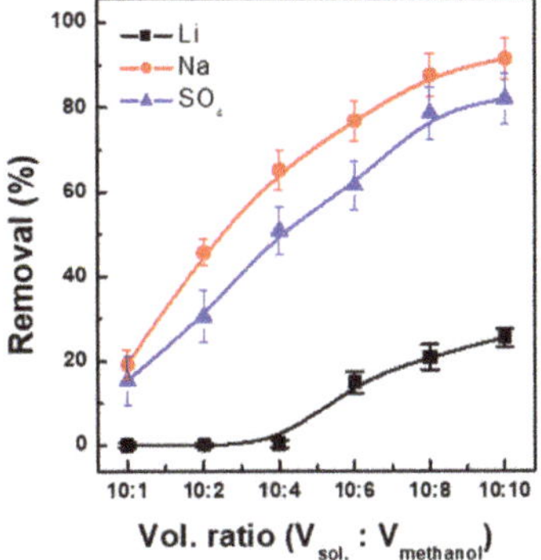

Figure 7. Effect of methanol addition on the sodium and sulfate removal (25 °C).

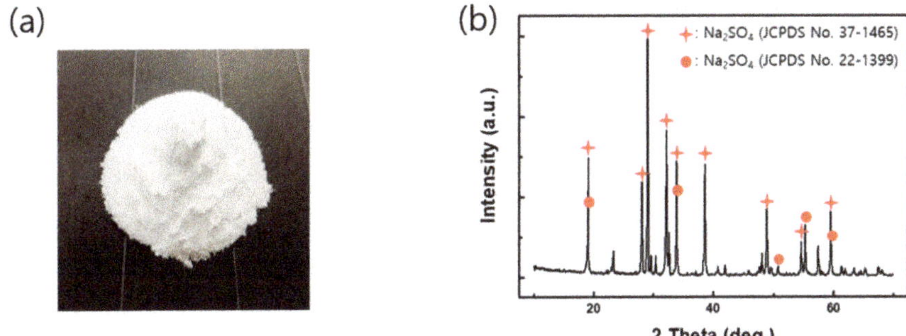

Figure 8. Photograph and result of XRD analysis of precipitate from Na_2SO_4 removal process by (**a**) precipitate and (**b**) XRD analysis of precipitate.

3.4. *Concentration of Lithium by MSC Electrodialysis after Sodium Sulfate Removal Process*

Sodium sulfate crystals were generated during concentration by ED using a raw material solution containing a large amount of sodium sulfate. Prior to the ED process, the MSC ED process was performed using a solution from which sodium sulfate was partially removed through a Na_2SO_4 removal process as a raw material. The first stage was repeated 20 times, the second stage eight times, the third stage five times, and the fourth stage three times. The concentrate solution from each stage was collected in a bottle and used as a feed solution in the next stage. Methanol was added to the raw material solution in Table 1 in a volume ratio of 0.4 to induce the reaction, and the solution from which sodium sulfate was removed was used as the raw material solution. Its composition is about 2.65 g/L Li, about 10 g/L Na, and about 71 g/L SO_4. In Figure 9, the concentration of lithium increased with the number of stages; it was concentrated to 10.02 g/L after four stages from an initial concentration of 2.65 g/L. Although the concentration of lithium was lower than that of the raw material solution, which had not undergone a Na_2SO_4 removal process, it showed a higher concentration even in the first stage, and the process time was shortened. Table 7 shows the concentration of lithium, sodium, and sulfate in each stage, the C_R (%) of lithium, and the process lead time. Compared to the initial concentration of lithium, the C_R (%) tended to increase, but compared to the previous stage, the rate tended to decrease gradually. This phenomenon occurred because water was migrated from the dilution chamber to the concentration chamber as shown in Table 8. In addition, it showed a tendency to increase the process lead time compared to the C_R (%) of lithium. However, compared to using a raw material solution that had not undergone a Na_2SO_4 removal process, a high C_R (%) is shown in a relatively short process lead time. If the concentration of components other than lithium is high in the raw material solution, it is considered efficient to perform ED after removal.

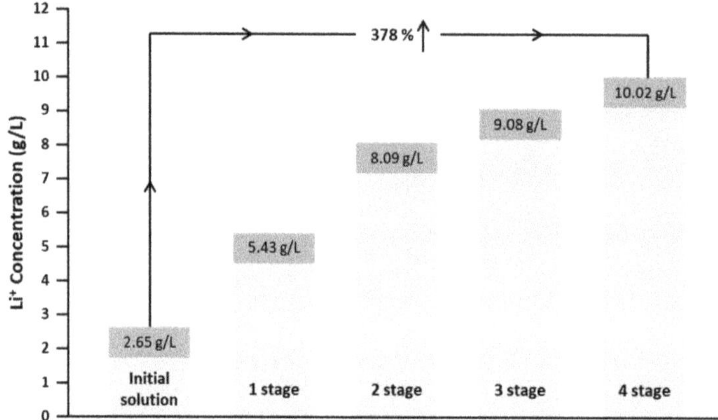

Figure 9. Effect of number of concentration stage on the lithium concentration after Na_2SO_4 removal process (10 V, 0.7 M Na_2SO_4, D:C [3:1]).

Table 7. Analytical results of the concentrated solution after Na_2SO_4 removal process and MSC process (10 V, 0.7 M Na_2SO_4, D:C [3:1]).

Stage	Concentration (g/L)			Concentration (%) Compared to Previous Stage Sol.	Concentration (%) Compared to Initial Sol.	Process Lead Time (min)
	Li	Na	SO_4			
Initial sol.	2.65	9.98	70.67	-	-	-
1st	5.43	20.67	142.54	204.91%	204.91%	150
2nd	8.09	30.09	208.21	148.99%	305.28%	210
3rd	9.08	36.92	260.78	112.24%	342.64%	260
4th	10.02	41.61	284.66	110.35%	378.11%	310

Table 8. Volume change of solution in chamber after Na_2SO_4 removal process and MSC process (10 V, 0.7 M Na_2SO_4, D:C [3:1]).

Stage	Volume Change of		Water Migration (%)
	Concentrated Sol.	Diluted Sol.	
1st	300 mL → 530 mL	900 mL → 670 mL	25.56%
2nd	300 mL → 810 mL	900 mL → 390 mL	56.67%
3rd	300 mL → 980 mL	900 mL → 220 mL	75.56%
4th	300 mL → 1020 mL	900 mL → 180 mL	80.00%

4. Conclusions

In this study, the valuable metals were recovered from the waste lithium secondary battery and a simulated solution was prepared with the same solution composition as the remaining filtrate. An experiment was performed to concentrate lithium from the simulated solution through a multi-stage concentration process using electrodialysis. The following conclusions were drawn:

(1) As the concentration of the electrode solution increased, the concentration ratio (%) of lithium increased and the process lead time decreased. When 0.7 M Na_2SO_4 was used as an electrode solution, the average flux during the process was high and the concentration ratio (%) of lithium increased, which was effective for concentration.

(2) In the case of multistage concentration electrodialysis using an initial raw material solution containing a large amount of sodium sulfate, sodium sulfate crystals were generated during the ED process. In addition, the concentration ratio (%) of lithium was low even after the third stage of ED, and a long process lead time was required.

(3) Sodium sulfate was precipitated by adding methanol to remove sodium from the raw material solution. As the volume of methanol increased, the sodium sulfate removal rate (%) increased, but the loss rate (%) of lithium also increased. When methanol as a raw material solution was added in a volume ratio of 0.4, sodium was removed by 65% and sulfate ion by 51%, and the loss rate of lithium was 0.3%.

(4) During multistage concentration (MSC) using electrodialysis with a solution from which sodium was partially removed as raw material, lithium was effectively concentrated to 10 g/L in the solution in four stages.

Author Contributions: Y.C., K.K., J.A., and J.L. carried out the experiments and analysis and interpretation of the data obtained. Y.C. and J.A. wrote and edited the paper. J.A. is the co-principal investigator of this work. All authors have read and agreed to the published version of the manuscript.

Funding: This study was supported by the Korea Institute of Energy Technology Evaluation and Planning (KETEP), funded by the Ministry of Trade, Industry and Energy, Republic of Korea (Project No. 20185210100050).

Conflicts of Interest: The authors declare no conflict of interest.

References

1. Yang, H.W.; Lee, J.H.; Jung, S.C.; Myung, S.T.; Kang, W.S.; Kim, S.J. Fabrication of Si/SiOx Anode Materials by a Solution Reaction-Based Method for Lithium Ion Batteries. *Korean J. Met. Mater.* **2016**, *54*, 780–786.
2. Yang, Y.M.; Loka, C.; Kim, D.P.; Joo, S.Y.; Moon, S.W.; Choi, Y.S.; Park, J.H.; Lee, K.S. Si-FeSi2/C Nanocomposite Anode Materials Produced by Two-Stage High-Energy Mechanical Milling. *Met. Mater. Int.* **2017**, *23*, 610–617. [CrossRef]
3. KIPOST (Korea Industry Post). Available online: https://www.kipost.net/news/articleView.html?idxno=203140 (accessed on 8 March 2020).
4. Park, J.G. Effect of Adding Carbonaceous Materials on Cathode Materials for Lithium Secondary Batteries. Master's Thesis, Kumoh National Institute of Technology, Gumi, Korea, 2007.
5. Ruiz, V.; Pfrang, A.; Kriston, A.; Omar, N.; Van den Bossche, P.; Boon-brett, L. A review of international abuse testing standards and regulations for lithium ion batteries in electric and hybrid electric vehicles. *Renew. Sustain. Energy Rev.* **2018**, *81*, 1427–1452. [CrossRef]
6. Hoshino, T. Development of technology for recovering lithium from seawater by electrodialysis using ionic liquid membrane. *Fus. Eng. Des.* **2013**, *88*, 2956–2959. [CrossRef]
7. Hoshino, T. Preliminary studies of lithium recovery technology from seawater by electrodialysis using ionic liquid membrane. *Desalination* **2013**, *317*, 11–16. [CrossRef]
8. Hoshino, T. Innovative lithium recovery technique from seawater by using world-first dialysis with a lithium ionic superconductor. *Desalination* **2015**, *359*, 59–63. [CrossRef]
9. Guo, Z.Y.; Ji, Z.Y.; Chen, Q.B.; Liu, J.; Zhao, Y.Y.; Li, F.; Liu, Z.Y.; Yuan, J.S. Prefractionation of LiCl from concentrated seawater/salt lake brines by electrodialysis with monovalent selective ion exchange membranes. *J. Clean. Prod.* **2018**, *193*, 338–350. [CrossRef]
10. Umeno, A.; Miyai, Y.; Takagi, N.; Chitraker, R.; Sakane, K.; Ooi, K. Preparation and adsorptive properties of membrane-type adsorbents for lithium recovery from seawater. *Ind. Eng. Chem. Res.* **2002**, *41*, 4281–4287. [CrossRef]
11. Lee, J.K.; Jeong, S.G.; Koo, S.J.; Kim, S.Y.; Ju, C.S. Solvent Extraction of Lithium Ion in Aqueous Solution Using TTA and TOPO. *Korean Chem. Eng. Res.* **2013**, *51*, 53–57. [CrossRef]
12. Ryu, T.G.; Shin, J.H.; Ghoreishian, S.M.; Chung, K.S.; Huh, Y.S. Recovery of lithium in seawater using a titanium intercalated lithium manganese oxide composite. *Hydrometallurgy* **2019**, *184*, 22–28. [CrossRef]
13. Lee, J.E. A Study of Lithium Concentration from Lithium-Containing Waste Solution by Electrodialysis. Master's Thesis, Daejin University, Pocheon, Korea, 2019.

14. Ahn, J.W.; Ahn, H.J.; Son, S.H.; Lee, K.W. Solvent Extraction of Ni and Li from Sulfate Leach Liquor of the Cathode Active Materials of Spent Li-ion Batteries by PC88A. *J. Korean Inst. Resour. Recycl.* **2012**, *21*, 58–64. [CrossRef]
15. Ahn, H.J.; Ahn, J.W.; Lee, K.W.; Son, H.T. Recovery of Li from the Lithium Containing Waste Solution by D2EHPA. *J. Korean Inst. Resour. Recycl.* **2014**, *23*, 21–27.
16. Kim, S.H. Non-Discharge System for Nickel Plating Rinse Waters by Electrodialysis. Ph.D. Thesis, Dankook University, Yongin, Korea, 2000.
17. Zhou, Y.; Yan, H.; Wang, X.; Wu, L.; Wang, Y.; Xu, T. Electrodialytic concentrating lithium salt from primary resource. *Desalination* **2018**, *425*, 30–36. [CrossRef]
18. Gmar, S.; Chagnes, A. Recent advances on electrodialysis for the recovery of lithium from primary and secondary resources. *Hydrometallurgy* **2019**, *189*, 1–12. [CrossRef]
19. Lee, J.E.; So, H.I.; Cho, Y.C.; Jang, I.H.; Ahn, J.W.; Lee, J.H. A Study on the Separation and Concentration of Li from Li-Containing Waste Solutions by Electrodialysis. *Korean J. Met. Mater.* **2019**, *57*, 656–662. [CrossRef]
20. Hasseine, A.; Menial, A.-H.; Korichi, M. Salting-out effect of single salts NaCl and KCl on the LLE of the systems (water + toluene + acetone), (water + cyclohexane + 2-propanol) and (water + xylene + methanol). *Desalination* **2009**, *242*, 264–276. [CrossRef]
21. Wang, C.; Lei, Y.D.; Wania, F. Effect of Sodium Sulfate, Ammonium Chloride, Ammonium Nitrate, and Salt Mixtures on Aqueous Phase Partitioning of Organic Compounds. *Environ. Sci. Technol.* **2016**, *50*, 12742–12749. [CrossRef]

Review

A Review of the Cyanidation Treatment of Copper-Gold Ores and Concentrates

Diego Medina * and Corby G. Anderson

Department of Mining Engineering, Kroll Institute for Extractive Metallurgy Colorado School of Mines 1500 Illinois St., Hill Hall 337, Golden, CO 80401, USA; cganders@mines.edu
* Correspondence: diegomedina@mymail.mines.edu; Tel.: +1-720-630-5159

Received: 21 May 2020; Accepted: 23 June 2020; Published: 5 July 2020

Abstract: Globally, copper, silver, and gold orebody grades have been dropping, and the mineralogy surrounding them has become more diversified and complex. The cyanidation process for gold production has remained dominant for over 130 years because of its selectivity and feasibility in the mining industry. For this reason, the industry has been adjusting its methods for the extraction of gold, by utilizing more efficient processes and technologies. Often, gold may be found in conjunction with copper and silver in ores and concentrates. Hence, the application of cyanide to these types of ores can present some difficulty, as the diversity of minerals found within these ores can cause the application of cyanidation to become more complicated. This paper outlines the practices, processes, and reagents proposed for the effective treatment of these ores. The primary purpose of this review paper is to present the hydrometallurgical processes that currently exist in the mining industry for the treatment of silver, copper, and gold ores, as well as concentrate treatments. In addition, this paper aims to present the most important challenges that the industry currently faces, so that future processes that are both more efficient and feasible may be established.

Keywords: gold cyanide leaching; sulfide minerals; SART process; cyanidation; activated carbon; metal–cyanide complex; copper ore; carbon in pulp (CIP), agitated tank; cyanide complexes

---

1. Introduction

The history of modern hydrometallurgy started with the discovery of how to obtain gold and silver from ores, on 19 October 1887, by John Steward MacArthur, who was recognized for having established the application of the cyanidation process. Gold production around the world readily doubled as a consequence of cyanidation's initial application within the mining industry. Following the first application of cyanidation in the recovery of gold, the hydrometallurgical industry has developed and grown according to the needs of the process and the mineral complexity of the ore deposits.

Hydrometallurgical processes can be defined as the leaching of a desired metal into a solution, followed by the concentration and purification of the pregnant solution, and finally, the recovery of the metal or its compounds. The processing of gold and silver ore by leaching is one of the most prominent examples of early hydrometallurgy-based processes.

Most of the gold extraction from ore is accomplished by the implementation of an alkaline cyanide leaching process. The chemical recovery of gold can be defined by two different operations: the oxidative dissolution of gold and the reductive precipitation of metallic gold from the solution. Cyanide is one of the most attractive lixiviants in the current industrial gold leaching process. During gold cyanidation, silver and copper are commonly present within the solution, which causes their metal ions to react with the cyanide (CN^-), thus forming complexes [1].

Cyanide is considered to be a hazardous compound because of its toxicity; there is currently environmental pressure by different groups around the world to ban the industrial use of cyanide.

Research on replacing cyanide as a lixiviant has been ongoing over the years, and has found that there are other potentially workable compounds, such as thiosulfate, thiourea, halides, various sulfide systems, ammonia, bacteria, natural acids, thiocyanate, nitriles, and combinations of cyanide with other compounds [1]. Many of these alternative gold processes are still in the early development stages. A key factor for the commercial success of these alternative lixiviants relates to the overall stability of the lixiviant and the gold complex in solution.

Currently, the mining industry faces the problem of separating these complex valuable minerals from the ore in which they reside. This paper outlines various options that hydrometallurgical processes offer for the treatment of these complex minerals, containing precious metals such as Cu, Ag, and Au.

2. Copper Mineral Complexity

The fact that the majority of copper ore deposits are complex has resulted in the better development of technologies able to extract precious metals more efficiently. Table 1 shows some of the sulfide minerals that can be found during the treatment of ores containing Cu, Ag, and Au ores.

Table 1. Main sulfide minerals for copper ores [2].

Sulfide Mineral	Element	Formula
Chalcopyrite	Copper	$CuFeS_2$
Chalcocite		Cu_2S
Covellite		CuS
Bornite		Cu_5FeS_4
Pyrite	Iron	FeS_2
Pyrholite		FeS
Argentite	Silver	Ag_2S

Copper sulfide deposits around the world are commonly associated with copper oxide minerals. Generally, copper oxide minerals do not respond to standard sulfide copper collectors and require the application of different flotation techniques [3]. The treatment of copper sulfide minerals containing a high percentage of oxide copper causes problems in the concentration process, decreasing the copper content. In order to increase the efficiency of the concentration process, it is necessary to treat the copper oxide component by leaching the grinded ore prior to the flotation, or by leaching the flotation tailings coming out from the concentrator [4,5].

3. Cyanidation of Complex Gold Ores and Concentrates

The cyanidation process has become one of the most used methods for the recovery of gold from ores. The use of cyanide leaching for gold recovery is based on gold's properties, whereby gold does not become oxidized at ordinary temperatures. Additionally, gold is not soluble in sulphuric, hydrochloric, or nitric acids, but can be dissolved in aqua regia (a mixture of nitric and hydrochloric acid). On the other hand, the most crucial fact about gold, in this case, is that it is soluble in dilute cyanide solutions. For this reason, cyanide is used as a lixiviant during the leaching process in order to perform the gold extraction through the use of this hydrometallurgical process [6]

Figure 1 shows that the hydrometallurgical process starts with the leaching agitators, where the slurry comes into contact with cyanide, oxygen, water, and lime, thereby enacting the leaching process. During the leaching process, other cyanide complexes, such as copper and silver sulfide minerals, are formed. The following cyanide complexes are formed during this stage of the cyanidation:

- Gold cyanide reaction:

$$4Au + 8NaCN + 2H_2O + O_2 \rightarrow 4NaAu(CN)_2 + 4NaOH \quad (1)$$

- Silver cyanide reaction:

$$4Ag + 8NaCN + 2H_2O + O_2 \rightarrow 4NaAg(CN)_2 + 4NaOH \quad (2)$$

In this case, the copper sulfide minerals can form complexes with cyanide, such as Cu $(CN)_2$, as the following reaction shows:

- Copper cyanide reaction:

$$4Cu + 8NaCN + 2H_2O + O_2 \rightarrow 4NaCu(CN)_2 + 4NaOH \quad (3)$$

The formation of copper and silver cyanide complexes affects the gold recovery in both the cyanide leaching process as well as the purification and refining stages [7]. These effects mainly interfere with the gold cyanide reaction and the carbon adsorption. Most of the copper minerals react rapidly with cyanide, forming multiple cyanide complexes. Table 2 shows the solubility of copper minerals in a cyanide solution. As can be shown, chalcopyrite is the copper mineral with the lowest percentage of copper dissolved and extracted, when compared to the other minerals [8].

Table 2. Solubility of copper minerals in 0.1% NaCN solutions. Reproduced and adapted from [8], with permission from Elsevier B.V., 2005.

Mineral	Formula	Percent Total Copper Dissolved [a]		g NaCN/ g Cu [b]	Extraction (% Cu) [b]
		23 C	45 C		
Azurite	$2Cu(CO)_3 \cdot Cu(OH)_2$	94.5	100	3.62	91.8
Malachite	$2CuCO_3(OH)_2$	90.2	100	4.48	99.7
Chalcocite	Cu_2S	90.2	100	2.76	92.6
Covellite	CuS	–	–	5.15	95.6
Native Copper	Cu	90	100	–	–
Cuprite	Cu_2O	85.5	100	4.94	96.6
Bornite	$FeS{\cdot}2Cu_2{\cdot}CuS$	70	100	5.13	96
Enargite	Cu_3AsS_4	65.8	75.1	–	–
Tetrahedrite	$(Cu,Fe,Ag,Zn)_{12}\,Sb_4\,S_{13}$	21.9	43.7	–	–
Chrysocolla	$CuSiO_3{\cdot}nH_2O$	11.8	15.7	–	–
Chalcopyrite	$CuFeS_2$	5.6	8.2	2.79	5.8

[a] Data after Hedley and Tabachnick (1958).[b] Data after Lower and Booth (1965). Cyanide consumption is expressed as g. NaCN/g of contained copper, data being generateed by leaching at room temperature for 6 h.

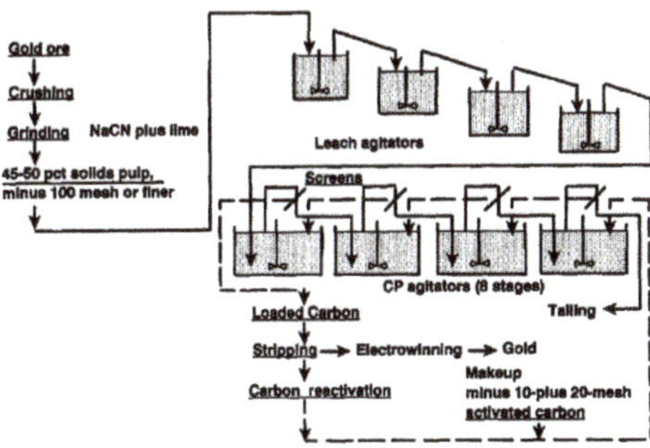

Figure 1. Diagram of gold cyanide leaching using activated carbon recovery. Reproduced and adapted from [9], with permission from PERGAMON, 1987.

There are high carbon-management and bullion-refining costs related to the interference of these cyanide complexes. The pH in the cyanide solution during the leaching plays an important role. The range of the cyano complexes depends on conditions such as the cyanide concentration and pH, as shown in Figure 2, where at an operational pH range of 10–11 there is a greater concentration of the complex $Cu(CN)_4^{3-}$ [8].

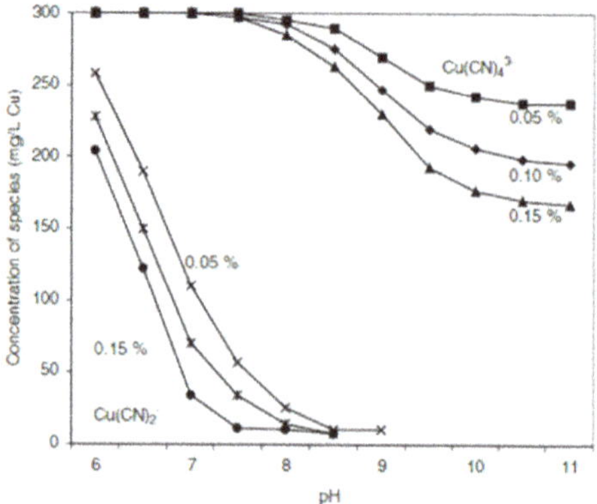

Figure 2. Comparison of cyanide species vs. pH. Reproduced and adapted from [8], with permission from Elservier B.V., 2005.

4. Cyanide Consumption

The formation of copper and silver cyanide complexes during the first stage of the leaching process affects the cyanide consumption for gold recovery. Consequently, the operational costs for cyanidation have increased significantly. As an approximation for cyanide, for every \$3600/t delivered to a mine site, it takes only 0.28 kg/t-ore of consumption to equate to \$1/t-ore or 1% of gold recovery [9]. Cyanide consumption remains one of the main economic considerations.

The behavior of a specific ore or concentrate in testing can be determined by performing rolled-bottle testing, or, alternatively, by testing in stirred vessels to measure the quantity of cyanide consumption per unit weight of ore. This value can be scaled up for engineering design purposes. The following factors affect cyanide consumption during the leaching process: The functions of ore mineralogy; Cyanide concentration; and Reaction kinetics [10].

A key factor to consider is the effects of the residence time and the pulp density during the cyanide leaching. The residence time determines and controls the reaction rate of the cyanide and oxygen at the surface of the free particles. The optimum time of the reaction for forming gold–cyanide complexes is reached within the first hours of the reaction [7].

The second important aspect to consider is the pulp density during the cyanide reaction. NaCN consumption decreases as the percentage of solids increases with different cyanide concentrations, meaning that if there is not an appropriate level of gold extraction from within the slurry, the cyanide consumption will be higher, as shown in Figure 3.

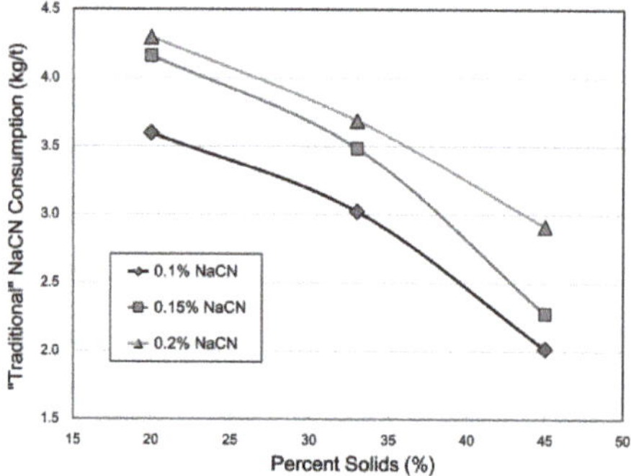

Figure 3. Consumption (kg/t) vs. percent solids (%) of NaCN. Reproduced and adapted from [7], with permission from Springer, 2015.

5. Gold Cyanidation in Copper Flotation Tailings

The production of copper concentrates from copper-gold sulfide minerals by froth flotation generally results in tailings with copper, silver, and gold values. The cyanidation of copper flotation tailings containing sulfides for gold recovery is an example of the formation of cyanide complexes with copper and silver during the leaching process [11,12]. Figure 4 shows the MLA Automated Minerology image of an ore sample from a copper mine in Mexico, containing chalcopyrite and bornite as copper sulfide minerals.

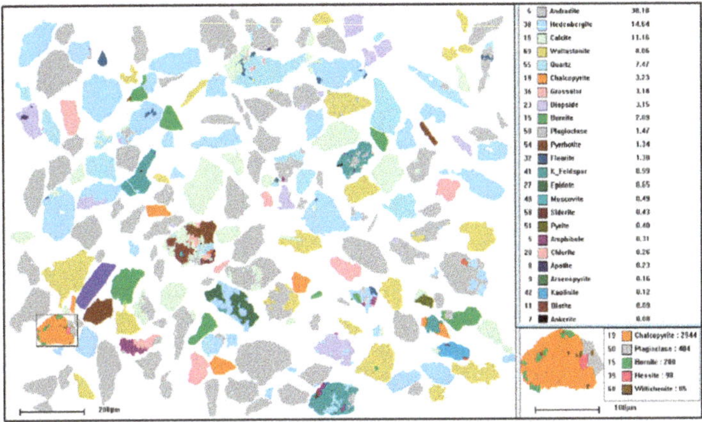

Figure 4. Image containing copper sulfide minerals. Reproduced and adapted from [13], with permission from the authors, 2020.

Studies were performed to analyze the gold recovery by cyanide leaching in the flotation tailings from this ore. There were considerable quantities of copper and silver in the tested tailings that could form cyanide complexes during the cyanide leaching. The beneficiation plant at this particular mine in Mexico produces a copper concentrate, however their flotation circuit only recovers around 75% of the gold; the rest is lost in the tailings [14,15].

Table 3 shows the results of the cyanide leaching performed on the flotation tailings of this Mexican beneficiation plant using a NaCN concentration of 0.5 mg/L. The copper and silver recovery percentage values were 33% and 35%, respectively, which means that most of the copper and silver values were lost in the leaching tailings, as shown in Table 4. The oxygen concentration was measured during the test to control the oxygen supply as it is needed for the gold leaching. [16]. These copper and silver values may represent a problem relating to the recirculation of the cyanide during leaching; for this reason, the implementation of a cyanide-cleaning detox process is advised [17].

The copper and silver that form cyanide complexes during the gold cyanidation can cause issues during the activated carbon adsorption, for example, by competing with the gold to be adsorbed, therefore requiring a higher free cyanide concentration [18]. In addition, the mineralogy consists of copper sulfide minerals, as previously mentioned. It is expected that the cyanide-barren solution retains a high sulfide content, which can be removed using the SART process discussed in this paper [19].

As mentioned, the process of treating large amounts of cyanide-contaminated effluents remains a challenge in the cyanidation process. The effluents contain free cyanide and metal–cyanide complexes that, in this specific case, would present as copper and silver complexes [20,21].

Table 3. Cyanide leaching results of copper-gold minerals. Reproduced from [13], with the permission from the authors.

Conditions				Assay (mg/kg)				Distribution %		
NaCN concentration (mg/L)	Time (h)	Dissolved Oxygen (mg/L)	pH	Au	Ag	Cu		Au	Ag	Cu
0.5	0.00	5.55	11.61	0.00	0.00	0.00		0.00	0.00	0.00
	2.00	4.87	11.60	0.11	0.45	85.31		12.00	7.20	10.73
	6.00	4.75	11.75	0.18	0.76	135.00		18.53	12.16	16.98
	17.00	4.34	11.62	0.21	1.24	186.80		22.11	19.84	23.49
	24.00	4.29	11.70	0.22	1.70	190.00		23.16	27.20	23.90
	32.00	3.80	11.71	0.23	2.10	198.00		24.21	33.60	24.90
	Recovery %			81.79	33.33	35.59	Head (Calc)	100.00	100.00	100.00

Table 4. Leaching tailings elemental analysis. Reproduced from [13], with the permission from the authors, 2020.

Sample	Au(mg/kg)	Ag(mg/kg)	Cu(%)
A	0.10	2.00	0.04
B	0.08	6.04	0.072

6. Alternatives Lixiviants to Cyanide

As mentioned in Section 1 there are alternative gold processes that utilize alternative lixiviants from cyanide. Table 5 shows the stability constants and standard reduction potentials for gold complexes. Clearly, the cyanide complex is more stable and inherently more selective than any other alternative reagent. For example, thiosulfate, thiourea, and bisulfide are several orders of magnitude less stable.

Table 5. Constants and standard reduction potentials for Au complexes at 25 °C. Reproduced and adapted from [4], with permission from Elsevier B.V., 2016.

Ligand	Au(I)or Au (II) Complex	Eo (V vs SHEa)	Stability Constants ß2 or ß4	pH Range
CN^-	$Au(CN)^{2-}$	−0.57	38.3	>9
$S_2O_3^{2-}$	$Au(S_2O_3)_2^{3-}$	0.17	28.7	8 to 10
$CS(NH_2)_2$	$Au(NH_2CSNH_2)_2^+$	0.38	23.3	<3
Cl^-	$AuCl_2^-$, $AuCl_4^-$	1.11,1	9.1, 25.3	<3
Br^-	$AuBr_2^-$, $AuBr_4^-$	0.98,0.97	12,32.8	5 to 8
I^-	AuI_2^-, AuI_4^-	0.58,0.69	18.6, 47.7	5 to 9
HS^-	$Au(HS)_2^-$	−0.25	29.9	<9
NH_3	$Au(NH_3)_2^+$	0.57	26.5	>9
Glycinate	$Au(NH_2CH_2COO)_2^-$	0.632	18	9
SCN^-	$Au(SCN)_2^-$, Au $SCN)_4^-$	0.66,0.66	17.1,43.9	<3
SO_3^{2-}	$Au(SO_3)_2^{3-}$	0.77	15.4	>4

The wide range of values of the stability constants for the gold complexes indicates that the standard reduction for the different gold ligand species varies by almost 2 V. Most reagents have a small operating window for effectively dissolving cyanide compared with cyanide, as shown in Figure 5. The high oxidizing potentials involved with some lixiviants lead to high reagent consumptions due to reaction with sulfide minerals contained in the ore and the oxidation of the reagent itself. A point to consider is that the adsorption of reagents and/or precipitation of gold onto some gangue minerals that can affect the overall gold recovery.

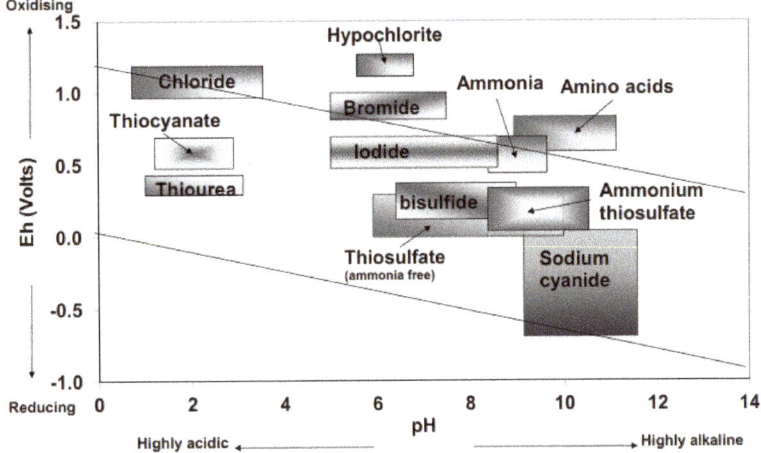

Figure 5. Diagram showing the typical range of operation of lixiviants. Reproduced and adapted from [22], with the permission from Elsevier B.V., 2016.

7. Carbon Adsorption

The formation of copper and silver cyanide complexes also affects the adsorption of gold during the CIP stage during the leaching. The metal-cyanide species adsorption occurs in a selective form, depending on the ionic diameters of the metal-cyanide complexes. The adsorption of metal-cyanides on activated carbon is selective, and the most potent adsorbed complex is $Au(CN)^{2-}$. On the other hand, silver-cyanide complex $Ag(CN)^{2-}$ adsorbs strongly but not as strongly as the gold–cyanide complex [23].

The copper-cyanide complexes $Cu(CN)^{2-}$ and $Cu(CN)_3{}^{2-}$ show different adsorption features, with the $Cu(CN)^{2-}$ complex adsorbing more strongly than the $Cu(CN)_3{}^{2-}$ complex [24]. The hydration is an essential aspect of carbon adsorption because it causes metal-cyanide complexes to increase their molecule diameter due to the fact that the metal ion is likely to have had a predetermined hydration that can affect the metal-cyanide carbon adsorption. The classification of those that are most strongly adsorbed to those that are the least absorbed is: Au > Ag > Cu, where the Ag and Cu have an effect on the gold recovery during the leaching process [23].

Maintaining a high free-cyanide concentration reduces the concentration of more adsorbed copper complexes. The increase of copper cyanide adsorption at low cyanide concentrations results in problems for gold adsorption, as previously mentioned [25].

8. Processing Options

Different options have been developed for the treatment of copper-gold ores. Approximately 20% of all gold deposits have significant copper mineralization associated with chalcopyrite, tetrahedrite, tennantite, bornite, and chalcocite. Most of the copper minerals, including copper oxides, carbonates, sulfides (with the exception of chalcopyrite), and native copper are highly soluble in cyanide solutions. As mentioned previously, the minerals that contain copper are problematic because during the gold cyanidation process, the copper also forms cyanide complexes, thereby consuming the cyanide [6,25].

The presence of copper-cyanide complexes creates competition with the gold during the activated adsorption; it also affects the electrowinning efficiency. Research has been undertaken relating to the treatment of copper-gold ores, as well as to the lowering of the effects of copper-cyanide complexes. Some of the processing options include [25,26]:

(1) Ore Segregation Technologies

Separating high-copper-containing ores selectively prior to the leaching process is an option in order to reduce the impact of the copper content. For example, the Red Dome Mine in Australia selectively mines and leaches ores containing less than 0.5% copper [25].

(2) Selective Leaching Technologies

Another processing option is to selectively leach the copper contained in the ores containing gold prior to the gold leaching. These processes are often associated to high reagent consumption and also to recovery issues of the leached copper. Studies have been undertaken on the application of copper/ammonia/cyanide processes for the treatment of copper-containing ores. These studies have shown that the addition of ammonia to the cyanide solution results in a lower cyanide consumption and a higher selectivity of gold leaching over copper. However, the rate of gold leaching is slower in this system when ammonia is used; in addition, it also represents an occupational health and environmental concern [24].

(3) Copper-Cyanide Destruction Technologies

This type of process uses SO_2 combined with air to produce cyanide oxidation to cyanite, whereby the copper is precipitated at the end of the reaction as copper hydroxide [27]. There are alternative processes for cyanide destruction such as alkaline chlorination, the use of hydrogen peroxide (Degussa process), the use of Caro's acid, electrochemical oxidation, biodegradation, the use of ultrasonic technology, or photolysis [5].

9. Sulfidization, Acidification, Recycling and Thickening (SART) Technology

Recently, new processes have been developed to treat the main issues previously mentioned about processing complex copper ores for gold recovery. One of these processes is the SART process. SGS Lakefield Group and Teck Corporation developed the SART process in the 1990s [26]. The benefit of having a SART process in the cyanidation process is that it breaks the base metal cyanide complexes, precipitates the metals as high-grade sulfide concentrates, and frees the cyanide for recirculation to the leaching process [28]. The SART process is described by the following sequence of unit operations:

Sulfidization and Acidification: During this stage, the cyanide solution is mixed with sodium hydrosulfide NaSH and H_2SO_4 to decrease the pH between 4–5 to form Cu_2S, using a precipitator reactor and thickener to form a Cu_2S precipitate as a co-product. Equations (4) and (5) show the reactions during this process [29]:

$$2NaCN + H_2SO_4 \rightarrow 2HCN\ (aq) + Na_2SO_4 \tag{4}$$

$$4Na_2Cu\ (CN)_3 + 2NaSH + 5H_2SO_4 \rightarrow 2Cu_2S(s) + 12HCN(aq) + 5Na_2SO_4 \tag{5}$$

Equation (4) shows the solution acidification, which promotes the dissociation of weak metal–cyanide complexes (WAD complexes), such as Cu and Ag metals. Equation (5) involves the precipitation of soluble metal ions formatting metallic sulfides such as Cu_2S. The Cu precipitation efficiency under standard process conditions is between 80% and 90%. [29].

Recycling: The remaining solution containing HCN is neutralized using CaO to reach a pH between 10 and 11 to form $CaSO_4$ (Gypsum) and to recycle the cyanide, respectively. The formed solid gypsum is then removed from the process by thickening and filtration, as shown on in Equations (6) and (7) [29]:

$$2HCN(aq) + Ca\ (OH)_2 \rightarrow Ca(CN)_2 + 2H_2O \tag{6}$$

$$H_2SO_4 + Ca(OH)_2 \rightarrow CaSO_4{\cdot}2H_2O\ (s) \tag{7}$$

Thickening: The gypsum is precipitated using a settler and a flocculent to separate the solution $Ca(CN)_2$ and the gypsum. The cyanide is recycled in the process, and the gypsum is precipitated for disposal. The overflow solution from the gypsum thickener is filtered and represents the final solution. This solution has a cyanide content, represented as Ca $(CN)_2$, that is equivalent to the free cyanide that is recycled in the cyanidation process [29].

These unit operations of the SART process describe the treatment of the cyanide solution after the stripping of the activated carbon. Figure 6 shows the general flowsheet of the SART process:

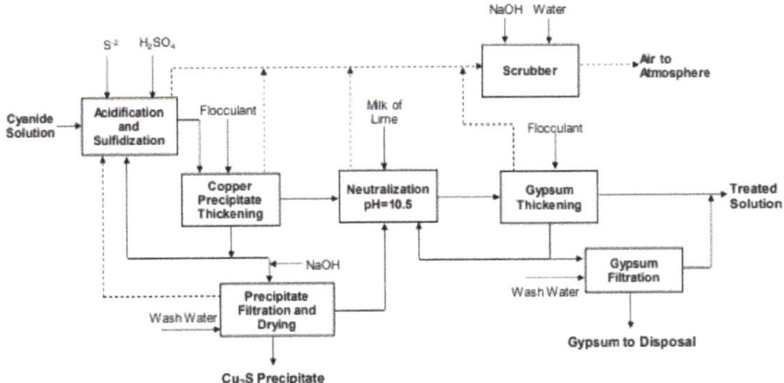

Figure 6. SART process flowsheet. Reproduced and adapted from [16], with permission from PERGAMON, 2020.

The SART process reduces the issues caused by the copper-cyanide complexes during the cyanide leaching process. It recycles the cyanide to diminish the cyanide consumption and provides operational cost savings. It avoids having free cyanide disposal, which can be a hazard for the environment [30]. Additionally, the formation of Cu_2S as a saleable product is one of the essential advantages of the SART process within cyanide leaching, because of the way it economically takes advantage of the copper content. The ideal performance of the SART process and thickening operation would be to produce a clean solution and to separate the solids from the treated solution in the thickener [31].

10. Conclusions

The challenges of treating ores and concentrates with copper, silver, and gold have increased significantly as mineralogies have become more complex. Hence, the technology for such treatment has been forced to become more efficient and innovative in order to face these current challenges. The application of the cyanidation process in conjunction with the SART process represents an innovative way to diminish the problems of having deleterious cyanide complexes involved in the process, by recovering marketable copper and silver sulfides. As a result, the carbon circuit can be made smaller and designed primarily for gold recovery. There are also environmental advantages in the application of these processes, including the reduction of dangerous chemical exposure to the environment.

Author Contributions: Conceptualization, D.M. and C.G.A.; methodology, D.M. and C.G.A.; D.M. and C.G.A.; validation, D.M. and C.G.A.; formal analysis, D.M. and C.G.A.; investigation, D.M. and C.G.A.; resources, D.M. and C.G.A.; data curation, D.M. and C.G.A.; writing—original draft preparation, D.M.; writing—review and editing, D.M.; visualization, D.M.; supervision, C.G.A.; project administration, C.G.A.; funding acquisition, C.G.A. All authors have read and agreed to the published version of the manuscript.

Funding: Research was funded by NEMISA, 470120.

Conflicts of Interest: The authors declare no conflict of interest.

References

1. Adams, M.D. *Advantages in Gold Ore Processing*, 1st ed.; Elsevier B.V.: Perth, Australia, 2005.

2. Twidwell, L.G. Montana College of Mineral Science and Technology Unit Processes in Extractive Metallurgy: Hydrometallurgy. 1970; p. 16. Available online: https://files.eric.ed.gov/fulltext/ED218136.pdf (accessed on 6 January 2020).
3. Lee, K.; Archibald, D.; McLean, J.; Reuter, M.A. Flotation of mixed copper oxide and sulphide minerals with xanthate and hydroxamate collectors. *Miner. Eng.* **2008**, *22*, 1–7. [CrossRef]
4. Sokic, M.D.; Milosevic, V.D.; Stankovic, V.D.; Matkovic, L.V.; Markovic, B.R. Acid leaching of oxide-sulfide copper ore prior the flotation—A way for an increased metal recovery. *Hemijska Industrija* **2015**, *69*, 454–458. [CrossRef]
5. McClelland, G.E.; McPartland, J.S. Metallurgical Comparisons from Testing to Production. *Adv. Gold Silver Process.* **1990**, *1*, 49–57.
6. Deschenes, G. Advances in the cyanidation of gold. *Dev. Miner. Process.* **2005**, *15*, 479–500.
7. Brittan, M.; Plenge, G. Estimating Process Design Gold Extraction, Leach Residence Time and Cyanide Consumption for High Cyanide-Consuming Gold Ore. *Miner. Metall. Process* **2015**, *32*, 111–120. [CrossRef]
8. Sceresini, B. Gold-copper ores. In *Advantages in Gold Ore Processing*, 1st ed.; Adams, M.D., Ed.; Elsevier B.V: Perth, Australia, 2005; pp. 789–821.
9. Hill Stephen, D. The carbon-in-pulp process. In *Precious Metals Recovery from Low-Grade Resources*; Bureau of Mines: Washington, DC, USA, 1986; pp. 40–43.
10. Breuer, P.L.; Rumball, J.A. Cyanide Measurement and Control for Complex Ores and Concentrates. In Proceedings of the Ninth Mill Operators Conference, Fremantle, WA, Australia, 19–21 March 2007; AusIMM: Victoria, Australia, 2007; pp. 249–253.
11. Estay, H. Designing the SART process—A review. *Hydrometallurgy* **2018**, *176*, 147–165. [CrossRef]
12. Thompson, P.; Runge, K.; Dunne, R. Sulfide Flotation testing. In *Mineral Processing and Extractive Metallurgy Handbook*, 1st ed.; Society for Mining, SME: Englewood, CO, USA, 2019; pp. 1029–1031.
13. Medina, D.; Anderson, C. Tailings Gold Recovery by Cyanide Leaching from Future Ores. Master' Thesis, Colorado School of Mines, Golden, CO, USA, 2020, (unpublished).
14. Xie, F.; Dreisinger, D.; Doyle, F. A review on recovery of copper and cyanide from waste cyanide solutions. *Miner. Process. Extr. Metall. Rev.* **2013**, *34*, 387–411. [CrossRef]
15. Markovic, Z.; Vusovic, N.; Milanovic, D. Old Copper Flotation Tailings Water Reprocessing. In Proceedings of the XXV International Mineral Processing Congress (IMPC) Proceedings, Brisbane, QLD, Australia, 6–10 September 2010; Australian Institute of Mining and Metallurgy: Brisbane, Australia, 2010; pp. 3825–3829.
16. Haque, K.E. The Role of Oxygen in Cyanide Leaching of Gold Ore. *CIM Bull.* **1992**, *85*, 31–38.
17. Anderson, C.G. Alkaline Sulfide Gold Leaching Kinetics. *Miner. Eng.* **2016**, *92*, 248–256. [CrossRef]
18. Barsky, G.; Swainson, S.J. Dissolution of Gold and Silver in Cyanide Solutions. *Trans. AIME* **1943**, *112*, 660–667.
19. Estay, H.; Minghai, G.K.; Gabriel, S.; Quilaqueo, M.; Barros, L.; Figueroa, R.; Troncoso, E. Optimizing the SART process: A critical assessment of its design criteria. *Miner. Process.* **2020**, *146*, 1–11. [CrossRef]
20. Zarate, G.E. Gold Tailings Processing by Heap Leaching. In *Small Mines Development in Precious Metals*; Society of Mining Engineers: Santiago, Chile, 1987; pp. 152–155.
21. Prasad, M.S.; Mensah, B.R.; Pizarro, R.S. Modern Trends in Gold Processing—Overview. *Miner. Eng.* **1991**, *4*, 1257–1277. [CrossRef]
22. Aylmore, M.G. Alternative Lixiviant to Cyanide for Leaching Gold Ores. In *Gold Ore Processing: Project Development and Operations*, 2nd ed.; Adams, M.D., Ed.; Elsevier B.V.: Amsterdam, The Netherlands, 2016; pp. 447–460.
23. Sayiner, B.; Acarkan, N. Effect of Silver, Nickel and Copper Cyanides on Gold Adsorption on Activated Carbon. *Physicochemical Prob. Miner. Process.* **2013**, *50*, 277–287.
24. Muir, D.M.; La Brooy, S.R.; Fenton, K. Processing copper-gold ores with ammonia or ammonia cyanide solutions. *World Gold* **1991**, *91*, 145–150.
25. Dai, X.; Simons, A.; Breuer, P. A review of copper cyanide recovery technologies for the cyanidation of copper containing gold ores. *Miner. Eng.* **2012**, *25*, 1–13. [CrossRef]
26. Littlejohn, P.; Kratochvil, D.; Hall, A. Sulfidisation-Acidification-Recycling-thickening for Complex Ores. In Proceedings of the World Gold, Brisbane, Australia, 26–29 September 2013; pp. 149–155.

27. Nicol, M.J.; Fleming, C.A.; Paul, R.L. The Chemistry of the Extraction of Gold. In *The Chemistry of Gold Extraction*; Marsden, J., House, I., Eds.; SME: Littleton, CO, USA, 2006; pp. 831–905.
28. Kratochvil, D.; Salari, D.; Avilez, T. SART Implementation at Heap Leach Operations in Mexico. In Proceedings of the 50th Annual Canadian Mineral Processors Operators Conference, Ottawa, ON, Canada, 24 January 2018; CIM: Ottawa, ON, Canada; pp. 1–13.
29. Estay, H.; Carvajal, P.; Hedjazi, F.; Zeller, T.V. The SART process experience in the Gedabel plant. In Proceedings of the 4th International Workshop on Process Hydrometallurgy, Santiago, Chile, 12–13 July 2012; Gecamin: Santiago, Chile; pp. 1–10.
30. Kevan, J.R.F.; Robert, D.H. Application of the SART Process to Heap Leaching. *SGS Miner. Serv. Tech. Bull.* **2008**, *51*, 1–12.
31. Estay, H.; Becker, J.; Carvajal, P.; Arriagada, F. Predicting HCN gas genertion in the SART process. *Hydrometallurgy* **2012**, *113*, 131–142. [CrossRef]

MDPI
St. Alban-Anlage 66
4052 Basel
Switzerland
Tel. +41 61 683 77 34
Fax +41 61 302 89 18
www.mdpi.com

Metals Editorial Office
E-mail: metals@mdpi.com
www.mdpi.com/journal/metals

www.ingramcontent.com/pod-product-compliance
Lightning Source LLC
LaVergne TN
LVHW071624170726
843515LV00009B/2471